Carbon Nanotubes and Their Applications

Pan Stanford Series on Carbon-Based Nanomaterials

Series Editors

Andrew Wee and Antonio H. Castro Neto

Titles in the Series

Published

Vol. 1
Carbon Nanotubes and Their Applications
Qing Zhang, ed.
2012
978-981-4241-90-8 (Hardcover)
978-981-4303-18-7 (eBook)

Forthcoming

Vol. 2
Chemistry of Graphene
Loh Kian Ping, ed.
2013

Vol. 3
2D Carbon: Fundamentals, Synthesis, and Applications
Wu Yihong, Shen Zexiang, and Yu Ting, eds.
2014

Pan Stanford Series on
Carbon-Based Nanomaterials — Volume 1

Carbon Nanotubes and Their Applications

edited by
Qing Zhang

PAN STANFORD PUBLISHING

Published by

Pan Stanford Publishing Pte. Ltd.
Penthouse Level, Suntec Tower 3
8 Temasek Boulevard
Singapore 038988

Email: editorial@panstanford.com
Web: www.panstanford.com

British Library Cataloguing-in-Publication Data
A catalogue record for this book is available from the British Library.

Carbon Nanotubes and Their Applications

ISBN 978-981-4241-90-8 (Hardcover)
ISBN 978-981-4303-18-7 (eBook)

Cover image courtesy: Chao Liu
Printed in the USA

Preface

Carbon nanotubes (CNTs) are a fantastic member of the carbon family. Their crystal structures are very close to graphite, belonging to sp^2-bonded carbon, rather than sp^3-hybridized carbon in diamond. Topologically, a single-wall CNT (SWCNT) can be constructed by rolling up a single layer of graphite or graphene along a certain direction into a tiny cylinder with a possible diameter from subnanometer to a few nanometers. Interestingly, the rolling-up direction and diameter or the chirality of an SWCNT determine its fundamental properties. Some SWCNTs have small energy bandgaps, showing semiconducting characteristics, whereas others do not have the bandgap and they are metallic ones. For CNTs with more than one cylindrical shell, the interactions between the shells must be taken into account. Indeed, different CNTs have different properties and application potentials. Until now, CNTs have not only attracted enormous research interest but also stimulated CNT-related applications and industrial development. This can be seen through the facts that more than 77,000 articles about CNTs have been published (ISI database, February 2012) and many CNT-relevant products are available on the market.

Extensive study of CNTs began in 1991 when very thin CNTs were observed by S. Iijima. Although tremendous progress has been made in understanding the fundamental properties of CNTs, optimizing CNT synthesis conditions and post-treatment processes, and exploring various potential applications, etc., are still very hot and active topics nowadays. The electrical and optical properties of SWCNTs are dominated by their chiralities. To obtain chirality-pure SWCNTs has been a dream for a long time. Precisely controlling CNT synthesis parameters and carefully selecting the catalysts used are confirmed to be the key factors that affect the chirality distribution of the grown CNTs. A large number of post-treatment processes have been reported to sort CNTs in terms of their semiconducting or metallic properties, their thickness and lengths, etc. The assembly of CNTs in desirable and controllable ways is another technical challenge. Many unique superior properties, say high electron mobility in electronic applications, high Young's modulus and yield strength in mechanic applications, etc., can be brought into play only when CNTs are uniformly oriented. Major advances in handling CNTs have been propelled by enormous interest in assembling CNTs. In contrast, when the orientation of CNTs is not a critical factor to be considered, CNT random network can be directly incorporated into the devices that could be printed on a soft substrate in a large scale at a low fabrication cost. SWCNTs can be regarded as a one-dimensional material that forms a perfect platform on which experimentalists could explore the mystery of physical puzzles observed only in a low-dimensional object. Super high sensitivity of CNTs, plus their high stability in most chemical environments and high compatibility to most biological materials, promotes CNTs to be utilized as the sensing elements in a wide variety of sensors for chemical and biological detections. Typically, CNT sensors are of much higher sensitivities than their counterparts made from other materials. However, how to improve the selectivity of CNT sensors is a big concern. CNT/polymers and CNT/metal composites have emerged as a class of new functional materials that are reinforced through embedded CNT network in favor of not only the composites' mechanical properties but also electrical properties. Besides, the study of CNTs and their composites in damping and viscoelastic properties is attracting more research attention. The investigation of these properties could lead to a feasible solution to dealing with the vibration damping in micro-/nanodevices. CNTs are not totally "black." Semiconducting CNTs are of direct bandgaps, and this greatly facilitates CNT-based optical and optoelectronic applications and opens the field of CNT nanophotonics.

This book collects 16 state-of-the-art chapters that cover the fundamental properties, relevant technologies and potential applications of CNTs. It gives an overview of the current status of the research and development activities of CNTs. It will be a valuable reference for scientists, researchers, engineers and students who wish to know more about CNTs. It will appeal to anyone involved in nanodevices, nanomaterials and nanofabrication.

Chapter 1 addresses the crystal structure and electronic band structure of SWCNTs. The density of states and singularities are presented to facilitate the discussion on the electronic and optical properties and basic phonon features of SWCNTs. Chapter 2 reviews various CNT synthesis techniques, especially chemical vapor deposition techniques, by which CNTs could grow at much lower temperatures compared with other commonly used methods. New approaches to efficiently growing high-quality CNTs at a large scale, macroscopically long CNTs, directed-grown CNTs, and chirality-pure CNTs are reviewed. The state-of-the-art synthetic assembly and post-synthetic assembly of CNTs are summarized in Chapter 3. It covers not only the assembly/alignment processes at a small scale for laboratory research purposes, but those for industrial production. Orientated CNTs are achievable in several post-synthetic treatments through electrical force, dielectrophoretic force, shear adhesion force, etc. Chapter 4 focuses on separating semiconducting CNTs from metallic ones. Several separation approaches with remarkably high capabilities are introduced in terms of the separation mechanisms and technical details. The electron transport properties and the electrical conductance of CNTs in response to electrical fields in the geometries of single-electron transistors, field-effect transistors, etc., are reviewed in Chapter 5. High carrier mobility is one of the key merits for CNT high frequency/speed electronic applications. The fabrication and high-frequency characterization of CNT field-effect transistors are summarized in Chapter 6. It has been demonstrated that tunneling field-effect transistors are capable of breaking through the smallest possible subthreshold swing for conventional metal–oxide–semiconductor field-effect transistors. In Chapter 7, the recent developments in CNT-based tunneling field-effect transistors are highlighted. High degree of CNT orientation may not be critical for some applications. A large category of CNT electronic devices are directly built on CNT random networks. Chapter 8 introduces several main techniques, including selective elimination or destruction of metallic CNTs and photolithography-assisted stripping to enhance the semiconducting behaviors of CNT networks. Metallic and multiwall CNTs have high electrical conductivities, and they also have extremely high aspect ratios. These make CNTs a very good material for electron field emitters. A comprehensive review on the recent development of CNT field emission technology for vacuum electronic applications in medical x-ray imaging and radiotherapy is presented in Chapter 9. There have been some arguments on ultrahigh temperature superconductivity of multiwall CNTs (MWCNTs). Chapter 10 presents several unusual magnetic experimental findings from Ni magnetic nanoparticles embedded in MWCNTs. The arguments on ultrahigh-temperature superconductivity of MWCNTs are revisited. The properties of CNTs are highly sensitive to the details how they interact with their surroundings. Based on these, a variety of CNT sensors have been developed. In Chapter 11, the electrocatalytic properties of CNTs and recent advances in CNT-based electrochemical biosensors are presented. In addition, CNT-based gas sensors and their sensing mechanisms are discussed in Chapter 12. Several measures to enhance the selectivity of CNT sensors are also highlighted. CNTs have excellent mechanical properties. The strength and weaknesses of applying CNTs and their composites to damping and viscoelastic applications are addressed in Chapter 13. Recent efforts have been devoted to spin and assemble CNTs into continuous lightweight and high-performance fibers. Various spinning methods, accompanied with different functionalities during spinning, are described in Chapter 14. The current status, future research focuses and challenges of several major CNT mechanical applications, including CNT nanomechanics and CNT actuators, are presented in Chapter 15. Finally, the origin of the optical nonlinearity of CNTs, the nonlinear effects, and potential CNT nonlinear applications are presented in Chapter 16.

I am thankful to all the authors for their substantial contributions to this book. I would like to express my gratitude to Pan Stanford Publishing for their kind support and assistance.

Qing Zhang
Singapore, February 2012

Contents

Chapter 1

Fundamental Properties of Carbon Nanotubes

Jianping Zou and Qing Zhang
Microelectronics Centre, School of Electrical and Electronic Engineering, Nanyang Technological University, 50 Nanyang Avenue, 639798, Singapore
jpzou@ntu.edu.sg, eqzhang@ntu.edu.sg

Carbon materials exist in multiple forms such as diamond, graphite, fullerenes, and carbon nanotubes (CNTs). These many different carbon materials arise because carbon atoms can form several hybridized orbitals and achieve relatively stable structures with different valence bonding. Geometrically, a CNT can be constructed by rolling a piece of graphene to create a seamless nanometer-scale cylinder. Depending on the number of the shells, CNTs can be grouped into single-wall CNTs and multi-wall CNTs. Single-wall CNTs can further be classified as semiconducting and metallic CNTs according to their chiralities. In this chapter, we summarize the atomic and electronic structures of CNTs and describe their basic properties that will be utilized in the following chapters for various applications.

Carbon Nanotubes and Their Applications
Edited by Qing Zhang

ISBN 978-981-4241-90-8 (Hardcover), 978-981-4303-18-7 (eBook)
www.panstanford.com

1.1 Bonding Between Carbon Atoms

The reason why carbon has many different structural forms is that carbon atoms can constitute multiple distinct types of valence bonds referred to as the hybridization of atomic orbitals.

Carbon is the sixth element of the periodic table and has the lowest atomic number of any elements in column IV of the periodic table. Each carbon atom has four valence electrons in $2s$, $2p_x$, $2p_y$, and $2p_z$ atomic orbitals, which are important in constituting covalent bonds in carbon materials. In comparison with the binding energy of the valence bonds, the energy difference between the $2p$ energy levels and the $2s$ energy level in carbon atom is small. Therefore, the wave functions for these four valence electrons can easily mix with each other, leading to the redistribution of these four electrons in the $2s$ and three $2p$ atomic orbitals so as to raise the binding energy of the carbon atom with its nearest neighbors. The mixture of $2s$ and $2p$ atomic orbitals is called the hybridization of orbitals, and the mixture of a single $2s$ electron with one, two, or three $2p$ electrons is called sp^n hybridization with $n = 1, 2, 3$ [1, 2]. Thus the $2s$ and $2p$ atomic orbitals can hybridize to form three possible hybridizations: sp, sp^2, and sp^3. The different bonding states induce certain structural arrangements, so that sp bonding gives rise to one-dimensional chain structures, sp^2 bonding to planar structures and sp^3 bonding to regular tetrahedral structures.

Each carbon atom has $(n + 1)$ σ bonds in sp^n hybridization. These σ bonds form a certain n-dimensional structural arrangements. For example, two σ bonds form a one-dimensional (1D) chain structure in sp hybridization. In sp^2 hybridization, three σ bonds form a planar structure in graphite. Because graphite has strong in-plane trigonal bonding, it is the stable ground state of carbon under ambient conditions [3, 4]. The sp^3 hybridization gives a regular three-dimensional (3D) tetrahedron called diamond structure.

In addition to the bulk phase of carbon, small carbon clusters have attracted much attention in the past 20 years [5]. Fullerenes and CNTs were discovered in 1985 [6] and 1991 [7], respectively. In fullerenes and CNTs, there exist planar local structures due to the sp^2 hybridization. Therefore, a CNT can be described as a graphene layer (a single layer of graphite) rolled into a cylindrical shape with axial symmetry.

As shown in Figure 1.1, CNTs usually exist in two kinds of forms: (1) single-wall CNT (SWCNT) consisting of only one graphene layer, and (2) multi-wall CNT (MWCNT) consisting of a nested coaxial array of SWCNTs, separated from one another by approximately 0.34 nm, slightly larger than the interlayer spacing of graphite.

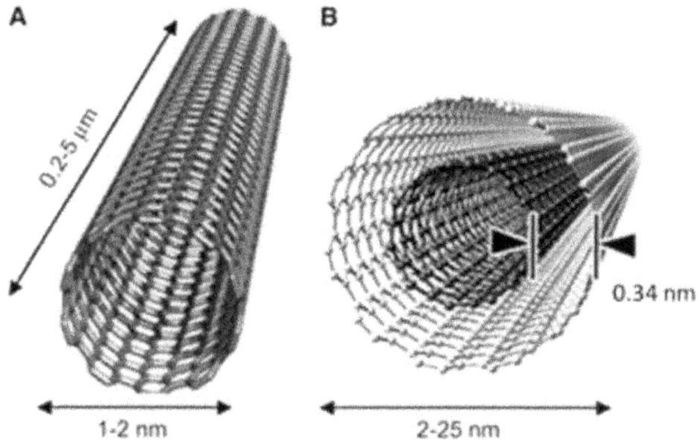

Figure 1.1 Schematic diagram of a single-wall carbon nanotube (SWCNT) (A) and a multi-wall carbon nanotube (MWCNT) (B) showing typical dimensions of length, diameter, and separation distance between two graphene layers in MWCNTs. Adapted from [8, 9].

1.2 Structure of a Single-Wall Carbon Nanotube

A single-wall carbon nanotube (SWCNT) can be described as a graphene layer rolled up to form a hollow cylinder [10]. The diameters of most of the SWCNTs are less than 2 nm. CNTs can be regarded as 1D nanostructure due to the high aspect ratio (i.e., the length over diameter).

Figure 1.2 shows the schematic diagram of a graphene lattice. The basis vectors $\boldsymbol{a}_1$ and $\boldsymbol{a}_2$ generate the graphene lattice. $|\boldsymbol{a}_1| = |\boldsymbol{a}_2| = a = \sqrt{3}\,a_{C\text{-}C}$, where $a_{C\text{-}C} = 0.142$nm is the nearest-neighbor distance between two carbon atoms. The vector $\boldsymbol{C}_h$ defines the circumference of a CNT. The graphene sheet is rolled up along the direction of $\boldsymbol{C}_h$ to form the CNT. Therefore, the vector $\boldsymbol{C}_h$ is called chiral vector of the CNT. Vector $\boldsymbol{T}$, called translation vector, is parallel to the nanotube axis, but normal to the chiral vector $\boldsymbol{C}_h$, corresponding to the first lattice point of the 2D graphene sheet through which the vector $\boldsymbol{T}$

passes. The rectangle generated by $\boldsymbol{C}_h$ and $\boldsymbol{T}$ (as shown in Figure 1.2) gives the unit cell of the CNT in real space.

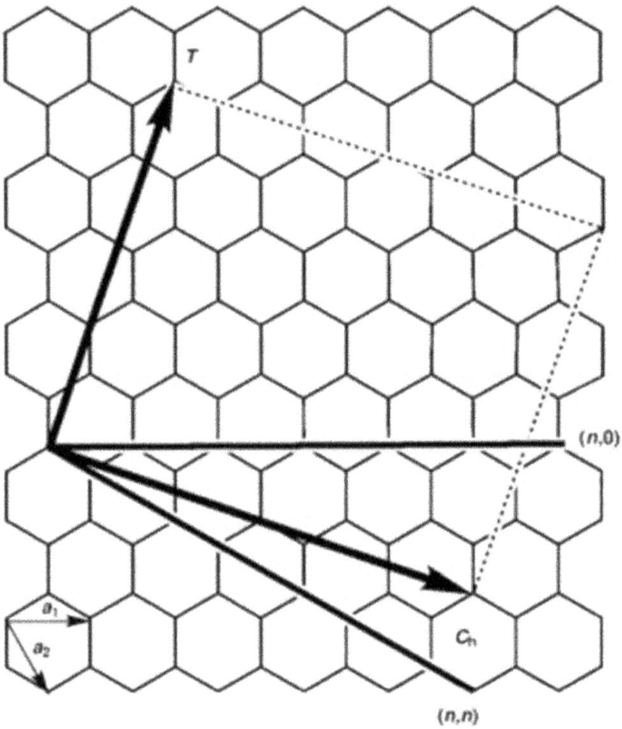

Figure 1.2 The unrolled CNT lattice. $\boldsymbol{a}_1$ and $\boldsymbol{a}_2$ are the lattice vectors of graphene, with $|\boldsymbol{a}_1| = |\boldsymbol{a}_2| = a = \sqrt{3}\, a_{C\text{-}C}$, where $a_{C\text{-}C} = 0.142$ nm is the carbon-carbon bond length. The chiral vector $\boldsymbol{C}_h$, defines the circumference of the CNT. The translation vector $\boldsymbol{T}$ is parallel to the CNT axis and is perpendicular to the chiral vector $\boldsymbol{C}_h$. Adapted from [8].

$\boldsymbol{C}_h$ can be defined using the basis vectors $\boldsymbol{a}_1$ and $\boldsymbol{a}_2$ (see Figure 1.2) of the graphene lattice:

$$\boldsymbol{C}_h = n\boldsymbol{a}_1 + m\boldsymbol{a}_2 \equiv (n, m) \tag{1.1}$$

where n and m are integers, and $0 \leq |m| \leq n$. (n, m) is called the chiral index or chirality of the CNT.

The diameter of the CNT, d_t, is given by $|\boldsymbol{C}_h|/\pi$:

$$d_t = |\boldsymbol{C}_h|/\pi = (a/\pi)\sqrt{n^2 + m^2 + nm} \tag{1.2}$$

As $\boldsymbol{C}_h$ can be defined with any possible combination of n and m, CNTs can be rolled from a graphene sheet in many ways. This suggests many possible structures for CNTs even though the CNTs are nanometer-size hollow cylinders. CNTs can be categorized as either achiral or chiral based on the primary symmetry. If the mirror image of a CNT is identical to the original one, it is an achiral CNT. Figure 1.3a, b shows two kinds of achiral nanotubes, i.e., armchair and zigzag nanotubes. These two kinds of achiral CNTs are named from the shape of the cross-sectional ring, as shown at the edge of the nanotubes in Figure 1.3a, b, respectively. Figure 1.3c shows the schematic diagram of a chiral nanotube. Chiral nanotubes show spiral symmetry and their mirror images are not identical to the original ones.

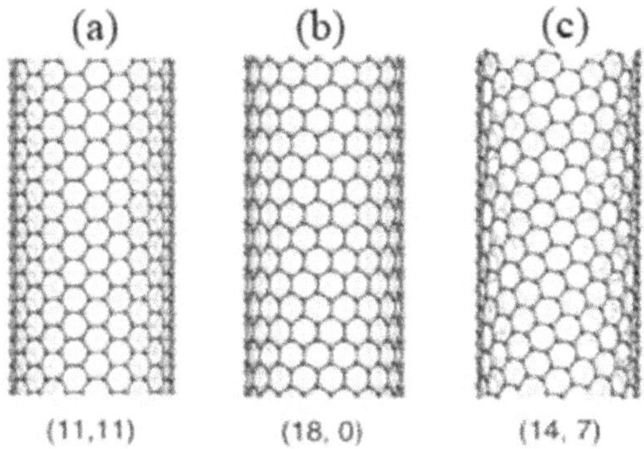

Figure 1.3 Classification of CNTs: (a) (11, 11) armchair CNT, (b) (18, 0) zigzag CNT, and (c) (14, 7) chiral CNT. Adapted from [9].

Two kinds of the chiral vector $\boldsymbol{C}_h$ deserve special mention. One is zigzag CNTs, whose chiral vectors lie purely along one of the two basis vectors, $\boldsymbol{C}_h = (n, 0)$, see Figure 1.3. The other is armchair CNTs, whose chiral vectors are along the direction exactly between the two basis vectors, $\boldsymbol{C}_h = (n, n)$. All other (n, m) CNTs with $n \neq m \neq 0$ correspond to chiral nanotubes.

1.3 Electronic Structure of Single-Wall Carbon Nanotubes

A SWCNT can be viewed as a graphene sheet, rolled up to form a 1D hollow cylinder. Thus, the band structure of a SWCNT roots from that of graphene. As shown in Figure 1.4a, the hexagonal arrangement of carbon atoms in real space generates the lattice structure of graphene. Isolated carbon atom has six electrons with orbital occupancy $1s^2 2s^2 2p^2$ and thus has four valence electrons in $2s$, $2p_x$, $2p_y$, and $2p_z$ atomic orbitals. When carbon atoms form graphene, three atomic orbitals, $2s$, $2p_x$, and $2p_y$, hybridize into three sp^2 orbitals in the plane of the graphene sheet, leaving an unsaturated p_z orbital, which remains perpendicular to the graphene sheet. This p_z orbital forms a delocalized π network out of the plane of the graphene sheet. The hybridized orbitals form σ bonds between the adjacent carbon atoms in the plane of the graphene sheet. Generally, the carriers near the Fermi level dominate electrical transport properties. In graphene, the delocalized π orbitals lie near the Fermi level, determining the electrical transport properties. Next we will discuss these electronic properties starting from a tight-binding model for graphene [11].

1.3.1 Graphene Electronic Structure

Graphene is a single layer of graphite, a 2D layered hexagonal lattice of carbon atoms. In 3D graphite, because the inter-layer separation of 0.34 nm is much larger than the nearest-neighbor distance ($a_{C\text{-}C}$ = 0.142 nm) between two carbon atoms, the intra-layer interactions are larger than that between two adjacent layers. Therefore, the electronic structure of 2D graphene can be viewed as a first approximation of that for 3D graphite.

In real space, graphene has a unit cell with two nonequivalent carbon atoms, A and B, as shown in Figure 1.4a. Figure 1.4b shows the reciprocal space representation of graphene with two unit vectors, $\boldsymbol{b}_1$ and $\boldsymbol{b}_2$, and high symmetry points (Γ, K, M). The directions of the unit vectors, $\boldsymbol{b}_1$ and $\boldsymbol{b}_2$, are rotated by 90° from the real space unit vectors, $\boldsymbol{a}_1$ and $\boldsymbol{a}_2$. The Brillouin zone of graphene shows high symmetry. The high symmetry points, Γ, K, and M are located at the center, the corner, and the center of the edge, respectively. The energy dispersion relations usually can be computed along the triangle ΓMK shown by the dashed lines in Figure 1.4b. Here we consider only π energy bands for graphene, because the delocalized π orbitals are the most important for determining the electrical transport properties of graphene. A tight-binding calculation can provide important insights for understanding the band structures of π electrons for graphene.

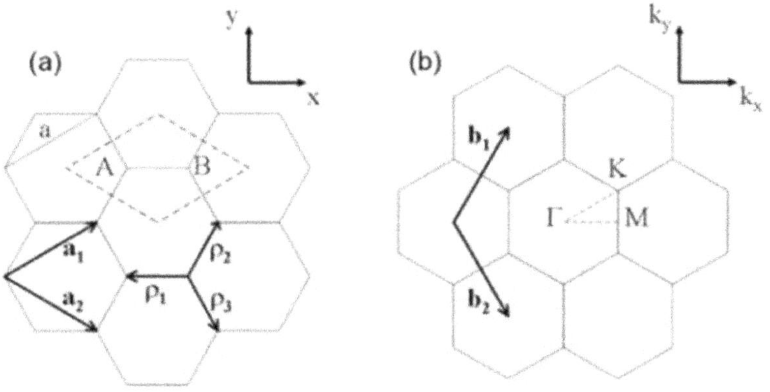

Figure 1.4 (a) A 2D graphene lattice in real space. The dashed rhombus region shows a unit cell with two carbon atoms (A and B). $\boldsymbol{a}_1$ and $\boldsymbol{a}_2$ are unit vectors with length of "a". Vectors $\boldsymbol{\rho}_i$ are defined to connecting the nearest neighbor carbon atoms. (b) Reciprocal space representation of a graphene lattice. $\boldsymbol{b}_1$ and $\boldsymbol{b}_2$ are two reciprocal lattice unit vectors. (Γ, K, M) are high-symmetry points in a Brillouin zone. Adapted from [10].

In order to obtain the π orbital band structure of the graphene, we start from the solution to the Schrödinger equation:

$$\boldsymbol{H}\Psi = E\Psi \tag{1.3}$$

where $\boldsymbol{H}$ is the Hamiltonian, E is the π electron energy of graphene, and Ψ is the total wave function. Since graphene is a periodic system, the total wave function commonly can be expressed by a linear combination of Bloch functions u_i with a periodicity of the lattice. A tight binding, Bloch function u_i is given by a linear combination of atomic wave functions. Because the π orbitals arise from the $2p_z$ orbital of carbon atom, we construct the Bloch function u_i for each atom from $2p_z$ orbitals of atom A and B in a unit cell as

$$u_{A(B)} = \frac{1}{\sqrt{N}} \sum_{A(B)} e^{i\boldsymbol{k}\cdot\boldsymbol{r}_{A(B)}} \varphi(\boldsymbol{r} - \boldsymbol{r}_{A(B)}), \tag{1.4}$$

where $\varphi(\boldsymbol{r})$ denotes the atomic orbital $2p_z$ wave function for an isolated carbon atom. The total wave function Ψ in Equation 1.3 is then a linear combination of u_A and u_B:

$$\Psi = C_A u_A + C_B u_B \tag{1.5}$$

By substituting Equation 1.5 into Equation 1.3, the Schrödinger equation can be written by a matrix form as follows:

$$\begin{pmatrix} \boldsymbol{H}_{AA} & \boldsymbol{H}_{AB} \\ \boldsymbol{H}_{BA} & \boldsymbol{H}_{BB} \end{pmatrix} \begin{pmatrix} C_A \\ C_B \end{pmatrix} = E \begin{pmatrix} \boldsymbol{S}_{AA} & \boldsymbol{S}_{AB} \\ \boldsymbol{S}_{BA} & \boldsymbol{S}_{BB} \end{pmatrix} \begin{pmatrix} C_A \\ C_B \end{pmatrix} \tag{1.6}$$

where $\boldsymbol{H}_{ij}$ and $\boldsymbol{S}_{ij}$ are called transfer integral matrices and overlap integral matrices, respectively, which are defined by

$$\boldsymbol{H}_{ij} = \langle u_i | \boldsymbol{H} | u_j \rangle, \boldsymbol{S}_{ij} = \langle u_i | u_j \rangle \tag{1.7}$$

If we assume that the atomic wave function is normalized, $\boldsymbol{S}_{AA} = \boldsymbol{S}_{BB} = 1$, $\boldsymbol{S}_{AB} = \boldsymbol{S}_{BA}{}^*$, then Equation 1.6 is simplified to

$$\begin{pmatrix} \boldsymbol{H}_{AA} - E & \boldsymbol{H}_{AB} - E\boldsymbol{S}_{AB} \\ \boldsymbol{H}_{BA} - E\boldsymbol{S}_{BA} & \boldsymbol{H}_{BB} - E \end{pmatrix} \begin{pmatrix} C_A \\ C_B \end{pmatrix} = \begin{pmatrix} 0 \\ 0 \end{pmatrix} \tag{1.8}$$

This matrix equation has a nontrivial solution only when

$$\begin{vmatrix} \boldsymbol{H}_{AA} - E & \boldsymbol{H}_{AB} - E\boldsymbol{S}_{AB} \\ \boldsymbol{H}_{BA} - E\boldsymbol{S}_{BA} & \boldsymbol{H}_{BB} - E \end{vmatrix} = 0 \tag{1.9}$$

where Equation 1.9 is called the secular equation. Due to the symmetry of the graphene lattice, atoms A and B cannot be distinguishable, leading to $\boldsymbol{H}_{AA} = \boldsymbol{H}_{BB}$ and $\boldsymbol{H}_{AB} = \boldsymbol{H}_{BA}{}^*$. Then, the energy eigenvalues E are given by

$$E = \frac{\boldsymbol{H}_{AA} \pm |\boldsymbol{H}_{AB}|}{1 \pm |\boldsymbol{S}_{AB}|} \tag{1.10}$$

$\boldsymbol{H}_{AA}$ (= $\boldsymbol{H}_{BB}$) can be calculated by substituting Equation 1.4 into Equation 1.7 as follows:

$$\boldsymbol{H}_{AA} = \frac{1}{N} \sum_{A} \sum_{A^*} e^{ik \cdot (r_A - r_{A^*})} \int \varphi^*(\boldsymbol{r} - \boldsymbol{r}_A) H \varphi(\boldsymbol{r} - \boldsymbol{r}_{A^*})\, d\tau = E_0, \tag{1.11}$$

where E_0 is the orbital energy of the 2p level.

The largest contribution to $\boldsymbol{H}_{AB}$ arise when atoms A and B are the nearest neighbors. Thus,

$$\boldsymbol{H}_{AB} = \frac{1}{N} \sum_{A} \sum_{B} e^{ik \cdot (r_A - r_B)} \int \varphi^*(\boldsymbol{r} - \boldsymbol{r}_A) H \varphi(\boldsymbol{r} - \boldsymbol{r}_B)\, d\tau$$

$$= \frac{1}{N} \sum_{j} e^{ik \cdot \rho_j} \int \varphi^*(\boldsymbol{r}) \boldsymbol{H} \varphi(\boldsymbol{r} - \boldsymbol{\rho}_j) \mathrm{d}\tau \tag{1.12}$$

where $\boldsymbol{\rho}_j$ is a vector connecting the nearest-neighbors between the A and B atoms (see Figure 1.4a). According to the coordinate system of the graphene in Figure 1.4a,

$$\boldsymbol{H}_{AB} = (e^{ik \cdot \rho_1} + e^{ik \cdot \rho_2} + e^{ik \cdot \rho_3}) \int \varphi^*(\boldsymbol{r}) \boldsymbol{H} \varphi\, (\boldsymbol{r} - \boldsymbol{\rho}_1) \mathrm{d}\tau$$

$$= t\left(e^{ik_x a/\sqrt{3}} + 2e^{-ik_x a/2\sqrt{3}} \cos\left(\frac{k_y a}{2}\right)\right), \tag{1.13}$$

where $t = \int \varphi^* (\boldsymbol{r})\boldsymbol{H}\varphi(\boldsymbol{r} - \boldsymbol{\rho}_1)\mathrm{d}\tau$, is called the transfer integral, which describes the strength of exchange interaction between the nearest-neighbor atoms.

The overlap integral matrices $\boldsymbol{S}_{AB}$ can be calculated by a similar method as used for $\boldsymbol{H}_{AB}$,

$$\boldsymbol{S}_{AB} = s\left(e^{ik_x a/\sqrt{3}} + 2e^{-ik_x a/2\sqrt{3}} \cos\left(\frac{k_y a}{2}\right)\right) \tag{1.14}$$

where $s = \int \varphi^* (\boldsymbol{r})\varphi(\boldsymbol{r} - \boldsymbol{\rho}_1)\mathrm{d}\tau$, is called the overlap integral between the nearest A and B atoms. Then, from Equations 1.11, 1.13, and 1.14, the energy dispersion in Equation 1.10 can be obtained as a function of $w(\boldsymbol{k})$:

$$E = \frac{E_0 \pm tw(\boldsymbol{k})}{1 \pm sw(\boldsymbol{k})} \tag{1.15}$$

In Equation 1.15, the negative signs give anti-bonding π^* orbitals (conduction bands of graphene), while the positive signs denote bonding π orbitals (valence bands of graphene). The function $w(\boldsymbol{k})$ is given by

$$w(\boldsymbol{k}) = \sqrt{1 + 4\cos\frac{\sqrt{3}k_x a}{2}\cos\frac{k_y a}{2} + 4\cos^2\frac{k_y a}{2}} \tag{1.16}$$

In Figure 1.5, the energy dispersion relations of graphene are plotted throughout the Brillouin zone. The inset shows the energy dispersion relations along the high symmetry directions in the reciprocal space with $E_0 = 0$. The upper branch of the energy dispersion curves denotes the π^* anti-bonding band, and the lower branch is the π bonding band. The remarkable feature of the energy dispersion of graphene is that the upper π^* anti-bonding band and the lower π bonding band meet at the six K points at the corners of the Brillouin zone, through which the Fermi energy passes. This suggests graphene is a zero bandgap semiconductor. The zero bandgap at the K points is determined by the symmetry of graphene lattice.

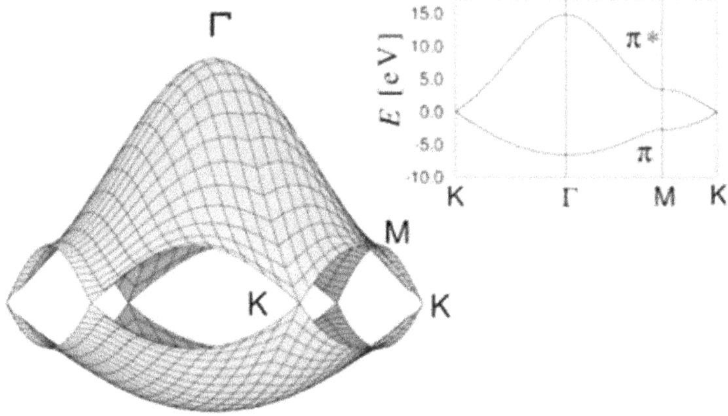

Figure 1.5 Surface plot of the energy dispersion for graphene throughout the whole Brillouin zone. The inset shows the energy dispersion along the high symmetry points of the triangle ΓMK defined in Figure 1.4b. The calculation is based on the parameters $E_0 = 0$, $t = -3.033$eV, and $s = 0.129$ [12]. Adapted from [10].

When the overlap integral $s = 0$, the lower π and upper π^* bands are symmetrical around $E = E_0$. In this case, the energy dispersion relations can be expressed as follows:

$$E = \pm t\left\{1 + 4\cos\frac{\sqrt{3}k_x a}{2}\cos\frac{k_y a}{2} + 4\cos^2\frac{k_y a}{2}\right\}^{1/2} \tag{1.17}$$

Equation 1.17 can be viewed as a simple approximation for the electronic structure of a graphene layer:

1.3.2 Band Structure of SWCNTs from Graphene

The electronic structure of a SWCNT can be simply constructed from that of graphene by imposing a periodic boundary condition along the circumferential direction of the SWCNT. There are two wave vectors related to a SWCNT with infinite length. One is the wave vector $\mathbf{k}_{||}$, which is along the nanotube axis and continuous for a nanotube of infinite length. The other one is the wave vector $\mathbf{k}_{\perp}$, which is along the $\boldsymbol{C}_h$ direction and quantized,

$$\mathbf{k}_{\perp} \cdot \boldsymbol{C}_h = \pi d_t k_{\perp} = 2\pi p \tag{1.18}$$

where d_t is the diameter of a SWCNT and p is an integer. Equation 1.18 describes the periodic boundary condition that $\mathbf{k}_\perp$ should satisfy. By cross-sectional cutting of the energy dispersion of graphene with the quantized values of allowed $k_\perp$ states, we can obtain the 1D band structure of SWCNTs. This method is called zone-folding scheme, as shown in Figure 1.6a. We can find each cross-sectional cutting gives a 1D subband of the SWCNTs. Therefore, the locations and the angles (with respect to the Brillouin surface of graphene) of the cutting lines determine the exact 1D band structures of SWCNTs. Particularly, those $k_\perp$ states that are closest to the K points determine the band structure near the Fermi level and they dominate the electrical transport properties. If the cutting line cuts across a K point of the 2D Brillouin zone as shown in Figure 1.6b, a zero energy gap arises. However, if the cutting line misses a K point as shown in Figure 1.6c, then the SWCNT has a finite energy gap between the valence and conduction bands. Therefore, two different kinds of SWCNTs can be expected depending on the chiral indices which determine the locations and the angles of the cutting lines. Metallic SWCNTs (m-SWCNTs) have a zero bandgap (see Figure 1.6b), and semiconducting SWCNTs (s-SWCNTs) have finite bandgaps (see Figure 1.6c).

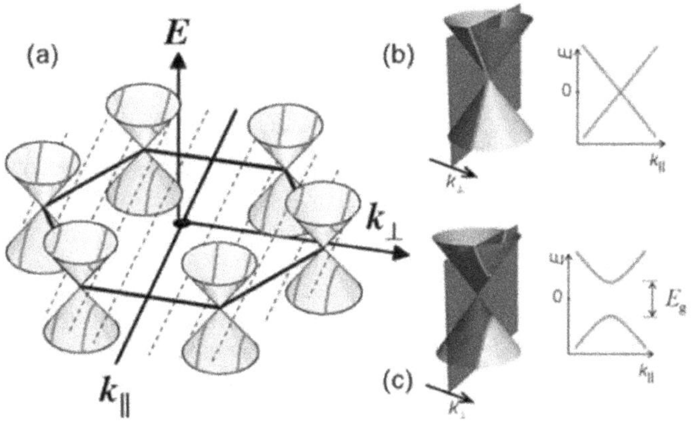

Figure 1.6 (a) Conic energy dispersions near six K points in the first Brillouin zone of graphene. The dashed cutting lines denote the allowed $k_\perp$ states in a SWCNT. The electronic structure of a SWCNT is achieved by cross-sectional cutting of the energy dispersion of graphene with the allowed $k_\perp$ states as indicated. 1D energy dispersions achieved by cross-sectional cutting for (b) a metallic SWCNT and (c) a semiconducting SWCNT. Adapted from [13].

The SWCNTs with the highest symmetry are usually the simplest cases for obtaining energy dispersion relations. For a zigzag $(n, 0)$ SWCNT, $\boldsymbol{C}_h = n\boldsymbol{a}_1$, the periodic boundary condition in Equation 1.18 can be expressed by

$$k_\perp na = 2\pi \mathrm{p} \tag{1.19}$$

If $n = 3q$, and q is an integer, there is an allowed $k_\perp$ which passes through a K point located at $(k_{||}, k_\perp) = (0, 4\pi/3a)$. By substitution,

$$k_\perp = \frac{2\pi p}{na} = \frac{3Kp}{2n} = \frac{Kp}{2q} \tag{1.20}$$

If the integer $p = 2q$, the allowed $k_\perp$ will pass through K points. This kind of zigzag SWCNTs $(n = 3q)$ are always metallic nanotubes with a zero bandgap as shown in Figure 1.6b.

If n is not a multiple of 3, we need to consider two cases. If $n = 3q + 1$, only when $p = 2q + 1$, there is a $k_\perp$, which is closest to K point (see Figure 1.6c):

$$k_\perp = \frac{2\pi p}{na} = \frac{2\pi(2q+1)}{(3q+1)a} = \frac{4\pi}{3a} + \frac{2}{3}\frac{\pi}{(3q+1)a} = \frac{4\pi}{3a} + \frac{2}{3d_t} \tag{1.21}$$

Likewise, if $n = 3q - 1$, the closest allowed $k_\perp$ corresponds to $p = 2q - 1$,

$$k_\perp = \frac{2\pi p}{na} = \frac{2\pi(2q-1)}{(3q-1)a} = \frac{4\pi}{3a} - \frac{2}{3}\frac{\pi}{(3q-1)a} = \frac{4\pi}{3a} - \frac{2}{3d_t} \tag{1.22}$$

For these two cases, the closest allowed $k_\perp$ misses K point by

$$\Delta k_\perp = \frac{2}{3d_t} \tag{1.23}$$

From Equation 1.23, we can find the minimum noncoincidence between an allowed $k_\perp$ and a K point is inversely proportional to the diameter of SWCNTs. Therefore, the bandgap E_g can be calculated from the slope of cones near K points as follows:

$$E_g = 2 \times \left(\frac{\partial E}{\partial k}\right) \times \frac{2}{3d_t} = 2\hbar v_F \left(\frac{2}{3d_t}\right) \approx 0.7eV / d_t(nm) \tag{1.24}$$

where $v_F = 8 \times 10^5$ m/s [14], is the Fermi velocity of electrons in graphene.

Similar argument shows the armchair (n, n) SWCNTs always have metallic behavior. Figure 1.7 shows the 1D dispersion relations for the (5, 5) armchair nanotube and (10, 0) zigzag nanotube, respectively.

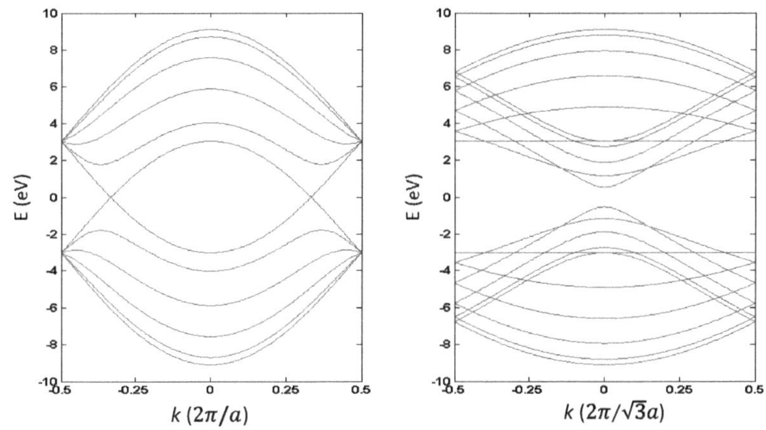

Figure 1.7 1D energy dispersion relations for (a) armchair (5, 5) and (b) zigzag (10, 0) carbon nanotubes (calculated from tight-binding model).

In summary, for a (n, m) SWCNT, when the difference between n and m is divided by 3 (i.e., $n - m = 3q + j$), the remainder j can be used to classify the SWCNTs into three different kinds [10, 15]:

(i) $j = 0$; metallic nanotube with a zero bandgap.
(ii) $j = -1$; semiconducting nanotube with a finite bandgap, $E_g \sim 0.7\text{eV}/d_t(\text{nm})$.
(iii) $j = +1$; semiconducting nanotube with a finite bandgap, $E_g \sim 0.7\text{eV}/d_t(\text{nm})$.

1.3.3 Effect of Curvature on Nanotube Bandgap

Since SWCNT can be viewed as a rolled up graphene layer, in the zone-folding scheme, a periodic boundary condition is applied to the graphene layer in order to obtain electronic structures of the SWCNT. Actually, the SWCNTs have certain curvature due to the cylinder shape. Therefore, we cannot treat π orbitals independently as in the graphene case because the π orbitals are not orthogonal to σ orbitals in the SWCNTs. The curvature of the SWCNTs mixes π and σ orbitals, leading to certain deviations in the electronic structures of nanotubes from the pure π orbital graphene picture. The smaller the nanotube diameter, larger the curvature and larger the deviations. All armchair nanotubes are always metallic regardless of the diameter. In contrast, a small bandgap is induced by curvature, exhibiting quasi-metallic behavior for the SWCNTs satisfying $n - m = 3q$. The small bandgap of quasi-metallic nanotubes is inversely proportional to the square of nanotube diameter. For example, a quasi-metallic nanotube with a diameter of 1 nm has a bandgap of only 40 meV, which is much smaller than that of a semiconducting nanotube with a comparable diameter. The simple π orbital graphene picture cannot describe $\sigma - \pi$ hybridization induced by the curvature of a nanotube. Therefore, small-diameter nanotubes usually require a more refined treatment such as the first principles calculations.

1.3.4 Density of States (DOS) in SWCNTs

The density of states (DOS) is an actual physical quantity measured in many experiments. In general, the DOS originated from a single, doubly degenerate 1D subband $\varepsilon(k)$ can be calculated as [16]

$$n(E) = \frac{\partial N(E)}{\partial E} = \frac{2}{l}\sum_i \int dk\delta(k - k_i)\left|\frac{\partial \varepsilon}{\partial k}\right|^{-1} \tag{1.25}$$

where k_i are the roots of the equation $E - \varepsilon(k_i) = 0$, $l = \int dk$ is the length of the 1D Brillouin zone, and $N(E)$ is the total number of electron states per unit cell at a given energy E. Equation 1.25 can be solved for SWCNTs as [16]

$$n(E) = \frac{2\sqrt{3}a_{c-c}}{\pi^2 rt}\sum_{m=-\infty}^{\infty} g(E, \varepsilon_m), \tag{1.26}$$

where

$$g(E, \varepsilon_m) = \begin{cases} |E|/\sqrt{E^2 - \varepsilon_m^2}, & |E| > |\varepsilon_m| \\ 0, & |E| < |\varepsilon_m| \end{cases} \tag{1.27}$$

It can be seen that $g(E, \varepsilon_m)$ shows a divergent van Hove singularity at the energy band edges ($|E| = |\varepsilon_m|$) for $|\varepsilon_m| \neq 0$, and that $g(E, 0) = 1$. Figure 1.8 shows the density of states for (11, 0) and (12, 0) nanotubes, respectively. Compared with bulk materials, the unique feature of the DOS of SWCNTs is the presence of singularities at the band edges. The presence of these singularities in the density of states has been verified by scanning tunneling microscopy of individual nanotubes [17].

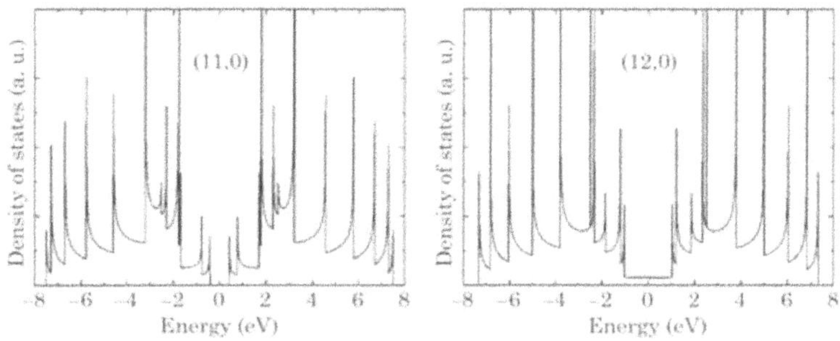

Figure 1.8 Calculated density of states for (11, 0) and (12, 0) carbon nanotubes by tight binding model. Adapted from [18].

1.4 Phonon Modes of Carbon Nanotubes

The electron–phonon interactions and thermal transport properties of a certain material are determined by its phonon spectrum. The phonon dispersion relations of CNTs can be understood based on zone-folding of the phonon spectrum of graphene, same as that used in treating the electronic structure. In this section, the phonon dispersion relations of graphene are presented first. Then, we discuss the basic phonon features of CNTs that will be useful in later chapters.

1.4.1 Phonon Dispersion Relations for Graphene

We begin with a method for calculating the phonon dispersion relations within a force constant model, in which the atoms are connected by springs with different spring constants. In the harmonic approximation, atomic displacements $\boldsymbol{u}_i$ around the equilibrium positions induce energy variation as follows:

$$E = \frac{1}{2} \sum_{i,j,\alpha,\beta} u_i^{\alpha} K_{ij}^{\alpha\beta} u_j^{\beta} \tag{1.28}$$

where the subscript i represents atom i in the lattice, and the Greek indices denote the x, y, z directions. $\boldsymbol{K}_{ij}$ represents the 3×3 force constant tensor between the atom *i* and the atom *j*. The equation of motion for the displacements of atoms is

$$M\frac{\partial^2 u_i^\alpha}{\partial t^2}=-\frac{\partial E}{\partial u_i^\alpha}=-\sum_{j,\beta} K_{ij}^{\alpha\beta} u_j^\alpha, \quad (1.29)$$

where M is the mass of the atoms.

The phonon modes arise from simultaneous displacements of all of the atoms with the same frequency ω. In a periodic system, in order to achieve the normal mode displacements, a Fourier transform of the displacement of the atom *i* with the wave number **k** can be expressed as

$$\hat{u}_i^\alpha(k,t)=\frac{1}{\sqrt{N_\Omega}}\sum_{r_i} u_i^\alpha e^{i(k\cdot r_i-\omega t)}, \quad (1.30)$$

where N_Ω is the atoms number in the considered volume and $\boldsymbol{r}_i$ is the original position of the atom *i*. The dynamical equation 1.29 can be expressed by the Fourier components as follows:

$$(M\omega^2 I-\Sigma_{j,\beta}K_{ij}^{\alpha\beta})\hat{u}_i^\alpha+\Sigma_{j,\beta}K_{ij}^{\alpha\beta}e^{ik\cdot\Delta R_{ij}}\hat{u}_j^\beta=0 \quad (1.31)$$

where $\boldsymbol{I}$ is a 3×3 unit matrix and $\Delta\boldsymbol{R}_{ij}=\boldsymbol{r}_i-\boldsymbol{r}_j$ is the relative coordinate of the atom *i* with respect to the atom *j*. The force constant tensor $\boldsymbol{K}_{ij}$ describes the coupling relations of the vibration of the atom *i* and the atom *j*.

Equation 1.31 can be written by defining a $3N \times 3N$ dynamical matrix $\boldsymbol{D}(\boldsymbol{k})$:

$$\boldsymbol{D}(\boldsymbol{k})\boldsymbol{u}_{\boldsymbol{k}}=0, \quad (1.32)$$

where the vector $\boldsymbol{u}_{\boldsymbol{k}}$ is a column vector with components ${}^t(\hat{u}_1^x,\hat{u}_1^y,\hat{u}_1^z,\hat{u}_2^x,\hat{u}_2^y,\hat{u}_2^z,\cdots\hat{u}_N^x,\hat{u}_N^y,\hat{u}_N^z)$ for a given $\boldsymbol{k}$, in which the superscript t represents the transpose of a row vector into a column vector, and vice versa. The dynamical matrix gives $3N$ equations for the N atoms in the first Brillouin zone. This matrix can be divided into small 3×3 matrices $D_{ij}^{\alpha\beta}$, which is denoted as

$$D_{ij}^{\alpha\beta}=(M\omega^2 I-\Sigma_{l'}K_{il'}^{\alpha\beta})\delta_{ij}\delta_{\alpha\beta}+\Sigma_l K_{il}^{\alpha\beta}e^{ik\cdot\Delta R_{il}}, \quad (1.33)$$

where the sum over l' is treated for all neighbor atoms from the atom *i* with $\boldsymbol{K}_{il'}\neq 0$, and the sum over l is treated for the equivalent atoms to the atom *j*.

In order to obtain the eigenvalues $\omega(\boldsymbol{k})$ and the eigenvectors $\boldsymbol{u}_k$, we need to solve the secular equation det$\boldsymbol{D}(\boldsymbol{k}) = 0$ for a given $\boldsymbol{k}$ vector. First, the force constant tensors $\boldsymbol{K}_{ij}$ need to be constructed. Although all atoms in the unit cell are coupled, we still can simplify the phonon modes calculation by considering the interaction to a few neighbor atoms. The calculations for graphene with considering the interactions up to the fourth neighbor atoms are consistent with the experimental data [10]. The force constant tensor between two atoms that lie on the x axis is given by

$$\boldsymbol{K}^n = \begin{pmatrix} \phi_R^n & 0 & 0 \\ 0 & \phi_{IPT}^n & 0 \\ 0 & 0 & \phi_{OPT}^n \end{pmatrix} \tag{1.34}$$

where ϕ_R^n, ϕ_{IPT}^n, and ϕ_{OPT}^n denote the force constant parameters in the radial (bond-stretching), in-plane and out-of-plane tangential (bond-bending) directions of the nth nearest-neighbors, respectively. Table 1.1 gives the force constant parameters achieved by fitting the 2D phonon dispersion relations over the Brillouin zone to that measured experimentally [19, 20].

Table 1.1 Force constant parameters for graphene up to the fourth neighbor, in units of eV/Å^2

First neighbor	Second neighbor	Third neighbor	Fourth neighbor
$\phi_R^1 = 22.81$	$\phi_R^2 = 5.50$	$\phi_R^3 = 1.88$	$\phi_R^4 = -1.20$
$\phi_{IPT}^1 = 15.31$	$\phi_{IPT}^2 = -2.02$	$\phi_{IPT}^3 = -3.28$	$\phi_{IPT}^4 = 1.43$
$\phi_{OPT}^1 = 6.14$	$\phi_{OPT}^2 = -0.25$	$\phi_{OPT}^3 = 0.094$	$\phi_{OPT}^4 = -0.36$

Adapted from [10].

The calculated phonon dispersion for 2D graphene using the force constant parameters given by Table 1.1 is shown in Figure 1.9. The phonon dispersion curves show the good agreement with experimental data points, indicating the force constant parameters by considering fourth-neighbor interactions are adequate for reproducing the phonon dispersion relations of 2D graphene.

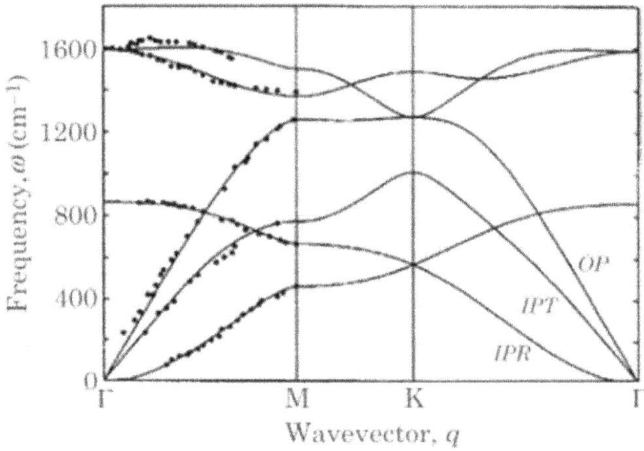

Figure 1.9 Phonon dispersion relations along high symmetry directions, for a 2D graphene, calculated by considering fourth-neighbor interactions. The experimental data points are shown by the black dots. Adapted from [21].

Since there are two carbon atoms per unit cell in graphene and each atom is of three degrees of freedom, there are six phonon branches in total dispersion relations. The three phonon dispersion branches arising from Γ point (see Figure 1.9) are acoustic modes corresponding to an out-of-plane, an in-plane tangential (bond-bending), and an in-plane radial (bond-stretching) mode, as indicated with increasing energy, respectively. The other three branches are optical modes: one out-of-plane mode and two in-plane modes.

1.4.2 Phonon Dispersion Relations for Carbon Nanotubes

The phonon dispersion of CNTs can be determined by zone folding the graphene phonon dispersion, which is the same as that for electrons to obtain the electronic structure [10]. The quantization condition is given by

$$\boldsymbol{k} \cdot \boldsymbol{C}_h = k_x C_{hx} + k_y C_{hy} = 2\pi p. \tag{1.35}$$

For a CNT with unit cell including N atoms, each of the $N/2$ values of p will cut across the six graphene phonon branches to generate $3N$ phonon branches, coincident with the three degrees of freedom for each atom.

Figure 1.10 shows the calculated phonon dispersion relation for a (10, 10) nanotube. Because each unit cell of a (10, 10) nanotube includes 40 atoms, we have 120 phonon branches. However, due to mode degeneracies, there are only 66 explicit phonon branches (12 modes are non-degenerate and 54 modes are doubly degenerate).

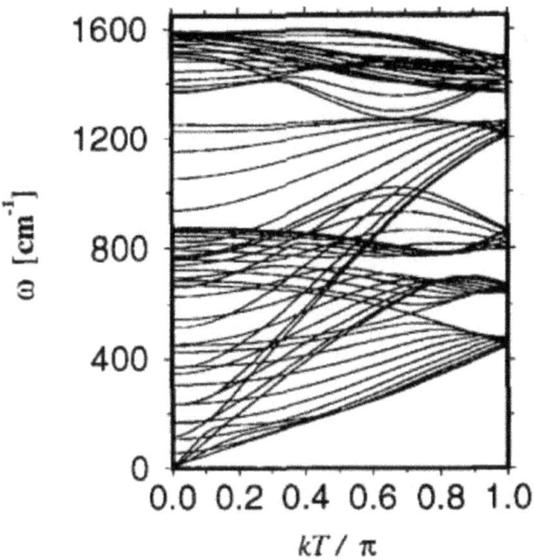

Figure 1.10 The calculated phonon dispersion curves for a (10, 10) armchair CNT. Adapted from [22].

Figure 1.11 shows the low-energy spectrum of the phonon dispersion for the (10, 10) CNT. There are four acoustic modes: two degenerate transverse modes (T) with the lowest energy, a twist mode (TW) and one longitudinal mode (L) with the highest energy. In addition to these four acoustic modes, there is an optical radial breathing mode (RBM) at 20 meV. The inset in Figure 1.11 shows the low-energy phonon density of states of the nanotube (solid line) and that of graphite (dashed line) and graphene (dot-dashed line). The phonon density of states of nanotube shows the presence of singularities at the band edges compared with graphene and graphite. Similar to the variations of the electronic structure, the curvature of CNTs with small diameter usually induces the corrections of the force constant. A detailed discussion of the phonon spectra of CNTs can be found in [22].

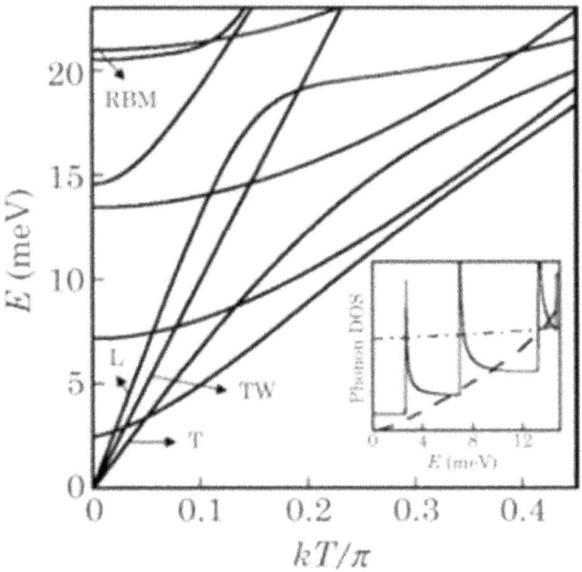

Figure 1.11 Phonon dispersion relations of the (10, 10) CNT near the Γ point. The inset shows the low-energy density of phonon states. Adapted from [22].

1.5 Optical Properties of Carbon Nanotubes

Recently, CNT opto-electronics has attracted more attentions because of the excellent opto-electronic properties of nanotubes. For example, s-CNTs are of direct bandgaps, which facilitate electronic transitions between the valence and conduction bands without involvement of phonons. An almost continuous optical response over a broad spectral range can be achieved by involving many nanotubes with different bandgaps. In addition, CNTs have low defect density, leading to less sensitive to nonradiative processes and significant improvement in the device efficiency that is not reached in traditional bulk materials. In this section, we will briefly discuss the optical properties of CNTs.

1.5.1 Selection Rules

We start with the simplest single-particle picture. Under optical radiation, the Hamiltonian of CNT is given by

$$\boldsymbol{H} = \frac{1}{2m}(\boldsymbol{p} + e\boldsymbol{A})^2 + U \tag{1.36}$$

where $\boldsymbol{p}$ is the electron momentum, m the electron mass, $\boldsymbol{A}$ the vector potential of the optical radiation and U indicates all other interactions. Given that the photon flux is not too large, the Hamiltonian of the system can be written to its first order in the magnetic vector potential as follows:

$$\boldsymbol{H} = \boldsymbol{H}_0 + \frac{e}{m}\boldsymbol{p} \cdot \boldsymbol{A} \tag{1.37}$$

where $\boldsymbol{H}_0$ is the Hamiltonian without the optical radiation. In Equation 1.37, the electron–photon interaction can be viewed as a perturbation induced by the optical radiation and is given by

$$\boldsymbol{H}_{\text{el-ph}} = \frac{e}{m}\boldsymbol{p} \cdot \boldsymbol{A} \tag{1.38}$$

The vector potential can be expressed by

$$\boldsymbol{A}(\boldsymbol{r}, t) = \hat{e}\left(\frac{2\hbar^2\sqrt{\mu}}{N\sqrt{\varepsilon}c\hbar\omega}I\right)^{1/2} e^{i(\boldsymbol{k}\cdot r - \omega t)} \tag{1.39}$$

where $\hat{e}$ is a unit vector in the direction of the light polarization, ε is the dielectric constant and μ the permeability of the CNT. I is the photon flux and ω the optical frequency. c is the light velocity. $\boldsymbol{k}$ is the optical wave vector, along the propagation direction which is normal to the electric and magnetic fields. According to Fermi's golden rule, the time-averaged transition probability between initial and final states can be expressed by

$$\Gamma_{i\to f} = \frac{2\pi}{\hbar}\delta(E_f - E_i - \hbar\omega)\left|\langle f|\boldsymbol{H}_{el-ph}||i\rangle\right|^2 \tag{1.40}$$

where the delta function suggests energy conservation during the transition. For band-to-band transitions induced by optical absorption, this means that an electron is excited from the valence band to the conduction band. In order to calculate the matrix elements, two special cases: light polarized parallel $(\hat{e} = \hat{z})$ and perpendicular $(\hat{e} = \hat{\phi})$ to the CNT axis are considered.

In a CNT the matrix elements corresponding to the electron–photon interaction can be described as

$$\langle q, J, c | \boldsymbol{H}_{\text{el-ph}} | q', J', v \rangle = \left(\frac{2\hbar^2 \sqrt{\mu}}{N\sqrt{\varepsilon} c\hbar\omega} I \right)^{1/2} \frac{e}{m} \langle q, J, c | (\boldsymbol{p} \cdot \hat{e}) e^{i\boldsymbol{k}\cdot\boldsymbol{r}} | q', J', v \rangle \tag{1.41}$$

where $|q, J, v\rangle$ denotes a wave function for an electron with axial wave vector q in valence subband J, and $|q, J, c\rangle$ is for the conduction band. The exponential term in Equation 1.41 can be approximated to be 1 because the nanotube radius is much smaller than the optical wavelengths. Therefore, the matrix elements corresponding to the electron–photon interaction can be written as

$$\langle f | \boldsymbol{H}_{\text{el-ph}} | i \rangle = \left(\frac{2\hbar^2 e^2 \sqrt{\mu}}{m^2 N\sqrt{\varepsilon} c\hbar\omega} I \right)^{1/2} \begin{cases} \langle q, J, c | p_z | q', J', v \rangle & \text{parallel polarization} \\ \langle q, J, c | p_\phi | q', J', v \rangle & \text{perpendicular polarization.} \end{cases} \tag{1.42}$$

According to the wave functions defined in the Section 1.3.1, the momentum matrix elements for the case of parallel polarization are

$$\langle q, J, C | p_z | q', J', v \rangle = \delta_{J,J'} \delta_{q,q'} \operatorname{Re}\Big(\sum_{i=1,2,3} \alpha_i e^{-iq\delta z i} e^{-iJ\delta\phi i} \Big) \tag{1.43}$$

where δz_i and $\delta\phi_i$ denote the axial and azimuthal coordinate differences between the positions of nearest-neighbor atoms. α_i is a coefficient related to spatial dependence of the atomic orbitals. In order to obtain Equation (1.43), we only consider the momentum matrix elements between the nearest-neighbor atoms. Equation 1.43 suggests the momentum ($q = q'$) and energy ($J = J'$) are conserved in the parallel polarization optical transition. Figure 1.12a shows the first two allowed band-to-band transitions for parallel polarization, i.e., E_{11} and E_{22} transitions.

The calculation of the matrix elements in the case of perpendicular polarization is particularly simple. Under the effect of the angular momentum operator, the quantum number J is changed by one unit:

$$\langle q, J, c | p_\phi | q', J', v \rangle \propto \langle q, J, c | q', J' \pm 1, v \rangle = \delta_{q,q'} \delta_{J,J'\pm 1} \tag{1.44}$$

where the ± signs correspond to right- or left-handed polarization, respectively. Figure 1.12b shows the first allowed optical transitions for perpendicular polarization, i.e., E_{12} and E_{21} transitions.

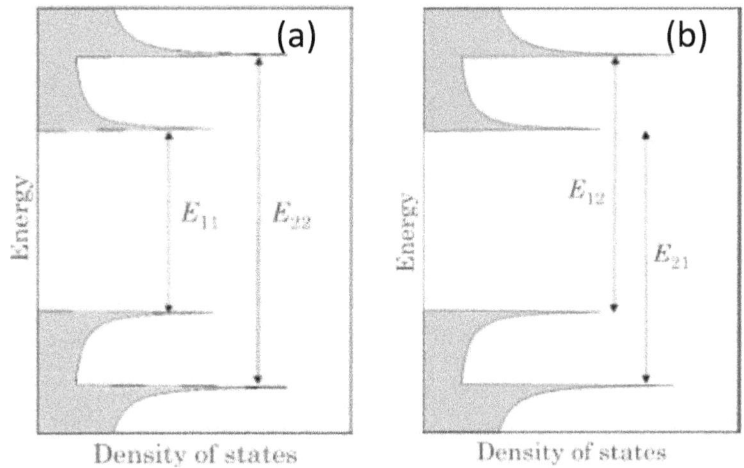

Figure 1.12 The first two band-to-band transitions for light polarized along the CNT axis (a) and perpendicular to the CNT axis (b). Adapted from [23].

1.5.2 Excitons

In general, the optical properties of a certain material can be well described by the single particle picture and band-to-band transitions. However, for 1D material such as CNTs, the optical properties mainly determined by many-body effects. Exciton is the most significant feature of many-body effects. The exciton is usually defined as electron–hole pair bound by the Coulomb interaction. Figure 1.13 shows a conceptual picture of exciton formation. An electron is excited across the bandgap by the absorption of a photon, leaving behind a hole in valence band. A hydrogen-like bound state can be formed between the excited electron and the hole through the Coulomb interaction. In an exciton, the electron and hole are separated by the exciton radius. The attractive Coulomb interaction potential can induce quantization of energy. The exciton binding energy can be defined as the difference between the free electron energy and the quantized energy levels.

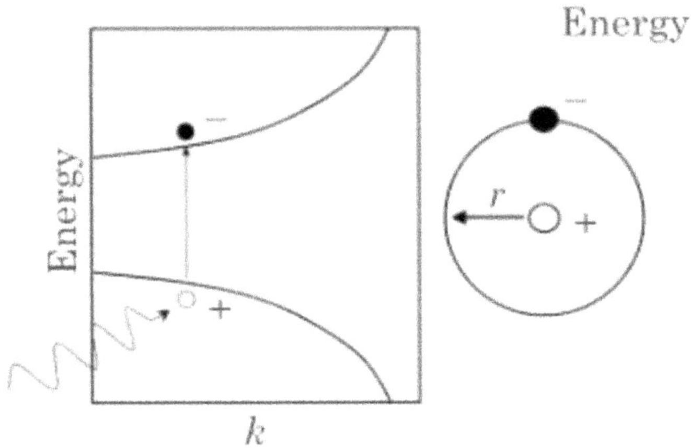

Figure 1.13 Schematic of exciton formation. With photon absorption, an electron is excited from the valence band to the conduction band. The electron can interact with the hole left in the conduction band by the Coulomb attractive interaction. The electron and the hole form a hydrogen-like bound state. Adapted from [23].

As we all know, the lowest energy level of a hydrogen atom can be expressed as

$$E_b^H = -\frac{m_0 e^4}{8h^2 r^2 \varepsilon_0^2} = -13.6 \text{ eV}, \tag{1.45}$$

where m_0 is the free electron mass, ε_0 is the permittivity of free space, and r is the Bohr radius. In solid materials, the effective mass is usually smaller than the free electron mass, and the screening of the Coulomb interaction usually gives rise to a much higher dielectric constant. Thus, as a hydrogen-like system, a simple model assuming the same exciton radius can be used to describe the binding energy of exciton:

$$E_b = -13.6\frac{m^*}{m_0}\left(\frac{\varepsilon_0}{\varepsilon}\right)^2 \tag{1.46}$$

In traditional semiconductors, due to the smaller effective mass and the strong screening of Coulomb interaction between electron and the hole, the binding energy of exciton is generally small. The excitons tend to be relevant only at temperatures much below room temperature. However, in CNTs, the electrostatic screening of Coulomb interaction is weak and the binding energy of excitons is remarkably increased. Theoretical work has predicted that the dielectric constant of s-SWCNT is approximately equal to 1 [24].

Both experiment and theory have proven the significance of excitons in CNTs. Figure 1.14 shows the calculated optical properties of a (8, 0) CNT. The solid line and the dashed line correspond to the calculations with and without electron–hole interactions, respectively. Peaks A, B, and C correspond to photon energies of 2.54, 2.66, and 3.7 eV, respectively, and they are caused by the band-to-band absorption. However, if the electron–hole interactions are considered, each band-to-band transition generates a series of sharp excitonic lines, labeled with A_1', A_2', A_3', B_1', B_2', C_1', C_2'. This suggests that the excitonic absorption can be very important in the optical spectra of s-SWCNTs.

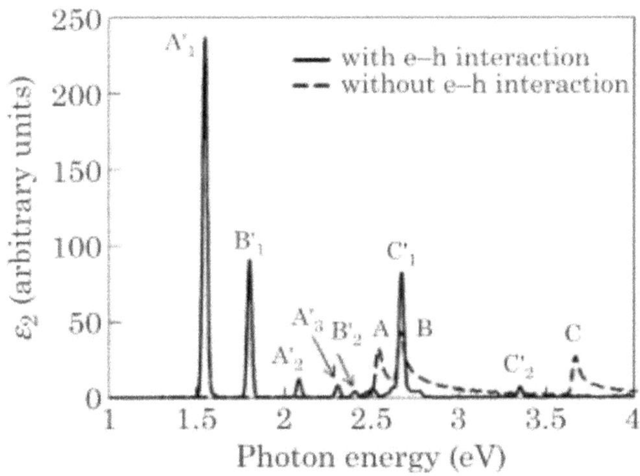

Figure 1.14 The calculated optical absorption spectra for a (8, 0) CNT with (solid line) and without (dashed line) the effect of the electron-hole interactions. Adapted from [25].

Owing to the strong electrostatic screening, excitons usually cannot form in metals. However, m-SWCNTs are exceptions. Excitons can form in m-SWCNTs with fairly large binding energies. Figure 1.15 shows the calculated optical spectra for (3, 3) and (5, 0) SWCNTs. Here the small-diameter (5, 0) nanotube is semi-metallic due to the strong curvature effects. Below the band-to-band peak, there is an exciton peak with the binding energy of 0.1 eV in the (3, 3) nanotube. It has been proven that the metallic screening induces the presence of a single bound exciton in the (3, 3) nanotube, causing an attractive delta-function potential with a single bound state along the nanotube axis [26]. In contrast, the (5, 0) nanotube shows limited influence of excitons on the optical spectra. Actually, in addition to attractive interactions, the electron–hole interaction also includes a repulsive interaction. In the (5, 0) nanotube, the repulsive interaction dominates the interaction, suppressing the appearance of bound exciton states [26].

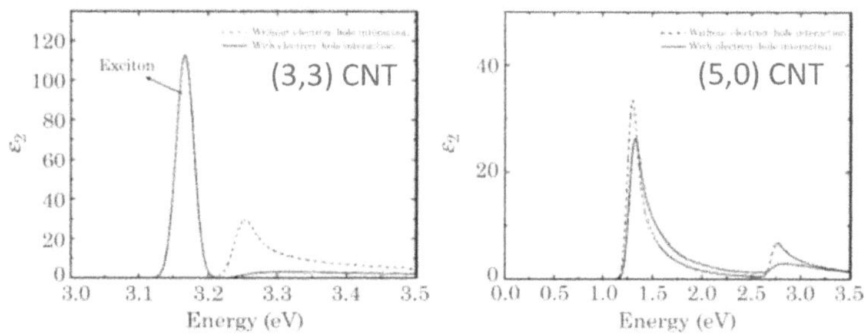

Figure 1.15 The calculated optical absorption spectra for (3, 3) and (5, 0) nanotubes with (solid lines) and without (dashed lines) electron–hole interactions. Adapted from [26].

Recently many experiments have provided obvious evidences for excitonic effects in SWCNTs [27, 28]. For example, two-photon experiments gave an experimental evidence for the excitonic picture. As shown in Figure 1.16, optical selection rules in a two-photon experiment are obviously different from single-photon excitation. The electrons can be excited up to the so-called 2*p* exciton or to the continuum. Then the excited states relax to the lowest energy exciton (1*s*) with emitting light. As shown in Figure 1.16a, the excitation and emission energies are different in the excitonic picture. While in the band-to-band picture as shown in Figure 1.16b, the excitation and emission energies are same. Figure 1.16c shows a color map of the measured emission intensity as a function of two-photon excitation energy. In the band-to-band picture, the excitation and emission energies are the same, and thus peaks corresponding to different SWCNTs should lie along a straight line with a slope of unity (solid red line in the figure). However, the observed peaks are on the line.

This suggests that the emission energy is much lower than the excitation energy due to the excitonic effect in SWCNTs.

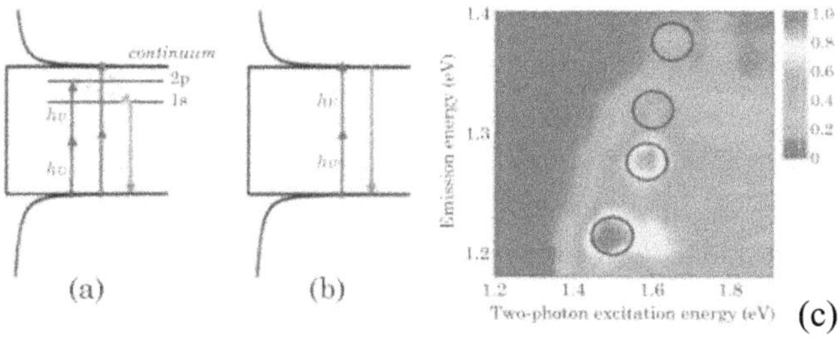

Figure 1.16 Schematic of two-photon excitation (blue lines) and succeeding fluorescence emission (red lines) in the exciton picture (a) and the band-to-band picture (b). (c) Contour plot of the measured one-photon fluorescence emission intensity as a function of the two-photon excitation energy. The solid red line describes equal excitation and emission energies, corresponding to the band-to-band picture. Adapted from [28].

1.6 Summary

In this chapter, we reviewed the electronic band structure of SWCNTs and the associated phonon and optical characteristics. Essential features of the nanotube band structure can be obtained by zone-folding of graphene. A SWCNT can be either metallic with linear dispersions and finite DOS near the Fermi level or semiconducting with a direct bandgap and zero DOS near the Fermi level, depending on its chirality and diameter. Owing to the 1D nature of SWCNTs, van Hove singularities are formed in the DOS. In addition, the phonon spectra of SWCNTs are addressed by applying zone folding scheme to graphene phonon dispersions. The fundamental optical properties of SWCNTs are also discussed in band-to-band transitions and excitonic picture, respectively.

References

1. Dresselhaus, M. S., Dresselhaus, G., Sugihara, K., Spain, I. L., Goldberg, H. A. (1988) *Graphite Fibers and Filaments*, Springer Series in Materials Science, Vol. 5 (Springer, Berlin, Heidelberg).
2. Kelly, B. T. (1981) *Physics of Graphite* (Applied Science, London).

3. Bundy, F. P. (1980). The *P*, *T* phase and reaction diagram for elemental carbon, *J. Geophys. Res.*, **85**, pp. 6930–6936.
4. Bundy, F. P., Bassett, W. A., Weathers, M. S., Hemley, R. J., Mao, H. K., and Goncharov, A. F. (1996). The pressure–temperature phase and transformation diagram for carbon, *Carbon*, **34**, pp. 141–153.
5. Dresselhaus, M. S., Dresselhaus, G., and Eklund, P. C. (1996) *Science of Fullerenes and Carbon Nanotubes* (Academic, New York).
6. Kroto, H. W., Heath, J. R., O'Brien, S. C., Curl, R. F., and Smalley, R. E. (1985). C_{60}: Buckminsterfullerene, *Nature (London)*, **318**, pp. 162–163.
7. Iijima, S. (1991). Helical microtubules of graphitic carbon, *Nature (London)*, **354**, pp. 56–58.
8. Hirsch, A. (2002). Funktionalisierung von einwandigen Kohlenstoffnanorohen, *Angew. Chem.*, **114**, pp. 1933–1939.
9. Iijima, S. (2002). Carbon nanotubes: past, present, and future, *Phys. B.*, **323**, pp. 1–5.
10. Saito, R., Dresselhaus, G., and Dresselhaus, M. S. (1998) *Physical Properties of Carbon Nanotubes* (Imperial College Press, London).
11. Wallace, P. R. (1947). The band theory of graphite, *Phys. Rev.*, **71**, pp. 622–634.
12. Painter, G. S., and Ellis, D. E. (1970), Electronic band structure and optical properties of graphite from a variational approach, *Phys. Rev. B.*, **1**, pp. 4747–4752.
13. Minot, E. (2004) *Tuning the Band Structure of Carbon Nanotubes*, Ph.D. thesis, Cornell University.
14. Kane, C. L., Mele, E. J., Lee, R. S., Fischer, J. E., Petit, P., Dai, H., Thess, A., Smalley, R. E., Verschueren, A. R. M., Tans, S. J., and Dekker, C. (1998). Temperature-dependent resistivity of single-wall carbon nanotubes, *Europhys. Lett.*, **41**, pp. 683–688.
15. Saito, R., Fujita, M., Dresselhaus, G., and Dresselhaus, M. S. (1992). Electronic structure of graphene tubules based on C 60 , *Phys. Rev. B*, **46**, pp. 1804–1811.
16. Mintmire, J. W., and White, C. T. (1998). Universal density of states for carbon nanotubes, *Phy. Rev. Lett.*, **81**, pp. 2506–2509.
17. Wildoer, J. W. G., Venema, L. C., Rinzer, A. G., Smalley, R. E., and Dekker, C. (1998). Electronic structure of atomically resolved carbon nanotubes, *Nature*, **391**, pp. 59–61.
18. Anantram, M. P., and Léonard, F. (2006). Physics of carbon nanotube electronic devices, *Rep. Prog. Phys.*, **69**, pp. 507–561.

19. Aizawa, T., Souda, R., Otani, S., Ishizawa, Y., and Oshima, C. (1990). Bond softening in monolayer graphite formed on transition-metal carbide surfaces, *Phys. Rev. B*, **42**, pp. 11469–11478.

20. Oshirna, C., Aizawa, T., Souda, R., Ishizawa, Y., and Sumiyoshi, Y. (1988). Surface phonon dispersion curves of graphite (0001) over the entire energy region, *Solid State Commun.*, **65**, pp. 1601–1604.

21. Ishi, R. A., Venkataraman, L., Dresselhaus, M. S., and Dresselhaus, G. (1993). Phonon modes in carbon nanotubules, *Chem. Phys. Lett.*, **209**, pp. 77–82.

22. Dresselhaus, M. S., and Eklund, P. C. (2000). Phonons in carbon nanotubes, *Adv. Phys.*, **49**, pp. 705–814.

23. Léonard, F. (2008) *The Physics of Carbon Nanotube Devices* (William Andrew, New York).

24. Léonard, F., and Tersoff, J. (2002). Dielectric response of semiconducting carbon nanotubes, *Appl. Phys. Lett.*, **81**, pp. 4835–4837.

25. Spataru, C. D., Ismail-Beigi, S., Benedict, L. X., and Louie, S. G. (2005). Excitonic effects and optical spectra of single-walled carbon nanotubes, *AIP Conf. Proc.*, **772**, pp. 1061.

26. Spataru, C. D., Ismail-Beigi, S., Benedict, L. X., and Louie, S. G. (2004) Excitonic effects and optical spectra of single-walled carbon nanotubes, *Phys. Rev. Lett.*, **92**, pp. 077402–077406.

27. Bachilo, S. M., Strano, M. S., Kittrell, C., Hauge, R. H., Smaley, R. E., and Weisman, R. B. (2002). Structure-assigned optical spectra of single-walled carbon nanotubes, *Science*, **298**, pp. 2361–2366.

28. Wang, F., Dukovic, G., Brus, L. E., and Heinz, T. F. (2005). The optical resonances in carbon nanotubes arise from excitons, *Science*, **308**, pp. 838–841.

Chapter 2

Synthesis of Carbon Nanotubes

San Hua Lim and Jianyi Lin
Institute of Chemical and Engineering Sciences, Singapore
lim_san_hua@ices.a-star.edu.sg, lin_jianyi@ices.a-star.edu.sg

This chapter reviews the recent advances in the synthesis of carbon nanotubes with special attention to chemical vapor deposition (CVD) methods. CVD methods are promising for synthesizing high quality and quantity of carbon nanotubes. Fluidized bed and floating catalyst CVDs enable continuous production of CNT to be scaled up. Plasma-enhanced CVD can greatly lower the CNT growth temperature and control the growth of aligned CNT at pre-determined areas for device applications. Improved CVD methods allow one to fabricate macroscopic assembly of carbon nanotubes. The chirality and electronic properties of nanotubes are more uniform using new growth strategies, purification and separation methods, which are useful for potential applications.

2.1 Introduction

Iijima [1991] reported the observation of crystallized and multiply layered tubular carbon nanostructures (multi-walled carbon

Carbon Nanotubes and Their Applications
Edited by Qing Zhang

ISBN 978-981-4241-90-8 (Hardcover), 978-981-4303-18-7 (eBook)
www.panstanford.com

nanotubes) in arc discharge soot in the early 1990s. The synthesis of single-walled carbon nanotube (SWCNT) was shortly demonstrated using a laser ablation technique [Thess *et al.*, 1996]. SWCNTs are visualized as seamlessly rolled-up sheets of graphenes and multi-walled carbon nanotubes (MWCNTs) consist of co-axial concentric tubes of SWCNTs. The molecular SWCNT is considered a new form of carbon, a sister material of C_{60} and a distinct form of carbon allotropes such as graphite and diamond. Hence carbon nanotubes have generated tremendous research interests to study its synthesis, characterizations, growth mechanism, chemical modifications and potential applications.

Nanotubes are ideal models for studying the physics and chemistry of one-dimensional (1D) solids. Carbon nanotubes exhibit extraordinary properties such as high tensile strength and excellent electrical and thermal conductivities [Ajayan, 1999]. The electronic properties of SWCNTs behave as either metallic or semiconducting depending on their diameters and chiralities [Saito *et al.*, 1992], while MWCNTs depend on the features of each coaxial carbon shell and conduction takes place within the basal plane of graphite [Bourlon *et al.*, 2004]. Theoretical studies predict that $\sim$1/3 of the SWCNTs are metallic and the remaining behaves as semiconducting nanotubes [Saito *et al.*, 1992]. The as-synthesized SWCNT samples contain a mixture of metallic and semiconducting nanotubes. Metallic SWCNTs are model systems for studying rich quantum phenomena such as ballistic transport, single-electron charging, Luttinger liquid, weak localization and quantum interference [Dekker, 1999]. Semiconducting SWCNTs are used to fabricate nanotube-based electronics such as transistors and logic, memory and sensory devices [Dai, 2002]. Therefore CNTs have great potential applications ranging over a wide field, including computer logic and memory, interconnect via, nano-actuators and nano-sensors, field emitters, polymer composites, energy storage, catalyst supports, and membranes.

Industrial application and device integration of CNTs require large quantities of bulk materials and scalable production. For example, the application of CNTs to polymer composites and energy storage requires high-quality CNTs at kilogram scale, which entails a scalable, simple, and relatively inexpensive synthesis method. The integration of CNTs into electronic devices requires synthesis methods that enable self-assembly and directed-growth of nanotubes. Electric arc discharge and laser ablation require very high energy input, replacement of spent carbon target, and re-evacuation of reactors, which pose difficulties to large-scale and continuous production of CNTs. In contrast, chemical vapor deposition (CVD) methods have gained enormous popularity due to their milder operation and flexible processing conditions, which have the greatest potentials to achieve continuous mass-production and integration of CNTs. In this chapter, in addition to the brief introduction to arc discharge and laser ablation methods, we will review and address mainly the challenges of CVD techniques so that CNTs may become a commercial reality. The challenges of nanotube synthesis are divided into five main areas: (1) an efficient synthesis method to produce high-quality nanotubes at large scales, (2) synthesis of macroscopic length carbon nanotubes, (3) controllable and directed growth of nanotubes, (4) uniform chirality and electronic property of SWCNTs, and (5) low-temperature synthesis of carbon nanotubes for electronic integration. We will also briefly discuss the purification and post-synthesis modification of single-walled carbon nanotubes, which are important for applications. The ultimate goal of chemical modification of SWCNTs is to achieve uniformity in chirality and electronic properties.

2.2 Arc Discharge

In electric arc discharge method, the arc is generated by passing direct currents of 50–100 A through opposing carbon anode and cathode in an inert (helium/argon) atmosphere. The high-temperature discharge vaporizes the carbon atoms, which re-condensate to form carbon nanotubes. MWCNTs produced by arc discharge are very straight due to their high crystallinity. For the growth of SWCNTs, transition metal catalysts are needed in the arc discharge. Journet *et al.* [1997] demonstrated large-scale production of SWCNTs using arc discharge method and achieved gram quantities. The yield of SWCNTs (70–90%) was improved by using a carbon anode containing 1 at.% yttrium and 4.2 at.% nickel as catalyst. However, the crude CNTs synthesized by arc discharge require extensive purification to remove the soot and catalytic metal particles.

Open-air synthesis of CNTs with welding arc torch and synthesis of MWCNTs in liquid nitrogen have been reported, in which low pressure or/and expensive inert gas can be avoided. These methods are promising routes for economic production of CNTs [Takikawa *et al.*, 2002; Antisar *et al.*, 2003].

2.3 Laser Ablation

High-quality SWCNTs were synthesized in yields of more than 70% using a laser ablation method [Thess *et al.* 1996]. Intense laser pulses were used to ablate a carbon target loaded with 0.5 at.% nickel and cobalt. Similar to arc discharge method, a metal catalyst is needed for the formation of SWCNTs. The carbon target was heated up to 1200°C in a tube furnace and the vaporized carbon atoms were carried downstream by an inert gas flow to form condensates (SWCNTs, soot and metal catalysts) at a cold finger. The SWCNTs self-organize into "ropes," which consist of 100–500 SWCNTs in a two-dimensional triangular lattice, due to van der Waals forces. Eklund *et al.* [2002] reported a high production rate of SWCNTs at ~1.5 g/hr using an ultrafast laser pulses at 20% power of a free electron laser. The production rate of SWCNTs is increased to ~45 g/hr if the power of the ultrafast laser pulses is running at maximum. A 2 KW continuous wave CO_2 laser ablation was used to produce SWCNT soot at a rate of 5 g/hr [Bolshakov *et al.*, 2002].

2.4 Chemical Vapor Deposition

A typical CVD method involves heating catalysts to suitable temperatures ~750–1200°C in a tube furnace and flowing hydrocarbon gases diluted in a process gas such as H_2, Ar or He for an extended time. The carbon nanotubes are synthesized over the catalysts. Catalytic nanoparticles of Fe, Co, or Ni or a combination supported on oxides is commonly used in CVD methods to grow CNTs [Terranova, *et al.*, 2006]. The general growth mechanism of CNTs in a CVD method involves the dissociation of hydrocarbons, dissolution and saturation of carbon atoms over the catalysts. Tubular carbon solids are precipitated from the super-saturated metal catalysts. In this book chapter, we will focus on improved CNT synthesis, which addresses the five challenges and might lead to the commercialization of CNT products.

2.4.1 Scalable Production of Carbon Nanotubes

2.4.1.1 Nano agglomerate fluidized (NAF) Process

The improvement of CVD methods shows that fluidized-bed reactors have greater commercial potential than packed bed reactors to produce CNTs at lower cost, higher purity, higher yield, and easier scalable production [See and Harris, 2007; Danafar *et al.*, 2009]. Fluidized bed reactors also exhibit excellent heat and mass transfer and have been widely used in industrial mass production.

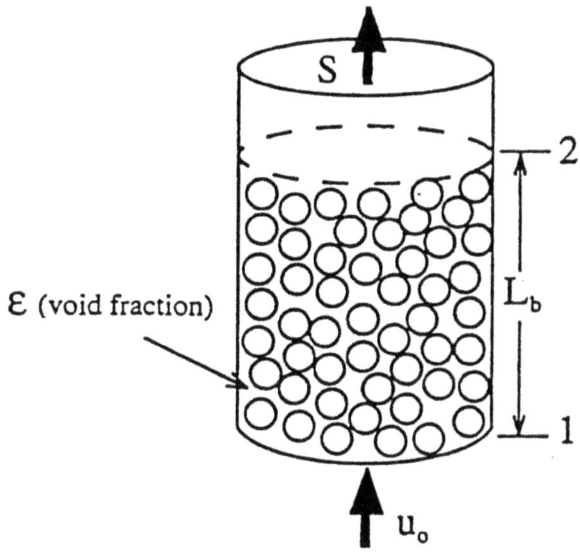

Figure 2.1 A schematic diagram showing a fluid flowing at a superficial velocity (U_o) through free-standing reactor of height (L_o) packed with spherical particulates (diameter, D_p).

Figure 2.1 depicts the flow of a fluid (gas/liquid) through a free-standing cylindrical reactor packed with spherical particulates (diameter D_p). The fluid flows through the packed bed at a superficial velocity (U_o), which is a measured upstream of the reactor entrance. A transition from a packed bed to a fluidized bed can occur only when the fluid superficial velocity reaches a minimum fluidization velocity, , which can be estimated from the equation below:

$$g(\rho_p - \rho) = \frac{150\mu(1-\varepsilon)U_{mf}}{\varepsilon^2 D_p^2} + \frac{1.75\rho U_{mf}^2}{\varepsilon^3 D_p} \tag{2.1}$$

where g, ρ_p, ρ, μ, ε, and D_p denote the standard gravity, particulate density, fluid density, fluid viscosity, void fraction of the bed and particulate diameter, respectively [Ruckenstein, 1967]. In other words, fluidization occurs when a pressurized fluid is intermixed with a solid particulate inside a bed reactor under suitable conditions such that the gravitational force of particulate bed is equal to the upward force of fluid flow, so that the solid particulates are lifted up and the fluidized bed exhibits fluid-like properties. The fluidized bed is considered to have a single bulk density. A particulate with a heavier density than the bed will sink, and vice versa. The bulk density of fluidized bed is altered by changing the fluid and solid fraction. The characteristic properties of fluidized beds include (i) high-surface-area contact between the fluids and solid particulates per unit bed volume, which greatly enhanced thermal transport inside the reactors, (ii) high relative velocities between the fluids and the solid particulates, (iii) high level of intermixing phase, and (iv) high frequency of particulate–particulate and particulate–wall collisions. The behavior of fluidized beds is classified according to Geldart's classification [Geldart, 1973] based on the density difference between the particulates (ρ_p - ρ), and the fluids and particulate size, D_p (see Table 2.1).

Table 2.1 Geldart classification of fluidized powders

Group A	The bed particle size is ~20–100 μm and density is typically ~1400 kg/m^3. After minimum fluidization and before the formation of bubbling, the particle density expands two to three times due to a decreased bulk density. Most powdered catalysts belong to group A.
Group B	The bed particle size is ~40–500 μm and density is typically ~1400–4500 kg/m^3. Bubbling appears at fluidization.
Group C	The bed particles are extremely fine (20–30 μm) and cohesive, and difficult to fluidize. May require the use of mechanical agitation
Group D	The particle size is >600 μm and produce deep spouting beds.

Fluidized bed reactors use fluidized catalysts to grow CNTs. At atomic level the growth of CNTs requires the use of Fe, Ni or Co catalysts or their alloys in nanometer scale. However, a single CNT or other well-dispersed nanoparticles can be easily entrained by the fluidizing gas. Fortunately, various CNT agglomerated structures can be formed as CNTs grow, and they can further coalesce into fractal clusters sized from several microns to several hundreds of microns.

Figure 2.2 shows the proposed formation of multi-agglomerate structures on the fluidized catalysts during CNT growth. The initial catalysts are composed of nanoparticle agglomerations. The growth of CNTs is nucleated by a base growth model, and the growing CNTs break up the catalyst agglomerations leading to increasing agglomerate size and decreasing density. The strong van der Waals interactions among the CNTs, catalysts and entanglement of CNTs result in large CNT agglomerations. As shown in Figure 2.2, the size of CNT agglomerates is ~400–500 μm and a bulk density of 40 kg/m^3, which is only 1/20 of the initial catalyst density. In the fluidized-bed reactor, the CNT-catalyst agglomerates must exist as low-density materials in order to maintain the fluidized state. In contrast, in a fixed-bed reactor, the reaction zone will be plugged up with CNT-catalyst agglomerates and caused significant heat and mass transfer problems. Therefore, the fluidized-bed reactor has the advantages to circumvent the issues of heat and mass transfer and allows continuous mass production of carbon nanotubes. The interaction between CNT agglomerates or other nanomaterials in a fluidized-bed reactor is more complicated than that between powders consisting of conventional small organic or inorganic molecules. The traditional facilities may have to be modified while dedicate selection of the catalyst and control of the operating condition are required for CNT production.

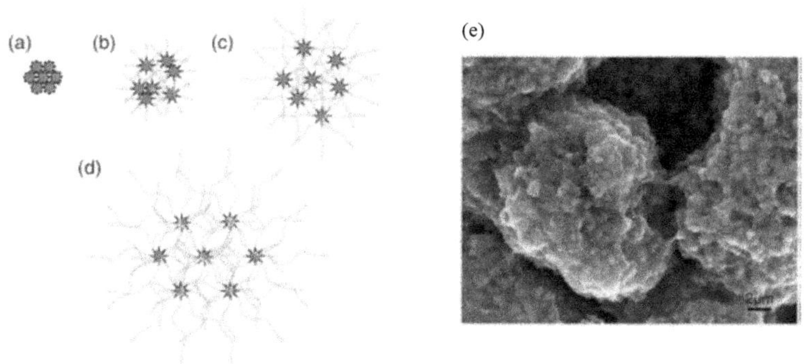

Figure 2.2 Mechanism of carbon nanotube agglomeration: (a) initial catalysts in the form of agglomerated nanoparticles, (b) breaking up of catalyst nanoparticles due to carbon nanotube growth, (c) formation of carbon nanotube sub-agglomerates, and (d) fully developed carbon nanotube agglomerates. (e) Scanning electron micrograph of MWCNT agglomeration. [Wei *et al.*, 2008]

Tsinghua University has pioneered a "Nano Agglomerate Fluidized" (NAF) process for large-scale and continuous production of MWCNTs [Wang *et al.*, 2002; Wei *et al.*, 2008]. Figure 2.3 shows the schematic setup of the fluidized-bed reactor and the pilot plant facility at Tsinghua University. Two big challenges in a stable NAF process are to control the agglomerate structure and to maintain a constant average agglomerate diameter. These can be done by careful selection of the catalyst and operating conditions. The NAF process reported by Wang *et al.* in 2002 used Fe nanoparticles supported on Al_2O_3 powders as the catalyst, which were fed into the reactor before the synthesis. Hydrocarbons having less than seven carbon atoms (such as methane, ethylene and propylene) were used as the carbon precursors [Wang *et al.*, 2002; Zhang *et al.*, 2008]. Nitrogen and hydrogen gases were used as the diluents. The flow rate of the gas mixture was as follows: ethylene/propylene = 5–10 m^3/h; N_2 = 0–0.5 m^3/h; and H_2 = 0–0.5 m^3/h. The reaction temperature was maintained at 600–900°C for the synthesis duration of 30–60 min under ambient pressure. The superficial gas velocity of ~0.08–2 m/s and the bed density of the reactor of ~20–800 kg/m^3 were applied during the NAF process so that the catalyst-CNT nano agglomerates were kept in a dense phase fluidization state. The NAF process is capable of continuous feeding of catalysts and reactants and continuous removal of the deposited CNTs from the reactor. The MWCNTs produced were ~4–100 nm in diameters and ~0.5–1000 μm in lengths. The size of CNT agglomerates ranged from ~1 to 1000 μm. The CNT agglomerates in NAF process were observed to behave like Geldart-A particles. A pilot plant facility operating with the NAF process has the capacity to synthesize MWCNTs at a production rate of 15 kg/h. The purity of MWCNTs was improved from 70–80% to > 99.9% after vacuum annealing at 2050 K.

The catalyst design has significant effects on the structures of CNTs. The application of different catalysts in NAF process may yield different types of CNTs. It was shown in 2008 that the addition of Mo-promoter could stabilize and prevent the sintering of Fe/(Co/Ni) supported on MgO catalysts at high temperature, which reduced diameter and size distribution of the MWCNTs. The as-synthesized MWCNTs had two to five walls with an outer diameter of 1–7 nm [Zhang *et al.*, 2008]. A hydrothermal treatment was reported to convert Fe/MgO to lamella-like $Fe/Mg(OH)_2$ catalysts, yielding double-walled CNTs [Ning *et al.*, 2007]. The presence of Na_2CO_3,

Na_3PO_4, and $CaCl_2$ additives was found to alter the activity of Ni/Cu/Al_2O_3 catalysts, which yield CNT with larger inner diameters while the outer diameters remain unchanged.

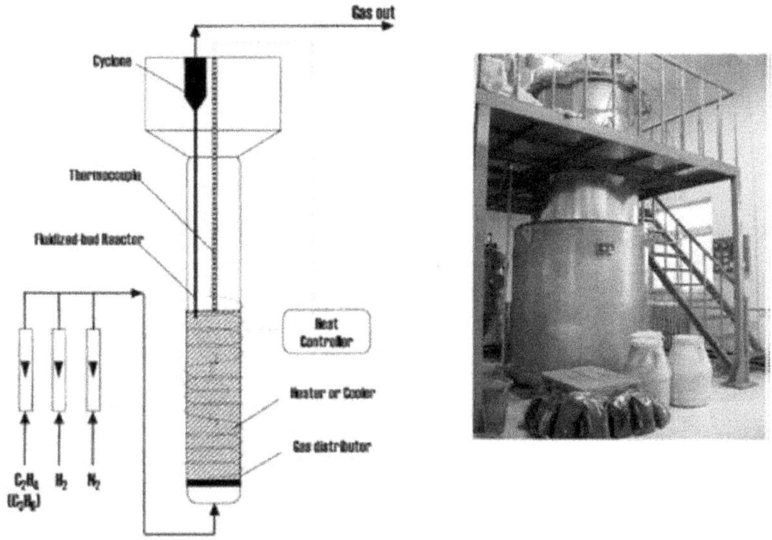

Figure 2.3 (a) Schematic representation of nano-agglomerate fluidized bed reactor for carbon nanotube production. [Wang *et al.*, 2002] (b) Pilot plant for carbon nanotube production in Tsinghua University. [Wei *et al.*, 2008]

2.4.1.2 HiPco process

A gas-phase high pressure carbon monoxide (HiPco) process was developed by Rice University in 2001, which could produce high-purity (97%) single-walled carbon nanotubes (SWCNT) in large scale at a rate of ~10.8 g/day or ~450 mg/h [Bronikowski *et al.*, 2001]. Figure 2.4 shows the schematic setup of the HiPco process, whereby SWCNTs are synthesized in high pressure (30–50 atm), high temperature (900–1100°C) flowing CO on clusters of iron catalysts formed *in situ* from iron pentacarbonyl, $Fe(CO)_5$. A fraction of the CO flow is passed through a liquid-filled $Fe(CO)_5$ bubbler and carried the CO/$Fe(CO)_5$ mixture into the hot reaction zone. These iron clusters act as nucleating and growing sites of SWCNTs. The production of SWCNTs in gas-phase is carried out by CO disproportionation (Boudouard reaction): $CO + CO \rightarrow C(SWCNTS) + CO_2$. The as-synthesized HiPco products consist of SWCNT ropes interspersed with small clusters of iron particles (2–5 nm), and the sidewalls of SWCNTs are free of amorphous carbon coating. The size and diameter distribution of SWCNTs are controllable by the CO pressure. The yield of ~70% can be achieved. Energy dispersive x-ray (EDX) spectroscopy revealed that the HiPco product consists of 97 mol% of carbon and 3 mol% of iron.

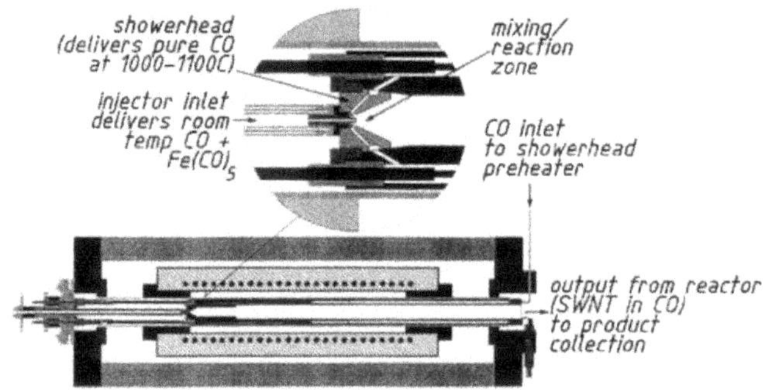

Figure 2.4 HiPco reactor, with the mixing/reaction zone show enlarged. [Bronikowski *et al.*, 2001]

2.4.1.3 CoMoCat process

Resasco *et al.* [2002] developed a scalable process (CoMoCAT) for the production of SWCNTs by catalytic disproportionation of CO on Co/Mo/SiO_2 solid catalysts. In the CoMoCAT process, high-quality SWCNTs with narrow distribution of tubular diameters are synthesized by CO disproportionation at 750–950°C in flow of pure CO at a pressure of 1–10 atm using fluidized-bed reactors. The production rate is ~0.25 g SWCNT/g catalyst in a couple of hours, with better than 80% selectivity towards SWCNT synthesis. The CoMoCAT process uses a low Co:Mo loading on silica supports (SiO_2 support with 6 nm average pore size and BET surface area of 480 m^2/g) as efficient catalysts for SWCNT production. The total loading of bimetallic Co-Mo in the catalyst is 2 wt% with a Co:Mo molar ratio of 1:3. The synergistic effect between Co and Mo enhances the performance of catalysts [Bachilo *et al.*, 2003]. The Co interacts with Mo to form a superficial Co molybdate-like species and the Mo is in the form of Mo(6+) oxidation state. During high-temperature reduction in hydrogen, the Co molybdate-like species remains as well dispersed Co^{2+} ion and resists sintering. The prevention of Co sintering is crucial to the selective synthesis of narrow diameter distribution of SWCNTs. As shown in Figure 2.5, SWCNTs synthesized by the CoMoCAT process has significantly narrower radial breathing modes (RBMs) than SWCNTs synthesized by the HiPco process. However, as the synthesis temperatures are higher than 750°C the diameter distribution of the as-synthesized SWCNTs becomes wider, which is due to the metal sintering and formation of large catalyst clusters. Figure 2.5 compares the near-infrared emission spectra of semiconducting SWCNTs synthesized by CoMoCAT and HiPco methods. Individually suspended SWCNTs show characteristic near-infrared emission that uniquely labeled the nanotubes' diameter and chirality using (n, m) notations. The CoMoCAT SWCNT samples are dominated by two structures, namely, (6, 5) and (7, 5), which account for 57% of the semiconducting nanotubes. In other words, the (6, 5) and (7, 5) semiconducting SWCNTs constitute ~38% of the total SWCNTs in the CoMoCAT. On the other hands, the near-infrared emission of HiPco SWCNT samples exhibit several types of semiconducting SWCNTs. The CoMoCAT process has great potential in obtaining SWCNT samples with better uniformity of electronic properties. Thus the CoMoCAT process has the potential to selectively grow SWCNT samples enriched with semiconducting (6, 5) and (7, 5) nanotubes.

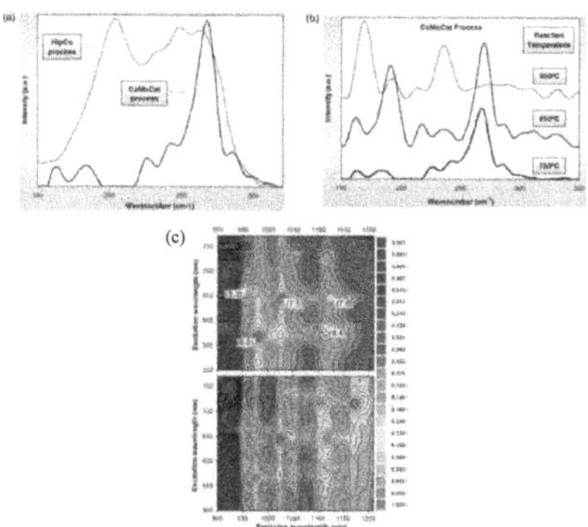

Figure 2.5 (a) Raman spectra of SWCNTs produced by CoMoCAT process at 750°C and by HiPco process respectively, obtained with an excitation laser of 532 nm. (b) Raman spectra of SWCNTs produced by CoMoCAT process at various temperatures. [Resasco *et al.*, 2002] (c) Contour plots of normalized fluorescence intensities for the CoMoCAT sample (top frame) and the HiPco sample (bottom frame). [Bachilo *et al.*, 2003] See also Color Insert.

2.4.1.4 Floating catalyst CVD method

Endo *et al.* [1988] have developed a CNT production system, in which the metal catalyst, carbon source, and gases are fed into a vertical furnace from the top end of the furnace [see Figure 2.6]. The CNTs grow while the catalyst particles are floating, gradually falling to the bottom of the furnace. This method is still used in a Japanese company, producing high-quality CNTs in large scale to the market.

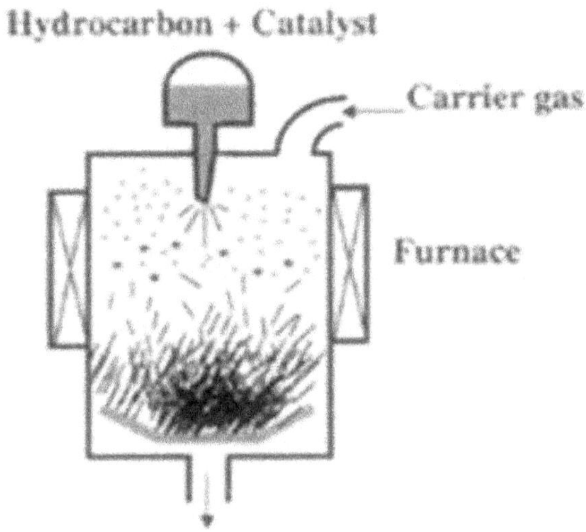

Figure 2.6 Illustration of floating method setup (left) and its growth model (right). The basic growth mechanism is the same as those of other CCVD methods, since the growth initiates from the catalyst. [Endo *et al.*, 1988]

2.4.2 Macroscopic Assembly of CNTs

Individual carbon nanotubes are useful to fabricate nanoscale electronic devices but have limited applications. However, macroscopic assemblies of carbon nanotubes will widen the scopes of application. For example surfactant-dispersed nanotubes could be post-processed with electric or magnetic field to obtain buckypapers, which were used directly as electrochemical electrodes [Li *et al.*, 2002; Fischer *et al.*, 2003]. Nanotube–acid and nanotube–polymer blends were drawn from solution to assembly CNT fibers [Shaffer *et al.*, 1998; Vigolo *et al.*, 2000]. The assembly of CNTs into sheets and fibers was due to the strong van der Waals interactions. However, it is better to obtain macroscopic assembly of CNT without the use of additives so that the pristine properties of nanotubes are preserved. Improved synthesis techniques enable the macroscopic assembly of CNTs into additive-free sheets and threads.

2.4.2.1 Direct synthesis of long CNT strands by floating catalyst CVD

Figure 2.7 shows the typical setup of a floating Catalyst CVD method employed to synthesize carbon nanotube strands. In packed or fluidized-bed reactors, the catalyst is supported on a substrate and the supported catalyst is loaded in the reactor. In contrast floating catalyst is injected into the reactor and formed *in situ* within the reaction zone during the pyrolysis of the carbon precursors. Metallocene and metal phthalocyanine are commonly used as iron-containing catalyst precursors, which are dissolved in organic solvents (also acting as carbon sources), for the synthesis of carbon nanotubes. The liquid precursor (organic solvent + dissolved catalyst) is injected together with hydrogen gas into the reaction zone of furnace to produce carbon nanotubes. A substrate may be place in the reaction zone to synthesize well aligned carbon nanotubes [Cao *et al.*, 2001].

Zhu *et al.* [2002] synthesized long strands of single-walled carbon nanotubes (~10–20 cm) in large yields using an enhanced floating catalyst CVD method. The n-hexane solution containing 18 mg/ml of ferrocene (as catalyst for SWCNT growth) and 0.4 wt% of thiophene (as promoter for SWCNT formation) was injected at 0.5 ml/min into a vertical tube furnace for pyrolysis at 1150°C using hydrogen (250 ml/min) as the carrier gas. The optimized floating catalyst CVD method yields ~0.5 g/hr of long SWCNT ropes. The lengths of macroscopic SWCNT ropes are ~10–20 cm and the diameters are ~0.3–0.5 mm. A single strand consists of well-aligned SWCNT bundles. These long SWCNT strands are mechanical robust and are easily handled and manipulated without breakage (see Figure 2.7 inset). Raman spectroscopic studies suggest that SWCNT strands are composed of a mixture of metallic and semiconducting nanotubes. Both Raman spectroscopy and x-ray diffraction indicate that the average diameter of individual SWCNTs is ~1.1 nm. The macroscopic electrical resistivity of long SWCNT strands was measured from room temperature to 5 K using a four-point probe.

The SWCNT strands exhibit a crossover from semiconducting to metallic behavior at ~90 K. At temperature range of 90–300 K, the resistivity of long SWCNT ropes is ρ = 5–7 × 10^{-6} ohm·m. The low values of resistivity strongly suggest that the conducting path along the entire length of long SWCNT strands is continuous. Young's modulus of SWCNT strand is estimated to be 49–77 GPa, which is five times Young's modulus of oriented SWCNT fibers and ribbons and 50 times of SWCNT buckypaper [Walters *et al.*, 1999; Park, *et al.*, 2009].

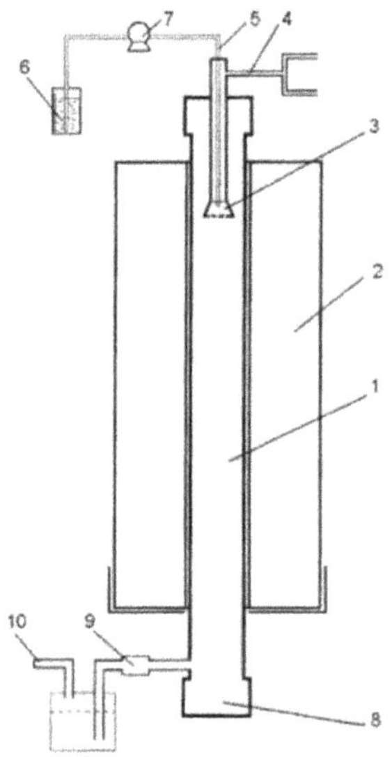

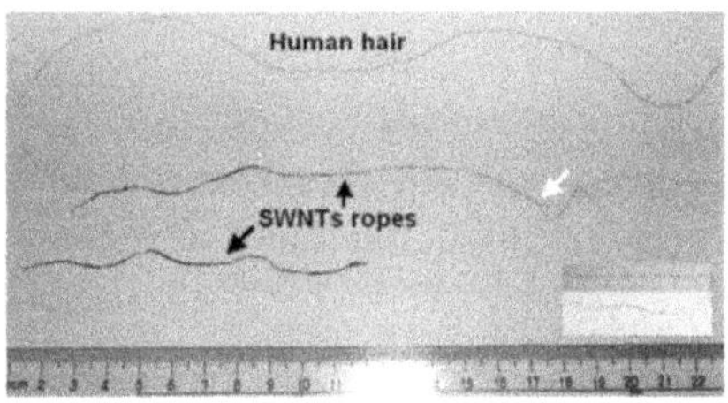

Figure 2.7 (a) Schematic of a floating catalyst chemical vapor deposition method used for the direct synthesis of long single-walled carbon nanotube strands. 1. Ceramic reaction chamber. 2. Electric furnace heating up to 1200°C. 3. Evaporator located at the upper part of the reactor. 4. Gas inlets. 5. Liquid (n-hexane + thiophene + ferrocene) inlet. 6. Liquid source vessel. 7. Liquid micro-pump. 8. Collector. 9. Filter. 10. Outlet. (B) Optical images showing a human hair and two strands of SWCNT. The lengths of SWCNT strands are ~ 20 and 10 cm, and have a diameter of the order 0.3 to 0.5 mm (white arrow indicates thinner ropes). The inset shows one straightened SWCNT strand and another tied in a knot. [Zhu *et al.*, 2002]

Li *et al.* [2004] spun fiber and ribbons of carbon nanotubes directly from the CVD hot zones using a floating catalyst method (see Figure 2.8). Carbon nanotube fibers were continuously drawn from the furnace by a rotating spindle and there is no length limitation for the spun CNT fiber. A mixture of ethanol, ferrocene (0.23 to 2.3 wt%) and thiophene (0.5 to 4.0 wt%) was injected at 0.08 to 0.25 ml/min from the top of the CVD furnace and H_2 carrier gas flowed at 400 to 1200 ml/min at temperatures of 1050–1200°C. Under optimized conditions, CNTs exist as aerogels in the hot zone, which appear as "elastic smoke" (see Figure 2.8), and the interactions among nanotubes impart certain degree of mechanical integrity to the aerogels so that it is possible to spin the aerogels into CNT fibers. Figure 2.9 shows the morphology of the spun MWCNT fibers with tubular diameters of 30 nm and an aspect ratio of ~1000. The purity of MWCNTs are ~85 to 95 wt% and well-aligned within the fibers. The spun CNT fibers are mechanically stable and it can be wound across two spindles. The electrical conductivity for the CNT fibers was measured ~$8.3 \times 10^5\ \Omega^{-1}m^{-1}$, which is slightly higher than typical carbon fibers. Young's modulus of twisted CNT fiber was determined to be ~0.1 to 1.0 GPa.

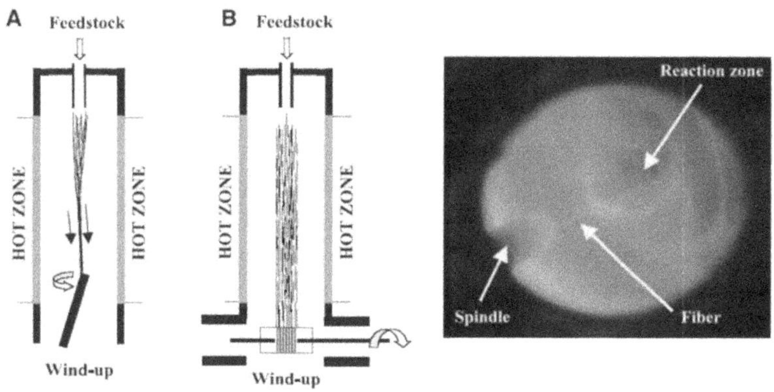

Figure 2.8 (A) Schematic of spinning process within the hot zone. (B) Schematic of the wind-up assembly that operates outside the furnace hot zone. A liquid feedstock (ferrocene + thiophene) is mixed with hydrogen and injected into hot zone. An aerogel of carbon nanotubes is formed at the hot zone, which is continuously wound out as a fiber or film. (C) Carbon nanotube fiber drawn from the aerogel onto the spindle. [Li *et al.*, 2004]

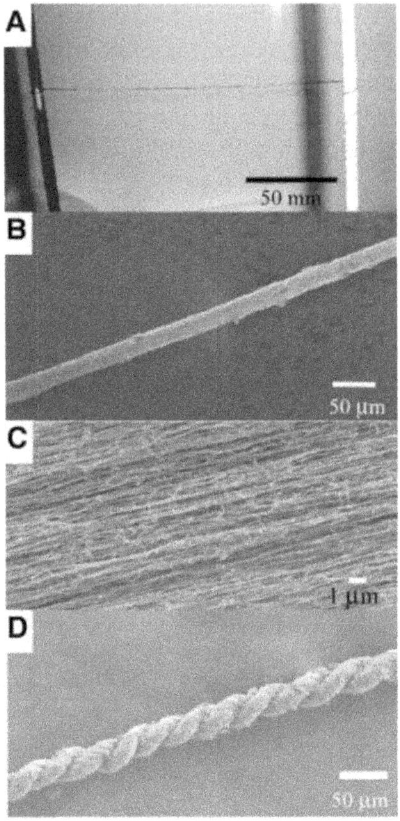

Figure 2.9 (A) Photographic image of carbon nanotube rope wound between tow spindles. (B and C) Scanning electron microscopic images of a fiber of multi-walled carbon nanotubes. (D) A permanently twisted carbon nanotube fiber after it is removed from the furnace. [Li *et al.*, 2004]

2.4.2.2 Spinning CNT fibers

Self-assembly of CNTs into yarns/sheets of up to meters long was achieved by continuous spinning from post-synthesized aligned carbon nanotube forests. Jiang *et al.* [2002] first demonstrated the spinning of pure CNT yarn by pulling out a bundle of CNTs from a CNT array several hundred micrometers high and grown on a silicon substrate. The continuous spinning of CNT yarns was very similar to the spinning of silk thread from a cocoon. Jiang *et al.* [2002] projected that 1 cm^2 of CNT arrays would generate ~10 m of CNT yarn (~200 μm wide). Similarly, Zhang *et al.* [2005] contacted the sidewall of MWCNT forest with an adhesive tape and initiated the automated spinning of MWCNT sheets of 5 cm wide at a rate of up to 10 m/min. However, the spinning of nanotube sheets depends on the structure of the MWCNT forests. Intermittent bundling of the MWCNT forests is crucial for the spinning process such that individual nanotubes are able to migrate from one bundle to another bundle. As shown in Figure 2.10A,B, the nanotube bundles are pulled from the entire sidewall of MWCNT forest so that the bundles are continuously joined together without breaking the CNT sheets. The as-produced MWCNT sheet was ~18 μm in thickness when drawn from a 245 μm-high forest. The measurement areal density and volume density of the MWCNT sheet were ~2.7 μg/cm^2 and 1.5 mg/cm^3, respectively. The as-produced MWCNT sheets are electrically conducting and highly anisotropic. The thickness and density of MWCNT sheets was increased to 50 μm and ~500 μg/cm^3 by adhering the sheet to a planar substrate and vertically immersing the sheet into ethanol and retracting the substrate along the nanotube alignment direction. The evaporation of ethanol densifies the MWCNT sheet ~360 folds. The MWCNT sheet acts as a planar incandescent light source of polarized UV, visible, and infrared for applications in sensors, infrared beacons, infrared imaging, and reference signals for device calibration. As shown in Figure 2.10E, the degree of polarization of the emitted radiation for a 2.5% stretched MWCNT sheets increased from 0.71 at 500 nm to 0.74 at 780 nm, which was much higher than previous reports. The polarization spectral radiance of MWCNT sheets were fitted with black-body radiation law and the degree of polarization did not greatly depend on the MWCNT sheet temperature for the observed temperature range from 1000–1600 K.

Although Jiang *et al.* [2002], Zhu *et al.* [2002], Li *et al.* [2004] and Zhang *et al.* [2005] have fabricated macroscopic size of carbon fabrics, the microtextural properties of the nanotube fabrics are different. The long SWCNT ropes synthesized by Zhu *et al.* [2002] are truly composed of SWCNTs with lengths that reach up to centimeter range. On the other hand, the long CNT ropes/sheets prepared by Li *et al.* [2004] and Zhang *et al.* [2002] are composed of many MWCNTs (lengths are hundreds of micrometers) that are weaved together to form threads/sheets. This also explains why spun CNT fibers have lower Young's modulus values than the long SWCNT ropes. However, the spinning of CNT yarn from nanotube forests has control over the dimensions of CNT yarns, particularly the lengths, which is useful for applications. The direct synthesis of centimeter long SWCNT rope using floating catalyst CVD has elements of randomness and limitation in controlling the desired length.

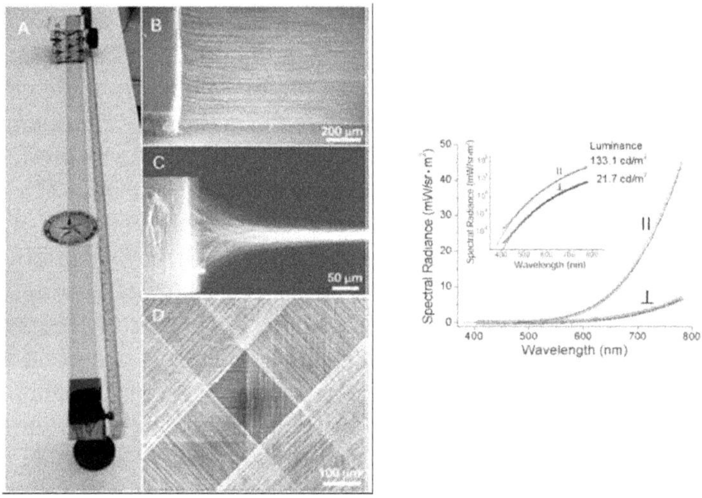

Figure 2.10 MWCNT forest conversion into sheets and assemblies of those sheets. (A) Photograph of a self-supporting 3.4 cm wide, meter long MWCNT sheet that has been hand drawn from a nanotube forest at an average rate of 1 m/min. Its transparency is illustrated by the visibility of the NanoTech Institute logo that is behind the MWCNT sheet. (B) SEM image, at 35° angle with respect to the forest plane, capturing a MWCNT forest being drawn into a sheet. (C) SEM micrograph showing the cooperative 90° rotation of MWCNTs in a forest to form a sheet. (D) SEM micrograph of a two-dimensionally by overlaying four nanotube sheets with a 45° shift in orientation between successive sheets. (E) Spectral radiance in directions parallel to (||) and perpendicular to (⊥) the draw direction of an as-drawn undensified MWCNT sheet after an added inelastic stretch in the initial draw direction of 2.5%. The luminance polarized parallel to the sheet draw direction for the direction for the indicated spectral range is 6.1 times higher than for the perpendicularly polarized luminance, which corresponds to a polarization factor of 0.72. The inset shows these data on a semi logarithmic scale. Underlying solid lines (largely obscured by coincidence with the data points) are data fits assuming black-body radiation with $T = 1410$ K. [Zhang *et al.*, 2005]

2.4.2.3 Super growth of aligned carbon nanotubes

The typical lengths of carbon nanotubes are from hundreds of nanometers to micrometers, regardless of synthesis methods. Extending the synthesis time does not infinitely increase the lengths of carbon nanotubes. The main reason is due to the poisoning of the catalysts, which have been coated with amorphous carbon. When the formation of amorphous carbon has been suppressed, "super growth" of carbon nanotubes with axial lengths in the range of millimeters and even centimeters will be achieved.

Hata *et al.* [2004] employed a water-assisted CVD method to synthesize impurity-free vertically aligned SWCNT forests with heights up to 2.5 mm in 10 min growth time (see Figure 2.11). High-resolution transmission electron microscopy reveals that the nanotubes are free from amorphous carbon and metal particles. The diameters of SWCNTs are estimated to be in the range of 1–3 nm. The amount of water vapor (20–500 ppm) must be controllably introduced into the $Ar/H_2/C_2H_4$ precursors in order to achieve super

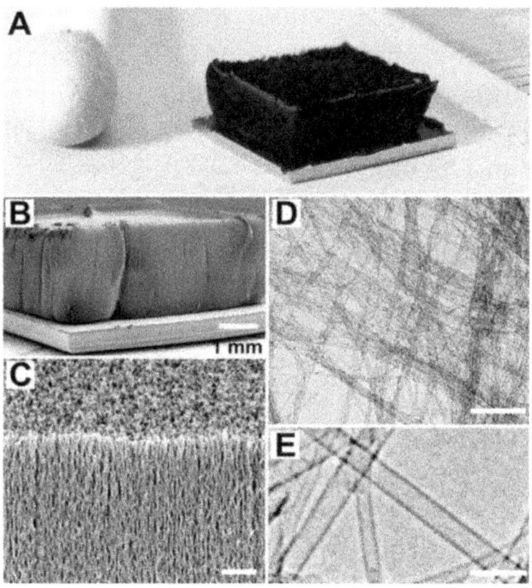

Figure 2.11 SWCNT forest grown with water-assisted CVD. (A) A 2.5 mm tall SWCNT forest synthesized on a 7 × 7 mm silicon wafer. A matchstick on the left and a rule with millimeter markings on the right are for size reference. (B) Scanning electron microscopy (SEM) image of the same SWCNT forest. (C) SEM and TEM images of the SWCNT forests. [Hata *et al.*, 2004]

growth of SWCNTs. The water vapor acts as a weak oxidizer and consumed the amorphous carbon coated on the catalyst particles during the SWCNT synthesis at temperature of 750°C. The removal of amorphous carbon is crucial to prolong and preserve the activity of the catalysts. Hata *et al.* [2004] claimed that water-assisted CVD growth of SWCNTs was not restricted by the choices of catalysts and substrates, but Noda *et al.* [2007] showed that Al_2O_3 support was necessary for the super growth of SWCNTs. The residual catalyst was only ~0.013% as the source of impurity and the carbon purity was greater than 99.98%. The water vapor significantly improves the efficiency of the catalysts and the weight ratio of SWCNT to catalyst is greater than 50000%, which is important for scaling up the production of vertically aligned carbon nanotubes. Futaba *et al.* [2006] showed that super-grown SWCNTs were shape-engineerable by using the zipping effect of liquids to fabricate very densely packed SWCNTs. These highly dense SWCNTs were made into a flexible heater and as electrochemical electrodes for supercapacitors.

The water-assisted CVD method had been extended to achieve super growth of vertically aligned MWCNTs (>1 mm) with a deposition time up to 12 h [Yun *et al.*, 2006]. Super growth of SWCNTs had also been achieved using hot filament CVD without the addition of water vapor. Xu *et al.* [2006] proposed that the generation of atomic hydrogen and rapid heating of the hot filament permit nucleation and sustained growth of SWCNTs.

The super growth of horizontally aligned CNTs has been demonstrated using improved CVD techniques. Wang *et al.* [2009] reported the water-assisted CVD synthesis of ultralong horizontally lying SWCNTs with axial lengths of 18.5 cm on clean Si substrates (see Figure 2.12). A CNT film was used as catalyst supports for the ~6 nm Fe-Mo nanoparticles so that the substrates remained clean. The ultralong SWCNTs showed uniform electric transport properties along the length, which is important for SWCNT-based field-effect transistors. A kite mechanism had been suggested to explain the super growth of horizontally lying SWCNTs [Huang *et al.*, 2004]. When the SWCNTs have grown sufficiently long, it is lifted off from the substrate by the carrier gas and continued to grow further and achieve super growth. As shown in Figure 2.13, 100 SWCNT-based field effect transistor (FET) device was constructed on a 3 cm individual SWCNT with a channel length of 2 μm using e-beam photolithography. The diameter of SWCNT was determined to be about 1.36 nm using atomic force microscopy and Raman spectroscopy. The source and drain electrodes are composed of 200 nm wide Cr (thickness 10 nm)/Pt (50 nm) and p-doped Si substrates as the back gate. The electrical properties of SWCNT-based FET devices were observed to be uniform along the 3 cm individual SWCNT. The maximum average current was ~17.7 μA at V_{gs} = −20 V and V_{ds} = 10 V. The on/off ratio was in the order of 10^6. The carrier mobility of SWCNT-based FET devices was estimated to be 0.14 pF/cm. The effective field effect mobility at the linear region of low bias conductance was ~800 cm^2/Vs at room temperature, which was higher than typical semiconductor devices with 2 μm channel length.

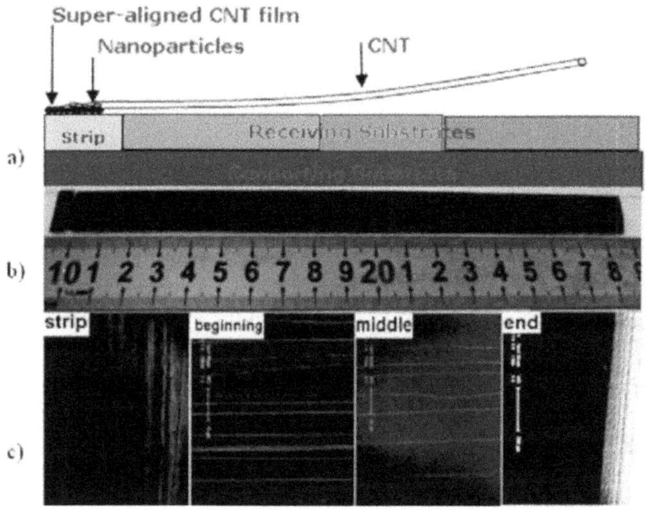

Figure 2.12 (a) Schematic of the growth method of ultralong CNTs. The catalyst were deposited on a superaligned CNT film, which was laid over a strip of silicon, and the catalyst nanoparticles were on the surface of the film. CNTs could be lifted up and caught by the laminar gas flow, and then, they grew in the direction parallel to the gas flow direction. Two or more pieces of SiO_2/Si substrates could be used to accept the CNT array simultaneously. The quartz glass was below the silicon chip as the support. (b) The optical images of an ultralong CNT array. (c) The SEM images of the strip and the beginning, middle, and end of the ultralong CNT array from left to right. The CNTs could cross the slit between the receiving substrates. We can see that some of the CNTs stopped growth and settled to the surface of the substrates during the growth process and became sparse at the end of the substrate. [Wang *et al.*, 2009]

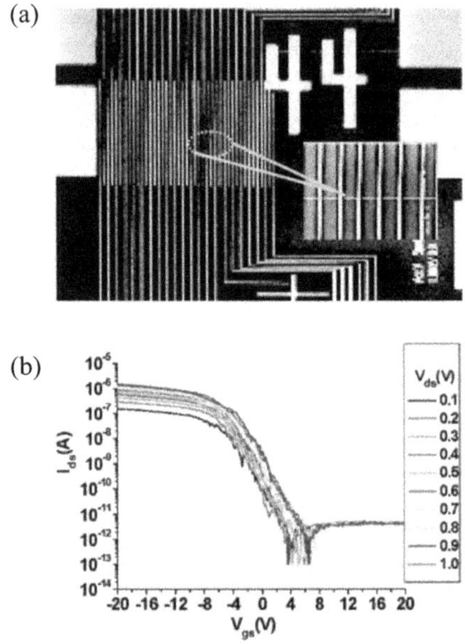

Figure 2.13 (a) The SEM image of FETs array made by photolithography and electron beam lithography. The inset is the enlarged view of the dashed circle. (b) Current (I_{ds}) versus gate voltage (V_{gs}) characteristics of ultralong SWCNT-FET for V_{ds} ranging from 0.1 to 1.0 V, showing that the on/off ratio is about 106 at this bias region. [Wang *et al.*, 2009] See also Color Insert.

2.4.3 Plasma-Assisted Growth of CNTs

2.4.3.1 Low-temperature synthesis

The synthesis of CNTs using chemical vapor deposition methods requires temperatures greater than ~750°C, which far exceeds the temperature limit of microelectronic fabrication of ~400–500°C. Plasma-enhanced chemical vapor deposition (PECVD) method might reduce the synthesis temperature of CNTs. The gas molecules are efficiently dissociated by the plasmatic energy and the synthesis of carbon nanotubes might occur at lower temperatures. Another advantage of using PECVD method is the presence of a built-in electric field in a plasma sheath, which will align the growing CNTs along the electric field lines. Large-scale Monte Carlo simulations [Tam and Ostrikov, 2008] of SWCNT synthesis show that PECVD methods are suitable for low-temperature synthesis and have two orders of magnitude higher growth rates than CVD methods. The delivery and redistribution of carbon adatoms between the catalysts and the nanotubes' bases are more efficiently controlled in the PECVD than CVD methods. The prevention of catalyst poisoning and formation of amorphous carbon are prevented due to plasma etching and the growth of uninterrupted ultra-long SWCNTs is possible. However, excessive plasma etchings in practical PECVD processes are undesirable for SWCNT growth.

The growth of CNTs using high-temperature CVD methods had been proposed to be a vapor–liquid–solid mechanism [Gavillet *et al.*, 2001]. Would the growth mechanism of CNTs change as the temperature is reduced? The catalyst for high-temperature CNT growth is in a liquid drop state and carbon species from the chemical vapor dissolved into it. Carbon nanotubes are precipitated from the supersaturated eutectic liquid. The activated energy for thermal CVD (~700°C) is reported to be ~1.2–1.8 eV [Lee and Park, 2002; Geohegan *et al.*, 2003]. A new growth mechanism of CNTs at low temperatures (< 400°C) is needed since the catalysts might remain as solids at very low temperatures. The activation energies for low-temperature plasma-assisted growth of CNTs were determined to be ~0.2–0.4 eV [Hoffmann *et al.*, 2004], which was similar to the activation energy of surface diffusion of carbon atoms on polycrystalline Ni (0.3 eV) [Mojica and Levenson, 1976]. Carbon diffusion on the catalyst surface was suggested as the rate-limiting step for low-temperature plasma-assisted growth of CNTs [Hoffman *et al.*, 2004].

With hindsight, excessively plasma etchings damages SWCNTs and favors the synthesis of multiwalled nanotubes, which are more resistant to etchings. The synthesis of SWCNTs requires the use of special plasma configurations such as remote plasma or point arc discharge and the substrates are minimally exposed to the plasma sheath. The plasma ion flux is very high (n_{ion} ~10^{10} cm^3) within the plasma sheath, and the ions energetically bombard SWCNTs at high energy of ~100 eV and causes C-C bond breakage. NH_x and H plasma radicals chemically etch the exterior surfaces of carbon nanotubes. The resistibility of the SWCNTs toward ion etching effects depends on the growth mechanism in a PECVD process. In tip-growth mechanism, the catalyst is at the tip of the vertically growing SWCNT and shields the nanotube walls from the ion etching effects. In a base-growth mechanism the catalyst is adhered to the substrate, and the uncapped tips of the vertically growing SWCNTs are easily destroyed by the impinging ions.

The operation modes of plasma are important to the PECVD synthesis of SWCNTs. Figure 2.14 shows the plasma modes of an atmospheric pressure radio frequency discharge (APRFD) reactor, which transited from a so-called α-mode (60 W) to γ-mode (100 W) as the input power is increased from 60 to 100 W [Nozaki *et al.*, 2007]. In the α-mode, emission layers are due to the spectra of CH at 432 nm and created near momentary cathode (I and II). The thickness of the plasma sheaths is ~630–910 μm and is formed between momentary cathode layers and electrodes. The potential drops drastically from the momentary cathode layers (I and II) towards the electrodes. At higher input power, the electric field strength in the plasma sheath also increases, causes more energetic ion bombardment and favors a transition to γ-mode. Intense ionization in the vicinity of electrodes occurs when the ion hit the electrodes, and secondary electrons are generated and accelerated into the plasma sheath by the electric field. In the γ-mode, the highly intense plasma spots induce plasma-heating of the catalysts and damage CNTs. Hence the synthesis of SWCNTs is greatly influenced by the modes of plasma and undesirable ion bombardment of substrate should be minimized.

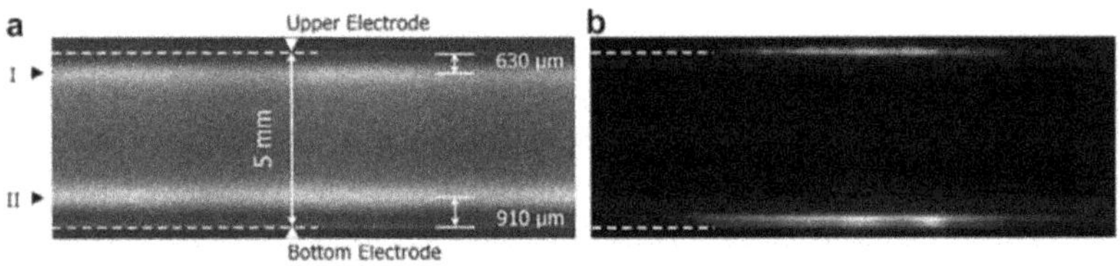

Figure 2.14 Emission distribution of CH (432 nm) in different operation regimes of APRFD. Emission intensity in the γ-mode is 10 times greater than that of α-mode regime: (a) α-mode (60 W); (b) γ-mode (100 W). [Nozaki *et al.*, 2007]

Luo *et al.* [2008] investigated the effects of ion bombardment and synthesized high-quality VA-SWCNTs in a plasma sheath of a capacitively coupled r.f. PECVD by optimizing the plasma input power (sheath voltage, V) and gas pressure. According to a simplified ion space-charge-limited model [Godyak *et al.*, 1991], the plasma ion flux (n_{ion}) and ion energy (E_{ion}) are qualitatively related to gas pressure (P) and sheath voltage (V) as follows:

$$E_{\text{ion}} \propto V^{4/5}P^{-1/2} \tag{2.2}$$

$$n_{\text{ion}} \propto VP^{3/4} \tag{2.3}$$

When the gas pressure (P) is fixed, the increment of plasma input power significantly increases the ion flux impinging SWCNTs, while the ion energy is moderately increased. On the other hand, for a fixed plasma input power, the plasma sheath varied with pressure as $V \propto P^{1/2}$. The plasma ion flux and energy can be rewritten as follow: $E_{\text{ion}} \propto P^{-1/10}$ and $n_{\text{ion}} \propto P^{5/4}$, which indicates that the ion etching effects are dominated by the ion flux. Thus reducing the incoming ion flux is essential to synthesize high-quality SWCNTs in a plasma sheath. Luo *et al.* [2008] also showed that the SWCNTs resisted ion etching at higher synthesis temperature (> 600°C).

The kinetics of reactive ion etching on the synthesis of SWCNTs was studied by Kato *et al.* [2008] studied in a parameter-controlled PECVD reactor. The time evolution of SWCNT graphitization due to reactive ion etching effects, and the SWCNT growth kinetics was monitored using Raman spectroscopy. The degree of graphitization of SWCNTs was assumed to be related to its tangential mode (I_G). For weak and negligible ion etching (low plasma power 40 W), the growth kinetics of SWCNTs in a PECVD method at 750°C was investigated to be very similar to a thermal CVD process, which can be expressed as follows:

$$I_G = I_o\left[1 - \exp\left[-\left(\frac{t_g - \Delta_t}{\tau_{gro}}\right)\right]\right] \tag{2.4}$$

where I_o, Δt, and τ_{gro} denote saturated tangential modes of SWCNTs, incubation time, and relaxation time of the growth respectively. To account for significant etching effect (high plasma power 100 W), Kato *et al.* [2008] proposed a new growth equation for the growth kinetics of SWCNTs:

$$I_G = I_o \exp\left[\frac{-t_g}{\tau_{\text{etc}}}\right]\left\{1 - \exp\left[\frac{-(t_g - \Delta t)}{\tau_{\text{gro}}}\right]\right\} \tag{2.5}$$

where τ_{etc} denotes relaxation time of the etching. The modified growth kinetics model of SWCNTs, which included the ion etching effects, agreed well with experimental results. The modified growth kinetics model of SWCNTs demonstrated that high radical densities in hydrocarbon plasma and H densities contributed significantly to the ion etching effects.

Zhang *et al.* [2006] observed that H and CH_3 ions in a methane plasma selectively etched metallic SWCNTs into hydrocarbon species, while semiconducting SWCNTs remained unmodified. The higher radius curvature and strain in the C-C bonding of small-diameter SWCNTs were also preferentially etched over larger ones. Theoretical studies had predicted that ~1/3 of as-synthesized SWCNTs were metallic and the remaining ~2/3 nanotubes were semiconducting. Li *et al.* [2004] and Qu *et al.* [2008] also reported the preferential etching of metallic SWCNTs during the PECVD synthesis steps. Therefore, PECVD methods yield mainly semiconducting SWCNTs.

To minimize the effects of reactive ion etching, which were inherent in PECVD processes, various strategies had been developed. For example, in a remote downstream PECVD process [Bae *et al.*, 2005], the plasma sheath was generated far away from the SWCNT substrates such that the ion etching effects were reduced. Kato *et al.* [2006] created a small opening (diameter 10 mm) in the center of the bottom electrode to diffuse the spatial distribution of plasma and placed the substrate placed below the opening. The diffused plasma reduced ion bombardment and promoted the growth of freestanding individual SWCNTs. Nozaki *et al.* [2008] had developed an atmospheric pressure radio frequency PECVD to synthesize VA-SWCNTs. The high collision frequency of the molecules at atmospheric pressure significantly reduced the ion etching of the SWCNTs.

2.4.3.2 Alignment of CNTs within plasma sheath

Merkulov *et al.* [2001] observed that the carbon nanofibers alignment was dependent on the location of the catalytic nanoparticles. Alignment of free-standing carbon nanofibers in a field was observed for tip growth model but not base growth model. The high polarizability of CNTs in an electric field assisted its directional growth. The non-alignment of carbon nanofibers is due to the bending of the nanofibers during synthesis. As shown in Figure 2.15, the alignment of carbon nanofibers is due to a feedback mechanism associated with a tensile-compressive stress generated at the catalyst–nanofibers interface [Merkulov *et al.*, 2001]. Neither ion bombardment nor electrostatic attraction plays an important role for the bending of carbon nanotubes. The carbon nanotubes would be synthesized vertically aligned when the axes of carbon nanotubes are growing perpendicularly to the substrates along the direction of the electric field and a uniform tensile stress occur across the catalyst–nanotubes interface (see Figure 2.15a,b). If there is a fluctuation in the carbon precipitation at the catalyst–nanotube interface, however, the carbon nanotubes would start to bend (Figure 2.15c,d). For tip growth model, the catalyst–nanotubes interface experienced a negative feedback, which equalized the fluctuation of the carbon precipitation and re-aligned the carbon nanotubes for vertical growth. When the catalyst–nanotube interface is attached to the substrate (base growth model), the tensile stress results in greater precipitation of carbon than the compressive stress and causes the carbon nanotubes to bend significantly. An unstable positive feedback occurs and results in non-aligned carbon nanotubes.

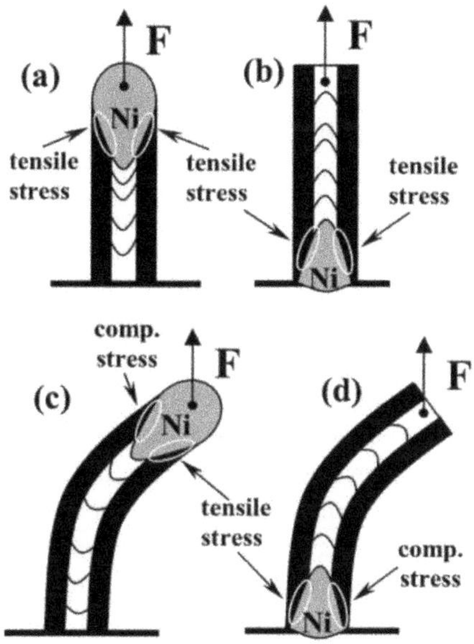

Figure 2.15 Bending of carbon nanofibers due to spatial fluctuations in carbon precipitation at the Ni catalyst/nanofiber interface. [Merkulov *et al.*, 2001]

However, the vertical alignment of dense CNT forests synthesized in base growth model has been observed in a PECVD process. The collective van der Waals interactions among the tubes (crowding effects) facilitate the alignment of the CNT forests.

When a conductive substrate is used to synthesize CNTs in a PECVD process, the electric field lines of the plasma sheath are always perpendicular to the substrate. On the other hand, when an insulating surface is deposited on top of the conducting substrate, plasma-induced surface charging occurs in the electric field of plasma sheath [Law *et al.*, 2007; Chai *et al.*, 2009]. The insulating surface gains an overall negative charge and repels the electron flux. In a steady-state plasma, the potential of the insulating surface (V_f) couples to the plasma potential (V_P) via the sheath: $V_f = V_P - V_{sh}$,where V_P and V_{sh} is the potential of the plasma and plasma sheath respectively. Law *et al.* [2007] used this plasma-induced surface charging phenomenon to redirect electric field of the plasma sheath to be horizontally across two adjacent electrodes and achieve horizontal growth of CNTs. As displayed in Figure 2.16, a "floating" electrode, which is separated from the substrate by an insulator, develops a floating potential V_{fe} with respect to the plasma potential. The potential of substrate (V_{fsub}) is the plasma potential at the wafer edges. Consequently, an electric field is developed across the electrodes due to the potential difference of V_{fe} and V_{fsub} (see Figure 2.16) and results in the synthesis of horizontally aligned SWCNTs.

Lin *et al.* [2004] tilted the substrates in a plasma sheath such that the electric field lines are aligned vertically from the plasma to the sample stage and synthesized inclined CNTs. The corners of microstructures in the proximity of the substrates distort the electric field and yield inclined CNTs. The synthesis of inclined CNTs with respect to the substrate has important applications such as AFM probe tips, microfluidic and channel valve or filter.

The growth direction of CNTs in a PECVD process is controllable by applying an external magnetic field to steer the direction of the ferromagnetic iron nano-catalysts at the tips of CNTs. An external magnetic field is applied to direct the alignment of CNTs instead of the built-in electric field within the plasma sheath. Ohmae *et al.* [2008] demonstrated the bending of CNTs due to varying external magnetic field direction (10 mT) during the CNT growth. CNT bundles with hook-, arch-, and ladder-shaped were synthesized observed when the direction of magnetic field were repeatedly changed. The iron nanoparticles were observed to be ellipsoidal in shape with the major axes were along the growth direction of CNTs. The iron nanoparticles followed the direction of magnetic field during growth and therefore the field controlled the alignment of CNTs. The self-bias of the PECVD (the electric potential between the plasma and substrate) was approximately –10 eV and generated a high electric field (10^3–10^4 V/m). The application of an external electric field to control the directional growth of CNT will be less effective than magnetic field.

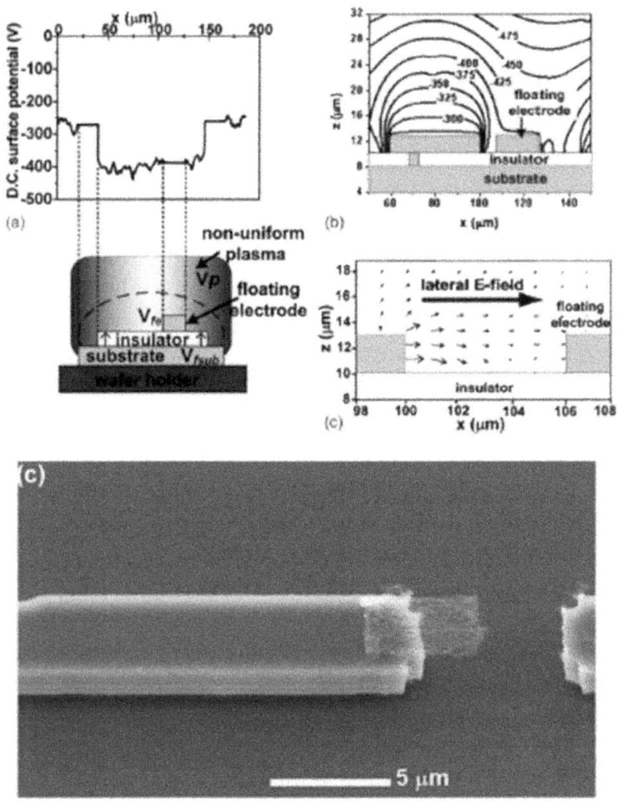

Figure 2.16 OOPIC PRO simulation showing (a) difference in surface potential between the substrate and the isolated electrode in a plasma, (b) equipotential lines due to charging encountered by the geometry, and (c) electric field vectors in the vicinity of the gap between the electrode pair in (b). (d) Horizontally directed growth of MWCNTs from the short/float electrode pair. [Law *et al.*, 2007]

2.4.3.3 Controlled growth of CNTs on a substrate with desired patterns

By careful deposition of the catalyst in predetermined areas via electron beam lithography or contact printing, aligned CNTs can be grown on the substrate with a desired pattern using PECVD. The type, diameter and length of the CNTs can be well controlled. This is particularly useful for device applications such as field emitter, flat panel display, and other vacuum microelectronics.

In addition to catalyst patterning, substrate rational design is also important, which can enhance catalyst-substrate interactions and control the catalyst particle size. Porous silicon obtained by electrochemical etching of n-type silicon wafers in hydrofluoric acid/methanol solutions is found to be an ideal substrate for this approach. Our group [Luo *et al.*, 2008] has developed a low-temperature PECVD by optimizing plasma input power and gas pressure, which enables us to controllably grow high-quality CNTs of various types at 450–500°C, using commercially available kitchen Al foil as the substrate. The Al foil is cheap and flexible. Fe or Co catalysts can be easily deposited on the substrate using simple technology such as dip coating, air-brush spraying or even pen brushing. The post-growth transfer of CNTs from Ai foil to other substrates can be easily performed by hot press or other techniques.

The growth of isolated nanotubes on specific surface sites on flat silicon oxide substrate has been developed by Kong *et al.* [1998]. This growth approach enables the integration of nanotubes into addressable structures. Our group [Lim *et al.*, 2010] has shown that the correct selection of the catalyst is crucial in this application. Under identical growth conditions the horizontal growth of SWCNTs is successful using a Co–Gd catalyst. The length of the horizontally aligned SWCNTs can be 1 mm. However, the growth of the horizontal SWCNTs on Co–Mn catalysts is self-terminated due to coiling, while no SWCNT growth occurs if Co–Mo catalysts are used.

2.5 Purification and Sorting of CNTs for Applications

2.5.1 Purification

As-synthesized CNTs are often contaminated with undesirable byproducts, which include metal nanoparticles, fullerenes, nanocrystalline graphite, and amorphous carbon coating the sidewalls of nanotubes. These impurities will interfere with CNTs properties and their applications. Efficient and mild purification methods, which remove impurities and do not significantly damage the carbon nanotubes, are desirable for fundamental studies and applications.

Traditional purification involves the use of strong oxidants such as HNO_3 and $KMnO_4/H_2SO_4$ and the controlled burning of CNTs in air to remove amorphous carbon and other graphitic impurities. However, the use of strong oxidants not only removes the impurities but also damages the nanotube substantially.

Feng *et al.* [2008] developed a mild three-step purification method that involves the selective oxidation of carbon species in hydrogen peroxide (H_2O_2) at room temperature. The as-synthesized CNTs were refluxed in concentrated HCl acid for 1 hr to remove metal catalysts and oxide supports. The CNTs were washed with distilled water and ethanol to remove the excess acid and dried at 90°C overnight. The resulting CNTs were dispersed in 30 wt% H_2O_2 by sonication and then stirred continuously in the H_2O_2 solution for an optimal period of 7 days at room temperature. The CNT were refluxed in concentrated HCl for 1 hr again to remove the residual catalysts. The H_2O_2 slowly but selectively etched away the carbonaceous impurities and oxidized metals within carbon nanoparticles, which could be removed by HCl treatment.

Harutyunyan *et al.* [2002] developed a scalable method for the purification of SWCNTs based on microwaving heating in air. The SWCNTs were heated at 500°C using microwave for 20 min in flowing dry air. The local temperature of the residual metal catalysts was substantially heated up by the microwaves and opened up the carbon shells encapsulating the metal catalysts. A mild 4 M HCl acid reflux for 6 hr removed most of the residual metal catalyst to lower than 0.2 wt%. The microwave heating of SWCNTs in air did not significantly damage the nanotubes due to the short process time.

2.5.2 Sorting of SWCNTs

2.5.2.1 Surfactant-assisted dispersion of SWCNTs

The size distribution and/or the electronic type of SWCNTs might have been narrowed using many possible synthetic strategies such as controlling the temperature, pressure, carbon feedstock and catalysts. However, the as-synthesized SWCNTs still exhibit heterogeneous properties, which fall short of stringent industrial application requirements. As a result, post-synthesis treatments are needed to further sort SWCNTs according to its lengths, diameters, chiralities, and electronic types [Hersam, 2008].

A typical surfactant-assisted dispersion of SWCNTs involves the interactions of nanotubes with active chemical agents and debundling of nanotube rope in a solvent using ultrasonication. The resulting SWCNT solution is centrifuged at high speed (~40000–100000 rpm) to obtain a supernatant that is composed of individually dispersed and functionalized SWCNTs. Pressurizing and forcing CNT dispersion through a special filtration unit forms a "buckypaper," which consists of 3D networks of nanotubes. By using the advantages of column chromatography (potassium polyacrylate as a stationary phase) and vacuum filtration large amount of SWCNTs can be purified and separated within hours. Dispersed SWCNTs can be controllably deposited onto patterned substrates for device application. Electronic devices based on purified SWCNTs have been shown to be vastly superior to those of circuits made from raw SWCNTs. In this application, SWCNT material dispersed in SDBS surfactant solution was flowed over a magnetic field gradient (~0.08 T/cm) to further remove the magnetic impurity. The magnetic purified suspension was deposited onto a degenerately doped oxidized silicon wafer, which was covered by a 3-aminopropyl triethoxysialne (APTS) monolayer. All these examples show that surfactant aided sonication is an essential purification for many important applications.

2.5.2.2 Separation of metallic from semiconducting SWCNTs

The exterior sidewalls of SWCNTs exhibit selective chemistry and vary as a function of electronic type, diameter and chirality, which have been exploited to sort out metallic and semiconducting SWCNTs. The rich π-electrons near the Fermi level in metallic SWCNTs offers charge-transfer process and bond formation. Aryl diazonium salts, dichlorocarbene and nitronium ions selectively form covalent bonds with metallic SWCNTs due to the rich π-electrons [Fu and Sun, 2003], while semiconducting SWCNTs do not participate in the reactions. However, covalent functionalization undesirably perturbs the intrinsic properties of SWCNTs and nitronium ions destroy small-diameter SWCNTs.

A milder functionalization of SWCNTs based on non-covalent chemistry, which is less destructive than covalent chemistry, has been preferred to sort SWCNTs. A selective and large-scale separation of metallic from semiconducting SWCNTs based on non-covalent amine chemistry has been demonstrated [Maeda *et al.* 2005]. Metallic SWCNTs are selectively functionalized with proplyamine and isoproplyamine in THF and the enrichment of metallic SWCNTs reaches up to 87%. Likewise, selective non-covalent functionalization of semiconducting SWCNTs has been achieved with porphyrin chemistry. The attachment of chiral porphyrin onto semiconducting SWCNTs also depends on the chirality of nanotubes. Therefore the chirality sorting of SWCNTs within semiconducting species is possible.

Non-covalent surfactant encapsulation of SWCNTs has been applied to sort the nanotubes according to electronic and diameter types [McDonald *et al.* 2006]. Two families of surfactants, namely anionic-alkyl amphiphiles (sodium dodecyl sulphate (SDS) and sodium dodecylbenzene sulphonate) and bile salts (sodium cholate (SC), sodium deoxycholate, and sodium taurodeoxycholate) readily interact with SWCNTs under ultrasonication condition. The separation of metallic from semiconducting surfactant-encapsulated SWCNTs has been achieved using dielectrophoresis method, which relies on the differences in the dielectric constants of SWCNT types [Krupke *et al.* 2003]. A most promising and scalable method called density gradient ultracentrifugation (DGU) sorts surfactant-encapsulated SWCNTs by diameters, bandgap, and electronic types based on subtle difference in the buoyant densities of nanotubes [Arnold *et al.* 2006]. In DGU process, an aqueous solution of non-ionic density gradient medium (iodixanol) is used to form density gradients in centrifuge tubes. When a particular type of SWCNT buoyant density matches the density gradient, that SWCNT species will form a buoyant layer in the centrifuge tube. Since the SWCNT buoyant density depends on its diameter, DGU method is capable of sorting SWCNTs. Sodium cholate (SC) is much more effective than sodium dodecyl sulphate (SDS) in separating SWCNTs into distinct buoyant layers using DGU method. As shown in Figure 2.17, SWCNTs in 1.1–1.6 nm diameter range are clearly fractionalized into several buoyant layers with the topmost layers (lowest buoyant density) consist of SWCNTs with smallest diameter. In addition, the topmost is also enriched with semiconducting SWCNTs based on the optical absorption spectra. Chiral enrichment of semiconducting SWCNTs is achieved by repeating the DGU process. After three cycles of DGU process, the as synthesized CoMoCAT SWCNTs, which are predominantly composed of (6, 5) and (7, 5) semiconducting tubes, are enriched with semiconducting (6, 5) tubes (mean diameter 0.76 nm) whereby >97% of the nanotubes' diameter is within 0.02 nm of the mean diameter. Although SDS does not separate SWCNTs in different buoyant fractions in DGU method, the addition of SDS to SC and judicious tuning of SC/SDS co-surfactant ratio greatly improve the separation of diameter and electronic type of SWCNTs. By varying the ratio of SDS/SC, the DGU-based separations of metallic (3:2 SDS/SC ratio) and semiconducting (1:4 SDS/SC ratio) SWCNTs are significantly improved. On the basis of optical absorption spectra (Figure 2.17), the orange layer is mainly composed of semiconducting SWCNTs, while the green layer just below the orange layer is enriched with metallic SWCNTs.

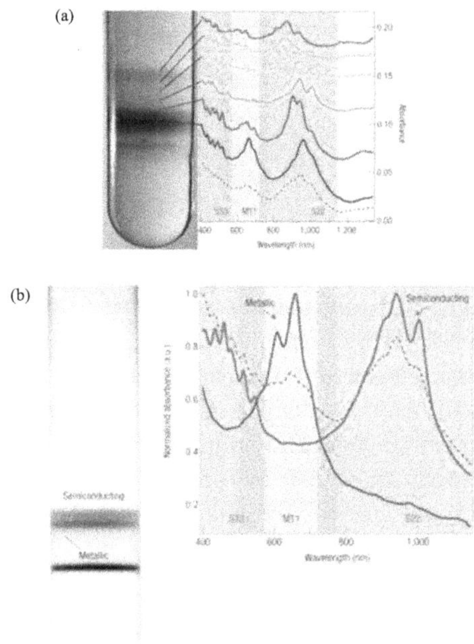

Figure 2.17 Separation of surfactant encapsulated SWCNTs into different buoyant layers using density gradient ultracentrifugation. (a) Sodium cholate encapsulated SWCNTs (diameter 1.1–1.6 nm) are sorted according to its diameter and electronic types. The species of SWCNTs within each buoyant layer are confirmed by optical absorbance spectra. (b) Optimization of separation by electronic types is carried using co-surfactant of sodium cholate and sodium dodecyl sulphate. [Arnold *et al.* 2006] See also Color Insert.

2.5.3 Modification of CNTs for Various Applications

Post-synthesis modifications of SWCNTs are often needed to adjust their properties to suit a particular application. SWCNTs are observed to disperse in *N*-methyl-2-pyrrolidone (NMP) solvent with the aid of sonication to form a stable CNT solution/ink [Beecher *et al.*, 2007]. The preparation of CNT inks is used in ink-jet printing for cheap and large area fabrication of plastic transistor, photovoltaic, organic light emitting diodes, and electrochemical electrodes. Du *et al.* [2002] employed electrophoretic deposition method to deposit MWCNT thin films on metallic substrates. The MWCNTs were dispersed in a mixture of ethanol and acetone solvents. Two electrodes placed 50 mm apart were immersed into the ethanol/acetone solvent and applied a constant voltage of 45 V. Strong hydrogen evolution was observed at the cathode and resulted in the deposition of porous MWCNT films with pore size ranging from 1 to 70 μm. The microstructure of the MWCNT films was greatly influence by the ratio of ethanol to acetone. These electrophoretic deposited MWCNT films exhibit excellent electrochemical supercapacitance.

2.6 Conclusions

In this chapter, an overview of various methods for the production of carbon nanotubes is presented with emphasis of chemical vapor deposition (CVD). For the continuous mass production of carbon nanotubes, both fluidized-bed CVD and floating catalyst methods are highly promising and widely adopted in industrial processes. Careful catalyst design and operating condition control are the two big challenges in these processes. The synthesis of dominant (6, 5) and (7, 5) semiconducting nanotubes depends on the selectivity of CoMoCAT process, while nano agglomerate fluidized-bed (NAF) process relies strongly on the control of the operating condition. Macroscopic assemblies of CNTs into sheet and yards can be obtained by spinning the yard from CNT forests or directly from the floating catalyst CVD furnaces. Super-growth of aligned SWCNTs can be realized by introducing water vapor into CVD process. The water vapor acts as a weak oxidizer and suppresses the formation of amorphous carbon and prevents catalyst poisoning. The application of plasma in CVD methods helps to greatly lower the growth temperature of CNTs. The selective synthesis of good quality SWCNTs is achieved by careful control of the operating condition when ion bombardment during the growth of CNT is minimized. The in-built electric field within the plasma sheath is manipulated advantageously to control the alignment of CNTs.

As-synthesized CNTs usually contain undesirable byproducts, including metal nanoparticles, nanocrystalline graphite, and amorphous carbon. These impurities will interfere with CNTs' properties and should be removed. A few efficient and mild purification methods that can remove the impurities but do not significantly damage the carbon nanotubes are summarized in the chapter, particularly suitable for fundamental studies. The separation of metallic and semiconducting nanotubes still requires post-synthesis/purification treatment. The most promising and scalable technique of density-gradient ultracentrifugation sorts carbon nanotube by diameter, bandgap, and electronic type using special surfactants. It will be a deep aspiration to develop a CVD technique that addresses the five challenges listed in the text at once.

References

Ajayan, P. M. (1999). Nanotubes from carbon. *Chem. Rev.* **99**, pp. 1787–1800.

Antisar, M. V., Marazzi, R., and Krsmanovic, R. (2003). Synthesis of multiwall carbon nanotubes by electric arc discharge in liquid environments. *Carbon* **41**, pp. 2393–2401.

Arnold, M. S., Green, A. A., Hulvat, J. F., Stupp, S. I., and Hersam, M. C. (2006). Sorting carbon nanotubes by electronic structure using density differentiation. *Nat. Nanotechnol.* **1**, pp. 60–65.

Bachilo, S. M., Balzano, L., Herrera, J. E., Pompeo, F., Resasco, D. E., and Weisman, R. B. (2003). Narrow (n,m)-distribution of single-walled carbon nanotubes grown using a solid supported catalyst. *J. Am. Chem. Soc.* **125**, pp. 11186–11187.

Bae, E. J., Min, Y., Kang, D., Ko, J., and Park, W. (2005). Low-temperature growth of single-walled carbon nanotubes by plasma enhanced chemical vapor deposition. *Chem. Mater.* **17**, pp. 5141–5145.

Beecher, P., Servati, P., Rozhin, A., Colli, A., Scardaci, V., Pisana, S., Hasan, T., Flewitt, A. J., Robertson, J., Hsieh, G. W., Li, F. M., Nathan, A., Ferrari, A. C. and Milne, W. I. (2007). Ink-jet printing of carbon nanotube thin film transistors. *J. Appl. Phys.* **102**, pp.043710-1-7.

Bolshakov, A. P., Uglov, S. A., Saveliev, A. V., Konov, V. I., Gorbunov, A., Pompe, W., and Graff, A. (2002). A novel CW laser-powder method of carbon single-wall nanotubes production. *Diamond Relat. Mater.* **11**, pp. 927–930.

Bourlon, B., Miko, C., Forro, L., Glattli, D. C., and Bachtold, A. (2004). Determination of intershell conductance in multiwalled carbon nanotubes, *Phys. Rev. Lett.* **93**, p. 176806.

Bronikowski, M. J., Willis, P. A., Colbert, D. T., Smith, K. A., and Smalley, R. E. (2001). Gas-phase production of carbon single-walled nanotubes from carbon monoxide via the HiPco process: A parametric study. *J. Vac. Sci. Technol. A.* **19**, pp. 1800–1805.

Cao, A., Ci, L, Wu, G., Wei, B., Xu, C., Liang, J., and Wu, D. (2001). An effective way to lower catalyst content in well-aligned carbon nanotube films. *Carbon* **39**, pp. 152–155.

Chai, Y., Xiao, Z., and Chan, P. C. H. (2009). Electron-shading effect on the horizontal aligned growth of carbon nanotubes. *Appl. Phys. Lett.* **94**, p. 043116.

Dai, H. (2002). Carbon nanotubes: synthesis, integration, and properties, *Acc. Chem. Res.* **35**, pp. 1035–1044.

Danafar, F., Fakhru'l-Rzai, A., Salleh, M. A. M. and Biak, D. R. A. (2009). Fluidized bed catalytic chemical vapor deposition synthesis of carbon nanotubes — A review, *Chem. Eng. J.* **155**, pp. 37–48.

Dekker, C. (1999). Carbon nanotubes as molecular quantum wires, *Phys. Today* **52**, pp. 22–28.

Du, C. S., Heldebrant, D., and Pan, N. (2002). Preparation and preliminary property study of carbon nanotubes films by electrophoretic deposition. *Mater. Lett.* **57**, pp. 434–438.

Eklund, P. C., Pradhan, B. K., Kim, U. J., Xiong, Q., Fischer, J. E., Friedman, A. D., Holloway, B. C., Jorgan, K., and Smith, M. W. (2002). Large-scale production of single-walled carbon nanotubes using ultrafast pulses from a free electron laser. *Nano Lett.* **2**, pp. 561–566.

Endo, M. (1988). Grow carbon fibers in the vapor phase. *CHEMTECH*. 18, pp. 568–576.

Feng, Y., Zhang H., Hou Y., McNicholas, T. P., Yuan, D., Yang, S., Ding, L., Feng, W., and Liu, J. (2008). Room temperature purification of few-walled carbon nanotubes with high yield, *ACS Nano* **2**, pp. 1634–1638.

Fischer, J. E., Zhou, W., Vavro, J., Llaguno, M. C., Gunthy, C., Haggenmueller, R., Casavant M. J., Walters, D. E., and Smalley, R. E. (2003). Magnetically aligned single wall carbon nanotube films: Preferred orientation and anisotropic transport properties. *J. Appl. Phys.* **93**, pp. 2157–2163.

Fu, K., and Sun, Y. (2003). Dispersion and solubilization of carbon nanotubes. *J. Nanosci. Nanotechnol.* **3**, pp. 351–364.

Futaba, D. N., Hata, K., Yamada, Hiraoka, T., Hayamizu, Y., Kakudate, Y., Tanaike, O., Hatori, H., Yumura, M., and Iijima, S. (2006). Shape-engineerable and highly densely packed single-walled carbon nanotubes and their application as supercapacitor electrodes. *Nat. Mater.* **5**, pp. 987–994.

Gavillet, J., Loiseau, A., Journet, C., Willaime, F., Ducatselle, F., and Charlier, J. C. (2001). Root-growth mechanism for single-wall carbon nanotubes. *Phys. Rev. Lett.* **87**, p. 275504.

Geldart, D. (1973). Types of gas fluidization. *Power Technol.* **7**, pp. 285–292.

Geohegan, D. B., Puretzky, A. A., Ivanov, I. N., Jesse, S., Eres, G., and Howe, J. Y. (2003). *In situ* growth rate measurements and length control during chemical vapor deposition of vertically aligned multiwall carbon nanotubes. *Appl. Phys. Lett.* **83**, pp. 1851–1853.

Gohier, A., Minea, T. M., Djouadi, M. A., and Granier, A. (2007). Impact of the etching gas on vertically oriented single wall and few walled carbon nanotubes by plasma enhanced chemical vapor deposition. *J. Appl. Phys.* **101**, pp. 054317.

Godyak, A., Piejak, R. B. and Alexandrovich, B. M. (1991). Electrical characteristics of parallel-plate RF discharges in argon. *IEEE Trans. Plasma Sci.* **19**, pp. 660–676.

Harutyunyan, A. R., Pradhan, B. K., Chang, J., Chen, G. and Eklund, P. C. (2002). Purification of single-walled carbon nanotubes by selective microwave heating of catalyst particles. *J. Phys. Chem. B* **106**, pp. 8671–8675.

Hata, K., Futaba, D. N., Mizuno, K., Namai, T., Yumura, M., and Iijima, S. (2004). Water-assisted highly efficient synthesis of impurity-free single-walled carbon nanotubes. *Science* **306**, pp. 1362–1364.

Hersam, M. C. (2008). Progress towards monodisperse single-walled carbon nanotubes. *Nat. Nanotechnol.* **3**, pp. 387–394.

Hoffmann, S., Kleinsorge, B., Ducati, C., Ferrari, A. C., and Robertson, J. (2004). Low-temperature plasma enhanced chemical vapor deposition of carbon nanotubes. *Diamond Relat. Mater.* **14**, pp. 1171–1176.

Huang, S. M., Woodson, M, Smalley, R., and Liu, J. (2004). Growth mechanism of oriented long single walled carbon nanotubes using "fast-heating" chemical vapor deposition process. *Nano Lett.* **4**, p. 1025.

Iijima, S. (1991). Helical microtubules of graphitic carbon. *Nature (London)* **354**, pp. 56–58.

Jiang, K., Li, Q., and Fan, S. (2002). Spinning continuous carbon nanotube yarns. *Nature* **419**, pp. 801.

Kato, T., Hatakeyama, R., and Tohji, K. (2006). Diffusion plasma chemical vapor deposition yielding freestanding individual single-walled carbon nanotubes on silicon-based flat substrate. *Nanotechnology* **17**, p. 2223.

Kato, T., and Hatakeyama, R. (2008). Kinetics of reactive ion etching upon single-walled carbon nanotubes. *Appl. Phys. Lett.* **92**, p. 031502.

Kong, J., Soh, H. T., Cassell, A. M., Quate, C. F., and Dai, H. (1998). Synthesis of individual single-walled carbon nanotubes on patterned silicon wafers. *Nature*. **395**, pp. 878–879.

Krupke, R., Hennrich, F., von Lohneysen, H., and Kappes, M. M. (2003). Separation of metallic from semiconducting single-walled carbon nanotubes. *Science* **301**, pp. 344–347.

Law, J. B. K., Koo, C. K., and Thong, J. T. L. (2007). Horizontally directed growth of carbon nanotubes utilizing self-generated electric field from plasma induced surface charging. *Appl. Phys. Lett.* **91**, p. 282447.

Lee, Y. T., and Park, J. (2002). Temperature-dependent growth of vertically aligned carbon nanotubes in the range 800–1100°C. *J. Phys. Chem. B* **106**, pp. 7614–7618.

Li, Y., Xu, C., Wei, B., Zhang, X., Zheng, M., Wu, D., and Ajayan, (2002). P. M. Self-organized ribbons of aligned carbon nanotubes. *Chem. Mater.* **14**, pp. 483–485.

Li, Y., Kinloch, I. A., and Windle, A. H. (2004). Direct spinning of carbon fibers from chemical vapor deposition synthesis. *Science* **304**, pp. 276–278.

Li, Y., Mann, D., Rolandi, M., Kim, W., Ural, A., Hung, S., Javey, A., Cao, J., Wang, D., Yenilmez, E., Wang, Q., Gibbons, J. F., Nishi, Y., and Dai, H. (2004). Preferential growth of semiconducting single-walled carbon nanotubes by a plasma enhanced CVD method. *Nano Lett.* **4**, p. 317.

Luo, Z., Lim, S., You, Y., Miao, J., Gong, H., Zhang, J., Wang, S., Lin, J., and Shen, Z. (2008). Effect of ion bombardment on the synthesis of vertically aligned single-walled carbon nanotubes by plasma-enhanced chemical vapor deposition. *Nanotechnology* **19**, p. 255607.

Lim, S. H., Lin, J., Luo, Z., and Shen, Z. (2010). Efficient growth of horizontally aligned single-walled carbon nanotubes by chemical vapor deposition over MgO-supported bimetallic Co-based catalysts. *J. Nanosci. Nanotechnol.* **10**, pp. 1-5.

Lin, C. C., Leu, I. C., Yen, J. H., and Hon, M. H. (2004). Sheath-dependent orientation control of carbon nanofibres and carbon nanotubes during plasma-enhanced chemical vapor deposition. *Nanotechnology* **15**, pp. 176–179.

Maeda, Y., Kimura, S., Kanda, M., Hirashima, Y., Hasegawa, T., Wakahara, T., Nakahodo, T., Lian, Y., Nakahodo, T., Tsuchiya, T., Akasaka, T., Lu, J., Zhang, X., Gao, Z., Yu, Y., Nagase, S., Kazaoui, S., Minami, N., Shimizu, T., Tokumoto, H., and Saito, R. (2005). Large-scale separation of metallic and semiconducting single-walled carbon nanotubes. *J. Am. Chem. Soc.* **127**, pp. 10287–10290.

McDonald, T. J., Engtrakul, C., Jones, M., Rumbles, G., and Heben, M. J. (2006). Kinetics of PL quenching during single-walled carbon nanotube rebundling and diameter-dependent surfactant interactions. *J. Phys. Chem. B* **110**, pp. 25339–25346.

Merkulov, V. I., Melechko, A. V., Guillorn, M. A., Lowndes, D. H. and Simpson, M. L. (2001). Alignment mechanism of carbon nanofibers produced by plasma-enhanced chemical vapor deposition. *Appl. Phys. Lett.* **79**, p. 2970.

Mojica, J. F., and Levenson, L. L. (1976). Bulk-to-surface precipitation and surface diffusion of carbon on polycrystalline nickel. *Surf. Sci.* **59**, pp. 447–460

Ning, G. Q., Liu, Y., Wei, F., Wen, Q., and Luo, G. H. (2007) Porous and lamella-like Fe/MgO catalysts prepared under hydrothermal conditions for high-yield synthesis of double-walled carbon nanotubes. *J. Phys. Chem. C* **111**, pp. 1969–1975.

Noda, S. Hasegawa, K., Sugime, H., Kakehi, K., Zhang, Z. Maruyama, S., and Yamaguchi, Y. (2007). Millimeter-thick single-walled carbon nanotube forests: Hidden role of a catalyst support. *Jpn. J. Appl. Phys.* **46**, pp. L399–L401.

Nozaki, T., Ohnishi, K., Okazaki, K., and Kortshagen, U. (2007). Fabrication of vertically aligned single-walled carbon nanotubes in atmospheric pressure non-thermal plasma CVD. *Carbon* **45**, pp. 364–374.

Ohmae, N. (2008). Shaping carbon nanotube bundles during growth using a magnetic field. *Carbon* **46**, pp. 544–546.

Park, J. G., Smithyman, Jesse, Lin, C., Cooke, A., Kismarahardja, A. W., Li, S., Liang, R., Brooks, J. S., Zhang, C., and Wang, B. (2009). Effects of surfactants and alignment on the physical properties of single-walled carbon nanotube buckypaper. *J. Appl. Phys.* **106**, p. 104310.

Qu, L., Du, F., and Dai, L. (2008). Preferential synthesis of semiconducting vertically aligned single-walled carbon nanotubes for direct use in FETs. *Nano Lett.* **8**, pp. 2682–2687.

Resasco, D. E., Alvarez, W. E., Pompeo, F., Balzano, L., Herrera, J. E., Kitiyanan, B., and Borgna, A. (2002). A scalable process for production of single-walled carbon nanotubes (SWNTs) by catalytic disproportionation of CO on a solid catalyst. *J. Nanopart. Res.* **4**, pp. 131–136.

Ruckenstein, E. (1967). Diffusion in a homogeneous fluidized bed. *Ind. Eng. Chem. Fundamen.* **6**, pp. 618–619.

Saito, R., Fujita, M., Dresselhaus, G., Dresselhaus, M.S., (1992). Electronic structure of chiral graphene tubules, *Appl. Phys. Lett.* **60**, pp. 2204–2206.

See, C. H., and Harris, A. T., (2007). A review of carbon nanotube synthesis via fluidized-bed chemical vapour deposition, *Ind. Eng. Res.* **46**, pp. 997–1012.

Shaffer, M. S. P., Fan, X., and Windle, A. H. (1998). Dispersion and packing of carbon nanotubes. *Carbon* **36**, pp. 1603–1612.

Takikawa, H., Ikeda, M., Hirahara, K., Hibi, Y., Tao, Y., Ruiz, P. A., Sakakibara, T., Itoh, S., and Iijima, S., (2002) Fabrication of single-walled carbon nanotubes and nanohorn by means of a torch arc in open air. *Physica B (Amsterdam, Neth.)* **323**, pp.1–4.

Tam, E., and Ostrikov, K., (2008) Plasma-controlled adatom delivery and (re) distribution: Enabling uninterrupted, low-temperature growth of ultralong vertically aligned single walled carbon nanotubes. *Appl. Phys. Lett.* **93**, p. 261504.

Terranova, M. L., Sessa, V., and Rossi, M., (2006). The world of carbon nanotubes: An overview of CVD growth methodologies, *Chem. Vap. Deposition* **12**, pp. 315–325.

Thess, A., Lee, R., Nikolaev, P., Dai, H. J., Robert, J., Xu, C. H., Lee, Y. H., Kim, S. G., Rinzler, A. G., Colbert, D. T., Scuseria, G. E., Tomanek, D., Fischer, J. E., and Smalley, R. E. (1996). Crystalline ropes of metallic carbon nanotubes, *Science* **273**, pp. 483–487.

Vigolo, B., Penicaud, A., Coulon, C., Sauder, C., Pailler, R,. Journet, C., Bernier, P. and Poulin, P. (2000). Macroscopic fibers and ribbons of oriented carbon nanotubes. *Science* **290**, pp. 1331–1334.

Walters, D. A., Ericson, L. M., Casavant, M. J., Liu, J., Colbert, D. T., Smith, K. A., and Smalley, R. E. (1999). Elastic strain of freely suspended single-walled carbon nanotube ropes. *Appl. Phys. Lett.* **74**, pp. 3803–3805.

Wang, Y., Wei, F., Luo, G., Yu, H., Gu, G. (2002). The large-scale production of carbon nanotubes in nano-agglomerate fluidized-bed reactor. *Chem. Phys. Lett.* **364**, pp. 568–572.

Wang. X., Li, Q., Xie, J., Jin, Z., Wang, J., Li, Y., Jiang, K., and Fan, S. (2009). Fabrication of ultralong and electrically uniform single-walled carbon nanotubes on clean substrates. *Nano Lett.* **9**, pp. 3137–3141.

Wei, F., Qian, W. Z., Yu, H., Wang, Y., Luo, G. H., Xu, G. H., and Wang, D. Z. (2008). The mass production of carbon nanotubes using nano-agglomerate fluidized bed reactor: A multiscale space-time analysis. *Powder Technol.* **183**, pp. 10–20.

Xu, Y., Flor, E., Schmidt, H., Smalley, R. E., and Hauge, R. H. (2006). Effects of atomic hydrogen and active carbon species in 1 mm vertically aligned single-walled carbon nanotube growth. *Appl. Phys. Lett.* **89**, p.123116.

Yun, Y., Shanov, V., Tu, Y., Subramaniam, S., and Schulz, M. J. (2006). Growth mechanism of long aligned multiwall carbon nanotube arrays by water-assisted chemical vapor deposition. *J. Phys. Chem. B* **110**, pp. 23920–23925.

Zhang, M., Fang, S., Zakhidov, A. A., Lee, S. B., Aliev, A. E., Williams, C. D., Atkinson, K. R., and Baughman, R. H. (2005). Strong, transparent, multifunctional, carbon nanotube sheets. *Science* **309**, pp. 1215–1219.

Zhang, G., Qi, P., Wang, X., Lu, Y., Li, X., Tu, R., Bangsaruntip, S., Mann, D., Zhang, L., and Dai, H. (2006). Selective etching of metallic carbon nanotubes by gas-phase reaction. *Science* **314**, pp. 974–977.

Zhang, Q., Yu, H., Liu, Y., Qian, W., Wang, Y., Luo, G., and Wei, F. (2008) Few walled carbon nanotube production in large-scale by nano-agglomerate fluidized-bed process. *NANO* **3**, pp. 45–50.

Zhu, H. W., Xu, C. L., Wu, D. H., Wei, B. Q., Vajtai, R., and Ajayan, P. M. (2002). Direct synthesis of long single-walled carbon nanotube strands. *Science* **296**, pp. 884–886.

Chapter 3

Assembly of Carbon Nanotubes Toward Practical Applications

Yehai Yan,[a] Jian Cui,[a] and Qing Zhang[b]
[a] *Key Laboratory of Rubber-Plastics of Ministry of Education, College of Polymer Science and Engineering, Qingdao University of Science and Technology, 53 Zhengzhou Road, 266042 Qingdao, China*
[b] *School of Electrical and Electronic Engineering, Nanyang Technological University 16 Nanyang Drive, 637722 Singapore, Singapore*
yhyan@qust.edu.cn

Remarkable size- and structure-dependent properties of carbon nanotubes (CNTs) are attractive for various potential applications, ranging from conducting wires in molecular devices to multifunctional fillers in nanocomposites, etc. To realize the potentials of CNTs in practical realities, some important barriers such as CNT cost, polydispersity in CNT type, and limitation in processing and assembly techniques should be hurdled beforehand. This chapter focuses on the topic of *Assembly of CNTs*, which is defined as the alignment and/or patterning of CNTs. To achieve the best possible performance of many CNT applications, especially electronic applications, people have to assemble CNTs into hierarchical arrays over large-scale areas with controllable shape, location, orientation, and density of the

Carbon Nanotubes and Their Applications
Edited by Qing Zhang

ISBN 978-981-4241-90-8 (Hardcover), 978-981-4303-18-7 (eBook)
www.panstanford.com

CNTs. Efforts from both the scientific and engineering viewpoints have been made to address this issue since discovery of CNTs. We review here the progress of CNT-assembly techniques in the last five years under the two rubrics of synthetic assembly and post-synthetic assembly. A few nanotechnology applications of such CNT assemblies are illustrated on occasion.

3.1 Introduction

Iijima's observation in 1991 of fullerene-like materials by high-resolution transmission electron microscopy (HRTEM) heralded the beginning of the CNT era [1]. The first observed CNTs were multi-wall carbon nanotubes (MWCNTs). Two years later, smaller-diameter, single-wall carbon nanotubes (SWCNTs) were independently discovered by Iijima's group [2] and Bethune's group [3]. The subsequent development of successful methods for the synthesis of large-quantity and high-quality CNTs has paved the way for worldwide R&D efforts. Currently, economical access to CNTs is less problematic. For example, Timesnano (http://www.timesnano.com, China) has marketed 20 nm-diameter MWCNTs (>95 wt% purity) at about $0.5 per gram and SWCNTs (>90 wt% purity) at about $50 per gram.

An SWCNT may be considered a single graphite sheet seamlessly rolled into a cylindrical and hollow tube, while a MWCNT is a collection of several concentric SWCNTs. The rolling-up phenomenon can be accomplished in several ways to match the dangling bonds at the join line. Consequently CNTs can have a variety of helical structures [4]. The graphite-like structure, intrinsic dimension (with a diameter down to 0.7 nm and a length up to several centimeters), and closed topology endow CNTs with not only in-plane properties of graphite such as high thermal [5] and electrical [6] conductivity, superior strength and stiffness [7], and strong chemical and thermal inertness [5], but also unusual electronic properties — depending on its helicity and diameter, a CNT may be metallic or semiconducting with a large bandgap [8, 9]. In addition, a large specific surface area endows CNTs with very high sensitivity. Due to these excellent properties, CNTs can be used potentially in many applications such as electronics, mechanical reinforcement, chemical and biological sensors, and microscope probe tips.

Since the discovery, CNTs have drawn wide and far attention. The exponential increase in patent filings and publications devoted to this intriguing material, on one hand, demonstrates ever-growing technological and academic interests and, on the other hand, indicate that there still exist many challenges toward the practical and bulk applications. Actually, in order to realize the potential of CNTs in practical realities, some important barriers such as CNT cost, polydispersity in CNT type, CNT quality and purity, and limitation in processing and assembly techniques should be resolved in advance. Even though each aspect mentioned above equally represents a very important research area in the CNT applications, all of them other than *Assembly of CNTs* fall outside of the scope of this chapter. In the following text, an advance in *Assembly of CNTs* is discussed in terms of necessity of assembling CNTs and assembly techniques.

3.2 Necessity of Assembling CNTs

A wealth of theoretical predictions and experimental verifications about CNTs have disclosed their remarkable size- and structure-dependent properties that are attractive for various potential applications, ranging from conducting wires in molecular devices to multifunctional fillers in nanocomposites. To achieve the best possible performance, many of these applications require assembly of CNTs into hierarchical arrays over large-scale areas with a controllable shape, location, orientation, and density of the CNTs. In what follows, the necessity of assembling CNTs is illustrated with some representative examples of applications.

3.2.1 Molecular Devices

Intrinsic small dimensions and remarkable electronic properties make CNTs promising building blocks for molecular devices, which may overcome fundamental physical and economic limitations of conventional Si-based VLSI (very large scale integration) fabrication techniques. For actualizing this objective, CNTs need to be assembled into hierarchical arrays over large-scale areas for them active components (such as transistors), which often need to be interfaced with other device components (such as electrodes). At the nanoscale, how to place CNTs at desired locations with targeted shapes, directions, and densities for further fabricating functional devices represents one of the longstanding challenges of CNT-based molecular devices [10], even though some exciting progresses have been made during the past three decades.

Representatively, Burke and coworkers have constructed a CNT-based radio receiver using a single CNT as the demodulator and have demonstrated a functioning radio that can pick up an amplitude-modulated signal generated in the lab and play music broadcast wirelessly across a room [11]. The demodulation functionality originates from the nonlinear current–voltage (I–V) characteristic of the CNT, and the demodulation effect can be maximized by properly biasing it. A similar work on the CNT radio performance was independently reported by Jensen et al. [12]. Thanks to the mechanical resonance characteristic, a single CNT can serve simultaneously as all essential components of a radio: antenna, tunable band-pass filter, amplifier, and demodulator. Recently, an entire CNT-based radio system has appeared with the SWCNT devices providing all of the key functions, including resonant antennas, fixed RF (radio frequency) amplifiers, RF mixers, and audio amplifiers [13]. Keys for actualizing these functions consist of assembling the SWCNTs in the channel regions of CNT based transistors with extremely linear configuration and high alignment levels, eliminating metallic SWCNTs, increasing semiconducting SWCNT density, and reducing the device dimensions in a proper way. The resultant radio was able to receive weak signals from a local radio station. This work represents an important first step toward the integrated nanosystems.

Dekker and coworkers produced the first SWCNT-based molecular field-effect transistor (FET) [14]. This device consists of a single semiconducting SWCNT bridging Pt electrodes and a Si substrate covered with a SiO_2 layer acting as a back gate. Its performance is comparable to existing devices in terms of switching speeds. Later, the same laboratory integrated such single-nanotube FETs onto a chip to demonstrate digital logic operations, such as inverter, logic NOR, static random-access memory cell, and ac (alternate current) ring oscillator [15].

Other examples of applications include field emitters [16–18], nanoelectrode arrays [19, 20], black body absorber [21], atomic force microscope or scanning tunneling microscope tips [22–24], and nanotweezers [25]. By comparing field emission measurement results of aligned CNTs with that of unprocessed CNT deposit, de Heer et al. concluded that aligned CNTs are required to produce higher field emission currents at low biased voltages [17]. To fabricate the nanoelectrode arrays, CNTs must be precisely positioned and well aligned [19]. In addition, the interspacing of the individual CNTs should be sufficiently larger than the diameter of the CNTs to make each of them function as an individual nanoelectrode. Fundamentally, aligned structure of the SWCNT forest plays a critical role in achieving black body behavior (a nearly constant and near-unity absorptivity of 0.98–0.99 across a wide spectral range of 0.2–200 μm) [21]. For probe tip and nanotweezer applications, it is clearly necessary to place a CNT at a prescribed position along a certain orientation.

3.2.2 CNT/Polymer Composites

CNTs have also been explored for use as reinforcement in composites; typical examples are polymer-based composites [26, 27]. As with traditional fillers, the main requirements for CNTs in a composite are to disperse them uniformly throughout the polymer matrix as well as to promote the CNT-matrix interfacial bonding that is required to achieve efficient load transfer across the interface. Additionally, to pursue the maximal reinforcement capability, CNTs should be oriented/aligned in matrices [4, 26, 27] in order to realize the advantage of their ultrahigh aspect ratios (length-to-diameter ratio), which can be 1000 or more. From geometric considerations, the difference between random orientation and perfect alignment is a factor of five in composite modulus [28]. This suggests that a reasonable degree of CNT alignment in the composite would play an important role of maximizing reinforcement.

In summary, to utilize CNTs in these applications and discover promising new applications, methods are urgently needed to assemble CNTs into hierarchical arrays over large-scale areas. Many efforts in two broad areas, which may be classified into synthetic and post-synthetic approaches, have been made to address the assembly of CNTs. Since the early work under this topic has been excellently reviewed by Dai et al. [10] and Yan et al. [29], this chapter will mainly focus on the progress of CNT-assembly techniques in the last five years.

3.3 Approaches to CNT Assembly

3.3.1 Synthetic Approaches

3.3.1.1 Perpendicularly aligned CNTs

The technique of chemical vapor deposition has been widely proved to be one of the most useful means to grow perpendicularly aligned CNTs on various substrates. Recently, Lu and coworkers reported an interesting work where controlled growth of CNTs with different alignments was achieved via a laser-assisted CVD process [30]. In this process, a Si wafer pre-patterned with ruthenium (Ru) electrodes was used as the substrate. In the subsequent CVD procedure, biasing voltages of different polarities were applied on the Ru electrodes to which tri-layer Al/Fe/Al catalysts were introduced. SEM observations showed that surface-bounded CNTs were obtained on the anodes, while perpendicularly aligned CNTs were produced on the cathodes. The growth alignment control was attributed to the movement of electrically charged catalyst nanoparticles under the influence of an external electric field. Besides the most commonly used Si or quartz substrate, the versatility of CVD technique makes it applicable for the CNT growth on various other substrates, such as single crystal magnesia [31], stainless steel [32], inconel [33], and even ceramic fibers [34].

A recent collaboration between Dai's group and Wang's group has exploited a low-pressure CVD process for growth of perpendicularly aligned MWCNT (PA-MWCNT) arrays that are dominated by a straight body segment but with a randomly entangled top layer [35]. After a small piece of such as-grown PA-MWCNT film (4 mm × 4 mm) was finger-pressed from the Si side onto a perpendicularly positioned glass slide, a 1480 g book was able to suspend from the glass surface, as shown in Fig. 3.1a. The calculated adhesion force parallel to the glass slide is 90.7 $N \cdot cm^{-2}$, almost 10 times that of a gecko foot, and much stronger than the normal adhesion force. That is to say, the array film is strongly bound along the shear direction and easily lifted in the normal direction. More important is that the array film could be repeatedly attached and detached from the glass surface with no decrease of the supported weight. Similar adhesion behaviors were also observed on other substrates such as poly(ethylene terephthalate) (PET) sheet, polytetrafluoroethylene

(PTFE) film, and rough sandpaper, demonstrating that the PA-MWCNT arrays could be used for fabrication of gecko-foot-mimetic dry adhesives.

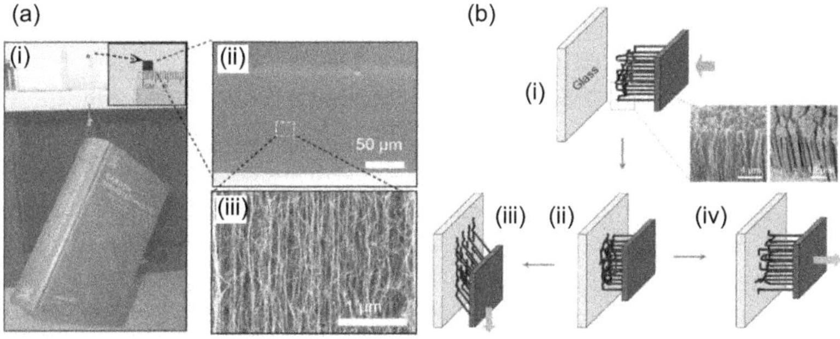

Figure 3.1 (a) A book of 1480 g in weight suspended from a glass surface with use of PA-MWCNTs supported on a Si wafer (i), and SEM images of the VA-MWCNT film under different magnifications (ii and iii), and (b) schematic diagrams for the morphological change of PA-MWCNT arrays during adhesion measurements: Preloading (i) (Inset: The structure similarity between the cross-sectional views of the VA-MWCNTs (left) and gecko's aligned elastic hairs (right)), attachment of the PA-MWCNTarray onto the glass substrate (ii), shear adhesion force stretching the nonaligned nanotubes on the substrate to form the line contact (iii), and normal adhesion force leading to the nonaligned CNTs point-by-point peel-off from the substrate (iv) [35]. See also Color Insert.

It is believed that the van der Waals force provides the adhesion force between the array film and the substrate, while the hierarchical structure of the array film (i.e., a straight aligning body along with a curly entangled top layer) is responsible for creating the anisotropic adhesion force (i.e., shear adhesion force much stronger than normal adhesion force). As shown in Fig. 3.1b, during the initial contact the top nonaligned MWCNTs adopt randomly distributed "line" contact with the substrate. To break such contact in the direction normal to the substrate, a very low force is sufficient because this breakage behavior is a "point-by-point" detaching process. When a shear force is applied on the array film, however, the nonalignment MWCNTs tend to align along the shear direction and the perpendicularly aligned MWCNTs trunks to tilt along the shear direction, causing a predominant aligned and more extensive line contact with the substrate surface and thus a much stronger van der Waals force between them. It is therefore reasonable to expect that the line contact detachment produces a much stronger shear adhesion force than the normal adhesion force governed by the point-by-point peel-off detachment.

Several parameters should be optimized for the synthesis of well-aligned CNTs by the CVD technique. The catalyst was found to be one of the most critical parameters. The thickness of catalyst film, namely, catalyst particle size usually determines both the alignment degree and the wall-number of CNTs [36–38]. In a water-assisted CVD process, Patole et al. observed that well-aligned CNTs with 2–4 walls were grown on an iron catalyst film ranging in thickness from 0.1 to 0.5 nm, less aligned CNTs with 4–10 walls on an iron catalyst film ranging in thickness from 0.6 to 1 nm, and misaligned CNTs with 8–16 walls when the thickness of the iron catalyst film is larger than 1nm [36]. Researchers found that, in a hot filament CVD process where a bias voltage of –500 V was applied between the substrate holder and the filament, simultaneous growth of randomly oriented and aligned CNTs occurred at the early growth stage (at that time, the catalyst particle sizes are relative small), but only aligned CNTs were produced when the catalyst particle sizes become big enough [33]. For a composite catalyst system, the deposition order influences the height of aligned CNT arrays, particularly evident during synthesis at elevated reaction pressures where carbon flux is high [39]. Besides the catalyst, other parameters including temperature, pressure, substrate surface roughness, and carbon flux composition were also investigated for a fundamental understanding of the synthesis of perpendicularly aligned CNTs. It should be especially pointed out that the synthesis of CNTs from a substrate is an intricate system engineering and each parameter mentioned above plays a certain role. The simultaneous investigation of two or more parameters in one synthesis process is thus absolutely necessary because it is much closer to the real world. Recently, Puretzk et al. [40] and Wood et al. [41] developed a model to support how temperature and carbon flux composition influence the growth rates and wall-number of the CNTs in the arrays. On the other hand, the experimental results revealed that an accurate balance between the catalyst and the carbon flux composite favors to gain a good alignment of CNTs [38].

3.3.1.2 Parallel aligned CNTs

It has been demonstrated in experiment that the application of electric fields in a CVD system can significantly promote the parallel growth of CNTs on substrates. In a recent work, Hongo et al. noticed that the CVD process under a local electrical field produced not only aligned SWCNTs but also many random ones [42]. Therefore, the controllability in the directional growth is still an open problem. To address this issue, the researchers inserted spacer layers between the catalyst for the CVD and the substrate, and investigated the spacer thickness dependence of the directional distribution width of the SWCNTs. They found that a thicker spacer layer resulted in better alignment of the SWCNTs. Typically, a sample having a 200 nm-thick spacer layer produced 7° of standard deviation along the electric field. These results thus revealed that suppression of the interaction between the substrate and the SWCNT tips is a key issue regarding the directional growth of SWCNTs through interaction between the SWCNTs and an electric field.

Liu's work has clearly shown that parallel aligned CNTs can be grown by controlling the gas flow direction in the CVD growth [43]. Recently, the mechanism for the gas-flow aligned growth of CNTs has been discussed. A detailed investigation elucidates that the alignment occurs only when a CNT exceeds a critical length, suggesting that short growth duration limits not only the CNT length but also the alignment [44]. In addition, it is also important to minimize the interaction between the growing CNTs and the substrate [45].This is a similar consideration that has already been mentioned in the electric-field aligned growth of CNTs. As an illustration [45], thin catalyst films were fabricated on the top of protruding patterns, and the subsequent gas-flow-assisted CVD growth produced the dense and well-aligned CNT arrays.

In contrast to perpendicular synthesis, parallel synthesis of CNTs received little attention due to its difficulty. However, this situation has been changed impressively with the recent appearance of epitaxial approach. The so-called "epitaxial approach" is to grow parallel aligned CNT arrays via the CVD process on single-crystal substrates, such as sapphire and quartz, without any external forces [46–56]. This approach has gradually emerged as a scalable way to produce massively aligned CNTs. Epitaxial growth, namely, surface-guided growth along step edges (step-templated growth) or lattice directions (lattice-directed growth) has been widely accepted as alignment mechanism [57–60]. In the case of lattice-directed growth, the driving force for the alignment was assumed to be anisotropic van der Waals interactions between the CNTs and the substrate. This has been experimentally verified by the $^{13}C/^{12}C$ isotope-labeling method together with the spatial Raman mapping technique [61]. Even if some evidence suggests that the lattice-directed growth is more appropriate [62], it is the fact that both growth modes coexist on a single substrate. A recent work further revealed that the SWCNT growth direction was sensitive to the surface geometry of sapphire, and competition occurred between the two growth modes [57]. Besides the growth alignment mechanism, another critical issue that should be clarified for better understanding the alignment phenomenon is why the catalyst can stay active for the growth of long SWCNTs with the strong interaction between the CNTs and the substrate. With this end in view, Yu et al. have developed a "raised-head" growth mechanism model [63] and suggested that the sapphire substrate is positively charged by Al-O dipoles under the growth environment, while the Fe catalyst is also positively charged owing to a contact potential at the interface of SWCNT and catalyst. The electrostatic force between the catalyst and the substrate keeps the catalyst from touching the substrate during the growth, thus ensuring the activity of the catalyst. From this model, any substrates not limited to the sapphire may be used in the parallel aligned growth of SWCNTs, provided that their surface atoms have a strong interaction with the SWCNTs in a certain crystallographic direction as well as they are positively charged in the growth environment.

Epitaxial approach makes it available to produce the perfectly aligned arrays of pristine, long, individual SWCNTs with perfectly linear geometries on a large scale. The obtained arrays have been examined as an effective thin-film semiconductor suitable for integration into transistors and other classes of electronic devices [64]. The parallel aligned SWCNT arrays were grown on a quartz wafer by the CVD of methane with the Fe catalysts (Fig. 3.2a). Then, the arrays were integrated into transistors by using standard photolithography to define source and drain electrodes (Ti/Pd = 1 nm/20 nm) on the SWCNT/quartz substrate. After spin-coating a 1.5 μm epoxy gate dielectric, the top gate electrodes (Ti/Au = 1 nm/20 nm) were defined again with photolithography. Fig. 3.2b,c give a schematic illustration of the transistor and SEM image of the channel region in such a device, respectively. Output currents (I_D) measured on more than 100 two-terminal test structures revealed that the devices not only provided high I_D but also possessed good device-to-device uniformity with a ~10% standard deviation in I_D, even though the incorporated

SWCNTs are electronically heterogeneous. Measurements on *p*- (made from as-grown SWCNT arrays) and *n*-channel (made from polyethylenimine-coated SWCNT arrays) transistors that involve as many as ~2100 SWCNTs revealed device-level mobilities and scaled transconductances approaching ~1000 $cm^2.V^{-1}.s^{-1}$ and ~3000 $S{\cdot}m^{-1}$, respectively. Once a transfer process was applied to these devices, they could be integrated onto unusual substrates such as flexible plastics (Fig. 3.2d). Additionally, by combining *n*- and *p*-channel transistors, CMOS and PMOS logic gates could be constructed. Collectively, these results represent a promising route to realize large-scale integrated CNT electronics using the parallel aligned SWCNT arrays grown by epitaxial approach.

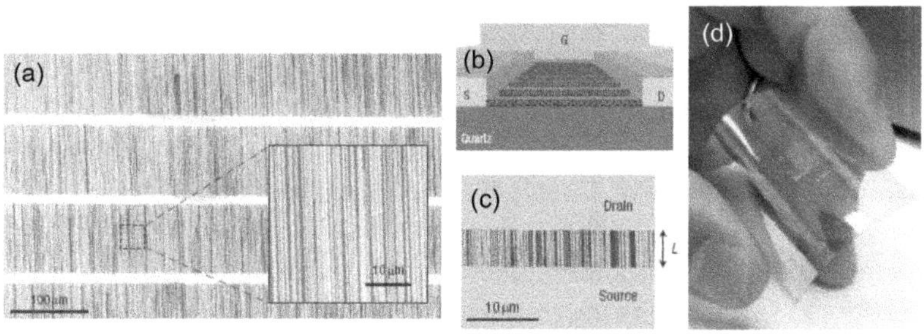

Figure 3.2 (a) SEM image of a pattern of perfectly aligned, linear SWCNTs formed by CVD growth on a quartz substrate (the bright horizontal stripes correspond to the regions of iron catalyst), (b) Schematic illustration of the layout of a type of transistor that incorporates aligned SWCNTs as the semiconductor, (c) SEM image of the channel region of such a device, and (d) optical image of an array of SWCNT transistors on a flexible plastic substrate (PET) [64]. See also Color Insert.

3.3.1.3 Perpendicularly aligned and patterned CNTs

It is well known that CVD synthesis strategies on catalytically patterned substrates can yield ordered CNT architectures [10, 29, 65]. Besides that, perpendicularly aligned CNTs can also grow on three-dimensional (3D) substrates. For instance, perpendicularly aligned CNTs have been grown on catalyst-coated bottoms of reactive-ion-etched (RIE) trenches in a Si substrate [66] as well as on a highly structured Si substrate surface and even underneath the suspended microstructure surface [67], which were fabricated by integration of three key processes: plasma surface treatment, self-assembled monolayer (SAM) coating, and contact displacement electroless plating.

Recently, Hart et al. proposed an interesting method to pattern the aligned CNTs on a planar substrate coated with a uniform thin film of Fe catalyst [68]. Before placing into the quartz tube furnace, the substrate was clamped against a microfabricated template. After the CVD growth, the CNT microforms were found to fill and conform to the template. This method allows growing arbitrary CNT microstructures with sloped and/or curved surfaces and nonorthogonal corners and it does not require patterning of the catalyst before growth. This technique is completely different from any other approaches mentioned above in which the catalyst pre-patterned on the substrates are normally required. Another method for producing perpendicularly aligned CNT patterns on a non-patterned substrate is to selectively prune the CNTs using the focused laser beam [69]. This process began with the growth of perpendicularly aligned CNTs by the PECVD of C_2H_2/H_2. The obtained sample was then placed in an optical microscope-focused laser beam system. When the laser beam was incident upon the sample, it caused the localized burning of CNTs. By moving the sample with respect to the laser beam, patterns of the desired design were then created on the CNT-covered substrate surface.

3.3.1.4 Parallel aligned and patterned CNTs

Based on the gas-flow-direction mechanism, 2D network and cross junction of CNTs have been fabricated on flat substrates [43, 70]. One method is to use a two-step CVD synthesis [43], the other bases on rotating the substrate *in situ* during the CVD synthesis [70].

Depending on the pre-patterned catalysts on the quartz wafer, epitaxial approach could also be used to form the parallel aligned SWCNT arrays with various geometries such as wavy lines, circles, and complex circuit-like patterns [71]. In principle, if an external force is properly applied in the epitaxial growth, the CNT crossbar architectures may be formed on a single-crystal substrate by simultaneous graphoepitaxy and field-directed growth. Such concept is depicted in Fig. 3.3a [72]. The CNTs originating from the catalysts lying on the bare sapphire grow along the nanosteps in the $[11\bar{2}0]$ direction, whereas those from the catalysts lying on the amorphous SiO_2 islands grow up free and will be aligned by the electric field that is applied perpendicular to the nanosteps. In order to actualize this concept, Pt electrodes and amorphous SiO_2 strips were successively patterned on the sapphire substrate using standard photolithography, and then the CVD growth of SWCNTs was performed in the presence of an external electric field. As presented in Fig. 3.3b,c, orthogonal self-assembly of SWCNT crossbar structures have been successfully fabricated. The SWCNT density along the $[11\bar{2}0]$ direction is mainly controlled by the catalyst concentration and the catalyst- and nanotube-substrate interactions. However, the SWCNT density along the electric field is determined by the number of free-grown SWCNTs originating from the SiO_2 strips.

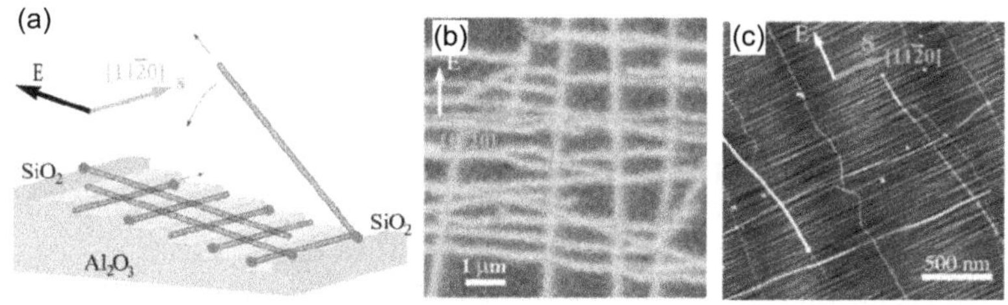

Figure 3.3 (a) Schematic representation of the orthogonal self-assembly of a SWCNT crossbar array by simultaneous graphoepitaxy and field-directed growth, and (b) SEM and (c) AFM images of dense SWCNT grids obtained by simultaneous graphoepitaxy and field-directed growth [72]. See also Color Insert.

3.3.2 Post-Synthetic Approaches

3.3.2.1 Perpendicularly aligned CNTs

As has been mentioned by Yan et al. [29], the attention on post-synthesis assembly of CNTs perpendicular to their substrate is much smaller than the assembly during synthesis. By far, this situation still remains and limited work has concerned this subject.

In one work, an electrophoretic deposition method has been used to form the perpendicularly aligned SWCNT films [73]. In this process, SWCNTs were first shortened and decorated with the acyl chloride groups. Then, the functionalized SWCNTs were dispersed in DMF to prepare a solution with the PH range of 7–7.5. Electrophoretic deposition was carried out in a quartz cell containing two electrodes (Au-coated Si/SiO_2 substrates pre-treated with (3-aminopropyl) triethoxysilane for 1 h in solution) separated by Teflon spacers.

After the deposition, the CNTs were found to stick perpendicular to the electrode when the gap size is more than about 5 mm. A post-treatment of ultrasonic irradiation with a reasonable duration further enhanced the alignment level of the CNTs in the deposited film.

In another work, Youn et al. demonstrated a novel magneto-evaporation method for perpendicular alignment of SWCNTs [74]. The used experimental setup is schematically shown in Fig. 3.4. This work starts with the fabrication of the SWCNT/Fe-oxide sample by reacting of shortened SWCNTs, iron-oleate complex, and oleic acid in octadecene. The obtained SWCNT/Fe-oxide sample was dispersed in DMF and then deposited onto an indium tin oxide (ITO) glass substrate by a spraying method. After evaporation of the SWCNT/Fe-oxide solution in the presence of a magnetic field generated by a magnet placed behind the ITO substrate, the sample was transferred into a vacuum-evaporator chamber and positioned between the electron-beam source and the poles of the magnet. This configuration rendered the magnetic field vector parallel to the electron-beam irradiation, thus enabling the simultaneous perpendicular alignment of the SWCNTs by the magnetic field and the fixation of the alignment by means of the direct evaporation of a Ti thin film.

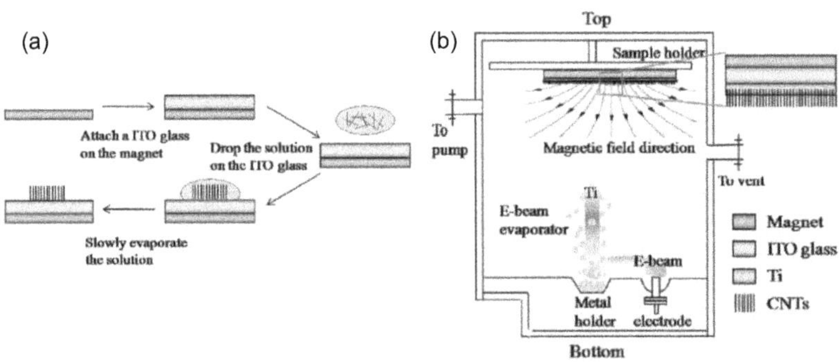

Figure 3.4 (a) Schematic representation of the fabrication of perpendicularly aligned CNTs, and (b) scheme of the magneto-evaporation alignment method. The perpendicularly aligned CNTs, placed on an ITO glass attached to a magnet, are inserted into an electron beam evaporator. The aligned CNTs are fixed to a Ti metal, which is evaporated using the electron beam [74]. See also Color Insert.

Although the CVD route is a favorable method for the fabrication of perpendicularly aligned CNT arrays, the involved high temperatures

(600–1200°C) in the growth process restrict the substrates to be high-melting-point semiconductors (e.g., Si) or insulators (e.g., quartz). A recently developed novel transfer technique has changed this situation and extended the applicable substrates to any other desired ones. For example, Zhang et al. have successfully realized the transfer of freestanding CNT films, which were separated from the SiO_2/Si substrate by HF etching, onto a Cu substrate with a thin polymer-binding layer at the Cu-CNT interface [75]. In a very similar way, Chai et al. have transferred CNT films onto Al and Cu substrates [76]. By using a low-melting-point tin-based solder, Zhu et al. have also reported the successful transfer of perpendicularly aligned CNTs grown on the Si substrate to the FR4 boards coated with Cu foil [77]. In addition, a low temperature transfer process was developed by using a conductive polymer composite, which consists of nano-size silver particles, micro-size silver flakes, and epoxy resin, as a transfer medium [78]. Sow and coworkers recently meliorated the transfer technique, endowing it with many desirable attributes including a simple and low-temperature process, substrate friendliness, strong adhesion, and satisfactory electrical contact [79]. For this end, PDMS was spin coated on the surface of various Pt-coated substrates, such as Cu foil, Fe foil, glass slide, paper, cloth, and flexible polymers. A piece of as-synthesized CNT array on Si substrate was inverted and pressed onto the surface of the PDMS-coated substrate, and subjected to a curing process of PDMS. After the curing, the Si substrate was lifted off and the CNTs were left on the target substrate.

3.3.2.2 Parallel aligned CNTs

Zhou and coworkers reported a simple polymer-mediated method for the parallel alignment of MWCNTs [80]. In this method, a composite was first prepared by casting a chloroform suspension of MWCNTs and poly(hydroxyaminoether) in a Teflon mold. Then, a constant force was applied on the strip-shape composite film to stretch it to a desired stretching ratio. TEM examination showed that the CNTs were aligned parallel to the stretching direction. Similarly, the parallel alignment of CNTs in other polymer matrices including polystyrene [81], gelatin [82], polyethylene [83], and epoxy resin [84] has been accomplished by mechanical stretching. Mechanical shearing was also used for realizing the alignment of CNTs in polymer composites [85, 86]. A representative example is the MWCNT-filled poly(furfuryl alcohol) composite prepared using the doctor blade technique [86]. Shear forces involved in the process are responsible for the preferential alignment of MWCNTs in the composite along the blading direction. Better alignment was achieved in composites with lower contents of MWCNTs.

Focusing on the basic idea of blown film extrusion, Lieber and coworkers have described another polymer-mediated method (namely, blown bubble film (BBF) method) for large-area, uniformly aligned and controlled-density CNT films [87]. The basic steps in this method (Fig. 3.5a) consist of (i) preparation of a homogeneous, stable and controlled concentration polymer suspension of CNTs by dispersing *n*-octadecylamine functionalized CNTs in tetrahydrofuran (THF) and then adding epoxy resin, (ii) expansion of a bubble from the polymer suspension, which has the desired viscosity of 15–25 Pa·s, using a circular die at a controlled direction and speed, and (iii) transfer of the bubble to substrates or open frame structures. In order to obtain consistent alignment of CNTs over large area and enable the overall alignment to be fixed during transfer, it is highly necessary to make the expansion along a defined direction, which was achieved by applying an external vertical force (*F*) on the bubble. Fig. 3.5b,c show the dark-field optical images of the transferred SWCNT- and MWCNT-BBFs where the CNTs are well aligned and uniformly dispersed over the 75 mm-diameter substrates. According to the authors' demonstration, the BBFs could be transferred to much larger substrates.

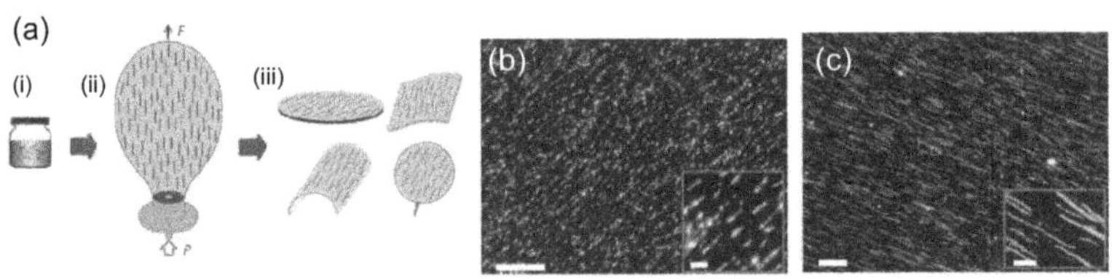

Figure 3.5 (a) Blown bubble film (BBF) process: (i) preparation of a CNT/epoxy suspension, (ii) bubble expansion over a circular die, and (iii) films transferred to substrates or open frame structures (black straight lines represent aligned CNTs embedded in the bubble film), and dark-field optical images of the transferred (b) SWCNT- and (c) MWCNT-BBFs- on a Si wafer (Scale bar in (b) represents 10 μm and in (c) 25 μm. Insets are high-resolution dark-field images, highlighting the alignment of individual nanotubes; the scale bar is 2 μm in (b) and 5 μm in (c)) [87].

Dispersing CNTs in polymer matrices and then spinning the composites into fibers has been proved to be an effective method to realize the alignment of CNTs in the fibers along the spinning direction. Driving force for the alignment originates from the shear force applied on the composite melt/solution during the injection step and the stretch force on the fiber in its collection step on a high-speed rotating drum. Both thermoplastic (e.g., poly(methyl methacrylate) [88], petroleum pitch [89], poly(*p*-phyenylene benzobisoxazole [90], cellulose [91], poly(vinyl acetate) [92], polycarbonate [93], nylon-6 [94, 95], and polyethelene [96]) and thermosetting resins (e.g., cyanate ester resin [97] and epoxy resin [98]) have been used as polymer matrices. The concerned spinning techniques included melt spinning [88, 89, 93–95], gel spinning [96], electrospinning [92], dry-jet wet spinning [90, 91], and reactive spinning [97, 98]. It should be emphasized that the spinning method for CNT alignment is not limited to the polymer composites. Actually, it has already been extended to ceramic composites, e.g., alumina/CNT composite [99].

Kimura and coworkers also developed a polymer-mediated method for parallel alignment of MWCNTs, in which the monomer solution of unsaturated polyester resin was mixed with MWCNTs by ultrasonication and then polymerized by adding a radical initiator in the presence of a constant magnetic field of 10 T [100]. TEM, magnetic susceptibility, electrical conductivity and dynamic mechanical measurements showed clear anisotropy of the resultant composite, indicating magnetic field-induced alignment of MWCNTs in the polymer matrix. By using this technique, Garmestani et al. [101, 102] and Steinert et al. [103] have realized the alignment of SWCNTs in epoxy resin matrix and poly(ethylene terephthalate) matrix, respectively. The alignment under a magnetic field occurs due to the cooperative effect of the magnetic torque exerted by the magnetic field directly on the CNTs and the hydrodynamic torque and viscous shear (i.e., drag forces) exerted on the CNTs by the polymer chains.

Electric field was also employed to assist the alignment of CNTs in polymer composites. To implement this process, CNTs were first dispersed ultrasonically in a liquid matrix (epoxy resin [104], methyl methacrylate monomer [105], or a urethane dimethacylate-based UV curable resin [106]), and then subjected to a constant electric field. Once the CNTs were aligned by electric field-induced dipolar interactions among them, the aligned CNTs were immediately immobilized by polymerizing the reactive monomers. Owing to the alignment of CNTs in matrices, the resultant composites showed significant anisotropic properties.

Liquid crystals (LCs) have been widely used as host for CNT dispersion and alignment. This is mainly based on the characters of LCs of fluidity and long-range orientation. The latter is particularly important because it enables the principal symmetry axis of the LC molecules spontaneously tending to align along the director where it is defined as the axis of average LC molecule orientation and can be easily reoriented by applying an external force. If CNTs are well dispersed, they will in general align with their long axes along the director so as to minimize distortions of the LC director field and thus the system's free energy [107].

The earliest attempts to use the LC self-organizing property for the alignment of CNTs were all performed on standard thermotropic nematic hosts such as 5CB and E7 [108–110]. Two recent reports showed that the reactive thermotropic nematic LC monomers could also be used for dispersing and aligning the CNTs [111, 112]. The alignment of CNTs was subsequently fixed by polymerizing the reactive monomers. A phenomenological theory has been developed to predict the alignment of CNTs in thermotropic nematic LCs [113, 114]. The phase ordering of the binary mixture was analyzed as a function of concentration of CNTs, temperature, and the LC-CNT coupling strength. The analyses suggest that the ordering degree of CNTs is enslaved by the properties of the LC host and can be tuned by changing the temperature or the CNT concentration. Recently, Jeong et al. demonstrated that the SWCNTs could be assembled into highly oriented plate with the help of a thermotropic smectic LC of 4'-hexyloxy-biphenyl-4-carboxylic acid ethyl ester [115].

With respect to thermotropic LCs having a limited dispersibility of CNTs, lyotropic LCs are more attractive to be used as hosts for CNT alignment due to their ability of involving well separated CNTs at high concentration. This situation is largely attributed to the fact that the standard surfactants used for preparing high-concentrated CNT dispersions are simultaneously available for making lyotropic LCs. One of the concerned lyotropic LCs is formed by adding ~5 wt% of decanol as co-surfactant into a ~25 wt% aqueous sodium dodecyl sulfate (SDS) solution [116]. By tuning the exact composition of the solution, either disc-micelle-type or rod-micelle-type nematic phase can be formed. As expected, the CNTs were aligned along the director for both micelle shapes. Triton X-100 is another most often used surfactant for making lyotropic LCs. The phase diagram of Triton X-100 indicates the formation of a hexagonal phase between 38 and 60 wt% in water at a temperature range of 273–303 K [117]. Recently, a lyotropic LC formed from a 50 wt% aqueous Triton X-100 solution has been successfully used for the SWCNT alignment [118]. Different from the abovementioned single-surfactant system, a catanionic-surfactant complex was used to form the lyotropic LC phase. A typical example is the binary system of sodium dodecyl benzenesulfonate (SDBS) and cetyltrimethylammonium bromide (CTAB) [119]. Why SDBS was chosen is because SDBS can produce a higher concentration of aqueous SWCNT dispersion compared with SDS. To produce the LC host phase, 28 wt% of CTAB was added into 1 wt% SDBS rather than increasing the SDBS concentration. One of the substantial advantages of this binary system is the greatly improved CNT concentration (up to 0.2 wt%) and sufficiently good alignment of CNTs. Other binary systems, such as Triton X-100/SDS [118] and polyoxyethylene/ethylammonium nitrate [120], were also developed to form the lyotropic LC phase for dispersing and aligning the CNTs.

Besides the use of LCs as host for CNT ordering, the CNTs themselves can form LC provided that the CNT concentration is high enough (according to the Onsager theory, it is about 0.5 wt% loading for SWCNTs and some ten times higher loading for MWCNTs [107]). The explored methods for preparing high-quality CNT dispersions at high loadings include dispersing oxidized MWCNTs in water [121–123], dispersing pristine SWCNTs in super acids [124–126], and dispersing CNTs in various solvents in the presence of dispersants such as biomolecules [127–130] and synthetic polymers [131, 132]. Once the CNT-based LCs were acquired, the macroscopic-scale uniform alignment of CNTs could be accomplished by applying the shear stress to control the director orientation.

Mechanical force can be applied on the bulk CNT samples for the parallel alignment. Homma and coworkers demonstrated that a tweezers tip with a diameter of about 30 μm could be used for achieving the alignment of SWCNTs on the substrates where they were grown [133]. The alignment direction was determined by the travel path of the tweezers tip. Liang and coworkers have developed a very simple mechanical-stretch method to align the CNTs in the randomly dispersed, commercially available sheets of mm-long MWCNTs [134]. Based on polarized Raman scattering tests, the near perfect alignment (more than 95% CNTs aligned along stretch direction) was obtained at a ~50% stretch ratio.

Lu predicted that SWCNTs should have an anisotropic magnetic susceptibility and that the lowest energy orientation for both metallic and semiconducting SWCNTs would be parallel to an impressed magnetic field [135]. In other words, application of a sufficiently strong magnetic field should align CNTs that are free to rotate in a suspension. Actually, the membrane consisting of highly aligned SWCNTs has been successfully fabricated by using a magnetic field of 7–25 T [136–138]. Field intensities of 7 T are commercially available and magnetic alignment is thus a feasible and accessible pathway for forming aligned SWCNTs. Even so, actualizing the alignment of CNTs in a much weaker magnetic field remains a challenge. As a response to this issue, Haddon and coworkers have realized MWCNT alignment in a magnetic field of H = 300 Oe, where the CVD-synthesized, perpendicularly aligned CNTs were Ni-capped via thermal evaporation before dispersion in DMF to form a suspension [139]. In another work, magnetic Fe_3O_4 nanoparticles were attached on the CNT surface with the assistance of a bi-functionalized carboxylic acid pyrene derivative, and such Fe_3O_4-decorated CNTs were aligned in a magnetic field of less than 250 mT [140]. Without any additional decoration of magnetic materials, Kordás et al. found that the aqueous carboxyl-functionalized MWCNTs could be aligned in a relatively weak magnetic field of 641 mT [141]. The magnetic response behavior of MWCNTs, which were grown by CVD of xylene/ferrocene, results from the ferromagnetic iron nanoparticles produced in the CNT growth and entrapped in the inner-tubular cavity of CNTs. Another possible contribution is from the defects introduced in the acid treatment of MWCNTs, which may act as localized magnetic moments.

Ajiki et al. have predicted that the difference between the magnetic susceptibility of SWCNT parallel to its axis and the perpendicular susceptibility is as great as one order of magnitude [142, 143]. However, this difference had not yet been evidenced in the experimental observations until Tumpane and coworkers reported their breakthrough work. The most likely reason is attributed to the effect of bundling, which may reduce the apparent differential susceptibility. Therefore, if once more soluble CNTs are available, one can expect to fully utilize the extreme difference in susceptibilities to achieve alignment of CNTs (without any magnetic decorations) at much lower field strength. With this idea in mind, Tumpane and coworkers have prepared water-soluble functionalized

SWCNTs through thermal condensation of paraformaldehyde and 2-[2-(*tert*-butoxycarbonyl)ethylamino]acetic acid in the presence of purified CNTs [144]. The linear dichroism tests showed that the functionalized and thus more soluble CNTs were well aligned in a very low magnetic field of 100 mT.

By employing a surface acoustic technique, Ebbecke et al. have prepared the aligned CNTs on the $LiNbO_3$ substrate [145]. By using a modified acoustic technique, Wen et al. have realized large-scale alignment of MWCNTs on other solid substrates (e.g., Si wafer) [146], not limited to piezoelectric materials (e.g., $LiNbO_3$ substrate) anymore.

Aligning CNTs between a pair of electrodes with an applied electric field, known as dielectrophoresis, is an attractive, practical, and efficient technique. To employ dielectrophoresis technique, a pair of electrodes have to be defined on the desirable substrates, through which a non-uniform electric field is applied. When CNT suspension is introduced to the electrodes, the electric field polarizes the CNTs, inducing effective dipole moments in each of the CNTs. Thus, the CNTs must rotate and align along the electric field under the influence of the induced dipole torque. In addition, the CNTs must move under a nonzero dielectrophoretic force caused by the nonuniform electric field. Simple simulations suggest that the time for the CNT rotation is much smaller than that for the CNT translation [147, 148].

The mechanism of the dielectrophoretic alignment of CNTs has been systematically studied by taking into consideration of various influence factors, such as electrode type, voltage, thermal noise, and already-deposited CNTs [149]. The results suggest that (i) the comb electrode has a higher efficiency on the alignment than the parallel one; (ii) already-deposited CNTs between the electrodes greatly alter the local electric field and screen off the later coming CNTs, resulting in the formation of randomly dispersed CNTs between parallel electrodes and guaranteeing only one CNT/bundle for each fingertip pair; (iii) the gap width of electrodes and the length of CNTs determine the screening distance that provides a way to estimate the density of aligned CNT; and (iv) the CNT array distribution is decided by the competition between the thermal noise, which can be suppressed by increasing the voltage, and the electric field strength.

Since dielectrophoretic alignment is based on the electric polarization induced in CNTs by the external electric field, variation in the electronic structures in a collection of CNTs will undoubtedly affect the movement of CNTs in the field through variation in the induced dipole moment. Pursuing this idea, Kappes and coworkers have achieved the effective separation of metallic and semiconducting SWCNTs via ac dielectrophoresis [150]. By using a novel radio frequency dielectrophoresis setup, Krupke and coworkers have fabricated the CNT network consisting of aligned metallic and randomly oriented semiconducting SWCNTs [151]. Chen et al. have realized simultaneous separation and alignment of SWCNTs in one dielectrophoretic process by using the multigap nanoelectrodes [152]. In a recent work, Padmaraj et al. have designed planar electrodes, which were built as metal–oxide–semiconductor nanogap capacitors with the substrate as one electrode and the patterned metal layer as the top electrode [153]. This design not only generates high electric fields but also enables the curved fringing field owing to the conductive substrate. As expected, the metallic SWCNTs were found to be aligned parallel to the electrodes. However, the perpendicular alignment of semiconducting SWCNTs was also observed. Such orthogonal alignment, which implies substantial separation, was attributed mainly to the unique electrode design.

The combination of dielectrophoresis method and other technique was able not only to realize the alignment of CNTs but also to endow the CNT assemblies with new characteristics. For example, the dielectrophoresis deposition together with fissure formation technique has been used to prepare clean SWCNT field emitters with increased adhesion to the substrate [154]. To this end, a SWCNT thin film was first deposited on a Ti substrate by dielectrophoresis from an aqueous mixture of SWCNTs and SDS, and then additional SDS layer was deposited on the SWCNT film from a 0.1 wt% SDS aqueous solution by the second dielectrophoresis. Next, the sample was fired at 350°C under a natural air environment. SEM observation showed that the SWCNTs with a clean surface were exposed in the fissures produced by firing. The electron field emission properties of the SWCNT emitters were improved significantly owing to the additional deposition of SDS layer. When the SWCNTs were cut by increasing the firing time, the field emission current density increased to some extent; however, their stability decreased considerably.

The Langmuir–Blodgett (LB) technique was also applicable for the alignment of CNTs. Most of the reported work related to covalently functionalized CNTs that were accomplished by an amidation reaction [25, 26, 155–158]. The functionalized CNTs were uniformly dispersed in various organic solvents and then spread on the water surface of a Langmuir trough. Finally, LB films were deposited on the substrates by either horizontal lifting or vertically dipping. Two mechanisms, resulting from compression and flow, are proposed to influence the alignment. Pristine SWCNTs could also be aligned by LB technique [159], in which the SWCNTs were noncovalently functionalized by poly(*m*-phenylenevinylene-*co*-2,5-dioctoxy-*p*-phenylenevinylene) (PmPV) to prepare the stable SWCNT solution in 1,2-dichloroethane.

With the assistance of an external force, the LBL technique has also applied for the alignment of CNTs. In a typical work, pristine SWCNTs were first noncovalently functionalized by an anionic polyelectrolyte PPE-SO_3^- consisting of a conjugated poly(*p*-phenylene ethynylene) (PPE) backbone and SO_3^--containing ionic side chains [160]. The functionalized SWCNTs were then dispersed in water to form an aqueous composite solution. For the LBL process, another aqueous solution of cationic PPE-$N(CH_3)_3^+$ was also prepared. The cleaned substrates were alternately dipped into the composite solution and the cationic solution. By carefully controlling the dipping direction as well as the dipping and withdrawing speeds, the polyelectrolyte/SWCNT composites were obtained with the SWCNTs aligned parallel to the substrate surface. Similarly, Liang and coworkers have also realized the alignment of SWCNTs in polyelectrolyte composites by a magnetic field-assisted LBL technique [161]. The SWCNT solution was prepared by dispersing the SWCNTs in an aqueous solution of 0.5 wt% anionic surfactant of sodium dodecylbenzenesulfonate (NaDDBS). The employed polyelectrolyte is poly(ethyleneimine) (PEI) that is positively charged in an aqueous solution when PH > 7.

Recently, Bao and coworkers have developed a spin-coating technique for the alignment of SWCNTs [162, 163]. To ensure the successful alignment, purified SWCNTs were dispersed in *N*-methyl pyrrolidone (NMP) to prepare the completely debundled SWCNT solution and the substrate surfaces were functionalized with amine- or phenyl-terminated silanes. After spin-coating, the alignment of SWCNTs was found radical from the center with the only unaligned area being the portion concentric with the vacuum chuck on the spin coater (~2 mm in diameter on a 12.7 mm-diameter wafer). The alignment degree could be tuned on the basis of spin-assembly conditions. By carefully depositing the solution onto the center of a spinning substrate, the alignment was obtained. But no alignment was observed by dispensing the solution followed by spinning the substrate. Additionally, increasing spin-coating rate significantly improved the alignment level. More important, this technique permits to fabricate the self-sorted SWCNT network, in which SWCNT chirality separation and simultaneous control of SWCNT density and alignment occur in one step. AFM and Raman examinations suggested that the enriched semiconducting SWCNTs were aligned parallel to the amine-functionalized substrate surfaces, while the enriched metallic SWCNTs parallel to the phenyl-functionalized ones. In short, the one-step, low-cost, and solution-based spin-coating technique provides a rather good alternative to prepare the aligned and sorted SWCNT networks (enriched with semiconducting/metallic SWCNTs) with a controlled SWCNT density. It, therefore, becomes practicable to fabricate high-performance and large-area thin-film transistors (TFTs) by using SWCNT networks on rigid or flexible substrate in an economic manner [164–166]. The TFT performance in detecting trace chemicals suggests that the TFT sensors can detect trace concentrations, down to 2 ppb, of dimethyl methylphosphonate and trinitrotoluene in aqueous solutions, showing a much higher sensitivity to analytes than those fabricated with the random and unsorted SWCNT networks [166].

Homma and coworkers have presented a simple contact printing process to transfer large-area random SWCNT films from the growth substrate to another surface and assemble the random SWCNT films into highly ordered arrays [167]. To perform this process, a flat receiver substrate was faced to a growth SiO_2 substrate (donor substrate) with an as-synthesized, randomly distributed SWCNT film. About 5 KPa downward pressure was then applied by hand. This made the SWCNTs sufficient contact with the receiver substrate surface to strongly interact with it through van der Waals interactions, eventually resulting in detachment of the SWCNTs from the growth substrate and direct transfer to the receiver substrate. When the contact process included directed sliding of the donor substrate, the random SWCNTs were aligned by a similar mechanism as in the tweezer-driven alignment.

Freestanding, highly dense, and parallel aligned CNT films could be fabricated directly from the as-synthesized, perpendicularly aligned CNT carpets through a mechanical rolling process (Fig. 3.6) [168]. In this method, a roller was used for the compression of the SWCNT carpets. The compression stems from an applied downward force by the hands during the rolling process in which the roller undergoes rotation and translation. The parallel aligned SWCNT films were then freed from the substrate by etching the catalyst with either 1 M HCl solution (wet chemical process) or H_2O vapor (dry process). The freestanding films, particularly those freed from the substrate by dry process, were found to be easily transferred to any arbitrary surface. In a followed work, such a film was transferred to a sapphire substrate for producing an optical polarizer [169]. The terahertz (THz) measurements showed that no absorption was noticed when the THz polarization was perpendicular to the alignment axis, but strong absorption took place when the polarization was parallel to the alignment direction. The polarization degree in terms of absorbance was measured to be 1 and the reduced linear dichroism was 3 over the frequency range of 0.1–1.8 THz. These observations suggest that this type of aligned SWCNT film performs as an ideal linear polarizer in the THz frequency range. On the basis of similar principle, parallel aligned CNT films were also fabricated by Hart and coworkers from the as-synthesized and perpendicularly aligned CNT arrays [170]. In their experiments, an elaborately designed force-controlled rolling machine rather than a hand-operated roller was employed to provide the compression, thus enabling the control of final thickness, packing density, and even alignment degree of the CNT films. The CNT films have been demonstrated to robustly transfer to other substrates, including PDMS and polyimide (PI) films.

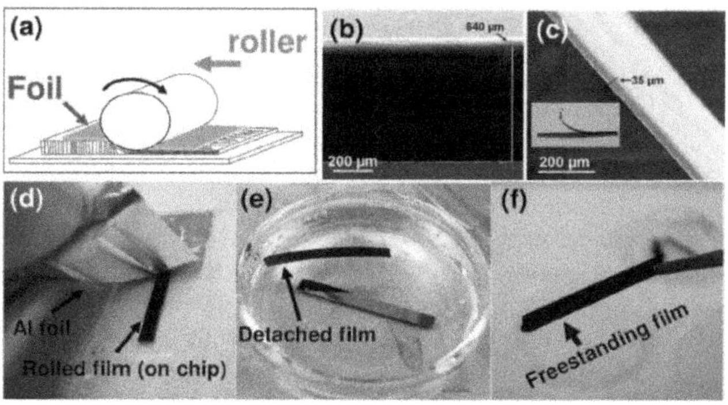

Figure 3.6 (a) Cartoon of the rolling process to form films of aligned SWCNTs, SEM images of the SWCNT carpets (b) before and (c) after the roll-over process, (d) image showing the Al foil detached from the carpet. Notice the lack of any transfer to the Al foil, (e) detachment of the film by catalyst etching in 1 M HCl solution, and (f) image of a freestanding film held by tweezers [168]. See also Color Insert.

3.3.2.3 Perpendicularly aligned and patterned CNTs

By using an ac dielectrophoresis, Ko et al. have also fabricated perpendicularly aligned SWCNTs on ITO glass substrates that were modified with SAM of 2-carboxyethyl phosphonic acid followed by covering a monolayer of monodispersed silica beads [171]. Perpendicularly aligned SWCNTs were found to be only implanted into the interstices of the silica beads, which function as supporting scaffold to prevent the SWCNTs from falling down and maintain them at low density. As a result, the SWCNT assembly exhibits enhanced field emission.

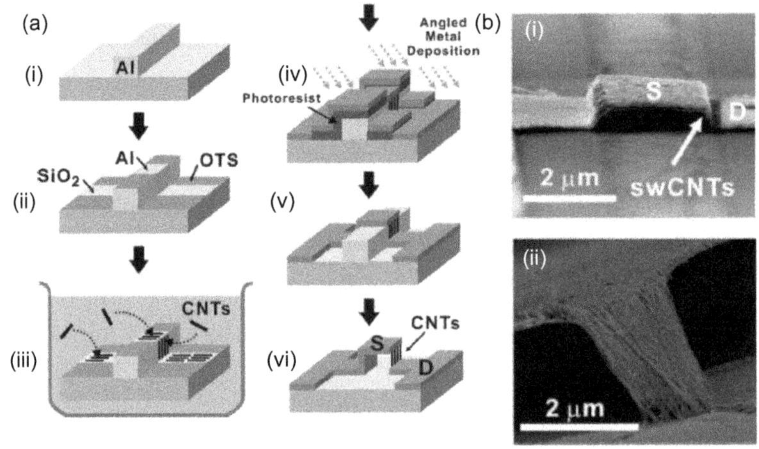

Figure 3.7 (a) Schematic diagram depicting the fabrication process: (i) formation of Al sacrificial structure on SiO_2 substrate, (ii) patterning of OTS SAM, (iii) selective absorption of the SWCNTs onto the unpatterned substrate surface, (iv) formation of the electrode patterns, (v) formation of the electrodes, and (vi) removal of the Al sacrificial structure, and (b) SEM images of (i) perpendicularly suspended SWCNT network junction and (ii) perpendicularly suspended SWCNT network stretched out by the residual stress in the electrode beam [172]. See also Color Insert.

By using only conventional microfabrication facilities, Lee et al. have fabricated the perpendicularly suspended and stretched SWCNT network junctions in large quantity via the directed assembly strategy [172]. The main procedures are given in Fig. 3.7a. Al sacrificial structure was first formed on SiO_2 substrate via photolithography and lift-off process. Then, octadecyltrichlorosilane (OTS) SAM was patterned via photolithography. When the patterned substrate was immersed in SWCNT solution (0.5 mg·ml^{-1} in 1,2-dichlorobenzene), the SWCNTs were selectively adsorbed onto the unpatterned substrate surface. Next, source and drain electrodes were fabricated via photolithography, angled metal deposition (this process permits a shadowed region on one perpendicular surface of the Al structure where the SWCNTs remains uncovered by the deposited metal), and lift-off process. Finally, perpendicularly suspended and stretched SWCNT structures were achieved by removing the Al sacrificial structure in base solution (Fig. 3.7b).

3.3.2.4 Parallel aligned and patterned CNTs

Various methods based on the self-assembly of CNTs on chemically modified surfaces to prepare parallel aligned and patterned CNTs have received great attention owing to the associated mild process conditions and the ability to produce well-defined structures.

By patterning the substrate surface with alternating polar (hydrophilic) and/or nonpolar (hydrophobic) SAM regions, CNTs have been found to self-assemble into the desired structures on planar substrates. For instance, Colbert and coworkers have demonstrated that individual SWCNTs can be precisely positioned on NH_2-functionalized nanolithographic patterns [173]. By vertically immersing glass substrates pre-patterned with hydrophobic and hydrophilic regions into an aqueous solution of acid-oxidized short SWCNTs, Zhou and coworkers have successfully fabricated ordered SWCNT structures [174]. Tsukruk and coworkers also reported the successful assembly of SWCNTs on the pre-patterned substrates where a SWCNT-SDS suspension was used as the SWCNT source and a casting process was used for densely packed SWCNT arrays, whereas a dip-coating process for low density SWCNT arrays [175, 176]. Wang et al. found when the polar and nonpolar patterns were respectively made from COOH-terminated and CH_3-terminated SAMs, SWCNTs preferred to strongly attract to the polar region and the SWCNTs with lengths larger than the dimensions of the COOH-SAM feature aligned along the boundaries between the polar and nonpolar regions [177]. By using these boundaries as affinity templates, the authors have demonstrated the ability to position, shape, and link μm-long SWCNTs. The constructed structures include SWCNT dot, ring, arc, letter, and ever more sophisticated structured thin films and continuous ropes, as shown in Fig. 3.8. Recently, Sharma et al. have developed a novel method for nanometer positioning, parallel alignment, and placement of individual SWCNTs using hydrodynamic forces in cylindrical droplets [178, 179]. The platform for performing this process is a patterned Au surface with alternating rectangular strips of polar and nonpolar SAMs. When a homogeneous thin film of SWCNT aqueous solution is deposited on the patterned substrate, the film segregates into the desired cylindrical droplets on the polar region. An internal hydrodynamic force developed in the droplet drying drives the SWCNTs to move toward the air–liquid–substrate interface. If the width of the polar SAM strips falls into the range of 175–950 nm, parallel alignment and placement of individual SWCNTs are observed along the centerline of the strips. For the polar strips with width larger than 3 μm, edge deposition with almost perfect parallel alignment of CNTs takes place. A transition/disordered regime exists when the strip width is about 2 μm. By using the microscale polar SAM patterns with gradient surface molecular density, Hong and coworkers have also realized the parallel alignment and placement of a-few-SWCNTs in the nanoscale regions at the center of the SAM patterns, but leaving the remaining regions empty [180]. Furthermore, Hong and coworkers found all of SWCNTs, double-walled carbon nanotubes (DWCNTs), and MWCNTs could be assembled onto the pre-patterned substrates, but smaller CNTs were found to have higher adsorption probability to polar surfaces than larger CNTs [181, 182].

For the self-assembly of CNTs on the SAM-patterned substrates, electrostatic interactions between them are widely accepted as the dominant driving force. However, the SAM patterns are usually made from the reactive molecules with alkane backbones, which are well known to be insulators. If the conjugated molecules are used to form the SAM patterns, it is undoubtedly favorable for the adsorption of CNTs on the charged substrate surface. On the other hand, the conductive characteristic of the conjugated molecules may decrease the contact resistance between the assembled CNTs and the electrodes. Based on these considerations, Hong and coworkers have chosen π-conjugated molecules of 2-mercaptonimidazole, 2-mercaptopyrimidine, and 4-mercaptopyridine to form the polar SAM patterns [183]. SWCNTs were found to selectively adsorb and precisely align onto the polar SAM regions from a 0.2 $mg \cdot ml^{-1}$ SWCNT solution in 1,2-dichlorobenzene. The conducting AFM test indicated that the assembled SWCNTs were electrically well-connected to the electrodes pre-patterned with the SAMs of 2-mercaptonimidazole.

The charged substrate surface with arbitrary patterns could also be created by charge writing with an atomic force microscope [184]. During its development in an aqueous solution of surfactant-stabilized SWCNTs, Coulomb attraction guides the positioning and alignment of the SWCNTs onto the charged patterns with a lateral resolution down to the diameter of a single SWCNT.

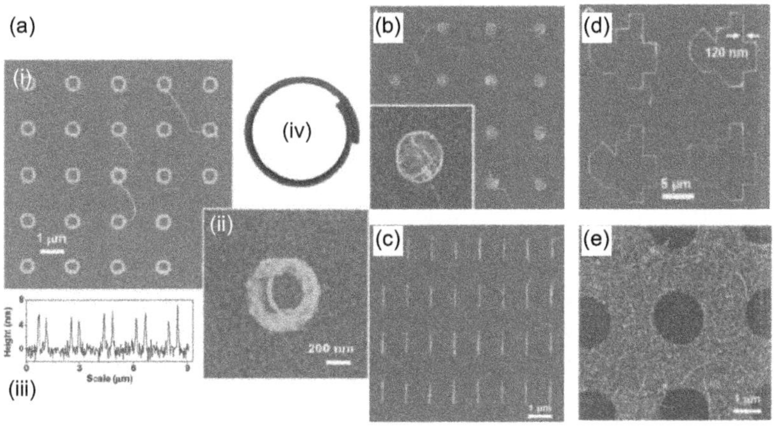

Figure 3.8 With nanopatterned affinity templates, SWCNTs were assembled into (a) rings and nanoletters [(i) AFM tapping mode topographic images, (ii) zoom-in AFM image, (iii) height profiles of SWCNT rings in a 5 × 5 array, and (iv) a molecular model of a coiled SWCNT], (b) 1 μm dots, (c) lines, (d) random line structures, (e) a network with 1.6 ± 0.1 μm-diameter pores [177].

By patterning the substrate surface with protruded polymer structures, highly organized SWCNT architectures in various dimensions, geometries, and densities could also be self-assembled within the trenches whose bottom surface (namely, substrate surface) were functionalized either by a simple $SF_6/O_2/Ar$ plasma treatment [185–187] or with an amine-terminated SAM [188–190]. After the assembly processing, the polymer structures used as a mask for selective deposition of CNTs at controlled locations were removed via a lift-off process. The narrowest lateral width of achieved SWCNT structures has been demonstrated to downsize to a few hundred nanometers [185, 187], which are limited only by the employed lithographic facilities and, in principle, can be reduced further down to lower sizes. By patterning the gate oxide (e.g., hafnium oxide (HfO_2))

surface with protruded SiO_2 structures, Tulevski and coworkers have realized the directed assembly of functionalized SWCNTs (*f*-SWCNTs) inside the HfO_2 bottomed trenches, allowing for the high yield (around 90%) and parallel fabrication of SWCNT device array [191]. In this process, the hydroxamic acid groups in *f*-SWCNTs selectively bind to basic metal oxides, thus directing the deposition of *f*-SWCNTs from the solution onto the HfO_2 surfaces, whereas the trench sidewalls induce the alignment of *f*-SWCNTs along the trench length. Since the post-treatment of 600°C-annealing removed the organic moieties to leave pristine SWCNTs behind, the resultant devices exhibit the electrical performance expected for SWCNT devices with no performance deterioration as a result of the assembling process.

Pristine SWCNTs could also be directly self-assembled from an aqueous solution into regular micron-wide strips on the bare and planar Si/SiO_2 substrate surface [192]. A slip-stick mechanism was proposed to be responsible for such self-assembly. As schematically shown in Fig. 3.9, when a planar substrate is immersed in the SWCNT solution a meniscus is formed at the contact line (solid-liquid-vapor interface). With the solvent evaporation, convective transport makes the aligned SWCNTs in the solution move to the contact line and deposit on the substrate. The surface roughness thus generated produces a frictional force (f), which together with the liquid surface tension (γ_f) pins the position of the contact line. As the evaporation proceeds, the capillary force (γ_L), which pulls the liquid inward, builds up and finally causes the contact line depinned. The contact line then jumps to a new position for developing a new SWCNT strip. The periodic stick-slip motion of the liquid eventually forms the regular CNT strips. The spacing between the adjacent strips depends mostly on the concentration of the SWCNT solution. A higher SWCNT concentration usually leads to more closely spaced SWCNT strips.

Yan et al. have developed a capillary-driven fluidic alignment method to fabricate the aligned SWCNT patterns with submicron feature size on various substrates over mm-scale surface areas [193]. Toward this end, a PDMS mould pre-patterned with an array of submicron-wide linear channels was brought to conformally contact with a flat substrate for the formation of an array of empty channels. When an aqueous SWCNT solution was deposited at the side of the mould that had open channels, the solution was drawn into the channels by the capillary effect, and the involved SWCNTs were aligned by the channels and flow geometry as they were drawn into and through the channels with the liquid. After filling the channels and evaporating the solvent, the PDMS mould was removed, and the SWCNT patterns with geometries defined by relief structures on the PDMS mould remained on the substrate surface. One of the key elements for the successful assembly process is to pre-align the SWCNTs before the solution entering the channels. Otherwise, the SWCNTs that are long compared with the channel width cannot flow into the channels and instead accumulate at the channel entrance, creating a log-jam to block the channel entrance. The adopted measures included (i) dropping the solution at a proper distance from the PDMS mould as well as tilting the substrate to assist the solution reaching the channels, (ii) using the horn-shaped channels that are tapered and varied the width from about 50 μm (at the channel section adjoining the feed port) to 1 μm (the narrowest section), and iii) developing a gas-assisted fluidic alignment method where two (pre- and post-capillary) gas flows were used to improve SWCNT alignment and significantly narrow the width of the SWCNT pattern down to about 20 nm [194]. Once a flat PDMS substrate was used in this process, the SWCNT-patterned substrate could be used as a stamp for further contact printing of aligned SWCNTs onto other flat substrate surface and even nonflat Au-electrode bearing surface [195]. In an independent work, Suh and coworkers have also reported a similar capillary-driven fluidic alignment method for assembling CNTs where a partially cured polyethylene glycol diacrylate (PEG-DA) mould was used to define an array of empty channels [196].

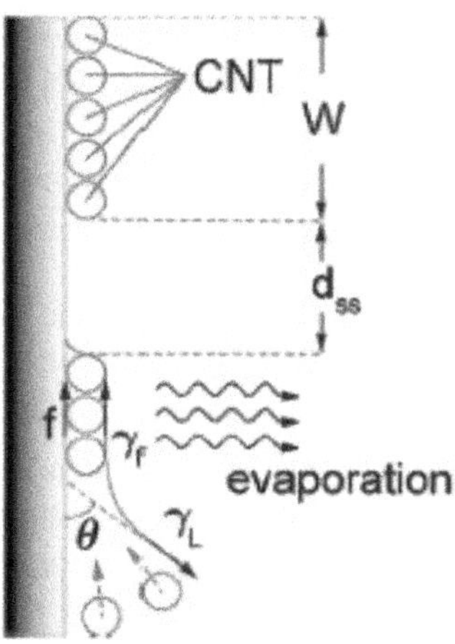

Figure 3.9 Cross-sectional schematic illustrating the slip-stick mechanism responsible for the self-assembly of SWCNTs [192].

In principle, most of the methods used for fabricating parallel aligned CNT arrays are simultaneously suitable to produce parallel aligned and patterned CNT arrays. One representative example is that once the aligned CNT monolayer was achieved on the substrate by a LB method, it could be further patterned and integrated into potential devices through removal of the unwanted CNTs using lithographic patterning techniques and O_2 plasma etching [159]. Another one is the gas-flow-based method. It has been demonstrated to create the orthogonal CNT network (by two orthogonal gas-flow-assisted depositions) with controlled CNT density (by varying the deposition cycles) [197]. Additionally, it was reported that the surface acoustic wave technique allowed the fabrication of aligned and patterned CNTs by using the pre-structured substrates [198, 199].

Besides for forming the parallel aligned CNTs, magnetic method can also be used for patterning the CNTs. For instance, Shim et al. have realized the precise self-assembly of one CNT at each Ni pattern on Au/Ti electrodes fabricated on the SiO_2/Si substrate [200]. In this work, pristine MWCNTs with one end capped with residual Fe catalyst were dispersed in an aqueous solution of 1 wt% SDBS and used as the starting material. A microfluidic channel was applied for guiding the CNT solution to the Ni pattern. When the CNT reaches the pattern, the Fe catalyst is entrapped by a magnetic force produced on the Ni pattern by applying an external magnetic field. The CNT is thus positioned. Finally, the shear force from the liquid flowing through the microchannel aligns the CNT parallel to the flow direction. Sufficiently small dimensions of the Ni pattern (200 nm-wide and 100 nm-thick) ensure the edge covered by the attached CNT, preventing the assembly of multiple CNTs at each pattern. By placing a field modulator in a strong magnetic field (5–12 T), Tsuda et al. have achieved the simultaneous alignment and micropatterning of CNTs in polymers and even in ceramic and silicone composites [201].

By use of a suitable design of electrode pairs, dielectrophoresis techniques can be employed to construct arrays of individual CNTs. The work by Krupke and his coworkers in this domain provides an illustration [202]. By application of an ac electric field (300 kHz and 2 V peak-to-peak voltage), each electrode pair was bridged by one SWCNT that was deposited from an aqueous SDBS solution with individually dispersed SWCNTs (Fig. 3.10). Further, by a two-step dielectrophoresis deposition, Diehl et al. have prepared wiring networks from SWCNTs [203]. And besides, recent development under this topic include (i) combining dielectrophoresis and dip coating processes to assemble SWCNTs at a fast rate (~1 min) over large areas (1.5×1.5 cm^2) with the feature size down to 80 nm [204], and (ii) combining dielectrophoresis and electrochemical deposition techniques to fabricate 1D wire devices [205].

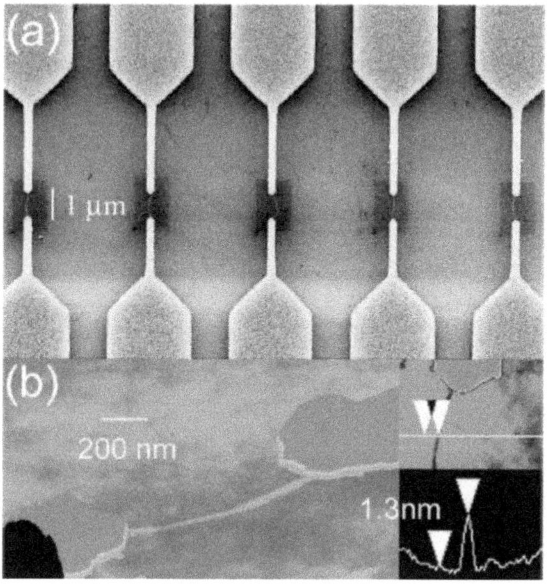

Figure 3.10 (a) SEM image of the electrode array having five adjacent devices with each electrode pair bridged by one SWCNT, and (b) AFM image of one such device (The height profile confirms the bridging by an individual SWCNT) [202]. See also Color Insert.

With the help of dielectrophoresis, SWCNTs have been assembled onto normal AFM probes to act as SWCNT-nanofibril AFM probes [206]. In this process, a 10 V ac (2 MHz) bias was applied between a working electrode (an AFM probe made from Si with Al coated on the reflex side and Pt on the tip side) and a counter electrode (a small hollow metal tube). A drop of SWCNT solution was used to form a droplet at the end of the metal tube, working as electrolyte. A positive dielectrophoretic force exerted on the SWCNTs drove them to align when precipitating out onto the probe, and the gradual withdrawal of the probe away from the solution consolidated the

SWCNTs into nanofibril. The diameter, length, and stiffness of such nanofibril probes could be tailored by varying the concentration of SWCNT solution, immersion time, pulling speed, and post-treatment conditions (dissolution in a SDS solution and vacuum annealing). Regrowth of a new SWCNT nanofibril from the side or at the end of a previously grown nanofibril was also demonstrated by repeating the dielectrophoresis deposition. The resultant SWCNT-nanofibril AFM probes are mechanically robust and electrically conductive and thereby can find various applications in biomechanics and electrochemical analysis of biological entities as well as deep trench structures.

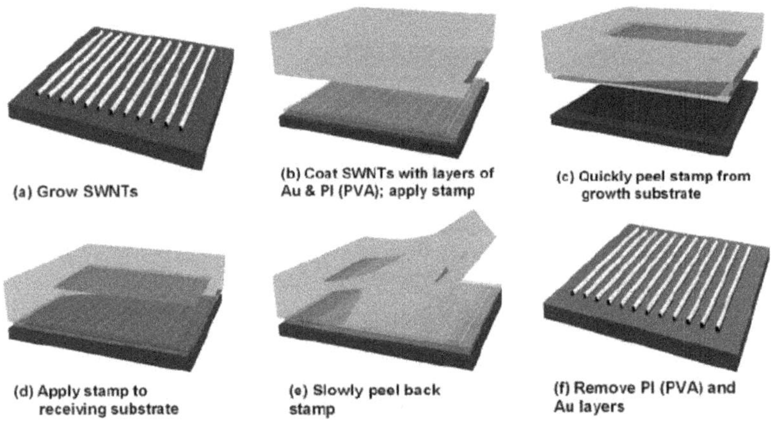

Figure 3.11 Schematic illustration of a transfer printing process: (a) patterned growth of SWCNTs on a quartz substrate by CVD, (b) deposition of a bilayer of Au/polymer (either PI or PVA), (c) removal of the SWCNT/Au/polymer layer with an PDMS stamp, (d) applying the stamp to a receiving substrate, (e) slowly peeling back the stamp and leaving the SWCNT/Au/polymer on the receiver, and (f) removing the polymer and Au by etching [207].

It is well known that PDMS has a much lower intrinsic surface energy than the substrates commonly used for CNT assembly, such as Si, SiO_2, mica, Au, and most polymeric substrates. On the basis of this property of PDMS, Rogers and coworkers have developed a methodology to form multilayer superstructures of large collections of SWCNTs configured in horizontally aligned arrays, random networks, and complex geometries of arrays and networks on various substrates [207]. As shown in Fig. 3.11, this process starts with parallel aligned and patterned growth of SWCNTs on a single-crystal quartz substrate by the CVD of ethanol, followed by deposition of a bilayer of Au/polymer (PI or polyvinyl alcohol (PVA)) on top of the SWCNTs as a carrier film, then application of a stamp made from a flat PDMS slab on the carrier, and removal of the SWCNT/Au/polymer layer with the PDMS stamp by quickly peeling it away. Applying the stamp to a receiving substrate and slowly peeling back the stamp leaves the SWCNT/Au/polymer on the receiver. Finally, the polymer/Au carrier was successively removed by O_2 reactive ion etching and wet etching with a commercial solution of Au-TFA. Besides the rigid and planar substrates, this process was also proved to be feasible for the transfer of SWCNT arrays to the curved surface of a glass tube and a thin sheet of Kapton plastic. More important, the process has the ability to transfer SWCNTs onto substrates that already support tubes and form well-defined multilayer superstructures of SWCNTs.

A direct transfer printing technique (namely, no stamp means) was also developed [208]. As depicted in Fig. 3.12, the perpendicularly aligned SWCNT patterns were first grown on a glass substrate (donor substrate). Then, the substrate with grown SWCNT patterns was faced down against a SiO_2 substrate (acceptor substrate) and slid around 1 mm to a certain direction with around 5 KPa of downward pressure. Finally, the parallel aligned SWCNT patterns were achieved on the acceptor substrate by removing the donor substrate. Recently, Hauge and coworkers reported another direct transfer printing technique — dry contact transfer printing technique [209]. As-grown patterns of perpendicularly aligned SWCNTs were undergone a post-growth H_2/H_2O vapor etch to release the chemical bonds between catalyst and SWCNTs, leaving the SWCNTs adhered to the growth substrate only by weaker van der Waals interactions. A rolling process was then used to transform the perpendicular alignment of SWCNTs to the parallel one. Finally, the transfer process was accomplished by applying an acceptor substrate on the top of the SWCNT layer. The principle for such transfer process is that van der Waals interaction between any acceptor substrate surface and the SWCNT layer is greater than the van der Waals interaction and bond strength of the SWCNT to the growth substrate.

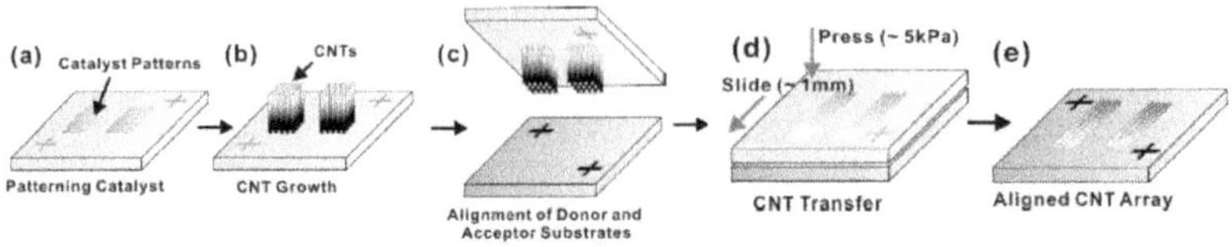

Figure 3.12 Schematic diagram depicting the direct transfer printing process of parallel aligned SWCNTs: (a) patterning catalyst film on a glass substrate, (b) perpendicular growth of SWCNTs, (c) placing the donor substrate with grown SWCNT patterns onto a SiO_2 substrate (acceptor substrate), (d) application of pressure and sliding of the donor substrate against the acceptor substrate, (e) formation of parallel aligned SWCNT patterns on the acceptor substrate after the removal of the donor substrate [208].

3.4 Conclusions

The realization of excellent potential of CNTs in their developed and potential applications requires a well-controllable and repeatable assembly of CNTs in various structures. The scientific and technical questions brought by such requirements have drawn the attention of a large number of researchers. Impressive progresses of CNT-assembly techniques have been made, especially in the last five years, under the two rubrics of synthetic assembly and post-synthetic assembly. Due to the applications of a series of new techniques, the CNTs have been successfully assembled into hierarchical arrays over large-scale areas on various substrates, such as semiconductors, metals, ceramics, plastics, and even papers. The proposed applications of such CNT assemblies include integrated electronic devices, sensors, nanoporous carbon membranes, nanocomposites, and so on. It is conceivable that we are approaching the point at which the available techniques will enable practical applications of CNTs. However, some important breakthroughs are still desired. The most highly anticipated applications of CNTs in molecular devices are still at the prototype level and the integration of CNT-based device elements into circuits remains a long-term goal, even though a few integrated nanosystems, e.g., a fully functional CNT-radio system, have been demonstrated. The new technologies that can economically fabricate in parallel complex large scale CNT-based circuit architectures with high efficiency and throughput approaching that of current semiconductor fabrication techniques are therefore in high demand. Gratifyingly, progress in this direction is rapid. If the current progress rate is any indication, it is fully possible in the near future to use CNT-based devices replacing digital CMOS ones made from silicon-based materials. Finally, it is undoubted that many new results will be presented while this chapter is in press.

References

1. Iijima, S. (1991). *Nature*, **354**, pp. 56–58.
2. Iijima, S., and Ichihashi, T. (1993). *Nature*, **363**, pp. 603–605.
3. Bethune, D. S., Kiang, C. H., de Vries, M. S., Gorman, G., Savoy, R., Vazquez, J., and Beyers, R. (1993). *Nature*, **363**, pp. 605–607.
4. Ajayan, P. M. (1999). *Chem. Rev.*, **99**, pp. 1787–1800.
5. Collins, P. G., and Avouris, P. (2000). *Sci. Am.*, **283**, pp. 62–69.
6. Dai, H., Wong, E. W., and Lieber, C. M. (1996). *Science*, **272**, pp. 523–526.
7. Qian, D., Wagner, G. J., Liu, W. K., Yu, M. F., and Ruoff, R. S. (2002). *Appl. Mech. Rev.* **55**, pp. 495–533.
8. Mintmire, J. W., Dunlap, B. I., and White, C. T. (1992). *Phys. Rev. Lett.*, **68**, pp. 631–634.
9. Hamada, N., Sawada, S., and Oshiyama, A. (1992). *Phys. Rev. Lett.*, **68**, pp. 1579–1581.
10. Dai, L. M., Patil, A., Gong, X. Y., Guo, Z. X., Liu, L. Q., Liu, Y., and Zhu, D. B. (2003). *Chemphyschem*, **4**, pp. 1150–1169.
11. Rutherglen, C., and Burke, P. (2007). *Nano Lett.*, **7**, pp. 3296–3299.
12. Jensen, K., Weldon, J., Garcia, H., and Zettl, A. (2007). *Nano Lett.*, **7**, pp. 3508–3511.
13. Kocabas, C., Kim, H., Banks, T., Rogers, J. A., Pesetski, A. A., Baumgardner, J. E., Krishnaswamy, S. V., and Zhang, H. (2008). *Proc. Natl. Acad. Sci. USA*, **105**, pp. 1405–1409.
14. Tans, S. J., Verschueren, A. R. M., and Dekker, C. (1998). *Nature*, **393**, pp. 49–52.
15. Bachtold, A., Hadley, P., Nakanishi, T., and Dekker, C. (2001). *Science*, **294**, pp. 1317–1320.
16. Rinzler, A. G., Hafner, J. H., Nikolaev, P., Lou, L., Kim, S. G., Tomanek, D., Nordlander, P., Colbert, D. T., and Smalley, R. E. (1995). *Science*, **269**, pp. 1550–1553.

17. de Heer, W. A., Chatelain, A., and Ugarte, D. (1995). *Science*, **270**, pp. 1179–1180.

18. Fan, S., Chapline, M. G., Franklin, N. R., Tombler, T. W., Cassell, A. M., and Dai, H. (1999). *Science*, **283**, pp. 512–514.

19. Li, J., Ng, H. T., Cassell, A., Fan, W., Chen, H., Ye, Q., Koehne, J., Han, J., and Meyyappan, M. (2003). *Nano Lett.*, **3**, pp. 597–602.

20. Tu, Y., Lin, Y. H., and Ren, Z. F. (2003). *Nano Lett.*, **3**, pp. 107–109.

21. Mizuno K., Ishii J., Kishida H., Hayamizu Y., Yasuda S., Futaba D. N., Yumura, M., and Hata, K. (2009). *Proc. Natl. Acad. Sci. USA*, **106**, pp. 6044–6047.

22. Dai, H. J., Hafner, J. H., Rinzler, A. G., Colbert, D. T., and Smalley, R. E. (1996). *Nature*, **384**, pp. 147–150.

23. Wong, S. S., Joselevich, E., Woolley, A. T., Cheung, C. L., and Lieber, C. M. (1998). *Nature*, **394**, pp. 52–55.

24. Xu, Z. W., and Fang, F. Z. (2009). *J. Vac. Sci. Technol. B*, **27**, pp. 1388–1393.

25. Kim, P., and Lieber, C. M. (1999). *Science*, **286**, pp. 2148–2150.

26. Shonaike, G. O., and Advani, S. G. (2003) *Advanced Polymeric Materials: Structure Property Relationships*, eds. Antonucci, V., Hsiao, K. T., and Advani, S. G. Chapter 11 "Review of Polymer Composites with Carbon Nanotubes" (CRC Press LLC, Boca Raton) pp. 1–41.

27. Coleman, J. N., Khan, U., and Gun'ko Y. K. (2006). *Adv. Mater.*, **18**, pp. 689–706.

28. Krenchel, H. (1964) *Fibre Reinforcement.* (Akademisk Forlag, Copenhagen, Denmark).

29. Yan, Y. H., Zhang, Q., and Chan-Park, M. B. (2006). *Small*, **3**, pp. 24–42.

30. Gao, Y., Zhou, Y. S., Xiong, W., Mahjouri-Samani, M., Mitchell, M., and Lu, Y. F. (2009). *Appl. Phys. Lett.*, **95**, pp. 143117(1–3).

31. Xiong, G. Y., Wang, D. Z., and Ren, Z. F. (2006). *Carbon*, **44**, pp. 969–973.

32. Masarapu, C., and Wei, B. (2007). *Langmuir*, **23**, pp. 9046–9049.

33. Yi, W., and Yang, Q. (2010). *Appl. Phys. A*, **98**, pp. 659–669.

34. Yamamoto, N., Hart, A. J., Garcia, E. J., Wicks, S. S., Duong, H. M., Slocum, A. H., and Wardle, B. L. (2009). *Carbon*, **47**, pp. 551–560.

35. Qu, L., Dai, L., Stone, M., Xia, Z., and Wang, Z. L. (2008). *Science*, **322**, pp. 238–242.

36. Patole, S. P., Alegaonkar, P. S., Shin, H.-C., and Yoo, J.-B. (2008). *J. Phys. D Appl. Phys.*, **41**, pp. 155311(1–6).

37. Chen, L.-C., Wen, C.-Y., Liang, C.-H., Hong, W.-K., Chen, K.-J., Cheng, H.-C., Shen, C.-S., Wu, C.-T., and Chen, K.-H. (2002). *Adv. Funct. Mater.*, **12**, pp. 687–692.

38. Signore, M. A., Rizzo, A., Rossi, R., Piscopiello, E., Luccio, T. D., Capodieci, L., Dikonimos, T., and Giorgi, R. (2008). *Diamond Relat. Mater.*, **17**, pp. 1936–1942.

39. Pint, C. L., Nicholas, N., Pheasant, S. T., Duque, J. G., Parra-Vasquez, A. N. G., Eres, G., Pasquali, M., and Hauge, R. H. (2008). *J. Phys. Chem. C*, **112**, PP. 14041–14051.

40. Puretzky, A. A., Geohegan, D. B., Jesse, S., Ivanov, I. N., and Eres, G. (2005). *Appl. Phys. A Mater. Sci. Process.*, **81**, pp. 223–240.

41. Wood, R. F., Pannala, S., Wells, J. C., Puretzky, A. A., and Geohegan, D. B. (2007). *Phys. Rev. B*, **75**, pp. 235446(1–8).

42. Hongo, H., Nihey, F., and Ochiai, Y. (2007). *J. Appl. Phys.*, **101**, pp. 024325(1–9).

43. Huang, S. M., Cai, X. Y., and Liu, J. (2003). *J. Am. Chem. Soc.*, **125**, pp. 5636–5637.

44. Reina, A., Hofmann, M., Zhu, D., and Kong, J. (2007). *J. Phys. Chem. C*, **111**, pp. 7292–7297.

45. Häffner, M., Heeren, A., Haug, A., Schuster, E., Sagar, A., Fleischer, M., Peisert, H., Burghard, M., Chassé, T., and Kern, D. P. (2008). *J. Var. Sci. Technol. B*, **26**, pp. 2447–2450.

46. Ismach, A., Segev, L., Wachtel, E., and Joselevich, E. (2004). *Angew. Chem. Int. Ed.*, **43**, pp. 6140–6143.

47. Han, S., Liu, X., and Zhou, C. (2005). *J. Am. Chem. Soc.*, **127**, pp. 5294–5295.

48. Ago, H., Nakamura, K., Ikeda, K., Uehara, N., Ishigami, N., and Tsuji, M. (2005). *Chem. Phys. Lett.*, **408**, pp. 433–438.

49. Ismach, A., Kantorovich, D., and Joselevich, E. (2005). *J. Am. Chem. Soc.*, **127**, pp. 11554–11555.

50. Kocabas, C., Hur, S. H., Gaur, A., Meitl, M. A., Shim, M., and Rogers, J. A. (2005). *Small*, **1**, pp. 1110–1116.

51. Ago, H., Uehara, N., Ikeda, K., Ohdo, R., Nakamura, K., and Tsuji, M. (2006). *Chem. Phys. Lett.*, **421**, pp. 399–403.

52. Kocabas, C., Pimparkar, N., Yesilyurt, O., Kang, S. J., Alam, M. A., and Rogers, J. A. (2007). *Nano Lett.*, **7**, pp. 1195–1202.

53. Arco, L. G.-D., Lei, B., Cronin, S., and Zhou, C. (2008). *Appl. Phys. Lett.*, **93**, pp. 123112(1–3).

54. Liu, X., Ryu, K., Badmaev, A., Han, S., and Zhou, C. (2008). *J. Phys. Chem. C*, **112**, pp. 15929–15933.

55. Xiao, J., Dunham, S., Liu, P., Zhang, Y., Kocabas, C., Moh, L., Huang, Y., Hwang, K.-C., Lu, C., Huang, W., and Rogers, J. A. (2009). *Nano Lett.*, **9**, pp. 4311–4319.

56. Rutkowska, A., Walker, D., Gorfman, S., Thomas, P. A., and Macpherson, J. V. (2009). *J. Phys. Chem. C*, **113**, pp. 17087–17096.

57. Ago, H., Imamoto, K., Ishigami, N., Ohdo, R., Ikeda, K., and Tsuji, M. (2007). *Appl. Phys. Lett.*, **90**, pp. 123112(1–3).

58. Ishigami, N., Ago, H., Imamoto, K., Tsuji, M., Iakoubovskii, K., and Minami, N. (2008). *J. Am. Chem. Soc.*, **130**, pp. 9918–9924.

59. Kocabas, C., Kang, S. J., Ozel, T., Shim, M., and Rogers, J. A. (2007). *J. Phys. Chem. C*, **111**, pp. 17879–17886.

60. Zhang, B., Hong, G., Peng, B., Zhang, J., Choi, W., Kim, J. M., Choi, J. Y., and Liu, Z. (2009). *J. Phys. Chem. C*, **113**, pp. 5341–5344.

61. Ago, H., Ishigami, N., Yoshihara, N., Imamoto, K., Akita, S., Ikeda, K., Tsuji, M., Ikuta, T., and Takahashi, K. (2008). *J. Phys. Chem. C*, **112**, pp. 1735–1738.

62. Ding, L., Yuan, D., and Liu, J. (2008). *J. Am. Chem. Soc.*, **130**, pp. 5428–5429.

63. Yu, Q., Qin, G, Li, H., Xia, Z., Nian, Y., and Pei, S.-S. (2006). *J. Phys. Chem. B*, **110**, pp. 22676–22680.

64. Kang, S. J., Kocabas, C., Ozel, T., Shim, M., Pimparkar, N., Alam, M. A., Rotkin, S. V., and Rogers, J. A. (2007). *Nature Nanotechnol.*, **2**, pp. 230–236.

65. Dresselhaus, M. S., Dresselhaus, G., and Avouris, P. (2000) *Carbon Nanotubes: Synthesis, Structure, Properties, and Applications*, eds. Dai, H. J. Chapter 3 "Nanotube Growth and Characterization" (Springer-Verlag, Heidelberg) pp. 29–53.

66. Sohn, J. I., and Lee, S. (2002). *Appl. Phys. A*, **74**, pp. 287–290.

67. Su, W.-S., Lin, C.-M., Chen, T.-H., and Fang, W. (2009). *J. Micromech. Microeng.*, **19**, pp. 105009(1–7).

68. Hart, A. J., and Slocum, A. H. (2006). *Nano Lett.*, **6**, pp. 1254–1260.

69. Lim, K. Y., Sow, C. H., Lin, J., Cheong, F. C., Shen, Z. X., Thong, J. T. L., Chin, K. C., and Wee, A. T. S. (2003). *Adv. Mater.*, **15**, pp. 300–303.

70. Hofmann, M., Nezich, D., Reina, A., and Kong, J. (2008). *Nano Lett.*, **8**, pp. 4122–4127.

71. Kocabas, C., Kang, S. J., Ozel, T., Shim, M., and Rogers, J. A. (2007). *J. Phys. Chem. C*, **111**, pp. 17879–17886.

72. Ismach, A., and Joselevich, E. (2006). *Nano Lett.*, **6**, pp. 1706–1710.

73. Kim, S.-K., Lee, H., Tanaka, H., and Weiss, P. S. (2008). *Langmuir*, **24**, pp. 12936–12942.

74. Youn, S. C., Jung, D.-H., Ko, Y. K., Jin, Y. W., Kim, J. M., and Jung, H.-T. (2009). *J. Am. Chem. Soc.*, **131**, pp. 742–748.

75. Zhang, G., Mann, D., Zhang, L., Javey, A., Li, Y., Yenilmez, E., Wang, Q., McVittie, J. P., Nishi, Y., Gibbons, J., and Dai, H. (2005). *Proc. Natl Acad. Sci. USA*, **102**, pp. 16141–16145.

76. Chai, Y., Gong, J., Zhang, K., Chan, P. C. H., and Yuen, M. M. F. (2007). *Nanotechnology*, **18**, pp. 355709(1–5).

77. Zhu, L., Sun, Y., Hess, D. W., and Wong, C.-P. (2006). *Nano Lett.*, **6**, pp. 243–247.

78. Jiang, H., Zhu, L., Moon, K.-S., and Wong, C. P. (2007). *Nanotechnology*, **18**, pp. 125203(1–4).

79. Zhu, Y., Lim, X., Sim, M. C., Lim, C. T., and Sow, C. H. (2008). *Nanotechnology*, **19**, pp. 325304(1–7).

80. Jin, L., Bower, C., and Zhou, O. (1998). *Appl. Phys. Lett.*, **73**, pp. 1197–1200.

81. Ichida, M., Mizuno, S., Kataura, H., Achiba, Y., and Nakamura, A. (2004). *Appl. Phys. A*, **78**, pp. 1117–1120.

82. Kim, Y., Minami, N., and Kazaoui, S. (2005). *Appl. Phys. Lett.*, **86**, pp. 073103(1–3).

83. Akima, N., Iwasa, Y., Brown, S., Barbour, A. M., Cao, J., Musfeldt, J. L., Matsui, H., Toyota, N., Shiraishi, M., Shimoda, H., and Zhou, O. (2006). *Adv. Mater.*, **18**, pp. 1166–1169.

84. Wang, Q., Dai, J., Li, W., Wei, Z., and Jiang, J. (2008). *Compos. Sci. Technol.*, **68**, pp. 1644–1648.

85. Camponeschi, E., Florkowski, B., Vance, R., Garrett, G., Garmestani, H., and Tannenbaum, R. (2006). *Langmuir*, **22**, pp. 1858–1862.

86. Lanticse, L. J., Tanabe, Y., Matsui, K., Kaburagi, Y., Suda, K., Hoteida, M., Endo, M., and Yasuda, E. (2006). *Carbon*, **44**, pp. 3078–3086.

87. Yu, G., Cao, A., and Lieber, C. M. (2007). *Nature Nanotechnol.*, **2**, pp. 372–377.

88. Haggenmueller, R., Gommans, H. H., Rinzler, A. G., Fisher, J. E., and Winey, K. I. (2000). *Chem. Phys. Lett.*, **330**, pp. 219–225.

89. Andrews, A., Jacques, D., Rao, A. M., Rantell, T., Derbyshire, F., Chen, Y., Chen, J., and Haddon, R. C. (1999). *Appl. Phys. Lett.*, **75**, pp. 1329–1331.

90. Zhou, C., Wang, S., Zhang, Y., Zhuang, Q., and Han, Z. (2008). *Polymer*, **49**, pp. 2520–2530.

91. Zhang, H., Wang, Z., Zhang, Z., Wu, J., Zhang, J., and He, J. (2007). *Adv. Mater.*, **19**, pp. 698–704.

92. Wang, G., Tan, Z., Liu, X., Chawda, S., Koo, J.-S., Samuilov, V., and Dudley, M. (2006). *Nanotechnology*, **17**, pp. 5829–5835.

93. Fornes, T. D., Baur, J. W., Sabba, Y., and Thomas, E. L. (2006). *Polymer*, **47**, pp. 1704–1714.

94. Mahfuz, H., Adnan, A., Rangari, V. K., Hasan, M. M., Jeelani, S., Wright, W. J., and DeTeresa, S. J. (2006). *Appl. Phys. Lett.*, **88**, pp. 083119(1–3).

95. Rangari, V. K., Yousuf, M., Jeelani, S., Pulikkathara, M. X., and Khabashesku, V. N. (2008). *Nanotechnology*, **19**, pp. 245703(1–9).

96. Ruan, S., Gao, P., and Yu, T. X. (2006). *Polymer*, **47**, pp. 1604–1611.

97. Che, J., and Chan-Park, M. B. (2008). *Adv. Funct. Mater.*, **18**, pp. 888–897.

98. Che, J., Yuan, W., Jiang, G., Dai, J., Lim, S. Y., and Chan-Park, M. B. (2009). *Chem. Mater.*, **21**, pp. 1471–1479.

99. Lu, P., Huang, Q., Jiang, D., Ding, B., Hsieh, Y.-L., Ovid'ko, I. A., and Mukherjee, A. (2009). *J. Am. Ceram. Soc.*, **92**, pp. 2583–2589.

100. Kimura, T., Ago, H., Tobita, M., Ohshima, S., Kyotani, M., and Yumura, M. (2002). *Adv. Mater.*, **14**, pp. 1380–1383.

101. Camponeschi, E., Vance, R., Al-Haik, M., Garmestani, H., and Tannenbaum, R. (2007). *Carbon*, **45**, pp. 2037–2046.

102. Garmestani, H., Al-Haik, M. S., Dahmen, K., Tannenbarm, R., Li, D. S., Sablin, S. S., and Hussaini, M. Y. (2003). *Adv. Mater.*, **15**, pp. 1918–1921.

103. Steinert, B. W., and Dean, D. R. (2009). *Polymer*, **50**, pp. 898–904.

104. Zhu, Y. F., Ma, C., Zhang, W., Zhang, R. P., Koratkar, N., and Liang, J. (2009). *J. Appl. Phys.*, **105**, pp. 054319(1–6).

105. Ma, C., Zhang, W., Zhu, Y., Ji, L., Zhang, R., Koratkar, N., and Liang, J. (2008). *Carbon*, **46**, pp. 706–720.

106. Park, C., Wilkinson, J., Banda, S., Ounaies, Z., Wise, K. E., Sauti, G., Lillehei, P. T., and Harrrison, J. S. (2006). *J. Polym. Sci. Part B Poly. Phys.*, **44**, 1751–1762.

107. Lagerwall, J. P. F., and Scalia, G. (2008). *J. Mater. Chem.*, **18**, pp. 2890–2898.

108. Lynch, M. D., and Patrick, D. L. (2002). *Nano Lett.*, **2**, pp. 1197–1201.

109. Dierking, I., Scalia, G., Morales, P., and Leclere, D. (2004). *Adv. Mater.*, **16**, pp. 865–869.

110. Dierking, I., Scalia, G., and Morales, P. (2005). *J. Appl. Phys.*, **97**, pp. 044309(1–5).

111. Cervini, R., Simon, G. P., Ginic-Markovic, M., Matisons, J. G., Huynh, C., and Hawkins, S. (2008). *Nanotechnology*, **19**, pp. 175602(1–10).

112. Bliznyuk, V. N., Singamaneni, S., Sanford, R. L., Chiappetta, D., Crooker, B., and Shibaev, P. V. (2006). *Polymer*, **47**, pp. 3915–3921.

113. Popa-Nita, V., and Kralj, S. (2010). *J. Chem. Phys.*, **132**, pp. 024902(1–8).

114. Schoot, P., Popa-Nita, V., and Kralj, S. (2008). *J. Phys. Chem. B*, **112**, pp. 4512–4518.

115. Jeong, H. S., Ko, Y. K., Kim, Y. H., Yoon, D. K., and Jung, H.-T. (2010). *Carbon*, **48**, pp. 774–780.

116. Lagerwall, J., Scalia, G., Haluska, M., Dettlaff-Weglikowska, U., Roth, S., and Giesselmann, F. (2007). *Adv. Mater.*, **19**, pp. 359–364.

117. Ahir, S. V., Petrov, P. G., and Terentjev, E. M. (2002). *Langmuir*, **18**, pp. 9140–9148.

118. Weiss, V., Thiruvengadathan, R., and Regev, O. (2006). *Langmuir*, **22**, pp. 854–856.

119. Scalia, G., Bühler, C., Hägele, C., Roth, S., Giesslmann, F., and Lagerwall, J. P. F. (2008). *Soft Matter.*, **4**, pp. 570–576.

120. Jiang, W., Yu, B., Liu, W., and Hao, J. (2007). *Langmuir*, **23**, pp. 8549–8553.

121. Zhang, S., Kinloch, I., and Windle, A. (2006). *Nano Lett.*, **6**, pp. 568–572.

122. Song, W., and Windle, A. (2005). *Macromolecules*, **38**, pp. 6181–6188.

123. Song, W., Kinloch, I., and Windle, A. (2003). *Science*, **302**, pp. 1363–1363.

124. Rai, P., Pinnick, R., Parra-Vasquez, A., Davis, V., Schmidt, H., Hauge, R., Smalley, R. E., and Pasquali, M. (2006). *J. Am. Chem. Soc.*, **128**, PP. 591–595.

125. Davis, V., Ericson, L., Parra-Vasquez, A., Fan, H., Wang, Y., Prieto, V., Longoria, J., Ramesh, S., Saini, R., Kittrell, C., Billups, W., Adams, W.,

Hauge, R., Smalley, R. E., and Pasquali, M. (2004). *Macromolecules*, **37**, pp. 154–160.

126. Ericson, L. M., Fan, H., Peng, H., Davis, V. A., Zhou, W., Sulpizio, J., Wang, Y., Booker, R., Vavro, J., Guthy, C., Parra-Vasquez, A. N., Kim, M. J., Ramesh, S., Saini, R. K., Kittrell, C., Lavin, G., Schmidt, H., Adams, W. W., Billups, W. E., Pasquali, M., Hwang, W. F., Hauge, R. H., Fischer, J. E., and Smalley, R. E. (2004). *Science*, **305**, pp. 1447–1450.
127. Moulton, S. E., Maugey, M., Poulin, P., and Wallace, G. G. (2007). *J. Am. Chem. Soc.*, **129**, pp. 9452–9457.
128. Badaire, S., Zakri, C., Maugey, M., Derre, A., Barisci, J., Wallace G., and Poulin, P. (2005). *Adv. Mater.*, **17**, pp. 1673–1676.
129. Lu, L., and Chen, W. (2010). *ACS Nano*, **4**, pp. 1042–1048.
130. Zamora-Ledezma, C., Blanc, C., Maugey, M., Zakri, C., Poulin, P., and Anglaret, E. (2008). *Nano Lett.*, **8**, pp. 4103–4107.
131. Lee, H. W., You, W., Barman, S., Hellstrom, S., LeMieux, M. C., Oh, J. H., Liu, S., Fujiwara, T., Wang, W. M., Chen, B., Jin, Y. W., Kim, J. M., and Bao, Z. (2009). *Small*, **9**, pp. 1019–1024.
132. Meuer, S., Braun, L., and Zentel, R. (2008). *Chem. Commun.*, **27**, pp. 3166–3168.
133. Liu, H., Chiashi, S., Ishiguro, M., and Homma, Y. (2008). *Nanotechnology*, **19**, pp. 445716(1–7).
134. Cheng, Q., Bao J., Park, J. G., Liang, Z., Zhang, C., and Wang, B. (2009). *Adv. Funct. Mater.*, **19**, pp. 3219–3225.
135. Lu, J. P. (1995). *Phys. Rev. Lett.*, **74**, pp. 1123–1126.
136. Walters, D. A., Casavant, M. J., Qin, X. C., Huffman, C. B., Boul, P. J., Ericson, L. M., Haroz, E. H., O'Connell, M. J., Smith, K., Colbert, D. T., and Smalley, R. E. (2001). *Chem. Phys. Lett.*, **338**, pp. 14–20.
137. Smith, B. W., Benes, Z., Luzzi, D. E., Fischer, J. E., Walters, D. A., Casavant, M. J., Schmidt, J., and Smalley, R. E. (2000). *Appl. Phys. Lett.*, **77**, pp. 663–665.
138. Casavant, M. J., Walters, D. A., Schmidt, J. J., and Smalley, R. E. (2003). *J. Appl. Phys.*, **93**, pp. 2153–2156.
139. Niyogi, S., Hangarter, C., Thamankar, R. M., Chiang, Y. F., Kawakami, R., Myung, N. V., and Haddon, R. C. (2004). *J. Phys. Chem. B*, **108**, pp. 19818–19824.
140. Marquez, F., Morant, C., Sanz, J. M., and Elizalde, E. (2009). *J. Nanosci. Nanotechnol.*, **9**, pp. 3810–3814.

141. Kordás, K., Mustonen, T., Tóth, G., Vähäkangas, J., Uusimäki, A., Jantunen, H., Gupta, A., Rao, K. V., Vajtai, R., and Ajayan, P. M. (2007). *Chem. Mater.*, **19**, pp. 787–791.

142. Ajiki, H., and Ando, T. (1994). *J. Phys. Soc. Jpn.*, **63**, pp. 4267.

143. Ajiki, H., and Ando, T. (1993). *J. Phys. Soc. Jpn.*, **62**, pp. 2470–2480.

144. Tumpane, J., Karousis, N., Tagmatarchis, N., and Nordén, B. (2008). *Angew. Chem. Int. Ed.*, **47**, pp. 5148–5152.

145. Strobl, C. J., Schaflein, C., Beierlein, U., Ebbecke, J., and Wixforth, A. (2004). *Appl. Phys. Lett.*, **85**, pp. 1427–1429.

146. Lim, W. P., Yao, K., and Chen, Y. (2007). *J. Phys. Chem. C*, **111**, pp. 16802–16807.

147. Li, J. Q., Zhang, Q., Peng, N., and Zhu, Q. (2005). *Appl. Phys. Lett.*, **86**, pp. 153116(1–3).

148. Peng, N., Zhang, Q., Li, J. Q., and Liu, N. Y. (2006). *J. Appl. Phys.*, **100**, pp. 024309(1–5).

149. Lu, Y., Chen, C., Yang, L., and Zhang, Y. (2009). *Nanoscale Res. Lett.*, **4**, pp. 157–164.

150. Krupke, R., Hennrich, F., Lohneysen, H. V., and Kappes, M. M. (2003). *Science*, **301**, pp. 344–347.

151. Blatt, S., Hennrich, F., Löhneysen, H. V., Kappes, M. M., Vijayaraghavan, A., and Krupke, R. (2007). *Nano Lett.*, **7**, pp. 1960–1966.

152. Chen, Z., Wu, Z., Tong, L., Pan, H., and Liu, Z. (2006). *Anal. Chem.*, **78**, pp. 8069–8075.

153. Padmaraj, D., Zagozdzon-Wosik, W., Xie, L.-M., Hadjiev, V. G., Cherukuri, P., and Wosik, J. (2009). *Nanotechnology*, **20**, pp. 035201(1–7).

154. Jung, S. M., Hahn, J., Jung, H. Y., and Suh, J. S. (2006). *Nano. Lett.*, **6**, pp. 1569–1573.

155. Kong, J., Franklin, N. R., Zhou, C. W., Chapline, M. G., Peng, S., Cho, K., and Dai, H. J. (2000). *Science*, **287**, pp. 622–625.

156. Li, J., and Zhang, Y. (2006). *Physica E*, **33**, pp. 235–239.

157. Li, J., and Zhang, Y. (2007). *Carbon*, **45**, pp. 493–498.

158. Jia, L., Zhang, Y., Li, J., You, C., and Xie, E. (2008). *J. App. Phys.*, **104**, pp. 074318(1–6).

159. Li, X., Zhang, L., Wang, X., Shimoyama, I., Sun, X., Seo, W.-S., and Dai, H. (2007). *J. Am. Chem. Soc.*, **129**, pp. 4890–4891.

160. Adachi, N., Fukawa, T., Tatewaki, Y., Shirai, H., and Kimura, M. (2008). *Macromol. Rapid Commun.*, **29**, pp. 1877–1881.

161. Tian, Y., Park, J. G., Cheng, Q., Liang, Z., Zhang, C., and Wang, B. (2009). *Nanotechnology*, **20**, pp. 335601(1–7).

162. LeMieux, M. C., Roberts, M., Barman, S., Jin, Y. W., Kim, J. M., and Bao, Z. (2008). *Science*, **321**, pp. 101–104.

163. Opatkiewicz, J. P., LeMieux, M. C., and Bao, Z. (2010). *ACS Nano*, **4**, pp. 1167–1177.

164. LeMieux, M. C., Sok, S., Roberts, M. E., Opatkiewicz, J. P., Liu, D., Barman, S. N., Patil, N., Mitra, S., and Bao, Z. (2009). *ACS Nano*, **3**, pp. 4089–4097.

165. Roberts, M. E., LeMieux, M. C., Sokolov, A. N., and Bao, Z. (2009). *Nano Lett.*, **9**, pp. 2526–2531.

166. Roberts, M. E., LeMieux, M. C., and Bao. Z. (2009). *ACS Nano*, **3**, pp. 3287–3293.

167. Liu, H., Takagi, D., Chiashi, S., and Homma, Y. (2010). *ACS Nano*, **4**, pp. 933–938.

168. Pint, C. L., Xu, Y.-Q., Pasquali, M., and Hauge, R. H. (2008). *ACS Nano*, **2**, pp. 1871–1878.

169. Ren, L., Pint, C. L., Booshehri, L. G., Rice, W. D., Wang, X., Hilton, D. J., Takeya, K., Kawayama, I., Tonouchi, M., Hauge, R. H., and Kono, J. (2009). *Nano Lett.*, **9**, pp. 2610–2613.

170. Tawfick, S., O'Brien, K., and Hart, A. J. (2009). *Small*, **5**, pp. 2467–2473.

171. Ko, Y. K., Geng, J., Jang, S.-G., Yang, S.-M., Jeong, T. W., Jin, Y. W., Kim, J. M., and Jung, H.-T. (2009). *Carbon*, **47**, pp. 1555–1560.

172. Lee, B. Y., Heo, K., Bak, J. H., Cho, S. U., Moon, S., Park, Y. D., and Hong, S. (2008). *Nano Lett.*, **8**, pp. 4483–4487.

173. Liu, J., Casavant, M. J., Cox, M., Walters, D. A., Boul, P., Lu, W., Rimberg, A. J., Smith, K. A., Colbert, D. T., and Smalley, R. E. (1999). *Chem. Phys. Lett.*, **303**, pp. 125–129.

174. Oh, S. J., Cheng, Y., Zhang, J., Shimoda, H., and Zhou, O. (2003) *Appl. Phys. Lett.*, **82**, pp. 2521–2523.

175. Tsukruk, V. V., Ko, H., and Peleshanko, S. (2004). *Phys. Rev. Lett.*, **92**, pp. 065502(1–4).

176. Ko, H., Peleshanko, S., and Tsukruk, V. V. (2004). *J. Phys. Chem. B*, **108**, pp. 4385–4393.

177. Wang, Y., Maspoch, D., Zou, S., Schatz, G. C., Smalley, R. E., and Mirkin, C. A. (2006). *Proc. Natl. Acad. Sci. USA*, **103**, pp. 2026–2031.

178. Sharma, R., Lee, C. Y., Choi, J. H., Chen, K., and Strano, M. S. (2007). *Nano Lett.*, **7**, pp. 2693–2700.

179. Sharma, R., and Strano, M. S. (2009). *Adv. Mater.*, 21, pp. 60–65.

180. Myung, S., Im, J., Huang, L., Rao, S. G., Kim, T., Lee, D. J., and Hong, S. (2006). *J. Phys. Chem. B*, **110**, pp. 10217–10219.

181. Rao, S. G., Huang, L., Setyawan, W., and Hong, S. (2003) *Nature*, **425**, pp. 36–37.

182. Im, J., Kang, J., Lee, M., Kim, B., and Hong, S. (2006). *J. Phys. Chem. B.*, **110**, pp. 12839–12842.

183. Im, J., Lee, M., Myung, S., Huang, L., Rao, S. G., Lee, D. J., Koh, J., and Hong, S. (2006). *Nanotechnology*, **17**, pp. 3569–3573.

184. Seemann, L., Stemmer, A., and Naujoks, N. (2007). *Nano Lett.*, **7**, pp. 3007–3012.

185. Xiong, X., Jaberansari, L., Hanm, M. G., Busnaina, A., and Jung, Y. J. (2006). *Small*, **3**, pp. 2006–2010.

186. Jaber-Ansari, L., Hahm, M. G., Somu, S., Sanz, Y. E., Busnaina, A., and Jung, Y. J. (2009). *J. Am. Chem. Soc.*, **131**, pp. 804–808.

187. Kim, Y. L., Li, B., An, X., Hahm, M. G., Chen, L., Washington, M., Ajayan, P. M., Nayak, S. K., Busnaina, A., Kar, S., and Jung, Y. J. (2009). *ACS Nano*, **3**, pp. 2818–2826.

188. Choi, K. H., Bourgoin, J. P., Auvray, S., Esteve, D., Duesberg, G. S., Roth, S., and Burghard, M. (2000). *Surf. Sci.*, **462**, pp. 195–202.

189. Lewenstein, J. C., Burgin, T. P., Ribayrol, A., Nagahara, L. A., and Tsui, R. K. (2002). *Nano Lett.*, **2**, pp. 443–446.

190. Ko, H., and Tsukruk, V. V. (2006). *Nano Lett.*, **6**, pp. 1443–1448.

191. Tulevski, G. S., Hannon, J., Afzali, A., Chen, Z., Avouris, P., and Kagan, C. R. (2007). *J. Am. Chem. Soc.*, **129**, pp. 11964–11968.

192. Engel, M., Small, J. P., Steiner, M., Freitag, M., Green, A. A., Hersam, M. C., and Avouris, P. (2008). *ACS Nano*, **2**, pp. 2445–2452.

193. Yan, Y. H., Li, S., Chan-Park, M. B., and Zhang, Q. (2006). *Nanotechnology*, **17**, pp. 5696–5701.

194. Li, S., Liu, N., Chan-Park, M. B., Yan, Y., and Zhang, Q. (2007). *Nanotechnology*, **18**, pp. 455302(1–7).

195. Li, S., Yan, Y., Liu, N., Chan-Park, M. B., and Zhang, Q. (2007). *Small*, **3**, pp. 616–621.

196. Kim, P., Baik, S., and Suh, K. Y. (2008). *Small*, **4**, 92–95.

197. Vichchulada, P., Zhang, Q., Duncan, A., and Lay, M. D. (2010). *ACS Appl. Mater. Interfaces*, **2**, pp. 467–473.

198. Seemann, K. M., Ebbecke, J., and Wixforth, A. (2006). *Nanotechnology*, **17**, pp. 4529–4532.

199. Ang, K. Y., Yao, K., Chen, Y., and Teo, S. L. (2009). *J. Nanosci. Nanotechnol.*, **9**, pp. 6523–6532.

200. Shim, J. S., Yun, Y.-H., Rust, M. J., Do, J., Shanov, V., Schulz, M. J., and Ahn, C. H. (2009). *Nanotechnology*, **20**, pp. 325607(1–7).

201. Tsuda, K., and Sakka, Y. (2009). *Sci. Technol. Adv. Mater.*, **10**, pp. 014603(1–6).

202. Vijayaraghavan, A., Blatt, S., Weissenberger, D., Oron-Carl, M., Hennrich, F., Gerthsen, D., Hahn, H., and Krupke, R. (2007). *Nano Lett.*, **7**, pp. 1556–1560.

203. Diehl, M. R., Yaliraki, S. N., Beckman, R. A., Barahona, M., and Heath, J. R. (2002). *Angew. Chem. Int. Ed.*, **41**, pp. 353–356.

204. Makaram, P., Somu, S., Xiong, X., Busnaina, A., Jung, Y. J., and McGruer, N. (2007). *Appl. Phys. Lett.*, **90**, pp. 243108(1–3).

205. Hong, S. H., Kang, M. G., Cha, H.-Y., Son, M. H., Hwang, J. S., Lee, H. J., Sull, S. H., Hwang, S. W., Whang, D., and Ahn, D. (2008). *Nanotechnology*, **19**, pp. 105305(1–5).

206. Wei, H., Kim, S. N., Zhao, M., Ju, S.-Y., Huey, B. D., Marcus, H. L., and Papadimitrakopoulos, F. (2008). *Chem. Mater.*, **20**, 2793–2801.

207. Kang, S. J., Kocabas, C., Kim, H.-S., Gao, Q., Meitl, M. A., Khang, D.-Y., and Rogers. A. J. (2007). *Nano Lett.*, **7**, 3343–3348.

208. Im, J., Lee, I.-H., Lee, B. Y., Kim, B., Park, J., Yu, W., Kim, U. J., Lee, Y. H., Seong, M.-J., Lee, E. H., Min, Y.-S., and Hong, S. (2009). *Appl. Phys. Lett.*, **94**, 053109(1–3).

209. Pint, C. L., Xu, Y. Q., Moghazy, S., Cherukuri, T., Alvarez, N. T., Haroz, E. H., Mahzooni, S., Doorn, S. K., Kono, J., Pasquali, M., and Hauge, R. H. (2010). *ACS Nano*, **4**, 1131–1145.

Chapter 4

Separation of Metallic and Semiconducting Single-Wall Carbon Nanotubes

Yuan Chen,[a] Andrew Keong Ng,[b] Shihe Bai,[a] Rongmei Si,[a] Li Wei,[a] and Qiang Wang[a]

[a] *School of Chemical and Biomedical Engineering, Nanyang Technological University Singapore, 637459, Singapore*

[b] *Institute for Infocomm Research, Agency for Science, Technology and Research, 138632, Singapore*

chenyuan@ntu.edu.sg

Single-wall carbon nanotubes (SWCNTs) can be either metallic (m) or semiconducting (s) depending on their atomic structures. Most practical applications of SWCNTs would require uniformly m– or s–SWCNTs. However, none of current synthesis methods can successfully produce SWCNTs with uniform metallicity. Intense research efforts have been devoted to post-synthetic separation of SWCNTs based on their metallicity. We begin this chapter by introducing the relation between SWCNT metallicity and their structures. Next, we describe separation approaches according to different electric conductivity and dielectric constants. We also discuss various chemical methods to augment the difference between

Carbon Nanotubes and Their Applications
Edited by Qing Zhang

ISBN 978-981-4241-90-8 (Hardcover), 978-981-4303-18-7 (eBook)
www.panstanford.com

m– and s– (m/s–) SWCNTs, followed by illustrating separation techniques for chemically modified SWCNTs. Last but not least, we review methodologies for evaluating abundances of SWCNTs with different metallicities.

4.1 Introduction

A SWCNT is a hollow carbon cylinder rolled up by a one-atom-thick planar graphene sheet. The atomic structure of a SWCNT, as depicted in Fig. 4.1, can be specified by a chiral index (*n*,*m*). Early studies applied the π-only band theory developed for graphite to describe electronic properties of SWCNTs [1]. Π-only tight binding calculations predicted that all infinite length armchair (*n*,*n*) SWCNTs are metallic; zigzag (*n*,0) and chiral (*n*,*m*) SWCNTs are metallic when (*n*-*m*)/3 is an integer; and other SWCNTs are semiconducting [2, 3]. These predications were shortly verified using scanning tunneling microscope (STM) [4,5]. SWCNTs have excellent electronic properties: m–SWCNTs can carry an electric current density ($\sim 4 \times 10^9$ A cm^{-2}) that is three orders of magnitude higher than a typical metal, such as copper or aluminum [6]; and the carrier mobility of individual s–SWCNTs ($\sim$10,000 cm^2 V^{-1} s^{-2}) is 10 times better than that of silicon [7]. Such exceptional electronic properties of SWCNTs inspire a huge range of applications, including conductive films, interconnects, transistors, logic gates, and sensors. Most of these applications would require homogeneous m/s–SWCNTs; however, none of current synthesis methods can successfully produce SWCNTs with identical metallicities. The lack of metallicity uniformity in SWCNT samples is one of major barriers toward their widespread applications. To overcome this barrier, researchers have exploited a number of selective synthesis and post-synthetic separation methods [8–10].

In chemistry and chemical engineering, a separation process is to transform a mixture of substances into separated products which differ in their chemical or physical properties. The performance of a SWCNT separation process can be evaluated on three aspects: (1) *Purity of separated products:* In the case of SWCNTs, the purity is the abundance of m/s–SWCNTs in separated products; (2) *Recovery rate:* It is the fraction of separated products in relation to starting materials. A high recovery rate means less SWCNTs would be discarded in the separation process. (3) *Scalability:* It is the ability to produce enough materials for intended applications. In this chapter, we focus on physical and chemical principles underlying different separation methods, and discuss the purity of separated products, recovery rate, and scalability.

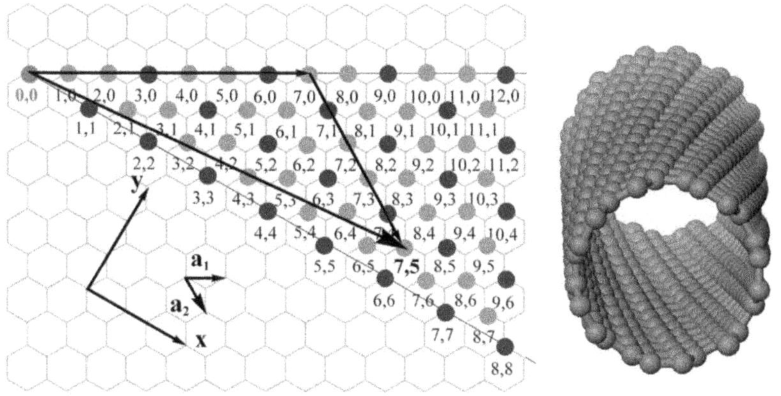

Figure 4.1 The atomic structure of a SWCNT can be considered as a graphene sheet rolled into a cylinder at a specific direction. Related to the origin point at (0,0), each point on the graphene sheet (the corner of each hexagon, where a carbon atom locates) can be identified by a pair of integers (n,m) $(0 \leq |m| \leq n)$. The rolling direction from the origin point towards any other points on the graphene sheet can be defined by a chiral vector, $C_h = n\mathbf{a}_1 + m\mathbf{a}_2$, where $(\mathbf{a}_1,\mathbf{a}_2)$ is a pair of lattice unit vectors of the graphene sheet. SWCNTs are known as armchair nanotubes when $n = m$, and as zigzag nanotubes when $m = 0$. Approximately 33% of SWCNTs are metallic (labeled in blue) and 67% of SWCNTs (labeled in grey) are semiconducting. The right portion of this figure shows the physical structure of a (7,5) SWCNT. See also Color Insert.

4.2 SWCNT Structure and Metallicity

The separation of SWCNTs relies on their differences in metallicity and other structure–dependent properties. Every carbon atom has four valence orbitals: 2*s*, $2p_x$, $2p_y$, and $2p_z$. In a graphene sheet, 2*s* orbital of a carbon atom mixes with its $2p_x$ and $2p_y$ orbitals, exhibiting a planar sp^2 orbital hybridization. Every carbon atom forms three σ and σ^* bonds with three neighboring in-plane carbon atoms at 120° internals. The remaining $2p_z$ orbital, which points out of the graphene sheet, interacts with $2p_z$ orbitals of neighboring carbon atoms to create delocalized π and π^* bands. Energy levels of σ and σ^* bands are far away from the Fermi level, thus they have limited influences on the metallicity of SWCNTs. In contrast, π and π^* bands lie close to the Fermi level, which determine the metallicity of SWCNTs [11]. As illustrated in Fig. 4.2, a graphene sheet is a semimetal because its π valence and π^* conduction bands degenerate at only six corners ($\boldsymbol{K}_{\mathrm{B}}$) of the hexagonal first Brillouin zone [12, 13]. When a graphene layer is rolled up in the direction of $\boldsymbol{C}_h$, it has periodic boundary conditions; hence, its electron wavevector $\boldsymbol{k}$ is quantized according to $\boldsymbol{C}_h \cdot \boldsymbol{k} = 2\pi q$ (q is an integer). In the π-only band theory, when $\boldsymbol{k}$ passes through a $\boldsymbol{K}_{\mathrm{B}}$ point of the graphene sheet, electron orbitals exist near the Fermi level, and therefore the SWCNT is metallic. Under such a condition, the (n,m) index satisfies $n - m = 3j$, where j is an integer. On the other hand, if $\boldsymbol{k}$ does not pass through a $\boldsymbol{K}_{\mathrm{B}}$ point, the SWCNT is semiconducting with a bandgap at Fermi level, and its $n - m = 3j + 1$ or $3j + 2$ [13]. Moreover, as shown in Fig. 4.2, after rolling up the graphene sheet, many electron states with similar energy exist near the energy minimum in subbands, giving rise to spikes in SWCNT electronic density of states, known as van Hove singularities [14].

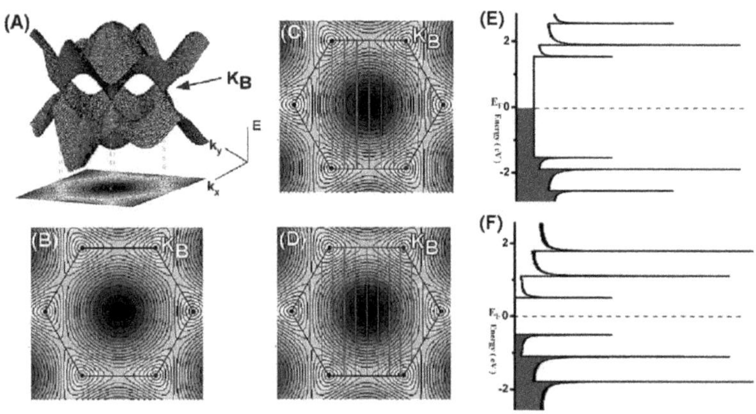

Figure 4.2 (A) Three-dimensional (3D) plot of π and π^* graphene energy bands and (B) its 2D projection. (C) Allowed 1D wavevectors for a metallic $(9,0)$ SWCNT. (D) Allowed 1D wavevectors for a semiconducting $(10,0)$ nanotube. (E) and (F) Electronic density of states of $(9,0)$ and $(10,0)$, respectively. The black hexagons define the first Brillouin zone of the graphene sheet, and black dots at hexagon corners are the $\boldsymbol{K}_B$ points. Straight lines represent allowed wavevectors, $\boldsymbol{k}$, within the first Brillouin zone. Modified from [12].

4.3 Electric Conductivity

The current-induced electrical breakdown can be considered a separation method, which directly utilizes different conductivities of SWCNTs. Carbon nanotubes (CNTs) can withstand a high current density (>10^9 A cm^{-2}); however, at high enough currents, CNTs would fail due to the rapid oxidation at current-induced defect sites [15, 16]. When SWCNTs are applied as an active layer in field effect transistors (FETs), the conductance of s-SWCNTs can be modulated by FET gate voltages, while that of m-SWCNTs cannot [17]. By varying gate voltages, carriers in s-SWCNTs can be depleted, and the current-inducted oxidation would only happen on current-carrying m-SWCNTs. Thus, m-SWCNTs are broken down without affecting s-SWCNTs. In order to obtain high-purity s-SWCNTs, the electric breakdown method would require that (1) SWCNTs be connected to electrodes independently, (2) SWCNTs be exposed to oxidizing environment, and (3) SWCNTs do not electrostatically shield each other. In an ideal condition, 100% purity s-SWCNTs can be obtained after removing all m-SWCNTs and a fraction of s-SWCNTs. However, in typical conditions, large SWCNT bundles or thick SWCNT films exist, and the above-mentioned three conditions may not be satisfied. Accordingly, a large amount of m/s-SWCNTs would be destroyed under high currents, and the recovery rate may be low. Moreover, complexities in FET fabrication also increase the difficulty of applying the current-induced electrical breakdown method for processing of bulk SWCNT samples; and this separation method is generally not scalable.

4.4 Dielectrophoresis

Besides conductivity, static dielectric constants (ε) of m/s-SWCNTs are significantly different. Static dielectric constants for s-SWCNTs of diameters from 0.8 to 1.4 nm are less than 5, whereas those for m-SWCNTs are many orders of magnitude larger attributed to their mobile carriers [18]. Consequently, if SWCNTs are exposed to an inhomogeneous alternating electric field, dielectrophoretic forces experienced by m-SWCNTs are significantly larger than those experienced by s-SWCNTs. Krupke *et al.* designed a microelectrode array for separating m-SWCNTs from a nanotube mixture. When a drop of SWCNT dispersion, containing sodium dodecyl sulfate (SDS)-dispersed individual nanotubes, was applied to the electrode array, m–SWCNTs were attracted toward electrodes, while s–SWCNTs remained in the dispersion [19]. SWCNTs can be regarded as long rods with their axis parallel to the electric field E at angular frequency ω. The dielectrophoretic forces (F_{DEP}) experienced by SWCNTs can be formulated as

$$F_{DEP} = \frac{\pi d^2 l}{8} \varepsilon_l \operatorname{Re}\left(\frac{\varepsilon_t^* - \varepsilon_l^*}{\varepsilon_l^* + (\varepsilon_t^* - \varepsilon_l^*)L} \right) \nabla E^2, \varepsilon^* = \varepsilon - i\frac{\sigma}{\omega},$$

where ε_t^* and ε_l^* are complex dielectric constants of SWCNTs and the surrounding liquid, respectively. ε and σ are the corresponding dielectric constants and conductivity. L is the depolarization factor determined by nanotube diameter d and length l [20]. Krupke *et al.* observed that surfactants could induce frequency dependent surface conductance on SWCNTs, which influences the actual F_{DEP}. To achieve good separation, the electric field strength has to be in the order of 10^6 V/m and at high frequency (>10^8 Hz) [20]. However, if the electric field is too large, s–SWCNTs would also be attracted to electrodes due to their transversal polarizability [21]. Several studies further demonstrated that the dielectrophoresis separation process depends not only on dielectric constants of SWCNTs, but also on nanotube aggregation, nanotube lengths, nanotube surface charges [22, 23], surfactants [24], solvents [25], electrode structures [26], electric field strength and frequency [27]. The initial microelectrode array can merely separate tiny amount of SWCNT samples. To increase the separation yield, Peng *et al.* proposed the dielectrophoresis field flow fractionation, in which SWCNT dispersion flew through a chamber with electrodes patterned at its bottom. By doing so, m–SWCNTs can be trapped on electrodes, and s–enriched SWCNT dispersion can be collected at the outlet of the chamber [28]. More recently, Shin *et al.* developed an H-shaped microfluidic channel to perform continuous extraction of m–SWCNTs [29]. However, it is difficult to precisely control all the above-mentioned parameters in a practical dielectrophoresis process. So far, the purity of SWCNTs produced by dielectrophoresis-based separation methods is not high. Moreover, the recovery rate is low because only a tiny fraction of SWCNTs in nanotube dispersions can be trapped on electrodes. Dielectrophoresis-based methods also require significant improvements in their scalability.

4.5 Chemical Reactivity

For chemists, the practical approach to achieve scalable separation of m/s–SWCNTs is to utilize their variations in chemical reactivity. The differences between m/s–SWCNTs can be amplified through chemical functionalizations. In this section, we discuss the electronic structural dependent chemical reactivity of SWCNTs, and subsequently explain principles of selective functionalization methods. Two most commonly used chemical approaches, namely covalent functionalization and non-covalent surfactant or polymer wrapping, are also detailed.

The chemical reactivity of SWCNTs was initially attributed to curvature-induced local strains induced by pyramidalization (the deformation of a trigonal planar molecule into a tetrahedral arrangement) and misalignment of π-orbitals among pairs of carbon atoms on SWCNT surfaces [30]. SWCNTs with small diameters have larger pyramidalization and severer π-orbital alignment; hence, they are usually more reactive. Reactivity of SWCNTs relies on SWCNT diameter and chirality due to different local strains. However, this curvature-induced reactivity concept only considers localized electronic structures. It is necessary to include the unique reactivity contributed by delocalized π electronic band structures of SWCNTs because the reactivity corresponding to their overall metallicity is more directly associated with the selective functionalization of m/s–SWCNTs.

Concepts of aromaticity, the highest occupied molecular orbital (HOMO), and the lowest unoccupied molecular orbital (LUMO), widely adopted in organic chemistry, can be used to explain the metallicity dependent reactivity of SWCNTs [31]. For instance, less aromatic or a smaller HOMO-LUMO gap normally indicates higher reactivity. The Clar aromatic sextet valence bond model suggests that only fully benzenoid SWCNTs are metallic and have smaller bandgaps, thereby possessing a higher reactivity than s–SWCNTs [32].

The frontier orbital theory provides a clearer picture to interpret the chemical reactivity of SWCNTs with different metallicities [33]. Frontier orbitals of s–SWCNTs are at the first van Hove singularities in valence and conduction bands, whereas those of m–SWCNTs locate at orbitals near the Fermi level [31]. When a molecule is adsorbed on a SWCNT, interactions between their occupied orbitals or between their empty orbitals become attractive if a charge transfer happens between their orbitals. These interactions may be significantly different between m/s-SWCNTs due to their distinct frontier orbital structures. M-SWCNTs exhibit stronger interactions with many molecules, while s-SWCNTs may have better interaction with some other molecules. Orbital interactions heavily depend on the possibility of charge transfer, which is determined by the relative position of Fermi levels between SWCNTs and molecules [34]. It has been found that frontier orbitals and HOMO-LUMO gaps of SWCNTs oscillate periodically with the increase of nanotube length, which suggests that a particular (n,m) type of SWCNTs may have different chemical reactivity when their lengths are different [35]. Furthermore, it is noteworthy that practical challenges in achieving selective SWCNT functionalization also often include a limited solubility in various solvents and the formation of nanotube bundles composing of both m/s-SWCNTs [36].

Table 4.1 Metallicity selective covalent functionalizations of SWCNTs

Ref.	Chemicals	Selectivity toward	Influencing parameters
[37–46]	Diazonium salts	M	Salt structures, external illumination, and pH
[47, 48]	Dichlorocarbene	M	Dichlorocarbene precursors
[49–51]	Nitronium ions	M	Small-diameter nanotubes
[52]	Osmium tetroxide	M	UV light
[53]	Plasma hydrocarbonation	M	Diameter sensitive
[54]	Fluorine gas	M	Small-diameter nanotubes
[55]	Fluorinated olefins	M	In FET device channels
[56–58]	Nitric and sulfuric acids	M	Microwave radiation
[59–61]	Hydrogen peroxide	S	UV light
[62]	Azomethine ylides	S	Soluble s-SWCNTs

Table 4.1 summarizes major covalent functionalization routes for m/s-SWCNT separation. We use the selective functionalization

of m-SWCNTs by diazonium salts [37] as a sample to illustrate key issues in developing an appropriate covalent functionalization method. First, the selectivity of a specific reaction often depends not only on SWCNT metallicity, but also SWCNT length, diameter and dispersion. Thus, it is necessary to choose chemical routes that are suitable for a particular type of SWCNT starting materials. Second, the selectivity can also be influenced by diverse reaction conditions, which should be precisely controlled in order to achieve good selectivity. For example, the selectivity of diazonium salts toward m–SWCNTs is dependent on the chemical structure of salts [40], the external illumination [44], and the pH of solvents [39, 40]. Third, a variety of separation techniques can be employed to further separate covalently functionalized SWCNTs. Electrophoresis is capable of separating SWCNTs with different electrophoretic mobilities because diazonium salts can induce negative charges on functionalized SWCNTs [42]. Extraction is another option when functionalized SWCNTs have a distinct solubility in different solvents. For instance, the diazonium salt of 4-heptadecafluorooctylaniline reacts with m– SWCNTs dispersed in water. This reaction produces fluoroderivatives, which can be extracted by perfluorohexane, leaving s–SWCNTs in the aqueous solvent [41]. Similarly, diazonium compounds with a long alkyl tail enable m–SWCNTs selectively soluble in tetrahydrofuran [45]. Ultracentrifugation can be applied to remove functionalized m-SWCNTs and nanotube bundles because of their large density as compared to un-functionalized s–SWCNTs [43]. Alternatively, covalent functionalization can be used to directly inactivate SWCNTs in thin-film devices, rather than removing them from a nanotube mixture [38, 46].

Table 4.2 Metallicity selective non-covalent functionalizations with SWCNTs

Ref.	Chemicals	Selectivity toward	Influencing parameters
[63, 64]	Propylamine/ isopropylamine	m	In tetraphydrofuran
[65]	Amine- and phenyl-terminated silanes	m/s	SWCNTs in a monolayer for FETs
[66–69]	Octadecylamine	s	Oxidized in an acidic solution

Ref.	Chemicals	Selectivity toward	Influencing parameters
[70, 71]	Bromine	m	Resulting in density differences
[72]	Large aromatic molecules	m	Simulation results
[73–76]	SDS, sodium dodecylbenzene sulphonate, sodium cholate	m/s	Sonication, cosurfactants
[77–79]	Porphyrin	s	Chiral handedness, diameter
[80]	Oligo-acene adducts		Large-diameter selective
[81–83]	Fluorine-based polymer	s	(*7,5*) nanotubes
[84–86]	ssDNA	s	DNA sequence and length
[87]	Au nanoparticles	s	(*6,5*) Nanotubes
[88, 89]	Flavin mononucleotide	s	(*8,6*) Nanotubes
[90]	Pentacenic/quaterrylene -based amphiphile	s	Armchair vs. zigzag
[91]	Heparin sodium salts	s	(*8,4*) Nanotubes
[92]	Solvents with electron-donating and –withdrawing groups	m/s	Aromatic and aliphatic backbones of solvents

As non-covalent functionalizations introduce fewer changes to SWCNTs, they are more desirable in practical applications where intrinsic properties of SWCNTs are needed. Various metallicity selective non-covalent interactions have been investigated. Table 4.2 summarizes major approaches explored so far, and these approaches can be broadly catalogued into two groups: small molecules and polymers. The metallicity selection mechanism of non-covalent functionalizations is still poorly understood. Several studies often cited explanations involving π-π stacking, charge-transfer interaction, and structural matching. As discussed earlier, SWCNTs have an extended π electron system, which can interact with other π electron systems through the π-π stacking. Due to the fact that m-SWCNTs have higher polarization than s–SWCNTs, they are expected to bind tighter in the π-π stacking with other molecules [72]. Since m-SWCNTs have higher tendency to form charge-transfer complexes in covalent functionalizations, they may also form a stronger interaction with physisorbed charge-transfer molecules in non-covalent functionalizations. However, the different binding affinity of m/s–SWCNTs alone is insufficient to explain some latest experiential results, especially those on the high selectivity toward specific (*n,m*) s–SWCNTs listed in Table 4.2. It has been recommended that the structural matching between SWCNTs and molecules or polymers could also be important [86]. For example, the single-stranded DNA (ssDNA) with different sequences and lengths can selectively wrap around chiral surfaces of different (*n,m*) SWCNTs [85, 86]. In addition, the initial dispersion-by-sonication step is necessary in most non-covalent functionalizations. It has been suggested that m/s–SWCNTs can be dispersed differently, causing distinct bundle sizes, charge densities or lengths. For instance, SDS can more effectively disperse m-SWCNTs, thus s–SWCNTs remain primarily in bundles [75]. In short, non-covalent interactions are significantly weaker than covalent bonding. Therefore, they are more susceptible to various experimental parameters, ranging from the structural variation of SWCNTs and molecules or polymers to environmental conditions, such as solvent, pH, temperature, and concentration. From a positive prospective, such a high susceptibility offers vast opportunities to develop new approaches for m/s–SWCNT separation methods.

4.6 Separation Techniques

Chemically modified SWCNTs with distinct density, surface charge or dimension can be readily separated using a variety of techniques, namely centrifugation, electrophoresis, chromatography, extraction, and adsorption. Some of these techniques have been introduced in the earlier section. Here, we only highlight three most successful separation techniques. Although a recent review has discussed their latest developments [10], we emphasize their fundamental principles, advantages, and current technological limitations.

Centrifugation, especially density gradient ultracentrifugation (DGU), separates SWCNTs via buoyant density. As exemplified in Fig. 4.3A, surfactants (SDS and sodium cholate) can disperse m/s–SWCNTs differently by non-covalent functionalizations

[93–95], leading to their density variation. A density gradient can be created in a centrifuge tube [96]. Surfactant-dispersed SWCNTs are inserted in the middle of the density gradient. Under a centrifugal field, SWCNTs move toward locations, where their densities match the local density of the gradient. The key advantages of DGU come from its tunability and scalability. DGU has a good sorting tunability because the density of SWCNTs can be simply changed using different covalent or non-covalent functionalization methods [43, 93, 94]. Moreover, the density gradient, centrifugal force and time can also be easily modified to yield different separation selectivities [97, 98]. The scalability and economic viability of DGU have already been reported in the literature. To date, DGU is the only separation technique that can deliver high-purity m/s–SWCNT commercial products (NanoIntegris). However, its technological limitations are also bounded by its scalability and economic viability. As there are only tiny density differences among dispersed SWCNT species, high centrifugal fields (about 200,000 *g*) are required to obtain good separation selectivities. It remains a challenge to realize large scale production at such a high centrifugal field [99]. Consequently, the cost of purified m/s–SWCNTs is still too high for widespread SWCNT applications.

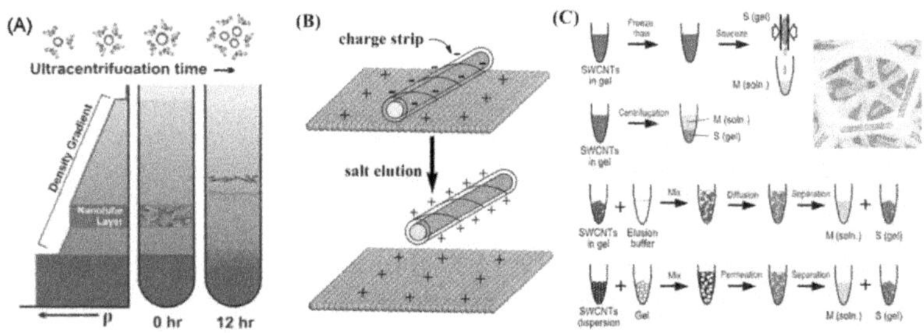

Figure 4.3 (A) Schematic of surfactant-encapsulated SWCNTs and ultracentrifuge tube in the DGU. Modified from [95]. (B) A DNA strand rolling up a nanotube and schematic of the anion exchange separation process [84, 86]. (C) A model of SWCNTs in agarose gel and schematic diagrams of m/s-SWCNT separation using agarose gel by freezing and squeezing, centrifugation, diffusion, and permeation. Modified from [101].

Chromatographic methods, especially ion-exchange chromatography (IEX), have demonstrated the most remarkable (*n*,*m*) selectivity in SWCNT separation. Twelve single-chirality samples were separated from HiPco SWCNTs [85, 86]. IEX separate ions or molecules based on their different charges. As rendered in Fig. 4.3B, ssDNA sequences self-assemble around SWCNTs, and carry negative charges to SWCNT surfaces [84]. SWCNT dispersion is then injected into an IEX column, which contains positively charged substrates composed of anion exchange resins. When the salt concentration of a SWCNT dispersion is low, negative charged SWCNTs are attracted to resin surfaces. With the increase of salt concentration, interactions between resin surfaces and SWCNTs are screened, and SWCNTs are eluted from the IEX column [84]. The separation selectivity can be influenced by surface charges induced by ssDNA, as well as the length of SWCNTs [100]. Since only ssDNA-dispersed SWCNTs can be efficiently separated by IEX, its key technological limitation is the high cost of ssDNA. A custom-made small number oligomer ssDNA with an acceptably robust dispersion would cost approximately $15 to $20/mg of dispersed SWCNTs [99]. On top of that, ssDNA would stick tightly to SWCNT surfaces after separation. If pristine SWCNTs are needed, appropriate methods for removing ssDNA without changing SWCNT properties are essential; unfortunately, such methods are still unavailable.

Recently, several agarose gel-based separation techniques have demonstrated promising results [101, 102]. Agarose gel electrophoresis is widely used to separate DNA or RNA with different lengths. Agarose is a linear polymer made up of repeating unit of agarobiose. It acts as a matrix to reduce the movement of molecules as they move under an electric field in electrophoresis. Since shorter molecules migrate more easily through pores of the gel, they move faster than longer ones. When SDS-dispersed SWCNTs are mixed together with an agarose gel, well-dispersed m-SWCNTs move quicker under an electric field or other forces, whereas s–SWCNTs in small bundles are trapped in the agarose gel [75]. The benefits of agarose gel based separation techniques are their low cost and ease of producing large amount of samples. On the other hand, the technological drawback is their relative low selectivity. The purity of separated m/s–SWCNTs by agarose gel based separation techniques is much lower than those obtained by DGU or IEX. Future developments may explore multiple stage separation approaches to further improve the purity of separated SWCNTs.

4.7 Metallicity Abundance Evaluation

Reliable metallicity abundance evaluation is a major step to objectively assess the performance of various separation techniques that aim to obtain high-purity m/s-SWCNTs. Current evaluation methodology can be catalogued into two types: individual counting-based and bulk sample based methods.

The composition of a m/s-SWCNT sample can be computed by counting numbers of individual SWCNTs with different metallicities. The metallicity of individual SWCNTs can be determined by STM [5], electron nano-diffraction [103, 104], voltage-contrast scanning electron microscope [105], electric force scanning probe microscope [106], and distinct performances of individual m/s-SWCNT FETs [107–109]. More recently, Weisman *et al.* proposed an evaluation technique based on atomic force microscope (AFM) and photoluminescence microscope [110]. The fraction of s-SWCNTs can be determined by comparing AFM images showing all SWCNTs with near-infrared photoluminescence images showing only s-SWCNTs. As listed in Table 4.3, results from this method can potentially serve as a useful reference in the absolute calibration of other abundance evaluation techniques.

Table 4.3 Compositions determined for various standard SWCNT samples [110]

Sample	Source	s-%	m-%	Error (%)
HiPco	Rice Univ.	62.9	37.1	0.5
CoMoCAT, standard grade	SWeNT Inc.	92.1	7.9	1.1
CoMoCAT, commercial grade	SWeNT Inc.	51.9	48.1	3.5
Laser ablation	SRC Inc./NASA-JSC	54.7	45.3	1.4
CVD growth	Honda Res. USA	15.4	84.6	2.6
HiPco, starting material	Northwestern Univ.	60.5	39.5	3.8
HiPco, s-enriched by DGU	Northwestern Univ.	96.0	4.0	0.6
HiPco, m-enriched by DGU	Northwestern Univ.	3.1	96.9	0.6

A common problem of individual counting-based methods is that they are usually time- and labor intensive. SWCNT samples need to be individually dispersed in solvents. Then, a very small fraction of the SWCNT dispersion is deposited on surfaces for microscope studies. SWCNTs may also rebundle together, when they are deposited on surfaces. Hence, only tiny fraction of SWCNTs can be directly measured under microscopes. The accuracy of individual counting-based methods depends strongly on procedures used in sample preparation.

Resonance Raman spectroscopy (RRS) [111] and ultraviolet-visible-near infrared absorption spectroscopy (UV-vis-NIR) [112] are the most widely used bulk sample based metallicity evaluation methods. RRS can measure either solid powders or liquid dispersions by radial breathing mode (RBM) or tangential vibrational modes (G band) of SWCNTs [111, 113]. RBM peaks correlate with different (*n,m*) species. The line shape of G–bands for m–SWCNTs is broader, followed a Breit–Wigner–Fano line shape, compared with that of s–SWCNTs at similar diameters [114]. RRS provides a relatively rapid tool to estimate relative contents of m/s–SWCNTs [19, 37, 84]. However, due to the strong resonance effect in SWCNT Raman peaks [115], performing a complete analysis of all (*n,m*) species would require continuous laser excitations to match resonance energies of every (*n,m*) species in a sample. Such continuous laser excitation systems are costly and not widely available, which limits the applicability of RRS. Moreover, SWCNT aggregation often induces changes in G-band peaks, which may interfere with the metallicity identification by RRS [116]. The measured RRS peak intensity also depends on the abundance of SWCNTs, as well as intrinsic photophysical parameters of particular (*n,m*) SWCNTs [117]. Several theoretical models have emerged to predict intrinsic photophysical parameters of different (*n,m*) species [118–121]. However, they need to be validated via well-characterized reference samples, which are presently unavailable.

UV-vis-NIR also reveals distinct optical transitions of m/s–SWCNTs [70, 115]. As UV-vis-NIR instruments are widely available in most research labs, they are frequently used to quantify m/s–SWCNTs [108, 122, 123]. However, there are several problems influencing their current usage. First, it is difficult to accurately deconvolute UV-vis-NIR spectra into contributions from different (*n,m*) SWCNTs because electron transitions of SWCNTs with different diameters may highly overlapped on their UV-vis-NIR spectra, especially for samples having a broad diameter distribution. Second, it is tedious to subtract background absorptions because background absorptions are composed of contributions from both carbon impurities and SWCNT defects. Third, good-quality absorption spectra can only be obtained when SWCNTs are individually dispersed, or presented in very small bundles [124]. Fourth, the sensitivity of UV-vis-NIR in differentiating high-purity m/s–SWCNTs is inadequate [125]. Fifth, similar to RRS, the measured absorption peak intensity depends on the abundance of SWCNTs, as well as intrinsic photophysical parameters of (n,m) SWCNTs [117].

The drawbacks of using RRS and UV-vis-NIR have led to increased interest in evaluating the metallicity of bulk SWCNT samples through the performances of SWCNT based thin-film FETs [93, 109]. SWCNT thin films fabricated from bulk samples serve as active layers in FETs. This popularity stems from the fact that FETs are basic components of many potential SWCNT applications, and that FET performance parameters (e.g., I_{on}/I_{off} ratio and mobility) are likely to be more sensitive than spectroscopic techniques in detecting the content of metallic nanotubes in a sample [125]. Nevertheless, FET performance parameters highly depend on the device structures: metal of electrodes, device channel width, channel length, thickness of oxide layers, contact-mode between SWCNTs and electrodes (top-gate or bottom-gate), annealing treatment history of devices, SWCNT density, aggregation, thickness and length, contact resistances among SWCNTs, and measurement conditions [126, 127]. All these parameters should be carefully considered when FET performance data are used to determine the metallicity of bulk SWCNT samples.

4.8 Summary

m/s–SWCNT separation is a complex process because the metallicity of SWCNTs is often coupled with many structural factors, such as nanotube diameter, length, and aggregation. A successful separation method needs to address not only the metallicity of SWCNT, but also SWCNT length sorting, dispersion, macroscopic assembling, and characterization methods. In recent years, several separation approaches have remarkably demonstrated the capability of obtaining high-purity m/s–SWCNTs. Commercial products are also now available. The remaining key challenge is to achieve an economical and large-scale production. The production capacity should be increased by many folds, while the production cost should be reduced to a small fraction of the current price. Unfortunately, no methods reviewed in this chapter could fulfill this requirement. Looking ahead, revolutionary methods are still waiting to be discovered, and the optimization of current separation techniques also offers hopes of achieving practical gains. The widespread applications of SWCNTs in real-world technologies are closer to us than ever before.

Acknowledgments

This work was supported by National Research Foundation, Singapore (NRFCRP2-2007-02 and NRF2010-POC001-021).

References

1. Wallace, P. R. (1947). The band theory of graphite, *Phys. Rev.*, **71**, pp. 622–634.
2. Dresselhaus, M. S.; Dresselhaus, G.; Saito, R. (1992). Carbon-fibers based on C-60 and their symmetry, *Phys. Rev. B Condens. Matter*, **45**, pp. 6234–6242.
3. Saito, R.; Fujita, M.; Dresselhaus, G.; Dresselhaus, M. S. (1992). Electronic-structure of graphene tubules based on C-60, *Phys. Rev. B Condens. Matter*, **46**, pp. 1804–1811.
4. Odom, T. W.; Huang, J. L.; Kim, P.; Lieber, C. M. (1998). Atomic structure and electronic properties of single-walled carbon nanotubes, *Nature*, **391**, pp. 62–64.
5. Wildoer, J. W. G.; Venema, L. C.; Rinzler, A. G.; Smalley, R. E.; Dekker, C. (1998). Electronic structure of atomically resolved carbon nanotubes, *Nature*, **391**, pp. 59–62.
6. Zhou, X. J.; Park, J. Y.; Huang, S. M.; Liu, J.; McEuen, P. L. (2005). Band structure, phonon scattering, and the performance limit of single-walled carbon nanotube transistors, *Phys. Rev. Lett.*, **95**, pp. 146805.
7. Dai, H. J.; Javey, A.; Pop, E.; Mann, D.; Kim, W.; Lu, Y. R. (2006). Electrical transport properties and field effect transistors of carbon nanotubes, *Nano*, **1**, pp. 1–13.
8. Banerjee, S.; Hemraj-Benny, T.; Wong, S. S. (2005). Routes towards separating metallic and semiconducting nanotubes, *J. Nanosci. Nanotechnol.*, **5**, pp. 841–855.
9. Hersam, M. C. (2008). Progress towards monodisperse single-walled carbon nanotubes, *Nat. Nanotechnol.*, **3**, pp. 387–394.

10. Liu, J.; Hersam, M. C. (2010). Recent developments in carbon nanotube sorting and selective growth, *MRS Bull.*, **35**, pp. 315–321.
11. Loiseau, A.; Launois, P.; Petit, P.; Roche, S.; Salvetat, J.-P. (2006) *Understanding Carbon Nanotubes from Basics to Applications*, 1st Ed. (Springer-Verlag, Berlin).
12. Ouyang, M.; Huang, J. L.; Lieber, C. M. (2002). Fundamental electronic properties and applications of single-walled carbon nanotubes, *Acc. Chem. Res.*, **35**, pp. 1018–1025.
13. Ouyang, M.; Huang, J. L.; Lieber, C. M. (2002). Scanning tunneling microscopy studies of the one-dimensional electronic properties of single-walled carbon nanotubes, *Annu. Rev. Phys. Chem.*, **53**, pp. 201–220.
14. Charlier, J. C.; Blase, X.; Roche, S. (2007). Electronic and transport properties of nanotubes, *Rev. Mod. Phys.*, **79**, pp. 677–732.
15. Collins, P. C.; Arnold, M. S.; Avouris, P. (2001). Engineering carbon nanotubes and nanotube circuits using electrical breakdown, *Science*, **292**, pp. 706–709.
16. Collins, P. G.; Hersam, M.; Arnold, M.; Martel, R.; Avouris, P. (2001). Current saturation and electrical breakdown in multiwalled carbon nanotubes, *Phys. Rev. Lett.*, **86**, pp. 3128–3131.
17. Stahl, H.; Appenzeller, J.; Martel, R.; Avouris, P.; Lengeler, B. (2000). Intertube coupling in ropes of single-wall carbon nanotubes, *Phys. Rev. Lett.*, **85**, pp. 5186–5189.
18. Benedict, L. X.; Louie, S. G.; Cohen, M. L. (1995). Static polarizabilities of single-wall carbon nanotubes, *Phys. Rev. B Condens. Matter*, **52**, pp. 8541–8549.
19. Krupke, R.; Hennrich, F.; von Lohneysen, H.; Kappes, M. M. (2003). Separation of metallic from semiconducting single-walled carbon nanotubes, *Science*, **301**, pp. 344–347.
20. Krupke, R.; Hennrich, F.; Kappes, M. M.; Lohneysen, H. V. (2004). Surface conductance induced dielectrophoresis of semiconducting single-walled carbon nanotubes, *Nano Lett.*, **4**, pp. 1395–1399.
21. Blatt, S.; Hennrich, F.; von Lohneysen, H.; Kappes, M. M.; Vijayaraghavan, A.; Krupke, R. (2007). Influence of structural and dielectric anisotropy on the dielectrophoresis of single-walled carbon nanotubes, *Nano Lett.*, **7**, pp. 1960–1966.
22. Hong, S.; Jung, S.; Choi, J.; Kim, Y.; Baik, S. (2007). Electrical transport characteristics of surface-conductance-controlled, dielectrophoretically separated single-walled carbon nanotubes, *Langmuir*, **23**, pp. 4749–4752.

23. Kim, Y.; Hong, S.; Jung, S.; Strano, M. S.; Choi, J.; Baik, S. (2006). Dielectrophoresis of surface conductance modulated single-walled carbon nanotubes using catanionic surfactants, *J. Phys. Chem. B*, **110**, pp. 1541–1545.

24. Kang, J.; Hong, S.; Kim, Y.; Baik, S. (2009). Controlling the carbon nanotube-to-medium conductivity ratio for dielectrophoretic separation, *Langmuir*, **25**, pp. 12471–12474.

25. Mendes, M. J.; Schmidt, H. K.; Pasquali, M. (2008). Brownian dynamics simulations of single-wall carbon nanotube separation by type using dielectrophoresis, *J. Phys. Chem. B*, **112**, pp. 7467–7477.

26. Lutz, T.; Donovan, K. J. (2005). Macroscopic scale separation of metallic and semiconducting nanotubes by dielectrophoresis, *Carbon*, **43**, pp. 2508–2513.

27. Burg, B. R.; Schneider, J.; Bianco, V.; Schirmer, N. C.; Poulikakos, D. (2010). Selective parallel integration of individual metallic single-walled carbon nanotubes from heterogeneous solutions, *Langmuir*, **26**, pp. 10419–10424.

28. Peng, H. Q.; Alvarez, N. T.; Kittrell, C.; Hauge, R. H.; Schmidt, H. K. (2006). Dielectrophoresis field flow fractionation of single-walled carbon nanotubes, *J. Am. Chem. Soc.*, **128**, pp. 8396–8397.

29. Shin, D. H.; Kim, J. E.; Shim, H. C.; Song, J. W.; Yoon, J. H.; Kim, J.; Jeong, S.; Kang, J.; Baik, S.; Han, C. S. (2008). Continuous extraction of highly pure metallic single-walled carbon nanotubes in a microfluidic channel, *Nano Lett.*, **8**, pp. 4380–4385.

30. Niyogi, S.; Hamon, M. A.; Hu, H.; Zhao, B.; Bhowmik, P.; Sen, R.; Itkis, M. E.; Haddon, R. C. (2002). Chemistry of single-walled carbon nanotubes, *Acc. Chem. Res.*, **35**, pp. 1105–1113.

31. Joselevich, E. (2004). Electronic structure and chemical reactivity of carbon nanotubes: a chemist's view, *ChemPhysChem*, **5**, pp. 619–624.

32. Ormsby, J. L.; King, B. T. (2004). Clar valence bond representation of pi-bonding in carbon nanotubes, *J. Org. Chem.*, **69**, pp. 4287–4291.

33. Hoffmann, R. (1988). A chemical and theoretical way to look at bonding on surfaces, *Rev. Mod. Phys.*, **60**, pp. 601–628.

34. Sumpter, B. G.; Jiang, D. E.; Meunier, V. (2008). New insight into carbon-nanotube electronic-structure selectivity, *Small*, **4**, pp. 2035–2042.

35. Matsuo, Y.; Tahara, K.; Nakamura, E. (2003). Theoretical studies on structures and aromaticity of finite-length armchair carbon nanotubes, *Org. Lett.*, **5**, pp. 3181–3184.

36. Lu, X.; Chen, Z. F. (2005). Curved pi-conjugation, aromaticity, and the related chemistry of small fullerenes (< C-60) and single-walled carbon nanotubes, *Chem. Rev.*, **105**, pp. 3643–3696.

37. Strano, M. S.; Dyke, C. A.; Usrey, M. L.; Barone, P. W.; Allen, M. J.; Shan, H. W.; Kittrell, C.; Hauge, R. H.; Tour, J. M.; Smalley, R. E. (2003). Electronic structure control of single-walled carbon nanotube functionalization, *Science*, **301**, pp. 1519–1522.

38. An, L.; Fu, Q. A.; Lu, C. G.; Liu, J. (2004). A simple chemical route to selectively eliminate metallic carbon nanotubes in nanotube network devices, *J. Am. Chem. Soc.*, **126**, pp. 10520–10521.

39. Usrey, M. L.; Lippmann, E. S.; Strano, M. S. (2005). Evidence for a two-step mechanism in electronically selective single-walled carbon nanotube reactions, *J. Am. Chem. Soc.*, **127**, pp. 16129–16135.

40. Doyle, C. D.; Rocha, J. D. R.; Weisman, R. B.; Tour, J. M. (2008). Structure-dependent reactivity of semiconducting single-walled carbon nanotubes with benzenediazonium salts, *J. Am. Chem. Soc.*, **130**, pp. 6795–6800.

41. Ghosh, S.; Rao, C. N. R. (2009). Separation of metallic and semiconducting single-walled carbon nanotubes through fluorous chemistry, *Nano Res.*, **2**, pp. 183–191.

42. Kim, W. J.; Usrey, M. L.; Strano, M. S. (2007). Selective functionalization and free solution electrophoresis of single-walled carbon nanotubes: separate enrichment of metallic and semiconducting SWNT, *Chem. Mater.*, **19**, pp. 1571–1576.

43. Lee, C. W.; Han, X. D.; Chen, F. M.; Wei, J.; Chen, Y.; Chan-Park, M. B.; Li, L. J. (2010). Solution-processable carbon nanotubes for semiconducting thin-film transistor devices, *Adv. Mater.*, **22**, pp. 1278–1282.

44. Nair, N.; Kim, W. J.; Usrey, M. L.; Strano, M. S. (2007). A structure-reactivity relationship for single walled carbon nanotubes reacting with 4-hydroxybenzene diazonium salt, *J. Am. Chem. Soc.*, **129**, pp. 3946–3954.

45. Toyoda, S.; Yamaguchi, Y.; Hiwatashi, M.; Tomonari, Y.; Murakami, H.; Nakashima, N. (2007). Separation of semiconducting single-walled carbon nanotubes by using a long-alkyl-chain benzenediazonium compound, *Chem. Asian J.*, **2**, pp. 145–149.

46. Banerjee, S.; Wong, S. S. (2004). Demonstration of diameter-selective reactivity in the sidewall ozonation of SWNTs by resonance Raman spectroscopy, *Nano Lett.*, **4**, pp. 1445–1450.

47. Hu, H.; Zhao, B.; Hamon, M. A.; Kamaras, K.; Itkis, M. E.; Haddon, R. C. (2003). Sidewall functionalization of single-walled carbon nanotubes

by addition of dichlorocarbene, *J. Am. Chem. Soc.*, **125**, pp. 14893–14900.

48. Kamaras, K.; Itkis, M. E.; Hu, H.; Zhao, B.; Haddon, R. C. (2003). Covalent bond formation to a carbon nanotube metal, *Science*, **301**, pp. 1501–1501.

49. An, K. H.; Park, J. S.; Yang, C. M.; Jeong, S. Y.; Lim, S. C.; Kang, C.; Son, J. H.; Jeong, M. S.; Lee, Y. H. (2005). A diameter-selective attack of metallic carbon nanotubes by nitronium ions, *J. Am. Chem. Soc.*, **127**, pp. 5196–5203.

50. An, K. H.; Yang, C. M.; Lee, J. Y.; Lim, S. C.; Kang, C.; Son, J. H.; Jeong, M. S.; Lee, Y. H. (2006). A diameter-selective chiral separation of single-wall carbon nanotubes using nitronium ions, *J. Electron. Mater.*, **35**, pp. 235–242.

51. Yang, C. M.; Park, J. S.; An, K. H.; Lim, S. C.; Seo, K.; Kim, B.; Park, K. A.; Han, S.; Park, C. Y.; Lee, Y. H. (2005). Selective removal of metallic single-walled carbon nanotubes with small diameters by using nitric and sulfuric acids, *J. Phys. Chem. B*, **109**, pp. 19242–19248.

52. Banerjee, S.; Wong, S. S. (2004). Selective metallic tube reactivity in the solution-phase osmylation of single-walled carbon nanotubes, *J. Am. Chem. Soc.*, **126**, pp. 2073–2081.

53. Zhang, G. Y.; Qi, P. F.; Wang, X. R.; Lu, Y. R.; Li, X. L.; Tu, R.; Bangsaruntip, S.; Mann, D.; Zhang, L.; Dai, H. J. (2006). Selective etching of metallic carbon nanotubes by gas-phase reaction, *Science*, **314**, pp. 974–977.

54. Yang, C. M.; An, K. H.; Park, J. S.; Park, K. A.; Lim, S. C.; Cho, S. H.; Lee, Y. S.; Park, W.; Park, C. Y.; Lee, Y. H. (2006). Preferential etching of metallic single-walled carbon nanotubes with small diameter by fluorine gas, *Phys. Rev. B Condens. Matter*, **73**, doi: 10.1103/PhysRevB.73.075419.

55. Kanungo, M.; Lu, H.; Malliaras, G. G.; Blanchet, G. B. (2009). Suppression of metallic conductivity of single-walled carbon nanotubes by cycloaddition reactions, *Science*, **323**, pp. 234–237.

56. Bergeret, C.; Cousseau, J.; Fernandez, V.; Mevellec, J. Y.; Lefrant, S. (2008). Spectroscopic evidence of carbon nanotubes' metallic character loss induced by covalent functionalization via nitric acid purification, *J. Phys. Chem. C*, **112**, pp. 16411–16416.

57. Qiu, H. X.; Maeda, Y.; Akasaka, T. (2009). Facile and scalable route for highly efficient enrichment of semiconducting single-walled carbon nanotubes, *J. Am. Chem. Soc.*, **131**, pp. 16529–16533.

58. Yu, B.; Hou, P. X.; Li, F.; Liu, B. L.; Liu, C.; Cheng, H. M. (2010). Selective removal of metallic single-walled carbon nanotubes by combined in situ and post-synthesis oxidation, *Carbon*, **48**, pp. 2941–2947.

59. Lu, J.; Lai, L.; Luo, G.; Zhou, J.; Qin, R.; Wang, D.; Wang, L.; Mei, W. N.; Li, G.; Gao, Z.; Nagase, S.; Maeda, Y.; Akasaka, T.; Yu, D. (2007). Why semiconducting single-walled carbon nanotubes are separated from their metallic counterparts, *Small*, **3**, pp. 1566–1576.

60. Miyata, Y.; Maniwa, Y.; Kataura, H. (2006). Selective oxidation of semiconducting single-wall carbon nanotubes by hydrogen peroxide, *J. Phys. Chem. B*, **110**, pp. 25–29.

61. Yudasaka, M.; Zhang, M.; Iijima, S. (2003). Diameter-selective removal of single-wall carbon nanotubes through light-assisted oxidation, *Chem. Phys. Lett.*, **374**, pp. 132–136.

62. Menard-Moyon, C.; Izard, N.; Doris, E.; Mioskowski, C. (2006). Separation of semiconducting from metallic carbon nanotubes by selective functionalization with azomethine ylides, *J. Am. Chem. Soc.*, **128**, pp. 6552–6553.

63. Maeda, Y.; Kanda, M.; Hashimoto, M.; Hasegawa, T.; Kimura, S.; Lian, Y. F.; Wakahara, T.; Akasaka, T.; Kazaoui, S.; Minami, N.; Okazaki, T.; Hayamizu, Y.; Hata, K.; Lu, J.; Nagase, S. (2006). Dispersion and separation of small-diameter single-walled carbon nanotubes, *J. Am. Chem. Soc.*, **128**, pp. 12239–12242.

64. Maeda, Y.; Kimura, S.; Kanda, M.; Hirashima, Y.; Hasegawa, T.; Wakahara, T.; Lian, Y. F.; Nakahodo, T.; Tsuchiya, T.; Akasaka, T.; Lu, J.; Zhang, X. W.; Gao, Z. X.; Yu, Y. P.; Nagase, S.; Kazaoui, S.; Minami, N.; Shimizu, T.; Tokumoto, H.; Saito, R. (2005). Large-scale separation of metallic and semiconducting single-walled carbon nanotubes, *J. Am. Chem. Soc.*, **127**, pp. 10287–10290.

65. LeMieux, M. C.; Roberts, M.; Barman, S.; Jin, Y. W.; Kim, J. M.; Bao, Z. N. (2008). Self-sorted, aligned nanotube networks for thin-film transistors, *Science*, **321**, pp. 101–104.

66. Chattopadhyay, D.; Galeska, L.; Papadimitrakopoulos, F. (2003). A route for bulk separation of semiconducting from metallic single-wall carbon nanotubes, *J. Am. Chem. Soc.*, **125**, pp. 3370–3375.

67. Ju, S. Y.; Utz, M.; Papadimitrakopoulos, F. (2009). Enrichment mechanism of semiconducting single-walled carbon nanotubes by surfactant amines, *J. Am. Chem. Soc.*, **131**, pp. 6775–6784.

68. Kim, S. N.; Luo, Z. T.; Papadimitrakopoulos, F. (2005). Diameter and metallicity dependent redox influences on the separation of single-wall carbon nanotubes, *Nano Lett.*, **5**, pp. 2500–2504.

69. Samsonidze, G. G.; Chou, S. G.; Santos, A. P.; Brar, V. W.; Dresselhaus, G.; Dresselhaus, M. S.; Selbst, A.; Swan, A. K.; Unlu, M. S.; Goldberg,

B. B.; Chattopadhyay, D.; Kim, S. N.; Papadimitrakopoulos, F. (2004). Quantitative evaluation of the octadecylamine-assisted bulk separation of semiconducting and metallic single-wall carbon nanotubes by resonance Raman spectroscopy, *Appl. Phys. Lett.*, **85**, pp. 1006–1008.

70. Chen, Z. H.; Du, X.; Du, M. H.; Rancken, C. D.; Cheng, H. P.; Rinzler, A. G. (2003). Bulk separative enrichment in metallic or semiconducting single-walled carbon nanotubes, *Nano Lett.*, **3**, pp. 1245–1249.

71. Park, N.; Miyamoto, Y.; Lee, K.; Choi, W. I.; Ihm, J.; Yu, J. J.; Han, S. W. (2005). Band gap sensitivity of bromine adsorption at carbon nanotubes, *Carbon*, **403**, pp. 135–139.

72. Lu, J.; Nagase, S.; Zhang, X. W.; Wang, D.; Ni, M.; Maeda, Y.; Wakahara, T.; Nakahodo, T.; Tsuchiya, T.; Akasaka, T.; Gao, Z. X.; Yu, D. P.; Ye, H. Q.; Mei, W. N.; Zhou, Y. S. (2006). Selective interaction of large or charge-transfer aromatic molecules with metallic single-wall carbon nanotubes: Critical role of the molecular size and orientation, *J. Am. Chem. Soc.*, **128**, pp. 5114–5118.

73. McDonald, T. J.; Blackburn, J. L.; Metzger, W. K.; Rumbles, G.; Heben, M. J. (2007). Chiral-selective protection of single-walled carbon nanotube photoluminescence by surfactant selection, *J. Phys. Chem. C*, **111**, pp. 17894–17900.

74. McDonald, T. J.; Engtrakul, C.; Jones, M.; Rumbles, G.; Heben, M. J. (2006). Kinetics of PL quenching during single-walled carbon nanotube rebundling and diameter-dependent surfactant interactions, *J. Phys. Chem. B*, **110**, pp. 25339–25346.

75. Moshammer, K.; Hennrich, F.; Kappes, M. M. (2009). Selective suspension in aqueous sodium dodecyl sulfate according to electronic structure type allows simple separation of metallic from semiconducting single-walled carbon nanotubes, *Nano Res.*, **2**, pp. 599–606.

76. Wei, L.; Wang, B.; Goh, T. H.; Li, L.-J.; Yang, Y.; Chan-Park, M. B.; Chen, Y. (2008). Selective enrichment of (6,5) and (8,3) single-walled carbon nanotubes via cosurfactant extraction from narrow (n,m) distribution samples, *J. Phys. Chem. B*, **112**, pp. 2771–2774.

77. Li, H. P.; Zhou, B.; Lin, Y.; Gu, L. R.; Wang, W.; Fernando, K. A. S.; Kumar, S.; Allard, L. F.; Sun, Y. P. (2004). Selective interactions of porphyrins with semiconducting single-walled carbon nanotubes, *J. Am. Chem. Soc.*, **126**, pp. 1014–1015.

78. Peng, X.; Komatsu, N.; Bhattacharya, S.; Shimawaki, T.; Aonuma, S.; Kimura, T.; Osuka, A. (2007). Optically active single-walled carbon nanotubes, *Nat. Nanotechnol.*, **2**, pp. 361–365.

79. Peng, X. B.; Komatsu, N.; Kimura, T.; Osuka, A. (2007). Improved optical enrichment of SWNTs through extraction with chiral nanotweezers of 2,6-pyridylene-bridged diporphyrins, *J. Am. Chem. Soc.*, **129**, pp. 15947–15953.

80. Tromp, R. M.; Afzali, A.; Freitag, M.; Mitzi, D. B.; Chen, Z. (2008). Novel strategy for diameter-selective separation and functionalization of single-wall carbon nanotubes, *Nano Lett.*, **8**, pp. 469–472.

81. Chen, F. M.; Wang, B.; Chen, Y.; Li, L. J. (2007). Toward the extraction of single species of single-walled carbon nanotubes using fluorene-based polymers, *Nano Lett.*, **7**, pp. 3013–3017.

82. Hwang, J. Y.; Nish, A.; Doig, J.; Douven, S.; Chen, C. W.; Chen, L. C.; Nicholas, R. J. (2008). Polymer structure and solvent effects on the selective dispersion of single-walled carbon nanotubes, *J. Am. Chem. Soc.*, **130**, pp. 3543–3553.

83. Nish, A.; Hwang, J. Y.; Doig, J.; Nicholas, R. J. (2007). Highly selective dispersion of singlewalled carbon nanotubes using aromatic polymers, *Nat. Nanotechnol.*, **2**, pp. 640–646.

84. Zheng, M.; Jagota, A.; Strano, M. S.; Santos, A. P.; Barone, P.; Chou, S. G.; Diner, B. A.; Dresselhaus, M. S.; McLean, R. S.; Onoa, G. B.; Samsonidze, G. G.; Semke, E. D.; Usrey, M.; Walls, D. J. (2003). Structure-based carbon nanotube sorting by sequence-dependent DNA assembly, *Science*, **302**, pp. 1545–1548.

85. Fantini, C.; Jorio, A.; Santos, A. P.; Peressinotto, V. S. T.; Pimenta, M. A. (2007). Characterization of DNA-wrapped carbon nanotubes by resonance Raman and optical absorption spectroscopies, *Chem. Phys. Lett.*, **439**, pp. 138–142.

86. Tu, X. M.; Manohar, S.; Jagota, A.; Zheng, M. (2009). DNA sequence motifs for structure-specific recognition and separation of carbon nanotubes, *Nature*, **460**, pp. 250–253.

87. Kato, Y.; Niidome, Y.; Nakashima, N. (2009). Efficient separation of (6,5) single-walled carbon nanotubes using a "nanometal sinker", *Angew. Chem. Int. Ed.*, **48**, pp. 5435–5438.

88. Ju, S. Y.; Doll, J.; Sharma, I.; Papadimitrakopoulos, F. (2008). Selection of carbon nanotubes with specific chiralities using helical assemblies of flavin mononucleotide, *Nat. Nanotechnol.*, **3**, pp. 356–362.

89. Ju, S. Y.; Kopcha, W. P.; Papadimitrakopoulos, F. (2009). Brightly fluorescent single-walled carbon nanotubes via an oxygen-excluding surfactant organization, *Science*, **323**, pp. 1319–1323.

90. Marquis, R.; Greco, C.; Sadokierska, I.; Lebedkin, S.; Kappes, M. M.; Michel, T.; Alvarez, L.; Sauvajol, J. L.; Meunier, S.; Mioskowski, C. (2008).

Supramolecular discrimination of carbon nanotubes according to their helicity, *Nano Lett.*, **8**, pp. 1830–1835.

91. Yan, L. Y.; Li, W. F.; Fan, X. F.; Wei, L.; Chen, Y.; Kuo, J. L.; Li, L. J.; Kwak, S. K.; Mu, Y. G.; Chan-Park, M. B. (2010). Enrichment of (8,4) single-walled carbon nanotubes through coextraction with heparin, *Small*, **6**, pp. 110–118.
92. Shin, H. J.; Kim, S. M.; Yoon, S. M.; Benayad, A.; Kim, K. K.; Kim, S. J.; Park, H. K.; Choi, J. Y.; Lee, Y. H. (2008). Tailoring electronic structures of carbon nanotubes by solvent with electron-donating and -withdrawing groups, *J. Am. Chem. Soc.*, **130**, pp. 2062–2066.
93. Arnold, M. S.; Green, A. A.; Hulvat, J. F.; Stupp, S. I.; Hersam, M. C. (2006). Sorting carbon nanotubes by electronic structure using density differentiation, *Nat. Nanotechnol.*, **1**, pp. 60–65.
94. Carvalho, E. J. F.; dos Santos, M. C. (2010). Role of surfactants in carbon nanotubes density gradient separation, *ACS Nano*, **4**, pp. 765–770.
95. Green, A. A.; Hersam, M. C. (2007). Ultracentrifugation of single-walled nanotubes, *Mater. Today*, **10**, pp. 59–60.
96. Ghosh, S.; Bachilo, S. M.; Weisman, R. B. (2010). Advanced sorting of single-walled carbon nanotubes by nonlinear density-gradient ultracentrifugation, *Nat. Nanotechnol.*, **5**, pp. 443–450.
97. Stiirzl, N.; Hennrich, F.; Lebedkin, S.; Kappes, M. M. (2009). Near monochiral single-walled carbon nanotube dispersions in organic solvents, *J. Phys. Chem. C*, **113**, pp. 14628–14632.
98. Yanagi, K.; Iitsuka, T.; Fujii, S.; Kataura, H. (2008). Separations of metallic and semiconducting carbon nanotubes by using sucrose as a gradient medium, *J. Phys. Chem. C*, **112**, pp. 18889–18894.
99. Fagan, J. A.; Becker, M. L.; Chun, J. H.; Nie, P. T.; Bauer, B. J.; Simpson, J. R.; Hight-Walker, A.; Hobbie, E. K. (2008). Centrifugal length separation of carbon nanotubes, *Langmuir*, **24**, pp. 13880–13889.
100. Zheng, M.; Semke, E. D. (2007). Enrichment of single chirality carbon nanotubes, *J. Am. Chem. Soc.*, **129**, pp. 6084–6085.
101. Tanaka, T.; Jin, H.; Miyata, Y.; Fujii, S.; Suga, H.; Naitoh, Y.; Minari, T.; Miyadera, T.; Tsukagoshi, K.; Kataura, H. (2009). Simple and scalable gel-based separation of metallic and semiconducting carbon nanotubes, *Nano Lett.*, **9**, pp. 1497–1500.
102. Yanagi, K.; Miyata, Y.; Kataura, H. (2008). Optical and conductive characteristics of metallic single-wall carbon nanotubes with three basic colors; Cyan, magenta, and yellow, *Appl. Phys. Express*, **1**, pp. 114001.

103. Gao, M.; Zuo, J. M.; Twesten, R. D.; Petrov, I.; Nagahara, L. A.; Zhang, R. (2003). Structure determination of individual single-wall carbon nanotubes by nanoarea electron diffraction, *Appl. Phys. Lett.*, **82**, pp. 2703–2705.

104. Qin, L. C.; Iijima, S.; Kataura, H.; Maniwa, Y.; Suzuki, S.; Achiba, Y. (1997). Helicity and packing of single-walled carbon nanotubes studied by electron nanodiffraction, *Chem. Phys. Lett.*, **268**, pp. 101–106.

105. Vijayaraghavan, A.; Blatt, S.; Marquardt, C.; Dehm, S.; Wahi, R.; Hennrich, F.; Krupke, R. (2008). Imaging electronic structure of carbon nanotubes by voltage-contrast scanning electron microscopy, *Nano Res.*, **1**, pp. 321–332.

106. Lu, W.; Xiong, Y.; Hassanien, A.; Zhao, W.; Zheng, M.; Chen, L. W. (2009). A scanning probe microscopy based assay for single-walled carbon nanotube metallicity, *Nano Lett.*, **9**, pp. 1668–1672.

107. Li, Y. M.; Mann, D.; Rolandi, M.; Kim, W.; Ural, A.; Hung, S.; Javey, A.; Cao, J.; Wang, D. W.; Yenilmez, E.; Wang, Q.; Gibbons, J. F.; Nishi, Y.; Dai, H. J. (2004). Preferential growth of semiconducting single-walled carbon nanotubes by a plasma enhanced CVD method, *Nano Lett.*, **4**, pp. 317–321.

108. Kim, W. J.; Lee, C. Y.; O'Brien, K. P.; Plombon, J. J.; Blackwell, J. M.; Strano, M. S. (2009). Connecting single molecule electrical measurements to ensemble spectroscopic properties for quantification of single-walled carbon nanotube separation, *J. Am. Chem. Soc.*, **131**, pp. 3128–3129.

109. Zhang, L.; Zaric, S.; Tu, X. M.; Wang, X. R.; Zhao, W.; Dai, H. J. (2008). Assessment of chemically separated carbon nanotubes for nanoelectronics, *J. Am. Chem. Soc.*, **130**, pp. 2686–2691.

110. Naumov, A. V.; Kuznetsov, O. A.; Harutyunyan, A. R.; Green, A. A.; Hersam, M. C.; Resasco, D. E.; Nikolaev, P. N.; Weisman, R. B. (2009). Quantifying the semiconducting fraction in single-walled carbon nanotube samples through comparative atomic force and photoluminescence microscopies, *Nano Lett.*, **9**, pp. 3203–3208.

111. Jorio, A.; Saito, R.; Hafner, J. H.; Lieber, C. M.; Hunter, M.; McClure, T.; Dresselhaus, G.; Dresselhaus, M. S. (2001). Structural (n,m) determination of isolated single-wall carbon nanotubes by resonant Raman scattering, *Phys. Rev. Lett.*, **86**, pp. 1118.

112. Hagen, A.; Hertel, T. (2003). Quantitative analysis of optical spectra from individual single-wall carbon nanotubes, *Nano Lett.*, **3**, pp. 383–388.

113. Rao, A. M.; Richter, E.; Bandow, S.; Chase, B.; Eklund, P. C.; Williams, K. A.; Fang, S.; Subbaswamy, K. R.; Menon, M.; Thess, A.; Smalley, R. E.;

Dresselhaus, G.; Dresselhaus, M. S. (1997). Diameter-selective Raman scattering from vibrational modes in carbon nanotubes, *Science*, **275**, pp. 187–191.

114. Pimenta, M. A.; Marucci, A.; Empedocles, S. A.; Bawendi, M. G.; Hanlon, E. B.; Rao, A. M.; Eklund, P. C.; Smalley, R. E.; Dresselhaus, G.; Dresselhaus, M. S. (1998). Raman modes of metallic carbon nanotubes, *Phys. Rev. B Condens. Matter*, **58**, pp. 16016–16019.

115. Kataura, H.; Kumazawa, Y.; Maniwa, Y.; Umezu, I.; Suzuki, S.; Ohtsuka, Y.; Achiba, Y. (1999). Optical properties of single-wall carbon nanotubes, *Synth. Met.*, **103**, pp. 2555–2558.

116. Heller, D. A.; Barone, P. W.; Swanson, J. P.; Mayrhofer, R. M.; Strano, M. S. (2004). Using Raman spectroscopy to elucidate the aggregation state of single-walled carbon nanotubes, *J. Phys. Chem. B*, **108**, pp. 6905–6909.

117. Tsyboulski, D. A.; Rocha, J.-D. R.; Bachilo, S. M.; Cognet, L.; Weisman, R. B. (2007). Structure-dependent fluorescence efficiencies of individual single-walled carbon nanotubes, *Nano Lett.*, **7**, pp. 3080–3085.

118. Jiang, J.; Saito, R.; Gruneis, A.; Chou, S. G.; Ge, G. S.; Jorio, A.; Dresselhaus, G.; Dresselhaus, M. S. (2005). Intensity of the resonance Raman excitation spectra of single-wall carbon nanotubes, *Phys. Rev. B Condens. Matter*, **71**, pp. 205420.

119. Oyama, Y.; Saito, R.; Sato, K.; Jiang, J.; Samsonidze, G. G.; Gruneis, A.; Miyauchi, Y.; Maruyama, S.; Jorio, A.; Dresselhaus, G.; Dresselhaus, M. S. (2006). Photoluminescence intensity of single-wall carbon nanotubes, *Carbon*, **44**, pp. 873–879.

120. Popov, V. N.; Henrard, L.; Lambin, P. (2004). Resonant Raman intensity of the radial breathing mode of single-walled carbon nanotubes within a nonorthogonal tight-binding model, *Nano Lett.*, **4**, pp. 1795–1799.

121. Reich, S.; Thomsen, C.; Robertson, J. (2005). Exciton resonances quench the photoluminescence of zigzag carbon nanotubes, *Phys. Rev. Lett.*, **95**, pp. 077402-077404.

122. Miyata, Y.; Yanagi, K.; Maniwa, Y.; Kataura, H. (2008). Optical evaluation of the metal-to-semiconductor ratio of single-wall carbon nanotubes, *J. Phys. Chem. C*, **112**, pp. 13187–13191.

123. Blackburn, J. L.; Barnes, T. M.; Beard, M. C.; Kim, Y. H.; Tenent, R. C.; McDonald, T. J.; To, B.; Coutts, T. J.; Heben, M. J. (2008). Transparent conductive single-walled carbon nanotube networks with precisely tunable ratios of semiconducting and metallic nanotubes, *Acs Nano*, **2**, pp. 1266–1274.

124. Bachilo, S. M.; Strano, M. S.; Kittrell, C.; Hauge, R. H.; Smalley, R. E.; Weisman, R. B. (2002). Structure-assigned optical spectra of single-walled carbon nanotubes, *Science*, **298**, pp. 2361–2366.

125. Wei, L.; Lee, C. W.; Li, L. J.; Sudibya, H. G.; Wang, B.; Chen, L. Q.; Chen, P.; Yang, Y. H.; Chan-Park, M. B.; Chen, Y. (2008). Assessment of (n,m) selectively enriched small diameter single-walled carbon nanotubes by density differentiation from cobalt-incorporated MCM-41 for macroelectronics, *Chem. Mater.*, **20**, pp. 7417–7424.

126. Avouris, P. (2004). Carbon nanotube electronics and optoelectronics, *MRS Bull.*, **29**, pp. 403–410.

127. Avouris, P.; Martel, R. (2010). Progress in carbon nanotube electronics and photonics, *MRS Bull.*, **35**, pp. 306–313.

Chapter 5

Electronic Applications of Single-Walled Carbon Nanotubes

Hong Li and Qing Zhang
Microelectronics Department, School of Electrical and Electronic Engineering, Nanyang Technological University, 50 Nanyang Avenue, 639798, Singapore
lihong@ntu.edu.sg

A single-wall carbon nanotube (SWCNT) can be viewed as a single layer of graphite (graphene) rolled up into a seamless cylinder. Differentiated by the direction of rolling up, an SWCNT can be a semiconducting SWCNT (s-SWCNT) or a metallic SWCNT (m-SWCNT). An S-SWCNT is a type of semiconducting material with bandgap inversely proportional to its diameter, whereas an m-SWCNT is generally considered a metallic wire. SWCNTs have attracted a lot of attention because of their simple structures and easily predictable electronic properties. The diameter-dependent energy gap makes s-SWCNTs attractive for various nanoelectronic devices. SWCNT based field-effect transistors (FETs) have the simplest geometry that consists of two metal electrodes connected by an s-SWCNT on the top of a conducting substrate capped by an insulating layer. Due to one-dimensional (1-D) nature of s-SWCNT, electron transport in the 1-D nanostructure is confined so that s-SWCNTs have an inherent advantage over bulk semiconducting materials to work as the key

Carbon Nanotubes and Their Applications
Edited by Qing Zhang

ISBN 978-981-4241-90-8 (Hardcover), 978-981-4303-18-7 (eBook)
www.panstanford.com

component in a single-electron transistor (SET). As SWCNTs have a very large surface-to-volume ratio, the conductance in SWCNTs is very sensitive to the electrostatic environment around the SWCNTs. The high sensitivity makes the s-SWCNT a perfect candidate for memory application. Because of these unique electronic properties, SWCNTs have been studied intensively in the past decade. The state-of-the-art electronic applications of the SWCNTs have been explored. In this chapter, we review the roles of SWCNTs in various types of semiconducting devices, such as FETs, SETs, and FET- and SET-based logic circuits, and memory devices.

5.1 Field-Effect Transistors and Logic Circuits

After an s-SWCNT is placed on top of a heavily doped Si substrate capped by a SiO_2 layer, a pair of metal electrodes is fabricated in contact with the SWCNT to complete the structure of a carbon nanotube field-effect transistor (CNTFET), in which the Si substrate acts as the back-gate electrode, the SiO_2 layer plays a role of back-gate dielectric, the two metal electrodes serve as the interchangeable source and drain contact, and the CNT works as the current channel. The gate electric field in a CNTFET at the s-SWCNT surface is enhanced due to the miniature dimension of the s-SWCNT. The enhanced electric field results in efficient gate control over the carrier concentration in the channel. As a result, the s-SWCNT channel can be effectively turned on and off by sweeping the gate voltage. The CNTFETs with Schottky contacts, ohmic contacts, as well as their relevant logic circuits will be introduced in this section. Moreover, the performance limit of CNTFETs will be discussed.

5.1.1 Schottky-Contact CNTFETs

The first CNTFET based on a single s-SWCNT was demonstrated in 1998, as shown in Figure 5.1a,b.[1] The CNTFET consisted of a 1.4 nm-thick s-SWCNT on a 300 nm-thick SiO_2 layer/highly doped silicon substrate. The SWCNT, SiO_2 layer and silicon substrate served as the conductance channel, gate dielectric and back gate, respectively. Since the work function of the SWCNT is 4.5 eV, smaller than that of the Pt electrodes, i.e., 5.7 eV, the Fermi level of Pt electrode was suggested to be aligned to the valence band of the SWCNT at the contacts. The SWCNT with a bandgap of ~0.6 eV was connected to the leads by tunneling contacts, labeled by the black vertical bar. Thus, a positive gate voltage $+V_{gate}$ could lead to an insulating state, while negative gate voltage $-V_{gate}$ could result in a metal-like conducting state, as shown in Figure 5.1c,d. Obviously, the SWCNT/Pt contact is very important to the performance of the transistor. Understanding the properties of the contact between 1-D SWCNT and bulk metal electrode is the first step for developing high-performance CNTFETs.

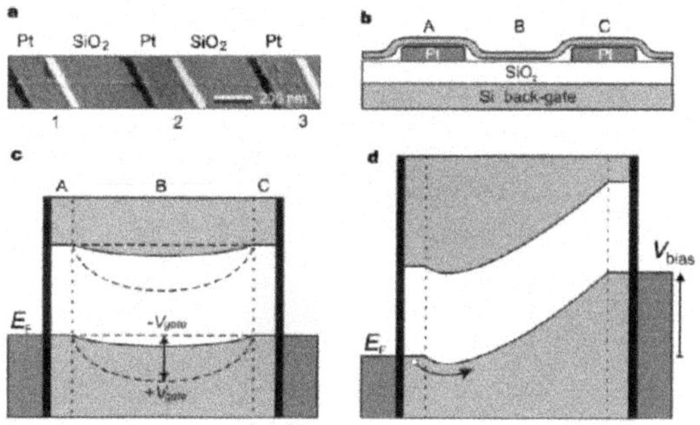

Figure 5.1 The first CNTFET based on a single s-SWCNT. (a) AFM image of the back-gated CNTFETs. (b) Schematic cross-sectional view of the CNTFET. (c) Energy band diagrams of the device at a positive (+V_{gate}) and negative (-V_{gate}) gate voltage, respectively, with the zero bias voltage. (d) Nonzero bias voltage V_{bias} [1].

Léonard *et al.* have done a systematic study on the SWCNT/metal contacts.[2,3] They found that "Fermi-level pinning" effect[4] is minor in SWCNT contacted to other bulk metals or semiconductors. As illustrated in Figure 5.2, metal-induced gap states (MIGS) are introduced in the semiconductor surface when the metal is brought to contact with the semiconductor. MIGS decays exponentially away from the interface. Charges in the semiconductor near the interface are balanced by image charges induced in the metal, leading to a dipole sheet formed at the two sides of the interface (see Figure 5.2a,b). The resulting electrostatic potential causes the energy band bending and E_F and E_N (the charge neutrality level) alignment, as depicted in Figure 5.2c. As a consequence, the length of the extra barrier due to Fermi level pinning in bulk contact and SWCNT contact are in the range of micrometers (L) and nanometers (L'), respectively. In the SWCNT/metal contact, electrons or holes can tunnel through the thin barrier easily, with a negligible effect of the Fermi level pinning. Léonard and coworkers showed quantitatively that the Fermi level pinning effect in SWCNT/metal contact is much less effective than that in the bulk semiconductor/metal contacts, as presented in Figure 5.3.[5] One can see that barrier widths at the SWCNT/metal contact are less than 10 nm at various MIGS levels.

Such thin barriers have negligible effects on the electron transport across the contacts. On the contrary, the barrier in a planar device is much thicker than that at the SWCNT/metal contacts (see the inset), easily thicker than 10 nm at various MIGS levels. Such a barrier is too thick for electrons to tunnel through; thus, this significantly affects the electron transport in the planar device.

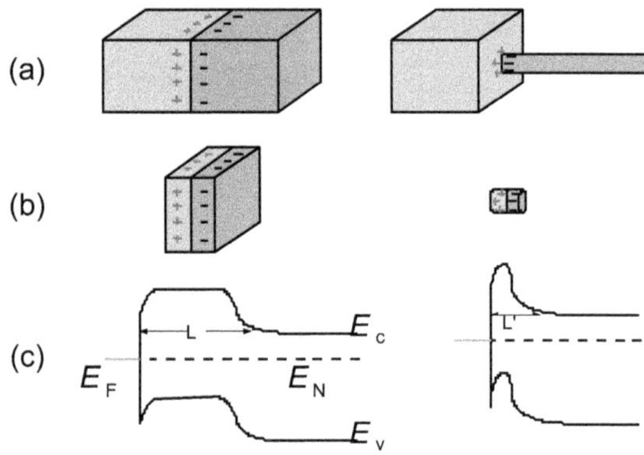

Figure 5.2 Schematic bulk semiconductor and SWCNT contact to bulk metal electrode, respectively. E_F is the Fermi level and E_N is the charge neutrality level. The charge distribution at the two sides of the interface is shown in (a) and (b). (c) Energy band bending at the contacts [3].

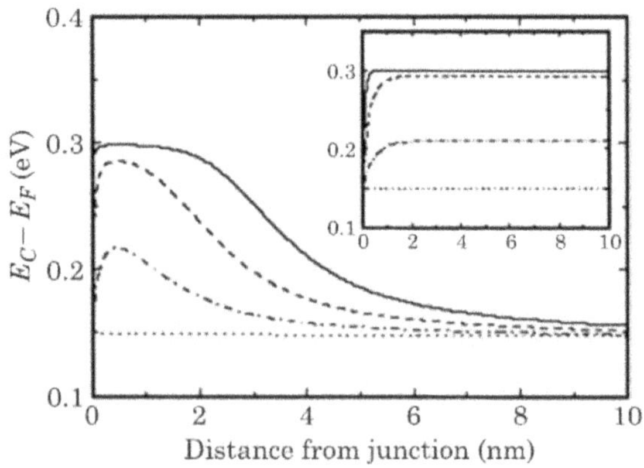

Figure 5.3 Calculated energy band-bending due to MIGS at the SWCNT/metal contact. The curves from top to bottom correspond to a decreasing density of MIGS. Inset: energy band-bending at the bulk semiconductor/metal contact with different MIGS (the curves correspond to those in the figure) [5].

To verify the concept that different metal work functions result in distinct energy barriers at metal/CNT contacts, we fabricate CNTFETs with aluminum (Al) and gold (Au) as the drain (D) and source (S) contact, respectively. We employed a self-alignment method to form a short SWCNT-channel, so that the influence of the defects in the SWCNT-channel is negligible and the electron transport could be ballistic. Figure 5.4a,b shows a schematic and an AFM image of a typical device (channel length ~90 nm), respectively. At the source contact (left electrode), the SWCNT-channel is sandwiched between SiO_2 (bottom) and Al layer (top). Thus, properties of the source contact depend on Al/SWCNT coupling. In contrast, at the drain side (right electrode), the SWCNT-channel contacts to the Au electrode, and then the contact properties should be dominated by Au/SWCNT coupling.

With the thermal emission transport model, the effective contact barrier height (activation energy E_a) can be estimated from the temperature-dependent I–V curves. The back-gate voltage V_{GS} dependent E_a is plotted in Figure 5.4c. The Arrhenius plot for obtaining E_a is shown in the inset of Figure 5.4c. It is found that E_a increases drastically when $V_{GS} < -10$ V. This could suggest that the flat-band condition is reached around $V_{GS} \sim -10$ V. In other words, the majority of electrons are thermally injected into the SWCNT-channel when $V_{GS} < -10$ V. But most of the electrons are likely to tunnel into the channel at $V_{GS} > -10$ V. We noted that the extracted activation energy values are smaller than expected values. The reason could be the imperfect contacts caused by absorbents or air gap between CNT and metals, resulting in an extra tunneling barrier. The tunneling barrier lowers the measured activation energy (detailed transport mechanisms will be present in a later part). More important, E_a at positive V_{DS} (right branches) are larger than those at the corresponding negative V_{DS} (left branches) by more than two times for various V_{GS}, as shown in Figure 5.4d. As Al and Au are connected as the D and S electrodes, respectively, electrons face the barrier at the Al/SWCNT (Au/SWCNT) contact at negative (positive) V_{DS}. Thus, the barrier height is expected to be larger at positive V_{DS} since Au with higher work function (~ 5.2 eV) could form a larger barrier height (for electrons) than that of Al (work function of ~ 4.3 eV) electrode. Our experimental results are qualitatively consistent with the aforementioned theoretical expectation.[5]

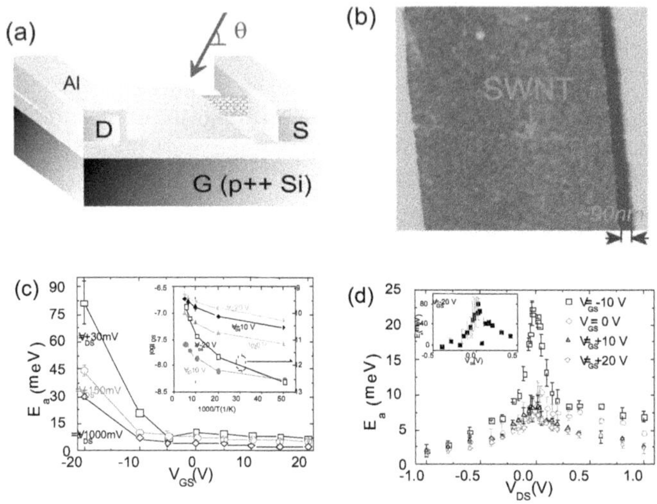

Figure 5.4 CNTFET with asymmetric metal contacts (Al and Au). (a) A schematic diagram of the device structure. (b) An AFM image of a ~90 nm-channel CNTFET. The arrows indicate SWCNTs. (c) E_a versus V_{GS} at V_{DS} = +30, +150 and + 1000 mV, respectively. Error bars are introduced due to linear fit of the Arrhenius plot for E_a estimation. Inset: log I_{DS} versus $1000/T$ plot at V_{DS} = +150 mV for V_{GS} = −20, −10, 0, +10, and +20 V, respectively. (d) E_a versus V_{DS} at V_{GS} = −10, 0, +10 and +20 V (from top to bottom), respectively. Inset: E_a versus V_{DS} at V_{GS} = −20 V (The large error bars are caused by the poor signal-to-noise ratio since the device is at "OFF" state with V_{GS} = −20 V.) [6].

The work function difference between the metal electrode and SWCNT results in a Schottky barrier (SB) at the contact.[7-9] Derycke *et al.* attributed the p-type conduction shown in a typical CNTFET to the adsorbed oxygen that modified the contact barriers.[10] Their experimental work evidenced that significant SBs existed at the CNT/metal contact. Heinze *et al.*[11] found that oxygen adsorption [the annealed (n-type) FET was exposed to oxygen for 2 min at P = 0 Torr (red), 10^{-4} Torr (orange), 5×10^{-4} Torr (light green), 5×10^{-3} Torr (dark green), 10^{-1} Torr (light blue), and in ambient (blue), from left to right respectively, as shown in Figure 5.5a] enhanced the hole conduction, but suppressed the electron conduction. However, the threshold voltage remained almost unchanged. In contrast, the potassium doping caused the p-to-n conduction type conversion (the curves from right to left correspond to increasing amounts of potassium, as depicted in Figure 5.5b). The threshold voltage of the CNTFET was found to change significantly. The origin of the differences is that the oxygen changes the work functions of the metal electrodes (especially Au), instead of doping the CNT channel like potassium deposition. As the SB height for holes is already reduced due to oxygen-induced metal work function change, the turn-on hole conductance is greatly enhanced. Similarly, the conductance for electrons is largely suppressed. Beyond the threshold voltage, the device is switched off due to the suppression of charge thermionic emission over the energy barrier in the center of the channel. Since the energy barrier is determined by the gate voltage and Fermi level of the CNT-channel and it does not depend on the SB height at the contact, the SB height variation at the CNT/metal contact due to

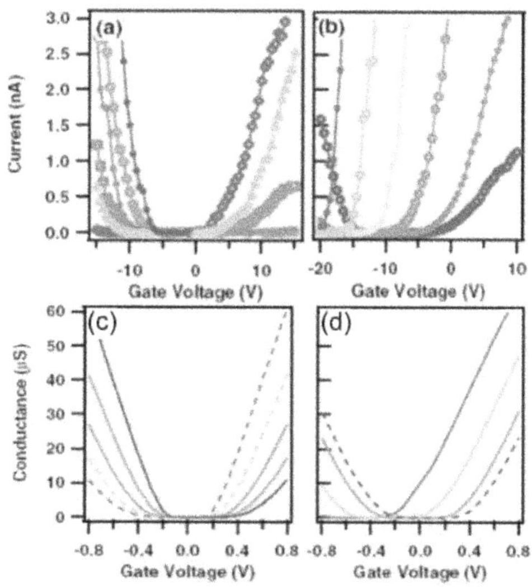

Figure 5.5 The experimentally measured effects of (a) oxygen adsorption [the annealed (n-type) FET was exposed to oxygen for 2 min at P = 0 Torr (red), 10^{-4} Torr (orange), 5 × 10^{-4} Torr (light green), 5 × 10^{-3} Torr (dark green), 10^{-1} Torr (light blue), and in ambient (blue), from left to right side, respectively.] (b) Potassium doping on CNTFETs (the curves from right to left side correspond to increasing amounts of potassium). (c) The calculated effects of metal work function variation [the work function of the metal electrode varied by −0.2 eV (red dashed), −0.1 eV (orange dashed), 0 eV (green), +0.1 eV (light blue), and +0.2 eV (blue), from left to right side, respectively.] (d) Effects of doping level change on CNTFETs (the doping atomic fraction is 10^{-3} (red), 5 × 10^{-4} (orange), and 10^{-4} (green), and 10^{-4} (blue dashed), from left to right side, respectively.)[10]. See also Color Insert.

oxygen adsorption could not modulate the threshold voltages. In contrast, potassium doping controlling the Fermi level position in the CNT-channel would alter the energy barrier in the center of the channel. Consequently, the threshold voltage is influenced by the degree of potassium doping.

Following the Schottky barrier transistor model, they calculated the channel current using Landauer formula as

$$I(V) = \frac{4e}{h}\int [F(E) - F(E + eV_D]P(E)dE \tag{5.1}$$

where e is the electron charge and h is the Planck constant. V_D is the drain voltage with respect to the source. $F(E)$ is the Fermi–Dirac distribution function. $P(E)$ is the electron tunneling probability through the Schottky contact. In order to describe the oxygen adsorption and potassium doping, they simulated two different cases (see Figure 5.5c,d): (1) The work function of the metal changes continuously (the work function of the metal electrode is changed by −0.2 eV (red dashed), −0.1 eV (orange dashed), 0 eV (green), +0.1 eV (light blue), and +0.2 eV (blue), from left to right side, respectively.) and (2) the doping level of the CNT varies successively [the doping atomic fraction is 10^{-3} (red), 5×10^{-4} (orange), and 10^{-4} (green), and 10^{-4} (blue dashed), from left to right side, respectively.] The work function of the metal enhances the turn-on conductance of one type of charge carrier and suppresses the other type of charge carrier. On the other hand, the doping level variation shifts the threshold voltage significantly (see Figure 5.5d). The simulation agreed well with the experimental observations, suggesting that (1) there is a significant SB existing at the CNT/metal contacts, and (2) oxygen modulates the metal work function, instead of the introducing doping into the CNT-channel.

The electron conduction mechanism in an SB-CNTFET could be tunneling or thermionic emission. Knowing the differences between these two mechanisms is crucial for understanding the operation of an SB-CNTFET. Appenzeller *et al.* suggested that tunneling and thermionic emission compete to dominate the transport mechanism in an SB-CNTFET.[12] Figure 5.6 depicts the simulated transfer curves [T = 300, 200, 150, 100, and 50 K (from top to bottom) at t_{ox} = 10 nm] of an SB-CNTFET with 0.3 eV SB height (ϕ_B = 0.3 meV). At negative gate voltages, the hole tunneling through the SB at the contact dominates the transport mechanism. On the contrary, thermionic emission is the main transport mechanism at more positive gate voltages and higher temperatures. This interpretation is supported by the I_d–$1/T$ plot shown in the inset (t_{ox} = 2 nm and ϕ_B = 150 meV at V_{gs} = −0.2, −0.1, and 0 V (from top to bottom)). Pure thermionic emission current can be formulated using

$$I_d \sim T^2 \exp\left(\frac{-q\phi_B}{k_B T}\right), \tag{5.2}$$

where T is the measurement temperature and K_B is the Boltzmann constant. From Equation 5.2, one can obtain

$$\ln(I_d) \sim 2\ln(T) + S\frac{1}{T}, \tag{5.3}$$

where $S = -q\phi_B/k_B$ is the slope of the $\ln(I_d)$ – $(1/T)$ curve. Thus, the straight dashed line suggests pure thermionic emission, while the flatter line indicates thermally assisted tunneling. The turning point (arrow pointing to the solid curve) separates the two temperature regions dominated by different transport mechanisms, respectively.

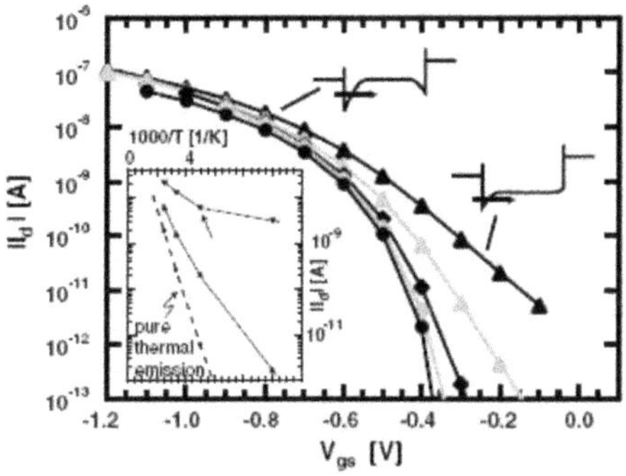

Figure 5.6 Simulated transfer characteristics of a SB-CNTFET at different temperatures (T = 300, 200, 150, 100, and 50 K from top to bottom) at t_{ox} = 10 nm and ϕ_B = 300 meV. Left lower inset: $|I_d|$–$1/T$ plot of another simulation with t_{ox} = 2 nm and ϕ_B = 150 meV at V_{gs} = −0.2, −0.1, and 0 V (from top to bottom). Right upper inset: schematic energy band diagrams at different V_{gs} [11].

With Equation 5.3, one can extract the barrier height ϕ_B from the slope of Arrhenius plot. Figure 5.7 summarizes the dependence of the extracted SB height on the back-gate voltage V_{gs} for CNTFETs with various oxide thickness t_{ox}, CNT diameter t_s and effective electron mass m^* [curve (i) with m^* = 0.1 m_0, t_{ox} = 2 nm, and t_s = 1.4 nm, curve (ii) at m^* = 1.0 m_0, t_{ox} = 2 nm, and t_s = 1.4 nm, curve (iii) with m^* = 1.0

m_0, t_{ox} = 10 nm, and t_s = 1.4 nm, and curve (iv) with m^* = 1.0 m_0, t_{ox} = 10 nm, and t_s = 5 nm]. What interests us is the meaning of different regions in the graph. All of the data points are calculated with assumption of ϕ_B = 150 meV. However, the extracted barrier height using the aforementioned method is not equal to 150 meV except for flat-band condition at point B, where V_{gs} = 0 V. In other words, the extracted barrier height is "true" at point B. Nevertheless, those data points in the shade area are with underestimated barrier height. The reason for underestimation is shown in the right upper inset. At negative V_{gs}, the energy bands shift upward, resulting in a thinner barrier for holes to tunnel through. Consequently, both tunneling and thermionic emission contribute to the measured current. In this case, the estimated barrier from the Arrhenius plot is smaller than the true barrier because the tunneling current has weaker temperature dependence than thermionic emission current. From A to B, there is no dependence of ϕ_B on V_{gs}, and true barrier height can be extracted from the thermal emission theory. From point B to C, any amount of change in V_{gs} leads to the same amount of variation in extracted ϕ_B. The reason is that only thermionic emission current exists. Moreover, there is negligible carrier accumulation in the channel. Thus, increase in V_{gs} directly changes into the same amount of increase in ϕ_B, suggesting that the slope of the line between B and C is unity. On the unity slope line, the extracted barrier height is true and V_{gs} corresponds to the flat-band voltage. The line from A to B to C indicates the thermal emission limit, which could be reached by increasing the gate oxide thickness, the semiconductor-channel size, or the effective carrier mass.

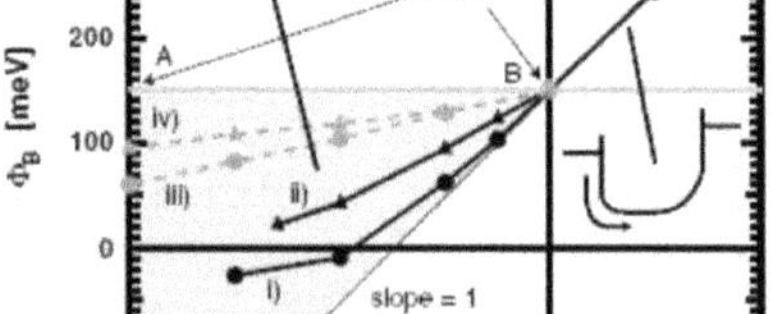

Figure 5.7 Calculated SB height versus V_{gs} for CNTFETs with various oxide thickness t_{ox}, CNT diameter t_s and effective electron mass m^*. (i) m^* = 0.1 m_0, t_{ox} = 2 nm, and t_s = 1.4 nm, (ii) m^* = 1.0 m_0, t_{ox} = 2 nm, and t_s = 1.4 nm, (iii) m^* = 1.0 m_0, t_{ox} = 10 nm, and t_s = 1.4 nm, (iv) m^* = 1.0 m_0, t_{ox} = 10 nm, and t_s = 5 nm [11].

5.1.2 Ohmic-Contact CNTFETs

The CNTFETs with Schottky contacts[13–16] show high contact resistances and small on-state conductance. Thus, improved contacts are favorable for most of the electronic applications. Various approaches in the ultimate goal of creating ohmic contact to SWCNT have been reported to be efficient in lowering the contact resistance.[17] Javey *et al.* reported the ohmic-contact CNTFET, and hence achieved the ballistic transport in the transistors.[18] Figure 5.8a shows scanning electron microscopy (SEM) and atomic force microscopy (AFM) images of two typical back-gated CNTFETs fabricated on the same SWCNT channel. For a long-channel (3 μm) CNTFET, the on-state conductance G_{ON} is well below the ballistic transport limit $4e^2/h$.[19] The room temperature on-state resistance is 60 KΩ, nearly one order of magnitude higher than the resistance of a ballistic m-SWCNT, 6.5 KΩ.[19,20] The on-state conductance exhibits metallic behavior (increase with decreasing temperature) between room temperature and

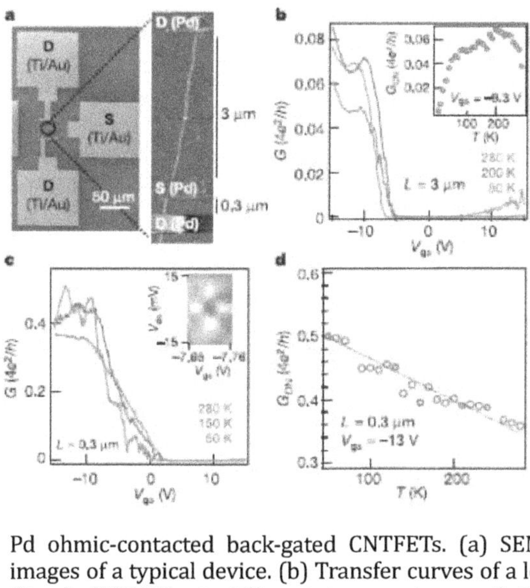

Figure 5.8 Pd ohmic-contacted back-gated CNTFETs. (a) SEM and AFM images of a typical device. (b) Transfer curves of a long-channel (3μm) CNTFET at different temperatures of 280 (green), 200 (blue) and 50 K (cyan), respectively. Inset: temperature-dependent on-state conductance at a fixed V_{gs} = –9.3 V. (c) Transfer curves of a short-channel (300 nm) CNTFET at various temperatures. Inset: differential conductance in V_{ds}–V_{gs} plane at 1.5 K. (d) Temperature-dependent (down to 50 K) on-state conductance of the short-channel CNTFET at fixed V_{gs} = –13 V [17]. See also Color Insert.

200 K, and then decreases upon further cooling (inset of Figure 5.8b). On the contrary, the on-state conductance of a short-channel (300 nm) CNTFET monotonically increases with decreasing temperature down to 50 K (Figure 5.8d). Interestingly, the device shows significant oscillation in conductance below 50 K. The conductance oscillation is caused by the Fabry–Perot interference of electron wave in the nanotube channel between the two contacts.[20] As depicted in the inset of Figure 5.8c, the color-coded conductance in the V_{ds}–V_{gs} plane shows Fabry–Perot interference-like pattern, where the bright and dark area correspond to $4e^2/h$ and $0.5 \times 4e^2/h$, respectively. This is evidence for ballistic hole transport in the SWCNT channel.

The SBs are sensitive to the contact metal work function due to the negligible Fermi level pinning effect at SWCNT/metal contacts as discussed previously.[2] Javey *et. al.* demonstrated the influences of SB variation on the electrical characterization of the CNTFET. Hydrogen exposure would decrease the work function of Pd electrode, resulting in higher SB height for holes and lower SB height for electrons. As a consequence, the p-type conductance decreases while the n-type conductance increases (Figure 5.9a). The corresponding energy

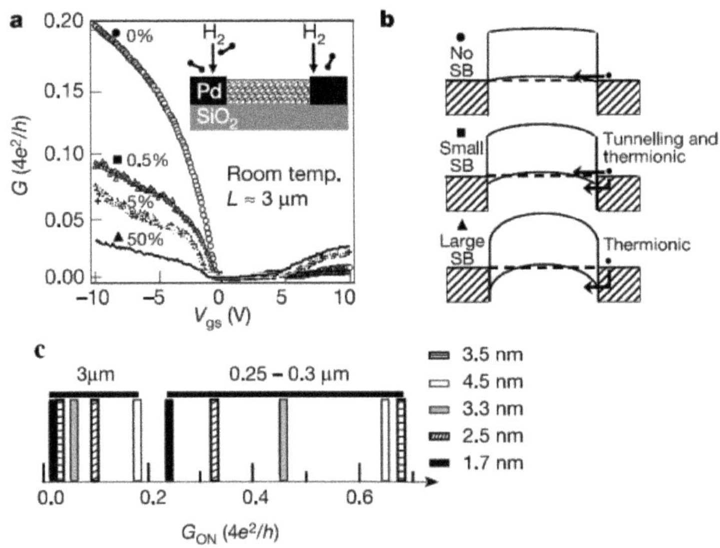

Figure 5.9 (a) Influences of hydrogen exposure on the SB height at the Pd/SWCNT contact (increased concentration of hydrogen from top to bottom). (b) Schematic energy band alignment at the Pd/SWCNT contact with different hydrogen concentrations. (c) Diameter-dependent on-state conductance [17].

band alignment is shown in Figure 5.9b. As discussed in the previous section, both tunneling and thermionic emission contribute to the hole transport if there are non-negligible SBs at the SWCNT/metal contacts. In comparison with Pt (work function of 5.7 eV), Pd has a lower work function of 5.1 eV. Thus, Pt should result in a negative barrier height for hole transport. Nevertheless, the author found that Pt/SWCNT contact shows non-ohmic behavior and lower on-state conductance than that of Pd/SWCNT contact experimentally. Thus a high work function of metal electrode is not necessary to achieve transparent contact for hole conduction in SWCNTs. The better contact between Pd and SWCNT is attributed to the better "wetting" interaction between them.[21] Moreover, the contact barrier also depends on the diameter of SWCNTs. For SWCNTs with diameter $d< 2$ nm, complete elimination of SBs is difficult.

As a summary, one can use the following three methods to fabricate ohmic-contact p-type CNTFETs:[22] (1) Pd as contact metal,[23–25] (2) short SWCNT-channel, and (3) thicker SWCNT with diameter > 2 nm typically. In addition, the transport in short channel CNTFET (less than 300 nm) is close to ballistic for all SWCNTs with various diameters.

5.1.3 CNTFET-Based Logic Circuits

In order to implement CNTFETs in logic circuits, top-gated geometry is necessary.[26–28] The first CNTFET-based logic circuit was demonstrated by Bachtold *et al.*[29] Aluminum passivated by alumina was used as the back-gate electrode. Due to the thin oxide layer (<10 nm), the gate coupling was significantly increased, resulting in effective switching on and off the CNTFET within very small back-gate voltage sweeping (±2 V). Thus, the gain of the inverter could reach up to 6. The inverter was constructed by connecting an off-chip 100 MΩ bias resistor and a p-type CNTFET in series. When the transistor was turned on with on-state resistance of 26 MΩ, the output voltage was about 20% of the bias voltage V_{DD}, corresponding to logical 0. When the transistor was switched off with off-state resistance of 1 GΩ, the output voltage was about 90% V_{DD}, resulting in the logical 1. Similarly, a logic NOR gate, static random access memory (SRAM) and a ring oscillator (3-stage inverters) were formed as shown in Figure 5.10. Two disadvantages of such logic circuits are as follows: (1) Constant power consumption by the bias resistor is always present. (2) It is difficult to scale down the logic circuit due to the size limitation of the off-chip bias resistor.

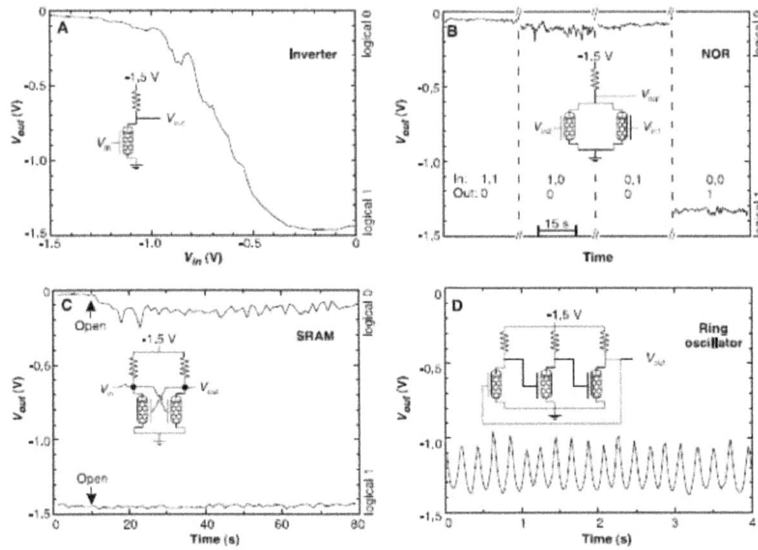

Figure 5.10 Output voltage versus the input voltage of an (a) inverter (the resistance is 100 MΩ). Output voltage versus time of a (b) NOR logic gate (the resistance is 50 MΩ), (c) SRAM (the resistance is 100 MΩ) and (d) ring oscillator (the three resistance are 100 MΩ, 100 MΩ, and 2 GΩ, respectively). Insets: schematics of the electronic circuits [28].

In order to avoid the disadvantages of the aforementioned logic circuits, n-type CNTFETs have to be fabricated. The first report on n-type doping of SWCNT was achieved by potassium doping, as shown in Figure 5.11a (The FET is initially p-type (curves 1–7). After seven doping cycles, no more current could be detected (open circles 8, 9). At higher doping levels, the device became n-type (curves 10-12)).[30] In addition to n-type doping, annealing in vacuum or inert gas also results in n-type conduction in SWCNT. Derycke *et al.* compared these two methods for n-type CNTFETs.[10] As shown in Figure 5.11b, the CNTFET showed p-type conduction in air and n-type conduction in vacuum. Moreover, the p-type conduction restored gradually when the device was exposed to oxygen. The origin of p-to-n conversion was the desorption of oxygen,[31,32] resulting in the modification of Schottky barrier. The main modification was due to shift of the work functions of the metal electrodes, instead of doping in the SWCNT, as discussed previously.

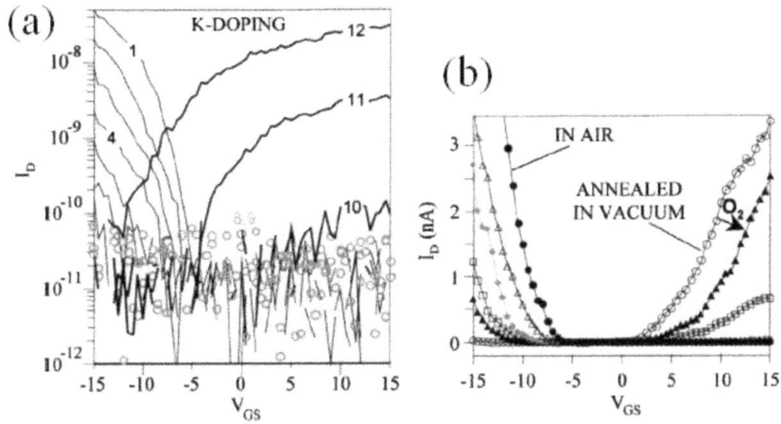

Figure 5.11 (a) Effect of various potassium doping levels on the transfer curves of a CNTFET. The transistor is initially p-type (curves 1–7). After seven doping cycles, no current is detectable (open circles 8, 9). At higher doping levels, the device becomes n-type (curves 10–12). (b) Effects of different oxygen concentrations on the transfer curves of a CNTFET. An annealed (n-type) FET has been exposed to oxygen for 2 min at $P = 0$ Torr (red), 10^{-4} Torr (orange), 5×10^{-4} Torr (light green), 5×10^{-3} Torr (dark green), 10^{-1} Torr (light blue), and in ambient (blue), from left to right side, respectively [9].

This method is not recommended for large-scale fabrication since the exposure to air will cause severe degradation of n-type conduction. A more reliable method is potassium doping, as shown in Figure 5.11a. In drastic contrast to oxygen exposure, the potassium doping changes the threshold voltage significantly. As discussed before, the origin behind this observation is believed to be the doping level variation, instead of modulation of energy band alignment at the metal/SWCNT junction (see Figure 5.12).

Another interesting method for fabricating n-type CNTFET is using low-work-function metals as contact electrodes.[33–35] Scandium (Sc) has been demonstrated to contact perfectly with SWCNT, resulting in ballistic electron transport in SWCNT.[36] Figure 5.13a,b shows the schematic and SEM image of a typical Sc-contacted CNTFET fabricated in a self-aligned manner. The SWCNT that spans the channel is with the length of 120 nm and diameter is 1.5 nm. The output characteristic of the device shows that the room temperature conductance can reach $0.3\ ' \ 4e^2/h$ and the saturation current can be as high as 25 μA (see Figure 5.13c). The maximum conductance and saturation current are close to those reported for that of p-type CNTFETs with ohmic contacts.[18] As shown in Figure 5.13d, the subthreshold swing S is 100 mV/dec with 15 nm-thick HfO_2. The other method to achieve n-type CNTFET include dielectric passivation,[37–39] chemical doping[40,41], etc.

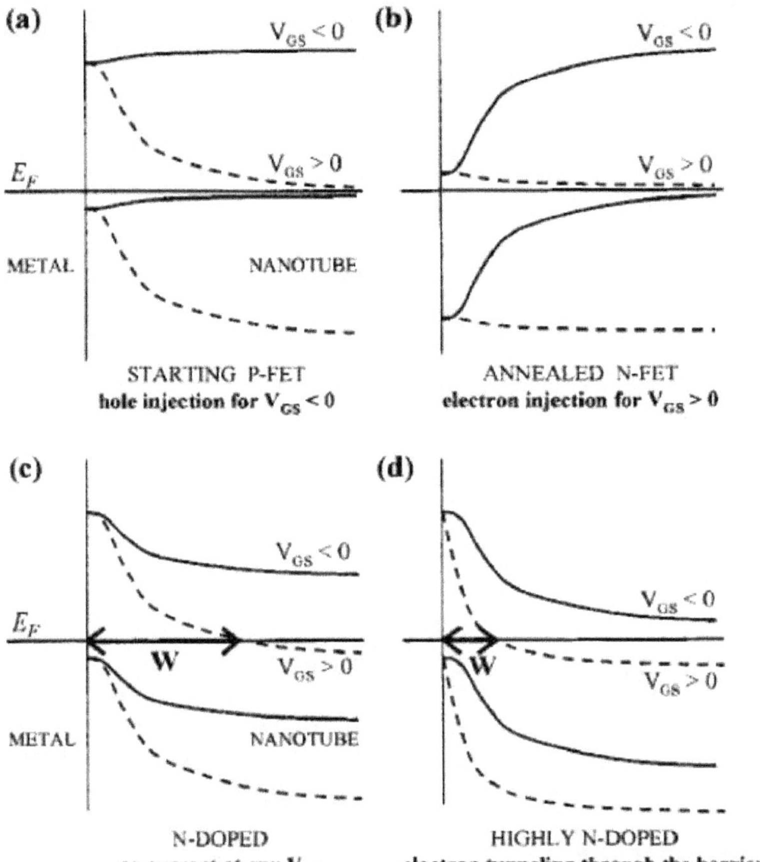

Figure 5.12 Schematic energy band diagrams at the metal/SWCNT contact in the absence of bias. (a) Energy bands at the contact of a p-type device in air, (b) Energy bands near the contact of the device after annealing in vacuum. (c) The energy bands at the contact of the device with medium doping level (curves 8 and 9 in Figure 5.11b). (d) The energy bands near the contact of the device with a high doping level [9].

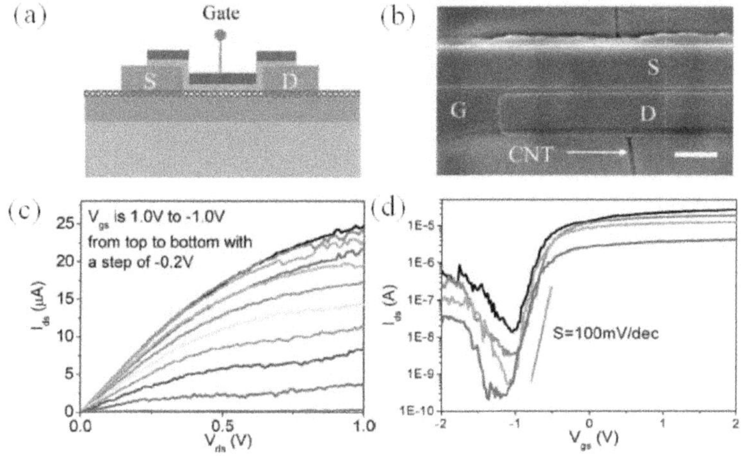

Figure 5.13 (a) Schematic of a self-aligned Sc-contacted CNTFET. (b) SEM image of a Sc-contact CNTFET. The scale bar is 1 μm. (c) Output characteristics of the device with V_{gs} ranging from 1.0 (top) to −1.0 V (bottom) with step of −0.2 V. (d) Transfer characteristics of a typical CNTFET with channel length of 120 nm and SWCNT diameter of 1.5 nm [35]. See also Color Insert.

Javey *et al.* integrated high-κ dielectric ZrO_2 into the CNTFETs and improved the performance of the logic gates significantly.[42] Figure 5.14a,b shows that the CNTFET has two gates, i.e., the global Si back-gate and the metallic partial top-gate. Thus, the device can be controlled through the CNT channel or the SB between the CNT/metal pads.[41,43,44] The SWCNT channel was buried between the ZrO_2 and SiO_2 layer, as illustrated in the transmission electron microscopy (TEM) images in Figure 5.14c,d. With the 8 nm-thick ZrO_2 as the gate dielectric, the electrostatic capacitance is $C_e \approx 2\pi\varepsilon_r\varepsilon_0 L/\ln(2h/r)$ = 1.1 fF, where ε_r = 25, h = 8 nm, and r = 1 nm are the dielectric constant of the ZrO_2 layer, length of the SWCNT channel, thickness of the ZrO_2 layer, and radius of the SWCNT-channel, respectively. The electrostatic capacitance is connected in series with the quantum capacitance $C_q \approx$ 0.8 fF, and the overall capacitance was 0.46 fF. With the improved gate capacitance, the subthreshold swing could be improved to 60–80 mV/dec and 90–100 mV/dec for p-type and n-type conduction, respectively.

As shown in Figure 5.15a, an inverter was formed by connecting a p-type and an n-type CNTFET. The n-type CNTFET was formed by

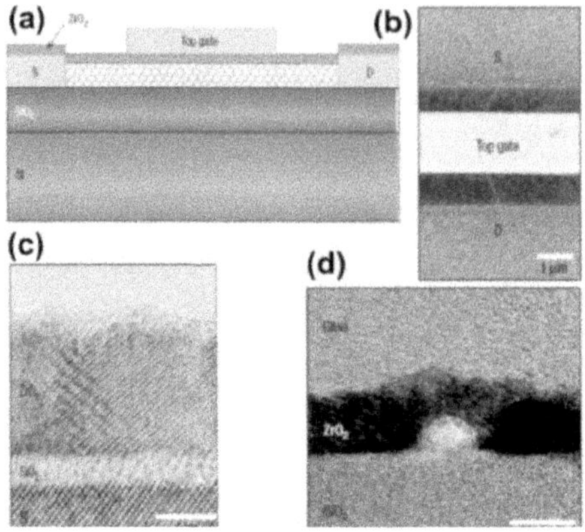

Figure 5.14 CNTFETs with high-κ gate dielectrics. (a) Schematic a top-gated CNTFET. (b) SEM image of the device. The scale bar is 1 μm. (c) and (d) are cross-sectional TEM images of the device. The scale bars are 5 nm [41].

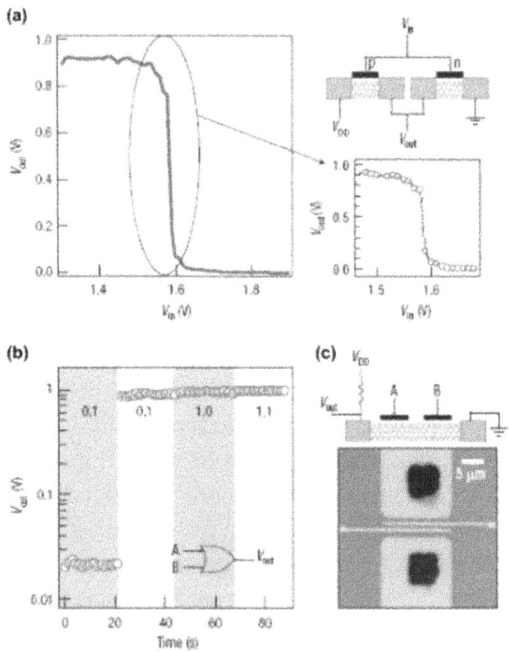

Figure 5.15 CNT-based logic gates. (a) Output characteristics of an inverter. Inset: schematic device structure and enlarged view of the transfer curve. (b) Output characteristics of an OR gate. Inset: symbol of an OR gate. (c) Schematic device structure and optical image of the device [41].

annealing the device in hydrogen at 400°C for 1 hour. Surprisingly, the voltage gain of the inverter could reach up to 60. The origin of the high voltage gain was a result of the high transconductance of the CNTFETs. This work demonstrated the high potential for SWCNT based logic circuits. In addition, a logic OR gate was demonstrated on a single SWCNT. Following this pioneer work, many efforts have been devoted to the application of SWCNT in logic circuits.[45]

5.1.4 Performance Limit of CNTFETs

The performance limit of a CNTFET is an interesting question that has been intensively discussed over recent years. Many efforts have been devoted to increase the gate coupling, reduce channel length, and improve subthreshold swing. A novel method to increase gate capacitance is to employ liquid gate in an electrochemical device configuration. The electrochemical CNTFET containing a MWCNT was first demonstrated by Krüger *et al.*[46] in 2001. Within a year, Rosenblatt *et al.* reported electrolyte-gated CNTFET based on a single s-SWCNT.[47] The optical and AFM images of the devices are shown in Figure 5.16a,b, respectively. As illustrated in Figure 5.16c, the gate voltage was applied to the silver (Ag) wire in the micropipette. A gate sweep between –0.9 V and 0.9 V was controlled to ensure there is negligible gate leakage current. An ion layer was formed on the surface of the SWCNT-channel in the NaCl solution under applied gate

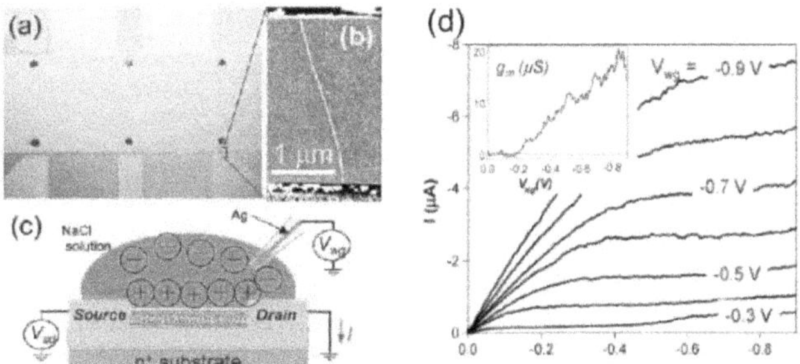

Figure 5.16 (a) Optical micrograph of the CNTFETs. (b) AFM image of a CNTFET. (c) Schematic setup of the measurement. A liquid gate voltage V_{wg} is applied to droplet through a thin silver wire. (d) Output characteristic of the CNTFET. The liquid gate voltage is in the range from –0.9 V to –0.3 V in 0.1 V steps (from top to bottom). Inset: the transconductance measured at V_{sd} = –0.8 V [46].

bias. The distance between this ionic layer and SWCNT-channel was in nanoscale range. Hence, the gate coupling was greatly enhanced in comparison with that of the back-gate or top-gate (solid).[48] As depicted in Figure 5.16d, the CNTFET can be effectively switched on and off when the gate voltage is varied between –0.3 and –0.9 V.

The increase in the current carrying capability requires the channel length to be as short as possible. In 2004, Javey *et al.* demonstrated a series of CNTFETs with short channels down to 10 nm, as shown in Figure 5.17a.[49] Previous reports have shown that the channel current saturates at 25 μA for a single SWCNT due to electron backscattering caused by optical or zone-boundary phonon emission.[18,50] It was found that the mean-free path of acoustic and optical phonon scattering is l_{ap} = 300 and l_{op}=15 nm, respectively. As a result, the low-bias conductance of an m-SWCNT shorter than 300 nm should be

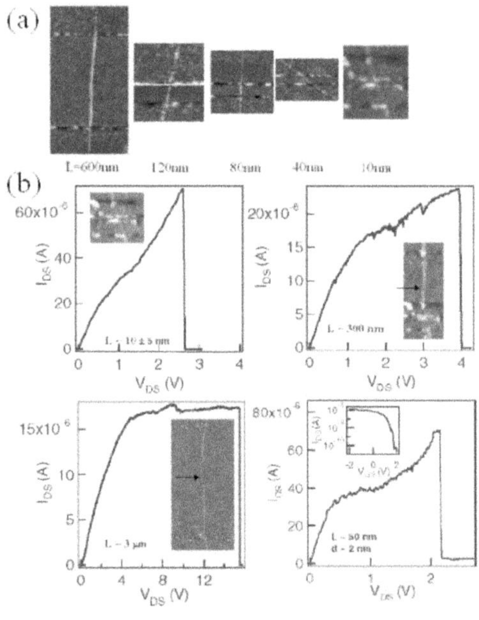

Figure 5.17 (a) AFM images of five CNTFETs that consist of individual SWCNTs with various channel length ranging from L = 600 to 10 nm. (b) Breakdown characteristics of CNTFETs with various channel lengths. The current channels of the first three samples are three sections of the same m-SWCNT with length of ~ 10 nm, ~300 nm and ~ 3μm, respectively. The back-gate oxide is 67 nm. The insets are the AFM image of the devices. The last sample has an s-SWCNT with length of 50 nm and the back-gate oxide of 10 nm. Insets: transfer curve of the device [48].

ballistic and independent of temperature. As depicted in Figure 5.17b, a 10 nm-long m-SWCNT was broken down at a bias of $V_{DS} = 2.6$ V (gate oxide thickness of 67 nm). A 300 nm-long segment from the same m-SWCNT was broken down at $V_{DS} = 4$ V and current $I_{DS} = 25$ μA, while a 3-μm-long segment was broken down at $V_{DS} = 13.8$ V and current $I_{DS} = 18$ μA. The flattening appearing before the breakdown point was caused by optical phonon scattering. Thus, the optical phonon scattering was not effective in the short-channel CNTFET ($L = 10$ nm), but it was very effective in the longer-channel CNTFETs ($L = 300$ nm and 3 μm). The reason is that the short-channel CNTFET is free of optical phonon scattering. Another notable feature in the breakdown characteristics was the drastic current increment before breakdown in the short-channel CNTFET at $V_{DS} = 1.5$ V. This was attributed to the conduction contrition from the second subband of the m-SWCNT. Similar observation was found in a 50 nm-long s-SWCNT (with oxide thickness of 10 nm). The mechanism of breakdown can be explained by high-field impact ionization ($E \sim 10^6$ V/cm).

A theoretical prediction shows that nearly ideal subthreshold slopes of ~63 mV/dec and maximum on/off ratios of 2.2×10^6 can be achieved in a 2 nm gate length CNTFET.[51] The simulated device was a coaxially gated, zero-SB CNTFET (for electron injection), as shown in Figure 5.18a. The source and drain underlaped, leading to a very low gate capacitance that resulted in Coulomb blockade of the ambipolar leakage current. The Coulomb blockade effect shifted the threshold voltage of the CNTFET because the hole-leakage current influenced. The threshold voltage of the CNTFET with $L = 30$ nm was the most negative, because it had the biggest source/drain underlap. Thus, the total gate capacitance was smaller as the fringing field coming from gate to source and gate to drain was weaker. Moreover, the tunneling barrier thicknesses for hole injection increased when the SWCNT length increased. As a result, hole leakage was suppressed, leading to a lower subthreshold swing and a higher current on/off ratio, but a lower on-state current. With 50 nm-long SWCNT, the subthreshold swing was reduced to 63 mV/dec and the current on/off ratio was increased to 2.2×10^6. The short-channel effect was notable in the device. The finite slope of the I_{DS}-V_{DS} curve in the saturation region was partially caused by the drain-induce barrier lowering (DIBL), though it was not significant. As displayed in the inset of Figure 5.18c, increasing V_{DS} from 0.2 (solid line) to 0.4 V (dashed line) would narrow the tunneling barrier, causing a tunneling current increase. Over the past few years, there have been continuous efforts on the fabrication and characterization of short-channel CNTFETs.[49,52,53] The most important conclusion is that the short-channel effect in the CNTFET is much less significant than the transistors based on other materials.

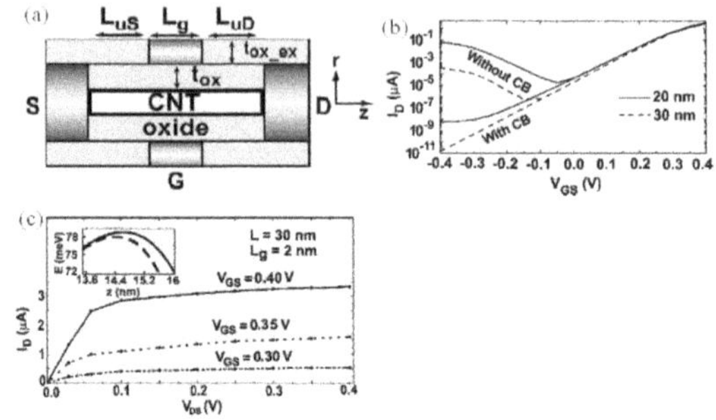

Figure 5.18 (a) Cross-sectional view of the CNTFET used in the simulation. The gate dielectric (ZrO_2) thickness is fixed at t_{ox} = 2 nm. (b) Transfer characteristics of a CNTFET with gate length L_g = 2 nm and V_{DS} = 0.4 V, and the SWCNT length of L = 20 and 30 nm, respectively. (c) Output characteristics of a CNTFET with L = 30 nm and L_g = 2 nm. Inset: conduction band profile at V_{GS} = 0.3 V and V_{DS} = 0.2 V (solid line) and 0.4 V (dashed line), respectively [50].

The limit of subthreshold swing S at room temperature in a conventional MOSFET is $k_B T \ln 10/q \approx 60$ mV/dec by scaling down the gate dielectric thickness and using high-κ materials. To beat this limit, a device having different switching mechanism is necessary. The first CNTFET involving band-to-band (BTB) tunneling showed S = 40 mV/dec.[54] As shown in Figure 5.19a, the gate dielectric of the CNTFET was about 4 nm-thick native aluminum, which ensured the extremely good gate coupling. Another important feature is that width of the gate is only 200 nm. The transfer curves of the device consisted of the p-type conduction branch (A) and the BTB tunneling conduction branch (B), as depicted in Figure 5.19b. Interestingly, the subthreshold swing S for the A branch was 65 mV/dec while that for the B branch was 40 mV/dec. The corresponding energy band diagrams are illustrated below the transfer curves. The A branch is interpreted by a thermal emission current as in conventional MOSFET. In contrast, the B branch can be understood by a bandpass-filter-like operation of the BTB tunneling. Because of the small transition region width (between the dash lines in Figure 5.19b), the BTB tunneling occurred easily in the CNTFET. Thanks to the small diameter of the SWCNT-channel and strong gate coupling, the tunneling barrier for holes to overcome before entering the conduction band was as thin as a few nanometers. This made it easy to achieve the BTB tunneling in CNTFETs. Simulations based on the aforementioned mechanisms gave results consistent with the experimental observation, as shown in Figure 5.19b. Following the first experimental demonstration, the study on this type of tunneling devices was furthered by many theoretical and experimental works.[44,55–58]

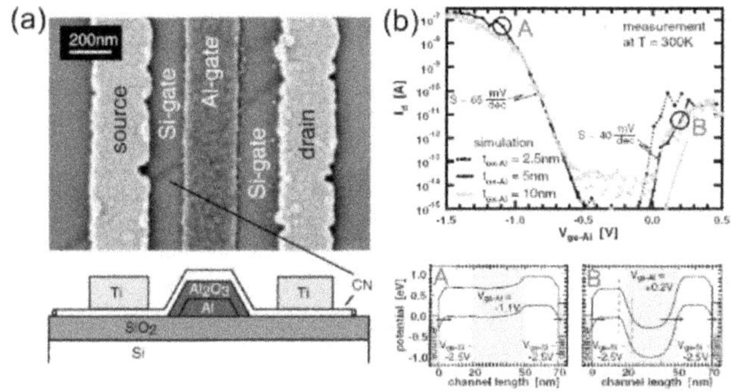

Figure 5.19 (a) The SEM image and schematic of a BTB tunneling CNTFET. (b) Experimental (gray solid circles) and simulated (symbol and line) transfer characteristics (V_{ds} = −0.5 V and $V_{gs\text{-}Si}$ = −3 V), and the corresponding energy band diagrams of the BTB tunneling CNTFET at point A and B of the transfer curves, respectively [53].

5.2 Single-Electron Transistors and Circuits

Single-electron/hole devices are very promising for future nanoelectronic devices with low power consumption, fast speed and ultra-high density integration. Various material systems such as Si/SiO_2[59,60] and GaAs[61] have been employed to fabricate single electron transistors (SETs). As shown in Figure 5.20, a typical SET consists of a quantum dot (QD) connected to the source and drain through tunneling capacitors (C_1 and C_2) and to the gate through a normal capacitor (C_g). The QD is so small that its energy levels become discrete below a certain temperature. Any potential variation will cause an electrostatic energy change in the QD. Electrons can only hop between the QD and source (drain) electrode, rather than the gate electrode. The total electrostatic energy of a QD with N electrons is calculated as $Q^2/2(2C) = (Ne)^2/(2C)$, where C is the total capacitance of the QD, which is the sum of the dot capacitance to source (C_1), drain (C_2) and gate (C_G), i.e., $C = C_1 + C_2 + C_G$. One regime that allows a detailed study of electron transport through an SET is the sequential tunneling limit, where electric current will flow when an electron can tunnel onto the QD from the source and subsequently off from it to the drain. When a state is available between the Fermi levels of the source electrode μ_S and drain electrode μ_D, the sequential tunneling process could occur through this state while the number of electrons on the QD keeps changing between N and $N + 1$. In this circumstance, a large current will flow and the device is at "ON" state. Otherwise, a very little current (due to a direct tunneling through the device if possible) may flow if there is no available state situated in between μ_S and μ_D. The total energy is obtained by

$$U(N+1)=\sum_{i=1}^{N} E_i + \frac{(Ne)^2}{2C},$$

where E_i is the chemical potential of the dot with i electrons, i.e., the energy of the orbital of the dot that the i-th electron would occupy if there were negligible electron–electron interaction, when N electrons are on the QD. The electrochemical potential, that is the minimum energy required for adding the N-th electron, is $\mu_N = U(N) - U(N-1) = E_N + (N - 1/2e)e^2/C$. Thus, if one more electron is added onto the QD, $\mu_{N+1} = \mu_N + e^2/C + \Delta E$ is needed and $\Delta E = E_{N+1} - E_N$ is a consequence of the quantized excitation spectrum of the dot. A single electron that tunnels onto the QD has to overcome the charge addition energy barrier of $E_{add} = E_C + \Delta E$, where $E_C = e^2/C$ is the charging energy.

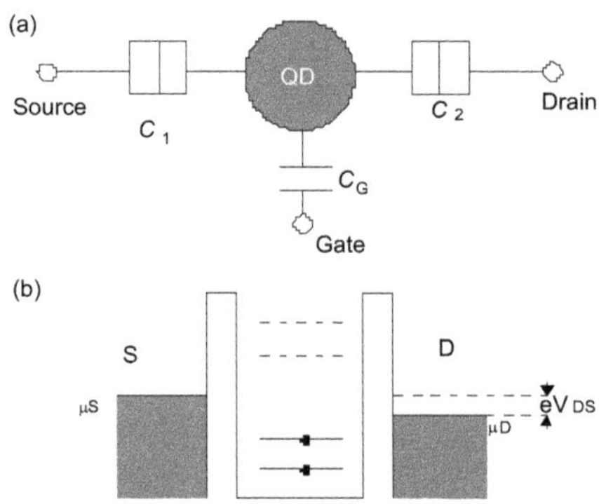

Figure 5.20 (a) A schematic of a single-electron transistor and its parameters. C_1 and C_2 are the tunneling capacitors, and C_g is the gate capacitor. (b) The corresponding energy band diagram of the device under bias voltage V_{DS}. Two unoccupied and two occupied energy states are illustrated in the figure.

In order to ensure the Coulomb blockade to be clearly observed, the lifetime (Δt) of an electron on the QD has to be long enough. This requirement means the tunnel capacitor (could be viewed as a capacitor connected to a resistor in parallel) must be large enough to prevent electron escaping from the QD. The lifetime Δt could be estimated by using RC-time constant: $\Delta t = RC$, where R is the resistance between the QD and source or drain electrode and C is the total capacitance to the QD. According Heisenberg's uncertainty relation, the lifetime must fulfill $\Delta E.\Delta t = \frac{e^2}{C}.RC > h$ in order to measure an energy difference of $\Delta E = E_C$. We thus get two necessary and sufficient conditions for observing the Coulomb blockade: (1) $R > \frac{h}{e^2} \sim 25.8\ \text{k}\Omega$, and (2) $E_C + \Delta E >> k_B T$.

5.2.1 SWCNT-Based SETs

An SWCNT has apparent inherent advantages over other materials for SET fabrication because of its tiny volume. The single-electron charging characteristics of an m-SWCNT were first reported in 1997.[62–64] As shown in Figure 5.21a, the CNT-based SET (CNTSET) consists of a 200 nm-long tube bundle as the "QD." The linear I–V curves at high temperatures become significantly nonlinear below 10 K. A conductance gap is observed at low temperatures, a manifest of Coulomb blockade effects. Interestingly, the transfer curve consists of a series of conductance oscillation peaks, which are separated by low-conductance regions. The period of the conductance peaks depends on the amount of energy required for the adding of one more electron to the dot. Moreover, Figure 5.21b shows that the peak amplitude increases as temperature decreases, and the peak width increases linearly with temperature T.

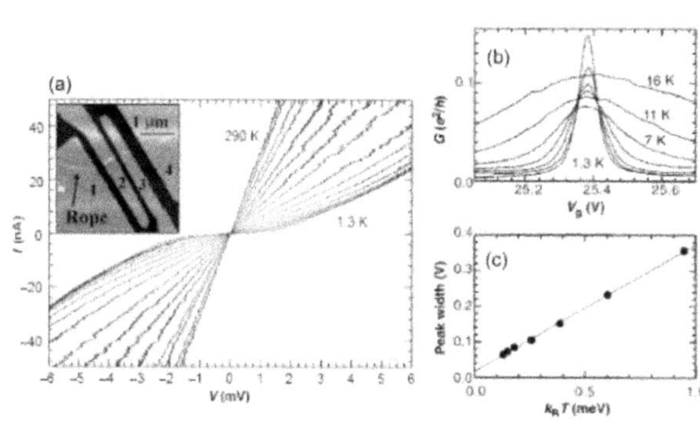

Figure 5.21 (a) Temperature-dependent I-V characteristics of a CNTSET from 1.3 to 290 K. Inset: The AFM image of the device. (b) Temperature dependence of a conductance oscillation peak from 1.3 to 16 K. (c) Width of the conductance oscillation peak as a function of temperature T [61].

The conductance oscillation (Coulomb oscillation) could be interpreted from the energy diagram shown in the inset of Figure 5.22. In the QD filled with N ($N = 2$ here) electrons, there is an energy gap $U + \Delta E$ for adding the (N +1)th electron into the QD. Above this energy gap, additional excited single-particle states of the dot correspond to adding the (N + 1)th electron to one of the states. At the conductance peak, the energy of the lowest empty state aligns with the Fermi level in the electrodes, and individual electrons can tunnel on and off the QD (point A). At the conductance valley, tunneling is suppressed because of the charging energy U (point B). When the source-drain bias V is increased so that the Fermi level of the right electrode moves below the energy of the highest filled state (the 3rd level here), an electron can tunnel on and off the QD again, leading to another conductance peak (point C). Further increasing the gate voltage would allow electrons residing in additional states (the second level here) to tunnel on and off the QD (point D). Within the extended CB model, the observed increased amplitude of the conductance peak with decreasing temperature means that the energy level separation is much larger than the thermal energy, i.e., $\Delta E >> k_B T$.

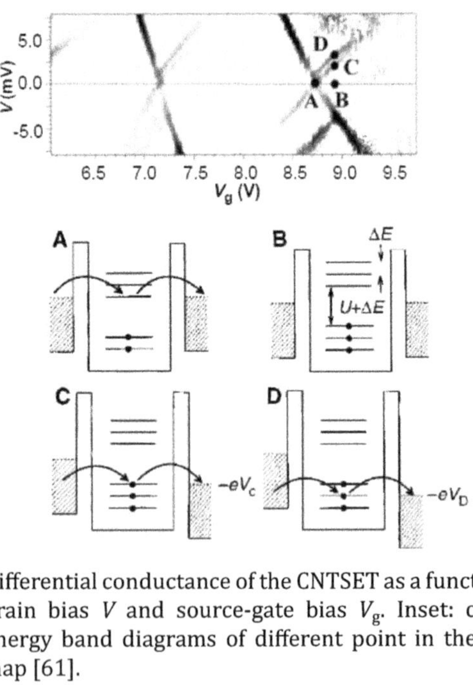

Figure 5.22 Differential conductance of the CNTSET as a function of source-drain bias V and source-gate bias V_g. Inset: corresponding energy band diagrams of different point in the conductance map [61].

For a real application of CNTSETs, the device must be workable at room temperatures. Postma *et al.* demonstrated a room-temperature CNTSET using a small m-SWCNT QD (~25 nm in length).[63] Figure 5.23a–f show the fabrication process of the CNTSET. The resistance of the m-SWCNT between two Au electrodes is about 50 KΩ. After imaged in tapping mode, the CNT is manipulated by the AFM tip that is firmly pressed against the substrate. Afterward, the AFM tip moves along the path indicated by the arrows in Figure 5.23a,b sequentially, a double-buckled m-SWCNT is formed, as depicted in Figure 5.23c,d. The QD is about 25 nm in length with a charging energy about 120 meV, much higher than the thermal energy at room temperature (26 meV). Thus, Coulomb oscillations and "diamonds" can be observed at 260 K, as shown in Figure 5.23e. The Coulomb oscillations persist until near room temperature, as depicted in Figure 5.23f.

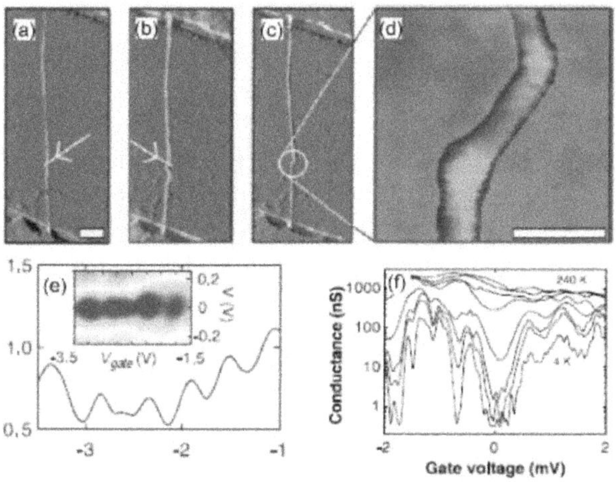

Figure 5.23 (a) to (d) AFM images of the SWCNT manipulated by an AFM tip to form a 25 nm-long QD. (e) Coulomb oscillation at 260 K. Inset: dI/dV vs. V_{gate} and V. (f) Temperature dependent Coulomb oscillations from 4 to 240 K [62].

Single-electron charging effects observed in CNTs can be divided into two categories: (1) classic regime ($\Delta E < k_BT$)[65,66] and (2) "quantum dot" regime ($\Delta E < k_BT$).[62,64,67–72] A detailed analysis of an example of "quantum dot" regime is shown in Figure 5.24.[72] The Coulomb oscillations were obtained below 50 K, implying that the tunneling resistance was significant and the SWCNT-rope was segregated into a "quantum dot." Figure 5.24b shows a single peak from (a) after zoom in. The height of the peak, plotted in

Figure 5.24c approaches $1/T$ dependence as T decreases while the area under it that is plotted in Figure 5.24d approaches a constant. These observations indicated that the transport occurred via one specific quantum transition for each Coulomb blockade peak. The observed peak sizes and spacings were randomly distributed, implying that there were multiple quantum dots involved in the transport. These quantum dots might be formed in parallel by several conducting tubes within a rope. On the other hand, the quantum dots could be formed in series within a single tube that is electrically broken up into segments by disorder.

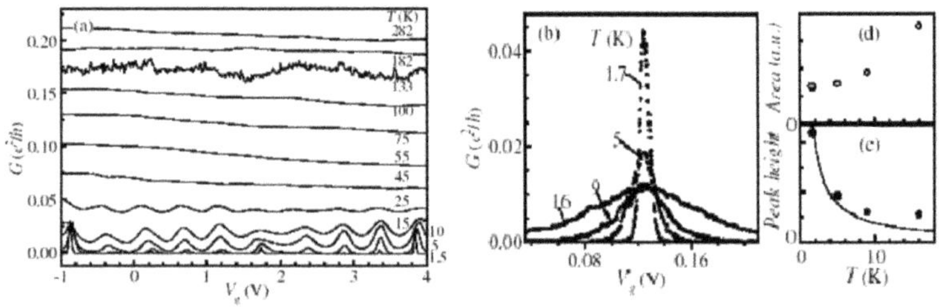

Figure 5.24 (a) Conductance G versus gate voltage V_g at temperatures ranging from 282 to 1.5 K for a metal-on-tube device. The device noises on the curves had been filtered from all traces except the one at 133 K, where its amplitude was at a maximum. (b) Profile of a Coulomb blockade peak in another device in the temperature range 16 and 1.7 K. (c) Peak heights versus T for the peak in (b). The solid curve shows $G \sim 1/T$. (d) peak area versus T [71].

Figure 5.25 shows the Coulomb "diamond" plot of a tube-under electrode device. At the peak points, the ground state energies for N and $N + 1$ electron on the dot are degenerate. The charging energy is $U = eV_{\text{max}}$, within the frame of constant-interaction model of the Coulomb blockade. The capacitance of an SWCNT is estimated as $C \sim \varepsilon_0 \varepsilon_r L$, and hence, $U = e^2/C \sim e^2/(\varepsilon_0 \varepsilon_r L) \sim 5$ [meV]/L [$^1\!/_4$m] is obtained, where $\varepsilon_r = 4$ for a SiO_2 (The geometrical factors that depend on the exact configuration of electrodes are difficult to estimate and thus neglected). $L \sim 0.6$ μm is obtained for $U = 8$ meV in this case. The extra white lines outside the "diamonds" in Figure 5.25c correspond to alignment of excited states. The level spacing expected for electrons with the appropriate dispersion confined in a 1-D box of length L can be estimated by $\Delta E = (dE/dk)(\Delta k/2)$ $= \hbar v_F \pi/(2L) = 1/L)$, where ΔE and L are in the unit of meV and μm, respectively. Hence $L \sim 0.5$ μm is obtained for a level spacing $\Delta E \sim 2$ meV here. Both estimations indicate that the segment of length around $L \sim 0.5$ μm, which is approximately the lithographically defined contact separation for the device. This finding implied that the evaporated metal strongly perturbed the underneath carbon tubes, but the electron states reached beyond the electrode region when the carbon nanotube is lying on the contacts electrodes.

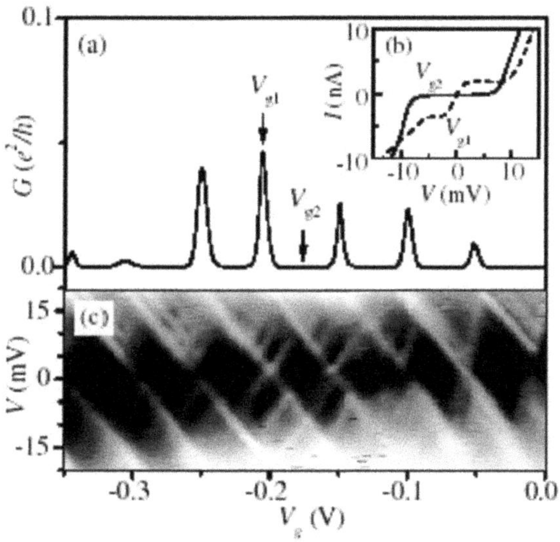

Figure 5.25 (a) G–V_g for a CNTSET at 4.2 K showing a regular series of Coulomb blockade peaks. (b) I–V characteristics taken at the center of a peak (V_{g1}) and in between peaks (V_{g2}). (c) Gray scale (bias spectroscopy) plot of dI/dV vs. V_g and V at 4.2 K in the same range of V_g (lighter color indicates more positive value) [71].

One can see that the CNT/metal contacts are critical for the characteristics of CNTSETs. Liang *et al.* have done systematic study on the transport spectroscopy of CNTSETs with different contact transparencies.[73] The transparency of the contact between an SWCNT and its electrodes is characterized by the gate-voltage-averaged conductance (G_{avg}), as shown in Figure 5.26. When $G_{avg} << G_0 = e^2/h$, the contact is the least transparent. The electron transport is governed by Coulomb blockade. The transport spectroscopy shows the information of energy-level quantization and spin filling. As G_{avg} approaches G_0, the transport spectroscopy shows high-order tunneling effect, such as Kondo resonance[74], inelastic co-tunneling, and four-shell electron filling. When $G_{avg} >> G_0$, the Coulomb blockade is suppressed and electron transport is ballistic and phase-coherent owing to the highly transparent contact.

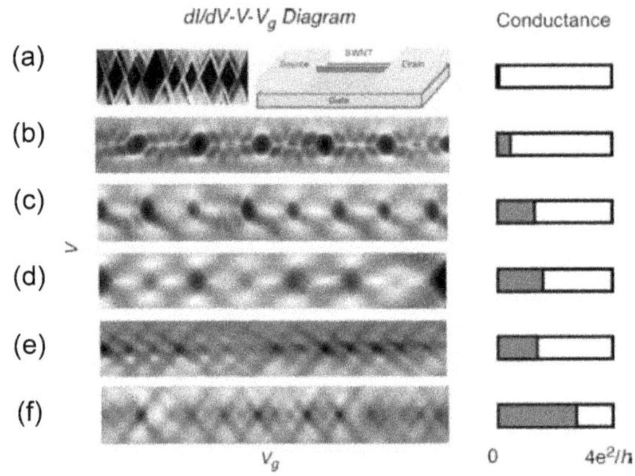

Figure 5.26 The differential conductance as a function of V_d and V_g from six different CNTSETs with decreasing resistance from (a) to (f). Left inset of (a): schematic device structure. Right inset of (a): color code of the conductance. Insets of (b) to (f): color code of the conductance [72]. See also Color Insert.

The single-electron tunneling with Coulomb blockade effect has been discussed in the previous section. Here, we focus on a CNTSET with more transparent contacts. Figure 5.27 depicts the transport spectroscopy of a CNTSET with $G_{avg} \approx e^2/h$. The Coulomb oscillations appeared in clusters of four peaks, suggesting that the electron addition to the SWCNT was with a four-electron periodicity. The four-electron periodicity was also observed in differential conductance mapping in V_d–V_g plane. Such a special pattern of conductance gap could be understood with the shell-filling model. An SWCNT has two spin-degenerate subbands. Thus, the quantization of these two subbands result in two sets of spin-degenerated electronic levels with the mean level spacing Δ within each set, as shown in the inset of Figure 5.27b,c. The energy mismatch between these two sets of electronic levels is labeled by δ in Figure 5.27d. Two possible cases are presented in Figure 5.27b. In the first case ($J + \delta U < \delta$, where J and δU are the exchange energy that favors spin alignment and excess Coulomb energy to put two electrons into a single level, respectively), two electrons of opposite spin fill each spin-degenerated level before filling higher energy level, leading to the sequence of ground state spins $S = 1/2$ (gap 1), $S = 0$ (gap 2), $S = 1/2$ (gap 3) and $S = 0$ (gap 4). In the second case ($J + \delta U > \delta$), the ground state for two electron is a spin triplet. In gap 2, the second spin-up electron is added to the lower level of the second set (yellow dot), leading to the total spin of 1. Similarly, in gap 4, the forth spin-up electron is added to the higher level of the first set and cause the total spin of 1. Thus, the ground state spins of the second case is $S = 1/2$ (gap 1), $S = 1$ (gap 2), $S = 1/2$ (gap 3) and $S = 1$ (gap 4).

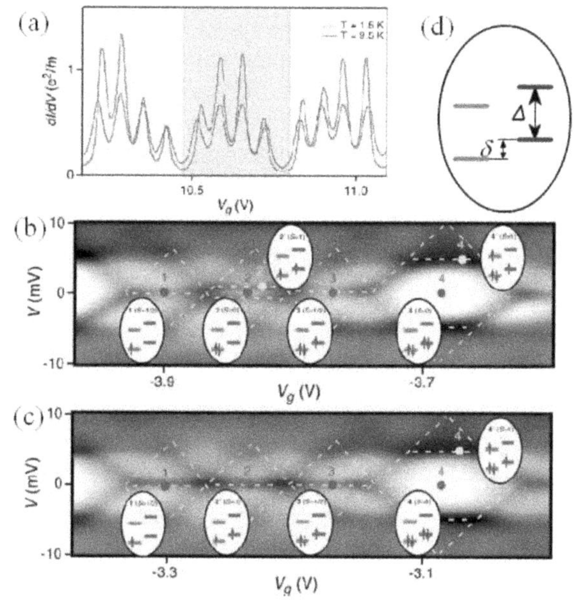

Figure 5.27 CNTFET with $G_{avg} \approx e^2/h$. (a) Coulomb oscillations at 1.5 and 9.5 K, respectively. (b) and (c) are the differential conductance as a function of V and V_g. The dashed lines delineate the conductance gap regions. Inset: schematic energy level filling in the SWCNT. (d) Energy level spacing Δ and the energy mismatch between the two sets of the electronic levels δ [72].

Comparing Figure 5.27b,c, one can see that the differential conductance at the zero bias $V = 0$ was always low (black) in gap 1 and 3 while it was always high (white) in gap 4. In contrast, the differential conductance could be low (Figure 5.27b) or high (Figure 5.27c) in gap 2. The black color at $V = 0$ is a signature of the Kondo resonance,[75] which is a result of the exchange interaction between a localized spin and the conduction electron in the metallic electrode.

It thus appeared when the electronic level of a quantum dot had nonzero spin. The electronic states in gap 2 could be spin singlet or triplet depending on the exact number of electron stored in the QD.

When either one or both contacts of the CNTSET have a conductance $G_{avg} > e^2/h$, the charges on the QD exhibit quantum fluctuation, as depicted in Figure 5.28a. Such a transport spectroscopy can be interpreted by interfacial electron scattering (at the metal contacts) coupled with ballistic electron transport in the SWCNT. The mesh of the crisscrossing dark lines originates from the quantum interference between electron waves multiply reflected between two SWCNT/metal contacts, analogous to light transmission in an optical Fabry–Perot cavity.[20] The bias voltage difference between two adjacent dark lines V_c was inversely proportional to the length of the SWCNT, implying that electron scattering only occurred at the SWCNT/metal contact while the electrons transport was ballistic inside the SWCNT. Figure 5.28b depicts the comparison between the experimental observation and the theoretical calculation. The agreement between these two results shows that the interpretation with Fabry–Perot model is correct and electron transport in the SWCNT is ballistic.

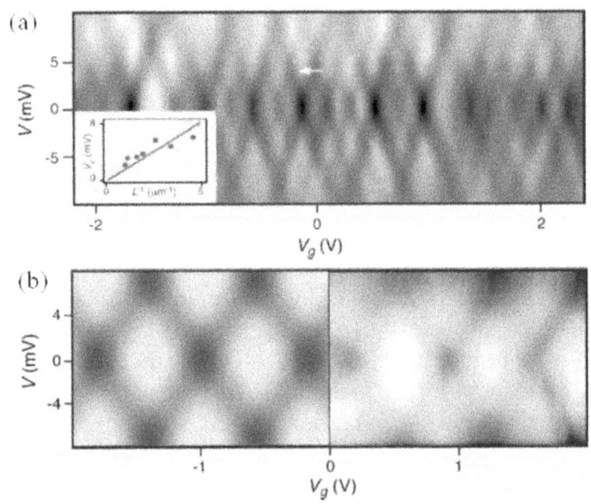

Figure 5.28 (a) Differential conductance in V–V_g plane of a CNTFET with $G_{avg} > e^2/h$. Inset: V_c versus L^{-1} of seven devices. (b) Comparison between calculated (left) and measured (right) transport spectroscopy [72].

Grove-Rasmussen *et al.* demonstrated open (high-transparence contact) and closed (low-transparence contact) quantum dot behavior in the same SWCNT.[76] As discussed previously, in a CNTSET with less transparent contacts, electron transport is governed by CB. The highly resistive contacts result in a long life time of single-electron states and electron tunnels through the device sequentially. Such a CNT can be viewed as a closed QD, in which the number of electrons is well defined (except at the charge-degeneracy point where single-electron tunneling occurs). On the contrary, highly transparent contacts enable the ballistic SWCNT to serve as an electron wave guide. This type of CNT is regarded as an open QD, in which the broad energy levels results in the observed resonances. In the transition region, the transparency of the contacts is between the aforementioned two cases. The number of electrons on the QD is still fixed, while high-order tunneling effects are possible. Figure 5.29a shows the conductance oscillations present on a wide gate voltage range. Broad oscillations at negative gate voltages (left) occurred accompanied with high conductance of hole transport. In contrast, sharp oscillations at positive gate voltages (right) appeared accompanied with low conductance of electron transport. The broad oscillations could be attributed to the Fabry–Perot interference, while the sharp oscillations are ascribed to the Coulomb blockade effect. Interestingly, at large positive gate voltages, the Coulomb oscillations were grouped into four peaks period (Figure 5.29b). Similar to the observations in Figure 5.27, the fourfold shell structure could be understood with the shell-filling model with high-order tunneling effects. The striking difference between the observation here and those in Figure 5.27 is that the different conductance oscillation patterns were found in the same device. The explanation is illustrated in Figure 5.29c. Hole transport faces small SB, resulting more transparent contact and higher conductance. Meanwhile, more transparent contacts enable the SWCNT channel act as an open QD. On the contrary, a relatively bigger SB to electron transport leads to less transparent contacts and close-QD behavior of the SWCNT is formed.

The detailed transport spectroscopy shows consistent features as the interpretation above (see Figure 5.30). The Fabry–Perot interference pattern is shown in Figure 5.30a; a mesh of the red lines (low-conductance regions) is a result of the interference of the hole wave reflected by the contacts back and forth. The level spacing can be extracted as (shown by the black arrow) $\Delta E \sim 4$ meV that was consistent with the length of the SWCNT $L = \hbar v_F \pi / \Delta E \approx 400$ nm. The Coulomb "diamonds" in Figure 5.30c appears as three small "diamonds" followed by a bigger "diamond." The origins of these patterns are the fourfold degenerate energy levels due to orbital and spin degrees of freedom. From transport spectroscopy shown in Figure 5.30c, the charge energy could be estimated as $U_c \sim 11$ meV, and the level spacing was estimated as $\Delta E \sim 4$ meV. Various CNTFETs with different geometries has also been demonstrated.[6,66,77,78]

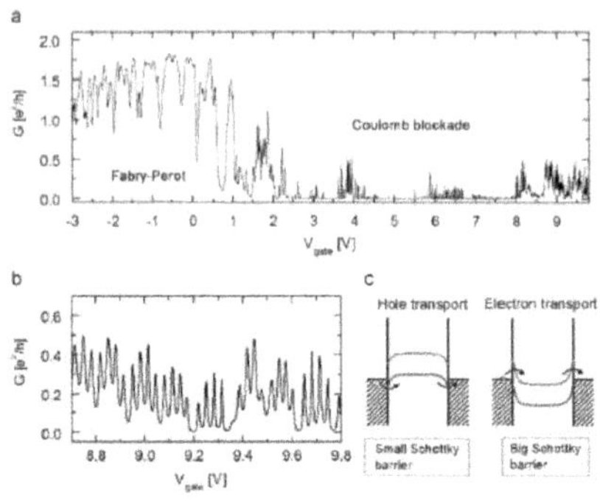

Figure 5.29 (a) Conductance oscillations in the transfer characteristic of a small bandgap s-SWCNT. (b) Zoom-in view of the transition region between Fabry–Perot interference oscillations and Coulomb oscillations. (c) Corresponding schematic band diagrams for the Fabry–Perot and Coulomb blockade circumstances [75].

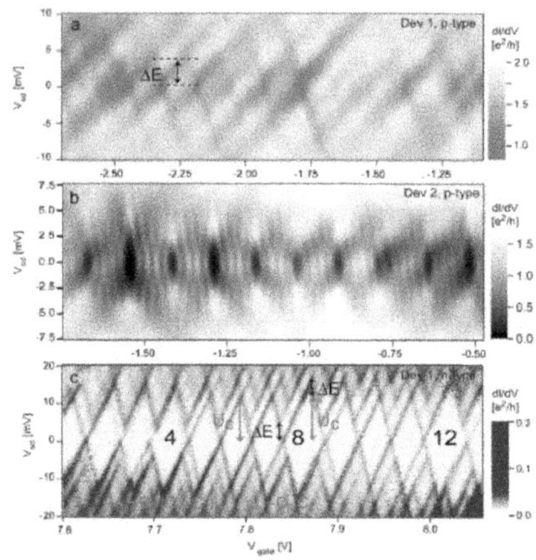

Figure 5.30 Conductance plot in the V_{sd}–V_{gate} plane for (a) Fabry–Perot interference region, where ΔE is energy level spacing. (b) Transition region where a single-hole transport is seen on top of the broad Fabry–Perot resonances. The four peaks with finite conductance in the valleys are caused by the Kondo effect and (c) Coulomb blockade region where the contact transparency is even lower. The fourfold degeneracy can be clearly observed and the charging energy and level-spacing energy can be extracted, as shown by the red and black arrows, respectively. The dashed red lines indicate the slope that is used to estimate the capacitances of the device. Insets: color code of the conductance [75].

SETs can also be employed to form logic circuits.[79,80] Figure 5.31 shows the operation of an inverter based on SETs. Similar to the CMOS inverter, two SETs connected in series shared the same gate that was the input terminal. Two side gates $V1$ and $V2$ (Figure 5.31b) were used to adjust the transfer curves of the SETs so that the Coulomb oscillation peak position of SET1 and SET2 were out of phase, as depicted in Figure 5.31c. In this case, only one of the SETs was switched on at a certain gate voltage. The output versus input characteristic of a SET inverter is presented in Figure 5.31d. The inverter operation was obtained, but with a small voltage gain (~0.6). The voltage gain could be improved by increasing the gate coupling through top-gate geometry with thinner gate dielectric, higher dielectric constant, etc.

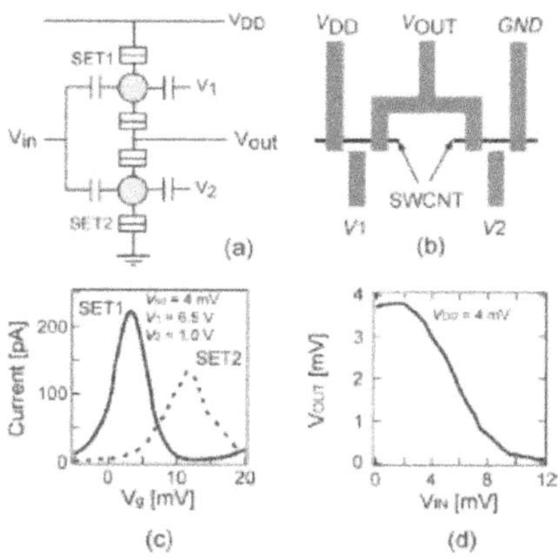

Figure 5.31 (a) Equivalent circuit of a single-electron inverter. (b) Schematic geometry of a single-electron inverter. (c) Coulomb oscillations of two SETs, SET1 and SET2. (d) Input–output curves of the fabricated inverter at bias V_{DD} = 4 mV [80].

A SET-based XOR gate is demonstrated in Figure 5.32. From the SEM image, one can see that the XOR gate consisted of a SET with two side gates V_{inA} and V_{inB}. The Coulomb oscillations due to the V_{inA} were adjusted through the common back-gate so that the center position of the two adjacent peaks were situated at V_{inA}=0. The application of V_{inB} (at a constant V_{inA}) shifted the peak by a half period (see Figure 5.32c). Under this condition, current flowed when either V_{inA} or V_{inB} was at "on" state, i.e., 125 mV in Figure 5.32c,d. The XOR operation was demonstrated at 1.5 K. When the temperature was increased to

40 K, the XOR operation persisted, but with a smaller logic swing. While the demonstrated XOR circuit had a current output, a voltage output could be realized with a load or a complementary device design.

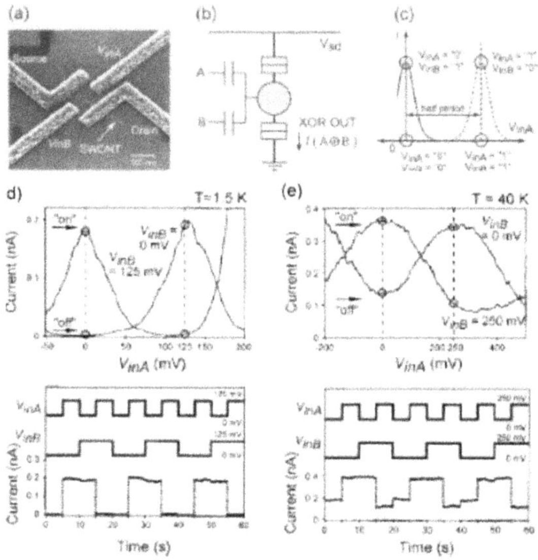

Figure 5.32 (a) SEM image of a SWCNT-based single-electron XOR logic gate. (b) Equivalent circuit of the single-electron XOR gate. (c) Operation principle of the XOR gate. When one of the two inputs (V_{inA} and V_{inB}) is "1" (on), the current (off) flows, but when both are "1" or "0", the current does not follow. Coulomb oscillation peak shift (upper) and the XOR operation (lower) at (d) 1.5 K and (e) 40 K [80].

5.3 Memory Devices

Fuhrer *et al.* reported the first CNTFET memory based on the hysteresis effects in 2002.[81] As shown in Figure 5.33b, a pronounced hysteresis loop ($\approx$ 6 V) was observed when the gate voltage was swept from –10 to +10 V and then swept back. The arrows in the figure indicate the sweeping directions of the gate voltage. The hysteresis loop is formed by the shift of threshold voltage of the CNTFET. The state of the device was read at the source-drain voltage (V_{sd}) of 500 mV and gate voltage (V_{g}) of –1 V. Writing and erasing operations were performed by pulses of gate voltage to ±8 V. A current (I_{sd}) of > 1 μA was switched, as shown in Figure 5.33c. After an initial decay (~50 s), I_{sd} remained constant. The hold time of the memory exceeded 5000 seconds. The memory effect was attributed to the charge storage in the dielectric layer just below the SWCNT channel. The electric field at the SWCNT/SiO_2 interface for V_g = 10 V was estimated as $E = V_g/[KR_t \ln(R_g/R_t)] \sim V_g/(23R_t)$,[82] where K = 3.9 was the dielectric constant of SiO_2, R_t = 1.35 nm was the SWCNT radius, and R_g = 500 nm was the dielectric thickness. Thus E > 0.3 V/nm and it was comparable to the breakdown field for SiO_2 (~ 0.25 V/nm). Therefore, electrons were easily injected into the dielectric from the SWCNT. The injected electrons would remain trapped in metastable states until the polarity is reversed. Moreover, it was reasonable to expect movement of the charges in the dielectric at such a high electric field of E = 0.3 V/nm. Another possible reason was that at such high electrical fields, the gas molecules around the SWCNT might experience corona discharge, and then the ionized gas could transport charges to the dielectric surface. In addition, it was also possible that the high electric field resulted in a rearrangement of adsorbents on the SWCNT such as oxygen. The interpretations above were proposed for the case in which gate voltage swept from positive to negative, when electrons were the major carriers from the beginning. These possible explanations were also valid for the case of sweeping gate voltage from negative to positive, i.e., holes were the major carriers initially.

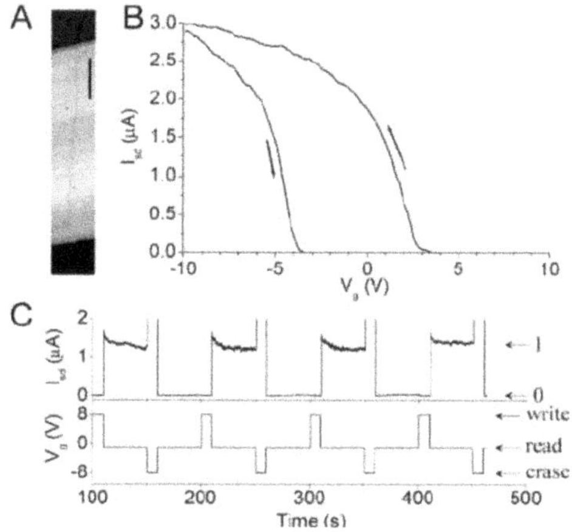

Figure 5.33 (a) AFM image of the CNTFET. The scale bar is 1 μm. b) I_{sd}–V_g curve of the CNTFET showing a hysteresis loop. (c) Four read/write cycles of the CNTFET memory at room temperature [81].

The amount of charges stored ΔQ could be obtained from the amount of threshold voltage shift, as $\Delta Q = Cg^* \Delta V_{th}$, where $C_g = 54$ aF was the gate capacitance determined directly from low-temperature Coulomb blockade measurements as discussed in the previous section, and $\Delta V_{th} = 6$ V was the shift of the threshold voltage. Hence, $\Delta Q = 3.24 \times 10^{-16}$ C was obtained, that is, around 2000 electrons had been trapped. The carrier mobility μ could also be determined from Figure 5.33b as $\mu = (L^2/C_g)(dG/dV_g)$,[83] where L is the channel length, and dG/dV_g for holes could be determined from the linear por as ~ 9000 cm^2/V·s (~ 450 cm^2/V·s in pure silicon). To summarize, CNTFET has the following merits: (1) high carrier mobility ($\mu \approx 9000$ cm^2/V·s for hole), (2) large geometrical capacitance, and (3) enhanced electric field due to cylindrical geometry ($E_{ox} \sim 0.3$ V/nm). Because of these merits, local charges affect the global conductance, as well as high charge injection capability. Therefore, the CNTFET can be an ideal charge-detecting device for nanoscale memory application.

Radosavljević and coworkers reported the nonvolatile memory (NVM) based on an n-type CNTFET.[84] The device was investigated in high vacuum to study the charge traps in bulk SiO_2. As shown in Figure 5.34, the hysteresis increased steadily as V_g-sweeping range increased. This indicated that the hysteresis originated from avalanching injection into bulk oxide traps. Taking Figure 5.34c as an example, after a sweep from –20 to +20 V, the threshold voltage moved toward a more positive value (–7 to 2 V), implying injection of electrons into oxide traps. Electrons were injected from the CNTFET channel, where the electric field was the enhanced due to the cylindrical device geometry. Figure 5.34d illustrates the processes for electrons injected from CNTFET channel. A memory cell containing an n-CNTFET and a 1 MΩ load resistor is shown in the inset of Figure 5.34e. Read ($V_{in} = 0$) and write ($V_{in} = +20$ or -20 V) were applied to input terminal (gate). Logical "1" ("0") was defined as $V_{out} = 1$ V (0 V). In order to write a "1" ("0") to the memory cell, V_{in} was switched rapidly to –20 V (+20 V) and then back to 0, so that the CNTFET was "on" ("off") at the read voltage ($V_g = 0$). The outputs as a function of time when a series of data bits were written are shown in Figure 5.34e.

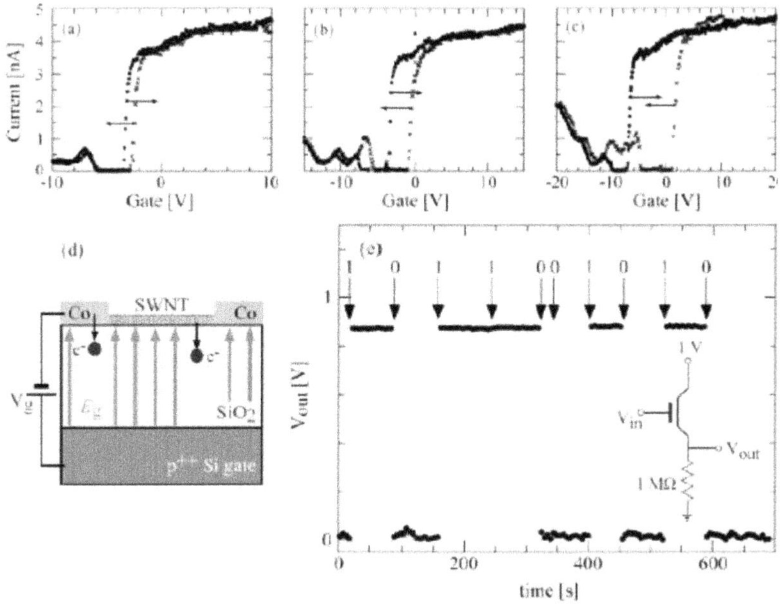

Figure 5.34 (a–c) I–V_g curves of a CNTFET in high vacuum with V_{ds} = 0.5 mV. The sweeping range of V_g is increased from (a) to (c). (d) Schematic of avalanche injection of electrons into bulk SiO_2 from CNTFET channel. (e) Real-time data of a nonvolatile molecular memory cell [84].

Cui *et al.* investigated the charge storage stability in CNTFETs.[85] They found the charges stored in CNTFETs were very stable for more than 12 days at room temperature. Figure 5.35 shows the memory effect at 84 and 12 K, respectively. It is noted that the curve measured at 12 K displayed many sharp peaks that could be caused by single-electron charging of the tube. Cui *et al.* also found that reproducible switching could be detected down to 5 K. The persistence of the hysteresis at low temperatures was not due to the mobile ions within the dielectric because the ion hopping would be frozen out at liquid-helium temperature (~ 4 K) where the associated activation energy was very high. They thus concluded that the mechanism for charge storage in dielectric was not sufficient to explain the memory effect observed in CNTFETs.

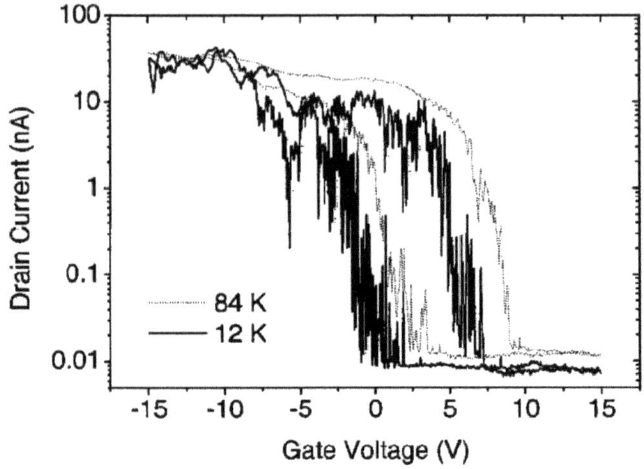

Figure 5.35 Drain current versus gate voltage in an annealed bundled-SWCNT FET at low temperatures (84 and 12 K) [85].

Kim and coworkers suggested that the hysteresis in CNTFET was caused by charges trapped by water molecules around the SWCNTs, such as SiO_2 surface-bound water proximal to the nanotubes.[86] As shown in the inset of Figure 5.36a, the CNTFET comprised fully suspended SWCNTs above the gate at a distance of 2 μm. The hysteresis loop at V_{ds} = 10 mV in ambient air for the device is shown in Figure 5.36a. As displayed in Figure 5.36b, the hysteresis quickly decreased to near zero immediately after it was placed in vacuum. The authors suggested that the charge trapping by water was an important cause of the hysteresis in CNTFETs exposed to the ambient environment. There were two types of water charge traps. Type 1 consisted of water molecules weakly adsorbed on the SWCNT surface, which could be easily removed by just pumping. Type 2 involved SiO_2 surface-bound water that was in very close proximity to the SWCNTs, as shown in Figure 5.36c. This finding implied that the SWCNT humidity sensors could be designed if a suitable water-sensitive coating was employed. Moreover, hysteresis was almost eliminated when the traps around SWCNT-bulk (suspended structure shown in Figure 5.36a) were removed. Thus, one could conclude, from this work, that the SWCNT-bulk instead of the contacts of the CNTFETs plays the dominant roles in the hysteresis.

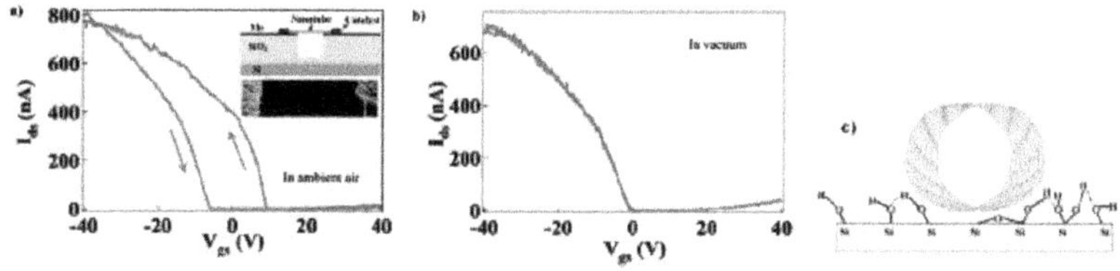

Figure 5.36 (a) Hysteresis of a suspended SWCNT FET exposed to ambient air at V_{ds} = 10 mV. Inset: A SEM image and a schematic structure of the device. (b) Eliminated hysteresis recorded with the same device after placing it in vacuum. (c) Schematic drawing that illustrates the water molecules hydrogen bonding to the Si–OH groups on the surface [86].

Besides the water molecules, other adsorbed gas molecules, e.g., ammonia, could also play a role of charge trapping centers.[87] Moreover, the passivation layer could also induce significant charge trapping in CNTFETs.[88] Choi *et al.* demonstrated the SWCNT-based NVM device using SiO_2–Si_3N_4–SiO_2 (ONO) layers as a storage node.[89] The structure of the CNT-flash memory is shown in Figure 5.37a. The memory device consisted of a top gate structure with a channel width of a few nanometers and the 14 nm-thick ONO layer that was embedded between the SWCNT and gate electrode. Figure 5.37b shows an SWCNT with diameter of 3 nm, length of 1 μm bridging a pair of Au/Ti electrodes. The SWCNT channel was completely covered by the ONO film to prevent atmosphere from influencing the electrical transport properties of the SWCNT. Two small hysteresis loops were observed at 5 K with different sweeping ranges of V_g. The hysteresis loops was suggested to be caused by charges trapped in ONO layer. Figure 5.37d shows a quantized increment of threshold voltage of 60 mV when the charging voltage was changed continuously. Each trap site containing charge in the ONO layer of the memory device was suggested to act like a quasi-quantized energy state for the flash memory.

Star and coworkers combined CNTFETs with a photosensitive polymer to fabricate optoelectronic memory devices.[90] Two types of photosensitive CNTFET devices were prepared. For the first type of devices, a composite of polymer and nanotubes was deposited in suspension form onto Si wafers with previously patterned electrical contacts. Their structure and performances are shown in Figure 5.38a,b. The UV illumination modulated the conductivity of the device. However, the modulation ($\Delta I_{SD}/I_{SD}$) was not efficient. This might be caused by the high contact resistance. Moreover, the channel of the device consisted of both m- and s-SWCNTs. For the second type of devices, a polymer layer was deposited over the contacts and SWCNTs by drop-casting of 0.05 μL of PmPV (0.1% wt) solution in $CHCL_3$ using a micropipet. Their structure and performances are shown in Figure 5.38c. The polymer itself did not significantly modify the device properties. However, when the device was illuminated with 365 nm light, the transfer characteristic shifted toward positive gate voltages. The threshold voltage change was 2 V. The devices recovered slowly, requiring hours to reach their original conductance, when the light was turned off. The memory operation of the devices in air at room temperature is shown in Figure 5.38d. The I_{SD} versus time response to UV illumination was measured under alternating V_G. The offset of V_G was 1 V, amplitude of 4 V and frequency of 1 KHz. Recall that the solution-deposited CNTFETs showed no optoelectronic memory behavior. In contrast, the polymer-coated CNTFETs showed a clear optoelectronic memory behavior. Therefore, it could be concluded that the proximity of SWCNTs to the silicon oxide substrate is necessary for long-term stable memories.

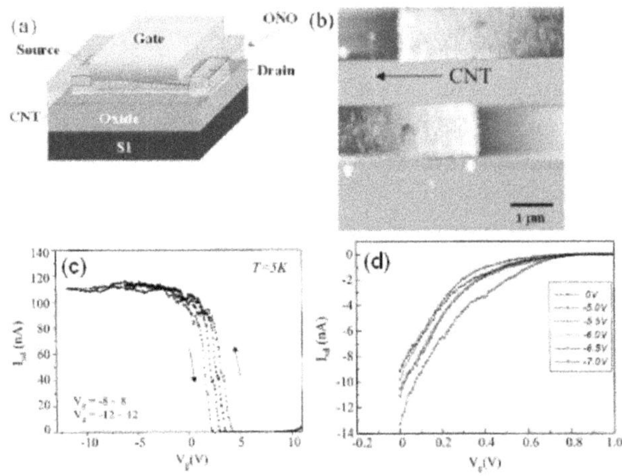

Figure 5.37 (a) Schematic structure of a CNT-based NVM with SiO_2–Si_3N_4–SiO_2 charge traps. (b) AFM image of the SWCNT attached between two metal electrodes. (c) I_{sd}–V_g at 5 K with V_{sd} = −0.9 V. Two loops correspond to V_g sweeping range of ±8 V (inner loop) and ±20 V (outer loop). d) I_{sd}–V_g after charging the SiO_2–Si_3N_4–SiO_2 storage node with different pulse amplitudes [89].

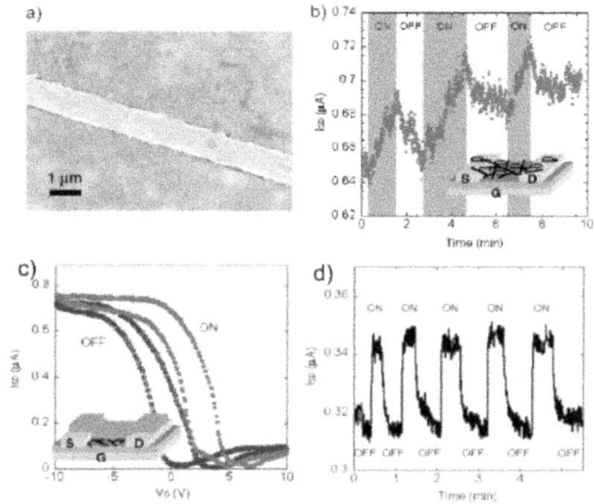

Figure 5.38 (a) SEM image of PmPV/CNT composite deposited on Si wafer with patterned Au electrodes. (b) Real-time record of current during UV illumination cycles in air. The shaded and unshaded regions mark the UV-on and –off periods, respectively. Inset: schematic device geometry. (c) I_{SD}–V_G in air with V_{SD} = 1 V at UV-off (blue curves) and UV-on (red curves) conditions, respectively. Inset: schematic device geometry. (d) Memory operation in air at room temperature [90].

The hysteresis was theoretically modeled by Robert-Peillard *et al.*[91] In their model, a CNT contacted to side electrodes lying on a SiO_2 layer. A model with full axial symmetry, instead of a planar one, was employed. The solution to the planar geometry was different from the axial one near the contacts only. The shift of the threshold voltage was due to the injection and the trapping/detrapping of carriers in the SiO_2 that were located within 3 nm around the CNT. These trapped electrons generated an additive potential to the external potential that was felt by the CNT charge carriers. The external potential on the CNT could be obtained as the sum of the gate voltage and the potential created by trapped charges in the oxide. The influences of two model parameters, sweeping rate and the range of the gate voltage ($-V_{GMax}$, V_{GMax}), on the hysteresis (H) were studied. As shown in Figure 5.39a, H increased linearly with the increase of V_{GMax} for large enough V_{GMax}. H would saturate at a value that is determined by the density of available traps. It is obvious in Figure 5.39b that the slower the V_G was swept, the larger the density of injected electrons. Although the simulated curves were not exactly the same as experimental observation, the simulated trends were qualitatively consistent with the experimental observation. In addition, other theoretical models were also proposed to describe the observed hysteresis.[92,93]

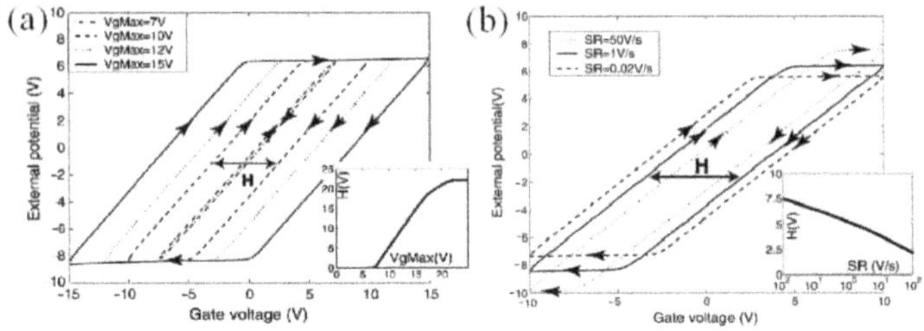

Figure 5.39 (a) Threshold voltage versus V_G calculated with various gate voltage ranges from 7 to 15 V (from inner to outer loop). The sweeping rate was 1 V/s. Inset: width of the hysteresis H as a function of V_{GMax}. (b) Threshold voltage versus V_G calculated for different SR. V_{GMax} = 10 V. Inset: semilog plot of the width of the hysteresis versus SR [91].

As a CNTFET can serve as a nanoscale charge sensor, it could be integrated into an ultimate miniaturized flash memory, i.e., the single-electron memory. As shown in Figure 5.40, the memory cell consisted of a normal CNTFET with deposited gold nanocrystals.[94] The gold nanocrystal acted as the storage nodes. Due to the Coulomb blockade effect, the number of electrons stored in the gold nanocrystal was quantized. The equivalent circuit is shown in Figure 5.40c. The gold nanocrystal was capacitively coupled to the SWCNT channel and the top-gate. For a 30 nm-diameter gold nanocrystal, the self-capacitance $C_{self} = 2\pi\varepsilon_0\varepsilon_r d$ = 5.3 aF, where the diameter d = 30 nm and permittivity of the surrounding medium (parylene C) ε_r = 3.2. The charging energy of the nanocrystal was estimated as $E_c = e^2/(2.C_{self})$ = 15 meV, equivalent to a temperature of 180 K. At a temperature well below this limit, Coulomb blockade dominated the electron tunneling processes onto/off the nanocrystal. The transport occurred at regularly spaced gate voltage separated by ΔV_w (Figure 5.40d), where $e\Delta V_w$ is the work done by the voltage source necessary to add or remove a charge quanta from the nanocrystal. The charge stored in the nanocrystal induced a local electrostatic field on the SWCNT-channel, which in turn modified the conductivity of the SWCNT-channel. In other words, the CNTFET acted as an electrometer probing the charges stored in the nanocrystal. The hysteresis ΔV_h observed in Figure 5.40d was quantitatively analyzed. The voltage separation ΔV_w and hysteresis ΔV_h depended on the coupling capacitances (Figure 5.40c) and the number of electrons n transferred into or out from the nanocrystal. This allowed us to confirm that the device was operating at the single-electron limit by evaluating n from the following equations: $\Delta V_w = ne/C_{gt}$ and $\Delta V_h = neC_{tc}/(C_{tc} + C_{gc})C_{tot}$. The total capacitance between the gate and device was $C_{tot} = C_{gc} + C_{tc}C_{gt}/(C_{tc} + C_{gt}) \approx C_{gc} + C_{gt}$ since C_{tc} was much larger than C_{gt} and C_{gc}. Thus, $n = \Delta V_h C_{gc}/[e(1 - \Delta V_h/\Delta V_w)]$ could be obtained. The number of electrons transferred was estimated to be between 1 and 8, which was much less than those of a conventional flash memory where thousands of electrons could be trapped and transferred.

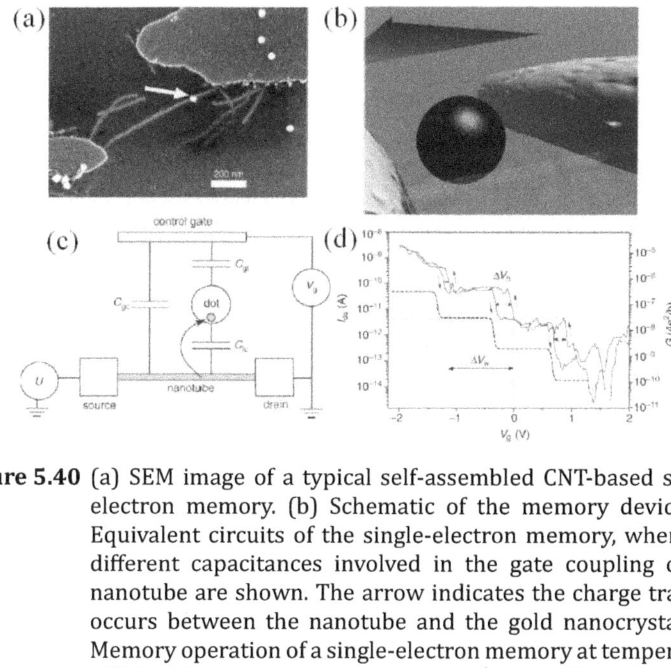

Figure 5.40 (a) SEM image of a typical self-assembled CNT-based single-electron memory. (b) Schematic of the memory device. (c) Equivalent circuits of the single-electron memory, where the different capacitances involved in the gate coupling of the nanotube are shown. The arrow indicates the charge transfer occurs between the nanotube and the gold nanocrystal. (d) Memory operation of a single-electron memory at temperature of 2 K with a low sweep rate of 10 mVs^{-1}. Dotted line (shifted for a clear view) was a fit for the Coulomb staircase of an electron box. The temperature for charging energy fit was 80 K [94].

In addition to the aforementioned CNT-based memories, much effort has been devoted to develop various types of memory devices. For instance, CNTs could be employed as charge storage nodes in NVM.[95] CNTs could be incorporated into metal-free silicon-molecule-nanotube memory devices.[96] CNT-based electromechanical switches could also serve as memory cells.[97–99]

5.4 Other Electronic Devices

In addition to the aforementioned applications, SWCNTs have also been used for constructing other electronic devices such as diodes. A CNT-based diode was formed by a p-n junction through chemical doping.[30] The p-n junction worked as a Zener diode due to the thin energy barrier in the CNT. However, chemical doping introduced defects into the intrinsic tube. Electrostatic doping in conjugation with splitting gate method also produced a p-n junction.[100,101] A p-n diode was demonstrated with SWCNT encapsulating Fe nanoparticles.[102] As discussed in the previous sections, Fermi level pinning effect is minor at CNT/metal contact, resulting in the metal work function dependent conduction type. Thus, a Schottky diode could be formed if different types of metals with high and low work functions were used as the source and drain electrodes, respectively.[6,103,104] CNT-based Schottky diodes could also be fabricated by contact engineering with chemical approach.[105] Moreover, CNT intermolecular and intramolecular junctions also show current rectification characteristics.[106–108] CNT-based chemical and biological sensors were also demonstrated for promising applications.[109–115]

5.5 Summary Remarks

SWCNTs have shown their unique electronic properties that make them perfect candidates as the current channels in many electronic applications. In fact, many prototypes of SWCNT electronic devices have been demonstrated with great application potentials. However, there are many technical challenges to be overcome before the SWCNT devices are used practically. First, large-scale integration of SWCNT devices is a paramount challenge. Second, the chirality and location of an SWCNT are still not controlled precisely. Third, the repeatability of the electrical performance of the SWCNT-based electronic devices is poor as the electrical and electronic properties of an SWCNT are largely affected by its metallic contacts and environmental conditions, etc.

References

1. Tans, S. J., Verschueren, A. R. M., and Dekker, C. Room-temperature transistor based on a single carbon nanotube. *Nature* **393**, 49–52 (1998).
2. Léonard, F., and Tersoff, J. Role of Fermi-level pinning in nanotube Schottky diodes. *Physical Review Letters* **84**, 4693 (2000).

3. Léonard, F., and Talin, A. A. Size-dependent effects on electrical contacts to nanotubes and nanowires. *Physical Review Letters* **97**, 026804 (2006).
4. Sze, S. M. *Physics of Semiconductor Devices*. 2nd edn (Wiley, 1981).
5. Léonard, F. *The Physics of Carbon Nanotube Devices* William Andrew, New York, (2009).
6. Li, H., Zhang, Q., and Marzari, N. Unique carbon-nanotube field-effect transistors with asymmetric source and drain contacts. *Nano Letters* **8**, 64–68, doi:10.1021/nl071905e (2007).
7. Léonard, F., and Tersoff, J. Novel length scales in nanotube devices. *Physical Review Letters* **83**, 5174 (1999).
8. Cui, X., Freitag, M., Martel, R., Brus, L., and Avouris, P. Controlling Energy-level alignments at carbon nanotube/Au contacts. *Nano Letters* **3**, 783–787, doi:10.1021/nl034193a (2003).
9. Freitag, M., *et al.* Imaging of the Schottky barriers and charge depletion in carbon nanotube transistors. *Nano Letters* **7**, 2037–2042, doi:doi:10.1021/nl070900e (2007).
10. Derycke, V., Martel, R., Appenzeller, J., and Ph, A. Controlling doping and carrier injection in carbon nanotube transistors. *Applied Physics Letters* **80**, 2773–2775 (2002).
11. Heinze, S., *et al.* Carbon nanotubes as Schottky barrier transistors. *Physical Review Letters* **89**, 106801 (2002).
12. Appenzeller, J., Radosavljevic, M., Knoch, J., and Avouris, P. Tunneling versus thermionic emission in one-dimensional semiconductors. *Physical Review Letters* **92**, 048301 (2004).
13. Martel, R., *et al.* Ambipolar electrical transport in semiconducting single-wall carbon nanotubes. *Physical Review Letters* **87**, 256805 (2001).
14. Appenzeller, J., *et al.* in *Electron Devices Meeting, 2002. IEDM '02. Digest. International.* 285–288.
15. Appenzeller, J., *et al.* Field-modulated carrier transport in carbon nanotube transistors. *Physical Review Letters* **89**, 126801 (2002).
16. Radosavljevic, M., Heinze, S., Tersoff, J., and Avouris, P. Drain voltage scaling in carbon nanotube transistors. *Applied Physics Letters* **83**, 2435–2437 (2003).
17. Dong, L., *et al.* Effects of local Joule heating on the reduction of contact resistance between carbon nanotubes and metal electrodes. *Journal of Applied Physics* **101**, 024320–024327 (2007).
18. Javey, A., Guo, J., Wang, Q., Lundstrom, M., and Dai, H. Ballistic carbon nanotube field-effect transistors. *Nature* **424**, 654–657 (2003).
19. White, C. T., and Todorov, T. N. Carbon nanotubes as long ballistic conductors. *Nature* **393**, 240–242 (1998).

20. Liang, W., *et al.* Fabry–Perot interference in a nanotube electron waveguide. *Nature* **411**, 665–669 (2001).
21. Zhang, Y., and Dai, H. Formation of metal nanowires on suspended single-walled carbon nanotubes. *Applied Physics Letters* **77**, 3015–3017 (2000).
22. Jing, G., Javey, A., Hongjai, D., and Lundstrom, M. in *Electron Devices Meeting, 2004. IEDM Technical Digest. IEEE International.* 703–706.
23. Mann, D., Javey, A., Kong, J., Wang, Q., and Dai, H. Ballistic Transport in Metallic Nanotubes with Reliable Pd Ohmic Contacts. *Nano Letters* **3**, 1541–1544, doi:10.1021/nl034700o (2003).
24. Chen, Z., Appenzeller, J., Knoch, J., Lin, Y.-M., and Avouris, P. The role of metal-nanotube contact in the performance of carbon nanotube field-effect transistors. *Nano Letters* **5**, 1497–1502, doi:10.1021/nl0508624 (2005).
25. Zhu, W., and Kaxiras, E. The nature of contact between Pd leads and semiconducting carbon nanotubes. *Nano Letters* **6**, 1415–1419, doi:10.1021/nl0604311 (2006).
26. Wind, S. J., Appenzeller, J., Martel, R., Derycke, V., and Avouris, P. Vertical scaling of carbon nanotube field-effect transistors using top gate electrodes. *Applied Physics Letters* **80**, 3817–3819 (2002).
27. Martel, R., Wong, H. S. P., Chan, K., and Avouris, P. in *Electron Devices Meeting, 2001. IEDM Technical Digest. International.* 7.5.1–7.5.4.
28. Yang, M. H., *et al.* Advantages of top-gate, high-k dielectric carbon nanotube field-effect transistors. *Applied Physics Letters* **88**, 113507–113503 (2006).
29. Bachtold, A., Hadley, P., Nakanishi, T., and Dekker, C. Logic circuits with carbon nanotube transistors. *Science* **294**, 1317–1320, doi:10.1126/science.1065824 (2001).
30. Zhou, C., Kong, J., Yenilmez, E., and Dai, H. Modulated chemical doping of individual carbon nanotubes. *Science* **290**, 1552–1555, doi:10.1126/science.290.5496.1552 (2000).
31. Shim, M., Back, J. H., Ozel, T., and Kwon, K.-W. Effects of oxygen on the electron transport properties of carbon nanotubes: Ultraviolet desorption and thermally induced processes. *Physical Review B* **71**, 205411 (2005).
32. Donghun, K., *et al.* Oxygen-induced p-type doping of a long individual single-walled carbon nanotube. *Nanotechnology* **16**, 1048 (2005).
33. Javey, A., Qian, W., Woong, K., and Hongjie, D. in *Electron Devices Meeting, 2003. IEDM '03 Technical Digest. IEEE International.* 31.32.31–31.32.34.
34. Nosho, Y., *et al.* Relation between conduction property and work function of contact metal in carbon nanotube field-effect transistors. *Nanotechnology* **17**, 3412 (2006).

35. Ding, L., *et al.* Y-contacted high-performance n-type single-walled carbon nanotube field-effect transistors: scaling and comparison with Sc-contacted devices. *Nano Letters* **9**, 4209–4214, doi:10.1021/nl9024243 (2009).

36. Zhang, Z., *et al.* Self-aligned ballistic n-type single-walled carbon nanotube field-effect transistors with adjustable threshold voltage. *Nano Letters* **8**, 3696–3701, doi:10.1021/nl8018802 (2008).

37. Kaminishi, D., *et al.* Air-stable n-type carbon nanotube field-effect transistors with Si_3N_4 passivation films fabricated by catalytic chemical vapor deposition. *Applied Physics Letters* **86**, 113115–113113 (2005).

38. Kojima, A., Chan, K.H., Kamimura, T., Maeda, M., and Matsumoto K. Air stable n-type top gate carbon nanotube field effect transistors with silicon nitride insulator deposited by thermal chemical vapor deposition. *Japanese Journal of Applied Physics* **44**, L328–L330 (2005).

39. Maehashi, K., *et al.* Electrical characterization of carbon nanotube field-effect transistors with SiN_x passivation films deposited by catalytic chemical vapor deposition. *Applied Physics Letters* **92**, 183111–183113 (2008).

40. Klinke, C., Chen, J., Afzali, A., and Avouris, P. Charge transfer induced polarity switching in carbon nanotube transistors. *Nano Letters* **5**, 555–558, doi:10.1021/nl048055c (2005).

41. Javey, A., *et al.* High performance n-type carbon nanotube field-effect transistors with chemically doped contacts. *Nano Letters* **5**, 345–348, doi:10.1021/nl047931j (2005).

42. Javey, A., *et al.* High-κ dielectrics for advanced carbon-nanotube transistors and logic gates. *Nature Materials* **1**, 241–246 (2002).

43. Wind, S. J., Appenzeller, J., and Avouris, P. Lateral scaling in carbon-nanotube field-effect transistors. *Physical Review Letters* **91**, 058301 (2003).

44. Javey, A., *et al.* Carbon nanotube field-effect transistors with integrated ohmic contacts and high-κ gate dielectrics. *Nano Letters* **4**, 447–450, doi:10.1021/nl035185x (2004).

45. Derycke, V., Martel, R., Appenzeller, J., and Avouris, P. Carbon nanotube inter- and intramolecular logic gates. *Nano Letters* **1**, 453–456, doi:doi:10.1021/nl015606f (2001).

46. Kruger, M., Buitelaar, M. R., Nussbaumer, T., Schonenberger, C., and Forro, L. Electrochemical carbon nanotube field-effect transistor. *Applied Physics Letters* **78**, 1291–1293 (2001).

47. Rosenblatt, S., *et al.* High performance electrolyte gated carbon nanotube transistors. *Nano Letters* **2**, 869–872, doi:10.1021/nl025639a (2002).

48. Minot, E. D., *et al.* Carbon nanotube biosensors: the critical role of the reference electrode. *Applied Physics Letters* **91**, 093507–093503 (2007).

49. Javey, A., *et al.* High-field quasiballistic transport in short carbon nanotubes. *Physical Review Letters* **92**, 106804 (2004).

50. Yao, Z., Kane, C. L., and Dekker, C. High-field electrical transport in single-wall carbon nanotubes. *Physical Review Letters* **84**, 2941 (2000).

51. Alam, K., and Lake, R. Performance of 2 nm gate length carbon nanotube field-effect transistors with source/drain underlaps. *Applied Physics Letters* **87**, 073104–073103 (2005).

52. Seidel, R. V., *et al.* Sub-20 nm short channel carbon nanotube transistors. *Nano Letters* **5**, 147–150, doi:10.1021/nl048312d (2004).

53. Chiu, P.-W., Chen, K.-J., and Lin, C.-Y. Fabrication and characteristics of ultrashort-channel carbon nanotube field-effect transistors. *Applied Physics Letters* **92**, 152111–152113 (2008).

54. Appenzeller, J., Lin, Y. M., Knoch, J., and Avouris, P. Band-to-band tunneling in carbon nanotube field-effect transistors. *Physical Review Letters* **93**, 196805 (2004).

55. Koswatta, S. O., Nikonov, D. E., and Lundstrom, M. S. in *Electron Devices Meeting, 2005. IEDM Technical Digest. IEEE International.* 518–521.

56. Koswatta, S. O., Lundstrom, M. S., and Nikonov, D. E. Band-to-band tunneling in a carbon nanotube metal–oxide–semiconductor field-effect transistor is dominated by phonon-assisted tunneling. *Nano Letters* **7**, 1160–1164, doi:10.1021/nl062843f (2007).

57. Koswatta, S. O., Lundstrom, M. S., Anantram, M. P., and Nikonov, D. E. Simulation of phonon-assisted band-to-band tunneling in carbon nanotube field-effect transistors. *Applied Physics Letters* **87**, 253107–253103 (2005).

58. Lu, Y., *et al.* DNA functionalization of carbon nanotubes for ultrathin atomic layer deposition of high-κ dielectrics for nanotube transistors with 60 mV/decade switching. *Journal of the American Chemical Society* **128**, 3518–3519, doi:10.1021/ja058836v (2006).

59. Koga, J., Ohba, R., Uchida, K., and Toriumi, A. in *Electron Devices Meeting, 2001. IEDM Technical Digest. International.* 7.1.1–7.1.4.

60. Takahashi, Y., Ono, Y., Fujiwara, A., and Inokawa, H. in *Solid-State and Integrated Circuits Technology, 2004. Proceedings. 7th International Conference on.* 624–629 vol.621.

61. Nakazato, K., Thornton, T. J., White, J., and Ahmed, H. Single-electron effects in a point contact using side-gating in delta-doped layers. *Applied Physics Letters* **61**, 3145–3147 (1992).

62. Bockrath, M., *et al.* Single-electron transport in ropes of carbon nanotubes. *Science* **275**, 1922–1925, doi:10.1126/science.275.5308.1922 (1997).

63. Postma, H. W. C., Teepen, T., Yao, Z., Grifoni, M., and Dekker, C. Carbon nanotube single-electron transistors at room temperature. *Science* **293**, 76–79, doi:10.1126/science.1061797 (2001).

64. Tans, S. J., *et al.* Individual single-wall carbon nanotubes as quantum wires. *Nature* **386**, 474–477 (1997).

65. McEuen, P. L., Bockrath, M., Cobden, D. H., Yoon, Y.-G., and Louie, S. G. Disorder, pseudospins, and backscattering in carbon nanotubes. *Physical Review Letters* **83**, 5098 (1999).

66. Li, H., and Zhang, Q. Tunable ambipolar Coulomb blockade characteristics in carbon nanotubes-gated carbon nanotube field-effect transistors. *Applied Physics Letters* **94**, 022101–022103 (2009).

67. Sapmaz, S., *et al.* Electronic excitation spectrum of metallic carbon nanotubes. *Physical Review B* **71**, 153402 (2005).

68. Moriyama, S., Fuse, T., Suzuki, M., Aoyagi, Y., and Ishibashi, K. Four-electron shell structures and an interacting two-electron system in carbon-nanotube quantum dots. *Physical Review Letters* **94**, 186806 (2005).

69. Tsuya, D., Suzuki, M., Aoyagi, Y., and Ishibashi, K. Characterization and estimation of tunnel barrier height in metallic single-wall carbon nanotube quantum dots. *Microelectronic Engineering* **82**, 196–200 (2005).

70. Cui, J. B., Burghard, M., and Kern, K. Room temperature single electron transistor by local chemical modification of carbon nanotubes. *Nano Letters* **2**, 117–120, doi:10.1021/nl015659x (2001).

71. Jarillo-Herrero, P., Sapmaz, S., Dekker, C., Kouwenhoven, L. P., and van der Zant, H. S. J. Electron-hole symmetry in a semiconducting carbon nanotube quantum dot. *Nature* **429**, 389–392 (2004).

72. Nygard, J., Cobden, D. H., Bockrath, M., McEuen, P. L., and Lindelof, P. E. Electrical transport measurements on single-walled carbon nanotubes. *Applied Physics A: Materials Science and Processing* **69**, 297–304, doi:10.1007/s003390051004 (1999).

73. Liang, W., Bockrath, M., and Park, H. Transport spectroscopy of chemical nanostructures: the case of metallic single-walled carbon nanotubes. *Annual Review of Physical Chemistry* **56**, 475–490, doi:10.1146/annurev.physchem.56.092503.141226 (2004).

74. Jarillo-Herrero, P., *et al.* Orbital Kondo effect in carbon nanotubes. *Nature* **434**, 484–488 (2005).

75. Glazman, L. K. a. L. Revival of the Kondo effect. *Physics World*, 33–38 (2001).

76. Grove-Rasmussen, K., JØgensen, H. I., and Lindelof, P. E. Fabry–Perot interference, Kondo effect and Coulomb blockade in carbon nanotubes. *Physica E: Low-Dimensional Systems and Nanostructures* **40**, 92–98 (2007).

77. Li, H., Zhang, Q., and Li, J. Carbon-nanotube-based single-electron/hole transistors. *Applied Physics Letters* **88**, 013508–013503 (2006).

78. Li, H., Zhang, Q., and Li, J. Interpretation of Coulomb oscillations in carbon-nanotube-based field-effect transistors. *Physical Review B* **73**, 235431 (2006).

79. Ishibashi, K., Tsuya, D., Suzuki, M., and Aoyagi, Y. Fabrication of a single-electron inverter in multiwall carbon nanotubes. *Applied Physics Letters* **82**, 3307–3309 (2003).

80. Ishibashi, K., Moriyama, S., Tsuya, D., Fuse, T. and Suzuki, M. Quantum-dot nanodevices with carbon nanotubes. *Symposium on Nanoscale Science and Technology*, 2005 AVS Meeting, Boston, USA (2005).

81. Fuhrer, M. S., Kim, B. M., Dürkop, T., and Brintlinger, T. High-mobility nanotube transistor memory. *Nano Letters* **2**, 755–759, doi:10.1021/nl025577o (2002).

82. P. M. Morse, and Freshbach, H. *Methods of Theoretical Physics* (McGraw-Hill, New York, 1953.)

83. McEuen, P. L., Fuhrer, M. S., and Hongkun, P. Single-walled carbon nanotube electronics. *Nanotechnology, IEEE Transactions on* **1**, 78–85 (2002).

84. Radosavljevic, M., Freitag, M., Thadani, K. V., and Johnson, A. T. Nonvolatile molecular memory elements based on ambipolar nanotube field effect transistors. *Nano Letters* **2**, 761–764, doi:10.1021/nl025584c (2002).

85. Cui, J. B., Sordan, R., Burghard, M., and Kern, K. Carbon nanotube memory devices of high charge storage stability. *Applied Physics Letters* **81**, 3260–3262 (2002).

86. Kim, W., *et al.* Hysteresis caused by water molecules in carbon nanotube field-effect transistors. *Nano Letters* **3**, 193–198, doi:10.1021/nl0259232 (2003).

87. Li, H., Zhang, Q., Peng, N., Liu, N., Lee, Y.C., Tan, O.K., Marazari, N., Thompson, C., V, Charge-trapping effects caused by ammonia in carbon nanotubes. *Journal of Nanoscience and Nanotechnology* **7**, 335–338 (2007).

88. Hong, L., *et al.* Global and local charge trapping in carbon nanotube field-effect transistors. *Nanotechnology* **19**, 175203 (2008).

89. Choi, W. B., *et al.* Carbon-nanotube-based nonvolatile memory with oxide–nitride–oxide film and nanoscale channel. *Applied Physics Letters* **82**, 275–277 (2003).

90. Star, A., Lu, Y., Bradley, K., and Grüner, G. Nanotube optoelectronic memory devices. *Nano Letters* **4**, 1587–1591, doi:10.1021/nl049337f (2004).

91. Robert-Peillard, A., and Rotkin, S. V. Modeling hysteresis phenomena in nanotube field-effect transistors. *Nanotechnology, IEEE Transactions on* **4**, 284–288 (2005).

92. Aravind, V., *et al.* Charge-injection-induced dynamic screening and origin of hysteresis in field-modulated transport in single-wall carbon nanotubes. *Applied Physics Letters* **89**, 162108 (2006).

93. Kar, S., *et al.* Quantitative analysis of hysteresis in carbon nanotube field-effect devices. *Applied Physics Letters* **89**, 132118–132113 (2006).

94. Marty, L., *et al.* Self-assembly of carbon-nanotube-based single-electron memories. *Small* **2**, 110–115, doi:10.1002/smll.200500148 (2006).

95. Ryu, S.-W., Huang, X.-J., and Choi, Y.-K. Vertically standing carbon nanotubes as charge storage nodes for an ultimately scaled nonvolatile memory application. *Applied Physics Letters* **91**, 063110–063113 (2007).

96. He, J., *et al.* Metal-free silicon-molecule-nanotube testbed and memory device. *Nat Mater* **5**, 63–68 (2006).

97. Rueckes, T., *et al.* Carbon nanotube-based nonvolatile random access memory for molecular computing. *Science* **289**, 94–97, doi:10.1126/science.289.5476.94 (2000).

98. Won Kang, J., *et al.* Nanoelectromechanical carbon nanotube memory analysis. *Physica E: Low-dimensional Systems and Nanostructures* **33**, 41–49 (2006).

99. Kang, J. W., Lee, J. H., Lee, H. J., Kwon, O. K., and Hwang, H. J. Electromechanical modeling and simulations of nanobridge memory device. *Physica E: Low-Dimensional Systems and Nanostructures* **28**, 273–280 (2005).

100. Ken, B., Nathan, G., and Paul, M. Transport in carbon nanotube p-i-n diodes. *Applied Physics Letters* **89**, 163121 (2006).

101. Lee, J. U., Gipp, P. P., and Heller, C. M. Carbon nanotube p-n junction diodes. *Applied Physics Letters* **85**, 145–147 (2004).

102. Li, Y. F., Hatakeyama, R., Shishido, J., Kato, T., and Kaneko, T. Air-stable p-n junction diodes based on single-walled carbon nanotubes encapsulating Fe nanoparticles. *Applied Physics Letters* **90**, 173127–173123 (2007).

103. Manohara, H. M., Wong, E. W., Schlecht, E., Hunt, B. D., and Siegel, P. H. Carbon Nanotube Schottky Diodes Using Ti-Schottky and Pt-ohmic Contacts for High Frequency Applications. *Nano Letters* **5**, 1469–1474, doi:10.1021/nl050829h (2005).

104. Yang, M. H., Teo, K. B. K., Milne, W. I., and Hasko, D. G. Carbon nanotube Schottky diode and directionally dependent field-effect transistor using asymmetrical contacts. *Applied Physics Letters* **87**, 253116–253113 (2005).

105. Kim, B.-K., *et al.* Carbon nanotube diode fabricated by contact engineering with self-assembled molecules. *Applied Physics Letters* **89**, 243115–243113 (2006).

106. Fuhrer, M. S., *et al.* Crossed nanotube junctions. *Science* **288**, 494–497, doi:10.1126/science.288.5465.494 (2000).

107. Yao, Z., Postma, H. W. C., Balents, L., and Dekker, C. Carbon nanotube intramolecular junctions. *Nature* **402**, 273–276 (1999).

108. Kim, D.-H., Huang, J., Shin, H.-K., Roy, S., and Choi, W. Transport phenomena and conduction mechanism of single-walled carbon nanotubes (SWCNTs) at Y- and crossed-junctions. *Nano Letters* **6**, 2821–2825 (2006).

109. Byon, H. R., and Choi, H. C. Network single-walled carbon nanotube-field effect transistors (SWCNT-FETs) with increased Schottky contact area for highly sensitive biosensor applications. *Journal of the American Chemical Society* **128**, 2188–2189, doi:10.1021/ja056897n (2006).

110. Heller, I., *et al.* Identifying the mechanism of biosensing with carbon nanotube transistors. *Nano Letters* **8**, 591–595, doi:doi:10.1021/nl072996i (2008).

111. Fushen, L., *et al.* Advances in bioapplications of carbon nanotubes. *Advanced Materials* **21**, 139–152 (2009).

112. Wang, S. G., *et al.* Multi-walled carbon nanotubes for the immobilization of enzyme in glucose biosensors. *Electrochemistry Communications* **5**, 800–803 (2003).

113. Wang, S. G., Zhang, Q., Yang, D. J., Sellin, P. J., and Zhong, G. F. Multi-walled carbon nanotube-based gas sensors for NH_3 detection. *Diamond and Related Materials* **13**, 1327–1332.

114. Wang, S. G., Wang, R., Sellin, P. J., and Zhang, Q. DNA biosensors based on self-assembled carbon nanotubes. *Biochemical and Biophysical Research Communications* **325**, 1433–1437 (2004).

115. Peng, N., Zhang, Q., Chow, C. L., Tan, O. K., and Marzari, N. Sensing mechanisms for carbon nanotube based NH_3 gas detection. *Nano Letters* **9**, 1626–1630, doi:10.1021/nl803930w (2009).

Chapter 6

Carbon Electronics for High-Frequency Applications

Henri Happy,[a] Laurianne Nougaret,[a] Nan Meng,[a] Emmanuelle Pichonat,[a] Vincent Derycke,[b] Dominique Vignaud,[a] and Gilles Dambrine[a]

[a] *IEMN UMR-CNRS 8520, Université Lille 1,Cité Scientifique, BP 60069, Avenue Poincaré, 59652 Villeneuve d'Ascq, France*

[b] *CEA Saclay, IRAMIS, Service de Physique de l'Etat Condensé (URA 2464), Laboratoire d'Electronique Moléculaire, 91191, Gif sur Yvette, France*

henri.happy@iemn.univ-lille1.fr

Carbon materials, including carbon nanotube (CNT) and graphene, exhibit a number of outstanding electronic and mechanical properties that make them very attractive for micro and nanoelectronic applications. In this chapter, our activities based on carbon materials are described. The first part of the study is related to single-wall carbon nanotube (SWCNT) devices. The studies developed concern fabrication and high-frequency (HF) characterization of carbon nanotube field effect transistors (CNTFET). As illustration, we have used high-purity solution with 99% of semiconducting tubes to fabricate CNTFET with intrinsic current-gain cut-off frequency of 80 GHz.

The second part is related to graphene material. Graphene used in this work was obtained by graphitization of silicon carbide

Carbon Nanotubes and Their Applications
Edited by Qing Zhang

ISBN 978-981-4241-90-8 (Hardcover), 978-981-4303-18-7 (eBook)
www.panstanford.com

(SiC) substrate, using thermal decomposition approach. One of our main objectives was to obtain large-area and high-quality material, in order to develop high-frequency graphene-based devices. To achieve high dynamic performance, a field effect transistor with an array of graphene nanoribbons connected in parallel was fabricated using e-beam lithography. The best intrinsic current gain cut-off frequency of 60 GHz and maximum oscillation frequency of 28 GHz were achieved.

6.1 High-Frequency Carbon Nanotube Field Effect Transistor

Because of their excellent electrical and mechanical properties, semiconducting SWCNTs are considered one of candidates for the "beyond complementary metal–oxide–semiconductor (CMOS)" era [1]. Using individual SWCNTs as a device channel, logic devices and circuits such as field effect transistors (FETs) [2–4], complementary metal–oxide devices [5–6], and ring oscillators [7] have been demonstrated. In the field of analog circuits, devices based on individual nanotube face two major problems: the high input impedance of an SWCNT and low drive currents in the "on" state. To circumvent these issues, devices based on an array of SWCNTs as the channel are required [8–12].

However, all nanotube synthesis methods invariably produce a mixture of both metallic and semiconducting nanotubes. As a consequence, the performance of network-based devices was limited by the presence of metallic nanotubes. This part describes strategies used to improve RF properties of CNTFETs, as well as device fabrication and characterization.

6.1.1 Conditions for High-Frequency Performance

Current gain cut-off frequency (f_t) and voltage gain (A_v) are some parameters that described the high-frequency properties of FET devices. These RF performances can be expressed, in an ideal case, by the following simplified expressions:

$$f_t = \frac{\alpha N g_{mo}}{2\pi(\alpha N C_g + C_p)}, \tag{6.1}$$

where N is a total number of tubes in the device channel; (αN) is the ratio of semiconducting tubes; g_{mo} the transconductance of a single

CNT; C_g the gate to channel capacitance of a single CNT (a few aF); and C_p the parasitic capacitance of the device structure (a few fF per unit length).

$$A_v = \frac{g_m}{g_d} = \frac{\alpha g_{mo}}{(1-\alpha) g_{do}}, \tag{6.2}$$

where $(1-\alpha)$ is the ratio of metallic nanotube and g_{do} the conductance of a single CNT.

Considering cut-off frequency, Equation (6.1) shows that by increasing the number of tubes in the channel, the influence of parasitic capacitance can be reduced, and consequently, the frequency can be improved.

In the same way, when considering g_{mo} equal to g_{do} (Equation 6.2), the voltage gain can be improved for high value of α, that means the proportion of semiconducting nanotubes is greater than the metallic ones. For example to obtain voltage gain of 10, it is necessary to have 90% of semiconducting nanotubes in the active channel of device. As a consequence, the performance of the CNTFET depends also on the proportion of metallic nanotubes in the device channel.

Therefore, high-density CNTFET using a high ratio of semiconducting nanotube is a best candidate for high-frequency performance.

The negative contribution of parasitic capacitances can be lowered also by improving the device geometry.

6.1.2 Device Fabrication

CNTFET can be fabricated using high-density SWCNT synthesized by two main techniques: (i) by growing horizontally aligned SWCNTs arrays with chemical vapor deposition (CVD) technique, (ii) or by using SWCNTs dispersed in solution. These synthesis methods invariably produce a mixture of both metallic and semiconducting nanotubes. Therefore strategies were developed to remove metallic nanotubes, while preserving semiconducting nanotube.

For CVD approach, the growth conditions are very important to reduce proportion of metallic CNT [13–15] and to increase nanotube density.

When using CNT in solution, a chemical route was chosen for sorting CNT and depositing them into devices. Progress was made for sorting CNT in solution. It's become possible to produce highly enriched semiconducting (99% of SC nanotubes) nanotube dispersions using density gradient ultracentrifugation [16].

To explore HF properties of CNTFET, the device structure chosen in our work is based on metal back gate device illustrated on Figure 6.1. It is a dual-gate device embedded in coplanar line access.

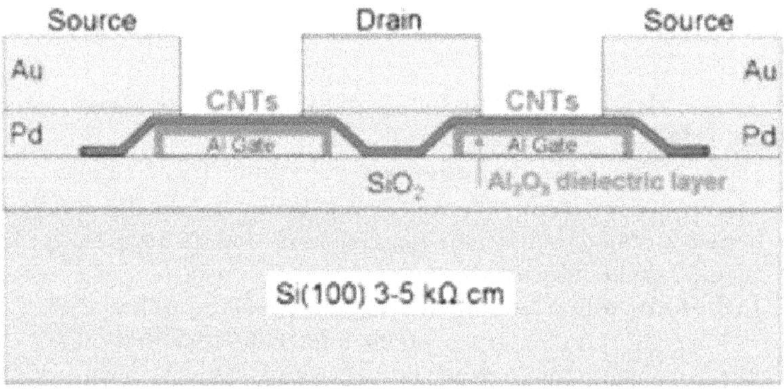

Figure 6.1 Cross section of the CNTFET structure with a dual gate [21].

Metallic back-gated CNTFETs were fabricated using standard microelectronic e-beam lithography combined with CNT deposition by wet self-assembly. High-resistivity silicon substrate (> 3 kΩ·cm) was used to avoid HF dielectric losses. A thick thermal oxide SiO_2 (260 nm) layer was grown prior to depositing coplanar access lines (titanium and gold). The metallic back-gate was fabricated with aluminum. A thin gate dielectric layer (Al_2O_3, about 2 nm) was obtained by the oxidation of aluminum with O_2 plasma. Then, the gate area was exposed to 3-aminopropyl triethoxysilane (APTS) vapor to form a sticky patch for which the CNs have a selective affinity when they are dispersed in an appropriate solvent (i.e., N-methyl-pyrrolidone) [17]. To improve the density of CNTs into the device channel, a dielectrophoresis [18], [19] approach can be associated to this surface treatment.

In this case, an ac electric field was applied between the access electrodes of the future source contacts. This electric field drives the deposition of the nanotubes. An important aspect of this technique was that not only the deposition can be made very dense, but it also results in nanotubes preferentially oriented along the field lines. Figure 6.2 shows a scanning electron microscopy (SEM) image of the active area after the DEP process. It is clear that the method results in a highly dense, mostly oriented and two-dimensional nanotube film. The very thin thickness of this film (a couple of nm) corresponding to a monolayer mainly is important since thick films would result in limited coupling to the gate due to the tube screening.

Figure 6.2 Scanning electron microscopy SEM images of the two-finger back-gate CNTFET active region: (a) after the dielectrophoresis process and (b) at the end of the fabrication process [21].

Duringourwork,threegenerationsofdeviceshavebeenfabricated, by applying one or two combinations of the previous techniques. For the first generation of devices, only the surface chemistry was used. It results in the density of tubes of about 10NTs/μm. (Figure 6.3a) [9, 20]. For the second generation of devices, the dielectrophoresis method is associated to the surface chemistry. Figure 6.3b shows the improvement of nanotube density in the device channel [21]. The first two device generations were fabricated using the conventional CNT solutions with one-third of metallic tubes and two-thirds of the semiconducting tube. For the third generation, we used a highly purified solution of nanotubes (99% of semiconducting NTs), and the same deposition technique as the second generation [22]. These three cases are illustrated in Figure 6.3.

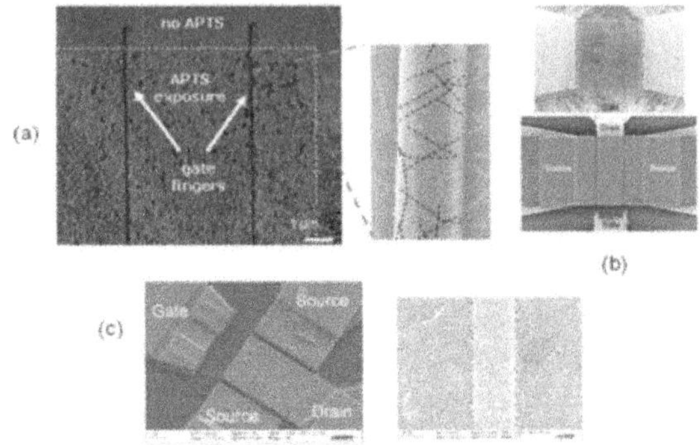

Figure 6.3 (a) First device generation: Illustration shows selective deposition of NTs with the surface chemistry. (b) Second device generation: Illustration of high-density of tube before and after deposition of source and drain accesses. Dielectrophoresis approach is associated to surface chemistry for tube deposition. (c) Third generation using enriched solution of semiconducting tubes.

6.1.3 Device Characterization

6.1.3.1 Direct current (dc) characterization

Figure 6.4 shows the I_{DS}–V_{DS} characteristic of CNFETs measured in air at room temperature, in the case of third generation. The device carries a high current (up to mA), in good correlation with the high density of nanotubes in the device channel. The devices do not show pinch-off, most likely as a result of small numbers of metallic nanotubes in the channel, which may be preferentially deposited by DEP.

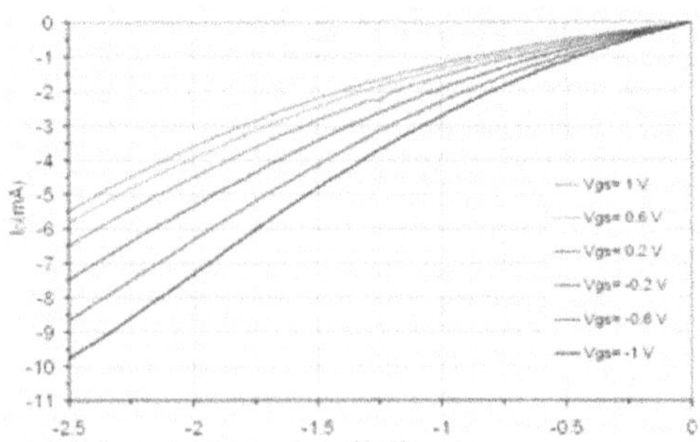

Figure 6.4 Measured dc I_{DS} vs V_{DS} at different V_{GS} — Third-generation device [22]. See also Color Insert.

Considering every device generation, we have defined, for a given V_{DS} bias, a factor of merit as the ratio of the effective contribution of semiconducting nanotubes to the total dc current versus the associated gate control bias range,

$$F_{\text{Merit}} = \frac{\Delta I_{\text{SC}}}{\Delta V_{\text{GS}}} = \frac{I_{\text{DS max}} - I_{\text{DS min}}}{\Delta V_{\text{GS}}} \tag{6.3}$$

This factor (which has the same units as transconductance) is related to the number of nanotubes per unit area and to the ratio of semiconducting and metallic nanotubes. We report in Table 6.1 characteristics of three device types, each having the same structure at similar bias points.

One can observe that F_{Merit} increases as the area density of semiconducting nanotubes increases (case I to case II) and that the best value is obtained in case III where the density of semiconducting nanotubes is maximized. Even though the nanotubes are randomly oriented, a significant improvement of dc performance is obtained. In the following, we show how it impacts radio-frequency (RF) performance.

Table 6.1 Characteristics of three device types

	Case I **(V_{DS} = 1 V)**	**Case II** **(V_{DS} = 1.5 V)**	**Case III** **(V_{DS} = 1.5 V)**
$I_{\text{DS max}}$	1.53 mA (V_{GS} = –2 V)	21 mA (V_{GS} = –2 V)	5.0 mA (V_{GS} = –1 V)
$I_{\text{DS min}}$	0.8 mA (V_{GS} = 2 V)	20 mA (V_{GS} = 0 V)	2 mA (V_{GS} = 1 V)
ΔV_{GS}	4 V	2 V	2 V
F_{Merit}	0.18	0.5	1.5
g_m (DC)	220 μS	500 μS	1.6 mS
g_d (DC)	0.8 mS	15 mS	4 mS
g_m/g_d	0.275	0.03	0.4

Note: Case I corresponds to an FET where conventional SWCNTs were deposited by selective surface chemistry. Case II corresponds to an FET where conventional SWCNTs were deposited by DEP. Case III corresponds to an FET where 99% pure semiconducting SWCNTs were deposited by DEP [22].

6.1.3.2 High-frequency characterization

Microwave measurements were performed using an Agilent E8361A PNA series network analyzer from 100 MHz to 55 GHz. The measurement and de-embedding techniques are described below.

To calculate extrinsic parameters, the influence of coplanar accesses is removed from the S parameters measurements using the specific structure illustrated in Figure 6.5(a). These extrinsic parameters are considered the actual high-frequency characteristics.

To calculate intrinsic parameters, an "open" test structure is used (Figure 6.5b) for de-embedding. This open structure is strictly identical to the active structure with CNTs except that no nanotube is deposited in the gate area. So all parasitic parameters close to the gate-electrode regions, fringing capacitances, gate-drain, and gate-source overlap capacitances, etc., are removed. With this technique, the contact resistance between access source and drain and array of nanotubes is not removed. It should be noted that test structures, which are passive, are fabricated on the same wafer as active structures (SWCNT FETs).

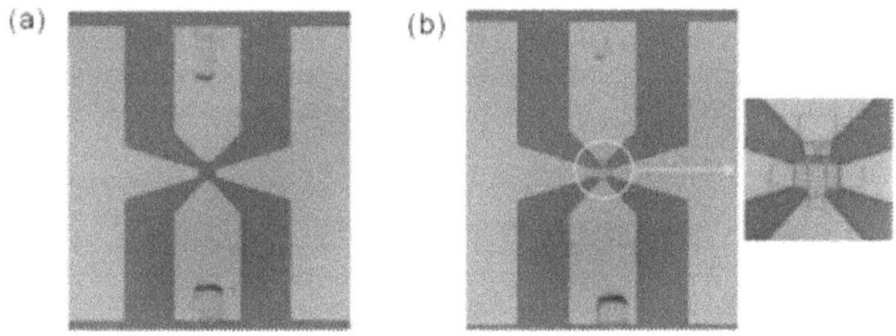

Figure 6.5 Structures used for de-embedding. (a) Structure used to extract extrinsic performance. (b) Structure used to extract intrinsic performance [22].

The current gain (H_{21}) obtained from this de-embedding procedure is plotted in Figure 6.6.

The high-frequency characteristics indicate significant improvement in terms of both extrinsic and intrinsic cutoff frequencies. For the best case, the extrinsic and intrinsic current gain cutoff frequencies are 15 and 80 GHz, respectively. Most important, this work shows that (i) increasing the CNT density in the device channel has improved device performance, (ii) for solution with 99% pure semiconducting nanotubes, the ratio between semiconducting nanotube current is versus metallic nanotube current is improved even if DEP deposition technique is used. Also, this work shows that the precise alignment of SWCNTs is not required to achieve high-frequency performance. This lifts a strong constraint in terms of fabrication processes, thus opening the route for the realistic application of nanotubes in randomly oriented networks.

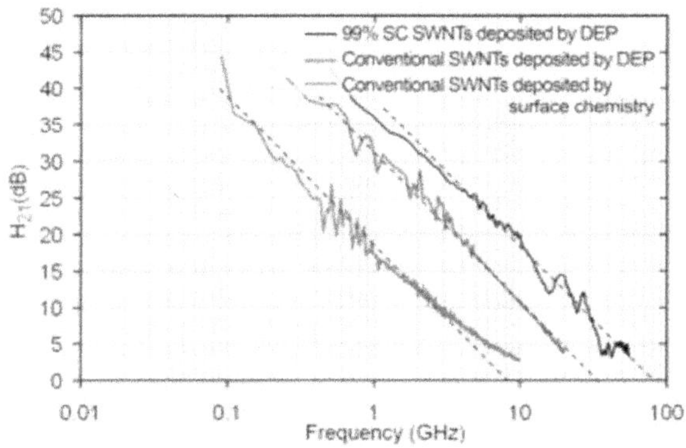

Figure 6.6 Current gain 10 log 10$|H_{21}|^2$, extracted from the S parameters for the three SWCNT FET cases. The dashed lines correspond to the ideal slopes of (20 dB/decade) for $|H_{21}|$. The respective f_t are 8 GHz, 30 GHz, and 80 GHz [22]. See also Color Insert.

6.2 High-Frequency Graphene Field Effect Transistor

Since the demonstration of the stability of graphene under ambient condition in 2004 [23], the research community has been attracted by its potentially high carrier mobility even at room temperature [23–25]. This makes graphene a promising candidate for the fabrication of future generation of high-frequency electronic devices, in particular field-effect transistors (FET) [23, 26–30]. Our work was oriented toward HF properties of graphene nanoribbon field effect transistors (GNRFET).

6.2.1 Graphene Synthesis

The multilayer graphene was obtained by thermal decomposition of Si-face semi-insulating on-axis SiC–4H (0001) substrates [31]. The surface preparation procedure, including a final smoothening ~1 hour-long Si exposure, has been already described in refs 32 and 33. Graphitization was then achieved by ultra-high vacuum annealing up to ~1400°C. We have investigated two samples with different graphene thickness, controlled by the heating times and temperature in the UHV chamber. The number of graphene layers is estimated to be about 10 and 5 for samples 1 and 2 respectively.

The layer properties of these two samples are reported in Table 6.2. The transport properties of graphene layers were studied by room temperature Hall Effect measurements. To this end, 100 μm × 100 μm Hall devices with metallic contacts were fabricated. The maximum mobility of 460 $cm^2/(V \cdot s)$ was obtained with a sheet carriers density of the order of 1×10^{13} cm^{-2}.

Table 6.2 Material properties of sample 1 and sample 2

	Sample 1	Sample 2
Graphene thickness (nm)	3.26	1.66
Number of graphene monolayer	~10	~5
Mobility (cm^2 V^{-1} s^{-1})	460	427
Carrier density (cm^{-2})	-1.2×10^{13}	-0.77×10^{13}

The full width at half-maximum (FWHM) Raman mapping for an area of 50 μm × 50 μm is shown in Figure 6.6a for sample 2. Good layer uniformity is observed, with FWHM about 55 cm^{-1} (Figure 6.6b).

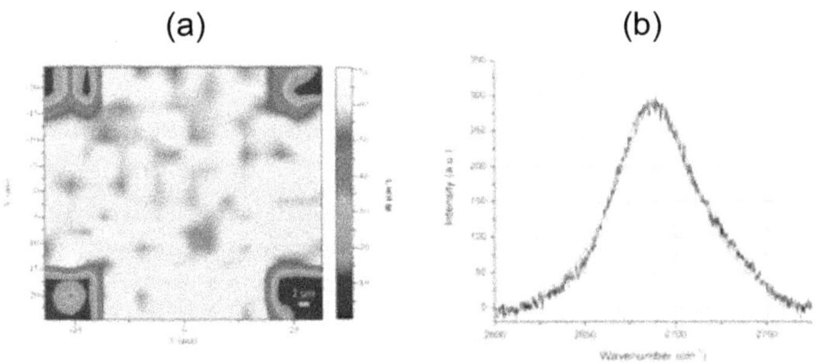

Figure 6.7 (a) FWMH Raman mapping for 50 μm × 50 μm area. (b) Raman spectrum for the zone indicated in (a) (ample 2, graphene thickness of 5 monolayers). See also Color Insert.

6.2.2 GNRFET Fabrication

Figure 6.8 shows scanning electron microscopy (SEM) image of the final device (Figure 6.8b), as well as the schematized cross section of the final structure (Figure 6.8c). The device channel is composed of arrays of parallel GNRs, defined by e-beam lithography. To achieve accurate ribbon width control, hydrogen silsesquioxane (HSQ) was used as mask material (Figure 6.8a). Graphene was then etched by O_2 RIE.

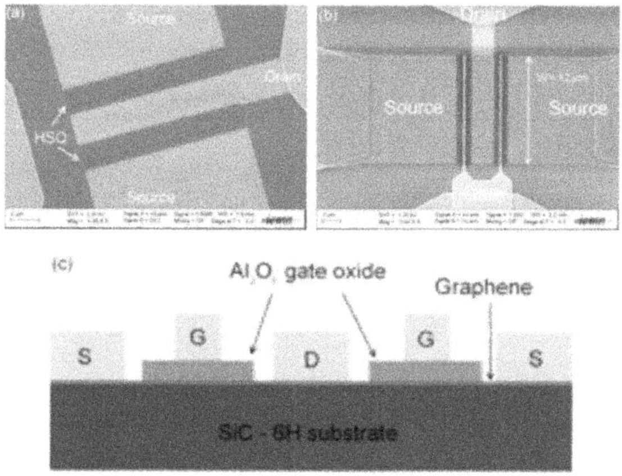

Figure 6.8 SEM image of graphene FET. (a) Array of e-beam resist used to etch graphene. (b) Final device structure with a top gate. (c) Cross section of device.

The source and drain contacts are realized with a standard lift-off process, from the evaporated Ni/Au (50 nm/300 nm) metals. The Al_2O_3 gate oxide was fabricated by oxidation of a thin aluminum layer (about 2 nm) deposited by e-beam evaporation. This was repeated twice to achieve a final oxide thickness of about 5 nm. The top gate with length L_g = 150 nm was defined by a lift-off process of Ni/Au 50 nm/300 nm. The photograph of the final device is shown in Figure 6.8b. The device gate width is 12 μm.

6.2.3 DC Characterization of GNRFET

Figure 6.9 shows the dc output characteristics (drain current versus drain voltage I_{ds}–V_{ds}) for devices on samples 1 and 2. The lower value of gate voltage was determined from the transfer characteristics measurement (drain current versus gate voltage I_{ds}–V_{gs}), for each individual device at V_{ds} = 1 V (inset of Figure 6.9a,b). For sample 1 with 40 ribbons of 100 nm width, the maximum drain current was 1.25 mA/μm at V_{ds} = 1 V and V_{gs} = 1 V. The maximum static transconductance (G_m) of 144 μS/μm was extracted at V_{ds}= 1 V and V_{gs}= −1 V. Nevertheless, the ratio between I_{on} and I_{off} was about 1.5 for sample 1. This may be related to the ribbon width, which is still too large to induce an energy gap [34–36], and also to the number of graphene layers [16]. To confirm this, the same device was fabricated on sample 2, which has about half the number of layers compared with sample 1; we obtained a maximum drain current of 1.1 mA/μm, $I_{on}/I_{off} \sim 3$, and a maximum transconductance of 265 μS/μm.

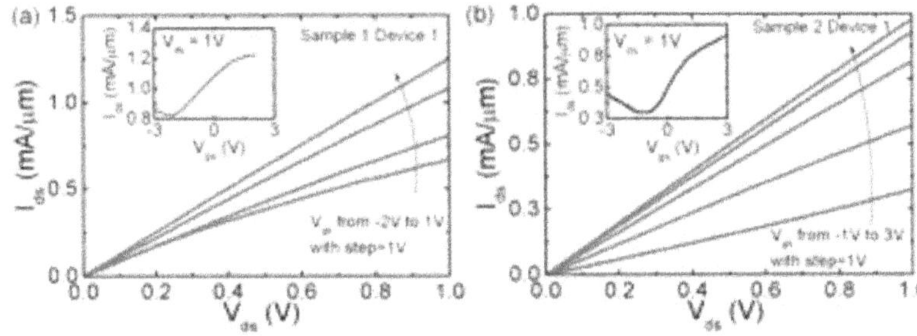

Figure 6.9 (a) DC output characteristics for Sample1_Device1 and (b) Sample2_ Device1. The inset curves show the observation of Dirac point. Both devices have 40 ribbons of 100 nm width.

On sample 1, electrical characterization of GNRFET at 300 K and 77 K was also carried out. Figure 6.10 shows the static output characteristic of this device. At 300 K, the drain current reaches a maximum value of 5 mA at drain voltage V_{ds} = 1 V and gate voltage V_{gs} = 1 V. At 77 K, with the previous bias conditions, the drain current increases. This result (Figure 6.10) illustrates the influence of traps in the structure. Possible explanations include either the contribution of localized edge states of graphene nanoribbons, induced by the structural disorder resulting from the etching process or surface states between the gate dielectric and the graphene.

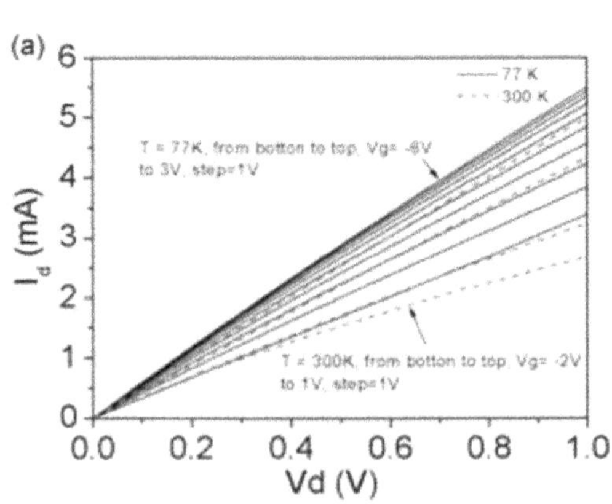

Figure 6.10 DC characteristics of GNRFET in two cases: (red) measurements at 300 K; (black): measurement at 77 K [37].

6.2.4 HF Characterization of GNRFET

Microwave measurements were performed from 10 MHz to 20 GHz. A common Line-Reflect-Match (LRM) calibration procedure was used. The de-embedding procedure is similar to the method described previously in the first part. Figure 6.11 gives the intrinsic current gain ($|h_{21}|_{intr}$) and unilateral gain (U) deduced from de-embedded S-parameters. In our best case, the GNRFETs are biased at V_{ds} = 3 V. For sample 2 (device 2), a current gain cut off frequency (f_t) of 60 GHz and a maximum oscillation frequency (f_{max}) of 28 GHz are obtained [38–39].

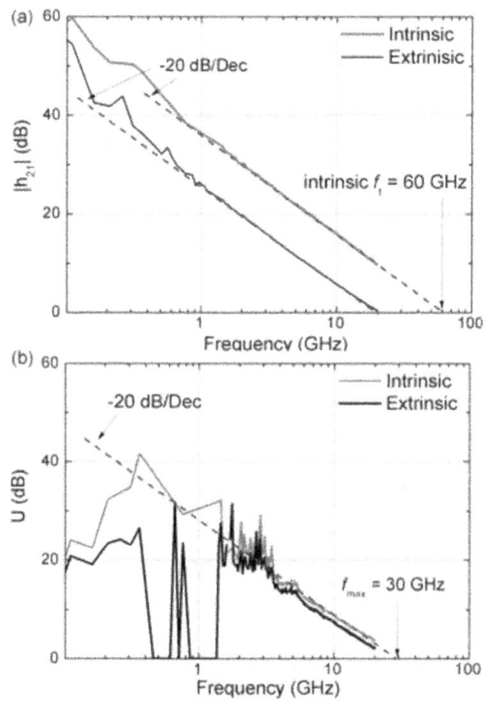

Figure 6.11 RF characteristics of this device under V_{ds}=3 V and V_{ds}=1 V. (a) Short circuit current gain $|h_{21}|$. (b) Unilateral gain (U). See also Color Insert.

In contrast with the change of current gain cut off frequency for extrinsic and intrinsic regimes, no significant difference are observed for f_{max}. This can be explained by using the simple expression of f_{max} given by

$$f_{\max} = \frac{f_t}{2\sqrt{(R_g + R_i)(g_{ds} + 2\pi f_t c_{pgd})}} \quad (6.4)$$

where R_g, R_i, and g_{ds} are the gate resistance, real part of input impedance due to non-quasi-static effect and output conductance, respectively, and C_{pgd} is gate-drain-coupled capacitance. Unlike current gain cut off frequency, which is impacted heavily by the parasitic capacitances, f_{max} is only slightly affected, i.e., f_{max} is not sensitive to the de-embedding procedure used in this work, which considers only the capacitive coupling.

Table 6.3 shows the results obtained on several devices.

Table 6.3 Analysis of devices' performance

	Sample 1		Sample 2		
	Device 1 (300 K)	Device 2 (77 K)	Device 1	Device 2	
	V_{ds}=1 V	V_{ds}=1 V	V_{ds}=1 V	V_{ds}=1 V	V_{ds}=3 V
Ribbon width (nm)	100	100	100	50	50
Ribbon number	40	40	40	120	120
Drive current (mA/μm)	1.25	1.3	1.1	2.1	6
G_m (μS/μm)	144	171	265	245	660
f_t (GHz)	10	15	20	30	60
f_{max} (GHz)	7	12	9	17	28

6.2.5 Conclusion

Carbon based devices are quickly becoming extremely interesting solutions for a wide variety of electronic devices and circuits. This work shows our contribution to the advancement of RF/microwave technology, based on carbon nanoelectronic devices.

Concerning the CNT field effect transport, we have clearly demonstrated the influence of nanotube density, and the influence of high-purity nanotube material on high-frequency characteristics. The main performance obtained with this device family are the extrinsic and intrinsic current gain cutoff frequencies of ~15 and ~80 GHz, respectively. Most important, this work shows that precise alignment of SWCNTs is not required to achieve high-frequency performance. This lifts a strong constraint in terms of fabrication processes, thus opening the route for realistic applications in radio frequency applications, using randomly oriented networks.

In the field of graphene material, even if this research field is quite new, it attracts the attention of the electron-device community, and the improvements of device performance are growing faster. Based on the graphene thermal growth on SiC, we have developed graphene nanoribbon field effect transistors with a good uniformity. The best transistor with gate length of 150 nm shows intrinsic cut-off frequency f_t of 60 GHz and maximum frequency of oscillation (f_{max}) of 28 GHz at V_{ds} = 3V. This value of f_{max} is the state of the art.

By improving material performances and device structures, it should be possible to reach a few hundred of GHz in the future.

References

1. ITRS (2007), Emerging Research Material chapter.
2. Tans, S. J., Verschuren, A.R.M., and Dekker, C. (1998), Room-temperature transistor based on a single carbon nanotube, *Nature (London)* **393**, pp. 49–52.
3. Martel, R., Schmidt, T., Shea, H. R., Hertel, T., and Avouris, P. (1998), Single- and multi-wall carbon nanotube field-effect transistors, *Appl. Phys. Lett.* **73**, pp. 2447–2449.
4. Martel, R., Derycke, V., Lavoie, C., Appenzeller, J., Chan, K., Tersoff, J., and Avouris, P. (201), Ambipolar electrical transport in semiconducting single-wall carbon nanotubes, *Phys. Rev. Lett.* **87**, 256805.
5. Javey, A., Wang, Q., Kim, W., and Dai, H. (2003), Advancements in complementary carbon nanotube field-effect transistors, *Tech. Dig. IEDM.,* 31.2.1–4.
6. Chen, C., Xu, D., Kong, E. S.-W., and Y., Zhang, *(2006),* Multichannel carbon-nanotube FETs and complementary logic gates with nanowelded contacts, *IEEE Electron Device Lett.,* **27**, 10, pp. 852–855.
7. Chen, Z., Appenzeller, J., Lin, Y.-M., Sippel-Aokley, J., Rinzler, A. G., Tang, J., Wind, S. J., Solomon, P. M., and Avouris, P. (2006), An integrated logic circuit assembled on a single carbon nanotube, *Science* **311**, 1735.
8. Kim, S., Choi, T.Y., Rabieirad, L., Jeon, J.-H., Shim, M., and Mohammadi, S. (2005), A poly-Si gate carbon nanotube field effect transistor for high frequency applications, *IEEE International Microwave Symposium Digest* (IMS, Long Beach, CA, 2005).
9. Bethoux, J.M., Happy, H., Derycke, V., Dambrine, G., Goffman, M., and Bourgoin, J.P. (2006), An 8-GHz ft carbon nanotube field-effect-

transistor for gigahertz range applications, *IEEE Electron Device Lett.* **27**, 8, pp. 681–683.

10. Wang, D., Yu, Z., McKernan, S., and Burke, P. J. (2007), Ultrahigh frequency carbon nanotube transistor based on a single nanotube, *IEEE Trans. Nanotechnol.* **6**, 4, 400–402.
11. Engel, M., Small, J. P., Steiner, M., Freitag, M, Green, A. A., Hersam, M. C., and Avouris, P. (2008), Thin film nanotube transistors based on self-assembled, aligned, semiconducting carbon nanotube arrays, *ACS Nano* **2**, 2445 (2008).
12. Cao, Q., Kim, H.-S., Pimparkar, N., Kulkarni, J. P., Wang, C., Shim, M., Roy, K., Alam, M. A., and Rogers, J. A. (2008), Medium-scale carbon nanotube thin-film integrated circuits on flexible plastic substrates, *Nature* **454**, pp. 495–502.
13. Kocabas, C., Dunham, S., Cao, Q., Cimino, K., Ho, X., Kim, H. S., Dawson, D., Payne, J., Stuenkel, M.,| Zhang, H., Banks, T., Feng, M., Rotkin, S.V., and Rogers, J. A. (2009), High-frequency performance of submicrometer transistors that use aligned arrays of single-walled carbon nanotubes, *Nano Lett.* **9**, 5, pp. 1937–1943.
14. Zhou, W., Rutherglen, C., and Burke, P. J. (2008), Wafer scale synthesis of dense aligned arrays of single-walled carbon nanotubes, *Nano Res.* **1**, pp. 158–165.
15. Ding, L., Tselev, A., Wang, J., Yuan, D., Chu, H., McNicholas, T. P., Li, Y., and Liu, J. (2009), Selective growth of well-aligned semiconducting single-walled carbon nanotubes, *Nano Lett.*, **9**, 2, pp. 800–805.
16. Arnold, M. S., Green, A. A., Hulvat, J. F., Stupp, S. I., and Hersam, M. C. (2006), Sorting carbon nanotubes by electronic structure using density differentiation, *Nat. Nanotechnol.* 1, 1, pp. 60–65.
17. Auvray, S., Derycke, V., Goffman, M., Filoramo, A., Jost, O., and Bourgoin, J. P. (2005), Chemical optimization of self-assembled carbon nanotube transistors, *Nano Lett.*, 5, 3, pp. 451–455.
18. Krupke, R., Hennrich, F., Weber, H. B., Kappes, M. M., and Löhneysen, H.V. (2003), Simultaneous deposition of metallic bundles of single-walled carbon nanotubes using ac-dielectrophoresis, *Nano Lett.*, 3, 8, pp. 1019–1023.
19. Boccaccini, A. R. , Cho, J., Roether, J. A., Thomas, B. J. C., Minay, E. J., and Shaffer, M. S. P. (2006), Electrophoretic deposition of carbon nanotubes, *Carbon* 44, pp. 3149–3160.
20. Bethoux J.M., Happy H., Siligaris A., Dambrine G., Borghetti J., Derycke V., Bourgoin J.P. (2006), Active properties of carbon nanotube field-

effect transistors deduced from S parameters measurements, *IEEE Trans. Nanotech.*, 5, 4, pp. 335–342.

21. Le Louarn, A., Kapche, F., Bethoux, J.-M., Happy, H., Dambrine, G., Derycke, V., Chenevier, P., Izard, N., Goffman, M. F., and Bourgoin, J.-P. (2007), Intrinsic current gain cutoff frequency of 30 GHz with carbon nanotube transistors, *Appl. Phys. Lett.* **90**, 233108.

22. Nougaret L., Happy H., Dambrine G., Derycke V., Bourgoin J.P., Green A.A., Hersam M.C. (2009), 80 GHz field-effect transistors produced using high purity semiconducting single-walled carbon nanotubes, *Appl. Phys. Lett.*, 94, 24, 243505.

23. Novoselov, K.S., Geim, A.K., Morozov, S.V., Jiang, D., Zhang, Y., Dubonos, S.V., Grigorieva, I.V., and Firsov, A.A. (2004), Electric field effect in atomically thin carbon film, *Science*, 306 (5696), pp. 666–669.

24. Morozov, S.V., Novoselov, K.S., Katsnelson, M.I., Schedin, F., Elias, D.C., Jaszczak, J.A., and Geim, A.K. (2008), Giant intrinsic carrier mobilities in graphene and its bilayer, *Phys. Rev. Lett.*, 100, 1, pp. 016602.

25. Bolotin, K., Sikes, K., Jiang, Z., Klima, M., Fudenberg, G., Hone, J., Kim, P., and Stormer, H. (2008), Ultrahigh electron mobility in suspended graphene, *Solid State Commun.*, **146**, 9/10, pp. 351–355.

26. Lin, Y., Dimitrakopoulos, C., Jenkins, K.A., Farmer, D.B., Chiu, H., Grill, A., and Avouris, P., 100-GHz transistors from wafer-scale epitaxial graphene (2010), *Science*, 327, 5966, p. 662.

27. Meric, I., Baklitskaya, N., Kim, P., and Shepard, K.L. (2008), RF performance of top-gated, zero-bandgap graphene field-effect transistors, *Proceedings of the IEEE International Electron Devices Meeting*, pp. 1–4.

28. Moon, J.S., Curtis, D., Hu, M., Wong, D., McGuire, C., Campbell, P.M., Jernigan, G., Tedesco, J.L., VanMil, B., Myers-Ward, R., Eddy, C., and Gaskill, D.K. (2009), Epitaxial-graphene RF field-effect transistors on Si-face 6H-SiC substrates, *Electron Device Lett.*, **30**,.6, pp. 650–652.

29. Meric, I., Han, M.Y., Young, A.F., Ozyilmaz, B., Kim, P., and Shepard, K.L. (2008), Current saturation in zero-bandgap, top-gated graphene field-effect transistors, *Nat. Nano*, **3**, 11, pp. 654–659.

30. Wu, Y., Lin, Y.-M., Bol, A. A., Jenkins, K. A., Xia, F., Farmer, D. B., Zhu, Y., and Avouris, P. (2011), High-frequency, scaled graphene transistors on diamond-like carbon, *Nature* **472**, pp. 74–78.

31. Berger, C., Song, Z., Li, T., Li, X., Ogbazghi, A.Y., Feng, R., Dai, Z., Marchenkov, A.N., Conrad, E.H., First, P.N., and De Heer, W.A. (2004), Ultrathin epitaxial graphite: 2D electron gas properties and a route

toward graphene-based nanoelectronics, *J. Phys. Chem. B*, **108**, 52, pp. 19912–19916.

32. Ferrer, F.J., Moreau, E., Vignaud, D., Godey, S., and Wallart, X. (2009), Atomic scale flattening, step formation and graphitization blocking on 6H- and 4H-SiC{0001} surfaces under Si flux, *Semicond. Sci. Technol.*, **24**, 12, p. 125014.
33. Starke, U., Bernhardt, J., Schardt, J., and Heinz, K. (1999), SiC surface reconstruction: relevancy of atomic structure for growth technology, *Surf Rev Lett*, **6**, 6, p. 1129–1141.
34. Han, M.Y., Ozyilmaz, B., Zhang, Y., and Kim, P. (2007), Energy band-gap engineering of graphene nanoribbons, *Phys. Rev. Lett.*, **98**, 20, pp. 206805.
35. Zhou, S., Siegel, D., Fedorov, A., Gabaly, F., Schmid, A., Neto, A.C., Lee, D., and Lanzara, A. (2008), Origin of the energy bandgap in epitaxial graphene, *Nat. Mater.*, **7**, 4, pp. 259–260.
36. Shemella, P., Zhang, Y., Mailman, M., Ajayan, P.M., and Nayak, S.K. (2007), Energy gaps in zero-dimensional graphene nanoribbons, *Appl. Phys. Lett.*, **91**, 4, pp. 042101.
37. Meng, N., Ferrer, F. J., Vignaud, D., Dambrine, G., and Happy, H. (2010) Influence of temperature on high frequency performance of graphene nano ribbon field effect, *Proceedings of the 68th Device Research Conference*, doi: 10.1109/DRC.2010.5551934, pp. 69–70
38. Meng, N., Ferrer, F. J., Vignaud, D., Dambrine, G., and Happy, H. (2010) 60 GHz current gain cut-off frequency graphene nanoribbon FET, *Int. J. Microw. Wireless Technol.*, 2, 5, pp. 441–444.
39. Meng, N., Ferrer, F. J., Vignaud, D., Dambrine, G., and Happy, H. (2011) Fabrication and characterization of epitaxial-graphene nano ribbon field-effect transistor , *IEEE Trans Electron Device.*, 58, 6, pp. 1594–1596.

Chapter 7

From Bulk TFETs to CNT-TFETs: Status and Trends

Chong Guan Low and Qing Zhang
Microelectronics Center, School of Electrical and Electronic Engineering, Nanyang Technological University, 639798 Singapore
lowc0023@e.ntu.edu.sg

This chapter presents the current status and trends of tunneling field effect transistors (TFET) and the advantages of carbon nanotube (CNT) based TFETs. It has been widely reported that TFET is of potential to be the next-generation device design structure as it can eliminate the scaling problems that exist in CMOS and improve the subthreshold swings. Performances of different TFETs designs, including CNTFETs, are discussed and compared with conventional transistors.

7.1 Introduction

The miniaturization of MOSFETs leads to fundamental physical limits that give rise to many problems such as leakage current, doping atom fluctuation etc. [1–3]. The off-current of conventional

Carbon Nanotubes and Their Applications
Edited by Qing Zhang

ISBN 978-981-4241-90-8 (Hardcover), 978-981-4303-18-7 (eBook)
www.panstanford.com

MOSFETs is exponentially increasing as a result of non-scalability of their threshold voltages. Leakage current has reached such a critical stage that power efficiency is of the major concern now. This has caused power crisis in the silicon chip industry [5]. To overcome this problem, researchers have been opting for new device design concepts that could bring the size of transistors down to 20 nm. Among all, tunnel field effect transistors (TFETs) have been drawing much attention and are expected to be the next-generation devices. As reported in ITRS 2005 [6], the TFET technology fits best into the Low Standby Power (LSTP) category due to its low off-current. On top of that, TFETs have excellent subthreshold swing of less than 60 mV/dec at 300 K, which is impossible to be achieved by conventional MOSFET technology. In TFETs, the conduction of carriers is caused by band-to-band-tunneling (BTBT) while, in MOSFETs, it relies on thermal emission of carriers over a channel barrier, which has the limit of $2.3{*}k_BT/q$ = 60 mV/dec at 300 K. The outstanding subthreshold swing and low leakage current make it an ideal device design to replace conventional MOSFET to realize high switching speed, high power efficiency, low operating current, and low leakage current.

However, the problems with TFETs have also been addressed. The on-current of TFETs is considerably low for high-speed circuit application. Recently, researchers have been reporting on various device designs to overcome this problem. One way is to use SiGe as the channel material to increase the on-current so that it is comparable to the on-current of MOSFETs. Considering the cost and compatibility with existing and matured silicon fabrication technology, standard CMOS fabrication and batch processing are preferred. In addition, the state-of-the-art device designs and the employment of new materials, such as multi-gate (MuG), high-k gate dielectric, channel and source/drain engineering, etc., can also be used to improve the on-current of TFETs.

7.2 Basic Structure

The basic structure of a typical TFET is a gated p-i-n diode. In this chapter, in order to be consistent with MOSFET technology, we assume that the source is made from p^+ silicon while the drain is made from n^+ silicon, so that when $V_{ds} > 0$ is applied, the p-i-n diode is reverse biased to realize n-TFET (consistent with n-MOSFET). The intrinsic silicon layer, which is the channel of TFETs, is sandwiched between the source and drain. Figure 7.1 shows a basic structure of n-TFETs with (a) single gate and (b) double gate (DG), respectively [7]. DG or multi-gate (MuG) TFETs are used to improve the gate control of carriers in the channel. This approach can be used to increase the on-current of the device [1].

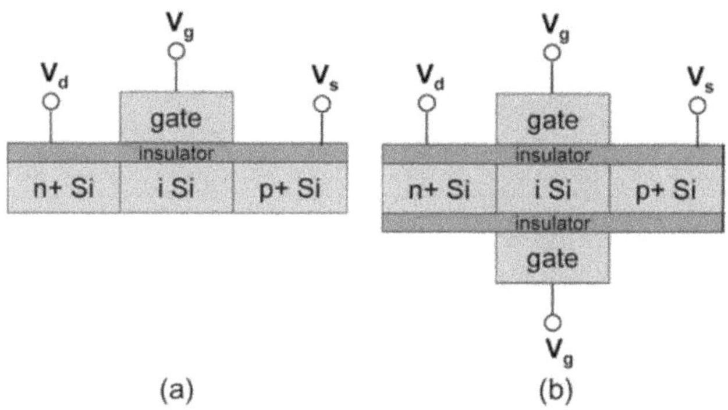

Figure 7.1 Basic structure of n-TFETs with (a) single gate and (b) double gate. Adapted from [7].

7.3 Device Operation and Issues of TFETs

7.3.1 Working Principle

Figure 7.2 shows the band diagram of an n-TFET with a positive voltage applied to the n^+ drain and the p^+ source is grounded [8]. The p-i-n structure is reverse-biased and it is worth mentioning that the reverse-biased p-i-n structure of TFET plays an important role in achieving low leakage current in off-state. Under a zero gate bias, as shown in Figure 7.2a, there is no current flow between the source and drain, as the energy barrier between the intrinsic and the p^+ region is so wide that the electrons cannot tunnel through it. Hence, the device is in the off-state. When a positive gate voltage is applied, as shown in Figure 7.2b, the energy barrier between the source and intrinsic channel becomes much narrower, increasing the probability of electrons tunneling through the barrier. Electrons in the valence band at the p^+ source will then tunnel through the energy barrier, causing a current from the drain to source. It is interesting to note that under a smaller gate voltage, the energy barrier is wider, resulting in a much smaller tunneling rate, as shown in Figure 7.2c.

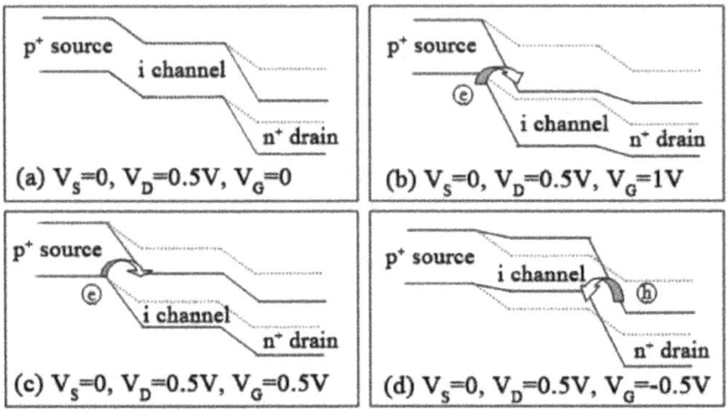

Figure 7.2 Band Diagrams of a TFET under (a) zero gate bias, (b) V_{DS} = 0.5 V and V_{GS} = 1 V (N-TFET), (c) V_{DS} = 0.5 V and V_{GS} = 0.5 V (N-TFET with small tunneling rate at source-channel), and (d) V_{DS} = 0.5 V and V_{GS} = −0.5 V (ambipolar conduction). Adapted from [8].

For a basic TFET with symmetric source and drain doping, ambipolar conduction can be observed, as shown in Figure 7.2d. In other words, the device will be turned on whenever a positive or negative gate voltage is applied. In this case, carrier tunneling takes place at either the source or the drain side. Due to the different working principle of TFETs compared with conventional MOSFETs, it is widely reported that a TFET is weakly dependent on temperature, which makes it an excellent device design for high-temperature applications.

7.3.2 Subthreshold Swing

Subthreshold swing/subthreshold slope is defined as the voltage required to increase or to decrease the source-drain current by one decade. It can be determined by the inverse of the slope of the log (I_{DS}) versus V_{GS} curve. For low-power logic applications, it is important to keep subthreshold swing as small as possible in order to completely turn off the device in the weak inversion regime. Ideally, when $V_{GS}<V_{th}$, there should be no current flow between the drain and the source so that the transistor can be completely turned off. In reality, for conventional MOSFETs, electrons with more energy tend to enter the channel, which gives rise to the so-called subthreshold leakage. This electron energy follows the Boltzmann distribution. For conventional MOSFETs, the current I_D in weak inversion of $V_{GS}<V_{th}$ (cutoff mode) can be expressed as the following formula:

$$I_{\mathrm{D}} \approx I_{\mathrm{D0}} e \frac{V_{\mathrm{GS}} - V_{\mathrm{th}}}{n V_{\mathrm{T}}}, \tag{7.1}$$

where $V_{\mathrm{T}} = k_{\mathrm{B}}T/\mathrm{q}$, I_{D0} = current at $V_{\mathrm{GS}} = V_{\mathrm{th}}$, and n is the slope factor. Subthreshold slope can be defined as

$$S = \frac{1}{\delta \log I_D / \delta V_G} \tag{7.2}$$

Hence,

$$S = \frac{1}{\dfrac{\delta \log I_{\mathrm{D}}}{\delta V_{\mathrm{G}}}}$$

$$= \frac{n k_{\mathrm{B}} T}{q \log e}$$

$$= \frac{n k_{\mathrm{B}} T \ln 10}{q}$$

$$= \frac{2.3 nkT}{q} [\mathrm{V/dec}] \tag{7.3}$$

From Equation 7.3, the smallest possible subthreshold swing for conventional MOSFETs is 2.3 kT/q when $n = 1$, which is around 60 mV/dec at 300 K.

The limitation of subthreshold swing of conventional MOSFETs remains to be the major problem when it comes to 20–30 nm node. The subthreshold swing is not only limited by diffusion current of the device in the weak inversion, but also limited by its dependence on temperatures. This is applicable to MOSFETs, which rely on thermal emission of carriers over a channel barrier. In contrast, the working principle of TFETs is based on tunneling, which allows subthreshold swing of smaller than 60 mV/dec, with less dependence on temperatures. The subthreshold swing of TFETs can be expressed as follows [7]:

$$S_{\mathrm{TFET}} = \frac{V_{\mathrm{GS}^2}}{5.75(V_{\mathrm{GS}} + \mathrm{Const})} \left[\frac{\mathrm{mV}}{\mathrm{dec}} \right] \tag{7.4}$$

The equation was derived based on the BTBT model as reported by Hurkx *et al.* [9]. As the gate voltage induces the band bending, the subthreshold swing of TFETs is dependent on the gate voltage. The constant in the equation is determined by the device dimensions and material parameters of the devices.

7.3.3 Off-Current, I_{off}

TFET has a very small off-current. Unlike MOSFETs, which suffer from short-channel effect (SCE), where the leakage current increases with the decrease of channel length, the leakage current in a TFET is small due to its built-in tunnel barrier and the reverse biased p-i-n structure. It is reported that the off-current of a TFET is dependent on temperature, as shown in Figure 7.3 [7]. This is because in TFETs, the off-current is mainly caused by the generation of carriers in a reverse biased junction. Hence, it can be observed that the off-current tends to increase when the temperature increases. However, this value remains sufficiently low since no BTBT exists under a reverse bias for the source and drain, if there is no gate voltage applied to shift the band diagram of the channel region.

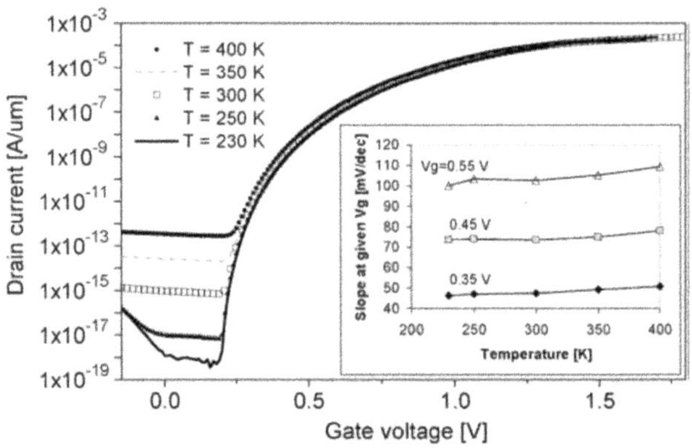

Figure 7.3 Temperature dependence of TFETs off-current. TFET off-current is caused by the thermal generation of carriers in a reverse biased junction. Hence, it increases with increasing temperature. The off-current of TFETs is sufficiently low compared with conventional MOSFETs. Adapted from [7].

7.3.4 On-current, I_{on}

It has been reported that TFETs tend to have a small on-current compared with conventional MOSFETs. As a result, the switching speed of TFETs is smaller than the switching speed of MOSFETs. Increasing the on-current of TFETs remains to be one of the major challenges for researchers in recent years. Various designs have been proposed and reported, for example, DG/MuG design, high-k gate dielectric etc, which will be discussed in next section. Based on Kane's model, an analytical expression for the TFET on-current I_{on} as a function of the gate voltage V_{GS} was first derived by Vandenberghe *et al.* in 2007 [10],

$$I_{on} \approx WLTe^{s'}\sqrt{V_{GS} - V_{onset}}\sqrt{V_{GS} - V_{onset}}\,, \tag{7.5}$$

where W and L are the width and length of the gate respectively, prefactor T can be expressed as

$$T = q\frac{A}{Bq^{3/2}}\frac{qN_a}{2\in_s}\sqrt{\frac{1}{E_g\gamma}}e^{-BqE_g\sqrt{2\in_S}/\sqrt{q^2Na}} \tag{7.6}$$

V_{onset}, the onset of BTBT in depletion region, can be expressed as

$$V_{onset} = V_{FB} + \frac{E_g}{q}\left(1 + 2t_{ox}\frac{\in_S}{\in_{OX}}\sqrt{\frac{q^2Na}{2Eg\in_S}}\right) \tag{7.7}$$

S' can be expressed as

$$S' = Bq\sqrt{\frac{2E_g\in_s 1}{qNa\,\gamma}} \tag{7.8}$$

where q is the electric charge, ε_s is the permittivity of semiconductor, A and B are the parameters depending on the effective mass of valence and conduction bands, E_g is the semiconductor bandgap, N_a is the acceptor doping concentration of the source. γ can be expressed as

$$\gamma = 1 + t_{ox}\frac{\in_S}{\in_{OX}}\sqrt{\frac{q^2N_a}{2E_g\in_S}} \tag{7.9}$$

The on-current is strongly dependent on the tunneling probability T at the energy barrier. It can be analytically calculated by using the following expression [11]:

$$T(E) \approx \exp\left(-\frac{4\sqrt{2m^*}E_g^{3/2}}{3q\hbar(E_g + \Delta\Phi)}\sqrt{\frac{\varepsilon_{si}}{\varepsilon_{OX}}t_{OX}t_{si}}\right) \tag{7.10}$$

where m^* is the effective carrier mass, E_g is the bandgap, $\Delta\Phi$ is the energy range over which tunneling can take place, and t_{ox}, t_{Si}, ε_{ox}, and ε_{Si} are the oxide and silicon film thickness and dielectric constants for the gate oxide layer and silicon, respectively. It can be seen that by decreasing the bandgap, decreasing the thickness of gate dielectric and using high-k dielectric, the tunneling rate will increase, resulting in a higher on-current. The bandgap is considered one of the most important component in determining high tunneling probability and high on-current as it determines the barrier between conduction and valence band. V_{onset} is also reduced for a smaller bandgap TFET as the voltage required to shift the conduction and valence band for BTBT is smaller, resulting in a larger on-current for the same gate voltage applied.

7.3.5 Ambipolar Conduction

As discussed above, the conduction of a TFET is based on carrier tunneling. When a positive gate voltage is applied, the valence band of the p^+ source is higher than the conduction band of intrinsic region, resulting in electrons tunneling at the source side, as shown in Figure 7.2 (b). It is the same for the drain side if a negative gate voltage is applied, causing the holes tunneling from the drain-end. The ambipolarity of a TFET can be observed if the source and drain have a symmetric doping. However, ambipolar conduction is not desired in most electronic applications as it increases the leakage current at the off-state. One approach to overcome this problem is to introduce a wider tunneling barrier at the drain-end and a thinner tunneling barrier at the source-end. As a result, the tunneling rate at the source-end is higher than the tunneling rate at the drain-end, hence achieving a high on-current and a low off-current. This can be done by introducing an underlapped drain, lower drain doping and high bandgap drain, which will be discussed in the next section. Figure 7.4 shows a typical ambipolar characteristic of a TFET [12].

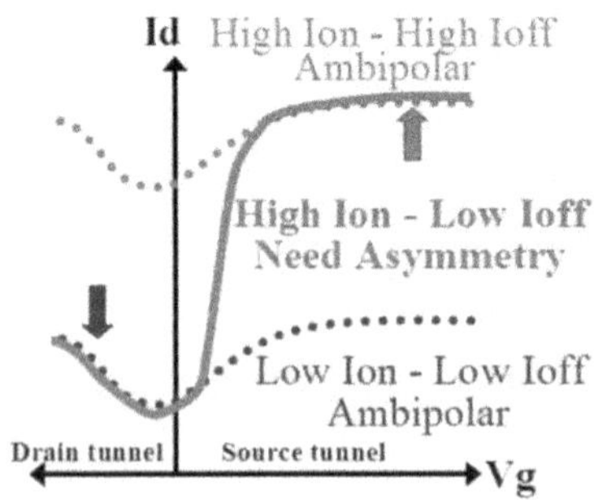

Figure 7.4 Symmetric TFETs exhibit ambipolar characteristics. In order to suppress the leakage current due to ambipolarity of TFETs, asymmetric source-drain can be used. Adapted from [12].

7.4 TFET Design Trends

7.4.1 Source/Drain Engineering

One way to eliminate the ambipolar conduction of TFETs is to adopt asymmetric source and drain structure. The transfer characteristics of symmetric and three different designs of asymmetric source-drain DG TFET are compared [12]. The three asymmetric strained-Ge (s-Ge) DG TFET configurations to be discussed are an underlapped drain, a low drain-doping, and a lateral heterostructure of a large-bandgap material, as shown in Figure 7.5 [12]. The general objective of adopting the asymmetric source and drain structure is to achieve a high tunneling rate at the source-end and a low tunneling rate at the drain-end in order to maintain a high on-current and a low off-current (leakage current).

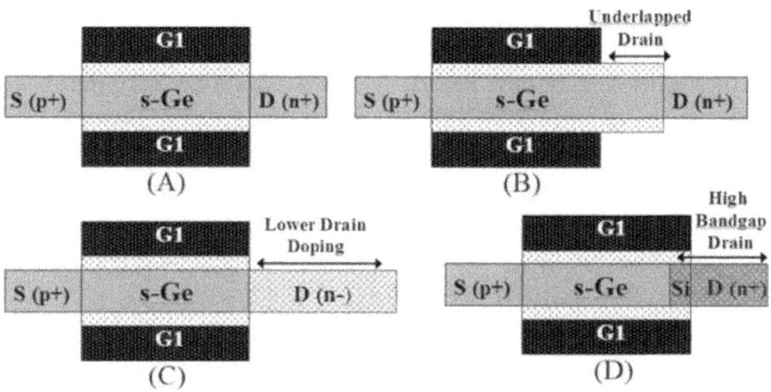

Figure 7.5 (a) Symmetric strained-Ge DG TFET structure with ambipolar characteristic. (b) Underlapped drain. (c) Lower drain doping. (d) Lateral heterostructure of large bandgap drain. Adapted from [12].

For the underlapped drain structure, the drain-end channel is not covered by the gate. In other words, only the source-end energy band is fully controlled by the gate. Contrary to MOSFET, a TFET can function with a gate that does not cover the whole channel. This flexibility allows the gate length of the TFET to be scaled down to 20 nm or smaller, without causing a significant short-channel effect. When a positive voltage bias is applied, the tunneling barrier at the source-end reduces, allowing the carriers to tunnel through it. For the drain-end channel, since the uncovered channel provides a longer depletion width, the electric field in this region is obviously smaller. Therefore, the tunneling rate can be reduced due to the large barrier width, leading to unipolar characteristics.

The concept for lower drain-doping is, in fact, similar to the case of the underlapped drain, which is to provide a longer depletion width to reduce the electric field in that region, so that the barrier width remains almost the same or sufficiently large when the gate voltage is applied. In this design, the drain length can also be increased to further reduce the electric field. Fundamentally, these two structural designs follow the same approach. However, Krishnamohan *et al.* show that, based on their simulation result, in order to achieve high on-current, low leakage current and excellent subthreshold swing of <60 mV/dec, the non-scaling drain need to be at least 30 nm [12].

Due to the restriction of the minimum length of drain, another approach, which is to increase the energy bandgap of the drain to decrease the tunneling rate, is introduced. This can be done by introducing a lateral heterostructure of doped large-bandgap material on the drain side [13]. As the channel material is strained Ge, the relatively large-bandgap material for the drain in this case can be, for example, silicon. By doing so, the drain length need not be increased. The relatively large bandgap of silicon, compared with Ge, provides a natural larger barrier to suppress the leakage current.

Since tunneling only takes place at the source/s-Ge interface, silicon drain can be extended to the channel region, which will then give a longer drain and hence reducing the electric field at the drain end. As shown in Figure 7.6, ambipolar conduction of TFETs can be greatly reduced by using the lateral heterostructure of s-Ge/Si [12]. It is interesting to note that the closer the s-Ge/Si interface to the p+ source, the lower leakage current is observed. However, it can be seen from Figure 7.6 that the on-current of the TFET is smaller when silicon is at the source side. This can be explained by the lower tunneling rate caused by the large-bandgap of silicon. The s-Ge/Si interface located at 2 nm from the source gives the best result of high on-current and almost no ambipolarity characteristic. Hence, lateral heterostructure is reported to be the most scalable approach to suppress ambipolarity of TFETs [12].

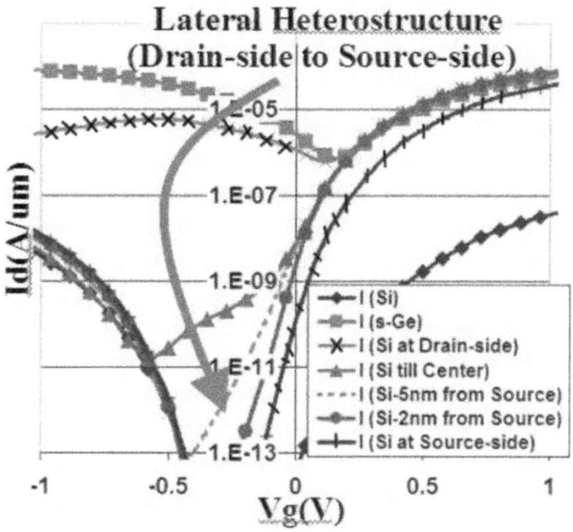

Figure 7.6 The effect of location of strained-Ge/Si interface on the leakage of the lateral heterostructure TFET. The S-Ge/Si interface located at 2·nm from the source gives the best result of high on-current and almost no ambipolarity characteristic. Adapted from [12]. See also Color Insert.

7.4.2 Channel Engineering

One way to increase the on-current of TFETs is to increase the tunneling rate at the source side, which is determined by several factors as discussed above. For this purpose, a narrow-gap semiconductor is used as the intrinsic channel region of TFETs as it can effectively enhance the tunneling rate at the source side due to the shorter tunneling distance and reduced barrier height.

Ge and InAs can be used as the channel material due to their narrow bandgaps [4]. Simulation results show that Ge and InAs TFETs have excellent performances in the switching delay and energy-delay product compared with Si TFETs. It is also reported that MOSFETs, unlike TFETS, do not meet the I_{ON}-I_{OFF} stipulation when it comes to low V_{DD} range of around 0.25 V. This provides another option for ultra-low standby power and high-speed application for low supply voltage range [4]. In Figure 7.7, it is shown that InAs TFET has the lowest switching delay compared with Ge TFET and Si TFET [4]. The device structures used in this simulation are based on DG and asymmetric source and drain doping for the purpose of increasing on-current and suppressing ambipolarity characteristic of the TFETs. It is reported that InAs TFETs show excellent performance with I_{ON}/I_{OFF} of > 4 × 10^4 and intrinsic delay of <1 ps under the condition of 0.25 V supply voltage, 30 nm gate length, 7 nm body thickness, and Gaussian doping profile with doping gradients of 2 nm/decade [4].

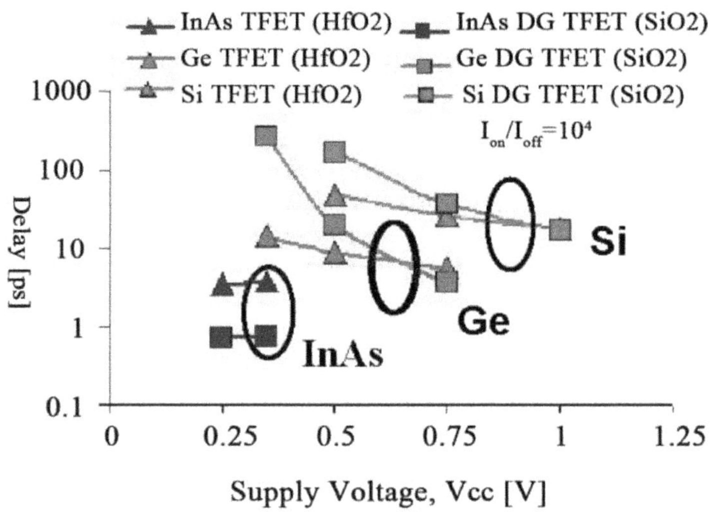

Figure 7.7 Intrinsic delay versus supply voltage of InAs TFET, Ge TFET, and Si TFET. It is shown that InAs TFET has the lowest delay among the devices. Adapted from [4]. See also Color Insert.

A DG s-Ge Heterostructure TFET has been reported to achieve a record high on-current of 300 μA/μm and 50 mV/dec subthreshold slope [12]. This on-current value is comparable to that of conventional MOSFET, and it can be attributed to the reduced tunneling resistance through the narrow bandgap of s-Ge. From the I_D–V_G curve of an experimental DG TFET with s-Ge heterostructure channel in Figure 7.8, it is seen that the on-current is around 300 μA/μm while the subthreshold slope is around 50 mV/dec [12].

Besides using small bandgap semiconductors as the intrinsic channel materials, nanowires or nanotubes such as CNT can also be used as the channel of TFETs due to its excellent 1-D ballistic transport. Further discussion of CNT based FETs will be given in the next section.

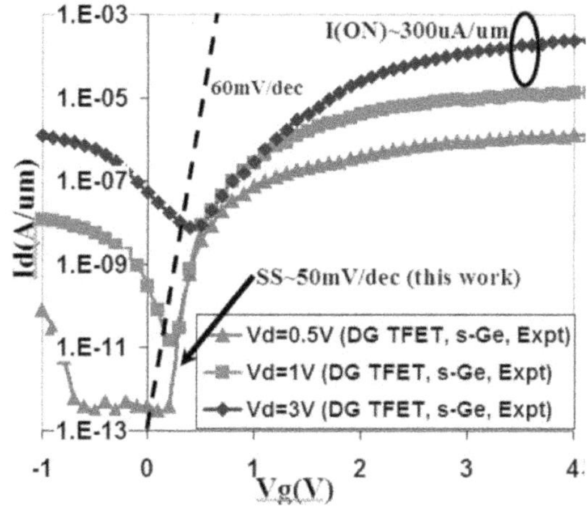

Figure 7.8 I_D versus V_G graph for experimental DG TFET with strained-Ge heterostructure channel, showing record high on currents of about 300 μA/μm and excellent subthreshold swing of 50 mV/dec. Adapted from [12]. See also Color Insert.

7.4.3 Double Gate (DG) / Multi-Gate (MuG)

The major problem for TFETs is the low on-current compared to conventional MOSFETs, which may limit the devices from high-speed applications. By using a DG or MuG to control the channel, one can expect the on-current to increase, as shown in Figure 7.9 [14]. The advantage of DG or MuG TFET can be realized by its double channel or multi-channel controlled by the DG or MuG, which in turn increases the current flows between the source and drain. It has been shown that a DG TFET has a comparable on-current with MOSFET [12].

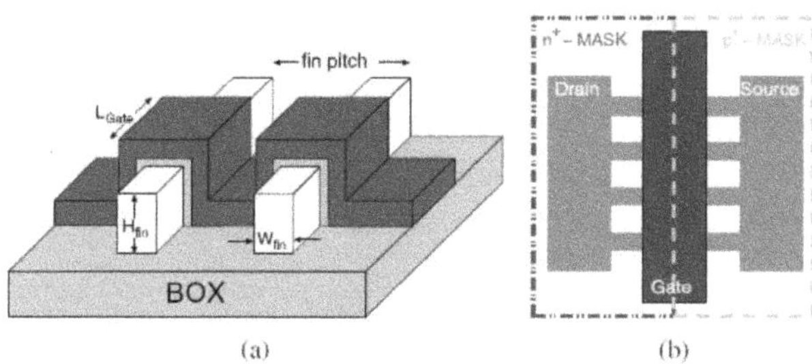

Figure 7.9 Schematic view of MuGFET structure and implementation. Adapted from [14].

However, the concept of combining MuG technology with TFETs is still not widely reported. Fulde *et al.* have reported on the performance of a SOI MuG TFET for the first time [14]. In their work, they combined the scaling potential of three-dimensional and tunneling devices, and showed that it is a promising candidate to replace planar technologies for 32 nm CMOS nodes. Their TFET is less sensitive to the channel length, allowing channel length scaling without affecting its performance and free from short-channel effect.

7.4.4 Gate Dielectric: Dielectric Constant and Thickness

It has been reported the possibility of using high-k material or reducing the thickness of gate dielectric to solve the major problem of TFETs with a low on-current. As discussed in previous section, the tunneling rate can be expressed as Equation 7.10. The thickness and dielectric constant of the gate dielectric layer play an important role to achieve high on-current as higher dielectric constant or small gate thickness tends to give a higher tunneling rate. This can be explained by the excellent gate control capability, where the gate has almost full control over the bands in the channel. Knoch *et al.* has reported the effect of quantum capacitance limit (QCL) to the gate delay [11, 15]. It has been found that the gate delay decreases greatly for a larger geometrical oxide capacitance. It was suggested that such QCL can be reached in the case of a wrap-gate device structure with a thin-gate dielectric thickness [11].

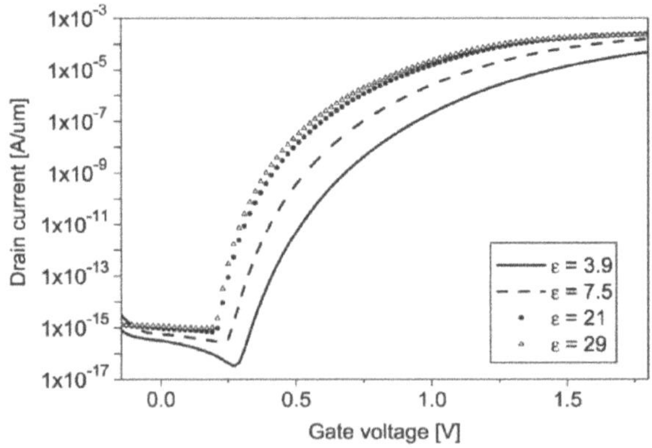

Figure 7.10 Drain current versus gate voltage for different dielectric constants. High-k dielectric tends to give higher on-current, with no significant change to the off-current. Adapted from [7].

Boucart *et al.* suggested that the on-current increases as the gate dielectric constant increases, based on their simulation results. This can be observed in Figure 7.10 [7]. Besides that, unlike MOSFET, the on-current of the TFETs does not increase proportionally with the increase in the gate capacitance. Instead, it increases exponentially. This can be explained by the exponential increase in the current caused by the reduction of the energy barrier width, as a result of improved coupling between the gate and the channel [7]. Figure 7.11 shows the relationship between the energy barrier width and gate voltage as discussed above [7].

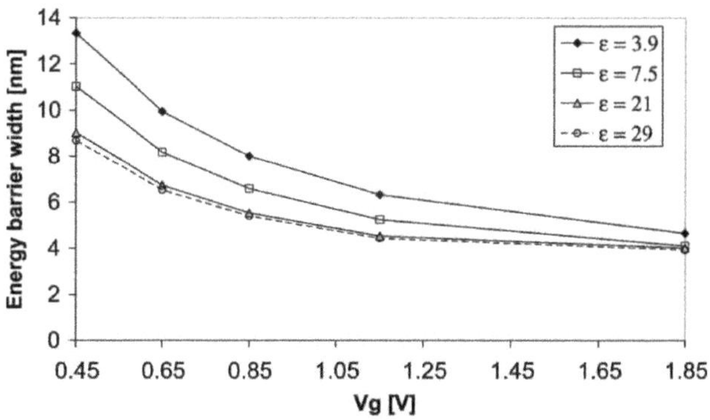

Figure 7.11 Energy barrier width (nm) for band-to-band tunneling versus gate voltage for different dielectric constant. The lower the dielectric constant is, the larger energy barrier width can be achieved. Adapted from [7].

While high-k gate dielectric can be used to improve the tunneling rate, low-k gate dielectric can also be used to reduce the leakage current in DG TFETs, as reported by Vadizadeh *et al.* [16]. As discussed in previous section, the basic approach to reduce the leakage current is to reduce the electric field near the drain-end by using underlapped drain or low doping drain, or high bandgap to suppress the carriers tunneling at the drain-end. Another way to realize high on-current and low off-current is by adopting the split gate structure as shown in Figure 7.12 [16]. For the section nearer to the source-end, a high-k dielectric is used to increase the tunneling rate of the TFET. For the drain-end, a low-k gate dielectric is used to reduce the tunneling rate, which is responsible for the leakage current. This approach increases the on-current and, at the same time, reduces the off-current, leading to a higher I_{on}/I_{off} ratio.

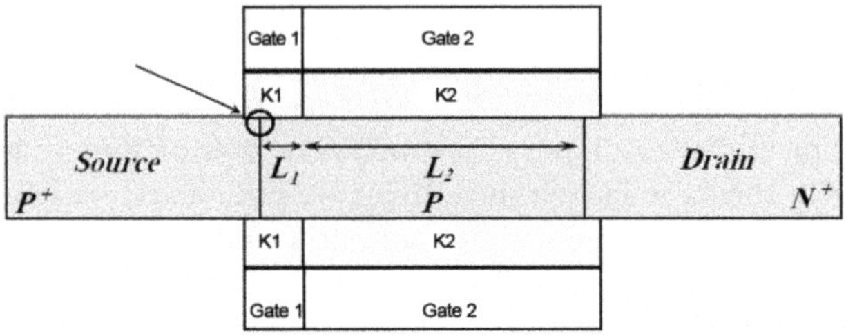

Figure 7.12 Schematic of a DG-TFET with two different dielectrics with different k values at source and drain sides. Adapted from [16].

Another concern is the leakage current at zero gate voltage. This is associated with the electric field at the barrier region, which depends on the metal–semiconductor work function when the gate voltage is zero. Using a low-k dielectric, the vertical electric field strength in the barrier region is reduced and hence the barrier width is increased. As a result, the tunneling leakage at zero gate voltage can be reduced. As reported, the off-current can be reduced up to 75% by adopting this technique [16]. Simulated off-current of TFETs under a different dielectric constant near the drain/channel end, K_2, is shown in Figure 7.13 [16]. In this simulation, the dielectric constant near the source/channel end, K_1, is 21. This high-k dielectric serves the purpose to increase the tunneling rate at source/channel end. In Figure 7.14, the energy band diagrams of $K_1 = K_2 = 21$ and $K_1 = 21$, $K_2 = 2.5$ are compared [16]. Based on the result, the barrier width in the drain/channel and source/channel interface increases with decreasing dielectric constant at drain/channel end, successfully suppressing the tunneling current in off-state.

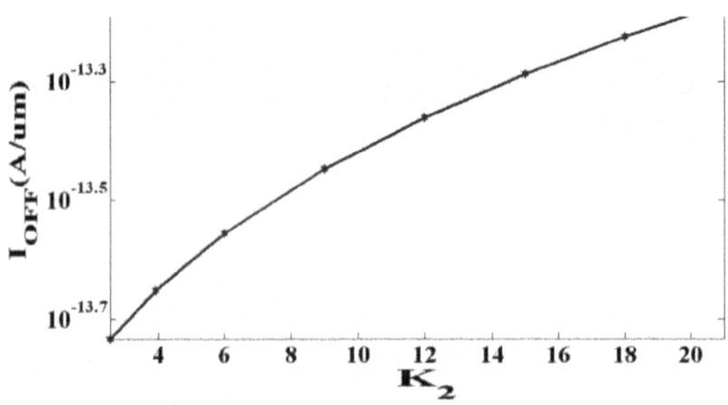

Figure 7.13 TFET off-current, I_{off}, versus gate dielectric constant at drain end, K_2. Low-k dielectric is used to suppress the leakage current, which in turn gives lower I_{off}. Adapted from [16].

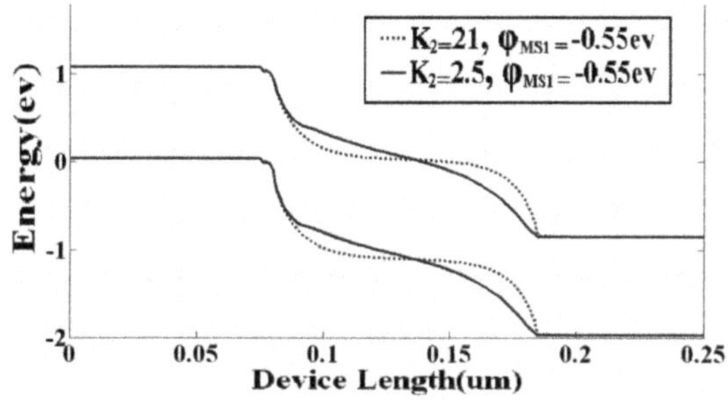

Figure 7.14 Comparison of energy band diagrams in OFF-state. The barrier width in the drain/channel and source/channel interface increases with decreasing dielectric constant at drain/channel end. Adapted from [16].

The effect of L_1 on I_{off} is also shown in Figure 7.15 [16]. It can be seen that the gate length does not have significant influence on the off-current. However, for the case of total gate length of 100 nm ($L_1 + L_2$=100 nm), when L_1 is greater than 90 nm (L_2 = 10 nm), the leakage current increases significantly. This is due to the Zener breakdown for this p-n$^+$ junction, which is caused by the high-k dielectric, resulting in more electrons tunneling through the drain-end barrier.

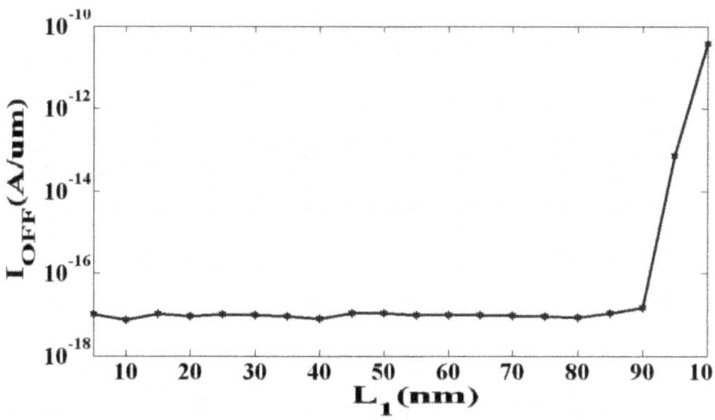

Figure 7.15 I_{OFF} as a function of L_1 for K_2=2.5. Total channel length, $L_1 + L_2$, is 100 nm. The increase of off-current for $L_1 > 90$ nm (or $L_2 < 10$ nm) is due to Zener breakdown. Adapted from [16].

7.4.5 Body Thickness

Silicon-on-insulator (SOI) and silicon-on-nothing fabrication technologies are the two commonly used techniques to realize DG thin-film devices. As reported by Boucart *et al.*, the thickness of the TFET influences the transfer curves, as shown in Figure 7.16 [7]. It can be observed that the off-current of the TFET decreases slightly as the body thickness decreases. On the other hand, the on-current of the TFET starts to decrease when the film thickness is smaller than 10 nm. This can be explained by the reduced cross-sectional area available for the current flow. In Figure 7.17, it can be seen that the optimum thickness for TFET with structure as shown in Figure 7.1 is around 7–8 nm [7]. However, the optimized thickness is affected by the supply voltage.

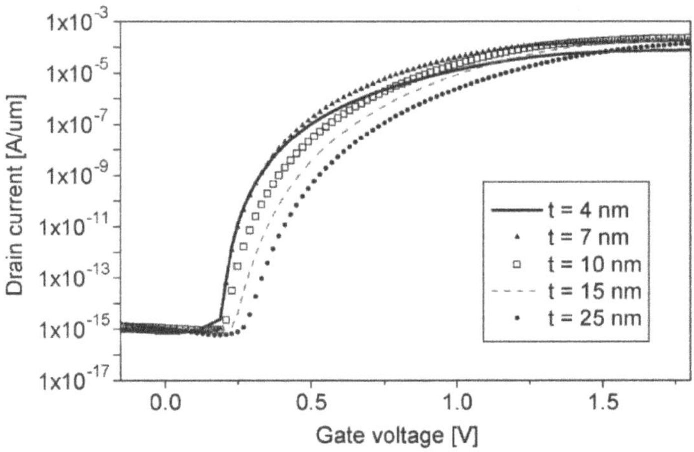

Figure 7.16 DG-TFET transfer characteristic for different silicon body thickness. The off-current of the TFET decreases slightly as the body thickness decreases. Adapted from [7].

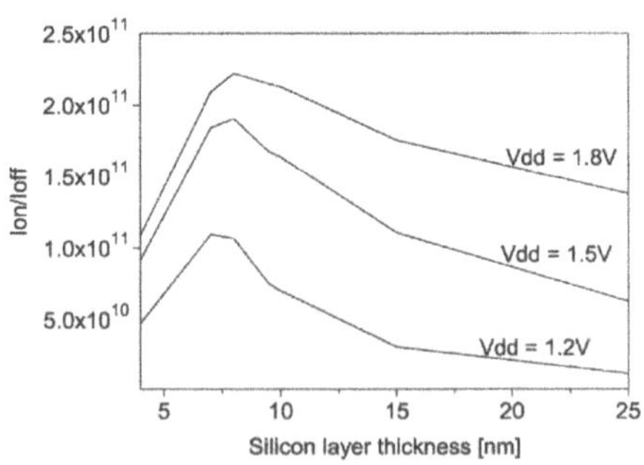

Figure 7.17 I_{ON}/I_{OFF} as a function of silicon body thickness for different V_{DD}. The optimum thickness is found to be around 7–8 nm. Adapted from [7].

7.5 Temperature Dependency of TFETs

It has been discussed that the TFET is an excellent device that has shown a reduced short-channel effect, outstanding subthreshold swing and low off-current. On top of that, unlike conventional MOSFETs, the working principle of TFETs makes them less sensitive to the operation temperatures. The weak dependency of TFETs on temperatures allows it to operate in a large temperature range without affecting their performance. This property of TFETs is significant especially for high performance and high-speed integrated circuits. Born *et al.* have reported on the high-temperature applications of TFETs [17]. Experimental transfer characteristics of the TFETs were compared at different temperatures ranging from 33 K to 373 K, as shown in Figure 7.18 [17]. Unlike conventional MOSFETs, the subthreshold swing does not change much for operation temperatures within 33 K to 373 K. It can be seen that there is a slight increase in the on-current, with increasing temperature. This could be due to the change of the bandgap of the channel material under different temperatures. The bandgap of a material at different temperatures can be expressed as follows [18]:

$$E_g(T) = E_g(0) - \frac{\alpha T^2}{T + \beta} \tag{7.11}$$

where α and β are material-dependent constants. The bandgap decreases with increasing temperature. As discussed in the previous section, the tunneling rate for a smaller bandgap channel material tends to be higher. Hence, a small decrease in the bandgap due to increasing temperature causes a slight increase in the drain current.

For a conventional MOSFET, the threshold voltage, V_T, shows a positive temperature coefficient in a range of 1–5 mV/K. This is caused by the shift of Fermi level with temperatures. The surface mobility decreases monotonically with increasing temperature. In addition, the subthreshold slope for a conventional MOSFET degrades under high temperatures, as shown in Figure 7.19 [17]. However, TFETs do not suffer from any of the aspects discussed above. Hence, TFETs can be designed to work over a wide range of temperatures without significant variation in their performance.

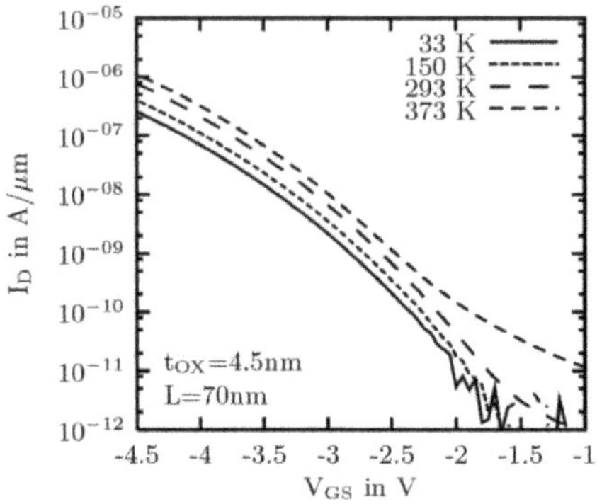

Figure 7.18 Transfer characteristic of TFET under different temperatures. The TFET performance is less sensitive to the operation temperatures. Adapted from [17].

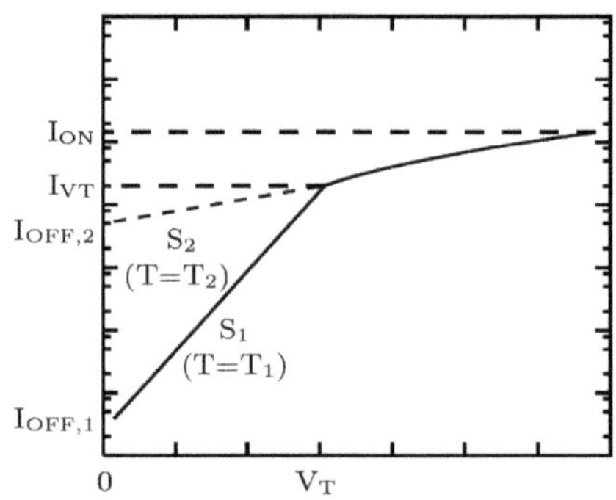

Figure 7.19 Transfer characteristic of conventional MOSFET under two different temperature of $T_2 > T_1$. The threshold voltage shows a positive temperature coefficient in a range of 1–5 mV/K. Adapted from [17].

7.6 Carbon Nanotube-Based Tunneling FETs

Appenzeller *et al.* reported a band-to-band tunneling effect in a carbon nanotube FET (CNTFET) [19]. They proposed, for the first time, a CNTFET whose working principle is based on tunneling effect in CNT and reported a subthreshold swing of 40 mV/dec for the tunneling current, as shown in Figures 7.20 and 7.21 [19]. The design concept is similar to other TFETs as discussed above, except that CNT is used to replace conventional silicon as the p-i-n structure. Instead of using a p-i-n structure, they simplified it by using a p-n-p structure, with the main objective to observe band-to-band tunneling phenomenon in the CNT. However, it is worth mentioning that p-i-n is still a preferred structure as there is only one tunneling barrier to overcome in order to turn on the device to achieve a higher on-current.

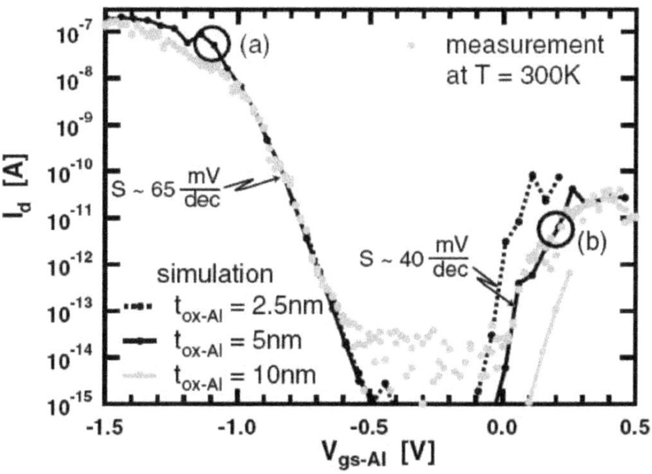

Figure 7.20 The experimental results from a CNTFET with controlled tunneling effect were reported by Appenzeller *et al.* The subthreshold swing was found to be as small as 40·mV/dec. Adapted from [19].

Another experiment was also carried out by them to differentiate the contributions of thermionic current and tunneling current of the device under different temperatures. As reported previously, temperature dependence of subthreshold swing can be used to distinguish the contribution of thermionic current and tunneling current as thermionic current is highly temperatures dependent [20, 21]. Based on this, the band-to-band tunneling effect in CNT was confirmed.

There are several advantages of CNTs being used as the active channel in the TFET. As reported by Appenzeller *et al.*, band-to-band-tunneling itself is not sufficient to achieve subthreshold swing of <60 mV/dec. One dimensionality of electronic transport and geometrical smallness of CNT are the main advantages of the CNT TFET to show a steep subthreshold swing [22]. It was found that the depletion lengths in a CNTFET with thin-gate dielectric can be so small that the tunneling current is significant [20].

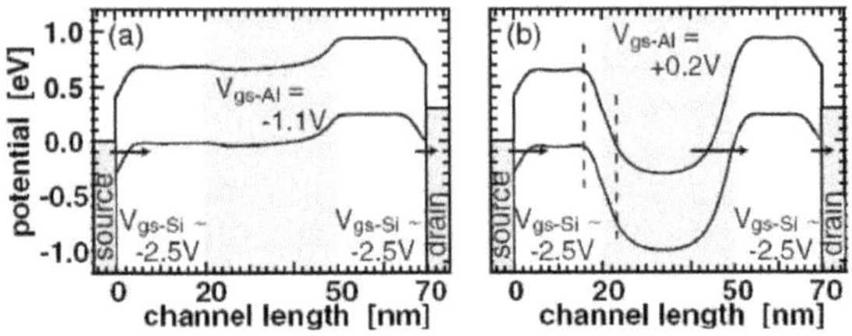

Figure 7.21 The band diagram of the device under two different operation modes, mainly contributed to thermionic current (left) and tunneling current (right). Adapted from [19].

Besides that, it can be observed from Equation 7.10 that, in order to achieve high tunneling rate, device parameters such as effective carrier mass, m^* and screening length (also called natural length or characteristic length), $\lambda=\sqrt{[(\varepsilon_{nt}/\varepsilon_{ox})d_{ox}d_{nt}]}$, where ε_{ox} and ε_{nt} are the dielectric constants of the gate oxide layer and CNT, d_{ox} is the thickness of the gate oxide and d_{nt} is the diameter of CNT, have to be as small as possible [11]. Screening length was originally derived for planar SOI MOSFETs and it characterizes the short-channel effect, with a smaller screening length being of a better short-channel effect immunity [23, 24]. In this case, the screening length is expected to be very small too, as the diameter of a semiconducting CNT is about 1 nm in general. Quantum capacitance limit can also be reached for one-dimensional nanowire/nanotube by introducing a tight gate control, such as a wrap gate. The wrap gate structure, as shown in Figure 7.22, allows the energy bands to have almost one-to-one response to the change of the gate potential; as a result, a TFET of high tunneling rate can be realized [11].

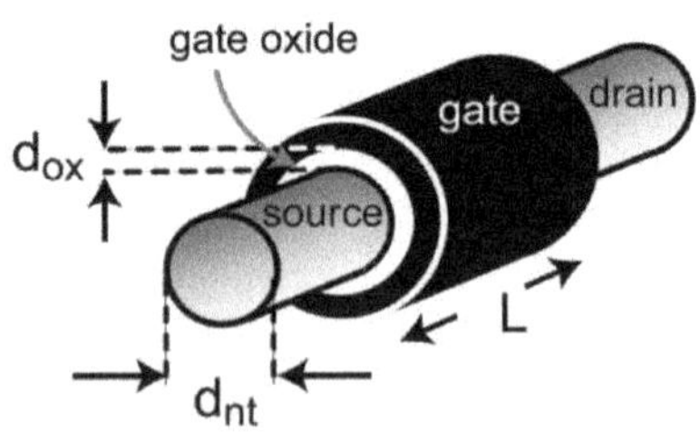

Figure 7.22 A wrap gate architecture is the ideal device geometric to realize ultimately scaled devices from the electrostatics point of view, especially for T-CNTFET as it gives almost one-to-one response to the change of gate voltage. Adapted from [11].

On the other hand, the small effective carrier mass can easily be achieved in CNTs. As reported, the effective mass of electrons in CNT can be as low as $0.108m_0$, depending on the chirality of the CNT [25]. The largest value of effective mass is $0.507m_0$ for chirality of (3,1), as listed in Table 7.1 [25]. These effective masses are much lower than the electron effective mass of silicon of $1.08m_0$. This property makes CNTs a suitable material to achieve high tunneling probability.

Table 7.1 Effective mass of electron in carbon nanotube of different chirality (adapted from [25])

(*n*,*m*)	**Effective mass for electrons**
(3,1)	$0.507\ m_0$
(3,2)	$0.222\ m_0$
(4,2)	$0.271\ m_0$
(4,3)	$0.175\ m_0$
(5,0)	$0.408\ m_0$
(5,1)	$0.159\ m_0$
(5,3)	$0.189\ m_0$
(6,1)	$0.255\ m_0$
(7,3)	$0.116\ m_0$
(9,2)	$0.099\ m_0$
(11,3)	$0.108\ m_0$

Knoch *et al.* have reported on the impact of the dimensionality on the performance of TFETs [22]. They compared the performance of bulk TFET and CNT-TFET, and concluded that a nanowire TFET, especially for a CNT TFET with one-dimensional transport and small geometry, has a high on-state current and a steep subthreshold slope. In contrast, for the case of planar bulk TFETs, there is a trade-off between high on-state current and steep subthreshold slope. This will be discussed in details in the following section.

The subthreshold swings of bulk TFET and CNT-TFET also depend on the position of the source/drain Fermi levels [22]. It was suggested that the Fermi level has to be as close to the conduction band as possible for the case of n-doped source/drain (or as close to the valance band for the case of p-doped drain/source). If the Fermi energy is too far away from the conduction band (or valence band) for the case of heavily doped source/drain, the bandgap does not cut off the low (or high) energy portion of the source/drain effectively [15, 22]. In this case, the band profile actually resembles that of a conventional FET, which makes it unable to realize subthreshold swing of <60 mV/dec. The band profile is shown in Figure 7.23 [22].

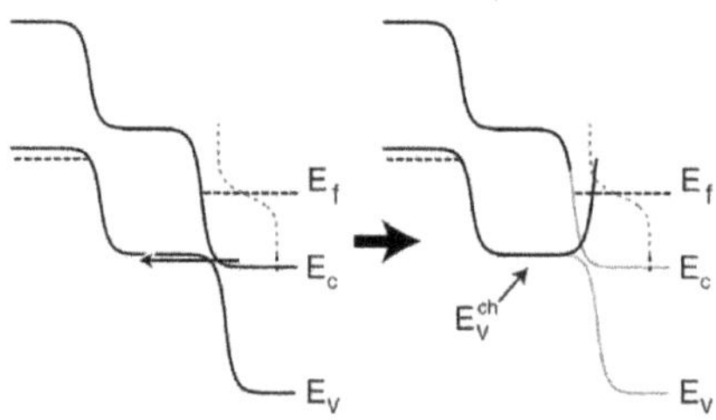

Figure 7.23 Band diagram of a TFET for a large Fermi level in n$^+$ source. It resembles a conventional FET with $E_V{}^{ch}$ as the potential barrier mediating thermionic emission. Adapted from [22].

However, for bulk TFETs, a heavily doped source/drain is always preferred in order to realize a thin tunneling barrier to achieve high on-state current and large carrier density. In other words, for the case of heavily doped source/drain, the Fermi level is not close to the conduction/valance band. As a consequence, there is always a trade-off between high tunneling rate and high on-current by doping the source/drain heavily, and steep subthreshold slope by bringing the Fermi level closer to the conduction/valance band.

As for nanowire TFET, such as CNT TFET, it is possible to realize steep subthreshold slope with the Fermi level as close to the conduction band as possible, at the same time maintaining a thin tunneling barrier, due to the one-dimensionality and excellent electrostatic gate control of the CNT. As discussed above, the depletion length of CNT can be so small that the tunneling current is significant [20]. Hence, one-dimensional materials, say CNTs, are superior materials for TFETs. The simulated transfer characteristics of the devices are shown in Figures 7.24 and 7.25 [22]. Based on the results, the smallest subthreshold swing for a planar bulk tunneling FET is 29 mV/dec, while 13 mV/dec can be achieved for a CNT-TFET [22]. Experimentally, Appenzeller *et al.* have already proved the argument by reporting on the lowest subthreshold swing of 40 mV/dec, by using CNT as the active channel of the TFET [19].

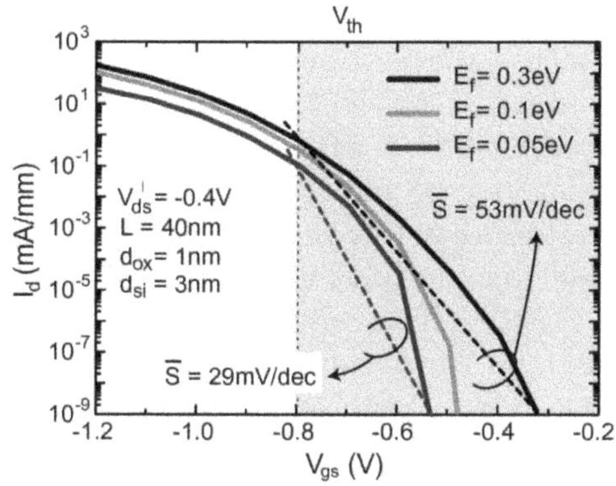

Figure 7.24 Transfer characteristics of planar bulk tunneling FETs for different source Fermi energies. Adapted from [22].

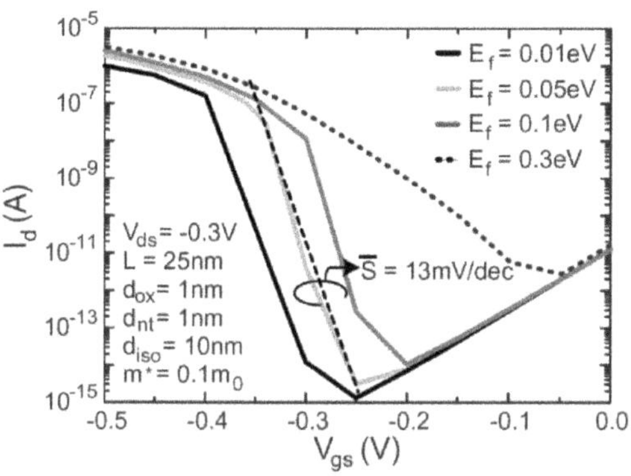

Figure 7.25 Transfer characteristics of a 1D tunneling CNTFET for different source Fermi energies. Adapted from [22].

In addition, CNT is a direct-bandgap semiconductor, which makes it a preferred choice over an indirect-bandgap semiconductor, such as conventional silicon. Koswatta *et al.* have also reported on a computational study of CNT p-i-n TFETs [26]. They pointed out that the direct bandgap allows band-to-band tunneling under ballistic transport in the CNT, unlike the indirect bandgap of silicon in where phonon emission or absorption has to be involved to conserve the transverse momentum of carriers [27]. Based on their simulation results, they showed that the current is mainly carried within the energy levels between the Fermi levels of the source and drain, which is a resonant process. This band pass filter behavior makes it possible for the CNT TFETs to be insensitive to temperatures. Figure 7.26 shows that the performance of a p-i-n CNT TFET is much better than an n-i-n CNT MOSFET in terms of subthreshold swing [26]. The on-current of CNT TFETs is lower than that of CNT MOSFET due to the tunneling barrier and quantum confinement effects in the CNT channel. Besides that, it is shown that a (13,0) CNT TFET has higher on-current than a (10,0) CNT TFET. This is due to a small bandgap of the (13,0) CNT as compared to (10,0) CNT, resulting in higher tunneling rate. The asymmetric source and drain doping of (13,0) CNT TFET shows a reduced ambipolar characteristic of the TFET in Figure 7.26 [26].

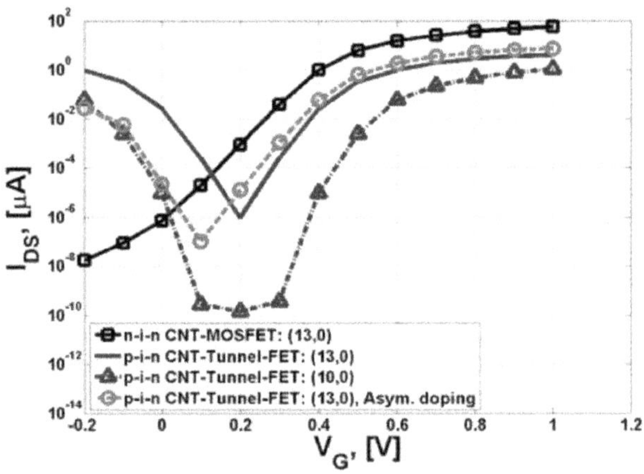

Figure 7.26 IDS vs VG for n-i-n CNT MOSFET and p-i-n CNT TFETs with different CNT bandgap and source-drain configuration. Adapted from [26].

7.7 Summary

Due to the outstanding subthreshold swing of TFETs compared with the conventional MOSFETs, TFETs have the potential to replace MOSFETs as the next-generation device structure. On the other hand, the downscaling of MOSFETs down to sub-20 nm node will be a challenging task as the power crisis of MOSFETs, another critical issue, is expected to be worse when the channel length is getting smaller. Due to the low off-current of TFETs, TFETs fit best in low standby power category. TFETs have been attracting much attention since then.

Semiconducting CNT is an excellent one-dimensional material. Due to the excellent electrical properties and one dimensionality of the CNT, tunneling phenomenon has been widely observed and reported. As the depletion width in CNT channel can be made very small and abrupt to facilitate BTBT, CNT can be a very good material to be used as the main component of TFET. The low effective mass of carriers in CNT makes it possible to realize ballistic transport in CNT to achieve high switching speed

One way to eliminate the ambipolarity of TFET is by introducing large-bandgap material for source/drain to suppress the off-state tunneling. Meanwhile, the on-state tunneling rate can be improved by using small bandgap material at the drain/source. The performances of CNT TFETs, especially their subthreshold swing and off-state current, are outstanding and significant. In addition, the direct bandgap of CNT is preferred over silicon as no phonon emission or absorption has to be involved to conserve the transverse momentum of carriers in the event of BTBT. All these make CNT-TFET a promising future device structure to realize logic circuit of high switching speed, low operating power, and low standby power.

References

1. Geppert, L., The amazing vanishing transistor act. *IEEE Spectr*, 2002. **39**(10): p. 28–33.
2. Chandrakasan, A.P., and R.W., Brodersen, minimizing power consumption in digital CMOS circuits. *Proc IEEE*, 1995. **83**(4): p. 498–523.
3. Horowitz, M., Scaling, power, and the future of CMOS technology. In *Device Research Conference, 2008*.
4. Mookerjea, S., and S. Datta. Comparative study of Si, Ge and InAs based steep subthreshold slope tunnel transistors for 0.25V supply voltage logic applications. In *Device Research Conference, 2008*.
5. Ieong, M., *et al.*, Silicon device scaling to the sub-10-nm regime. *Science*, 2004. **306**(5704): p. 2057–2060.
6. Semiconductor Industry Association (SIA), *International Technology Roadmap for Semiconductors* (*ITRS*), 2005. [Online]. Available: http://www.itrs.net/.
7. Boucart, K., and A.M. Ionescu, Double-gate tunnel FET with high-k gate dielectric. *IEEE Trans Electron Devices*, 2007. **54**(7): p. 1725–1733.

8. Chen, Z.X., *et al.*, Demonstration of tunneling FETs based on highly scalable vertical silicon nanowires. *IEEE Electron Device Lett, 2009.* ***30****(7): p. 754–756.*

9. Hurkx, G.A.M., D.B.M. Klaassen, and M.P.G. Knuvers, A new recombination model for device simulation including tunneling. *IEEE Trans Electron Devices*, 1992. **39**(2): p. 331–338.

10. Vandenberghe, W.G., *et al.*, Analytical model for a tunnel field-effect transistor. In *Electrotechnical Conference, 2008. MELECON 2008. The 14th IEEE Mediterranean*. 2008.

11. Knoch, J., and J. Appenzeller, Tunneling phenomena in carbon nanotube field-effect transistors. *Phys Status Solidi* (A), 2008. **205**(4): p. 679–694.

12. Krishnamohan, T., *et al.*, Double-gate strained-Ge heterostructure tunneling fet (TFET) with record high drive currents and <<60mV/dec subthreshold slope. In *Electron Devices Meeting, 2008. IEDM 2008. IEEE International*. 2008.

13. Mayer, F., *et al.*, Impact of SOI, $Si_{1-x}Ge_xOI$ and GeOI substrates on CMOS compatible Tunnel FET performance. In *Electron Devices Meeting, 2008. IEDM 2008. IEEE International*. 2008.

14. Fulde, M., *et al.*, Fabrication, optimization and application of complementary Multiple-Gate Tunneling FETs. In *Nanoelectronics Conference, 2008. INEC 2008. 2nd IEEE International*. 2008.

15. Knoch, J. Optimizing tunnel FET performance: Impact of device structure, transistor dimensions and choice of material. In *VLSI Technology, Systems, and Applications, 2009. VLSI-TSA '09. International Symp on*. 2009.

16. Vadizadeh, M., and M. Fathipour. Using low-k oxide for reduction of leakage current in Double Gate Tunnel FET. In *10th International Conference on Ultimate Integration of Silicon, 2009. ULIS 2009*. 2009.

17. Born, M., *et al.*, Tunnel FET: a CMOS device for high temperature applications. In *25th International Conference on Microelectronics, 2006*. 2006.

18. Sze, S.M., *Physics of Semiconductor Devices*. 1981.

19. Appenzeller, J., *et al.*, Band-to-band tunneling in carbon nanotube field-effect transistors. *Phys Rev Lett*, 2004. **93**(19): p. 196805.

20. Appenzeller, J., *et al.*, Field-modulated carrier transport in carbon nanotube transistors. *Phys Rev Lett*, 2002. **89**(12): p. 126801.

21. Appenzeller, J., *et al.*, Tunneling versus thermionic emission in one-dimensional semiconductors. *Phys Rev Lett*, 2004. **92**(4): p. 048301.

22. Knoch, J., S. Mantl, and J. Appenzeller, Impact of the dimensionality on the performance of tunneling FETs: Bulk versus one-dimensional devices. *Solid State Electron,* 2007. **51**(4): p. 572–578.

23. Yan, R.H., A. Ourmazd, and K.F. Lee, Scaling the Si MOSFET: from bulk to SOI to bulk. *IEEE Trans Electron Devices,* 1992. **39**(7): p. 1704–1710.

24. Suzuki, K., *et al.*, Scaling theory for double-gate SOI MOSFET's. *IEEE Trans Electron Devices,* 1993. **40**(12): p. 2326–2329.

25. Marulanda, J.M., and A. Srivastava. Carrier density and effective mass calculations for carbon nanotubes. In *Integrated Circuit Design and Technology, 2007. IEEE International Conference on ICICDT '07.* 2007.

26. Koswatta, S.O., D.E. Nikonov, and M.S. Lundstrom. Computational study of carbon nanotube p-i-n tunnel FETs. In *Electron Devices Meeting, 2005. IEDM Technical Digest. IEEE International.* 2005.

27. Sedlmaier, S., *et al.*, Phonon assisted tunneling in gated p-i-n diodes. *Mater Science Eng B*, 2002. **89**(1-3): p. 116–119.

Chapter 8

Transistors Based on Carbon Nanotube Random Networks

Lain-Jong Li[a] and Chun-Wei Lee[b]

[a] *Institute of Atomic and Molecular Sciences, Academia Sincia, Taipei, 11529, Taiwan*

[b] *School of Chemical and Biomedical Engineering, Nanyang Technological University, 16 Nanyang Drive, 637722 Singapore*

lanceli@gate.sinica.edu.tw, leecw@ntu.edu.sg

Single-wall carbon nanotube (SWCNT) networks have been potentially envisioned as one of promising semiconducting materials for transistors in logic circuit applications. The advantage of using SWCNT random networks as field-effect transistors (FET) is the ease of fabrication, which practically makes cost-effective solution processable and printable electronics feasible. It is challenging to fabricate SWCNT FET on a large scale with reasonably high mobility and full yield of semiconductor devices owing to the co-existence of metallic (M) and semiconducting (S) tubes in its networks or films. This chapter will introduce several major approaches, including specific synthesis of S-SWCNTs, selective elimination or destruction of M-SWCNTs, and photolithography-assisted stripping to intentionally improve the semiconducting characteristics from its networks. Meanwhile, tremendous efforts have been devoted

Carbon Nanotubes and Their Applications
Edited by Qing Zhang

ISBN 978-981-4241-90-8 (Hardcover), 978-981-4303-18-7 (eBook)
www.panstanford.com

to sort out M- and S-SWCNTs for preparing S-SWCNT rich ink for solution-based fabrication, including surfactant wrapping, aromatic extraction, amine reaction, selective oxidation, dielectrophoresis, bundled tubes removal, DNA or polymer wrapping, density gradient ultracentrifugation (DGU) and gel-based separation techniques. In addition, methods such as selective modification of SWCNTs with diazonium reagents or organic free radicals are adopted to successfully demonstrate full semiconductor device yield and reasonably high carrier mobility. Fundamental issues which limit the performance of SWCNT networks are also discussed.

8.1 Introduction

SWCNT, which is considered a rolled-up hollow cylinder of graphite monolayer with ~1 nm diameter, are widely regarded to be an outstanding candidate for field-effect transistors (FETs) owing to its unique properties. Based on their roll-up chirality, the electronic structure of SWCNTs is either metallic (M) or semiconducting (S). Due to the extremely high aspect ratio and outstanding nearly defect-free electronic structure, electronic transporting property of SWCNTs is ballistic, resulting in superior field-effect behavior on semiconducting SWCNTs and its intrinsic carrier mobility can achieve about 100,000 cm^2/Vs. Owing to its outstanding electrical properties, SWCNTs have been widely anticipated as one of excellent potential candidates for electronic applications. However, the variations of chirality and diameter in SWCNT assembly lead to inconsistence of electrical properties from single tube devices. In addition, the integration of single-tube devices into a circuit is another technological hurdle for realizing its electronic applications. Therefore the development of FETs based on SWCNT networks is intensified with tremendous efforts from researchers. There are several remarkable advantages from SWCNTs devices based on random networks such as ease of production, high reproducibility with consistent electrical properties and high scalability as well as high compatibility with conventional Si technology. Hence, Random SWCNT networks or thin films are considered as ideal semiconducting materials for low-cost carbon based electronics. Despite the great promise, realization of low-cost carbon-based electronics still remains difficult owing to inevitably technological challenges on device fabrication. There are several major challenges on using SWCNT random networks or thin films for low-cost high-performance electronic applications, such as the following:

(1) Coexistence of M- and S-SWCNTs leads to high off-current, resulting in low on-off current ratio in FET configuration.
(2) Relatively low effective field-effect mobility from SWCNT-FETs is widely attributed to the contact resistance of inter-nanotube junctions existing within SWCNT networks or thin films.
(3) Low yield in SWCNT-FET production as well as compatibility issue with conventional processing method have to be carefully addressed for making commercial applications.

Despite being widely regarded as the most promising candidate for realizing low-cost high-performance carbon-based electronics, FETs based on SWCNT networks or thin films exhibit the issues mentioned above, which require tremendous research efforts and innovatively strategic approaches to resolve them. Thus, the intensive studies on understanding the above issues arising from SWCNT-FETs are essential from various aspects such as fundamental researches and technological challenges in order to achieve decent performance electronics.

8.2 Structure of Single-Wall Carbon Nanotubes

8.2.1 Geometrical Structure

SWCNT is widely depicted as a rolled-up hollow cylinder of graphene sheet, which is assembled with carbon atoms in hexagonal structure. SWCNTs can be identified by graphene lattice vectors owing to highly similar microscopic structure between graphene and SWCNT. Its circumferential vector ($\vec{C_h}$) is used to index its structure as illustrated in Figure 8.1.[1] The definition of the chiral vector is a vector for bonding two specific atoms on the graphene sheet. Bonding carbon atoms via sp^2 hybridization forms a tube called as SWCNT. In General, a (n,m) SWCNT is denoted from its chiral vector $\vec{C_h} = n\vec{a_1} + m\vec{a_2}$, where $\vec{a_1}$ and $\vec{a_2}$ are unit vectors of the graphite primitive lattice. The angle between $\vec{C_h}$ and $\vec{a_1}$ defines as chiral angle θ varying from 0° to 30° (as illustrated in Figure 8.1.)

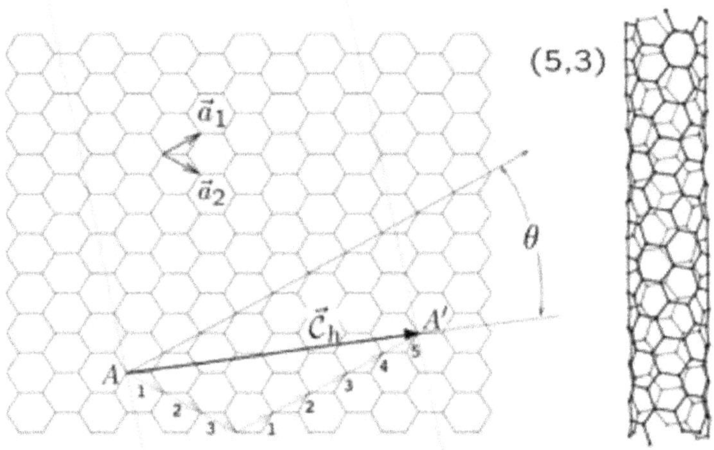

Figure 8.1 Graphene sheet with primitive lattice vectors a_1 and a_2. The chiral vector $\vec{C}_h = 5\vec{a}_1 + 3\vec{a}_2$ represents a possible direction for rolling up a two-dimensional graphene sheet into a hollow cylinder. The tube axis is perpendicular with $\vec{C}_h$. The chiral angle θ is defined as the angle between $\vec{C}_h$ and $\vec{a}_1$. The physical structure of (5,3) SWCNT is illustrated on the right. Reproduced from ref. 1.

The nanotube diameter (d_t) can be estimated from the following equation:

$$d_t = \left|\vec{C}_h\right| / \pi = \frac{a}{\pi}\sqrt{n^2 + m + m^2} \tag{8.1}$$

where a is the lattice constant of graphite and its value is equivalent to $a = \sqrt{3} \times a_{cc}$, where a_{cc} = 1.42 Å, the C–C bond length. According to its indices, nanotubes can be classified as zigzag, armchair, and chiral tubes. The zigzag tube possesses (n, 0) index and (θ = 0°), where its circumferential vector lies along one of the two vectors showing zigzag pattern along its circumference. While the armchair tube possesses m = n and (θ = 30°) expressing armchair pattern along its circumference. Others are called chiral tubes with (n, $m \neq n$ $\neq$ 0) and (0° < θ >30°).

8.2.2 Electronic Structure

The bonding structure of SWCNT, similar to graphene, is planar sp^2 hybridized. An individual carbon atom in graphene sheet possesses

4 valence orbitals (2s, $2p_x$, $2p_y$, and $2p_z$ orbitals). The in-plane σ (bonding or occupied) and σ* (antibonding or unoccupied) orbitals are formed by combining s, p_x and p_y orbitals. The interaction of P_z orbital with the neighboring P_z orbitals forms delocalized π (bonding) and π* (antibonding) orbitals. Similarly, the carbon atoms in SWCNT are bonded by in-plane σ bonds which are responsible for binding energy between them. A delocalized π networks across a SWCNT are responsible for its electronic properties. The conjugating π electrons also contribute to the formation of SWCNT bundles through van der Waals forces. Therefore, it is necessary to study the electronic structure of graphene in order to understand the electronic properties of SWCNT.

8.2.2.1 Tight binding calculation from 2-D graphite to nanotube

The two-dimensional graphite system has been widely depicted by tight binding model, which is also adopted to discuss on SWCNT by Saito and Dresselhaus.[2] It provides a fundamental study on understanding electronic properties of SWCNT. A unit cell of hexagonal graphene lattice consists of two atoms (A and B). The lattice basis vectors ($\vec{a}_1$ and $\vec{a}_2$) in the graphene sheet and the corresponding reciprocal-lattice basis vectors ($\vec{b}_1$ and $\vec{b}_2$) in the Brillion zone are shown in Figure 8.2a,b, respectively.

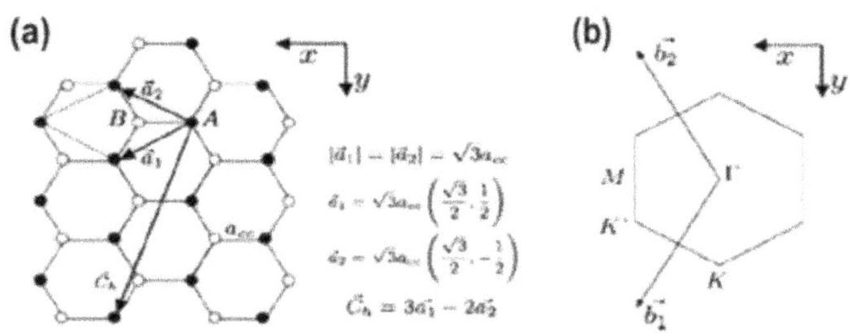

Figure 8.2 Brillouin zone and reciprocal space. (a) Basis vectors in the hexagonal lattice of graphene. (b) Brillion zone. The corresponding reciprocal basis vectors read $\vec{b}_1 = b(1/2, \sqrt{3}/2)$ and $\vec{b}_2 = b(1/2, -\sqrt{3}/2)$, with $b = 4\pi / a\sqrt{3}$. Reproduced from ref. 2.

A carbon atom at the position r_s has an unsaturated π orbital described by the wavefunction $\chi_{rs}(r)$.[3] Based on tight binding model, the P_z orbitals only interact with the nearest-neighbor carbon atoms. By solving the corresponding Schrödinger equation with applying the Bloch theorem, the wave function of the carbon atoms interaction can be expressed as

$$\langle \chi_{rA}|H|\chi_{rA}\rangle = \langle \chi_{rB}|H|\chi_{rB}\rangle = 0 \tag{8.2}$$

$$\langle \chi_{rA}|H|\chi_{rB}\rangle = \langle \chi_{rB}|H|\chi_{rA}\rangle = \gamma\delta_{rA-rB=a} \tag{8.3}$$

where a is a vector connecting nearest-neighbors between the A and B sublattices and the on-site interaction energy is zero.

In order to calculate the electronic structure, the Bloch wavefunction of each sublattices is expressed as

$$\phi_{sk}(r)=\sum_{r_s} e^{ik\cdot r_s}\chi_{r_s}(r), \tag{8.4}$$

where s represents the sublattices (A or B) and r_s is a set of data related to sublattices. The total wavefunction can be described as the linear combination of these two functions as

$$\Psi_k(r)=\frac{1}{\sqrt{2}}[\phi_{Ak}(r)+\lambda_k\phi_{Bk}(r)] \tag{8.5}$$

The Hamiltonian matrix elements are written as

$$\langle \phi_{Ak}|H|\phi_{Ak}\rangle = \langle \phi_{Bk}|H|\phi_{Bk}\rangle = 0, \tag{8.6}$$

$$\langle \phi_{Ak}|H|\phi_{Bk}\rangle = \langle \phi_{Bk}|H|\phi_{Ak}\rangle^* = H_{AB} = \gamma\sum_{a} e^{ik\cdot a} \tag{8.7}$$

To obtain the Schrodinger equation in a matrix form:

$$\begin{pmatrix} E & -H_{AB} \\ -H_{AB}^* & E \end{pmatrix}\begin{pmatrix} 1 \\ \lambda_k \end{pmatrix} \tag{8.8}$$

After solving this matrix, the solution becomes:

$$E=\pm|H_{AB}|=\pm\lambda\sqrt{1+4\cos\left(\frac{3}{2}k_y a\right)\cos\left(\frac{\sqrt{3}}{2}k_x a\right)+4\cos^2\left(\frac{\sqrt{3}}{2}k_x a\right)} \tag{8.9}$$

This is the band structure of graphene as function of k_x and k_y, which can be plotted as shown in Figure 8.3. The valence and conduction bands actually connect at the corners of the Brillouin zone, resulting that graphene is a zero-gap semi-metal. Reproduced from ref. 3.

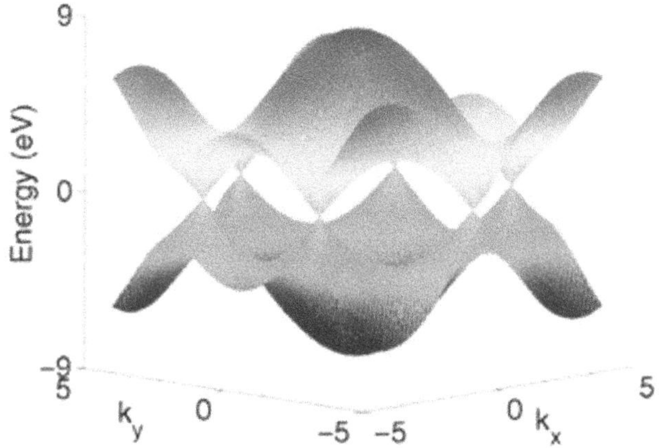

Figure 8.3 Electronic structure of graphene calculated from tight-binding model only consisting of π electrons. Reproduced from ref. 3. See also Color Insert.

To reveal the electronic structure of SWCNTs, the electron wavevector quantization along the circumferential direction is presented as

$$k \cdot C = k_x C_x + k_y C_y = 2\pi t \tag{8.10}$$

where C is the wrapping vector of the tube and t is an integer. A relation between k_x and k_y defining lines in the (k_x, k_y) plane is shown in equation 8.10. The intersection point of these lines with graphene band structure is determined by value of C_x, C_y and t. Hence, various SWCNTs exhibit different electronic structures because of the difference in the set of C_x, C_y and t. From the band structure perspective, SWCNT exhibits semiconducting electronic structure as there is no intersection between quantized wavevector lines and Fermi points. Whereas metallic SWCNT is obtained as there is intersection between quantized wavevector lines and Fermi points as shown in Figure 8.4. The band structure of semiconducting and metallic zigzag CNTs is also illustrated in Figure 8.4.

In general, the determination of (n,m) metallic SWCNTs is governed by $n\text{-}m = 3p$, where p is an integer. The (n,m) SWCNTs with $n\text{-}m \neq 3p$ and $n\text{-}m = 3p$ exhibit semiconducting and metallic characteristic, respectively. The transport properties of SWCNTs are determined by the states near the K point. For example, the states along line closest to the K point represent the conduction band and valence band in a semiconducting tube. For metallic tubes, the states intercept with the K point. It is worth noting that the relation between bandgap and diameter[19,20] has been depicted as $E_g \approx \gamma a/R$.

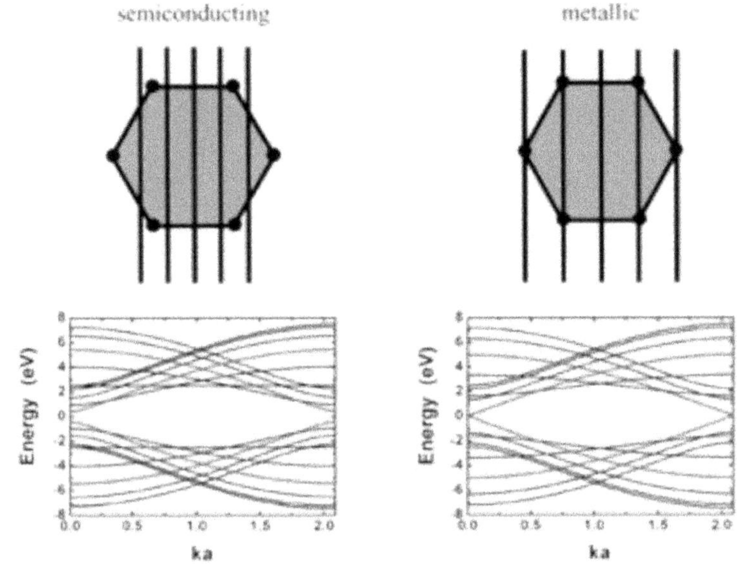

Figure 8.4 Illustration of allowed wavevector lines leading to semiconducting and metallic CNTs. The band structures of semiconducting and metallic zigzag CNTs. Reproduced from ref. 3.

8.2.2.2 Density of states (DOS) of SWCNT

DOS of SWCNTs has been used to describe the electronic properties of SWCNTs as

$$D(E) = \frac{\sqrt{3}a^2}{2\pi R} \sum_i \int dk\delta(k - k_i) \left| \frac{\partial \varepsilon}{\partial k} \right|^{-1} \tag{8.11}$$

Where $\varepsilon(k_i)$ = E. The calculated DOS of SWCNTs exhibits one-dimensional van Hove singularities as the band edges as illustrated in Figure 8.5. Experimentally, scanning-tunneling microscopy is adopted for verifying the presence of these singularities in DOS. In general, SWCNT behaves as a metal due to the existence of DOS at Fermi energy level. Whereas there is no DOS at Fermi energy level for semiconducting tubes. Hence, SWCNTs are either semiconducting or metallic based on their roll-up chiralities.

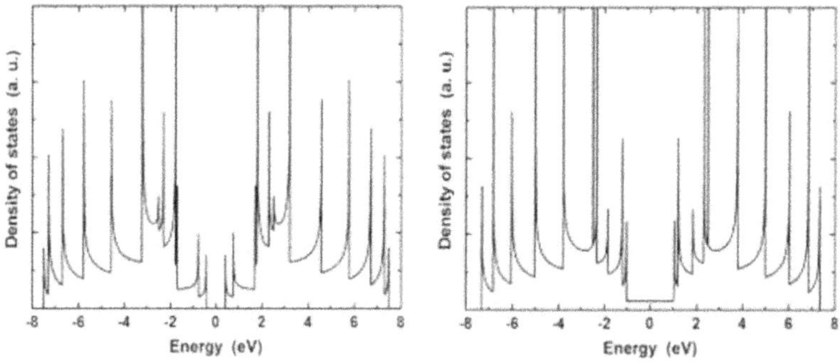

Figure 8.5 Density of states for (11,0) semiconducting and (12,0) metallic CNTs showing van Hove singularities from tight–binding model. Reproduced from ref. 3.

8.2.3 Individual Single-Wall Carbon Nanotube-Based Transistor (SWCNT-FET)

Surprisingly, the electrical performance of the prototype SWCNT-FET has been demonstrated superior to the best silicon-based metal-oxide-semiconductor FET (MOSFET).[4,5] Indeed, it is reasonable from material science perspective due to the following reasons. First, owing to high aspect ratio of SWCNT the scattering probability of carriers is low. Thus, short-channel SWCNT-FET operates in the ballistic transport regime. Moreover, a passivation layer, technologically required in conventional MOSFETs, is not necessary at the nanotube and gate dielectric interface due to absence of dangling bond at its surface. Based on the above merits, it attracts tremendous efforts from nanotube researchers to develop SWCNT-FET and to fully utilize its unique properties.

Initially, the Si backgate configuration with noble metal as electrodes is used for SWCNT-FETs. Despite showing fascinating electrical properties, such device configuration exhibits large contact resistances and poor electrical characteristic of SWCNT-FET resulting in underperforming its intrinsic electrical properties.[6,7] With efforts on improvement of device performance from nanotube researchers, there is a remarkable achievement in term of device performance.[4] The top-gate configuration has been established in conventional MOSFET and proved superior to Si backgate configuration. Dai's group has demonstrated top-gate SWCNT-FET with Pd electrodes and Al top-gate separated with HfO_2 high-k dielectric.[8] The high-performance enhancement-mode semiconducting SWCNT-FETs exhibit a high on-off ratio (up to 10^5), and 80 mV/decade subthreshold swings can be achieved as shown in Figure 8.6.

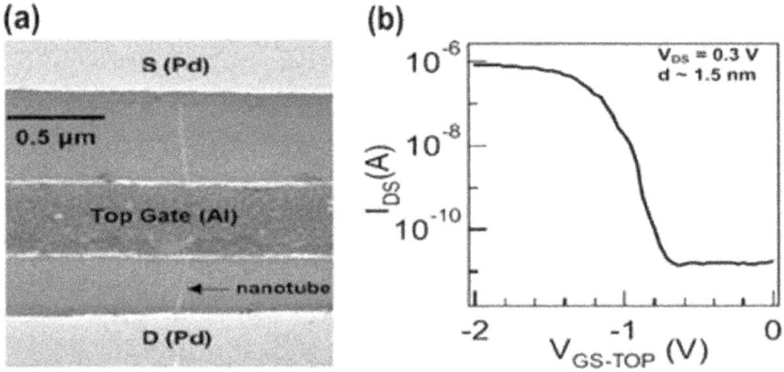

Figure 8.6 (a) A SEM image of top-gated SWCNT-FET and (b) its transfer characteristic. Reproduced from ref. 8.

Theoretically, it is possible to observe an ambipolar characteristic from SWCNT-FETs with appropriate energy-level alignment between metal electrode and nanotube. In Figure 8.7, an ambipolar SWCNT-FET is observed likely because the Fermi level of metal electrode aligns with the midgap of nanotube, leading to large Schottky barrier energies for both electron and hole injection. Surprisingly, output curve (I_{ds}–V_{ds}) is ohmic at room temperature as shown in Figure 8.7, indicating that no barrier exists at nanotube/metal interface.[9] Based on the temperature-dependent measurements, the observation of electron and hole accumulation region suggests that the barrier height for both electron and hole injection are shallow. Thus, electron or hole carriers can tunnel through the thin junction at the nanotube/metal interface with only thermal excitation, and this is known as a tunneling mechanism. In addition, SWCNT is highly sensitive to the surrounding environment which could alter its electrical characteristics. For instance, a SWCNT-FET fabricated with Co metals as electrodes and annealing under hydrogen gas condition exhibited air-stable and ambipolar characteristic,[10] suggesting that the electrical behavior of SWCNT-FET is significantly affected from metal electrodes and adsorbates. Intrinsically, the decent electrical performance of SWCNT-FETs, such as high on-off ratio, could only be demonstrated from semiconducting SWCNT species. Therefore, sorting SWCNT is extremely crucial to realize its potential electronic applications, which is discussed in the following section.

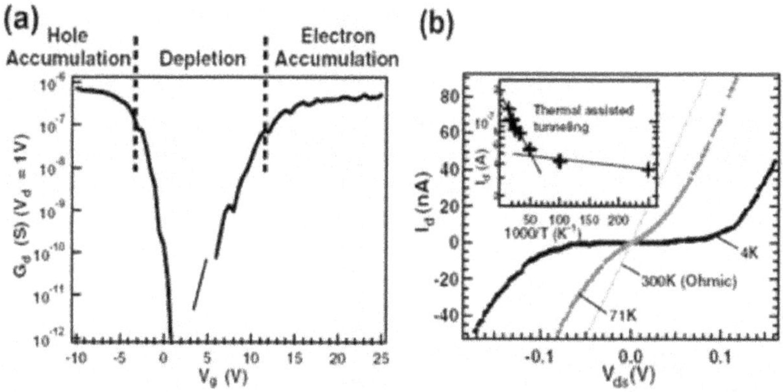

Figure 8.7 (a) A plot for transfer characteristic of an ambipolar SWCNT-FET with indicating the hole accumulation, depletion and electron accumulation region. (b) A plot for output characteristic of an ambipolar SWCNT-FET with various temperatures. Ohmic behavior is observed at room temperature suggesting small energy barrier. (Inset) An Arrhenius plot for this device indicates occurrence of thermal assisted tunneling. Reproduced from ref. 9.

8.3 Sorting Carbon Nanotubes

The selection of SWCNTs with desired properties is critically important for electronic applications. Hence, the enrichment of S-SWCNTs from the raw nanotubes is the major challenge for realizing the electronic applications based on SWCNT-FET. Despite tremendous efforts on the development of SWCNT growth, the selective SWCNTs growth technique for effective producing identical population of SWCNTs with desired properties is still not yet realized. Thus, development of effective and innovative solutions is essential for overcoming this issue. Recently, the development of post-synthesis sorting techniques and selective growth methods for assembly of SWCNTs with desired properties have been intensively investigated. [11–13]

To accomplish optimal performance of SWCNT network based FET in electronic applications, populations of SWCNTs ideally should be monodispersed in terms of electronic type (S-SWCNTs). Besides enriching S-SWCNTs from pristine SWCNTs, the evaluation of an effective SWCNT sorting technique should consider the subsequent several factors: (1) high scalability technique for fulfilling the demand of increase in SWCNTs production, (2) high chemical stability for retaining its outstanding electronic transport properties after introducing treatments for SWCNT sorting, (3) high reproducibility of sorting technique for further improving in purity level, and (4) high affordability for realizing incorporation into target applications. Several popular effective SWCNTs sorting technique such as electrophoresis, polymer wrapping, DNA wrapping and selective functionalization of SWCNT will be introduced in the following section.

8.3.1 Electrophoresis

Owing to the similar dimension and size between SWCNTs and many biomolecules, researchers have borrowed established separation techniques, for instance, electrophoretic separation, for sorting out SWCNTs in terms of their physical and electronic properties. Despite the high aspect ratio of SWCNTs, they are still relatively mobile in gel or capillary with the aid of a direct current (d.c.) and such merit is widely used for electrophoretic separation of SWCNTs. The gel and capillary electrophoresis techniques have been demonstrated as an effective separation technique on SWCNTs in terms of diameter and length.[14–17] The selectivity on electronic type of SWCNTs was not possible until 2003, when Krupke, R. et al. demonstrated that alternating current (a.c.) dielectrophoresis technique can successfully separate SWCNTs based on electronic type owing to the differences in dielectric constant between M- and S-SWCNTs.[18] Figure 8.8 illustrates the experimental setup for dielectrophoresis. A drop of SWCNT dispersion was introduced on a micoelectrode array and M-SWCNTs have greater tendency in moving towards high electric field, resulting in selective depositing on the substrate between micoelectrodes via applying a.c. Furthermore, SWCNT alignment is achievable via a.c. dieletrophoresis as shown in Figure 8.8b. It has been reported that the efficiency of sorting SWCNTs based on electronic type can be significantly enhanced by optimizing the electric filed intensity as well as its frequency.[19, 20] In addition, neutralization of the surface charges on S-SWCNTs via the aid of surfactant encapsulation can further improve the separation selectivity for SWCNTs.[21] Despite the presence of an effective SWCNT sorting technique, the relatively low throughput hinders its potential for scale-up production. Therefore, many other approaches are also investigated.

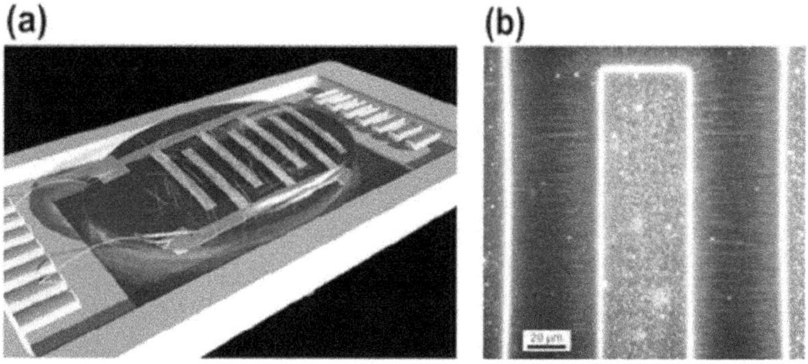

Figure 8.8 (a) A schematic illustration of the typical experimental setup for dielectrophoresis of a drop of SWCNTs solution using a microelectrode array. (b) From dark-field microscope image during dielectrophoresis, the Rayleigh scattered light indicates SWCNTs alignment between microelectrodes with the electric field. Reproduced from ref. 18.

8.3.2 Polymer Wrapping

Conventional surfactants, such as sodium dodecyl sulphate (SDS)[22] and sodium dodecylbenzene sulphonate (SDBS)[23] are utilized to disperse SWCNTs based on the energetically favorable interaction between the highly hydrophobic SWCNTs and the aliphatic chain of these surfactants in aqueous phases. The surfactant can be envisaged to closely wrap on nanotube surfaces, forming a micelle system and stabilizing the dispersion. Similarly, conjugating polymers with suitable functionality may also impose such effect for stabilizing the polymer-SWCNT suspensions.[24, 25] The backbone or side-chains of these aromatic polymers favorably wrap around SWCNTs, resulting in certain degree of selectivity in terms of either SWCNT chirality or its diameter.[26, 27] Interestingly, Nicholas and coworkers have demonstrated that poly(9,9-dioctylfluorenyl-2,7-diyl) (PFO) exhibits particularly high selectivity in wrapping SWCNTs with specific diameter and chiral angles.[28] Particularly, the CoMoCAT SWCNTs with small diameter was used to investigate the selectivity of SWCNT via PFO wrapping. From optical absorption spectra and photoluminescence excitation (PLE) maps as shown in Figure 8.9a,b, PFO has preferentially select SWCNTs with large chiral angle. Besides high selectivity of large chiral angle SWCNTs, enrichment of (7,5) species and content reduction on (6,5) species in CoMoCAT SWCNTs via PFO wrapping also suggest highly selective on specific diameter range.

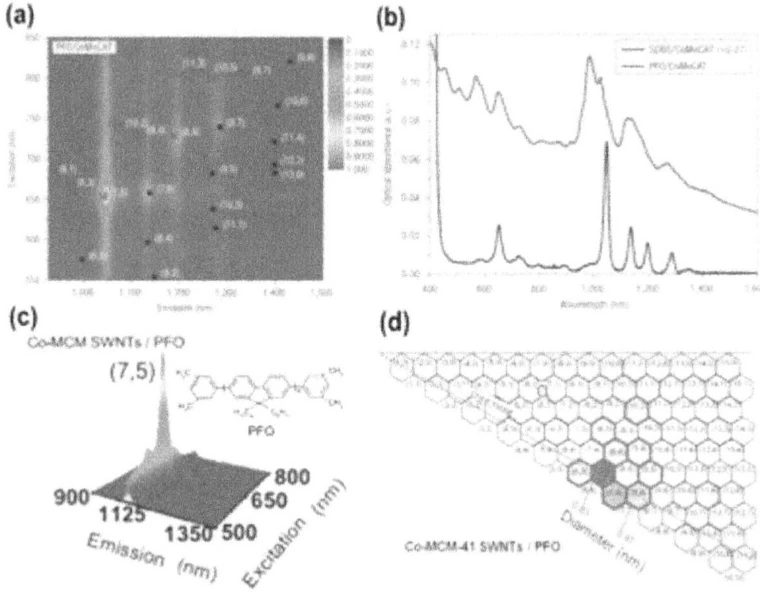

Figure 8.9 (a) Highly resolved PL spectra of PFO-assisted SWCNT dispersion from the CoMoCAT process. (n,m) chiral indices are indicated. The emission is dominated by a very prominent (7,5) peak. (b) Absorbance spectra comparing CoMoCAT SWCNTs dispersed in aqueous SDBS surfactant solution (red lines) and in toluene solutions of the polymer PFO (black lines). Reproduced from ref. 28. (c) PLE mappings for the Co-MCM SWCNTs dispersed in PFO/toluene. Inset presents schematic molecular structure of PFO. (d) Chirality map showing the wrapping preference of PFO. Reproduced from ref. 29.

The similar research work has been also explored by Li's group. Specific enrichment of (7,5) species (79% of the S-SWCNT ensemble) can be achieved via PFO polymer wrapping from the smaller diameter SWCNTs with a narrow-diameter distribution produced by chemical vapor deposition based on CO disproportionation using Co-MCM-41 catalysts (Co-MCM SWCNTs) as shown in Figure 8.9c.[29] The experimental result further corroborates the sensitivity of PFO for large chiral angle and specific diameter for smaller diameter SWCNTs. The strong sensitivity of PFO with specific diameter is attributed the formation of self-assembled intermolecular structures. Moreover, high selectivity of specific semiconducting species with large chiral angle and insensitivity of metallic species suggests the possible electronic interaction between specific polymers and nanotubes. Recently, sorting of SWCNTs by electronic type (metallic/semiconducting) and diameters has been reported via polymethacrylate with pendant atomatic funcational groups.[44] Therefore, better separation may be obtained by fine tuning the polymer wrapping approach. It is noteworthy pointing out that for electronic applications, the removal of these polymers is critically important, which may remain as a challenge at this moment.

8.3.3 DNA Wrapping

The most popular sorting approach in biochemistry field is chromatographic method. Indeed, ion-exchange chromatography (IEX) of DNA-encapsulated SWCNTs hybrids system has recently been shown as a decent approach for sorting SWCNTs in terms of diameter or electronic type.[30] Despite the encouraging results from IEX sorting process, variation in SWCNT length complicates the sorting process.[31] Therefore, it is necessary to sort the SWCNTs length and ensure the effectiveness of the IEX process in sorting diameter and electronic type. Hence, size exclusion chromatography (SEC) is widely used for sorting SWCNTs length before carrying out IEX sorting process.[32–34] The success of desired SWCNT length and chiral angle separation has been achieved by combining SEC and IEX sorting process on narrow diameter SWCNTs.[30] However, such approach on DNA-SWCNT hybrid system is highly sensitive to SWCNT diameter distribution and only effective for SWCNTs with small diameters (<1.2nm). Hence, it is essential to understand the role of the DNA sequence used for SWCNT sorting and the dynamic interaction between DNA and SWCNTs. The DNA sequence and length govern the effectiveness of SWCNT sorting. DNA length must be comparable with the sorted SWCNT length. Experimentally, the sorting of DNA-SWCNT hybrid has been shown generally possible by using the DNA sequence of

$(GT)_n$, where n is 10 to 45.[35] A further detailed study by Zheng, M. and coworkers has demonstrated the DNA sequence specificity in SWCNT separation, where they reported that by using IEX with specific DNA sequence wrapping one can purify each single-chirality S-SWCNT species from a synthetic mixture of SWCNTs.[36] Figure 8.10a summarizes a list of DNA sequences enabling chromatographic purification of single chirality S-SWCNTs. Figure 8.10b reveals that optical absorption spectra of all purified (n,m) S-SWCNT species after

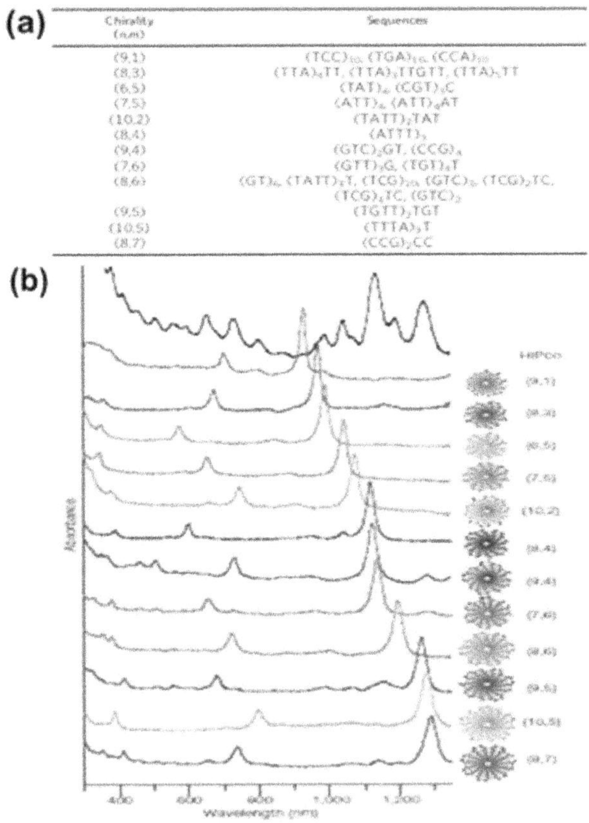
(a)

Chirality (n,m)	Sequences
(9,1)	$(TCC)_{10}$, $(TGA)_{10}$, $(CCA)_{10}$
(8,3)	$(TTA)_4TT$, $(TTA)_3TTGTT$, $(TTA)_5TT$
(6,5)	$(TAT)_4$, $(CGT)_3C$
(7,5)	$(ATT)_4$, $(ATT)_4AT$
(10,2)	$(TATT)_2TAT$
(8,4)	$(ATTT)_3$
(9,4)	$(GTC)_2GT$, $(CCG)_4$
(7,6)	$(GTT)_3G$, $(TGT)_4T$
(8,6)	$(GT)_6$, $(TATT)_2T$, $(TCG)_{10}$, $(GTC)_3$, $(TCG)_2TC$, $(TCG)_4TC$, $(GTC)_2$
(9,5)	$(TGTT)_2TGT$
(10,5)	$(TTTA)_3T$
(8,7)	$(CCG)_2CC$

Figure 8.10 (a) Table of DNA sequence versus SWCNT chirality. (b) Optical absorption spectra and atomic structures of purified semiconducting SWCNTs and the starting HiPCO mixture. The structure of each purified SWCNT species (viewed along the tube axis) and its (n,m) notation are given at the right side of the corresponding spectrum. Reproduced from ref. 36. See also Color Insert.

undergoing chromatographic purification and the starting material HiPCO SWCNTs for comparison. Owing to the extremely high purity of specific purified semiconducting (n,m) species obtained, the third lowest electronic transition of semiconducting species (S33) can be clearly observed in the respective optical absorption spectra. Despite demonstrating an astonishing finding on single-chirality purification, the structural basis for the observed DNA sequence specificity in SWCNT purification is still under investigation. It is exceptionally important to find out the DNA sequence specificity for further purifying wider range of semiconducting as well as metallic SWCNT species.

8.3.4 Selective Functionalization

Despite the perfect hexagonal geometrical structure of SWCNT surface, it is still widely anticipated that the high specificity reaction with reactive chemical molecules such as diazonium reagent with hexagonal benzene structure. The exploration of chemical selectivity in terms of SWCNT electronic type, diameter and/or chiral handedness always remains as a promising approach for obtaining SWCNTs with uniform desired properties to target applications. Based on different strategies for chemistry on SWCNTs, they basically can be classified into different major categories in terms of chemistry nature as follow: (1) covalent sidewall chemistry; (2) covalent chemistry at defect sites or open ends; (3) non-covalent surfactant encapsulation; (4) non-covalent polymer wrapping, which has been extensively discussed in the previous Section 8.3.2; (5) presence of molecules into the SWCNT interior[37] as schematically illustrated in Figure 8.11. Owing to high reactivity and ease of accessibility, chemistry on SWCNT exterior sidewalls is widely regarded a most effective approach for sorting SWCNTs. For instance, functionalization of SWCNTs with diazonium reagents reveals high selectivity on M-SWCNTs by covalent sidewall chemistry.[38,39] The proposed mechanism on selectivity of M-SWCNTs is the availability of electrons near the Fermi level in M-SWCNTs, facilitating the reaction, and resulting in preferable formation of covalent bonds between M-SWCNTs and diazonium reagents. Covalent functionalization of M-SWCNT with diazonium reagents under highly favorable condition leads to suppression of M-SWCNTs conductance in thin-film electronic device owing to the change in the electronic properties of M-SWCNTs.[40,41]

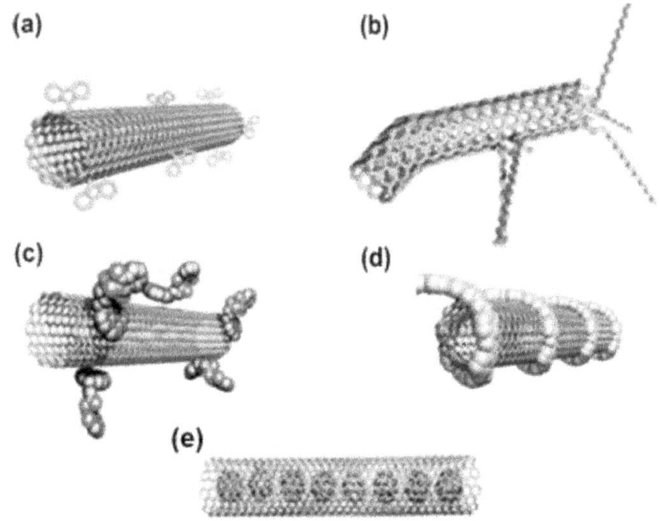

Figure 8.11 A schematic illustration of several selective chemical functionalization approaches for sorting out SWCNTs. (a) Covalent sidewall chemistry. (b) Covalent chemistry at defects or open ends. (c) Non-covalent surfactant encapsulation. (d) Non-covalent polymer wrapping. (e) Molecular insertion into the SWCNT interior. Reproduced from ref. 37.

In addition, dichlorocarbene covalent chemistry has been reported as an effective method for converting M-SWCNTs to semiconductors. The preferential attachment of dichlorocarbene on preferential M-SWCNTs via covalent bonding causes splitting of band structure at its Fermi level, resulting in semiconductor behaviors.[42] Furthermore, achievement on diameter selectivity of SWCNTs has been demonstrated via direct protonation using binary sulphonic acid mixtures due to great sensitivity of geometric strain or Fermi energy from protonation.[43] Normally, there is no significant difference between M- and S-SWCNTs in terms of physical properties resulting in difficulties in identification and separation via conventional separation techniques. By utilizing the discrete functionalization of specific electronic type of SWCNTs, it is possible to sort SWCNTs according to electronic type with the aid of conventional separation techniques. One approach is the cycloaddition of azomethine ylides derived from trialklyamine-N-oxides labeled with polycyclic aromatics for functionalizing S-SWCNTs,[44] where the sorting of functionalized S-SWCNTs from pristine SWCNTs has been demonstrated. Recently, the cycloaddition of fluorinated olefins has been explored. Interestingly, this study suggests the possibility to suppress metallic conductivity without degrading semiconducting tubes.[45]

Knowing that SWCNTs possess unique properties, it is crucial to retain these unique properties for future applications. Hence, selective non-covalent chemistries are preferable owing to the smaller alteration on the unique properties of SWCNTs upon functionalization. The selective non-covalent functionalization from amines derivatives such as proplyamine and isopropylamine has shown that M-SWCNTs can be preferentially functionalized.[46, 47] By contrast, S-SWCNTs exhibiting relatively greater extent of interaction with octadecylamine compared to M-SWCNTs species.[48, 49] In addition, the ease of charge-transfer complex formation between bromine[50] and charge-transfer aromatics[51] with M-SWCNTs owing to carrier abundance paves a way for sorting tubes with electronic type. It is noted that common surfactants (sodium dodecyl sulphate, sodium dodecylbenzene sulphonate and sodium cholate) have not only exhibited the marvelous ability of dispersing SWCNTs in aqueous phases, but also experimentally confirmed with certain degree of diameter selectivity due to their distinct interaction with SWCNTs.[52] Therefore, the role of conventional surfactants has recently recalled attention from researchers.

Owing to the remarkable merit of selective non-covalent chemistry, it has motivated researchers for exploring the possibility of sorting electronic type SWCNTs while retaining its incredible unique properties. With enormous efforts, another effective approach on functionalization of S-SWCNTs has been demonstrated based on porphyrin chemistry via utilizing discrete interaction between porphyrin and S-SWCNTs.[53] Besides revealing the effectiveness of sorting electronic types of SWCNTs, the SWCNTs chiral handedness selectivity has recently demonstrated by using chiral diporphyrin molecules.[54] By utilizing the non-covalent interaction between polymers such as fluorine-based polymers and SWCNTs, the fluorine-based polymers not only wrap SWCNTs but also exhibit high selectivity for SWCNT diameter and electronic type particular in small-diameter SWCNTs[28, 29] as discussed in Section 8.3.2. Moreover, it has been demonstrated that some solvents consisting of aromatic backbones and electron-withdrawing groups exhibit high affinity to M-SWCNTs owing to the existence of electrons close to the Fermi level of M-SWCNTs.[55]

8.4 Transistors Based on As-Grown SWCNT Networks

In Section 1.2.3, high-performance individual SWCNT-FET has been demonstrated. Despite their good electrical performance for FETs, a precise positioning and alignment of SWCNTs for transistor fabrication still remains as a main technological challenge. In contrast, the formation of SWCNT networks is relatively easy, for example, by drop-casting, spin-coating, or inkjet-printing. The density of the SWCNT networks determines its electrical characteristics. Owing to the coexistence of M- and S-SWCNTs in the networks, metallic characteristic dominates when the density of the M-SWCNTs exceeds the percolation threshold. Snow and coworkers have primarily demonstrated that as-grown SWCNT network FET exhibits p-type semiconductor characteristics with a mobility of ~10 cm^2/V s and on-off ratio up to 5 orders at relatively low nanotube networks density (~1 μm^{-2}) [56] as illustrated at Figure 8.12. The possibility of forming metallic path increases with increasing network density, resulting in metallic characteristic. Despite obtaining devices with

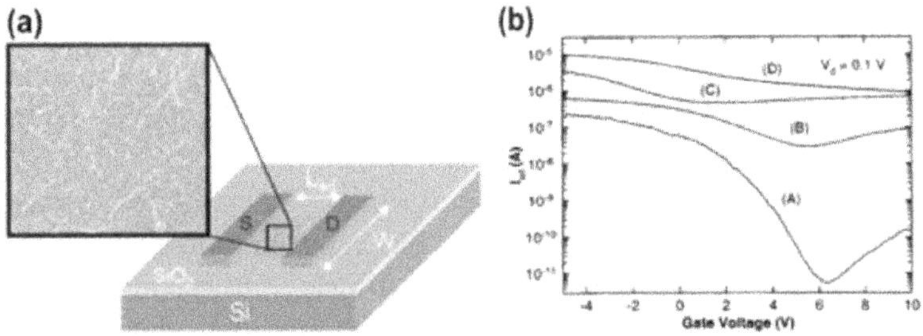

Figure 8.12 (a) Schematic of the back-gate device geometry, where SWCNT network was directly grown on Si substrate with SiO_2 as gate dielectric. Titanium was thermally evaporated on Si substrate as source and drain electrodes. An enlarge 5 μm × 5 μm AFM image represents as-grown SWCNT network on channel region. (b) Log(I_{sd}) vs V_g at V_d = 0.1 V for four devices from four different growths, corresponding to curves (A)–(D), each with a channel width of 35 μm and a channel length of 10 μm. The gate dependence indicates that the networks behave like a p-type semiconducting film with field-effect mobilities ranging from 7 to 270 cm^2/Vs. Only the network corresponding to curve (A) can be gated off. Reproduced from ref. 56.

semiconducting characteristic at appropriate nanotube network density, its electrical performance is inferior to single-tube device likely owing to the inter-tube contact resistance within the network. Therefore, there is a pool of researchers working on the development of well-aligned SWCNTs array which avoids inter-tube contact issue and retains intrinsic electrical properties of SWCNTs.

8.4.1 Aligned SWCNTs from CVD Growth

Well-aligned array of SWCNTs can be visualized as multiplying of numerous individual nanotubes. Therefore, the electrical performance of aligned SWCNT array is expected to be superior to random networks. There is a remarkable breakthrough on well-aligned arrays of SWCNT growth by Roger's group. They have demonstrated high-performance electronics based on dense, perfectly aligned arrays of SWCNTs from CVD growth.[57] The exceptional merits of using these arrays are large current outputs, relatively large active areas and small variation in device-to-device properties. The dense, perfectly aligned SWCNT arrays shown in Figure 8.13a not only preserve the intrinsic properties of SWCNTs but also eliminate the drawback of percolating transport pathways and inter-nanotubes overlap junctions.[56, 58, 59] In addition, the well-aligned SWCNT array FETs exhibiting p-channel characteristics in ambient condition could be effectively converted to n-channel characteristics after coating with polyetherimide (PEI0 on top of the FET devices, where their electrical characteristics are shown in Figure 8.13b. Interestingly, thousands of fabricated devices with 5 μm channel length possess average 1,000 $cm^2 V^{-1} s^{-1}$ and 3,000 $S m^{-1}$ in their mobility and transconductance, together with exceptionally high output current up to 1 A. Figure 8.13d,e reveals that the mobility increases but the transconductance decreases with the channel length. Despite highly impressive electrical performance, developments on precise control growth and large-scale production of these dense, perfectly aligned arrays of SWCNTs still require massive efforts and research in order to implement in commercial applications. The major fundamental issue is the coexistence of M- and S-SWCNTs in the dense, perfectly well-aligned SWCNT arrays, resulting in low on-off ratio, and low yield of semiconductor devices.

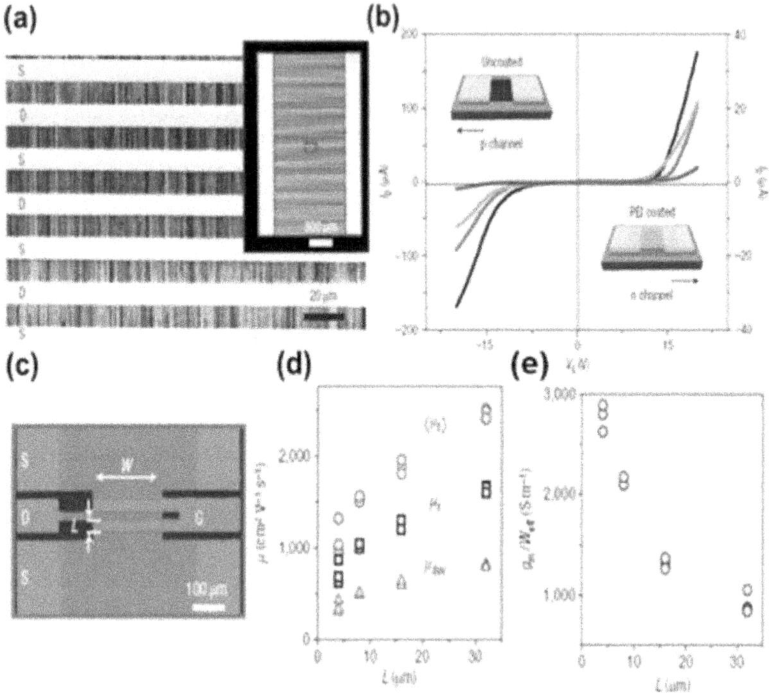

Figure 8.13 (a) optical (inset) and SEM images of dense, perfectly aligned arrays of SWCNTs. (b) A plot for transfer characteristic of p- and n-channel transistors from aligned arrays of SWCNTs and PEI-coated arrays of SWCNTs, respectively. (c) Optical micrograph of high performance top-gated transistor with L = 32 um and W = 200 um. (d) A plot for expressing mobilities as function of channel length by using parallel plate (μ_{dev}, blue triangle), rigorous (μ_r, black square) model, together with average mobilities per tube (μ_t, red circles). (e) A plot for scaled transconductances (g_m/W_{eff}) as function of channel length. Reproduced from ref. 57.

8.4.2 Random SWCNTs from CVD Growth

Fundamentally, high off-current from FET device based on as-grown SWCNT network or thin film is due to coexistence of M and S-SWCNTs. Recently, Roger's group has strategically exploited post-grown treatment on the SWCNTs grown by CVD to improve on-off ratio. Basically, a sub-monolayer, random SWCNTs networks grown by CVD on the substrate as presented was cut into desire strips geometries by combining phase-shift lithography and reactive-ion etching for preventing formation of metallic conductive paths as presented in Figure 8.14a. The electrical characteristic is favorably dominated by semiconducting paths. Such networks, serve as semiconducting materials, were fabricated into integrated circuits

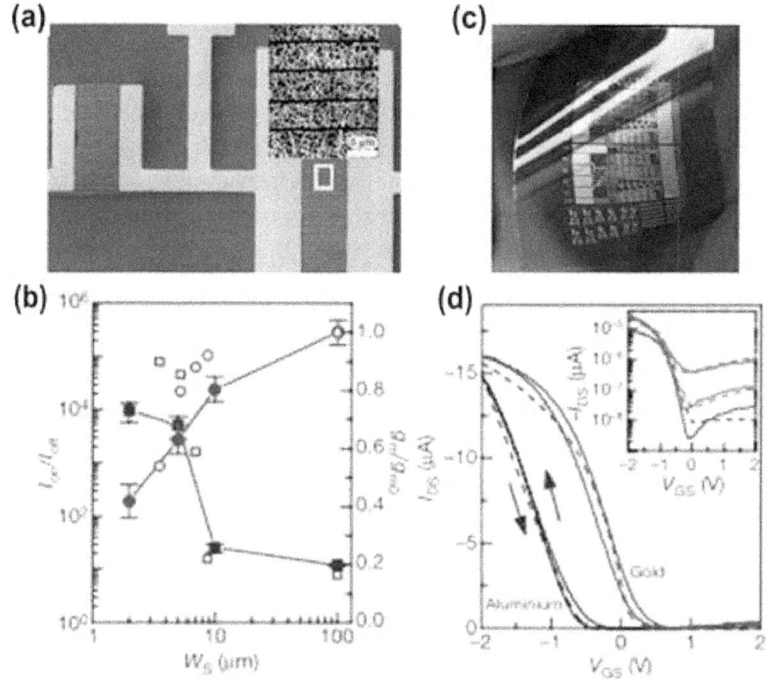

Figure 8.14 (a) A SEM image of part of SWCNTs circuit with presence of SWCNTs strips in channel region. The inset is magnified view of the network strips corresponding to the highlighted channel region. (b) A photograph of a collection of SWCNT transistors and circuits on a flexible plastic substrate (PI). (c) A plot of the measured (filled) and stimulated (open) data on the on-off ratio and normalized transconductance (g_m/g_{mo}, where g_{mo} represents the response without strips, blue) of devices with 100 μm channel length as function of strips width. (d) Measured (solid) and stimulated (dashed) transfer curve of depletion-mode (blue) and enhancement-mode (black) transistors with L = 100 μm and W = 200 μm. Inset shows its transfer curve with various V_{DS} = −0.5 V (navy), −2 V (green), and −5 V (magenta). Reproduced from ref. 60.

with exceptionally decent electrical performance. The electronic properties of the medium-scale integrated circuits consisting of nearly 100 carbon-based transistors are with high mobilities (80 $cm^2 V^{-1} s^{-1}$), low subthreshold slopes (140 V dec^{-1}), low operating voltage (<5 V), high on-off ratio up to 5 orders, and relatively good switching speed in KHz range with ~100 μm channel length device geometries [60] as shown in Figure 8.14. Besides having outstanding electrical performance, the fabricated integrated circuits on plastic substrates possess good mechanical flexibility, high uniformity and good reproducibility. The exceptional performance characteristic from sub-monolayer, random SWCNT networks suggests that it is possible to use random SWCNT networks in commercial electronic applications such as flexible integrated circuits. It is noted that this approach in principle is feasible for scale-up production but it would require delicate photolithography processes, which may still retard the investments from flexible and macroelectronic industry.

8.5 Solution Processable SWCNT Networks

This section focuses on several methods that can improve the electrical performance of SWCNT network FET devices.

8.5.1 Ultracentrifugation

Ultracentrifugation is another fascinating post-synthetic sorting approach which is scalable and compatible with Si conventional processing technology. In conventional ultracentrifugation, SWCNTs are sorted in a constant density medium based on their sedimentation coefficient. It is crucial to consider buoyant density and molecular weight for further enhancing SWCNT sorting efficiency by conventional ultracentrifugation since it greatly depends on the sedimentation coefficient. Thus, the achievement of sorting SWCNTs electronic type either metallic or semiconducting tubes via conventional ultracentrifugation has been reported through strategic utilization of the specific chemical interaction between added molecules and specific electronic type of SWCNTs to differentiate their sedimentation coefficient.[50, 61]

Amazingly, Hersam's group has adopted a sorting technique known as density gradient ultracentrifugation (DGU), which basically exploits a gradient density medium to create density gradient within the centrifuge tube and SWCNTs are exclusively separated by their buoyant density after ultracentrifugation.[62] Theoretically, SWCNTs diffuse to their respective isopynic points, in which SWCNT density matches the density of the gradient within the gradient via ultracentrifugation resulting in accomplishment of SWCNTs separation. According to their buoyant density, SWCNTs are differentiated in several layers in a centrifuge tube. A hydrodynamic model is adopted to describe the motion of SWCNTs and surfactant encapsulation surrounding SWCNTs during DGU.[63]

Sorting SWCNTs by diameter via DGU separation technique was primarily demonstrated in DNA-SWCNT hybrids system.[62] Despite high selectivity on SWCNT diameter, this system fundamentally has several issues such as irreversible wrapping DNA surrounding SWCNTs obstructing recovery of its intrinsic properties, high cost of DNA limiting the potential to be a promising separation technique in future applications, and limitation on inability to disperse large-diameter (>1.2 nm) SWCNTs. Therefore, it is crucial to search a replacement for DNA in dispersing SWCNTs, and the most widely used candidate is a surfactant such as sodium dodecyl sulfate (SDS), sodium dodecylbenzenesulphonate (SDBS) and sodium cholate (SC). The pioneer research group on discovery of DGU separation technique has reported that the achievement of SWCNTs diameter and electronic sorting via DGU can be realized with the aid of SC surfactant encapsulation and co-surfactant system mixtures of SC and SDS (4:1 wt % ratio), respectively[64] as shown in Figure 8.15. For the surfactant assisted DGU, it is speculated that the sorting of SWCNTs is likely due to the non-uniform binding of the two surfactants as a function of the SWCNT polarizability. The fabricated CNT network FET (CN-FET) devices with transfer-print technique possess impressive electrical characteristic with its mobility 20 $cm^2\ V^{-1}\ s^{-1}$ and on-off ratio greater than 4 orders,[64] respectively, as shown in Figure 8.15.

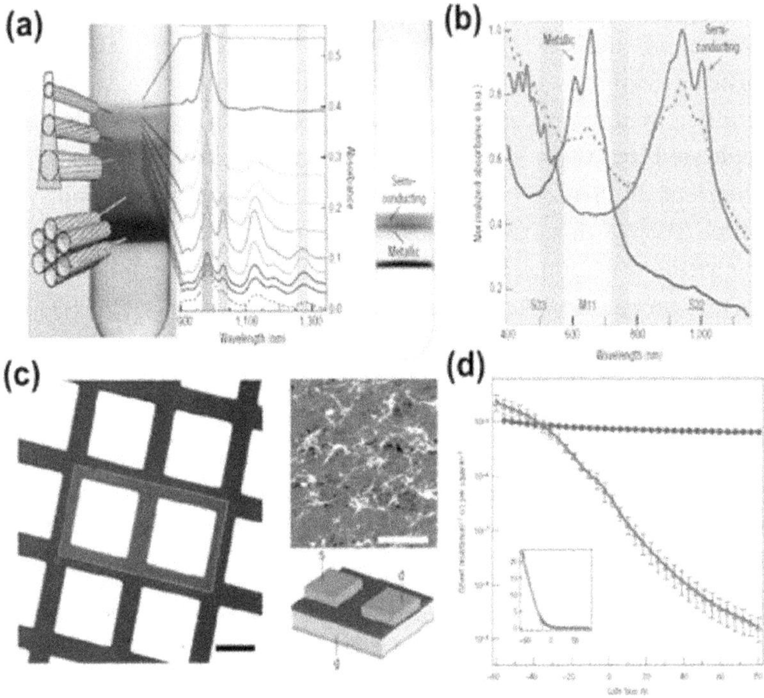

Figure 8.15 (a) An image of SWCNT sorting with diameter by density gradient ultracentrifugation. The optical absorbance spectra of different fractions indicate the selectivity of SWCNTs diameter. (b) Sorting SWCNT with electronic type can be achieved by density gradient ultracentrifugation with co-surfactant system. The optical absorbance spectra confirm the sorting by electronic type from analyzing the unique optical transitions of semiconducting SWCNTs (S22 and S33) and metallic SWCNTs (M11). (c) A representative atomic force microscopy image of random SWCNTs network and a schematic geometry of CN-FET with SiO_2 dielectric. (d) A plot for the transfer characteristic of a semiconducting CN-FET (red) and a metallic CN-FET (blue) in log scale. The inset shows the linear scale transfer curve of a semiconducting CN-FET. Reproduced from ref. 64.

8.5.2 De-Bundling

High tendency of aggregation of SWCNT tubes in bundles is due to substantial van der Waals interaction forces (950 me/nm), [65] which results in difficulty of obtaining individual SWCNT in its dispersion. Since the discovery of SWCNT in 1991, dispersibility of SWCNT commonly remains as a major issue. With tremendous efforts on finding the suitable candidates for dispersing SWCNT in solution forms, it can be achieved by using surfactants or polymers to stabilize SWCNTs dispersion via reducing its surface energy.[66, 67] In the conventional method, SWCNT is well-dispersed in aqueous solutions with assistance of surfactant encapsulation via ultrasonication and then followed by ultracentrifugation for obtaining individual dispersed SWCNTs. Recently, Ziegler's group has reported the removal of bundled SWCNTs in its dispersion via organic-aqueous interfacial trapping purification (ITP).[68] The proposed mechanism for removing bundled SWCNTs via interfacial trapping is likely due to reduction in free energy induced by the change of wetting properties. Thus, bundled SWCNTs show great tendency of diffusion to the toluene-water interface, which allows further separation of these bundled tubes.

Recently, Lee and coworkers have demonstrated that SWCNT bundles removal via ITP can further improve the electrical performance of CN-FET devices.[69] Enrichment of S species has accomplished via DGU before removing bundled tubes via ITP. The layer 1 (top layer after DGU) was extracted out and subsequently carried out by ITP for bundle removal. The typical transfer characteristic of CN-FET fabricated from layer 1+ITP is revealed at Figure 8.16a. The histogram of device percentage with various on/off ratios demonstrates that the on/off ratios of the device fabricated from layer 1+ITP is improved with comparison to layer 1 as illustrated at Figure 8.16b. Figure 8.16c further corroborates that the on/off ratio of the CN-FET devices is enhanced by ~ 1 order of magnitude after removing bundled tubes via ITP. Combining DGU (to enrich S species) and ITP (to remove bundled tubes), the best solution-processed CN-FET devices with representative effective mobility $\sim$1.5 cm^2/Vs and on/off ratio $\sim 10^4$ can be achieved at a promising yield without applying electrical field to selectively destruction of M-SWCNTs as discussed later.

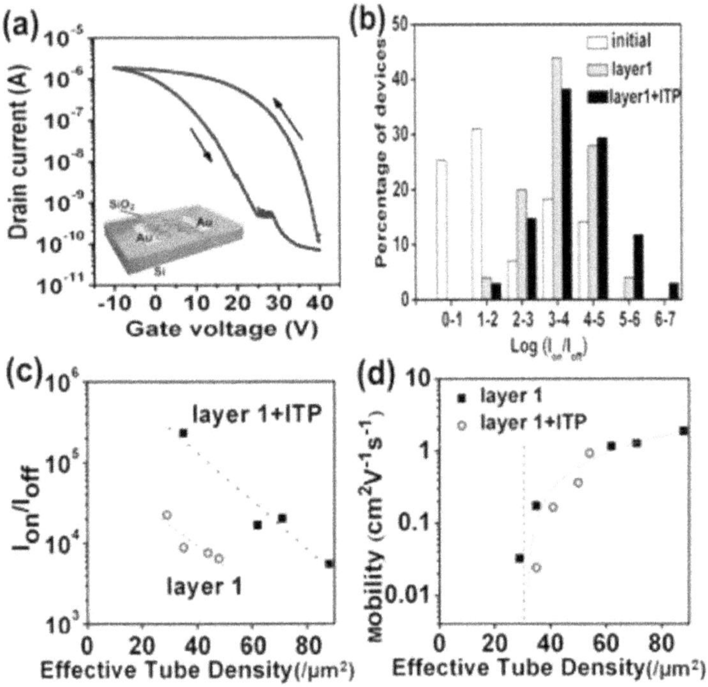

Figure 8.16 (a) The typical transfer characteristics, I_d vs. V_g for a CN-FET drop-cast from the layer 1+ITP solution. (b) The histogram of device percentage with various on/off ratios for the devices made from the initial, layer 1 and layer1+ITP solutions respectively. (c) The on-off ratio, (d) the mobility, and (c) vs. tube density of for the devices fabricated from layer 1 and layer + ITP. The red dot and black square represent layer 1 and layer 1+ITP respectively. Reproduced from ref. 69.

8.5.3 Selective Destruction of Metallic Tubes

In addition to S-SWCNT enrichments, another strategic approach is the selective destruction of metallic nanotubes via chemical functionalization such as selective reaction of diazonium reagents with M-SWCNTs, functionalization via highly reactive organic radical species as well as the cycloaddition of fluorinated olefins with M-SWCNTs. In principle, the conductivity of M-SWCNTs significantly diminishes after selective destruction of M-SWCNTs in the networks, which accomplishes decent electrical device performance with reasonably high mobility while retaining low off-current. In addition, such kinds of approaches generally are solution processable which potentially offer high scalability on device fabrication. Hence, selective destruction of M-SWCNTs approaches has been widely investigated and utilized for achieving decent device performance.

Richard Smalley and coworkers have primarily reported the highly selective reaction of diazonium reagents with M-SWCNTs.[39] The selectivity is likely due to the existence of electrons near the Fermi level in M-SWCNTs to initiate the reaction, which subsequently stabilizes a charge-transfer transition state, resulting in bond formation. Several research groups have adopted this method to selectively suppress the conductivity of M-SWCNTs to realize decent semiconductor devices. Recently, achievement of full semiconductor device yield has been demonstrated via combining selective functionalization of SWCNTs with 4-bromobenzene diazonium tetrafluoroborate (BBDT) compounds and DGU for eliminating most of M-SWCNTs species and impurities.[70] Spectroscopically, the optical absorption spectrum reveals suppression of peaks corresponding to M-SWCNTs species suggesting preferentially selective reaction of BBDT with M-SWCNTs as illustrated at Figure 8.17a. The device performance shown in Figure 8.17c experimentally corroborates the necessity of combining functionalized SWCNT with BBDT and DGU methods in order to achieve full semiconductor device yield. The reported best device performance possesses a reasonably high effective mobility ~10cm^2/Vs with an on-off ratio higher than 5000 as plotted in Figure 8.17d.

Radical species have been widely used to react with C=C bonds on the sidewalls of SWCNTs.[39, 71, 72] Recently, a cost-effective and facile solution processable method for obtaining semiconductor devices based on SWCNT networks via chemically modifying SWCNTs with free radicals generated by an organic radical initiator, 1,1′-azobis(cyanocyclohexane) (ACN) has been demonstrated.[73] The optical absorption spectrum of the pristine SWCNTs and the chemically modified SWCNT with ACN shown in Figure 8.18a suggest the preferential suppression of M- and smaller-diameter SWCNTs. In addition, comparison on the Raman spectra of the pristine and ACN-modified SWCNTs further corroborates the effectively preferential suppression of metallic tubes in SWCNT ensemble based on the notable reduction of the peak at ~1544 cm^{-1} corresponding to the longitudinal G$^-$ band from M-SWCNTs species (Figure 8.18b).

Furthermore, the D band enhancement of SWCNT after modification suggests the destruction of graphitic lattice induced by sidewall functionalization of SWCNT with free radicals. Figure 8.18d reveals a significant improvement in on/off ratio after ACN modification. Semiconductor devices based on the ACN modified SWCNT ensembles with high on/off ratio and high mobility (up to 10 cm^2/Vs) are achievable. Interestingly, the modified M-SWCNTs do not need to be separated from S-SWCNTs in the networks due to that the M-SWCNTs are electrically silent after radical modification.

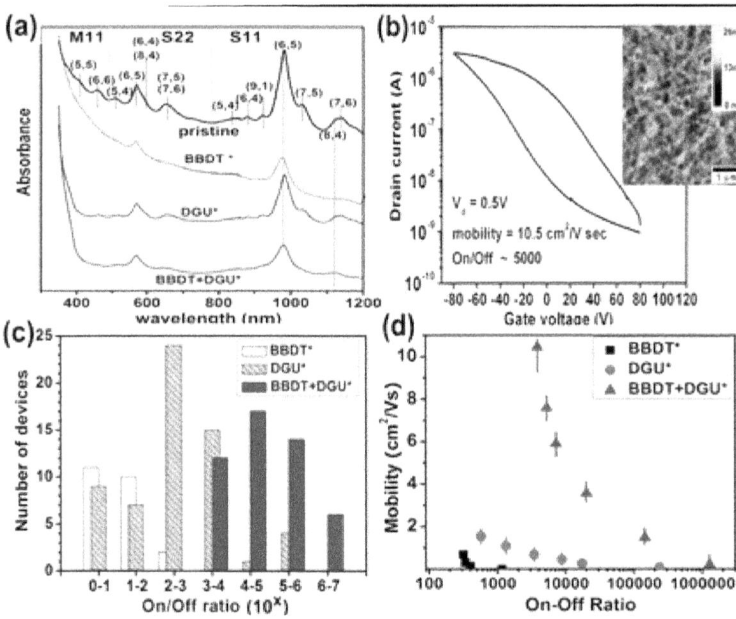

Figure 8.17 (a) Optical absorption spectra of pristine SWCNTs, the SWCNTs after treated with only BBDT (labeled as BBDT*), with only DGU (labeled as DGU*) and those with both BBDT and DGU (labeled as BBDT+DGU*). (b) Transfer characteristics for bottom-gated FETs drop-cast from the BBDT+DGU* solution. The channel length and width are 25 μm and 25 μm, respectively. Inset shows the AFM image for the SWCNTs in the channel region. (c) Histogram of on-off ratios and (d) the trend of on-off ratio with device mobility for devices constructed from BBDT*, DGU* and BBDT+DGU*. The number *x* in *x*-axis represents 10^x. Reproduced from ref. 70.

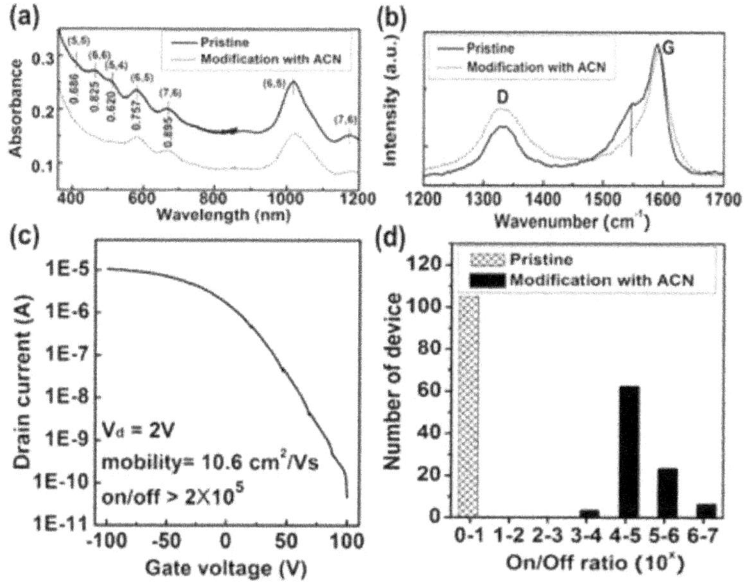

Figure 8.18 (a) The optical absorption spectra and (b) Raman spectra of the pristine and ACN-modified SWCNTs. The diameter (nm) for each tube is also indicated below the chiral index. (c) Transfer characteristics of the bottom-gated thin film transistor drop-cast from ACN-modified SWCNT suspensions. (d) The histogram of the on/off ratios of the devices fabricated using the pristine and functionalized SWCNTs. Reproduced from ref. 73.

Recently, cycloaddition of fluorinated olefins such as perfluoro-2(2-flurosulfonylethoxy) propyl vinyl ether (PSEPVE) and perfluoro (5-methyl-3,6-dioxanon-1-ene) (PMDE) has been reported to deactivate M tubes.[45] Figure 8.19b reveals the AFM images of pristine and functionalized SWCNTs with fluorination levels (c = 0.018 and 0.038), indicating that the concentration of the fluorination level is crucial for obtaining optimized electrical characteristics of SWCNT network FETs. Its optimized fluorination level ($c \sim 0.018$) under controlled conditions is used for fabricating high electrical performance TFT devices. As plotted in Figure 8.19c,d, the fabricated TFT devices from PSEPVE-SWCNTs and PMDE-SWCNTs exhibit mobilities of 10 $cm^2\ V^{-1}\ s^{-1}$ and 104 $cm^2V^{-1}s^{-1}$, respectively, and both with an on/off ratio up to 5 orders. It is noted that there is still no information about the reproducibility or device yield for this approach.

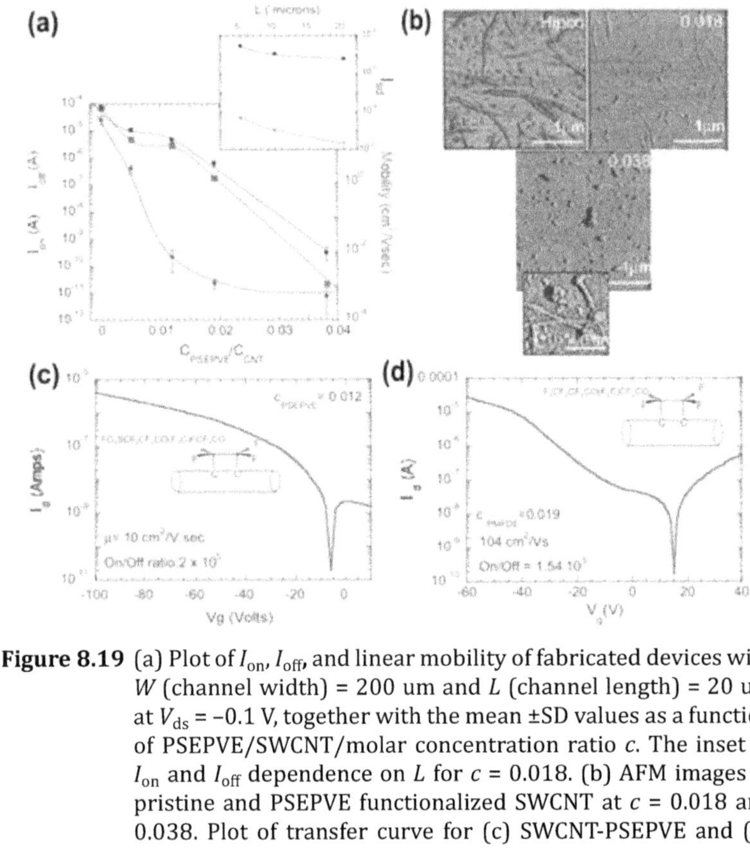

Figure 8.19 (a) Plot of I_{on}, I_{off}, and linear mobility of fabricated devices with W (channel width) = 200 um and L (channel length) = 20 um at V_{ds} = −0.1 V, together with the mean ±SD values as a function of PSEPVE/SWCNT/molar concentration ratio c. The inset of I_{on} and I_{off} dependence on L for c = 0.018. (b) AFM images of pristine and PSEPVE functionalized SWCNT at c = 0.018 and 0.038. Plot of transfer curve for (c) SWCNT-PSEPVE and (d) SWCNT-PFMDE TFTs at c = 0.018 and V_{ds} = −0.1 V and −0.01 V, respectively. Reproduced from ref. 45.

8.6 Conclusions

This chapter starts with the introduction on geometrical and electronic structures of SWCNTs, followed by the detailing several methods for sorting CNTs based on diameter or electronic types. Some issues and FET device fabrication with SWCNT networks are also discussed. The SWCNT network FETs exhibit many applications such as gas sensors or biosensors. Readers who are interested in this research area may refer to a review article by Fu et al.[74] It is apparent that to achieve acceptable electrical performance for electronic applications the following two approaches have to work hand-in-hand: (1) growth of SWCNT with narrower chirality distribution or desired species and (2) cost-effective methods for CNT sorting or M-tube deactivating. The remaining challenge is to develop an effective approach for achieving full semiconductor device yield and reasonably high carrier mobility, which shall allow the scale-up production for flexible and macroelectronic applications.

References

1. Charlier, J.C., Blase, X., and Roche, S. Electronic and transport properties of nanotubes. *Rev. Mod. Phys.* **79**, 677–732 (2007).
2. Dresselhaus, M.S., Dresselhaus, G., and Saito, R. *Physical Properties of Carbon Nanotubes*. Imperial College Press, London, 259 (1998).
3. Anantram, M.P., and L'eonard, F. Physics of carbon nanotube electronic devices. *Rep. Prog. Phys.* **69**, 507–561 (2006).
4. Javey, A. et al., High-κ dielectrics for advanced carbon nanotube transistors and logic gates. *Nat. Mater.* **1**, 241–246 (2002).
5. Wind, S.J., Appenzeller, J., Martel, R., Derycke, V., and Avouris, P. Vertical scaling of carbon nanotube field-effect transistors using top gate electrodes. *Appl. Phys. Lett.* **80**, 3817–3819 (2002).
6. Tan, S.J., Verschueren, A., and Dekker, C. Room-temperature transistor based on a single carbon nanotube. *Nature* **393**, 49–52 (1998).
7. Hyongsok T. Soh, H.T., et al. Integrated nanotube circuits: controlled growth and ohmic contacting of single-walled carbon nanotubes. *Appl. Phys. Lett.* **75**, 627–629 (1999).
8. Javey, A., et al. Carbon nanotube field-effect transistors with integrated ohmic contacts and high-κ gate dielectrics. *Nano Lett.* **4**, 447–450 (2004).
9. Martel, R., et al. Ambipolar electrical transport in semiconducting single-wall carbon nanotubes. *Phys. Rev. Lett.* **87**, 256805–256808 (2001).
10. Radosavljivic, M., Freitag, M., Thadani, K.V., and Johnson, A.T. Nonvolatile molecular memory elements based on ambipolar nanotube field effect transistors. *Nano Lett.* **2**, 761764 (2002).
11. Krupke, R., and Hennrich, F. Separation techniques for carbon nanotubes. *Adv. Eng. Mat.* **7**, 111–116 (2005).
12. Banerjee, S., Hemraj-Benny, T., and Wong, S.S. Routes towards separating metallic and semiconducting nanotubes. *J. Nanosci. Nanotechnol.* **5**, 841–855 (2005).

13. Haddon, R.C., Sippel, J., Rinzler, A.G., and Papadimitrakopoulos, F. Purification and separation of carbon nanotubes. *Mater. Res. Bull.* **29**, 252–259 (2004).
14. Heller, D.A., et al. Concomitant length and diameter separation of single-walled carbon nanotubes. *J. Am. Chem. Soc.* **126**, 14567–14573 (2004).
15. Vetcher, A.A., et al., Fractionation of SWNT/nucleic acid complexes by agarose gel electrophoresis. *Nanotechnology* **17**, 4263–4269 (2006).
16. Doorn, S.K., et al. High resolution capillary electrophoresis of carbon nanotubes. *J. Am. Chem. Soc.* **124**, 3169–3174 (2002).
17. Doorn, S.K., et al. Capillary electrophoresis separations of bundled and individual carbon nanotubes. *J. Phys. Chem. B* **107**, 6063–6069 (2003).
18. Krupe, R., Hennrich, F., Lohneysen, H.V., and Kappes, M.M. Separation of metallic from semiconducting single-walled carbon nanotubes. *Science* **301**, 344–347 (2003).
19. Krupke, R., Hennrich, F., Kappes, M.M., and Lohneysen, H.V. Surface conductance induced dielectrophoresis of semiconducting single-walled carbon nanotubes. *Nano Lett.* **4**, 1395–1399 (2004).
20. Krupke, R., Linden, S., Rapp, M., and Hennrich, F. Thin films of metallic carbon nanotubes prepared by dielectrophoresis. *Adv. Mater.* **18**, 1468–1470 (2006).
21. Hong, S., Jung, S., Choi, J., Kim, Y., and Baik, S. Electrical transport characteristics of surface-conductance-controlled, dielectrophoretically separated single-walled carbon nanotubes. *Langmuir* **23**, 4749–4752 (2007).
22. Nish, A., and Nicholas, R.J. Temperature induced restoration of fluorescence from oxidised single-walled carbon nanotubes in aqueous sodium dodecylsulfate solution. *Phys. Chem. Chem. Phys.* **8**, 3547–3351 (2006).
23. Matarredona, O., et al. Dispersion of single-walled carbon nanotubes in aqueous solutions of the anionic surfactant NaDDBS. *J. Phys. Chem. B* **107**, 13357–13367 (2003).
24. O' Connell, M.J., et al. Reversible water-solubilization of single-walled carbon nanotubes by polymer wrapping. *Chem. Phys. Lett.* **342**, 265–271 (2001).
25. Rice, N.A., Soper, K., Zhou, N., Merschrod, E., and Zhao, Y. Dispersing as-prepared single-walled carbon nanotube powders with linear conjugated polymers. *Chem. Commun.*, 4937–4939 (2006).

26. McCarthy, B., et al. A microscopic and spectroscopic study of interactions between carbon nanotubes and a conjugated polymer. *J. Phys. Chem. B* **106**, 2210–2216 (2002).

27. Yang, M., Koutsos, V., and Zaiser, M. Interactions between polymers and carbon nanotubes: A molecular dynamics study. *J. Phys. Chem. B* **109**, 10009–10014 (2005).

28. Nish, A., Hwang, J.-Y., Doig, J., and Nicholas, R.J. Highly selective dispersion of single-walled carbon nanotubes using aromatic polymers. *Nat. Nanotechnol.* **2**, 640–646 (2007).

29. Chen, F., Wang, B., Chen, Y., and Li, L.-J. Toward the extraction of single species of single-walled carbon nanotubes using fluorene-based polymers. *Nano Lett.* **7**, 3013–3017 (2007).

30. Zheng, M., and Semke, E.D. Enrichment of single chirality carbon nanotubes. *J. Am. Chem. Soc.* **129**, 6084–6085 (2007).

31. Zheng, M., et al., DNA-assisted dispersion and separation of carbon nanotubes. *Nat. Mater.* **2**, 338–342 (2003).

32. Duesberg, G.S., Muster, J., Krstic, V., Burghard, M., and Roth, S. Chromatographic size separation of single-wall carbon nanotubes. . *Appl. Phys. A* **67**, 117–119 (1998).

33. Huang, X.Y., McLean, R.S., and Zheng, M. High-resolution length sorting and purification of DNA-wrapped carbon nanotubes by size-exclusion chromatography. . *Anal. Chem.* **77**, 6225–6228 (2005).

34. Arnold, K., Hennrich, F., Krupke, R., Lebedkin, S., and Kappes, M.M. Length separation studies of single walled carbon nanotube dispersions. *Phys. Status Solidi B* **243**, 3073–3076 (2006).

35. Zheng, M., et al. Structure-based carbon nanotube sorting by sequence-dependent DNA assembly. *Science* **302**, 1545–1548 (2003).

36. Tu X, Manohar, S., Jagota, A., and Zheng, M. DNA sequence motifs for structure-specific recognition and separation of carbon nanotubes. *Science* **460**, 250–253 (2009).

37. Hirsch, A. Functionalization of single-walled carbon nanotubes. *Angew. Chem. Int. Ed.,* **41**, 1853–1859 (2002).

38. Campidelli, S., Meneghetti, M., and Prato, M. Separation of metallic and semiconducting single-walled carbon nanotubes via covalent functionalization. *Small* **3**, 1672–1676 (2007).

39. Strano, M.S., et al., Electronic structure control of single-walled carbon nanotube functionalization. *Science* **301**, 1519–1522 (2003).

40. An, L., Fu, Q.A., Lu, C.G., and Liu, J. A simple chemical route to selectively eliminate metallic carbon nanotubes in nanotube network devices. *J. Am. Chem. Soc.* **126**, 10520–10521 (2004).

41. Balasubramanian, K., Sordan, R., Burghard, M., and Kern, K.A. Selective electrochemical approach to carbon nanotube field-effect transistors. *Nano Lett.* **4**, 827–830 (2004).

42. Hu, H., et al., Sidewall functionalization of single-walled carbon nanotubes by addition of dichlorocarbene. *J. Am. Chem. Soc.* **125**, 14893–14900 (2003).

43. Ramesh, S., et al., Diameter selection of single-walled carbon nanotubes through programmable solvation in binary sulfonic acid mixtures. *J. Phys. Chem. C* **111**, 17827–17834 (2007).

44. Menard-Moyon, C., Izard, N., Doris, E., and Mioskowski, C. Separation of semiconducting from metallic carbon nanotubes by selective functionalization with azomethine ylides. *J. Am. Chem. Soc.* **128**, 6552–6553 (2006).

45. Kanungo, M., Lu, H., Malliaras, G.G., and Blanchet, G.B. Suppression of metallic conductivity of single-walled carbon nanotubes by cycloaddition reactions. *Science* **323**, 234–237 (2009).

46. Maeda, Y., et al., Large-scale separation of metallic and semiconducting single-walled carbon nanotubes. *J. Am. Chem. Soc.* **127**, 10287–10290 (2005).

47. Maeda, Y., et al., Dispersion and separation of small-diameter single-walled carbon nanotubes. *J. Am. Chem. Soc.* **128**, 12239–12242 (2006).

48. Chattopadhyay, D., Galeska, L., and Papadimitrakopoulos, F. A route for bulk separation of semiconducting from metallic single-wall carbon nanotubes. *J. Am. Chem. Soc.* **125**, 3370–3375 (2003).

49. Kim, S.N., Luo, Z.T., and Papadimitrakopoulos, F. Diameter and metallicity dependent redox influences on the separation of single-wall carbon nanotubes. *Nano Lett.* **5**, 2500–2504 (2005).

50. Chen, Z.H., et al., Bulk separative enrichment in metallic or semiconducting single-walled carbon nanotubes. *Nano Lett.* **3**, 1245–1249 (2003).

51. Lu, J., et al., Selective interaction of large or charge-transfer aromatic molecules with metallic single-wall carbon nanotubes: Critical role of the molecular size and orientation. *J. Am. Chem. Soc.* **128**, 5114–5118 (2006).

52. McDonald, T.J., Engtrakul, C., Jones, M., Rumbles, G. and Heben, M.J. Kinetics of PL quenching during single-walled carbon nanotube rebundling and diameter-dependent surfactant interactions. *J. Phys. Chem. B* **110**, 25339–25346 (2006).

53. Li, H.P., et al., Selective interactions of porphyrins with semiconducting single-walled carbon nanotubes. *J. Am. Chem. Soc.* **126**, 1014-1015 (2004).

54. Peng, X., et al., Optically active single-walled carbon nanotubes. *Nat. Nanotech.* **2**, 361–365 (2007).

55. Shin, H.-J., et al., Tailoring electronic structures of carbon nanotubes by solvent with electron-donating and - withdrawing groups. *J. Am. Chem. Soc.* **130**, 2062–2066 (2008).

56. Snow, E.S., Novak, J.P., Campbell, P.M. and Park, D. Random networks of carbon nanotubes as an electronic material. *Appl. Phys. Lett.* **82**, 2145–2147 (2003).

57. Kang, S.J., et al., High-performance electronics using dense, perfectly aligned arrays of single-walled carbon nanotubes. *Nat. Nanotechnol.* **2**, 230–236 (2007).

58. Zhou, Y., et al., p-Channel, n-Channel Thin Film Transistors and p-n Diodes Based on Single Wall Carbon Nanotube Networks. *Nano Lett.* **4**, 2031–2035 (2004).

59. Hu, L., Hecht, D.S., and Gruner, G. Percolation in transparent and conducting carbon nanotube networks. *Nano Lett.* **4**, 2513–2517 (2004).

60. Q. Cao, Q., et al., Medium scale carbon nanotube thin film integrated circuits on flexible plastic substrates. *Nature* **454**, 495–500 (2008).

61. Wei, L., et al., Selective enrichment of (6,5) and (8,3) single-walled carbon nanotubes via cosurfactants extraction from narrow (n,m) distribution samples. *J. Phys. Chem. B* **112**, 2771–2774 (2008).

62. Arnold, M.S., Stupp, S.I., and Hersam, M.C. Enrichment of single-walled carbon nanotubes by diameter in density gradients. *Nano Lett.* **5**, 713–718 (2005).

63. Nair, N., Kim, W.-J., Braatz, R.D. and Strano, S.M. Dynamics of surfactant-suspended single-walled carbon nanotubes in a centrifugal field. *Langmuir* **24**, 1790–1795 (2008).

64. Arnold, M.S., Green, A.A., Hulvat, J.F., Stupp, S.I. and Hersam, M.C. Sorting carbon nanotubes by electronic structure using density differentiation. *Nat. Nanotechnol.* **1**, 60–65 (2006).

65. Thess, N., et al., Crystalline ropes of metallic carbon nanotubes. *Science* **273**, 483–487 (1996).
66. O'Connell, M.J., et al., Band gap fluorescence from individual single-walled carbon nanotubes. *Science* **297**, 593–596 (2002).
67. Moore, V.C., Strano, M.S., Haroz, E.H., Hauge, R.H., and Smalley, R.E. Individually suspended single-walled carbon nanotubes in various surfactants. *Nano Lett.* **3**, 1379–1382 (2003).
68. Wang, R.K., Reeves, R.D., and Ziegler, K.J. Interfacial trapping of single-walled carbon nanotube bundles. *J. Am. Chem. Soc.* **129**, 15124–15125 (2007).
69. Lee, C.W., et al., Toward high-performance solution-processed carbon nanotube network transistors by removing nanotube bundles. *J. Phys. Chem. C* **112**, 12089–12091 (2008).
70. Lee, C.W., et al., Solution-processable carbon nanotubes for semiconducting thin-film transistor devices. *Adv. Mater.* **22**, 1278–1282 (2009).
71. McIntosh, D., Khasbashesku, V.N., and Barrera, E.V. Benzoyl peroxide initiated in situ functionalization, processing, and mechanical properties of single-walled carbon nanotube polypropylene composite fibers. *J. Phys. Chem. C* **111**, 1592–1600 (2007).
72. Peng, H.-Q., Reverdy, P., Khabashesku, V.N., and Margrave, J.L. Sidewall functionalization of single-walled carbon nanotubes with organic peroxides *Chem. Commun.*, 362–363 (2003).
73. Zhao, J., et al., Solution-processable semiconducting thin-film transistors using single-walled carbon nanotubes chemically modified by organic radical initiators. *Chem. Commun.*, 7182–7184 (2009).
74. Fu, D., and Li, L.-J. Label-free electrical detection of DNA hybridization using carbon nanotubes and graphene. *Nano Rev.* **1**, 5354 (2010).

Chapter 9

Carbon Nanotube Field Emission Technology for Vacuum Electronic Device Applications

Sigen Wang
Department of Radiation Oncology and Department of Physics and Astronomy, University of North Carolina, Chapel Hill, NC 27599, USA
sgwang@email.unc.edu, or sgwang88@yahoo.com

Carbon nanotube (CNT) is a relatively new carbon allotrope discovered in 1991 by Iijima *et al.* A CNT is a honeycomb lattice rolled into a cylinder. In 1995, four years after the discovery of CNT, field emission from CNTs was reported by three groups at low operating temperature, low-turn-on fields, and high current densities. From 1998 onward, the perspective to use CNTs in field emission devices such as field emission display, x-ray source, and microwave source spurred on efforts worldwide. In this chapter, we briefly review the recent development of carbon nanotube field emission technology for future vacuum electronic device applications in medical x-ray imaging and radiotherapy, high-power microwave and flat-panel displays.

Carbon Nanotubes and Their Applications
Edited by Qing Zhang

ISBN 978-981-4241-90-8 (Hardcover), 978-981-4303-18-7 (eBook)
www.panstanford.com

9.1 Introduction

In 1991, Iijima *et al.* [1] discovered CNT, a relatively new carbon allotrope. A CNT is a honeycomb lattice rolled into a cylinder. It comprises either a single graphene shell [1] termed single-wall CNT (SWCNT) or multiple concentric graphene shells termed a multiwall CNT (MWCNT) [2]. The diameter of CNTs is from subnanometer to a hundred nanometers and the length of CNTs can be up to hundreds of microns. CNTs are an extremely interesting new area of research with many potential future applications. CNTs have been envisioned as one of the promising materials for field emitters due to their unique structure and properties. With graphene layers parallel to the filament axis, CNTs (single-wall metallic-type or multiwall) exhibit high electrical conductivity at room temperature. They also have a high aspect ratio (length/diameter), which can be as large as 10^4–10^5 and a whisker-like shape. In addition, CNTs can be very stable emitters, even at high temperatures. Purcell *et al.* [3] demonstrated that an MWCNT emitter could be heated by its emission current up to 2000 K and remain stable, unlike metal emitters, which suffer from thermal runaway. For an MWCNT, its resistance decreases with temperature, which suppresses ohmic heat generation, and surface diffusion is less likely in the strong C–C covalent bonds. These unique characteristics of CNTs make them remarkable field emitters.

The power of CNTs as electron field emitters was apparent from the first three reports of extremely low turn-on fields and high current densities in 1995 [4–6]. Only a few studies followed during the next two years [7]. From 1998 on, the perspective to use CNTs as field emission devices spurred efforts worldwide to develop applications such as CNT field emission displays [8,9], lighting elements [10], x-ray sources [11,12] and high-power microwave sources [13,14], etc.

In this chapter, we briefly review the recent development of CNT field emission technology for future vacuum electronic device applications in x-ray sources for medical imaging and radiotherapy systems, high-power microwave devices and flat-panel displays.

9.2 Carbon Nanotube Field Emission

Electron field emission, also known as Fowler–Nordheim quantum mechanical tunneling, is the process of releasing electrons from metal surface via barrier penetration in the presence of a high electric field, as illustrated in Figure 9.1.

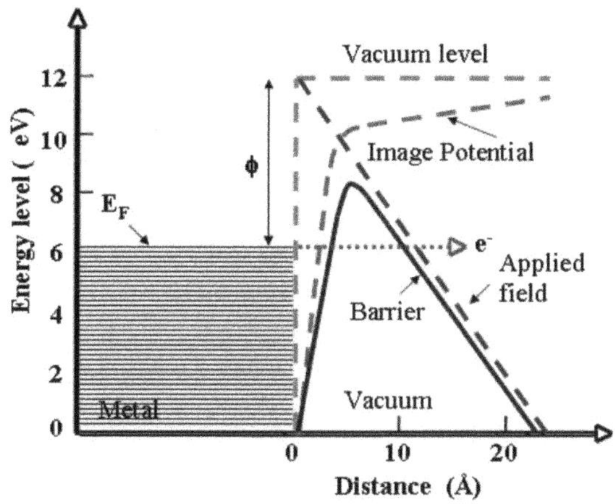

Figure 9.1 Schematic illustration of field emission (quantum tunneling effect). When an electric field is applied to the metal, electrons can escape from the metal surface.

Unlike the escaping over a potential barrier as with photoemission and thermionic processes, field emission is purely a quantum mechanical phenomenon. The theory of electron field emission was formally published in 1928 by Fowler and Nordheim [15]. Initially, little progress was made with this theory outside the realm of pure science. However, this certainly changed in 1937 when Erwin Muller first used this phenomenon on the experimental front and devised a new way to do microscopy [16]. Muller was successfully able to use the extracted electrons from a field-emitting metal tip to create an image of the tip revealing its surface structure.

The Fowler–Nordheim model [15] for field emission describes the electron current density J emitting from a surface into vacuum as a function of the applied field E. The model parameters that govern the emission are the material (effective work function, ϕ), the area available for emission Ae and a geometric enhancement factor β, which describes how electric fields can be enhanced by protrusions from the emitting surface. $B1$ and $B2$ are known fundamental constants, I is the pre-break down current, and d is the gap distance [17]. Emission current density J with local surface field E is given by

$$J = (B_1 E^2/\phi) \exp(-B_2 \phi^{3/2}/E) \quad (9.1)$$

Under the assumption that there is one major site of field emission and its emitting area denoted by Ae and with $E = \beta(V/d)$ accounting for enhancement where V is the applied voltage, the total pre-breakdown current I as $I = JAe$ in logarithmic form is given by

$$I = (AeB_1\beta^2V^2/(\phi d^2)) \exp(-dB_2\phi^{3/2}/(\beta V)) \tag{9.2}$$

The complete Fowler–Nordheim result is given by

$$\log(I/V^2) = \log((1.54 \times 10^{-6}Ae\beta^2 10^{4.52\phi\,\wedge -0.5})/(\phi d^2)) - ((2.84 \times 10^9 d\phi^{1.5})/\beta)(1/V) \tag{9.3}$$

The Fowler–Nordheim equation states that the emission current (I) increases exponentially with increasing voltage (V). According to the above equations, the emission current is strongly dependent on the following three factors: (a) the work function of the emitter surface, (b) the radius of curvature of the emitter apex and (c) the emission area.

For a metal with a flat surface, however, the threshold field is impractically high and is typically around 10^4 V/μm. All the electron field emitters rely on field enhancement factor β at the sharp tips/protrusions. One way to fabricate sharp field emitters is a lithography process [18]. It is immediately apparent that a field emitter could be more power-efficient than a thermionic emitter, which requires heating. Field emission sources also offer several attractive characteristics such as instantaneous response to electric field variation, resistance to temperature fluctuation, and a high degree of focus ability in electron optics due to their sharp (0.2–0.3 eV) energy spread. Furthermore, the field emitted electrons are confined to a narrow cone angle along the electrical field direction, where thermal electrons are randomly distributed. However, due to the high electric field experienced at the tips of the materials during field emission, the metal atoms often diffuse or electromigrate, causing failure and thermal runaway. As a result, the conventional field emitters of Spindt tips have not been used in practical devices since they in general suffer from high turn-on field, low emission current, high cost and poor stability with limited lifetime.

CNT is a unique form of carbon filament/fiber in which the graphene walls roll up to form tubes, with diameters typically 1–50 nm and lengths of a few microns. It has high electrical and thermal conductivity, very high tensile strength, and a high aspect ratio. This makes it a very good material for field emitters. Figure 9.2a illustrates how a CNT emits electrons under the influence of a large electric field.

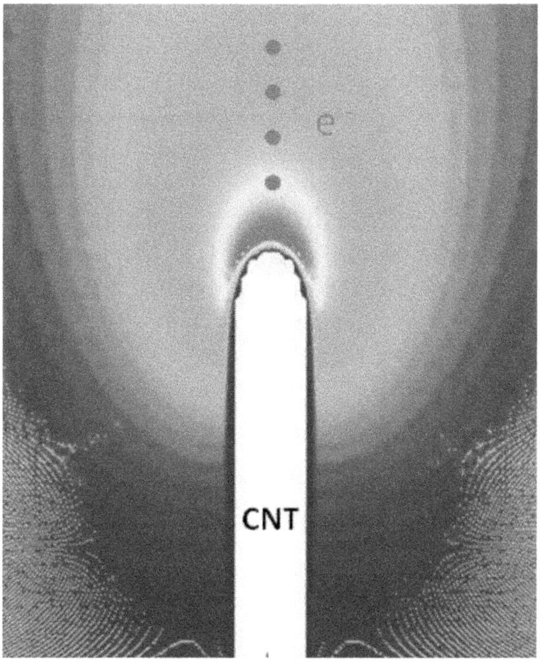

(a)

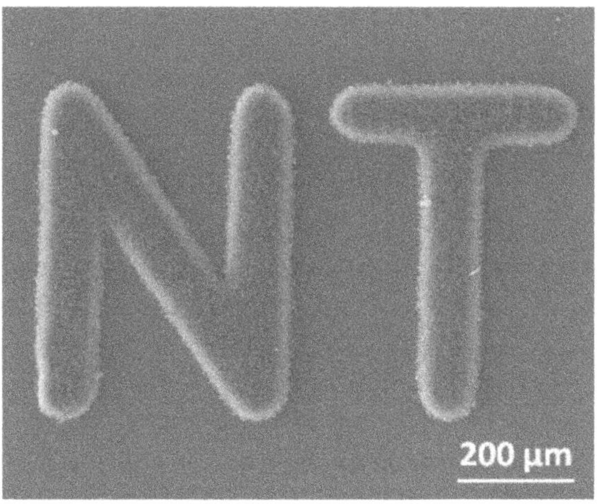

(b)

Figure 9.2 (a) Schematic illustration of field emission from a CNT; and (b) a scanning electron microscopy (SEM) image of patterned MWCNTs [20].

The mechanism of electron emission from CNTs was confirmed to be a field emission process [19]. The CNT behaves as a sharp metallic tip. It has been well documented that CNTs have better field emission properties than other sharp metal based emitters. Besides, they can provide great flexibility in cathode configuration, such as curved array, linear array, 2 dimension (2D) array, and other desired shapes of patterns, as shown in Figure 9.2b [20].

CNTs have already been applied in various field emission applications. In the following we will review the applications of CNT field emission technology in vacuum electronic devices. We will focus on CNT based medical x-rays, high-power microwave and flat-panel displays.

9.3 Applications of Carbon Nanotube Field Emission in Vacuum Electronic Devices

9.3.1 Carbon Nanotube Field Emission x-Ray Sources for Medical Imaging and Radiotherapy Applications

Conventional x-ray tubes use the thermionic electron source. In the thermionic electron source, a metallic filament is heated to a high temperature and its electrons gain sufficient energy to overcome the work function of the metal. In contrast, CNT technology-based x-ray tubes use CNT field emission electron sources.

The CNT based x-ray tubes have several advantages over conventional thermionic x-ray tubes: (a) the life span can potentially be prolonged; (b) the size and power consumption can be reduced significantly; (c) focused electron beams with very small energy spread and programmable pulse width and repetition rate can be easily realized [21]; and (d) the small size and low power consumption characterizations can lead to portable x-ray sources for on-site medical and industrial applications. Furthermore, CNT field emission x-ray technology offers two properties that are lacking in the thermionic x-ray sources: temporal modulation and spatial configuration (the measure of how closely lines can be resolved in an image). This makes CNT field emission x-ray technology ideal for development of novel medical imaging and irradiation systems.

9.3.1.1 Carbon nanotube field emission x-ray tubes

Our group led by Dr. Otto Zhou at the University of North Carolina is among the first to demonstrate a CNT based field emission x-ray source that is capable of producing sufficient flux for imaging human extremities as shown in Figure 9.3 [12, 22]. The structure of the first generation CNT field emission x-ray tube developed is schematically illustrated in Figure 9.4 [12]. It has a triode-type structure with a CNT cathode, a gate electrode and a metal target housed in a vacuum tube with a beryllium (Be) window.

The electrons are extracted from the CNT cathode by applying a gate voltage between the cathode and gate. X-rays are produced when the emitted electrons are accelerated and bombard on the copper target. The tube current and the acceleration voltage are independently controlled. By replacing the direct current (DC) gate voltage with an amplified signal from a function generator, electron emission and thus x-ray radiation with programmable waveform are easily produced.

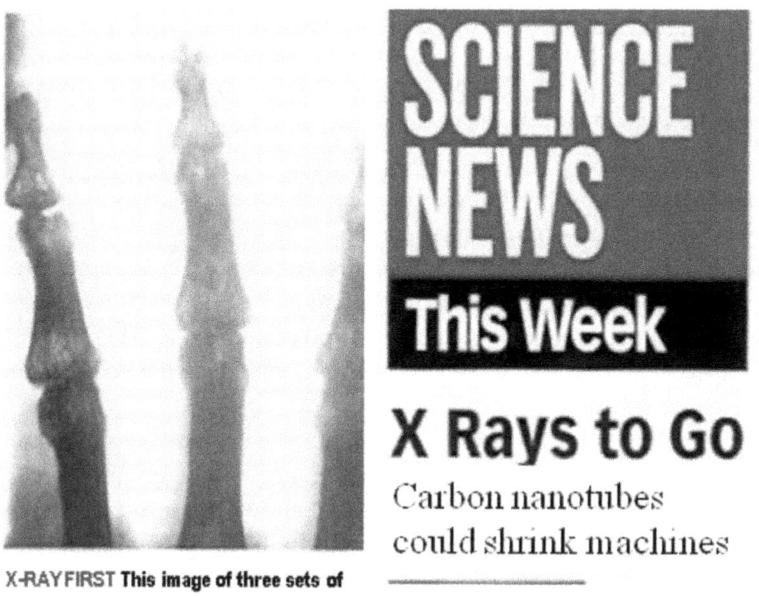

Figure 9.3 An x-ray image of a humanoid hand taken using the first generation CNT field emission x-ray tube [12,22].

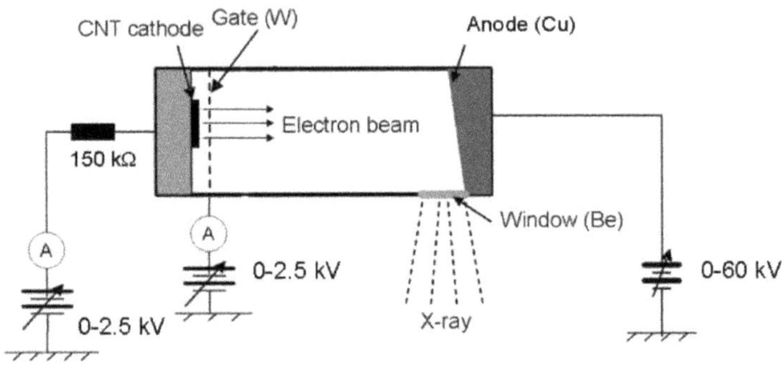

Figure 9.4 A schematic drawing of the CNT field emission x-ray source [12].

CNT based micro-focus field emission x-ray tube with high spatial resolution, temporal resolution, and stable emission has also been demonstrated in our group [21]. This micro-focus x-ray source comprised of a CNT cathode, a gate electrode, an electrostatic focusing unit, and a molybdenum anode with 9° tilting angle. The entire x-ray source was housed in a vacuum chamber at 1×10^{-7} Torr base pressure. Field emitted electrons were extracted from the CNT cathode by applying a bias voltage on the gate electrode and were subsequently focused by the focusing lenses before reaching the anode. The flux generated by this x-ray tube has a 30 μm spatial resolution, which is higher than those used in the typical micro-computed tomography (micro-CT) imaging protocols with a fixed-anode thermionic x-ray source operating at a comparable resolution where the current is less than 0.1 mA at 40 kV (peak). The combined high spatial and temporal resolutions of the CNT field emission micro-focus x-ray source are highly attractive for dynamic tomography imaging.

Since the initial report, the design of the x-ray tube has undergone several modifications and the performance characterizations have been significantly improved. Figure 9.5 shows the latest developed compact CNT field emission x-ray tube whose body dimension is only 150 mm × 70 mm × 70 mm [23]. It has one CNT pixel inside and can easily be adapted to a commercial gantry system. It is capable of delivering as much as 3 mA current stably at about 1600 V gate voltage.

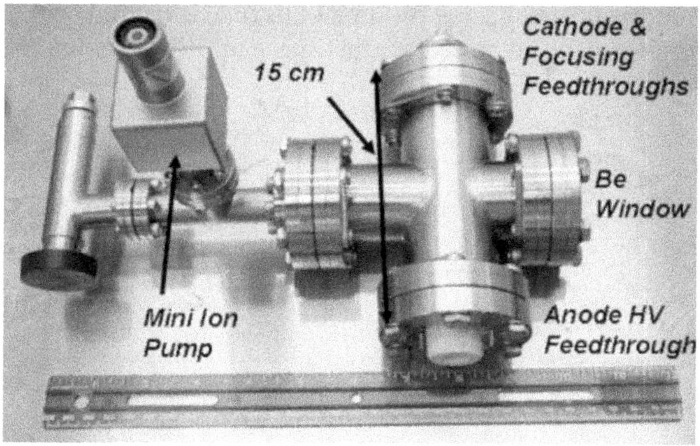

Figure 9.5 An image of the compact CNT field emission micro-focus x-ray tube with a body dimension of 150 mm × 70 mm × 70 mm [23].

Figure 9.6 shows the structure of a CNT field emission micro focus x-ray source. Inset of Figure 9.6 shows an SEM image of the CNTs deposited on the cathode surface, the CNT cathode can deliver the maximum power allowed by the x-ray anode and provide long-term stability [24].

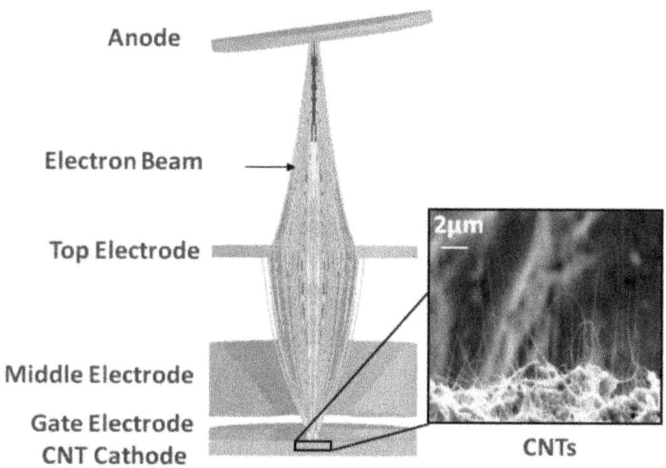

Figure 9.6 Structure of a CNT field emission micro focus x-ray source, which consists of CNT cathode, gate electrode, focusing electrode and a stationary anode. Inset: an SEM image of CNTs on cathode surface [24].

Another distinguishing feature of this tube is that it could easily be programmed and pulsed at a fast speed in various waveforms, as shown in Figure 9.7.

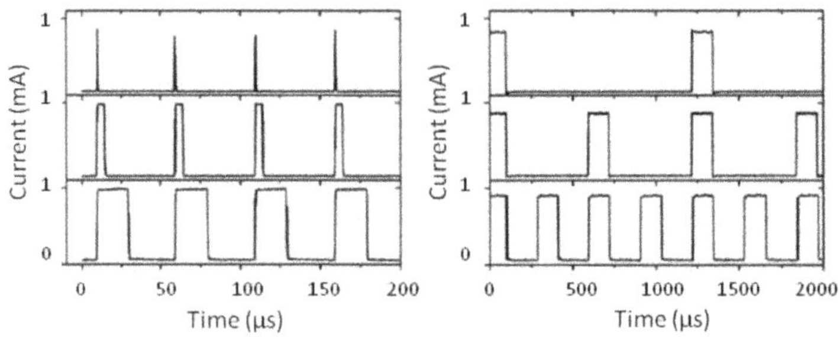

Figure 9.7 Various waveforms of the field emission current showing that the CNT field emission x-ray tube could easily be programmed and pulsed at a fast speed.

9.3.1.2 Carbon nanotube field emission single-beam dynamic micro-CT scanner

The invention of computed tomography (CT) is considered to be the greatest innovation in the field of radiology since the discovery of x-rays. The fundamental concept of CT is to reconstruct the cross-sections of the internal structure of an object from multiple projections taken from different viewing angles [25].

CT scanners are widely used for diagnostic medical imaging and security screening. With CT technique, it is possible to obtain the three dimensional distribution of x-ray attenuation coefficients within an object in a completely non-destructive way. Today, CT has become an important probing and diagnostic tool for investigations in biological, medical, material, and industrial applications.

Micro-CT employs a new adaptive geometry to provide non-destructive three dimension (3D) microscopy of the internal structures of small objects such as small animals with a high spatial resolution at an unprecedented speed. The micro-CT machine is much smaller in design compared to the human version of the x-ray tomography systems (CAT-scans) used in medicine. It works in exactly the same way as the CAT-scans but with much finer resolution and faster scan speed to obtain good quality image. In general, there are two types of micro-CT scanner setups. In one setup, the x-ray source and detector are typically stationary during the scan while the sample/animal rotates. The second setup, much more like a clinical CT scanner, is gantry based where the animal/specimen is stationary in space while the x-ray tube and detector rotate around. These scanners are typically used for small animals (in-vivo scanners), biomedical samples, foods, microfossils, and other studies for which minute detail is desired. The first x-ray microtomography system was conceived and built by Jim Elliott in the early 1980s. The first published x-ray microtomographic images were reconstructed slices of a small tropical snail, with pixel size about 50 micrometers [26].

Conventional micro-CT scanners use conventional x-ray tubes. CNT field emission single-beam dynamic micro-CT scanners developed in our group use a compact CNT field emission x-ray tube as shown in Figure 9.5. We have developed a number of prototype CNT field emission micro-CT scanners. Figure 9.8a shows an image of our most current micro-CT scanner. Figure 9.8b shows its internal structure, which consists of a custom compact CNT field emission x-ray tube operated at 50 kVp and 2 mA anode current. The tube has a tungsten anode, internal electron focusing and a 100 μm effective focal spot size. The CsI based flat-panel detector (C7940DK-02, Hamamatsu Photonics, Japan) was mounted opposite the x-ray source on a Huber 430 goniometer (Huber Diffraktiontechnik, Germany) to allow for rotation of the source-detector pair around the animal. During imaging, the animal was placed horizontally in a custom-made mouse bed, which is attached to a homemade positioning stage. The positioning stage has three-axis linear translation to ensure the object be positioned exactly at the isocenter and aligned with the axis of rotation. The positioning stage has a long travel distance (200 mm) in the *z*-axis (i.e., the axis of rotation), so that the bed can be translated completely outside the system field of view (FOV). A computer interface was developed under LABVIEW (National Instruments, Austin, TX) to automate the image acquisition process.

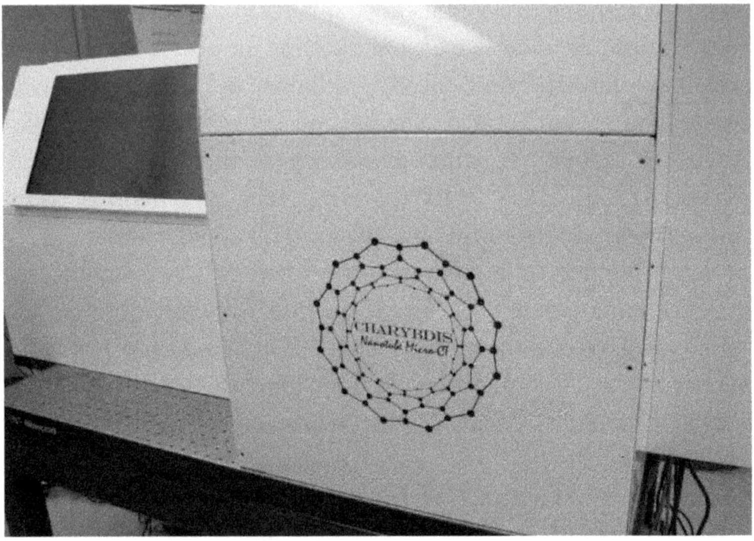

(a)

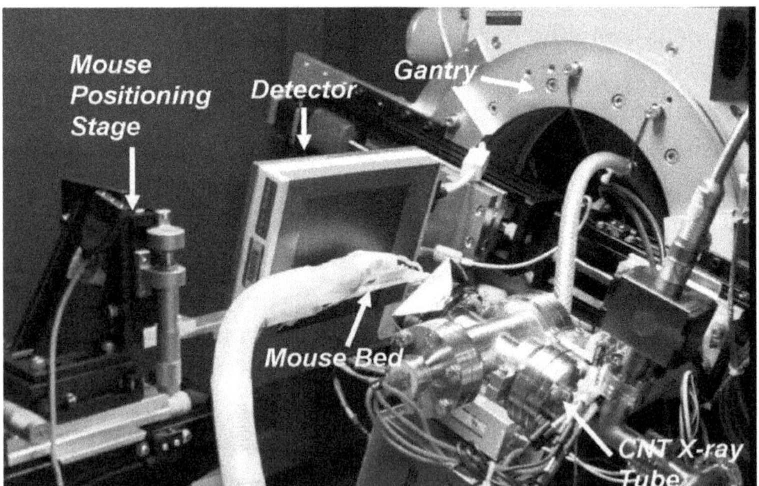

(b)

Figure 9.8 The desktop *in vivo* micro-CT scanner (a) that consists of the CNT field emission microfocus x-ray tube (b), also see Figure 9.5, a flat-panel detector, a small-diameter rotating gantry, and a stationary mouse bed.

In terms of micro-CT performance, the CNT x-ray tube has many advantages over the conventional x-ray tube. It could easily be programmed and pulsed at a fast speed in various waveforms, so that the CNT micro-CT can obtain gated imaging with a free breathing animal. When coupling x-ray pulses with the physiological signal from mouse breathing and flat-panel detector, images were captured only at the same desired phase in multiple breathing cycles and, therefore, the motion blur was greatly reduced. Figure 9.9 shows images of a wild-type mouse at 0 msec (a, c) and 55 msec (b, d) after the R wave. More encouraging, with a higher flux micro-focus tube, cardiac gated imaging can also be obtained [23].

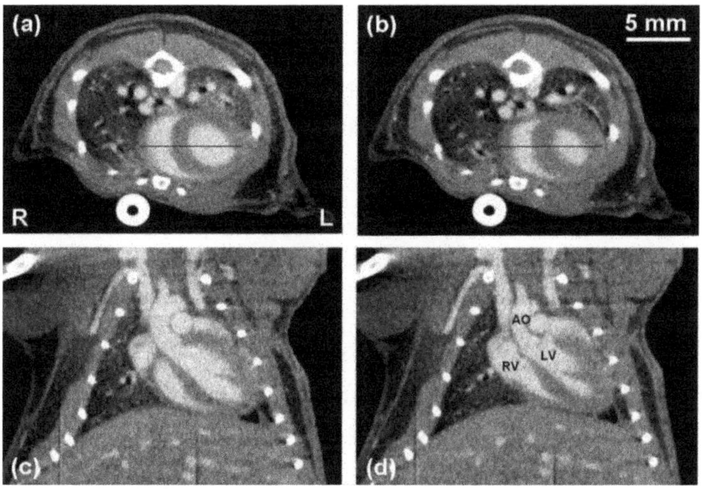

Figure 9.9 CNT field emission micro-CT images of a wild-type mouse at 0 msec (a, c) and 55 msec (b, d) after the R wave [23].

9.3.1.3 Carbon nanotube field emission multi-beam stationary micro-CT scanner

Most commercial CT and micro-CT systems use a single x-ray source and detector, which are mounted on a gantry, and rotated around the object [27,28]. In such a design, the temporal resolution and scanning speed are limited by the x-ray tube peak power, the gantry rotation speed, and the detector frame rate. In order to overcome these challenges to obtain high quality diagnostic images, a fast scanning speed, and a higher temporal resolution, researchers started to study alternative solutions and multiple x-ray sources were proposed [29]. One of the methods is to pack more x-ray sources to cover sufficient projection angles so that mechanical movement of the gantry is no longer needed. However, this method has not been realized because of the engineering difficulties to program and pack multiple thermionic x-ray sources together.

CNT field emission x-ray sources would offer a great alternative solution to overcome the disadvantages suffered by conventional x-ray tubes mentioned above, especially the programmability and flexibility in source design. It is well known that CNTs can provide great flexibility in cathode configuration, such as curved array, linear array, and 2D array. This makes CNT field emission x-ray source technology feasible for multi-beam x-ray source (Multiplexing). Meanwhile, CNT field emission technology results in x-ray beam with a high temporal resolution and a high spatial resolution. Using multiplexing x-ray source, a simultaneous collection of multiple images is available. This feature makes the stationary CT design feasible.

In terms of the multiplexing concept, we proposed a design of CNT field emission multi-beam stationary micro-CT scanner, as shown in Figure 9.10a. Figure 9.10b shows a photo of a CNT field emission multi-beam stationary micro-CT scanner [30]. The scanner has three main components (see Figure 9.10b): a multibeam x-ray source sealed inside a vacuum chamber, a rotary object holder, and a flat-panel x-ray detector for image collection.

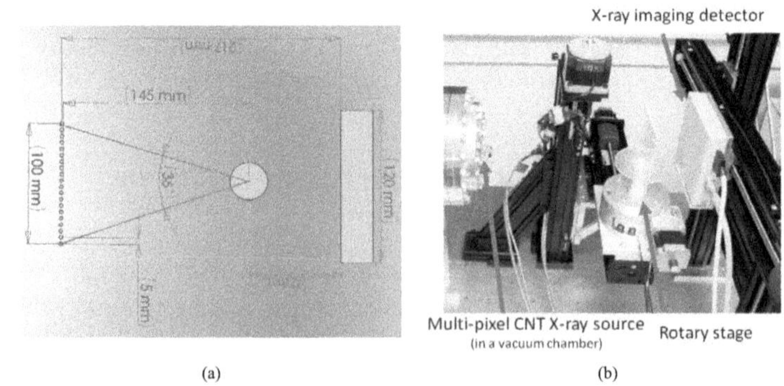

Figure 9.10 (a) The geometric configuration and (b) the prototype of the CNT field emission multi-beam stationary micro-CT scanner [30].

The x-ray source comprises 20 discrete CNT cathode pixels. Each CNT pixel has its own electronic control circuitry and can be addressed independently. The multi-beam x-ray source array can be programmed to generate a scanning x-ray beam. The micro-CT scanner using this source array is able to capture up to 20 projection images from different viewing angles without any mechanical movement of the system. After one linear scan from 20 pixels, the object would rotate 36° to the next imaging position until it finishes a full 360° rotation within only 10 steps.

Figure 9.11a shows an image of a cathode array, the 20 CNT cathode pixels (black oval dots) were deposited on a glass substrate. Figure 9.11b shows a structure of an individual x-ray pixel source. On the top of the CNT cathode, there is a gate electrode for electron extraction and a focusing electrode for electron beam formation. The molybdenum anode is supported by a copper anode holder for better heat dissipation. The entire anode assembly is tilted at a 12° angle. When applying a sufficient positive electric field on the extraction gate electrode, an array of electron beams from CNTs is extracted and accelerated towards and bombards the anode and generates an array of x-ray beams.

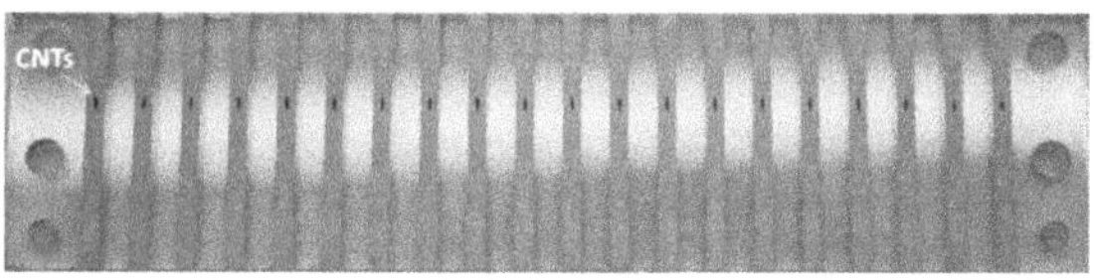

(a)

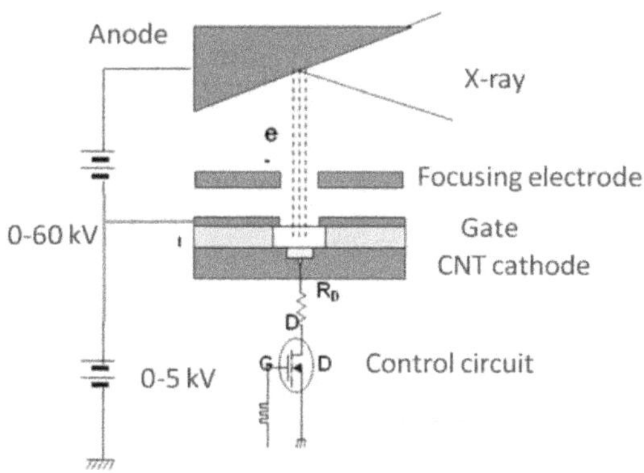

(b)

Figure 9.11 (a) An image of a 20-pixel CNT cathode (black oval dots) array deposited on a glass substrate and (b) structure of one individual micro-focus x-ray source [30].

Figure 9.12 shows field emission *I–V* curves from the 20 CNT cathode pixels. The emission current measured from individual cathode is ~5 mA. The emission current from these pixels has a good consistency.

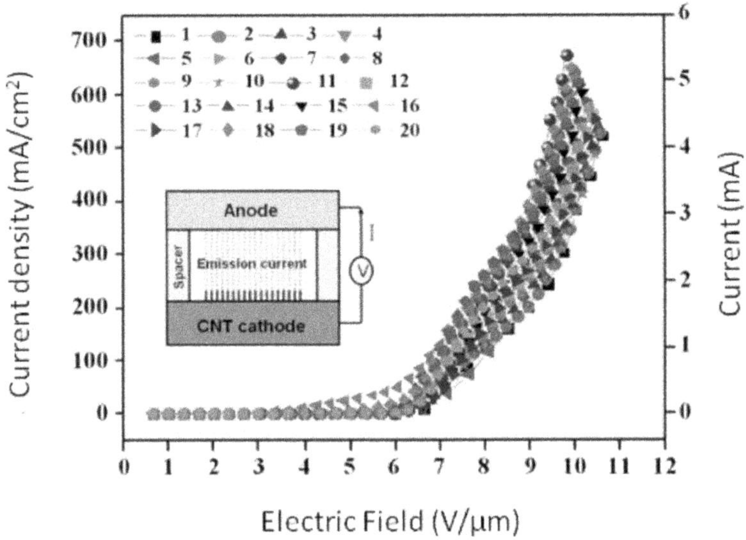

Figure 9.12 Field emission *I–V* curves from 20 different CNT cathode pixels [30]. See also Color Insert.

A mouse was imaged using this scanner system. During imaging, in order to make sure that the detector was able to capture all the projection images generated by the multi-beam source array, the distance between the imaging object (mouse) and x-ray detector was reduced to 3 cm. The source-to-object distance (SOD) was set at 17.5 cm and the mouse was aligned with central line of the 20-beam x-ray source array. The source-to-detector distance (SDD) was 20.5 cm for the scanning.

Figure 9.13 shows three sets of projection images generated by the CNT x-ray pixel 1, 6, 13 and 16. Each set of images correspond to a certain rotation position of the mouse (0°, 112°, and 224° respectively). The projection images demonstrated a good bone delineation, soft tissue contrast and reasonable spatial resolution.

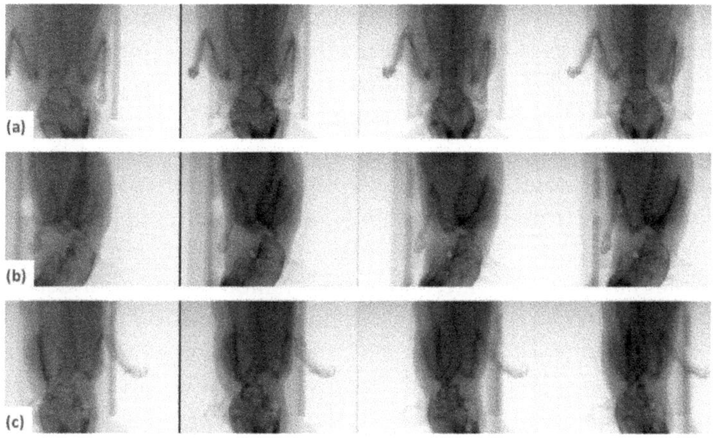

Figure 9.13 Projection images were recorded at (a) 0°, (b) 112°, and (c) 224° using the prototype of the CNT field emission multi-beam stationary micro-CT scanner [30].

9.3.1.4 Carbon nanotube field emission multi-beam micro-radiotherapy system

Today, many cancer patients receive state-of-the-art radiotherapy in which the radiation dose distribution is highly conformal around the tumor and normal tissue around tumor is largely spared. However, traditional research irradiators such as Cesium irradiators and x-ray irradiators lack many of the capabilities found in today's clinical practice, such as spatial resolution and temporal resolution. Recent advances in both genetically engineered mouse modeling and cancer biomarkers promise better preclinical studies to understand human cancers and cancer biomarkers [31]. These and other advances in cancer research demand new research tools with high spatial and temporal resolution for preclinical study before clinical trials. For small animal irradiation, there is a need for high-resolution and image–guided micro-radiotherapy (micro-RT) technology that can deliver radiation treatment similar to human radiotherapy but at mouse scales. Currently, there are several micro-radiotherapy systems developed in different research labs using Ir-192 isotope [32] or conventional x-ray tube [33] as the radiation sources.

Our group developed a CNT field emission multi-pixel micro-RT system. Figure 9.14 illustrates the conceptual design of the multi-pixel x-ray micro-RT system. After the electron pixel beams are generated from the CNT cathode pixels, they are accelerated towards the target under 80–100 kV voltage and produce x-rays in the target. The x-rays produced are then collimated into individual x-ray pixel beams.

This novel multi-pixel x-ray beam array micro-RT system is capable of electronic field shaping and intensity modulation. None of the existing micro-RT systems is capable of intensity-modulated treatment at a mouse scale, and intensity-modulated radiotherapy is widely used in human prostate and head and neck cancer treatment today. In this micro-RT system, CNT field emission x-rays are used as the irradiation source. Figure 9.15 shows an image of a 5 × 10 electron beam array incident on a phosphor screen at an applied voltage of 1400 V. It can be seen that 50 electron beams are roughly uniform. The emission uniformity of the cathode chip is critical for micro-RT application, as it determines the uniformity of the generated x-ray beams, which again dominate a predictable and consistent radiation dose to tumors [34].

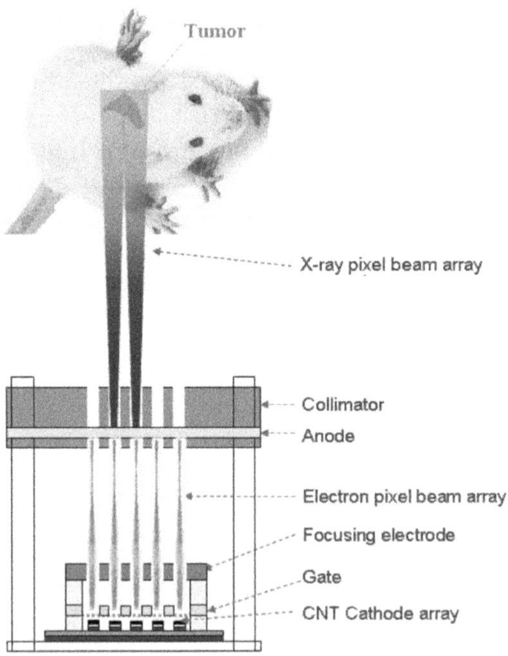

Figure 9.14 Schematic illustration of the CNT field emission multi-pixel x-ray beam array micro-radiotherapy system.

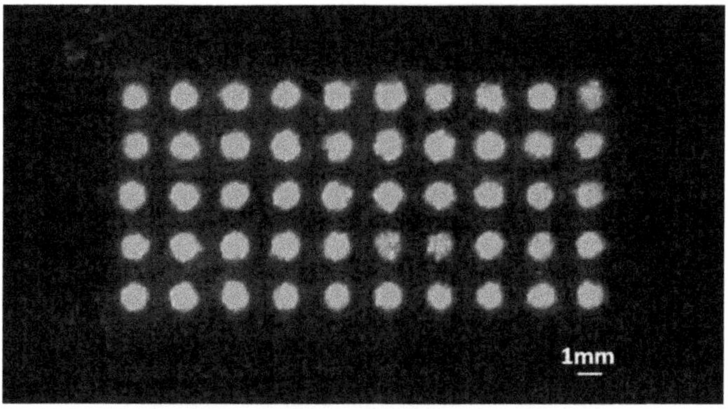

Figure 9.15 Electron field emission image from a 5 × 10 pixel CNT cathode array at an applied voltage of 1400 V [34].

Individual pixel control is necessary for defining arbitrary beam shapes and intensity distributions. The individual pixel control performance was tested for the 5 × 10 pixel cathode array chip. Figure 9.16a shows an image in which only 17 cathodes of the 50 pixel cathode chip were addressed with a voltage of 1400 V. It can be seen that electron beams were only observed from the selected 17 cathodes to form an electron pattern of "NCI", indicating that this cathode chip can individually control each pixel. With the cathode array chip and transmission anode of tungsten, the x-ray beams were produced by using the prototype micro-RT system schematically illustrated in Figure 9.14. Figure 9.16b shows the corresponding dose image delivered to a Gafchromic EBT film. Each x-ray beam size is ~2 mm. The dose rate of each x-ray beam was calibrated and a dose rate of 1.24 Gy/min at the center of the irradiated object was obtained at an emission current of 3.0 mA [35]. The ability of individual irradiation pixel control is one of the key advantages of the CNT field emission micro-RT over the other micro-RT systems. It enables the micro-RT to electronically form arbitrary shapes and intensity distribution of the irradiation field.

The CNT micro-RT has an ability to electronically, not mechanically, form arbitrary field shapes and intensity modulation patterns for very small radiation fields. The inset of Figure 9.17 shows that out of 50 pixel beams only 6 irradiation beams are electronically "turning on." Different pixel beams can deliver different dose and dose rate by individual field emission control [35, 36]. Figure 9.17 illustrates a mouse image generated by our CNT field emission micro-CT and the Monte Carlo simulation generated dose distribution from an irregular field formed by 6 pixel beams of the micro-RT system.

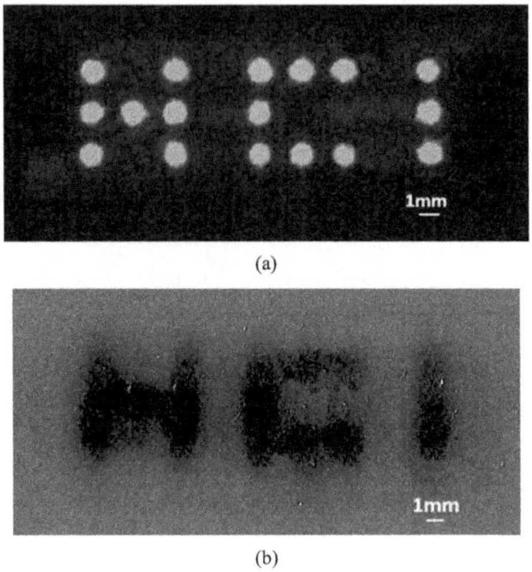

Figure 9.16 Test of electron and x-ray beams of individual pixel beam control: (a) 17 pixel electron beams (out of a 5 × 10 pixel cathode array) incident on a phosphor screen and (b) the corresponding 17 x-ray pixel beams to a Gafchromic EBT film to form a pattern of "NCI" [34].

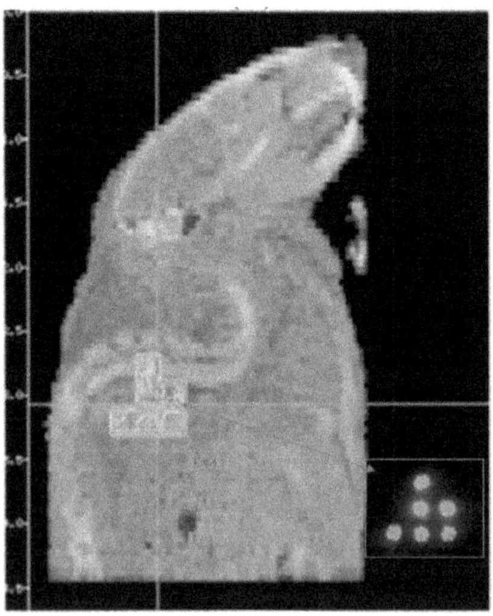

Figure 9.17 Mouse image from our CNT field emission based micro-CT and the Monte Carlo simulated dosimetry from a 6-pixel beams formed irregular radiation field. Inset: Illustration of individual pixel beam field emission control for the proposed micro-RT — 6 out of 50 pixel cathodes were turned on [36].

Figure 9.18 displays the intensity modulation ability of the x-ray pixel beam array [37]. Because each of the x-ray pixel beams is individually controlled, the micro-RT system can irradiate any spatial or intensity modulation pattern simply by activating a given subset of the x-ray pixel beams electronically.

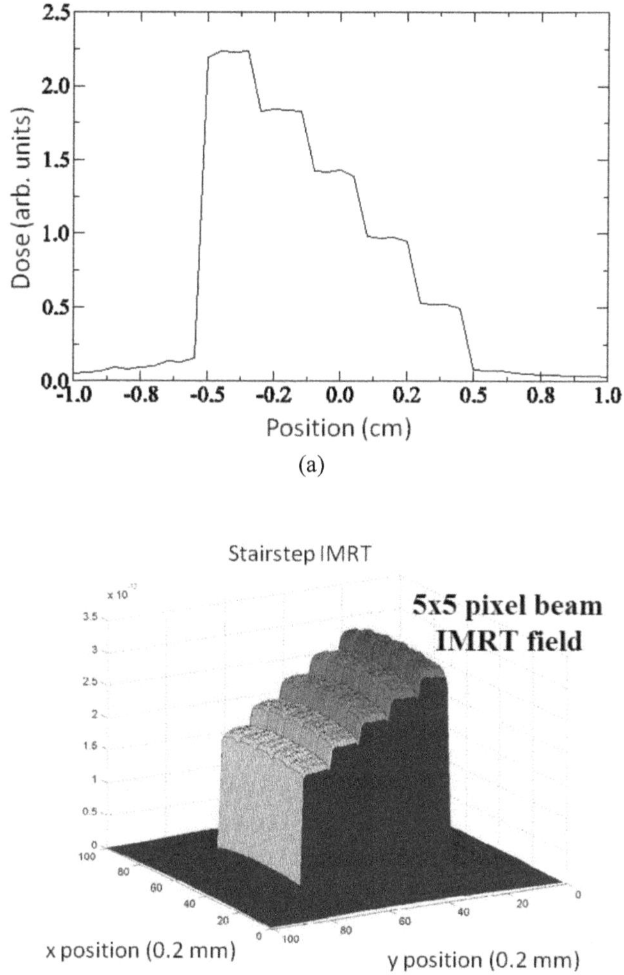

(a)

(b)

Figure 9.18 Illustration of the intensity modulation ability of the x-ray pixel beam array.

Currently we are developing an image guided radiotherapy system (micro-CT-RT) in which the CNT field emission micro-RT is integrated with the CNT field emission micro-CT. Figure 9.19a shows the design of the system and Figure 9.19b shows an image of the developing system. This forms a high spatial and temporal resolution image-guided and intensity-modulated irradiation system that is analogous to state-of-the-art clinical image-guided radiotherapy systems for human patients at the mouse scale [35]. The system is undergoing calibration and testing.

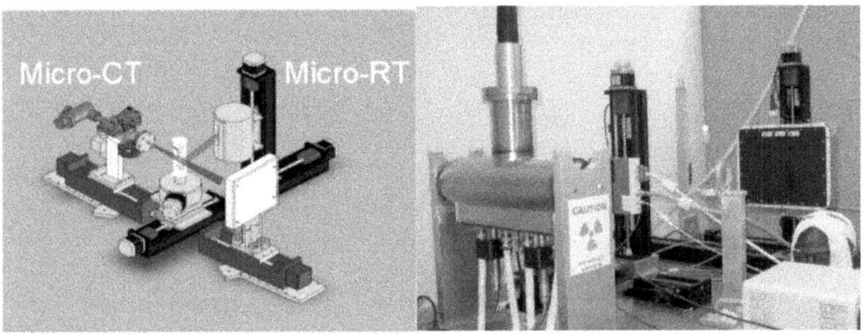

(a) (b)

Figure 9.19 (a) Design of our proposed CNT micro-CT-RT system for real time imaged-guided intensity-modulated radiation therapy (IMRT) and treatment response observation in small animals; and (b) an image of the CNT micro-CT-RT system under development in our lab [35].

9.3.2 Carbon Nanotube Field Emission High-Power Microwave Devices

There has been considerable interest recently in high-power microwave (HPM) sources for communication, strategic and scientific applications. High-power microwave sources are being developed that provide peak powers over 1 GW at frequencies on the order of 1 GHz in short pulses, typically on the order of 100 ns. High-power microwaves are electromagnetic radiation spanning the frequency range of approximately 1 GHz to over 100 GHz with 100 GW microwave power. The main requirements for the field emitters for high-power microwave (or radio frequency (RF) amplifier) application are high current density, high trans-conductance and low array capacitance [38].

Today, the high-power microwave devices use thermionic emitters, which are bulky and heavy. More problematically, such high-power microwave devices are not able to be switched on or off instantaneously due to heating of the emitter.

CNT emitters have attracted interest as an alternative to thermionic emitters in high-power microwave devices [13,14,39]. Teo *et al.* [14] reported a microwave diode that uses a cold-cathode electron source consisting of CNTs and that operates at high frequency and at high current densities. Figure 9.20a shows a representation of the equivalent electrical circuit. Figure 9.20b shows images of CNT cathodes. The CNT array consisted of uniform individual CNT spaced at a distance approximately twice their height to minimize screening from adjacent CNTs. The cathode had an active area of 0.5×0.5 mm^2, which corresponded to 2500 CNTs. The device was operated at 1.5 GHz using various RF-input powers and 3.2 mA of current was achieved at an applied RF electric field of 29 MV/m, which was equivalent to an average current density of 1.3 A/cm^2. In the output wave form the peak current corresponded to 30 mA and a current density of 12 A/cm^2 [14].

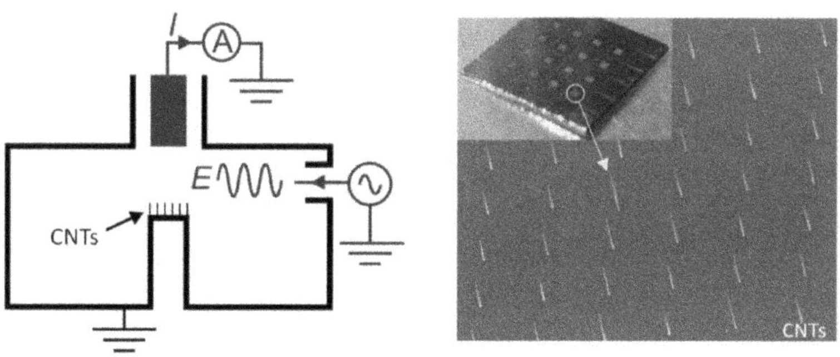

(a) (b)

Figure 9.20 (a) Representation of the equivalent electrical circuit, where E is the applied electric field and I is the emitted current; and (b) image of CNT cathodes [14].

The experimental results show that the CNT cathode already delivers average- and peak-current densities that are similar to those used in present-day microwave transmission devices. Due to their small size and ability to generate and modulate the beam directly on demand without the need for high temperatures, CNT cathodes hold promise for a new generation of light weight, efficient and compact microwave devices for telecommunications in satellites or spacecrafts. Due to the power requirement of high-power microwave tubes, tremendous amounts of electrons are needed to extract from the cathodes and bombard on the anodes with very high kinetic energy. However, overheating, out-gassing, and formation of secondary electrons of the anode cause serious problems so that more efforts are still needed to overcome these challenges.

9.3.3 Carbon Nanotube Field Emission Displays

A field emission display (FED) is a flat-panel display technology that uses large-area field electron sources to provide electrons that strike colored phosphor to produce a color image. In a general sense, an FED consists of a matrix of cathode ray tubes, each tube producing a single sub-pixel, grouped in threes to form red-green-blue (RGB) pixels. FEDs combine the advantages of cathode ray tubes (CRTs), namely their high contrast levels and very fast response times, with the packaging advantages of liquid crystal display (LCD) and other flat-panel technologies. They also offer the possibility of requiring less power, about half that of an LCD system. An FED has several advantages over other technologies including higher brightness, power efficiency, wider operating temperature and viewing angle. Operation principle of an FED is similar to a cathode ray tube. Instead of a single hot cathode, which consists of several emitters for each pixel resulting in large redundancy, the electric field is applied between the emitters and the anode (or gate electrode) to generate electrons, which create an image upon interaction with the phosphor screen. Phosphor coating is applied to a transparent anode facing the emitters and sub-pixels are designed to generate the primary colors.

Since CNT field emission display was first reported by Wang *et al.* [8] and Choi *et al.* [9], CNT field emission display has become a very interesting area of research. Figure 9.21 shows a CNT FED with SWCNT emitters. The CNT FED was a diode type with screen printed CNTs as the cathode. The FED had a brightness of 1800 cd/m^2 at 4 V/μm with stable electron emission. Since the first demonstration of CNT FED, several prototype displays with triode structure have been fabricated. In 2002, Chung *et al.* [40] reported a prototype of a CNT-FED with a gate structure, an active area of 12.7 cm in diagonal, in full color and at 100 Hz. In 2005, researchers at Motorola successfully demonstrated a prototype of fully operational 5 inch, 3.3 mm thick FED based on CNTs (directly grown on glass substrates. Motorola terms it a nano-emissive display (NED)). Recently, Choi *et al.* reported that a double gated CNT field-emitter array was utilized for FED, which has minimized beam spread with uniform emission [41].

However, several challenges remain in the development of CNT FED. The major issues in a FED operation include emission uniformity and stability and degradation of vacuum level due to the out-gassing of phosphors (light emitting materials) and other device components. At this stage, CNT FED faces huge competition from LCD and organic light-emitting diodes (OLED) based technologies. Since FEDs have lower ultimate cost base and better power efficiency though, it may be a viable option in the near future for manufacturing over 45–60 inch displays.

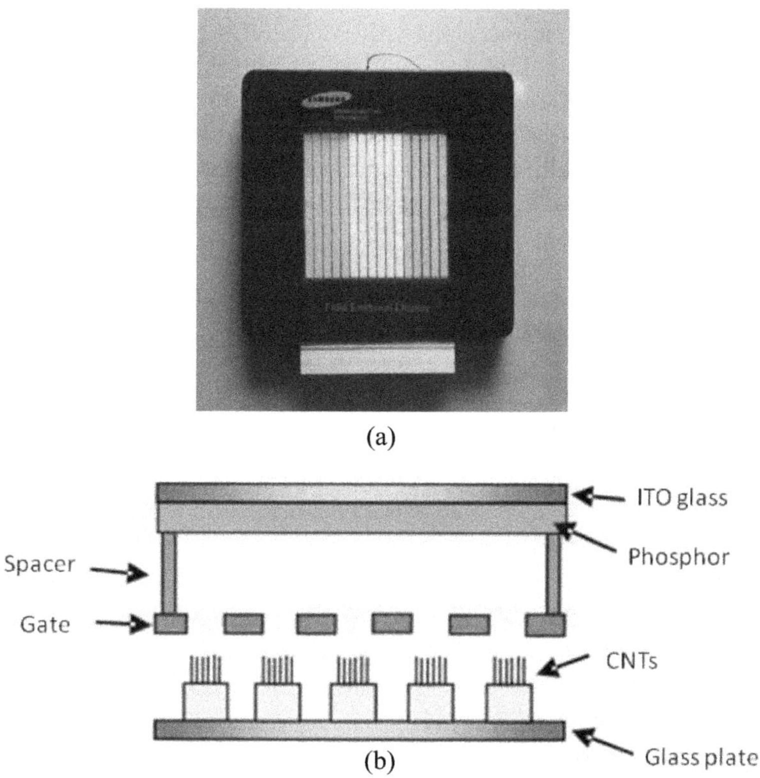

Figure 9.21 (a) SWCNT-FED with red, green, and blue phosphor columns; and (b) FED schematic [9]. See also Color Insert.

9.4 Summary and Future Directions

The unique structural and electrical properties of CNTs make them ideal electron field emitters for vacuum electronics applications. In recent years, tremendous progress has been made in the development of vacuum electronic devices.

The CNT based "cold" cathode technology enables the fabrication of matrix addressable single or multipixel x-ray sources to not only replace the hot filament x-ray sources but also add new capabilities and functionality, which are used for medical imaging and radiotherapy systems, high-power microwave source and flat-panel displays. These devices (systems) show great promise in commercial and research practical applications in the future. The CNT cold-cathode x-ray and microwave technology can potentially lead to portable and miniature medical x-ray and microwave devices.

However, to commercialize these applications, some issues, including vacuum stability, emission uniformity, and highly reproducible cost-efficient nanofabrication processes, should be further improved. With fast development in the area, we believe that in the near future more and more CNT field emission vacuum electronic devices can be found in the commercial market.

References

1. Iijima, S. (1991). Helical microtubules of graphitic carbon, *Nature*, **354**, pp. 56–58.
2. Iijima, S., and Ichihashi, T. (1993). Single shell carbon nanotubes of one nanometer diameter, *Nature*, **363**, pp. 603–605.
3. Purcell, S.T., Vincent, P., Journet, C., and Binh, V.T. (2002). Hot nanotubes: stable heating of individual multiwall carbon nanotubes to 2000 K induced by the field-emission current, *Phys. Rev. Lett.*, **88**, pp. 105502.
4. Chernozatonskii, L.A., Gulyaev, Y.V., Kosakovskaja, Z.J., Sinitsyn, N.I., Torgashov, G.V., Zakharchenko, Y.F., Fedorov, E.A., and Valchuk, V.P. (1995). Electron field-emission from nanofilament carbon-film, *Chem. Phys. Lett.*, **233**, pp. 63.
5. de Heer, W.A., Chatelain, A., and Ugarte D. (1995). A carbon nanotube field-emission electron source, *Science*, **270**, pp. 1179.
6. Rinzler, A.G., Hafner, J.H., Nikolaev, P., Lou, L., Kim, S.G., Tomanek, D., Nordlander, P., Colbert, D.T., and Smalley, R.E. (1995). Unraveling

nanotubes-field-emission from an atomic wire, *Science*, **269**, pp. 1550.

7. de Jonge, N., and Bonard, J.M. (2004). Carbon nanotube electron sources and applications, *Philos. Trans. R. Soc. Lond. A Math. Phys. Eng. Sci.*, **362**, pp. 2239–2266.
8. Wang, Q.H., Setlur, A.A., Lauerhaas, J.M., Dai, J.Y., Seelig, E.W., and Chang, R.P.H. (1998). A nanotube-based field-emission flat panel display, *Appl. Phys. Lett.*, **72**, pp. 2912–2913.
9. Choi, W.B., Chung, D.S., Kang, J.H., Kim, H.Y., Jin, Y. W., Han, I.T., Lee, Y.H., Jung, J.E., Lee, N.S., Park, G.S., and Kim, J.M. (1999). Fully sealed high-brightness carbon-nanotube field-emission display, *Appl. Phys. Lett.*, **75**, pp. 3129–3131.
10. Saito, Y., and Uemura, S. (2000). Field emission from carbon nanotubes and its application to electron sources, *Carbon*, **38**, pp. 169–182.
11. Sugie, H., Tanemura, M., Filip, V., Iwata, K., Takahashi, K., and Okuyama, F. (2001). Carbon nanotubes as electron source in an x-ray tube, *Appl. Phys. Lett.*, **78**, pp. 2578–2580.
12. Yue, G.Z., Qiu, Q., Gao, B. Cheng, Y., Zhang, J., Shimoda, H., Chang, S., Lu, J.P., and Zhou, O. (2002). Generation of continuous and pulsed diagnostic imaging x-ray radiation using a carbon-nanotube-based field-emission cathode, *Appl. Phys. Lett.*, **81**, pp. 355–357.
13. Jensen, K.L., Abrams, R.H., and Parker, R. K. (1998). Field emitter array development for high frequency applications, *J. Vac. Sci. Technol. B*, **16**, pp. 749–753.
14. Teo, K.B.K., Minoux, E., Hudanski, L., Peauger F., Schnell J.P., Gangloff, L., Legagneux, P., Dieumegard, D., Amaratunga, G.A.J., and Milne, W.I. (2005). Microwave devices — Carbon nanotube cold cathodes, *Nature*, **437**, pp. 968.
15. Fowler, R.H., and Nordheim, L.W. (1928). Electron emission in intense electric fields, *Proc. R. Soc. Lond. A*, **119**, pp. 173–181.
16. French A.P., and Taylor, E. F. (1979) *An Introduction to Quantum Physics.* (CRC Press), pp. 87–413.
17. Latham, R.V. (1995) *High Voltage Vacuum Insulation: Basic Concepts and Technological Practice* (Academic Press, London), pp 26–227.
18. Brodie, I., and Spindt, C.A. (1992). Vacuum microelectronics, *Adv. Electron. Electron. Phys.*, **83**, pp. 1–106.
19. Bonard, J.M., Salvetat, J.P., Stockli, T., Forro, L., and Châtelain, A. (1999). Field emission from carbon nanotubes: perspectives for applications and clues to the emission mechanism, *Appl. Phys. A*, **69**, pp. 245–254.

20. Wang, S.G., Sellin, P.J., Lian, J., Ozsan, E., and Chang, S. (2009). Improvement of electron field emission in patterned carbon nanotubes by high temperature hydrogen plasma treatment, *Curr. Nanosci.*, **5**, pp. 54–57.

21. Liu, Z., Yang, G., Lee, Y., Bordelon, D., Lu, J., and Zhou, O. (2006). Carbon nanotube based microfocus field emission x-ray source for microcomputed tomography, *Appl. Phys. Lett.*, **89**, pp. 103111.

22. Zhou, O., and Lu, J.P. (2003) U.S. 6,553,096.

23. Cao, G., Burk, L.M., Lee, Y.Z., Calderon-Colon, X., Sultana, S., Lu, J., and Zhou, O. (2010). Prospective-gated cardiac micro-CT imaging of free-breathing mice using carbon nanotube field emission x-ray, *Med. Phys.* **37,** pp. 5306–5312.

24. Calderon-Colon, X., Geng, H., Gao, B., An, L., Cao, G., and Zhou, O. (2009). A carbon nanotube field emission cathode with high current density and long-term stability, *Nanotechnology*, **20,** pp. 325707.

25. Hsieh, J. (2009) *Computed Tomography: Principles, Design, Artifacts, and Recent Advances,* 2nd Ed. (Wiley), pp. 1–16.

26. Elliott, J.C., and Dover, S. D. (1982). X-ray microtomography, *J. Microscopy*, **126**, pp. 211–213.

27. Paulus, M.J., Gleason, S.S., Kennel, S.J., Hunsicker, P.R., and Johnson, D. K. (2000). High resolution X-ray computed tomography: an emerging tool for small animal cancer research, *Neoplasia*, **2**, pp. 62–70.

28. Postnov, A.A., Haese, P.C.D., Neven, E., De Clerck, N.M., and Persy, V.P. (2009). Possibilities and limits of X-ray microtomography for in vivo and ex vivo detection of vascular calcifications, *Int. J. Cardiovasc. Imaging*, **25**, pp. 615–624.

29. Robb, R.A., Hoffman, E.A., Sinak, L.J., Harris, L.D., and Ritman, E.L. (1983). High-speed three-dimensional X-ray computed tomography: The dynamic spatial reconstructor, *Proc. IEEE*, **71(3),** pp. 308–319.

30. Peng, R., Zhang, J., Calderon-Colon, X., Wang, S.G., Sultana, S., Chang, S., Lu, J., and Zhou O. (2009). Stationary micro-CT scanner using a distributed multi-beam field emission x-ray source: a feasibility study, *SPIE,* **7258**, pp. 725847.

31. Whitfield, M.L., George, L.K., Grant, G.D., and Perou, C.M. (2006). Common markers of proliferation, *Nat. Rev. Cancer,* **6,** pp. 99–106.

32. Stojadinovic, S., Low, D.A., Vicic, M., Mutic, S., Deasy, J.O., Hope, A.J., Parikh, P. J., and Grigsby, P. W. (2006). Progress toward a microradiation therapy small animal conformal irradiator, *Med. Phys.,* **33(10),** pp. 3834–3845.

33. Deng, H., Kennedy, C.W., Armour, E., Tryggestad, E., Ford, E., McNutt, T., Jiang, L., and Wong, J. (2007). The small-animal radiation research platform (SARRP): dosimetry of a focused lens system, *Phys. Med. Biol.*, **52,** pp. 2729–2740.

34. Wang, S.G., Calderon-Colon, X., Peng, R., Schreiber, E.C., Zhou, O., and Chang, S. (2011). A carbon nanotube field emission multipixel x-ray array source for microradiotherapy application, *Appl. Phys. Lett.*, **98**, pp. 213701.

35. Wang, S., Schreiber, E., Calderon-Colon, X., Peng, R., Sultana, S., Zhang, J., Zhou, O., and Chang, S. (2009). An image-guided micro-radiotherapy system based on carbon nanotube field emission technology for small animal irradiation for cancer research, *Med. Phys.*, **36**, pp. 2798.

36. Wang, S.G., Liu, Z.J., Sultana, S., Schreiber, E., Zhou, O., and Chang. S. (2007). A novel high resolution micro-radiotherapy system for small animal irradiation for cancer research, *BioFactors,* **30(4),** pp. 265–270.

37. Schreiber, E. C., and Chang, X.S. (2009). Monte Carlo simulation of an x-ray pixel beam microirradiation system, *Radiat. Res.*, **171,** pp. 332.

38. Seelaboyina, R., and Choi, W. (2007). Recent progress of carbon nanotube field emitters and their application, *Recent Patents Nanotechnol.*, **1(3),** pp. 1–7.

39. Verma, P., Gautam, S., Pal, S., Kumar, P., Chaturvedi, P., Rawat, J.S.B.S., Chaudhary, P.K., Harsh, Basu, P.K., and Bhatanagar, P.K. (2008). Carbon nanotube-based cold cathode for high power microwave vacuum electronic devices: A potential field emitter, *Defence Sci. J.*, **58**, pp. 650–654.

40. Chung, D.S., Park, S.H., Lee, H.W., Choi, J.H., Cha, S.N., Kim, J.W., Jang, J.E., Min, K.W., Cho, S.H., Yoon, M.J., Lee, J.S., Lee, C.K., Yoo, J.H., Kim, J.M., Jung, J.E., Jin, Y.W., Park, Y.J., and You, J.B. (2002). Carbon nanotube electron emitters with a gated structure using backside exposure processes, *Appl. Phys. Lett.,* **80**, pp. 4045–4047.

41. Choi, Y.C., Jeong, K.S., Han, I.T., Kim, H.J., Jin, Y.W., Kim, J.M., Lee, B.G., Park, J.H., and Choe, D.H. (2006). Double-gated field emitter array with carbon nanotubes grown by chemical vapor deposition, *Appl. Phys. Lett.*, **88**, pp. 263504.

Chapter 10

Novel Magnetic and Electrical Properties of Carbon Nanotubes: Consistent with Ultrahigh Temperature Superconductivity

Guo-meng Zhao
Department of Physics and Astronomy, California State University at Los Angeles
Los Angeles, CA 90032, USA
gzhao2@exchange.calstatela.edu

We present detailed magnetic properties of Ni magnetic nanoparticles embedded in multiwall carbon nanotubes (MWCNTs). The measured room-temperature saturation magnetization for the nickel nanoparticles is about three times larger than the expected value from the nickel concentration determined independently from inductively coupled plasma mass spectrometer and high energy synchrotron x-ray diffractometer. What is more intriguing is that the Curie–Weiss constant above the Curie temperature of nickel is enhanced by a factor of 12.2. We show that the moment enhancement factor is about two orders of magnitude larger than that predicted from a magnetic-proximity effect. Alternatively, the giant moment enhancement can be naturally explained if MWCNTs are ultra-high temperature superconductors. There is also independent evidence

Carbon Nanotubes and Their Applications
Edited by Qing Zhang

ISBN 978-981-4241-90-8 (Hardcover), 978-981-4303-18-7 (eBook)
www.panstanford.com

of ultrahigh temperature superconductivity in MWCNTs. The measured room-temperature diamagnetic susceptibility of pure MWCNTs for the magnetic field parallel to the tube-axis direction agrees quantitatively with the expected diamagnetic Meissner effect. Because of a finite number of transverse conduction channels in ultra-thin superconducting tubes, quantum phase slips are significant and the on-tube resistance is not expected to be zero below the mean-field superconducting transition temperature. Nonetheless, the room-temperature on-tube resistivity has been found to be indistinguishable from zero for many individual MWCNTs. We further show that the temperature dependencies of the resistivity in individual single-wall carbon nanotubes (SWCNTs) are inconsistent with ballistic electrical transport mechanism but can be quantitatively explained in terms of quantum phase slips in quasi-one-dimensional superconductors.

10.1 Introduction

Graphene is a sheet of carbon atoms distributed in a honeycomb lattice and is the building block for carbon-based materials such as graphite and carbon nanotubes (CNTs). The massless relativistic Dirac fermions in graphene are a result of its unique electronic structure, characterized by conical valence and conduction bands that meet at a single point in momentum space. The integer quantum Hall effect recently observed in graphene [Novoselov *et al.* (2005); Zhang *et al.* (2005)] is related to the Dirac fermions. More intriguingly, there were reports of ultrahigh-temperature superconducting behaviors in carbon films [Antonowicz (1974); Lebedev (2004)], CNTs [Zhao and Wang (2001); Zhao (2004, 2006)], graphite [Kopelevich *et al.* (2000)], and graphite–sulfur composites [Da Silva *et al.* (2001); Moehlecke *et al.* (2004)]. Highly oriented pyrolithic graphite (HOPG) was shown to display either a partial superconducting or a ferromagnetic-like response to an applied magnetic field even at temperatures well above room temperature [Kopelevich *et al.* (2000)]. In addition to the observation of unusual high-temperature ferromagnetism in the carbon-based materials [Mombru *et al.* (2005); Esquinazi *et al.* (2003); Cervenka *et al.* (2009)], there was a report of extra magnetic moment induced in graphite due to a large magnetic proximity effect between graphite and magnetic nanoparticles [Coey *et al.* (2002)]. Similarly, high-temperature magnetic data of multiwall carbon nanotube (MWCNT) mat samples embedded with Fe and/or Fe_3O_4 nanoparticles [Zhao and Beeli (2008)] indicated that the room-temperature saturation magnetizations of the magnetic nanoparticles embedded in the MWCNTs are enhanced by a factor of about 2–3 as compared with what they would be expected to have for free magnetic nanoparticles. Recently, the study has been extended to nickel nanoparticles embedded in MWCNTs [Wang *et al.* (2010)]. High-energy synchrotron x-ray diffraction spectrum shows that the sample contains about 0.43% nickel that has face-centered-cubic structure and is ferromagnetic. Magnetic measurements show that the room-temperature saturation magnetization of the nickel nanoparticles embedded in the MWCNTs is enhanced by a factor of about 3.4 as compared with that expected for free (unembedded) nanoparticles [Wang *et al.* (2010)]. What is more intriguing is that the Curie–Weiss constant above the Curie temperature is enhanced by a factor of 12.2. The observed enhancement in the saturation magnetization is about two orders of magnitude larger than that predicted from the magnetic proximity effect [Wang *et al.* (2010)]. The colossal enhancement in the Curie–Weiss constant is even more difficult to explain in terms of the magnetic proximity effect [Wang *et al.* (2010)]. Alternatively, it is possible that these moment enhancements might arise from the interplay between ferromagnetism of magnetic nanoparticles and ultrahigh temperature superconductivity in MWCNTs [Wang *et al.* (2010)].

The existence of ultrahigh-temperature superconductivity in the carbon-based materials is not accidental. The unique electronic structures of the carbon-based materials make them ideal for high-temperature superconductivity. Several theoretical models based on different types of interactions predict high-temperature superconductivity in quasi-one-dimensional (quasi-1D) and/or quasi-two-dimensional (quasi-2D) electronic systems. Alexandrov and Mott [Alexandrov and Mott (1995)] demonstrated that strong electron–phonon coupling can lead to the formation of intersite bipolarons and that the Bose-Einstein condensation of the bipolarons can explain high-temperature superconductivity in cuprates. Little [Little (1964)] proposed that high-temperature or room-temperature superconductivity could be realized by exchanging high-energy excitons in quasi-1D systems. Lee and Mendoza showed that superconductivity as high as 500 K can be achieved through a pairing interaction mediated by undamped acoustic plasmon modes in quasi-1D systems [Lee and Mendoza (1989)]. High-temperature superconductivity can also occur in a multi-layer electronic system due to an attraction of charge carriers in the same conducting layer via exchange of virtual plasmons in neighboring layers [Cui and Tsai (1991)]. If the plasmon-mediated pairing mechanisms are relevant, one should be able to find high-temperature superconductivity in quasi-one-dimensional and/or multi-layer systems such as cuprates, CNTs, and graphite. In contrast to the mechanisms based on the attractive interactions between electrons by virtually exchanging phonons, excitons, and/or plasmons, an exotic model based on resonating-valence-bond (RVB) theory originally proposed by Anderson [Anderson (1987)] even predicts ultrahigh temperature d-wave superconductivity in heavily doped graphene [Black-Schaffer and Doniach (2007)].

In order to confirm the existence of superconductivity, it is essential to provide two important signatures: the diamagnetic Meissner effect and the sharp resistive transition to the zero resistance state. However, these two essential signatures are less obvious in a quasi-one-dimensional superconducting wire that has a finite number of transverse conduction channels. Due to large superconducting fluctuations, the resistive transition in quasi-1D superconductors could be very broad. Because of the finite number of transverse channels, the four-probe resistance never goes to zero even though the on-wire resistance approaches zero. Because the magnetic penetration depth may be far larger than the transverse dimension, the diamagnetic Meissner effect might be negligible, which makes it difficult to distinguish between the diamagnetic Meissner effect and orbital diamagnetism. Therefore, it is not straightforward to unambiguously identify quasi-1D superconductivity in ultra-thin wires such as CNTs because their magnetic and electrical properties are not typical of bulk superconductivity.

In this chapter, we will present the detailed magnetic properties of MWCNTs embedded with Ni nanoparticles [Wang *et al.* (2010)]. Magnetic measurements were carried out using a quantum design vibrating sample magnetometer (VSM). Inductively coupled plasma mass spectrometer and high-energy synchrotron x-ray diffractometer were used to accurately determine the impurity concentration. Scanning electron microscope (SEM) and transmission electron microscopy (TEM) were used to characterize MWCNTs and magnetic nanoparticles embedded. In section 10.2, we will present the experimental results for MWCNTs embedded with Ni nanoparticles. In section 10.3, we will identify the diamagnetic Meissner effect in the magnetic field parallel to the tube axis up to room temperature for aligned MWCNTs that are physically separated and have negligibly small magnetic impurities. The magnitude of the Meissner effect is in quantitative agreement with the predicted penetration depth expected from the measured carrier density. In section 10.4, we will provide the evidence of negligible on-tube resistances at room temperature in individual MWCNTs, and rule out the possibility of ballistic transport as a mechanism for the negligible on-tube resistances at room temperature. Alternatively, the temperature dependencies of the resistivity of individual single-wall carbon nanotubes (SWCNTs) can be quantitatively explained in terms of quantum phase slips in quasi-one-dimensional superconductors. In section 10.5, we will give concluding remarks and discuss possible microscopic mechanisms for high-temperature superconductivity in CNTs.

10.2 Magnetic Properties of Nickel Nanoparticles Embedded in MWCNTs

Purified MWCNT mat samples (Lot No. TS0636) from SES Research of Houston were synthesized by chemical vapor deposition under catalyzation of nickel nanoparticles. Some of nickel nanoparticles got into the nanotubes during the nanotube growth. After purification of as-grown MWCNT mat samples, most nickel nanoparticles were removed except for those embedded inside the innermost shells. The morphology of the mat sample can be seen from scanning electron microscopy image shown in Figure 10.1a. The SEM image was taken by a field emission scanning electron microcopy (FE-SEM, Hitachi S-4800) using an accelerating voltage of 3 kV. One can see that the outer diameters of these MWCNTs are in the range of 30–50 nm and centered around 40 nm. The mean inner diameter of the MWCNTs is about 10 nm, as seen from the transmission electron microscopy images (Figure 10.1b,c) recorded by FEI Tecnai F20 with an accelerating voltage of 200 kV. The nickel nanoparticles sit inside the innermost shells near the ends of the tubes, as labeled by A, B, C, and D in Figure 10.1c. Some nickel nanoparticles are connected to form a continuous chain (see a location labeled by D in Figure 10.1c).

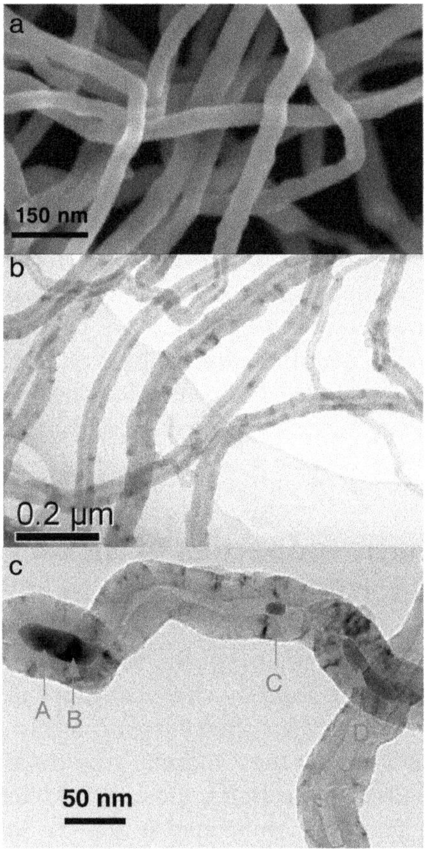

Figure 10.1 (a) SEM image of a Ni-filled MWCNT mat sample. (b) TEM image of a MWCNT mat sample. (c) TEM image of a selected MWCNT filled with nickel nanoparticles labeled by A, B, C, and D. After [Wang *et al.* (2010)].

The total metal-based impurity concentrations of the mat sample can be determined from the composition analysis of the residual of the sample, which was obtained by burning off carbon-based materials in air. A Perkin-Elmer Elan-DRCe inductively coupled plasma mass spectrometer (ICP-MS) was used to analyze the composition of the residual. From the weight (1.33%) of the residual and the ICP-MS analysis, we obtain the metal-based magnetic impurity concentrations by weight: Ni = 0.476%, Fe = 0.00907%, and Co = 0.0133%.

In order to quantitatively understand the magnetism of the Ni-filled MWCNTs, it is essential to determine the concentration of the ferromagnetic nickel phase. We can achieve this goal by performing high-energy synchrotron x-ray diffraction (XRD) experiment. Figure 10.2a shows a synchrotron XRD spectrum for a virgin Ni-filled MWCNT sample. The XRD spectrum was taken on a high-energy synchrotron x-ray beam-line 11-ID-C at the Advanced Photon Source, Argonne National Laboratory, using monochromatic radiation with a wavelength of λ = 0.1078 Å. The major peaks in the spectrum of Figure 10.2a correspond to the diffraction peaks of MWCNT [Reznik *et al.* (1995)] and the face-centered cubic (fcc) phase of Ni. In particular, the (002) diffraction peak of MWCNT is seen at 2θ = 1.815° and the Ni (311) peak at 2θ = 5.815°. Figure 10.2b,c displays the expanded views of the MWCNT (002) and Ni (311) peaks, respectively. The solid red line in Figure 10.2b is the fitted curve by the sum of a Gaussian and a Lorentzian function, which takes into account both domain size broadening and strain broadening [Reznik *et al.* (1995)]. The fit with the sum of two Gaussian or two Lorentzian functions is not excellent. The solid red line in Figure 10.2c is the fitted curve by a Gaussian function that is consistent with particle-size broadening [Reznik *et al.* (1995)]. The integrated intensity of the Ni (311) peak is found to be 0.84% of the intensity of the MWCNT (002) peak. Using the calculated standard intensities of the graphite (002) and Ni (311) peaks and assuming that the intensity of MWCNT (002) peak is the same as that of graphite (002), we find that the ferromagnetic fcc nickel concentration is 0.43% (by weight), which is slightly lower than the total Ni concentration (0.476%) inferred from ICP-MS. This implies that the ferromagnetic fcc nickel is the dominant phase while the concentrations of other nonmagnetic nickel-based phases are too small to be seen in the XRD spectrum.

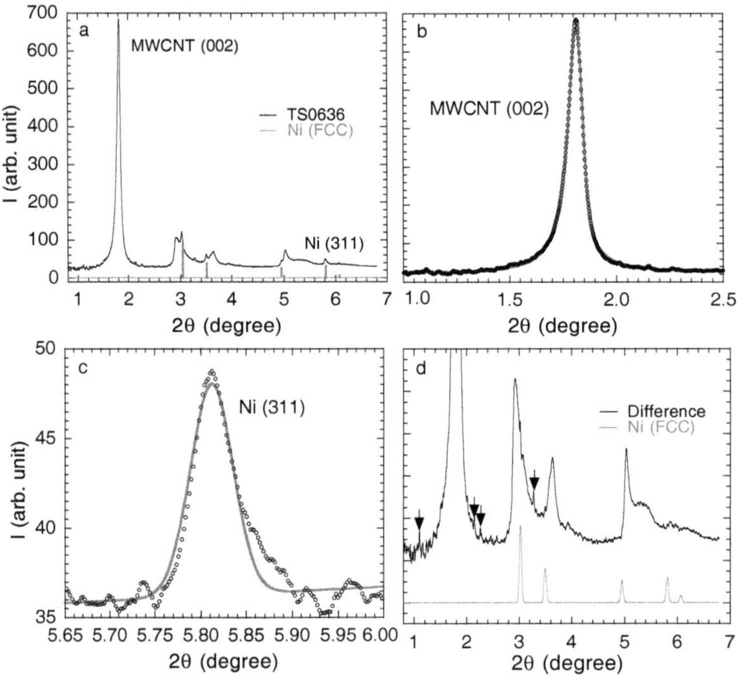

Figure 10.2 (a) High-energy synchrotron x-ray diffraction (XRD) spectrum for a virgin Ni-filled MWCNT sample (TS0636) and the standard spectrum for fcc nickel. (b) The expanded view of the MWCNT (002) peak. (c) The expanded view of the Ni (311) peak. (d) The expected XRD spectrum of the fcc Ni (red line) based on the nickel concentration (0.43%) and the difference spectrum (blue line), which is obtained by subtracting the Ni spectrum from the spectrum of TS0636 in Figure 10.2a. After [Wang *et al.* (2010)].

In order to check the reliability of our inferred ferromagnetic nickel concentration based on the Ni (311) peak, we show, in 10.2d, the expected XRD spectrum of the fcc Ni with the concentration of 0.43% (red line) and the difference spectrum (blue line), which is obtained by subtracting the Ni spectrum from the spectrum of TS0636 in Figure 10.2a. The difference spectrum shows no observable residual of any peaks of the fcc nickel, implying that the inferred Ni concentration is indeed reliable. Furthermore, all the peaks except for some peaks indicated by arrows in the difference spectrum agree with the peaks observed in pure MWCNTs [Reznik *et al.* (1995)]. The extra peaks indicated by the arrows should be associated with other impurity phases.

Since magnetic properties of nanoparticles depend strongly on the particle size, it is important to determine the average diameter d of the ferromagnetic Ni nanoparticles embedded in MWCNTs. We can determine d from the peak width of the XRD spectrum. The full width at half maximum of the Ni (311) peak is found to be 0.0556° from the Gaussian fit in Figure 10.2c. Using the Scherrer equation: $d = 0.89\ \lambda/(\beta \cos\theta)$ and with $\beta = 0.0511°$ (after correcting for the instrumental broadening), we calculate d = 11 nm, in good agreement with the average inner diameter of the tubes (see Figure 10.1b).

With the information of the average diameter (11 nm) of the ferromagnetic Ni nanoparticles and its concentration of 0.43% in our Ni-filled MWCNT sample, we would expect the room-temperature saturation magnetization (M_s) to be 0.14 emu/g from the known M_s = 30-32 emu/g for pure fcc Ni nanoparticles with d = 11-12 nm [Chen and Hsieh (2002)]. If we use the upper limit of M_s = 55 emu/g for bulk Ni, the expected M_s for our Ni-filled MWCNT sample would be 0.24 emu/g. Figure 10.3 shows magnetization versus magnetic field for our Ni-filled MWCNT mat sample at 320 K. The magnetization was measured using a Quantum Design vibrating sample magnetometer. The linear field dependence of the magnetization with a negative slope at H > 10 kOe is due to the diamagnetic contribution. The linear extrapolation to $H = 0$ yields $|\chi_{dia}| = 4.2 \times 10^{-6}$ emu/g and M_s = 0.47 emu/g. The measured M_s value is a factor of 3.4 larger than the expected value (0.14 emu/g) from the measured Ni concentration and average diameter. Thus, there is a giant magnetic moment enhancement of the Ni nanoparticles when they are embedded inside the innermost shells of MWCNTs. Using the measured nickel concentration (0.43%), the enhanced saturation magnetization ΔM_s is calculated to be about 76.7 emu per gram of nickel or 683 emu per cubic centimeter of nickel.

We have also found similar moment enhancement of Fe nanoparticles (d = 44 nm) embedded in MWCNTs with a mean inner diameter of about 50 nm and a mean outer diameter of about 70 nm. The concentration of the ferromagnetic Fe is found to be about 0.21% by weight while the saturation magnetization at room temperature is about 0.95 emu/g. Then ΔM_s is calculated to be about 273 emu per gram of iron or 2154 emu per cubic centimeter of iron. In contrast, we have found a negligible moment enhancement of Fe_3C

nanoparticles (d = 14 nm) embedded outside MWCNTs with a mean inner diameter of about 7 nm and mean outer diameter of about 15 nm. The detailed results will be published elsewhere.

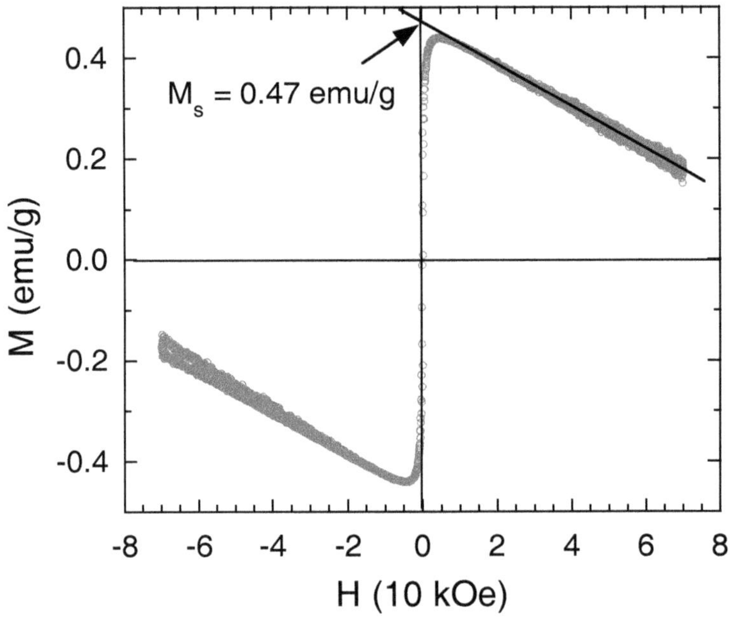

Figure 10.3 Magnetic hysteresis loop of the Ni-filled MWCNT mat sample at 320 K. After [Wang *et al.* (2010)].

If the observed giant magnetic moment enhancement arises from the magnetic proximity effect (MPE), then as proposed in [Coey *et al.* (2002)], there should exist two distinctive ferromagnetic-like transitions. Figure 10.4a shows zero-field-cooled (ZFC) and field-cooled (FC) magnetic susceptibilities of another virgin Ni-filled MWCNT sample in a field of 100 Oe. We clearly see that there is only one ferromagnetic transition at about 630 K. For comparison, Figure 10.4b shows the ZFC and FC susceptibilities for pure nickel powders (obtained from Shanghai Chemical Regent Co., China). The mean diameter of the Ni crystallites (subparticles) is calculated to be about 80 nm from the width of the XRD (311) peak. Within the experimental uncertainty, the Curie temperature of the pure nickel powders is close to that of the nickel nanoparticles embedded in MWCNTs.

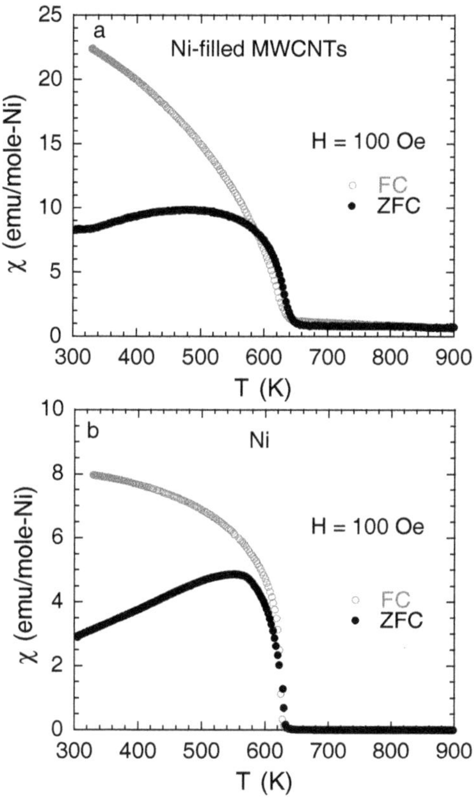

Figure 10.4 (a) ZFC and FC susceptibilities of a virgin Ni-filled MWCNT mat sample in a field of 100 Oe. (b) ZFC and FC susceptibilities for pure nickel powders. After [Wang *et al.* (2010)].

In Figure 10.5a, we plot the expanded view of the FC susceptibility just above the Curie temperature for the Ni-filled MWCNT mat sample. The solid line is a fitted curve by a Curie–Weiss law: $\chi = \chi_0 + C/(T - T_C)$ with $T_C = 625.7$ K and $C = 5.95$ emuK/mole-Ni, where χ_0 represents any background and other paramagnetic contributions, which should be nearly temperature independent within the narrow temperature interval. The Curie–Weiss constant corresponds to an effective moment of 6.95 μ_B per Ni atom (where μ_B is the Bohr magneton). In Figure 10.5b we show the expanded view of the FC susceptibility just above the Curie temperature for the pure nickel nanoparticles. The solid line is a fitted curve by the Curie–Weiss law with $T_C = 626.7$ K and $C = 0.486$ emuK/mole-Ni. The Curie–Weiss constant corresponds to an effective moment of 1.99 μ_B per Ni atom.

The Curie–Weiss constant for the nickel nanoparticles embedded in MWCNTs is enhanced by a factor of 12.2 compared with what would be expected for free nickel nanoparticles. It is interesting that the effective moment is enhanced by a factor of 3.49, which is close to the enhancement factor (3.4) for the saturation magnetization.

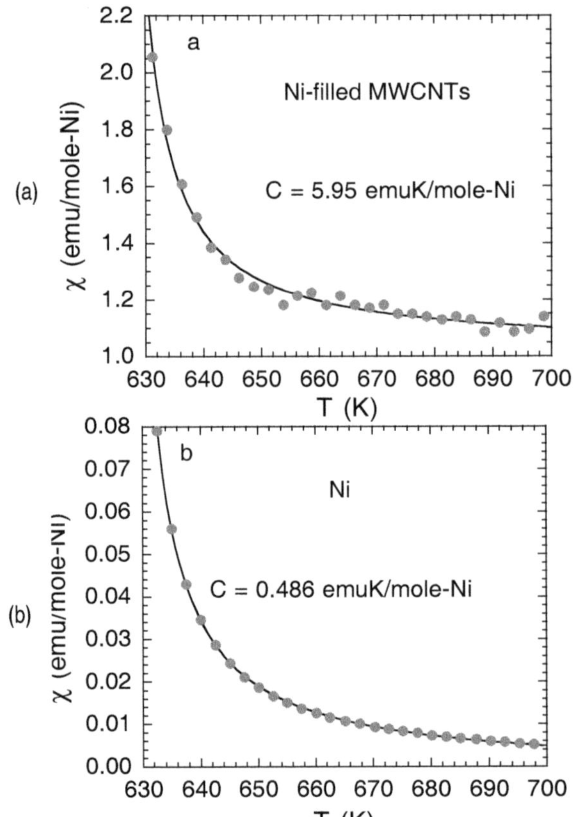

Figure 10.5 (a) The expanded view of the FC susceptibility just above the Curie temperature for the Ni-filled MWCNT mat sample. The solid line is a fitted curve by a Curie–Weiss law: $\chi = \chi_0 + C/(T - T_C)$ with T_C = 625.7 K and C = 5.95 emuK/mole-Ni. (b) The expanded view of the FC susceptibility just above the Curie temperature for the pure Ni nanoparticles. The solid line is a fitted curve by a Curie–Weiss law: $\chi = \chi_0 + C/(T - T_C)$ with T_C = 626.7 K and C = 0.486 emuK/mole-Ni. The Curie–Weiss constant for the nickel nanoparticles embedded in MWCNTs is enhanced by a factor of 12.2 compared with what would be expected for free nickel nanoparticles.

Now we turn to discuss the origin of the giant magnetic moment enhancement of the nickel nanoparticles embedded in MWCNTs. One possibility is that this enhancement arises from a large magnetic proximity effect [Coey *et al.* (2002); Cespedes *et al.* (2004)]. We consider a simple case where our ferromagnetic nanoparticles have a cylindrical shape with both length and diameter equal to d and the curve surface of the cylinder contacts with the innermost shell of a MWCNT (this is the most favorable case for the proximity effect). The curved surface area is equal to πd^2 and the total number of the contact carbon is $\pi N_c d^2$, where N_c is the number of carbon per unit area and equal to $3.82 \times 10^{15}/\text{cm}^2$ [Wallace (1947)]. If the induced magnetic moment is $m\mu_B$ per contact carbon atom, then the induced saturation magnetization normalized to the volume of the ferromagnetic nanoparticle is $\Delta M_s = 4N_C m\mu_B/d = 1420(m/d)$ emu/cm^3 (here d is in units of nm). Using the measured $\Delta M_s = 683$ emu/cm^3 and $d = 11$ nm for ferromagnetic nickel nanoparticles, we find that $m = 5.3$, which is a factor of 53 larger than the value (~0.1) calculated using density function theory [Cespedes *et al.* (2004)]. For ferromagnetic iron nanoparticles with $d = 44$ nm, the measured $\Delta M_s = 2154$ emu/cm^3. This implies $m = 66.7$, about three orders of magnitude larger than the value predicted from the magnetic proximity effect. Therefore, the magnetic proximity model is unlikely to explain such a giant magnetic moment enhancement.

Alternatively, it is possible that a strong diamagnetic tube could enhance the extrinsic magnetic moment of a (single-domain) magnet embedded inside it. If the tube were a perfect diamagnet, the "poles" of the magnet would be extended further apart (to the length of the tube) without changing their strength, thus giving an extrinsic enhancement to the magnetic moment. This is because the perfect diamagnetism of the tube prevents the magnetic field lines of the magnet from leaking out through the wall of the tube. This scenario can also explain why there is negligible moment enhancement when Fe_3C nanoparticles are embedded outside of the tubes.

The plausibility of this interpretation depends on whether MWCNTs show strong diamagnetism when the magnetic field is applied in the tube-axis direction in which the orbital diamagnetism is negligibly small [Lu (1995)]. If MWCNTs are ultrahigh temperature superconductors, they will exhibit strong diamagnetism. The observation of superconducting-like hysteresis loops in HOPG at 400 K should be a good indication of local superconductivity well above room temperature [Kopelevich *et al.* (2000)]. Similarly, there is also compelling evidence for ultrahigh temperature superconductivity in MWCNTs (see a review article [Zhao (2004)]). A recent theoretical work [Black-Schaffer and Doniach (2007)] predicts ultrahigh temperature *d*-wave superconductivity in well-doped graphene based on RVB theory originally proposed by Anderson [Anderson (1987)]. A similar model for layered cuprates [Lee *et al.* (2006)] predicts that an optimal superconducting transition temperature $T_c \simeq 0.14J/k_B$ (where J is the antiferromagnetic exchange energy and k_B is the Boltzmann constant). In graphene, $J \simeq t \simeq 3.0$ eV [Black-Schaffer and Doniach (2007)], so the optimal T_c should be about 0.42 eV$/k_B \simeq 5000$ K, in quantitative agreement with the numerical calculation [Black-Schaffer and Doniach (2007)]. Very recent large-scale quantum Monte Carlo simulations of correlated Dirac fermions on a honeycomb lattice (realized in graphene) have confirmed the existence of a short-range RVB liquid [Meng *et al.* (2010)]. If the RVB theory for superconductivity is relevant, ultrahigh temperature superconductivity can be realized in the MWCNTs where sufficient doping is realized by charge-transfer between ferromagnetic nanoparticles and MWCNTs. The special morphology (granular nature) in the mat samples, and the presence of magnetic nanoparticles can also promote the paramagnetic Meissner effect.

10.3 Identification of the Diamagnetic Meissner Effect in Pure MWCNTs

One of the most important signatures of superconductivity is the existence of the diamagnetic Meissner effect. Therefore, it is essential to provide evidence for the existence of the diamagnetic Meissner effect to prove ultrahigh temperature superconductivity. However, the diamagnetic Meissner effect may be less visible because the outer diameters of the tubes may be much smaller than the magnetic penetration depth. Further, the orbital diamagnetic susceptibility in the magnetic field perpendicular to the graphite sheet is large, making it difficult to separate the diamagnetic Meissner effect from the large orbital diamagnetic susceptibility. Fortunately, the orbital diamagnetic susceptibility of CNTs in the magnetic field parallel to the tube axes is predicted to be very small at room temperature [Lu (1995)]. This makes it possible to extract the diamagnetic Meissner effect from the measured susceptibility in the parallel field.

Figure 10.6 shows the temperature dependence of the orbital diamagnetic susceptibility of an SWCNT in the parallel magnetic field (solid circles), which was calculated based on the tight-binding approximation [Lu (1995)]. Here $E_g = \gamma_0 a_{C-C}/r$, a_{C-C} (= 0.142 nm) is the bonding length, r is the radius of the tube, and γ_0 (= 2.6 eV) is the tight-binding transfer matrix element [Lu (1995)]. What is remarkable is that the temperature dependence of the orbital susceptibility can be excellently fitted by an equation (solid line):

$$\frac{\chi_{||}^{orb}(T)}{r} = \frac{\chi_{||}^{orb}(0)}{r}[1 - 1.242\sqrt{E_g/k_BT}\exp(-0.283E_g/k_BT)]. \quad (10.1)$$

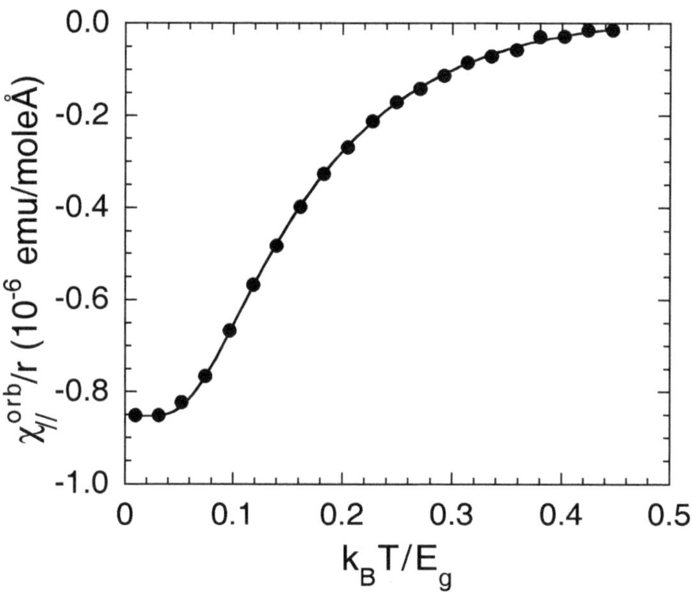

Figure 10.6 The calculated temperature dependence of the orbital diamagnetic susceptibility of an SWCNT in the parallel magnetic field. The calculation is based on the tight-binding approximation [Lu (1995)].

From Eq. 10.1, we can determine the orbital diamagnetic contribution for a tube or shell. To calculate the orbital diamagnetic susceptibility for a MWCNT nanotube comprising several concentric shells, one needs to replace E_g and r in Eq. 10.1 by the averaged $< E_g >$ and $< r >$. Since both $\chi_{||}^{orb}$ and the mass of each shell are proportional to r, the average $< r >$ should be $(2/3)r_{out}$ (where r_{out} is the outer radius of a MWCNT). Considering the fact that $\chi_{||}^{orb}$ is proportional to γ_0^2 [Kotosonov and Kuvshinnikov (1997)], we finally have

$$\chi_{||}^{orb}(T) = -7.0\times10^{-9}\gamma_0^2 r_{out}\left[1-1.52\sqrt{\frac{a_{C-C}\gamma_0}{r_{out}k_B T}}\exp\left(-\frac{0.425a_{C-C}\gamma_0}{r_{out}k_B T}\right)\right], \quad (10.2)$$

where $\chi_{||}^{orb}$, γ_0, and r_{out} are in units of emu/g, eV, and Å, respectively.

Now we consider the diamagnetic Meissner effect for a superconducting MWCNT in the magnetic field parallel to the tube axis. We are particularly interested in the case where the magnetic penetration depth is larger than the outer radius of the tube. In this case, the diamagnetic susceptibility due to the Meissner effect is given by

$$\chi_{||}^{S}(T) = \frac{r_{out}^2}{32\pi\lambda_\theta^2(T)}, \quad (10.3)$$

where $\lambda_\theta(T)$ is the penetration depth when carriers move along the circumferential direction (or the field is parallel to the tube axis). For a macroscopic sample consisting of a macroscopic number of MWCNTs, the r_{out}^2 in the above equation should be replaced by $<r_{out}^2>$, the average of r_{out}^2. In the low temperature limit: $k_B T \leq 0.2\Delta_{min}(0)$ [where $\Delta_{min}(0)$ is the minimum value of the superconducting gap at zero temperature], the penetration depth follows the following expression:

$$\lambda_\theta(T) = \lambda_\theta(0) + \lambda_\theta(0)\sqrt{\pi\Delta_{min}(0)/2k_B T}\,\exp[-\Delta_{min}(0)/k_B T]. \quad (10.4)$$

Combining Eqs. 10.3 and 10.4, we can readily show that

$$\chi_{||}^{S}(T) = \chi_{||}^{S}(0)\,(1-2\sqrt{\pi\Delta_{min}(0)/2k_B T}\,\exp[-\Delta_{min}(0)/k_B T]). \quad (10.5)$$

The total diamagnetic contribution of a superconducting MWCNT is the sum of $\chi_{||}^{S}(T)$ and $\chi_{||}^{orb}(T)$, that is,

$$\chi_{||}(T) = \chi_{||}^{S}(0)\,(1-2\sqrt{\pi\Delta_{min}(0)/2k_B T}\,\exp[-\Delta_{min}(0)/k_B T]) -7.0\times10^{-9}\gamma_0^2 r_{out}\left[1-1.52\sqrt{\frac{a_{C-C}\gamma_0}{r_{out}k_B T}}\exp\left(-\frac{0.425a_{C-C}\gamma_0}{r_{out}k_B T}\right)\right] \quad (10.6)$$

Figure 10.7 shows the temperature dependence of the parallel-field susceptibility for pure MWCNTs, which are physically separated and aligned [Chauvet *et al.* (1995)]. The outer diameters of the tubes are 10 ± 5 nm and the lengths are on the order of 1 μm [Chauvet *et al.* (1995)]. It is apparent that the diamagnetic susceptibility is significant up to 265 K. If these MWCNTs are ultrahigh-temperature superconductors with $\Delta_{min} \geq 100$ meV, the data should be consistent with Eq. 10.6. The solid line in Figure 10.7 is the best fitted curve by Eq. 10.6. It is striking that the fit is excellent. The fitting parameters are the following: $\chi^{S}_{||}(0) = -(7.6 \pm 0.2) \times 10^{-6}$ emu/g, $\Delta_{min} = 124 \pm 14$ meV, $\gamma_0 = 2.80 \pm 0.09$ eV, and $r_{out} = 62.9 \pm 0.7$ Å. The value of Δ_{min} = 124 meV justifies the temperature range for the fitting. If we use the BCS relation between the gap and superconducting transition temperature: $T_c = \Delta(0)/1.76k_B$, we find that $T_c \geq 800$ K. The value of r_{out} = 62.9 Å is consistent with the average outer radius of 50 ± 25 Å, which was directly measured by TEM [Chauvet *et al.* (1995)]. The value of γ_0 = 2.80 eV is in quantitative agreement with both theory and experiment.

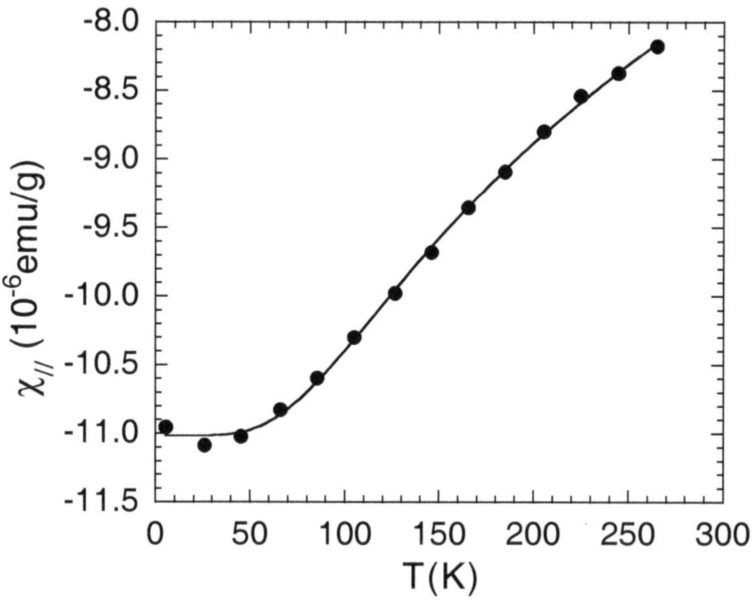

Figure 10.7 Temperature dependence of the susceptibility for physically separated and aligned MWCNTs in a magnetic field parallel to the tube axes. The data are extracted from [Chauvet *et al.* (1995)].

Now we would like to check if the parameter $\chi_{||}^{S}(0) = -(7.6 \pm 0.2) \times 10^{-6}$ emu/g is consistent with the expected Meissner effect. If we assume that the outer radii of MWCNTs are equally distributed from 23 to 103 Å with $< r_{out} > = 63$ Å (in agreement with that inferred from the best fit above), we find $< r_{out}^2 > = 4502$ Å^2. With the weight density of 2.17 g/cm^3 [Qian *et al.* (2000)] and $\chi_{||}^{S}(0) = -7.6 \times 10^{-6}$ emu/g, we calculate $\lambda_\theta(0) \simeq 1648$ Å using Eq. 10.3. This value of the penetration depth corresponds to $n/m_\theta^* = 1.04 \times 10^{21}$/cm$^3m_e$, where n is the carrier density, m_θ^* is the effective mass of carriers along the circumferential direction. If we take $m_\theta^* = 0.012m_e$, typical for graphite [Bayot *et al.* (1989)], we estimate $n = 1.25 \times 10^{19}$/cm^3, in quantitative agreement with the Hall effect measurement [Baumgartner *et al.* (1997)] which gives $n = 1.6 \times 10^{19}$/cm^3. It is worthy of noting that the inferred magnetic penetration depth is far larger than the outer radii of MWCNTs, which justifies Eq. 10.3. This also ensures that the Hall effect in the superconducting state is the same as that in the normal state.

A CNT should behave like graphene when the electron mean free path is shorter than the circumference of the tube [Schonenberger *et al.* (1999)]. In graphene, the effective mass of carriers is given by $m^* = \sqrt{\pi n_{2D}}\hbar/v_F$ [Novoselov *et al.* (2005)], where n_{2D} is the sheet carrier density and v_F is the Fermi velocity. Using $\hbar v_F = 1.5a_{C\text{-}C}\gamma_0 = 5.96$ eVÅ and taking $n = 1.6 \times 10^{19}$/cm^3, we find that $m^* = 0.018\ m_e$. This leads to $n/m^* = 0.89 \times 10^{21}$/cm$^3m_e$, very close to what we have inferred from the susceptibility data. Therefore, the susceptibility data of the aligned MWCNTs are in quantitative agreement with ultrahigh temperature superconductivity.

10.4 Electrical Transport and Quantum Phase Slips

The phenomenon of superconductivity depends on the coherence of the phase of the superconducting order parameter. The phase coherence of the superconducting order parameter leads to the zero-resistance state. For three-dimensional bulk systems, the transition to the zero-resistance state occurs right below the mean-field superconducting transition temperature T_{c0} such that the resistive transition is very sharp and the transition width is negligibly small. In contrast, the resistive transition in quasi-1D superconductors is broad because of large superconducting fluctuations. A quantum theory to describe the resistive transition in quasi-1D superconductors was developed by Langer, Ambegaokar, McCumber, and Halperin [Langer and Ambegaokar (1967)] over 30 years ago. The theory is based on thermally activated phase slips (TAPS), which cause the resistance to decrease to zero exponentially. This theory was tested quantitatively on a superconducting SWCNT bundle [Zhao (2005)]. In addition to the thermally activated phase slips, there also exist quantum phase slips due to a finite number of transverse channels [Zaikin *et al.* (1997); De Martino and Egger (2004)], which prevent ultra-thin wires or tubes from being true superconductors with absolutely zero resistance. Since any CNT has a finite number of transverse channels N_{ch}, we should not expect to see a zero-resistance state at room temperature even if its T_{c0} is significantly higher than 300 K. However, the on-tube resistance at room temperature should be negligibly small or significantly lower than the normal-state resistance if N_{ch} is not too small.

Several independent groups have observed negligible on-tube resistances in many individual MWCNTs [Frank *et al.* (1998); De Pablo *et al.* (1999); Poncharal *et al.* (2002); Urbina *et al.* (2003)]. These results have been tentatively explained in terms of ballistic transport at room temperature. Room-temperature on-tube resistivity (resistance per unit length) has also been measured for many individual SWCNTs (d = 1.4 nm) with two transverse channels [Kajiura *et al.* (2005)]. The resistivity shows a large variation from 496 Ω/μm to 645 kΩ/μm [Kajiura *et al.* (2005)]. The variation in the resistivity was attributed to the variation in the impurity scattering [Kajiura *et al.* (2005)]. In terms of the generally accepted ballistic transport mechanism, the smallest resistivity 496 Ω/μm should be the upper limit of the resistivity caused by electron–phonon scattering. Here we will show that electron–phonon scattering at room temperature is significant and the expected resistivity for a 1.4 nm SWCNT is at least 9.7 kΩ/μm. This is about 20 times larger than the observed upper limit: 496 Ω/μm. We further show that the on-tube resistivity values for many individual MWCNTs are too small to be explained by the ballistic transport mechanism. Alternatively, the data can be quantitatively explained in terms of quantum phase slips in quasi-1D ultrahigh-temperature superconducting CNTs.

With neglect of impurity scattering, the room-temperature on-tube resistivity ρ_{on} for a metallic single-wall carbon nanotube (m-SWCNT) is predicted to be nearly independent of the chirality of the tube [Suzuura and Ando (2002)]. For armchair tubes with the chiral angle $\eta = \pi/6$, the resistivity ρ_A is caused only by a twist mode. For zigzag tubes with $\eta = 0$, the resistivity ρ_z arises from scattering by a radial breathing mode. For other metallic tubes, the resistivity is simply equal to $\rho_A \sin^2 3\eta + \rho_z \cos^2 3\eta$ [Suzuura and Ando (2002)]. At high temperatures, ρ_A and ρ_z have the same magnitude so that the high-temperature resistivity is the same for m-SWCNTs with the same diameter. The expression for ρ_A is given by [Suzuura and Ando (2002)]

$$\rho_A = \frac{R_Q}{2} \frac{3k_B T \kappa^2 \beta^2}{\mu_T \pi a^2 d}, \tag{10.7}$$

where $R_Q = h/2e^2 = 12.9$ kΩ, $a = 2.46$ Å, $\kappa = 0.369$, μ_T is the shear modulus and equal to 57.4 kg/s^2 in graphite [Suzuura and Ando (2002)], $\beta = (a_{C\text{-}C}/\gamma_0)(d\gamma_0/da_{C\text{-}C})$. The deformation potential $d\gamma_0/da_{C\text{-}C}$ has been calculated to be 7.8 eV/Å based on the density functional approximation (DFA) [Lazzeri *et al.* (2005)]. It is remarkable that the predicted deformation potential is very close to that (7.21 eV/Å) determined from graphite phonon dispersion of the highest optical branch near Γ point [Lazzeri *et al.* (2006)]. This leads to $\beta = 3.65$ with $\gamma_0 = 2.6$ eV and $a_{C\text{-}C} = 1.42$ Å. Using this unbiased β value, we can calculate ρ_A for an armchair tube with $d = 1.37$ nm, which is displayed in Figure 10.8. For comparison, we also include ρ_z in Figure 10.8 for a zigzag tube with $d = 1.37$ nm. It is clear that ρ_A increases linearly with temperature while ρ_z increases quasi-linearly with temperature. At room temperature, ρ_A and ρ_z are close to each other and equal to about 11 kΩ/μm. This corresponds to an inelastic mean-free path of about 586 nm.

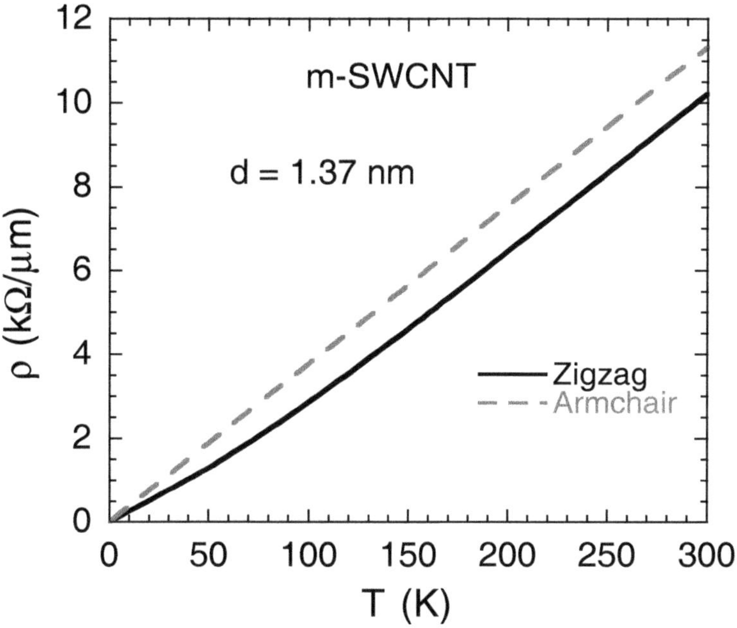

Figure 10.8 Temperature dependencies of the resistivity calculated for armchair and zigzag SWCNTs.

In Figure 10.9, we show the temperature dependencies of the resistances for a m-SWCNT, which are hole doped by applying gate voltages (V_g) of −10 V and 2 V, respectively. The diameter d of the tube is about 1.5 nm and the length L is about 200 nm. The data are taken from [Kong *et al.* (2001)]. In both doping levels, the resistance does not increase linearly with temperature. Instead, the temperature dependencies can be well fitted (solid lines) by the equation: $R(T) = R_0 + AT^p$. The best fit yields p = 1.58 ± 0.12 and 2.30 ± 0.18 for V_g = −10 V and 2 V, respectively. The temperature dependencies are incompatible with the parameter-free calculation (Figure 10.8). The values of the exponent p in both cases are well off from the theoretical value of about 1. Using L = 200 nm, the temperature dependent resistivity at room temperature ρ_T(300K) is calculated to be 74.2 kΩ/μm at V_g = 2 V and 19.3 kΩ/μm at V_g = −10 V. These resistivity values are significantly larger than the theoretically predicted value of about 10 kΩ/μm.

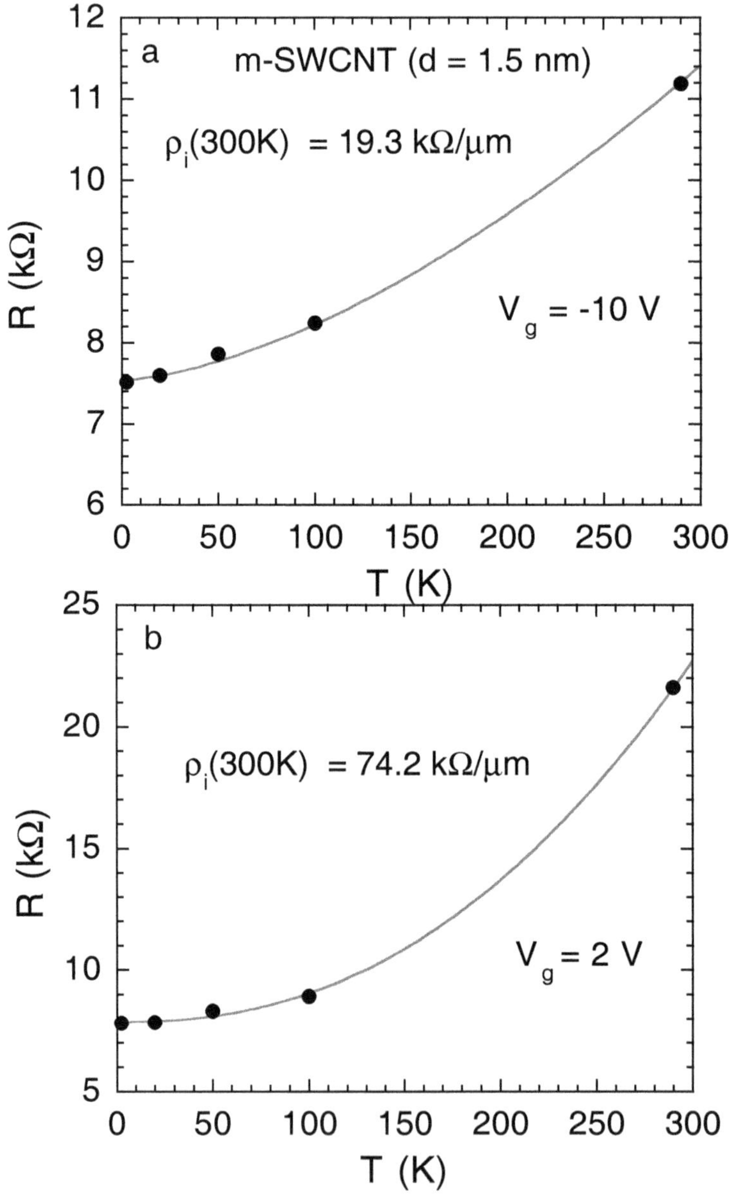

Figure 10.9 Temperature dependencies of the resistance for a m-SWCNT, which are hole doped by applying gate voltages of –10 V and 2 V, respectively.

Similar electrical transport data have been obtained [Mann *et al.* (2003)] for another m-SWCNT with d = 1.7 nm and L = 4000 nm. Figure 10.10 shows the resistance as a function of temperature at V_g = −10 V. The data can also be fitted by a power law with p = 1.42 ± 0.11, which is very close to that (1.58 ± 0.12) found for the shorter tube at V_g = −10 V. The temperature dependent resistivity at room temperature ρ_T(300K) is calculated to be 3.72 kΩ/μm, which is much lower than those found for the shorter tube. The large variation of the temperature dependent resistivity and the power-law temperature dependence of the resistivity are in sharp contrast with the predicted curves (Figure 10.8) on the basis of electron–phonon scattering.

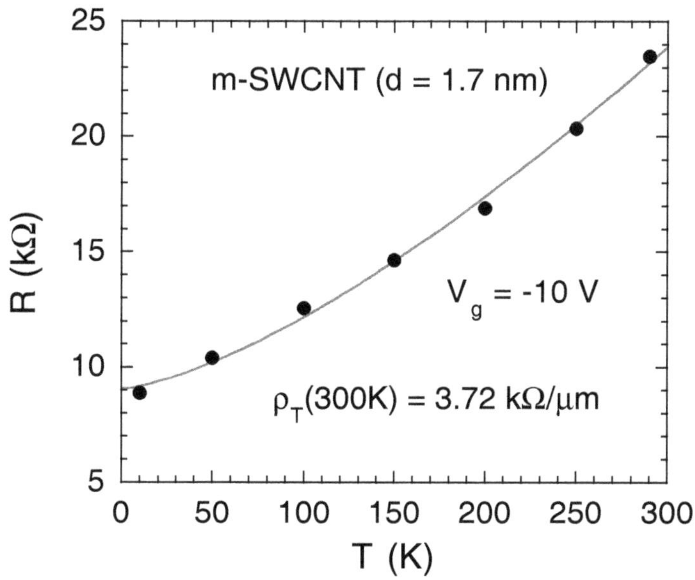

Figure 10.10 Temperature dependence of the resistance for a m-SWCNT with d = 1.7 nm and L = 4000 nm, which are hole doped by applying gate voltages of −10 V.

If ρ_T(300K) = 74.2 kΩ/μm at V_g = 2 V (corresponding to a low hole doping level) for the 1.5 nm tube were caused by electron–phonon scattering, the inelastic mean-free path of this tube would be 87 nm at room temperature. Then ρ_T(300K) = 7.42 kΩ/μm for a metallic tube with d = 15 nm. This would imply that the room-temperature on-tube resistivity ρ_{on}(300K) > 7.42 kΩ/μm for d =15 nm and 5.57 kΩ/μm for d = 20 nm.

Poncharal *et al.* [Poncharal *et al.* (2002)] made extensive transport studies on over 50 individual MWCNTs with diameters ranging from 5 to 25 nm. These MWCNTs, which protrude from unprocessed arc produced nanotube fibers and are contacted with liquid metals, show very small ρ_{on} values at room temperature (<92 Ω/μm). For example, ρ_{on} is found to be 14 Ω/μm for a MWCNT with $d \simeq 20$ nm, which is over two orders of magnitude smaller than what would be expected. Therefore, these electrical transport data cannot be consistently explained if these tubes were not phase-incoherent room-temperature superconductors.

On the other hand, we can consistently explain the electrical transport data in terms of phase-incoherent ultrahigh temperature superconductivity in these doped CNTs. Because these tubes are ultra thin and have a finite number of transverse channels, both quantum phase slips (QPSs) and thermally activated phase slips are inevitable. Thermally activated phase slips lead to broadening of the resistive transition just below T_{c0} while quantum phase slips give rise to a finite resistivity at any finite temperature below T_{c0} [Zaikin *et al.* (1997)]. A QPS is a topological vortex-like excitation of the superconducting phase field where superconducting order is locally destroyed and a phase cannot be defined.

Well below T_{c0} (e.g., $T < 0.3T_{c0}$), the dimensionless phase-stiffness parameter $\mu \propto 1/\lambda(T)$ [Zaikin *et al.* (1997)] is nearly independent of temperature, where $\lambda(T)$ is the magnetic penetration depth. The intrinsic on-wire (on-tube) resistance R_{tube} of a superconducting wire or tube is given by [De Martino and Egger (2004)]:

$$R_{tube} = \frac{R_Q \pi L}{2x_0}\left(\frac{\pi \exp(-S_c^*)\Gamma(\mu/2)}{\Gamma(\mu/2+1/2)}\right)^2 \left(\frac{L_T}{x_0}\right)^{3-2\mu}$$
$$\times \int_0^\infty du \frac{2/\pi}{1+u^2}\left|\frac{\Gamma(\mu/2+iuL_T/2L)}{\Gamma(\mu/2)}\right|^4, \qquad (10.8)$$

where x_0 is the QPS core size, S_c^* is the effective QPS action, and L_T is the thermal length defined as $L_T = \hbar c_0/\pi T$ (where c_0 is the Mooij-Schon velocity that is very close to v_F in m-SWCNTs [De Martino and Egger (2004)]). For $L/L_T >> 1$, the u-integral approaches unity, and hence $R_{tube} \propto T^{2\mu-3}$, while for $L/L_T << 1$, the u-integral is proportional to T so that $R_{tube} \propto T^{2\mu-2}$. Similar formulas have also been derived for the diffusive wires having a finite mean-free path l in the normal state [Zaikin *et al.* (1997)].

Considering electrical current dissipation in the QPS core, the effective core action S_C^* becomes the sum of the superconducting condensation energy and the dissipation energy of currents lost inside the QPS core during a phase-slip event [Zaikin *et al.* (1997)]. Adding the contribution of the electromagnetic fields outside the core and minimizing the action with respect to x_0 and τ_0 (where $\tau_0 = x_0/c_0$), S_C^* happens to be three times the superconducting condensation energy [Zaikin *et al.* (1997)], that is, $S_C^* = 3S\tau_0 x_0 \Delta^2(0)N(0)/2$, where S is the cross-sectional area, $N(0)$ is the density of states at the Fermi level, and $\Delta(0)$ is the superconducting gap at zero temperature [Zaikin *et al.* (1997)]. Considering $c_0 \simeq v_F$ and the 3D conductivity $\sigma = e^2N(0)v_F l$ for m-SWCNTs, S_C^* and x_0 can be re-written as

$$S_C^* = \frac{3}{\pi}\left[\frac{l\Delta(0)}{2\hbar v_F}\right]^{2/3}, \tag{10.9}$$

and

$$x_0 = \left[\frac{(\hbar v_F)^2 l}{2\Delta^2(0)}\right]^{1/3}. \tag{10.10}$$

For two probe measurements, the total resistance $R(T) = R_0 + R_{tube}$, where $R_0 = R_c + R_t$, R_c is the contact resistance, and R_t is the tunneling resistance [Schonenberger *et al.* (1999)]. In the true superconducting state ($R_{tube} = 0$) at $T = 0$, the tunneling resistance due to the Andreev reflection is given by [Blonder *et al.* (1982)]

$$R_t = \frac{R_Q(1+2Z^2)^2}{4N_{ch}}, \tag{10.11}$$

where Z is a dimensionless barrier strength. In the normal state, $R_t = R_Q(1 + Z^2)/2N_{ch}$. When $Z = 0$, the tunneling resistance is $R_Q/4N_{ch}$ in the superconducting state while it is $R_Q/2N_{ch}$ in the normal state. When $Z < 0.455$, R_t in the superconducting state is smaller than $R_Q/2N_{ch}$, expected for a normal wire with perfect contacts. When $Z = 0.555$, $R_t = 1.305R_Q/2N_{ch}$, for both the superconducting and normal states. Therefore, a wire must be in the superconducting state if $R_0 < R_Q/2N_{ch}$.

Now we compare the QPS theory with the resistance data of SWCNTs that are known to exhibit quasi-1D superconductivity. Investigation of the magnetic and transport properties of the smallest SWCNTs embedded in a zeolite matrix revealed that at temperatures below 15 K, 4 angstrom tubes with a semiconducting chirality index (5,0) exhibit phase incoherent superconductivity [Tang *et al.* (2001)]. The superconducting behavior of the tubes can be well described by Ginzburg-Landau equation in 1D systems [Tang *et al.* (2001)]. The magnetic data are consistent with T_{c0} = 15 K and the zero-temperature coherent length ξ_0 = 4.2 nm. Using the standard equations: $\xi_0 = 0.85\sqrt{l\xi_{BCS}}$ and $\xi_{\mathrm{BCS}} = 0.18\,\hbar v_{\mathrm{F}}/k_B T_{c0}$, and taking $\hbar v_F$ = 2.40 eVÅ (see discussion below), we calculate l = 5.88 Å. Several band-structure calculations [Barnett *et al.* (2005); Lyakutti *et al.* (2006)] on (5,0) tube based on the local density approximation consistently show that the Fermi level crosses two subbands with three conducting channels (one subband is non-degenerate and another subband is doubly degenerate [Barnett *et al.* (2005)]). Because of strong curvature effect, the conduction and valance subbands of the semiconducting-chirality tube are overlapped, making it metallic. With N_{ch} = 3, we calculate the normal-state on-tube resistivity $\rho_{tube} = R_Q/N_{ch}l$ = 7313 kΩ/μm. Then, the normal-state on-tube resistance is calculated to be 365.7 kΩ for the SWCNT with L = 50 nm.

In Figure 10.11a, we plot the two-probe resistance as a function of temperature for the 4 angstrom SWCNT with L = 50 nm. The data are taken from [Tang *et al.* (2001)] and rescaled to ensure the on-tube resistance at 15 K to be 365.7 kΩ. It is apparent that the resistance increases with temperature and flattens out at about 14 K, suggesting that $T_{c0} \simeq$ 15 K, in agreement with magnetic and tunneling data [Tang *et al.* (2001)].

The temperature dependence of the resistance of the superconducting tube would not be consistent with superconductivity on the basis of the bulk signature of superconductivity. Since the tube is a quasi-1D superconductor, the temperature dependence of the resistance should be consistent with Eq. 10.8. In Figure 10.11b, we show the expanded view of the resistance below 6 K. The solid line is the calculated result using μ = 1.73, l = 5.88 Å, $\Delta(0) = 1.76k_BT_{c0}$, and $\hbar v_F$ = 2.40 eVÅ (the same value used above for calculating l). It is remarkable that the calculated curve is in excellent agreement with the data. In other words, the data can be well fitted by Eq. 10.8 with the fitting parameters: μ = 1.73 and $\hbar v_F$ = 2.40 eVÅ.

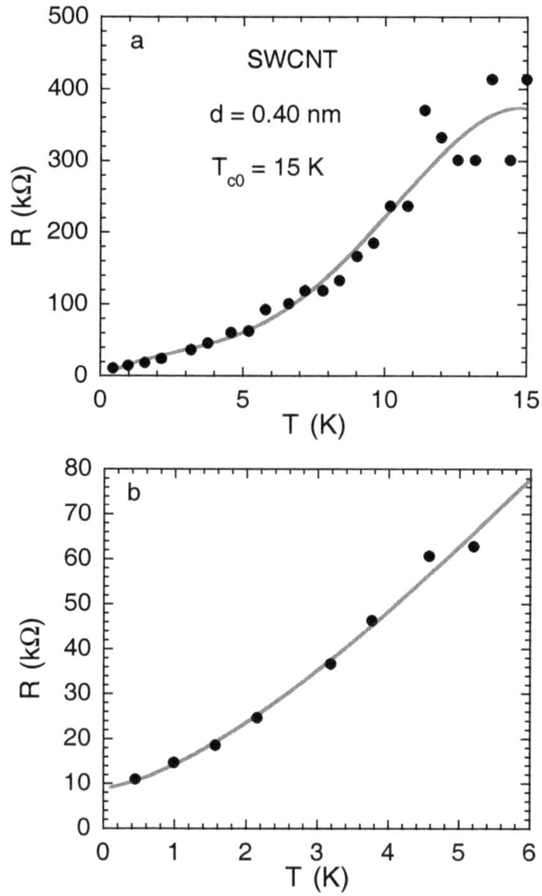

Figure 10.11 (a) Two-probe resistance as a function of temperature for a 4 angstrom SWCNT with L = 50 nm. The data are taken from [Tang *et al.* (2001)] and rescaled to ensure the on-tube resistance at 15 K to be 365.7 kΩ. (b) The expanded view of the resistance below 6 K. The solid line is the calculated result using Eq. 10.8 with $\mu = 1.73$, $l = 5.88$ Å, $\Delta(0) = 1.76k_BT_{c0}$, and $\hbar v_F = 2.40$ eVÅ.

We can compare the inferred Fermi velocity with that [Barnett *et al.* (2005)] predicted from LDA calculations. Figure 10.12 shows the calculated band dispersion near the Fermi level $E_F = 0$ for (5,0) tube. There are two subbands labeled by A and B, which have the different values of the bare Fermi velocity v_F. The average value of $\hbar v_F$ is

3.0 eVÅ, which is slightly larger than the inferred value of 2.40 eVÅ. This quantitative agreement between the data and Eq. 10.8 justifies the QPS theory.

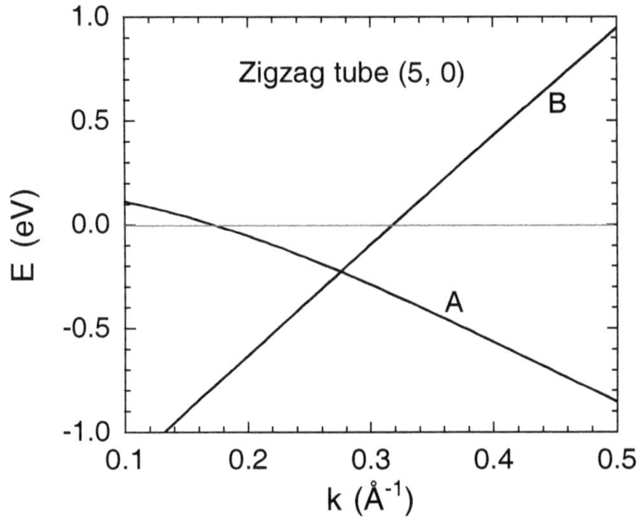

Figure 10.12 The calculated band dispersion near the Fermi level $E_F = 0$ for a zigzag tube with a chirality index (5,0) [Barnett *et al.* (2005)]. There are two subbands labeled by A and B. Band B is non-degenerate while band A is doubly degenerate [Barnett *et al.* (2005)], suggesting three conduction channels at E_F.

Now we explain the resistance data for the long m-SWCNT with $L = 4000$ nm in terms of Eq. 10.8. In this case $L >> L_T$ down to 5 K, and the u-integral is simply equal to 1. Then $R_{tube} = AT^p$ with $p = 2\mu - 3$. The power law fit above has shown that $p = 1.42$, which implies $\mu = 2.2$. The coefficient $A = 0.00446$ kΩK$^{-1.42}$ is consistent with the following parameters: $\Delta(0)$ = 105 meV [Zhao (2004)], $\hbar v_F$ = 5.3 eVÅ, and l = 384 Å. The A value or the room-temperature resistivity is very sensitive to the mean free path l. For example, when l is reduced from 384 Å to 188 Å, the A value increases by three orders of magnitude. This can naturally explain why the on-tube resistivity can be varied from 496 Ω/μm to 645 kΩ/μm (varied by over three orders of magnitude). If we keep all the other parameters and only vary l, we can easily show that the room-temperature on-tube resistivity can vary from 496 Ω/μm to 645 kΩ/μm when l changes from 424 Å to 253 Å.

It has been shown that the impurity scattering is much less effective for the linearly dispersed metallic subband than for the parabolic-like subbands [Seri and Ando (1997)]. The calculated mean-free path for the first parabolic-like subband is about four times shorter than for the linearly dispersed subband when they have the same density of states [Seri and Ando (1997)]. The density of states for the linearly dispersed subband is constant and given by $N(0) = \sqrt{3}\, a_{\mathrm{C-C}} / (\pi^2 \gamma_0 d)$ per eV per C atom per spin [Mintmire and White (1998)]. For the 1.7 nm m-SWCNT, $N(0)$ is calculated to be 0.00565 per eV per C atom per spin. This is a factor of 14 smaller than that (0.08 per eV per C atom per spin [Barnett *et al.* (2005)]) for the (5,0) tube where the Fermi level crosses the parabolic-like subbands. This implies that the mean-free path l for the 1.7 nm m-SWCNT should be a factor of about 56 times longer than that for the 0.4 nm SWCNT if they have similar disorder. Then l for the 1.7 nm metallic tube should be about 330 Å, in good agreement with that (384 Å) inferred from the QPS theory.

10.5 Concluding Remarks

It is well known that copper-based perovskite oxides rightly enjoy consensus as high-temperature superconductors on the basis of two signatures: the resistive transition and the Meissner effect. Here we have provided magnetic and electrical evidences for ultrahigh temperature superconductivity in CNTs. The giant magnetic moment enhancement found for the magnetic nanoparticles embedded inside the innermost shells of MWCNTs cannot be explained by the magnetic proximity effect. But rather the result can be naturally explained in terms of the interplay between magnetism of the magnetic nanoparticles and ultrahigh temperature superconductivity in MWCNTs. The diamagnetic susceptibility of pure MWCNTs for the field parallel to the tube axes agrees quantitatively with the predicted penetration depth from the measured carrier density. Furthermore, bundling of individual MWCNTs into closely packed bundles leads to a large enhancement in the diamagnetic susceptibility, which can be naturally explained by the Josephson coupling among the tubes in the bundles [Zhao and Beeli (2008)]. Because of a finite number of transverse conduction channels, both quantum and thermally activated phase slips are important and the on-tube resistance will never go to zero below the mean-field superconducting transition temperature. Nonetheless, the room-temperature on-tube resistance has been found to be indistinguishable from zero for many individual MWCNTs. Our parameter-free calculations based on the inelastic electron–phonon scattering suggest that the m-SWCNTs with the same diameter should have nearly the same inelastic mean-free path at room temperature. But experimental data show a very large variation in the room-temperature resistivity and in some cases the inferred inelastic mean-free path is about two orders of magnitude larger than the value predicted from electron–phonon scattering. In contrast, the temperature dependencies of the resistivity for m-SWCNTs can be quantitatively explained by the quantum phase slip theory. The very large variation of room-temperature on-tube resistivity in many individual SWCNTs is the natural consequence of the quantum phase slips that depend very strongly on the normal-state resistivity.

There are also other independent evidences for ultrahigh temperature superconductivity in both SWCNTs and MWCNTs [Zhao and Wang (2001); Zhao (2004, 2006)]. Some resistivity data of MWCNTs and SWCNTs show quite broad superconducting transitions above room temperature and can be well explained [Zhao (2006)] in terms of thermally activated phase slip theory developed by Langer, Ambegaokar, McCumber, and Halperin. Raman data and tunneling spectra of SWCNTs consistently show single particle excitation gaps in the range of 100-200 meV [Zhao (2006)]. This would imply that T_{c0} = 600-1200 K. The tunneling spectra of some MWCNTs also indicate a gap of about 200 meV, which is too large to be consistent with the expected semiconducting gap for semiconducting-chirality tubes [Zhao (2006)].

Although electron–phonon coupling in graphite and related materials is strong and the phonon energy is high (> 100 meV), the calculated electron–phonon coupling constants for various carbon-based materials [Barnett *et al.* (2005); Park *et al.* (2008)] are small due to low density of states. This implies that electron–phonon interaction alone should be insufficient to explain ultrahigh temperature superconductivity in CNTs, graphite, and carbon films. Although the RVB theory [Anderson (1987); Black-Schaffer and Doniach (2007)] might be able to explain ultrahigh temperature superconductivity in heavily-doped graphene and MWCNTs, it does not predict ultrahigh temperature superconductivity at low doping. We speculate that strong electron–electron correlation of the relativistic Dirac fermions may lead to a huge enhancement of electron–phonon coupling. Indeed, the electron–phonon coupling constant has been calculated to be about 0.04 for graphene and graphite [Park *et al.* (2008)] while the coupling constant determined by angle-resolved photoemission spectroscopy is about 1.0 [Sugawara *et al.* (2007)]. The enhancement factor is over one order of magnitude, similar to the case in strongly correlated cuprates. The strongly enhanced electron–phonon coupling along with strong coupling to the high-energy acoustic plasmons inherent in quasi-1D and 2D electronic systems [Lee and Mendoza (1989); Cui and Tsai (1991)] is the key to achieve ultrahigh temperature superconductivity. In order to take further advantage of strong electron–electron correlation, the order parameters in doubly-degenerate bands near K and K' points might be of opposite signs (node-less d-wave). Another important factor to influence superconductivity in CNTs is the strong long-range Coulomb interaction, which can completely destroy superconductivity if it is not well screened by substrates and/or electrodes [De Martino and Egger (2004); Zhao (2006)]. More extensive experimental and theoretical investigations are required to understand the pairing mechanism of ultrahigh temperature superconductivity in carbon-related materials.

References

Alexandrov, A. S., and Mott, N. F. (1995). *Polarons and Bipolarons*, World Scientific, ISBN-981022298X, Singapore.

Anderson, P. W. (1987). The resonating valence bond state in La_2CuO_4 and superconductivity. *Science*, **235**, 4793 (March 1987), 1196–1198.

Antonowicz, K. (1974). Possible superconductivity at room temperature. *Nature* (London), **247**, 5440 (February 1974), 358–360.

Barnett, R., Demler, E., and Kaxiras, E. (2005). Electron–phonon interaction in ultrasmall-radius carbon nanotubes. *Phys. Rev. B*, **71**, 3 (January 2005), 035429–035450.

Baumgartner, G., Carrard, M., Zuppiroli, L., Bacsa, W., De Heer, W. A., and Forro, L. (1997). Hall effect and magnetoresistance of carbon nanotube films. *Phys. Rev. B*, **55**, 11 (March 1997), 6704–6707.

Black-Schaffer, A. M., and Doniach, S. (2007). Resonating valence bonds and mean-field d-wave superconductivity in graphite. *Phys. Rev. B*, **75**, 13 (April 2007), 134512–134521.

Blonder, G. E., Tinkham, M., and Klapwijk, T. M. (1982). Transition from metallic to tunneling regimes in superconducting microconstrictions: Excess current, charge imbalance, and supercurrent conversion. *Phys. Rev. B*, **25**, 7 (April 1982), 4515–4532 (1982).

Bayot, V., Piraux, L., Michenaud, J.-P., and Issi, J. -P. (1989). Weak localization in pregraphitic carbon fibers. *Phys. Rev. B*, **40**, 6 (August 1989), 3514–3523.

Cervenka, J., Katsnelson, M. I., and Flipse, C. F. J. (2009). Room-temperature ferromagnetism in graphite driven by two-dimensional networks of point defects. *Nat. Phys.*, 11 (October 2009), 840–844.

Cespedes, O., Ferreira, M. S., Sanvito, S., Kociak, M., and Coey, J. M. D. (2004). Contact induced magnetism in carbon nanotubes. *J. Phys. Condens. Matter*, **16**, 10 (February 2004), L155–162.

Chauvet, O, Forro, L., Bacsa, W., Ugarte, D., Doudin, B and. de Heer, W. A. (1995). Magnetic anisotropies of aligned carbon nanotubes. *Phys. Rev. B*, **52**, 10 (September 1995), R6963–6966.

Chen, D. H and. Hsieh, C. H. (2002). Synthesis of nickel nanoparticles in aqueous cationic surfactant solutions. *J. Mater. Chem.*, **12**, 8 (June 2002), 2412–2415.

Coey, J. M. D., Venkatesan, M., Fitzgerald, C. B., Douvalis, A. P and. Sanders, I. S. (2002). Ferromagnetism of a graphite nodule from the Canyon Diablo meteorite. *Nature* (London), **420**, 6912 (November 2002), 156–159.

Cui, S. M., and Tsai, C. H. (1991). Plasmon theory of high-T_c superconductivity. *Phys. Rev. B*, **44**, 22 (December 1991), 12500–12510.

Da Silva, R. R., Torres, J. H. S., and Kopelevich, Y. (2001). Indication of superconductivity at 35 K in graphite-sulfur composites. *Phys. Rev. Lett.*, **87**, 14 (September 2001), 147001–147003.

De Martino, A., and Egger, R. (2004). Effective low-energy theory of superconductivity in carbon nanotube ropes. *Phys. Rev. B*, **70**, 1 (July 2004), 014508–014517.

De Pablo, P. J., Graugnard, E., Walsh, B., Andres, R. P., Datta, S., and Reifenbergera, R. (1999). A simple, reliable technique for making electrical contact to multiwalled carbon nanotubes. *Appl. Phys. Lett.*, **74**, 2 (January 1999), 323–325.

Esquinazi, P., Spemann, D., Hohne, R., Setzer, A., Han, K. H., and Butz, T. (2003). Induced magnetic ordering by proton irradiation in graphite. *Phys. Rev. Lett.*, **91** (November 2003), 227201–227204.

Frank, S., Poncharal, P., Wang, Z. L., and de Heer, W. A. (1998). Carbon nanotube quantum resistors. *Science*, **280**, 5370 (June 1998), 1744–1746.

Ginzburg, V. L. (1982). *High-Temperature Superconductivity* (Plenum Publishing Corporation, New York.) ISBN-0306109700, New York.

Gong, W., Li, H., Zhao, Z., and Chen, J. (1991). Ultrafine particles of Fe, Co, and Ni ferromagnetic metals. *J. Appl. Phys.*, **69**, 8 (April 1991), 5119–5121.

Klug H. P., and Alexander, L. E. (1974). *X-Ray Diffraction Procedures for Polycrystalline and Amorphous Materials*, 2nd Edition (Wiley Interscience, New York, 1974) p. 689.

Kajiura, H., Nandyala, A., Coskun, U. C., Bezryadin, A., Shiraishib, M., and Ata, M. (2005). Electronic mean free path in as-produced and purified single-wall carbon nanotubes. *Appl. Phys. Lett.*, **86**, 12 (March 2005), 122106–122108.

Kong, J., Yenilmez, E., Tombler, T. W., Kim, W., Dai, H.-J., Laughlin, R. B., Liu, L., Jayanthi, C. S., and Wu, S. Y. (2001). Quantum interference and ballistic transmission in nanotube electron waveguides. *Phys. Rev. Lett.*, **87**, 10 (March 2001), 106801–106804.

Kopelevich, Y., Esquinazi, P., Torres, J. H. S., and Moehlecke, S. (2000). Ferromagnetic- and superconducting-like behavior of graphite. *J. Low Temp. Phys.*, **119**, 5 (April 2000), 691–702.

Kotosonov, A. S., and Kuvshinnikov, S. V. (1997). Diamagnetism of some quasi-two-dimensional graphites and multiwall carbon nanotubes. *Phys. Lett. A*, **229**, 5 (June 1997), 377–380.

Langer, J. S., and Ambegaokar, V. (1967). Intrinsic Resistive Transition in narrow superconducting channels. *Phys. Rev.*, **164**, 2 (Dec. 1967), 498–510.

Lazzeri, M., Piscanec, S., Mauri, F., Ferrari, A. C., and Robertson, J. (2005). Electron transport and hot phonons in carbon nanotubes. *Phys. Rev. Lett.*, **95**, 23 (Nov. 2005), 236802–236805.

Lazzeri, M., Piscanec, S. , Mauri, F. , Ferrari, A. C., and Robertson, J. (2006). Phonon linewidths and electron–phonon coupling in graphite and nanotubes. *Phys. Rev. B*, **73** (April 2006), 155426–155431.

Lebedev, S. G. (2004). Particle irradiation for verification of superconducting-like behavior in carbon arc films. *Nucl. Instr. Meth.*, **A521**, 1 (March 2004), 22–29.

Lee, Y. C., and Mendoza, B. S. (1989). Possible high-T_c superconductivity in thin wires. *Phys. Rev. B*, **39**, 7 (March 1989), 4776–4779.

Lee, P. A., Nagaosa, N., and Wen, X.-G. (2006). Doping a Mott insulator: Physics of high-temperature superconductivity. *Rev. Mod. Phys.*, **78**, 1 (January 2006), 17–85.

Little,W. A. (1964). Possibility of Synthesizing an Organic Superconductor. *Phys. Rev.*, **164**, 6A (June 1964), A1416–1424.

Loss, D., and Martin, T. (1994). Wentzel-Bardeen singularity and phase diagram for interacting electrons coupled to acoustic phonons in one dimension. *Phys. Rev. B*, **50**, 16 (October 1994), 12160–12163.

Lu, J. P. (1995). Novel magnetic properties of carbon nanotubes. *Phys. Rev. Lett.*, **74**, 7 (February 1995), 1123–1126.

Lyakutti, K., Bodapati, A., Peng, X.-H., Keblinski, P., and Nayak, S. K. (2006). Electronic band structure, electron-phonon interaction, and superconductivity of (5,5), (10,10), and (5,0) carbon nanotubes. *Phys. Rev. B*, **73**, 3 (January 2006), 035413–035417.

Mann, D., Javey, A., Kong, J.,Wang, Q. Dai, H.-J. (2003). Ballistic Transport in Metallic Nanotubes with Reliable Pd Ohmic Contacts. *Nano Lett.*, **3**, 11 (October 2003), 1541–1544.

Meng, Z. Y., Lang, T. C., Wessel, S., Assaad, F. F., and Muramatsu, A. (2010). Quantum spin liquid emerging in two-dimensional correlated Dirac fermions. *Nature* (London), **464**, 7290 (April 2010), 847–851.

Mintmire, J.W., and White, C. T. (1998). Universal Density of States for Carbon Nanotubes. *Phys. Rev. Lett.*, **81**, 12 (September 1998), 2506–2509.

Moehlecke, S., Kopelevich, Y., and Maple, M. B. (2004). Interaction between superconducting and ferromagnetic order parameters in graphite-sulfur composites. *Phys. Rev. B*, **69**, 13 (April 2004), 134519–134523.

Mombru, A. W., Pardo, H., Faccio, R., de Lima, O. F., Leite, E. R., Zanelatto, G., Lanfredi, A. J. C., Cardoso, C. A., and Araujo-Moreira, F. M. (2005). Multilevel ferromagnetic behavior of room-temperature bulk magnetic graphite. *Phys. Rev. B*, **71**, 10 (March 2005), 100404–100407(R).

Novoselov, K. S., Geim, A. K., Morozov, S. V., Jiang, D., Katsnelson, M. I., Grigorieva, I. V., Dubonos, S. V., and Firsov, A. A. (2005). Two-dimensional gas of massless Dirac fermions in graphene. *Nature* (London), **438**, 7065 (November 2005), 197–200.

Park, C.-H., Giustino, F., Cohen, M. L., and Louie, S. G. (2008). Electron–phonon interactions in graphene, bilayer graphene, and graphite. *Nano Lett.*, **8**, 12 (November 2008), 4229–4233.

Poncharal, P., Berger, C., Yi, Y., Wang, Z. L., and de Heer, W. A. (2002). Room temperature ballistic conduction in carbon nanotubes. *J. Phys. Chem. B*, 106, 47 (November 2002), 12104–12118.

Qian, D., Dickeya, E. C., Andrews, R., and Rantell, T. (2000). Load transfer and deformation mechanisms in carbon nanotube-polystyrene composites. *Appl. Phys. Lett.*, **76**, 20 (May 2000), 2868–2870.

Reznik, D., Olk, C. H., Neumann, D. A., and Copley, J. R. D. (1995). X-ray powder diffraction from carbon nanotubes and nanoparticles. *Phys. Rev. B*, **52**, 1 (July 1995), 116–124.

Schönenberger, C., Bachtold, A., Strunk, C., Salvetat, J.-P., and Forro, L. (1999). Interference and interaction in multi-wall carbon nanotubes. *Appl. Phys. A*, **69**, 3 (August 1999), 283–295.

Seri, T., and Ando, T. (1997). Boltzmann Conductivity of a Carbon Nanotube in Magnetic Fields. *J. Phys. Soc. Jap.*, **66**, 169–173.

Sugawara, K., Sato, T., Souma, S., Takahashi, T., and Suematsu, H. (2007). Anomalous quasi-particle lifetime and strong electron–phonon coupling in graphite. *Phys. Rev. Lett.*, **98**, 3 (January 2007), 036801–036804.

Suzuura, H., and Ando, T. (2002). Phonons and electron-phonon scattering in carbon nanotubes. *Phys. Rev. B*, **65**, 23 (May 2002), 235412–235426.

Tang, Z. K., Zhang, L. Y., Wang, N., Zhang, X. X., Wen, G. H., Li, G. D., Wang, J. N., Chan, C. T., and Sheng, P. (2001). Superconductivity in 4 Angstrom Single-Walled Carbon Nanotubes. *Science*, **292**, 5526 (June 2001), 2462–2465.

Urbina, A., Echeverra, I., Perez-Garrido, A., Daz-Sanchez, A., and Abellan, J. (2003). Quantum conductance steps in solutions of multiwalled carbon nanotubes. *Phys. Rev. Lett.*, **90**, 10 (March 2003), 106603–106606.

Wallace, P. R. (1947). The band theory of graphite. *Phys. Rev.*, **71**, 9 (May 1947), 622–634.

Wang, J., Beeli, P., Ren, Y., and Zhao, G. M. (2010). Giant magnetic moment enhancement of nickel nanoparticles embedded in multiwalled carbon nanotubes. *Phys. Rev. B*, **82**, 19 (November 2010), 193410–193413.

Zaikin, A. D., Golubev, D. S., van Otterlo, A., and Zimanyi, G. T. (1997). Quantum Phase Slips and Transport in Ultrathin Superconducting Wires. *Phys. Rev. Lett.*, **78**, 8 (February 1997), 1552–1555.

Zhang, Y.-B., Tan, Y.-W., Stormer, H. L., and Kim, P. (2005). Experimental observation of the quantum Hall effect and Berry's phase in graphene. *Nature* (London) **438**, 7065 (November 2005), 201–204.

Zhao, G. M., and Wang, Y. S. (2001). Possible superconductivity above 400 K in carbon-based multiwall nanotubes. arXiv:cond-mat/0111268v2.

Zhao, G. M. (2004). Arguments for quasi-one-dimensional room-temperature superconductivity in carbon nanotubes, In: *Molecular Nanowires and Other Quantum Objects*, Alexandrov, A. S.; Demsar, J., and Yanson, I. K. (Ed.), 95-106, Nato Science Series, Kluwer Academic Publishers, ISBN-978-1-4020-2068-1, The Netherlands.

Zhao, G. M. (2005). Quantitative test of a quantum theory for the resistive transition in a superconducting single-walled carbon nanotube bundle. *Phys. Rev. B*, **71**, 11 (March 2005), 113404–113407.

Zhao, G. M., and Beeli, P. (2008). Observation of an ultrahigh-temperature ferromagnetic-like transition in iron-contaminated multiwalled carbon nanotube mats. *Phys. Rev. B*, **77**, 24 (June 2008), 245433–245438.

Chapter 11

Carbon Nanotube-Based Biosensors

Liang Su, Wenzhao Jia, and Yu Lei
Department of Chemical, Materials and Biomolecular Engineering, University of Connecticut, 191 Auditorium Road, Storrs, CT 06269, USA
liang.su@engr.uconn.edu, ylei@engr.uconn.edu

Carbon nanotubes (CNTs) have been widely employed as the key components in various biosensors due to their unique properties, such as very large surface to volume ratio, high sensitivity to their surface adsorbates, and good compatibility with most chemical and biological compounds. In this chapter, the electrocatalytic properties of CNTs followed by the functionalization of CNTs and the construction of CNT-modified electrodes will be discussed. Furthermore, we will focus on recent advances in CNT-based electrochemical biosensors.

11.1 Introduction

Carbon nanotubes (CNTs) have been captivating researchers in a great variety of scientific fields since the tremendously impacted *Nature* paper about "helical microtubules of graphitic carbon" published by Dr. Sumio Iijima in 1991 [1]. Interestingly, the first possible observation of multi-wall carbon nanotubes (MWCNTs)

Carbon Nanotubes and Their Applications
Edited by Qing Zhang

ISBN 978-981-4241-90-8 (Hardcover), 978-981-4303-18-7 (eBook)
www.panstanford.com

was documented in 1952 followed by a huge amount of recurrent investigations into the nanometer-scale carbon filaments before the well-known "discovery" of MWCNTs in 1991 [2].

Nevertheless, it was undoubtedly true that the occurrence of CNTs in 1991 exceedingly advanced the perception of nanometer-scale materials and extensively stimulated the research interest in CNTs. In 1992, Ebbesen and Ajayan developed the method for mass production of CNTs by arc-discharge technique [3], which effectively narrowed the gap between demand and supply. In 1993, single-wall carbon nanotubes (SWCNTs) were first synthesized in the vapor phase by the catalysis of iron [4] or cobalt [5], which convincingly enriched the conception of CNTs. Since then, the exploration of CNTs reached at an unprecedented level and quickly superseded fullerene as the hottest subject of carbon materials **(Figure 11.1)**.

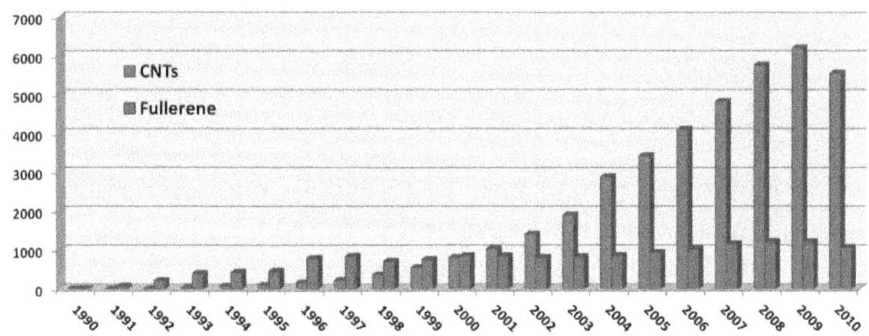

Figure 11.1 The comparison between the number of publications (research articles and review articles) about CNTs and fullerene from 1990 to 2010. See also Color Insert.

In this chapter, the electrocatalytic properties of CNTs will be discussed followed by the functionalization of CNTs and the construction of CNT-modified electrodes. Furthermore, we will focus on recent advances in CNT-based electrochemical biosensors.

11.2 Electrochemical Properties of CNTs

11.2.1 Structures and Electronic Properties of CNTs

SWCNTs can be visualized as a graphene sheet rolling up into a seamless, cylindrical shape with a diameter of 1–2 nm [4]. Hamada et al. developed a schematic notation to illustrate the helicity of

CNTs [6]. As shown in **Figure 11.2A,** the orientation of the wrapping axis can be specified by the roll-up vector (n,m), resulting in three types of CNTs: zigzag ($m = 0$), armchair ($m = n$), and chiral ($m \neq n \neq 0$) [7]. Moreover, when $n - m = 3l$ (l is an integer), CNTs are expected to be metallic. Otherwise ($n - m \neq 3l$), they are predicted to be semiconducting [8]. As shown in **Figure 11.2B,** MWCNTs, actually, consist of multiple concentric hollow cylinders with an interlayer distance of about 0.34 nm [1]. CNTs can also be classified as a division of fullerene in the form of tube, the growth of which is initiated (nucleation) from a hemi-fullerene dome (e.g., C_{80}) [9] and could be terminated (closure) at another hemi-fullerene dome [10]. Therefore, CNTs can be described as a long cylinder made up of the hexagonal honeycomb lattice of carbon, bound by two pieces of hemi-fullerenes at the ends [7].

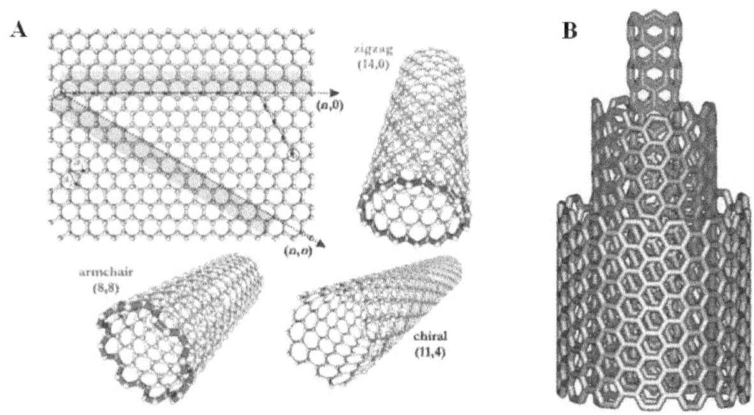

Figure 11.2 (A) Roll-up of a graphene sheet leading to Figure 11.2 three different types of CNT. (B) Structure of a MWCNT made up of three shells of differing chirality. Reprinted with permission from K. Balasubramanian, M. Burghard. *Small* 2005, 1, 180. Copyright 2005 Wiley-VCH. See also Color Insert.

In graphene, carbon atoms connect to each other through typical sp^2 hybrid orbitals in a trigonal plane with a symmetric π-electron conjugation. However, in the case of CNTs, the electron cloud is distorted due to the curvature of the tube. The σ–π rehybridization in CNTs results in more wave function localizing outside the tube [11], which makes CNTs more electrochemically active [12]. Furthermore, in contrast to graphene, the carbon atoms in CNTs are nonplanar (pyramidalized) and some π-orbitals are misaligned. The pyramidalization of carbon atoms and the misalignment of π-orbitals induce local strain widespread on CNTs [13]. The intrinsic high reactivity of CNTs can also be attributed to the tendency to relieve this strain [14]. In addition, since the curvature, the pyramidalization, and the misalignment inversely change with the scale of the tubes, an increased reactivity is expected with decreasing the diameter of CNTs [13].

11.2.2 Electrochemical Properties of CNTs

CNTs have been successfully applied in the detection of many analytically important chemical or biochemical compounds, such as hydrogen peroxide (H_2O_2) [15] and β-nicotinamide adenine dinucleotide (NADH) [16]. The outstanding electrocatalytic ability of CNTs are reflected by the enhanced electron transfer (decreased overpotential), the amplified current response (high sensitivity and low detection limit), and the improved anti-fouling effect (stable response). However, intact CNTs with perfect tubular structures can be hardly applied in the electrocatalysis. In fact, different kinds of "defects" on CNTs are responsible for their remarkable electrochemical properties.

11.2.2.1 Edge-plane defects

Before discussing the electrocatalytic activity of CNTs, let's have a look at another widely used carbon material, highly ordered pyrolytic graphite (HOPG). As shown in **Figure 11.3A,** HOPG consists of a series of graphene sheets parallel to each other with an interlayer distance of 0.335 nm [17]. According to the position of different planes, HOPG has been applied in electrochemistry in forms of basal plane pyrolytic graphite (BPPG) electrode and edge plane pyrolytic graphite (EPPG) electrode. Compton et al. calculated the standard electron transfer constants for BPPG and EPPG and predicted that the electron transfer on the edge plane was much faster than that on the basal plane [18]. Meanwhile, the same group also designed an inventive experiment to provide the convincing evidence that edge-plane sites were electrochemically active, while basal-plane sites were effectively inert [19].

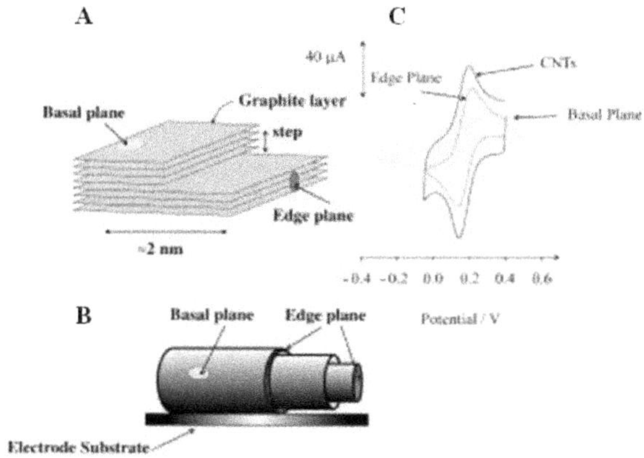

Figure 11.3 (A) Schematic representation of a crystal of highly ordered pyrolytic graphite where the layers of graphite have an interlayer spacing of 3.35 Å. (B) A single MWCNT on an electrode surface where the edge-plane-like sites are shown at the end of the tube and along the tube axis. (C) The difference in the voltammetric response for the reduction of $K_3Fe(CN)_6$ in an aqueous solution using basal-plane and edge-plane pyrolytic graphite electrodes. Note the identical response for the CNT-modified electrode compared with the edge-plane pyrolytic graphite electrode. Reprinted with permission from C. E. Banks, A. Crossley, C. Salter, S. J. Wilkins, R. G. Compton. *Angew. Chem. Int. Ed.* 2006, 45, 2533. Copyright 2006 Wiley-VCH.

Similar to HOPG, CNTs also possess edge-plane sites on the end and basal-plane site on the wall **(Figure 11.3B)**, the electrochemical properties of which are expected to be distinct according to the discussion above. Compton et al. have shown that CNT-modified or EPPG-modified electrode behaved identically toward the electron transfer of $Fe(CN)_6^{3-/4-}$ **(Figure 11.3C)** [20]. The analogous situations were obtained in response to epinephrine [21] and ascorbic acid [22] with the same electrodes. In order to directly compare the activity of the open end with the side wall, an imaginative strategy was developed by electrodepositing MoO_2 on MWCNTs [23]. Since MoO_2 could be selectively coated on electrochemically active sites, the resultant MoO_2 nanoplugs only on the ends of CNTs conclusively indicated the higher electrochemical activity of these regions **(Figure 11.4A)**. Furthermore, Gooding et al. provided first-hand evidence to demonstrate the more favorable electrochemical property of open ends than that of side walls. As shown in **Figure 11.4B**, the potential difference between anodic and cathodic peaks (ΔE_p) for aligned CNTs (72 mV) was smaller than that for randomly dispersed CNTs (105 mV), suggesting a better electron transfer kinetics with vertically aligned CNTs (more edge sites) than that with randomly dispersed CNTs (more basal sites) [24]. In the case of randomly dispersed CNTs, bamboo-like CNTs with more edge-plane sites on the surface exhibited a more favorable activity for the surface absorbed electroactive compounds (guanine and adenine) than regular SWCNTs [25].

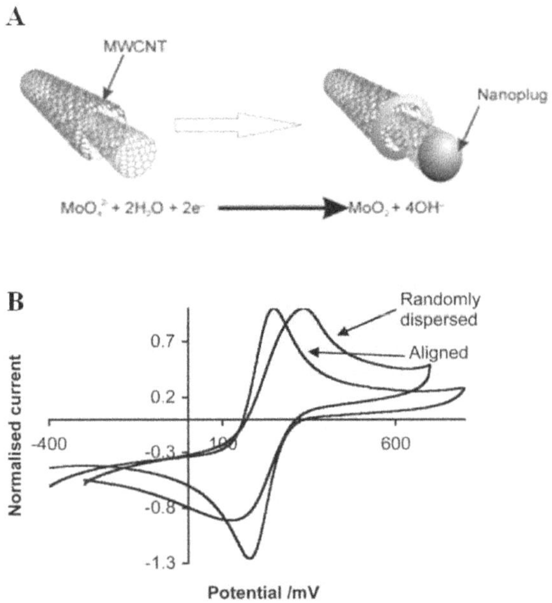

Figure 11.4 (A) A schematic representation of the structure of MWCNTs, showing the electrochemical deposition of MoO_2 at edge-plane-like defect sites along the walls of the MWCNTs and at the ends of the tubes. Reprinted with permission from K. Jurkschat, S. J. Wilkins, C. J. Salter, H. C. Leventis, G. G. Wildgoose, L. Jiang, T. G. J. Jones, A. Crossley, R. G. Compton. *Small* 2006, 2, 95. Copyright 2006 Wiley-VCH. (B) Cyclic voltammograms of SWCNT modified electrodes where the tubes were cut for 2 hours and either randomly dispersed or vertically aligned. CVs are recorded relative to an Ag/AgCl reference electrode in 1 mM $K_3Fe(CN)_6$ in a background of 0.05 M KH_2PO_4 and 0.05 M KCl at pH 7.0 at 100 mV s^{-1}. The y-axis shows the current normalized scale relative to the anodic peak current. Reprinted with permission from A. Chou, T. Bocking, N. K. Singh and J. J. Gooding. *Chem. Commun.* 2005, 842. Copyright 2005 The Royal Society of Chemistry.

11.2.2.2 Oxygenated species and dopant defects

The existence of oxygen-containing functional groups on the surface of carbon electrodes, including carbonyls, phenolic OH, lactones, ethers, and carboxylates, have been under extensive study [26]. Among all these oxygenated species, carboxylic group has been widely used as a surface-modified functional group on CNTs. Gooding et al. reported the unambiguous evidence for the improved electron transfer rate of $Fe(CN)_6^{3-/4-}$ by CNTs, which could be ascribed to the increased oxygen-containing defects substantiated by X-ray photoelectron spectroscopy (XPS) [24]. In addition, oxygenated species have been found very useful in the covalent modification of CNTs, which will be discussed in next section. However, it is difficult to compare the weight of the effect of oxygenated species with that of edge-plane sites [27].

In contrast to the oxygen-containing functional groups "protruding" from the surface of CNTs, doped CNTs are obtained by replacing carbon atoms in the hexagonal network by other elements [28]. Due to their similar radii to that of carbon, boron (one electron less than carbon) and nitrogen (one electron more than carbon) atoms have reasonably better opportunity to enter the lattice of CNT through the atomically substitutional doping [29]. Therefore, the doped CNTs tend to serve as an electron donor (for B) or acceptor (for N), which could greatly facilitate electrochemical reactions [30]. For example, Dai et al. demonstrated the superior electrocatalytic ability of nitrogen-doped CNTs toward the oxygen reduction reaction (ORR) in alkaline solution [31]. Furthermore, in the context of biosensors, both boron-doped CNTs (BCNTs) and nitrogen-doped CNTs (NCNTs) have shown to be advantageous to undoped CNTs in the oxidation of NADH [32] and H_2O_2 [33], respectively.

11.2.2.3 Electrocatalysis of CNTs

CNTs have shown favorable electrochemical properties to promote the electron transfer and/or facilitate the redox reaction of various electroactive compounds. However, controversial issues with regard to the function of CNTs in the improved electrochemical reaction still exist [34]. For example, Compton et al. claimed that, in some cases, the electrocatalytic performances of cheaper carbon materials, such as EPPG [17] or even graphite powder [35], were comparable with that of CNTs. They also challenged the enhanced electrochemical reactions of H_2O_2 [36] and hydrazine [20] catalyzed by CNTs, which were almost taken for granted. Another ambiguity of CNTs lies in the roles of side walls, ends, and oxygenated species in the electrocatalysis [37]. In order to address this uncertainty, Dai et al. developed a series of ingenious experiments which revealed that different electrochemical probes ($K_3[Fe(CN)_6]$, NADH, cysteine, H_2O_2, oxygen, and ascorbic acid) showed diverse electrosensitivity to the ends, sidewalls, and oxygen-containing surface groups on CNTs [38]. These findings could facilitate the design and development of novel CNT-based electrodes for various task-oriented applications.

11.3 Functionalization of CNTs

As mentioned before, pristine CNTs have very limited electrocatalytic activity. The production of CNTs will also generate a lot of impurities, such as transition metal catalysts and amorphous carbonaceous species, which may have undesired effect on the electrocatalysis of CNTs. In addition, due to the strong van der Waals interactions, CNT usually entangles with each other, which makes them difficult to be dissolved in most solvents and further manipulated. Moreover, in the context of biosensing, favorable surface properties of CNTs are indispensable for the successful immobilization of biomolecules. However, intact CNTs are not readily accessible to the integration of biological compounds. In order to overcome aforementioned limitations, chemical functionalization of CNTs is the most widely used strategy in the application of CNT-based electrochemical biosensors. As a matter of fact, functionalization can be viewed as a bridge between physics and chemistry of CNTs. A number of significant review articles have been published to elaborately summarize the chemical modifications of CNTs [39–48]. In order to avoid redundancy, we will only concentrate on the functionalization of CNTs toward the application of electrochemical biosensors.

11.3.1 Oxidation Treatment

The oxidation-treatment of CNTs has been unanimously applied as the starting pointing in the fabrication of CNT-based electrochemical biosensors. The pioneering work by Smalley et al. reached a milestone in the application of CNTs [49]. They first revealed that the nanoscale impurities in CNTs could be effectively removed by refluxing the raw sample of SWCNTs in 2.6 M nitric acid for 45 hours and resuspending the precipitate in an alkaline solution (pH 10) with surfactant followed by another filtration procedure with a cross-flow filtration system. Subsequently, the purified but still tangled SWCNT ropes could be efficiently separated with each other and cut into 100–300 nm lengths by a prolonged sonication of the nitric acid–treated SWCNT rope material in a mixture of concentrated sulfuric and nitric acids (3:1 v/v, 98% and 70%, respectively) at 40°C for 24 hours. They also found that such purified and shortened SWCNTs could form stable colloidal aqueous solution with the assistance of anionic surfactants such as sodium dodecyl sulfate (SDS) or nonionic surfactants such as Triton X-100, which could greatly facilitate the further manipulation of SWCNTs.

Since then, a lot of research has been conducted to modify the existing oxidation methods. For example, in order to decrease the carbon loss, slow down the etching rate, and minimize the sidewall damage, a controllable oxidative cutting of SWCNTs was achieved by stirring raw SWCNTs in Piranha solutions (4:1 v/v, 96% H_2SO_4 and 30% H_2O_2, respectively) at room temperature [50]. In another work, a great population of carbonyl group and improved integrity of carbon skeleton were obtained by sequentially sonicating MWCNTs in 3 M nitric acid for 2 hours and 30% H_2O_2 for 2 hours [51].

Such oxidation-treatment of CNTs also resulted in the opening of caps in the ends as well as the formation of holes on the sidewalls with the concomitant introduction of carboxylate *in situ* [46]. The presence of these carboxylic groups can not only dramatically increase the electrocatalytic ability and strongly prevent the aggregation of CNTs, but also enable CNTs to be further modified by other (bio)molecules of interest through the covalent functionalization.

11.3.2 Covalent Functionalization

Taking advantage of the surface carboxylic group, CNTs can be functionalized by polymer, such as polyethlyeneimine, polyamidoamine, and poly-L-lysine, or biomolecules, such as protein, enzyme, and DNA, via the well-established carbodiimide chemistry

[48]. As shown in **Figure 11.5**, the amidation between the carboxylic group on the CNT and the amino group on the functional moiety can be activated by a carbodiimide at room temperature in the presence of N-hydroxybenzotriazole (HOBt), N-hydroxysuccinimide (NHS), or N-hydroxysulfosuccinimide (NHSS), serving as the additive to increase yields and decrease side reactions. The most commonly used carbodiimide includes dicyclohexylcarbodiimide (DCC), N,N′-diisopropylcarbodiimide (DIC), and 1-ethyl-3-(3-dimethylaminopropyl)carbodiimide (EDC, also EDAC).

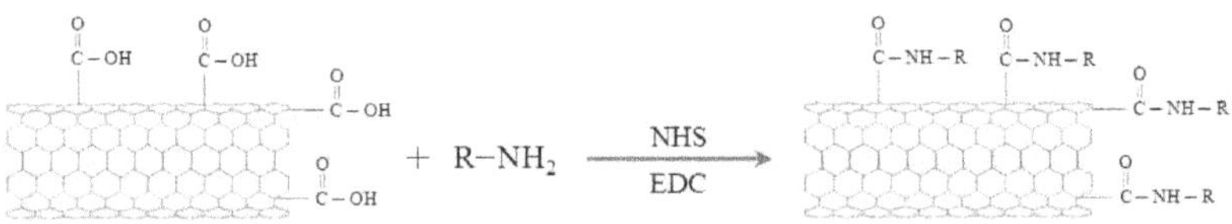

Figure 11.5 Amidation reaction of oxidized CNTs via carbodiimide chemistry.

According to this mechanism, Sun et al. reported the covalent attachment of bovine serum albumin (BSA) on sidewalls of both SWCNTs and MWCNTs and 90% of the total proteins still retained their activities in the resultant conjugates [52]. Lin et al. described the covalent immobilization of glucose oxidase (GOx) on the exposed end of acid-treated CNT nanoelectrode using EDC + NHSS [53]. Willner et al. reconstituted GOx holoenzyme by linking the apoenzyme to the covalently bound FAD on the end of SWCNTs [54]. Similarly, carbodiimide chemistry has also been applied to covalently binding other amine-terminated biomolecules, such as nucleic acid [55] and antibody [56], onto CNTs for the construction of ultrasensitive DNA- and immuno-sensors, respectively.

CNTs can also be covalently functionalized by different polymers via carbodiimide chemistry. Poly(ethyleneimine) (PEI) [57] and poly (vinyl alcohol) (PVA) [58] were grafted to CNTs via diimide-activated amidation reactions to dramatically enhance the solubility of CNTs. As a promising polymeric material in drug delivery, polyamideamine (PAMAM) dendrimer was successfully introduced to the surface of oxidized SWCNTs facilitated by DCC [42]. Poly-L-lysine (PLL), which could promote the cell adhesion, was also covalently bound to SWCNTs to provide a biocompatible matrix for further modification [59].

11.3.3 Non-Covalent Functionalization

The functionalization of CNTs can also be realized by the non-covalent attachment of various compounds in order to greatly improve the solubility of CNTs in aqueous system with the conservation of the integrity of structural and electronic properties of CNTs [60, 61]. In addition, it has been demonstrated the direct, non-covalent adsorption of a variety of metalloproteins and enzymes on CNTs with the preservation of enzyme activities, which provides a convenient route for the biofunctionalization of CNTs [62].

As a typical polynuclear aromatic compound, pyrene has been under intensive investigation in the non-covalent functionalization of CNTs via π–π stacking interaction [63] **(Figure 11.6)**. The application of various pyrene-based compounds, such as ionic pyrene–ammonium amphiphiles [64] and nonionic pyrene-β-cyclodextrin [65] has been demonstrated in the solubilization of CNTs. Recently, Ajayan et al. devised a non-covalent strategy as an alternative to the commonly used oxidative acid treatment to introduce carboxylic group on the surface of CNTs using pyrenecarboxylic acid [66]. Bertozzi et al. developed biocompatible SWCNTs non-covalently functionalized by pyrene–glycodendrimer derivatives, which could not only dramatically improve the solubility but also greatly mitigate the cytotoxicity of CNTs [67]. In another research, a water-soluble protein scaffold was synthesized by the non-covalent attachment of α-pyrene-ω-aminooxypolyethyleneglycol on the surface of SWCNTs [68]. Furthermore, the seminal work by Dai et al. provided a general and versatile approach to modify CNTs with biomolecules combining the advantages of both covalent and non-covalent interactions [69]. A bifunctional molecule, 1-pyrenebutanoic acid, succinimidyl ester, was non-covalently attached on the surface of CNTs via π–π stacking, followed by the covalent immobilization of biomolecules through carbodiimide chemistry. Furthermore, a proof-of-principle glucose biosensor was constructed by immobilizing glucose oxidase on the surface of an individual SWCNT with the aid of 1-pyrenebutanoic acid, succinimidyl ester [70].

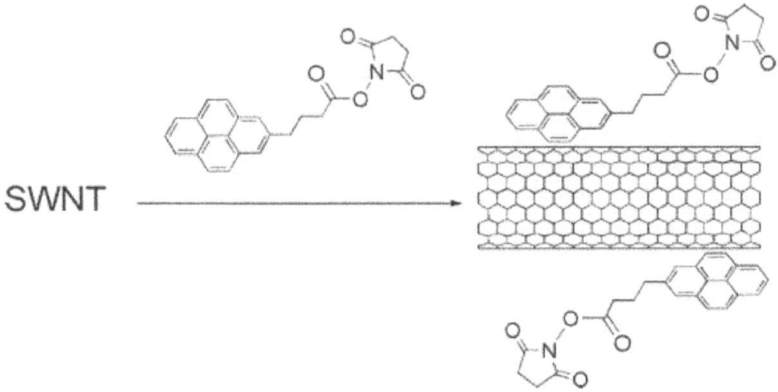

Figure 11.6 Interactions between CNTs and pyrene derivatives. Reprinted with permission from D. Tasis, N. Tagmatarchis, A. Bianco, M. Prato. *Chem. Rev.* 2006, 106, 1105. Copyright 2006 American Chemical Society.

The non-covalent modification of CNTs with amphiphilic molecules (surfactants) provides a non-destructive method to effectively solubilize CNTs in the solvent [71]. The hydrophobic moieties of surfactants are absorbed on the surface of CNTs via hydrophobic effect; the hydrophilic ends are involved in the interaction with water. Typical surfactants in the application of biosensors can be divided into two categories: charged surfactant such as sodium dodecyl sulfate (SDS) and cetyltrimethyl ammonium bromide (CTAB), and neutral surfactants such as Triton and Tween 20. Similarly, imidazolium-based (such as 1-butyl-3-methyl imidazolium) ionic liquids can also interact with CNTs via van der Waals force to facilitate their dispersion without the influence on the electronic property of CNTs [72].

The solubilization of CNTs can also be achieved under the influence of macromolecular wrapping agents without impairing physical properties of CNTs. In this regard, polymers have been widely used to facilitate the dispersion of CNTs toward the construction of CNT-based biosensors. Wang et al. reported the stable solution of both SWCNTs and MWCNTs in ethanol containing perfluorosulfonated polymer Nafion [15]. This finding opens up the opportunity for using CNT in a wide range of biosensors and nanoscale electronic devices. Another enzyme–CNT composite was explored as an active and stable surface material with high enzyme activity by dispersing α-chymotrypsin and SWCNTs in toluene in the presence of poly(methyl methacrylate) (PMMA) [73]. Moreover, in order to improve the specificity of the attachment of enzyme, Dai et al. reported the non-covalent modification of amine-terminated poly(ethylene glycol) (PEG) which could prevent the non-specific binding of protein, and Triton (surfactant) which could enhance the adsorption of PEG, on the surface of SWCNTs [74]. The specific binding of streptavidin was demonstrated in the presence of biotin covalently attached with the amino group on PEG. According to this idea, highly specific electronic biosensors for detecting clinically important biomolecules were advanced with specific receptors for the target proteins [75].

CNTs can also be effectively dispersed in aqueous system assisted by different biomacromolecules such as chitosan (CHIT) [76], starch [77], hyaluronic Acid [78], protein [79], peptide [80], and DNA [81]. The resultant CNT-based matrices, which can be further modified with other functional biomolecules, are very attractive in medical and biological applications due to their excellent biocompatibility. For example, Bertozzi et al. engineered CNTs by non-covalently attaching C_{18} lipid-terminated mucin (natural glycoprotein) mimic with α-N-acetylgalactosamine residues to realize the molecular recognition of CNTs [82]. The water-soluble, functionalized CNTs could be interfaced with cells via specific carbohydrate receptors without any toxicity to the cell [83].

11.4 Construction and Advantage of CNT-Modified Electrode for Biosensors

On the basis of the scientific studies of CNTs, we will proceed the discussion about the engineering of CNTs by integrating them with or converting them to an electrochemical transducer in the application of biosensors. According to its arrangement, CNT-based electrochemical transducer can be generalized into three categories: (i) non-oriented CNT electrode, (ii) oriented CNT array electrode, and (iii) individual CNT nanoelectrode.

11.4.1 Non-Oriented CNT Electrode

11.4.1.1 Drop-casting electrode

Drop-casting method is the most straightforward and versatile technique to modify an existing electrode such as glassy carbon electrode (GCE) [15], platinum electrode (PTE) [84], gold electrode (AUE) [85], indium tin oxide (ITO) [86], and screen printed carbon electrode (SPCE) [87]. Generally, acid-treated CNTs are first randomly dispersed in aqueous system or organic solvent such as N,N-dimethylformamide (DMF), and then casted on a working electrode. The integration of enzyme can be realized either by mixing enzyme with CNTs beforehand or by subsequently immobilizing enzyme on the CNT-coated surface via cross-linking, electropolymerization, or electrostatic adsorption. CNT–Nafion system [15] and CNT–CHIT [88] system are the most commonly used matrices in this category due to their excellent biocompatibility and large surface area for the enzyme loading. In addition, Nafion is also well known for its anti-fouling and discriminative properties[15].

Another advantage of the drop-casting method is reflected by its flexibility for the incorporation of other components, such as metal nanoparticles (NPs) [89], redox mediator [90], conducting polymer (CP) [91], and ionic liquid (IL) [92], in order to further improve the sensing performance. Moreover, under this regime, sandwich-like layer-by-layer (LBL) biosensors can be facilely constructed using self-assembled CNT–enzyme composites mediated by redox polymer such as poly[(vinylpyridine)Os(bipyridyl)$_2$Cl$^{2+/3+}$] (PVP-Os) [93] or polyelectrolyte such as poly(diallyldimethylammonium chloride) (PDDA)[94].

11.4.1.2 Composite paste electrode

Motivated by the commonly used carbon (graphite) pasted electrodes (CPE), CNTs are also used to prepare CNT paste electrode (CNTPE). SWCNTs or MWCNTs are first mixed with a binder in a certain ratio and the composite is then firmly packed into an electrode cavity. The advantages of CNTPE over CNT drop-casting electrode are the robust CNT packing and the readily renewable electrode surface.

The first CNTPE adopted bromoform as the binder and was successfully applied in the examination of the electrooxidation of dopamine [95]. Similarly, CNTPE was also constructed by mixing HNO_3-treated SWCNTs with mineral oil [96]. Compared to the one with conventional CPE, the two-electron oxidation of dopamine to dopaminedquinone exhibited an excellent reversibility in cyclic voltammetry using the resultant CNTPE. Wang et al. reported a MWCNTs–Teflon composite electrode with enhanced mechanical strength [97]. The advantageous electrocatalytic activity toward H_2O_2 and NADH along with minimal surface fouling made the MWCNTs–Teflon electrode extremely attractive in the construction of oxidase- or dehydrogenase-based electrochemical biosensors. Interestingly, a binderless CNT paste enzyme electrode was developed by direct mixing MWCNTs and glucose oxidase using a pestle in a mortar [87]. H_2O_2 can be reduced at −0.1 V (vs. Ag/AgCl), which allowed the highly selective detection of glucose without any interference from the coexisting ascorbic acid, acetaminophen, and uric acid. Recently, Musameh et al. designed a CNTPE with IL, *n*-octylpyridinum hexafluorophosphate (OPFP) as the binder [98]. Besides the favorable electrocatalytic ability toward H_2O_2 and NADH with high sensitivity and low overpotential, the major advantage of the resultant composite electrode lay in its extremely low capacitance and background current compared to other electrodes using CNTs or IL.

11.4.2 Oriented CNT Array Electrode

In contrast to the non-oriented CNT electrode, oriented CNT array electrode with well-organized spatial arrangement perfectly instantiates the structural advantage of CNTs characterized by high aspect ratio. Furthermore, since the open ends of CNTs are more electroactive than side walls **(Figure 11.4B)**, vertical aligned CNT electrode is expected to possess much higher electrocatalytic ability per unit area than randomly dispersed CNT electrode does. In this regard, Gooding et al. have demonstrated that the electron transfer between AUE and ferrocene via vertically aligned SWCNT was 40 times faster than that via randomly dispersed SWCNTs [99] **(Figure 11.7)**. Moreover, it has been shown that the electron transfer rate constant varied inversely with the average length of CNTs [54,99]. Generally, oriented CNT electrode can be fabricated by self-assembly or *in situ* growth.

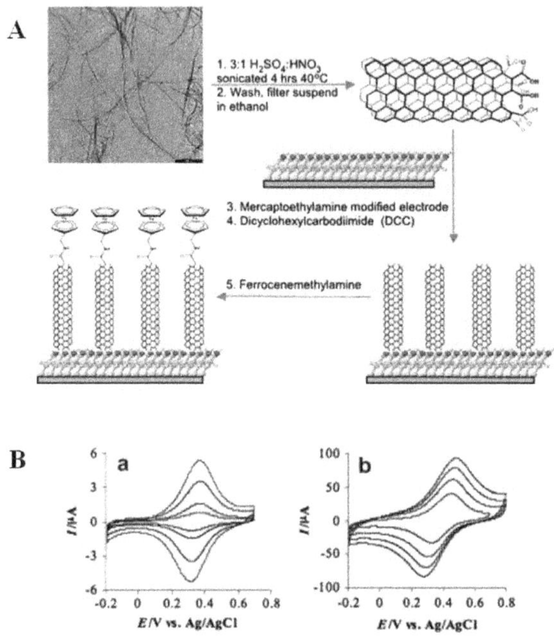

Figure 11.7 (A) Schematic representation of the steps involved in preparing a Ferrocene-vertically aligned SWCNT-Mercaptoethylamine-AUE. (B) Cyclic voltammograms of ferrocenemethylamine at (a) 6 h–cut vertically aligned SWCNT modified AUE at scan rates of 0.1, 0.2, 0.5 and 1 V s^{-1} and (b) 6 h–cut randomly dispersed SWCNT modified AUE at scan rates of 0.1, 0.2, 0.3 and 0.4 V s^{-1} in increasing peak current. CVs were obtained in 0.1 M $HClO_4$. Reprinted with permission from J. J. Gooding , A. Chou, J. Liu, D. Losic, J. G. Shapter, D. B. Hibbert. *Electrochem. Commun.* 2007, 9, 1677. Copyright 2007 Elsevier B.V.

11.4.2.1 Vertically aligned SWCNT nanowires produced by self-assembly

Sufficient chemical shortening process is indispensable in the fabrication of oriented SWCNTs by self-assembly, which not only introduces the carboxylic group to the open ends for the further treatment, but also cuts the long SWCNTs into shorter pieces so as to make them “stand” on the substrate [100]. As shown in **Figure 11.8**, pre-shortened SWCNTs can be subsequently attached on the surface of a substrate via a covalent bond or a non-covalent interaction to form an ordered orientation. In the first route, a monolayer of aminothiol (such as cysteamine) assembles on the surface of gold substrate through the formation of Au–S bond. The exposed amino group is then applied to linking carboxylic-terminated SWCNTs through the formation of the amide bond [101] . In the other approach, the surface of a substrate (such as glass, silicon wafer, etc.) is first functionalized by ferric hydroxide at elevated pH, followed by the attachment of oxidized SWCNTs via electrostatic interaction [102].

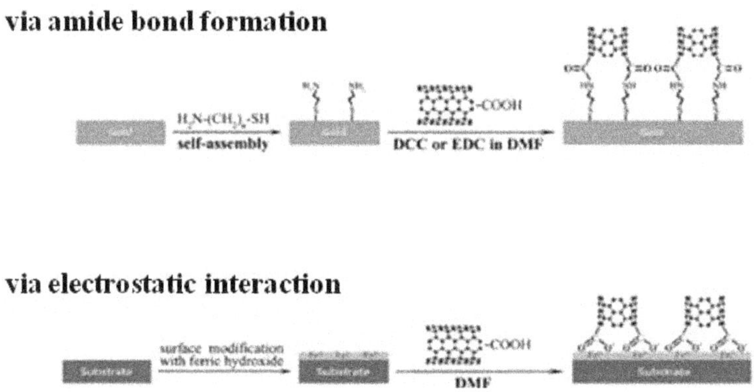

Figure 11.8 Schematic representation of the strategies to construct vertically aligned SWCNTs by self-assembly. Reprinted with permission from P. Diao, Z. Liu. *Adv. Mater.* 2010, 22, 1430. Copyright 2010 Wiley-VCH.

Self-assembled, oriented CNT electrode is well known for functioning as an efficient molecular wire to facilitate the electron transfer which is the fundamental mechanism for numerous biochemical reactions, such as respiration and photosynthesis. Many redox enzymes/proteins have been identified to be able to efficiently transport electron to their natural electron acceptor (e.g., oxygen). However, due to the shielding of redox active sites by glycoprotein shells, the pathway of electron transfer from unmodified electrode to redox active centers is usually blocked [60, 103, 104]. In order to realize direct electron transfer (DET) of redox enzymes/proteins on the electrode, a variety of strategies for effectively "wiring" the redox molecule have been developed [54, 105, 106]. Furthermore, with these sophisticated configurations, the direct electrochemistry has become a powerful tool to inquire into the kinetic and molecular aspects of the catalytic mechanism of redox enzymes [107].

The DET between redox molecules and CNTs has been under extensive study in order to develop miniaturized biosensors without electron mediator [54, 101, 108]. The direct electrochemistry of redox proteins facilitated by CNTs was first reported by Hill et al [109] in 1997. The groundbreaking work of Rusling et al. and Gooding et al. revealed that aligned SWCNT array behaved electrically to conduct electrons from the electrode surface to an iron protoporphyrin IX active center of myoglobin [108], horseradish peroxidase [108], or microperoxidase-11 [101] covalently attached on the end of SWCNTs. Furthermore, an inventive extension of attaching a biomolecule on

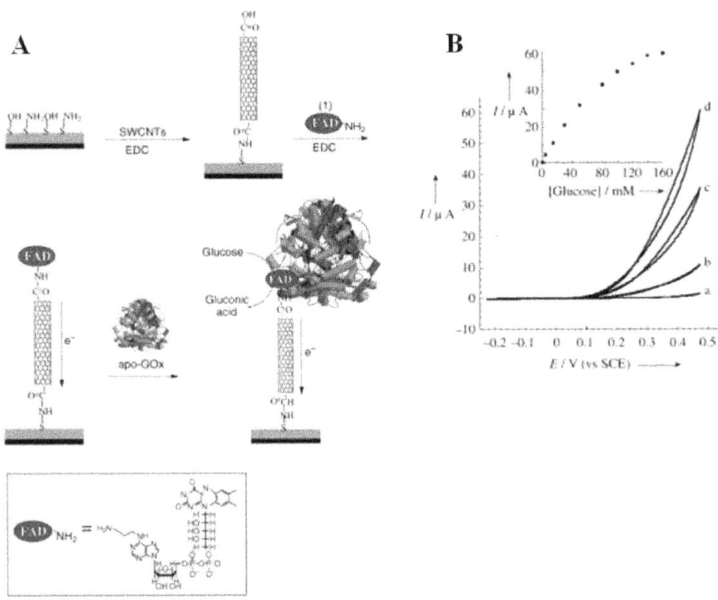

Figure 11.9 (A) Assembly of the SWCNT electrically contacted glucose oxidase electrode. (B) Cyclic voltammograms corresponding to the electrocatalyzed oxidation of different concentrations of glucose by the GOx reconstituted on the 25 nm long FAD-functionalized CNTs assembly: (a) 0 mM glucose. (b) 20 mM glucose, (c) 60 mM glucose, (d) 160 mM glucose. Data recorded in phosphate buffer, 0.1 M, pH 7.4, scan rate 5 $mV s^{-1}$. Inset shows the calibration curve corresponding to the amperometric responses of the reconstituted GOx/CNT (25 nm) electrode (at E = 0.45 V) in the presence of different concentrations of glucose. Reprinted with permission from F. Patolsky, Y. Weizmann, I. Willner. *Angew. Chem. Int. Ed.* 2004, 116, 2165. Copyright 2004 Wiley-VCH.

the end of SWCNT was to "plug" the SWCNT nanowire inside the biomolecule. The proof-of-principle experiment by Willner et al. demonstrated the aligned reconstitution of apo-enzyme plus redox active center by SWCNTs [54]. As shown in **Figure 11.9A**, instead of the active GOx, the prosthetic group was first coupled to the carboxylic group at the free edges of the vertical SWCNTs on the AUE. Apo–GOx was then reconstituted with the FAD–SWCNTs to give the holoenzyme, which still possessed good catalytic ability toward the oxidation of glucose **(Figure 11.9B)**. The electrons could be transported along a distance greater than 150 nm and the rate of electron transfer was controllable by the length of SWCNTs.

11.4.2.2 CNT Nanoelectrode ensembles produced by *in situ* growth

Well-aligned, uniformly distributed CNT array with controllable diameter and length of individual CNT can directly grow on the substrate by chemical vapor deposition [110]. Specifically, Ren et al. reported the fabrication of low-site density CNT nanoelectrode arrays [111] **(Figure 11.10)**. Briefly, Ni nanoparticles, serving as the starting sites for the growth of CNTs, were randomly electrodeposited on a Cr-coated silicon substrate. The size and the density of the Ni nanoparticles were regulated by the amplitude and the duration of the pulse current. The plasma-enhanced chemical vapor deposition of CNT was carried out by the gaseous mixture of acetylene and ammonia at 650°C. After that, a thin layer of SiO_2 was coated on the surface of CNTs as well as the silicon substrate followed by covering another layer of insulated epoxy resin between individual CNT electrodes to provide the mechanical support to CNTs. Finally, the resultant nanoelectrodes were polished to expose the tips of CNTs with a diameter of 100 nm and an average spacing of 10 μm. Lin et al. further investigated the electrochemical behavior of the nanoelectrode ensembles (NEEs) in this pattern [53]. The sigmoidal shape of the cyclic voltammogram for $Fe(CN)_6^{3-/4-}$, which was a typical behavior of the nanoelectrode, indicated that there was no overlap between the diffusion layers of individual electrodes, which was ascribed to the large spacing between CNTs compared with the small diameter of CNTs. In addition, extremely low background current was secured by the insulated epoxy layer on the top of the substrate.

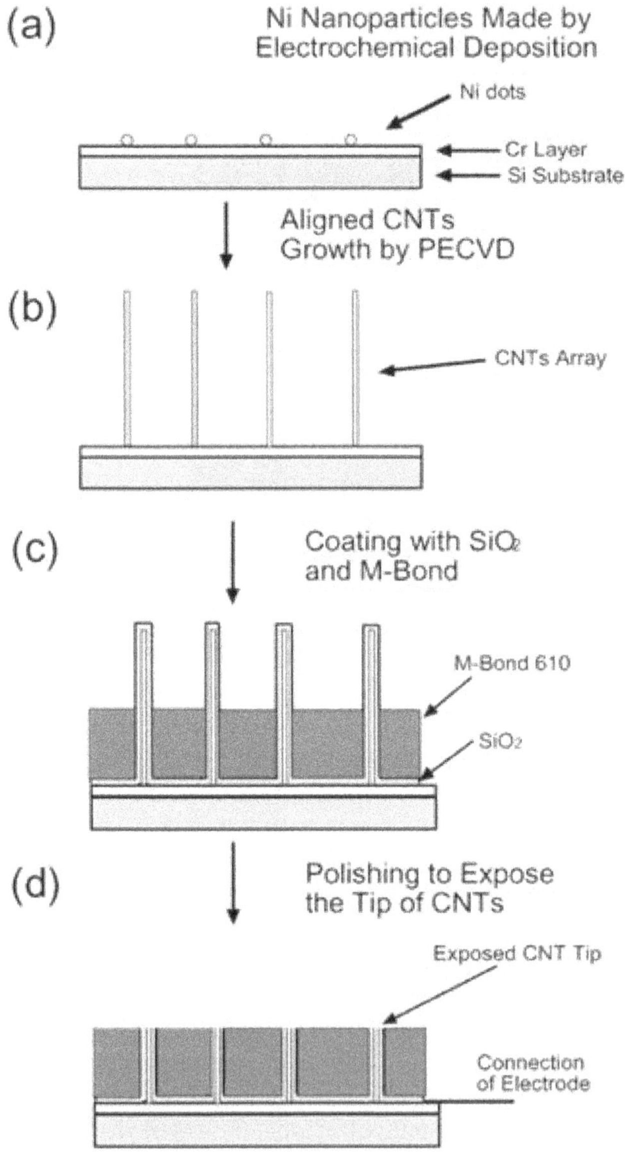

Figure 11.10 Fabrication scheme of the NEAs. (a) electrodeposition of Ni nanoparticles; (b) aligned carbon nanotube growth; (c) coating of SiO_2 and M-Bond (epoxy resin); and (d) polishing to expose CNTs. Reprinted with permission from Y. Tu, Y. Lin, Z. F. Ren. *Nano Lett.* 2003, 3, 101. Copyright 2003 American Chemical Society.

Self-assembled CNT arrays may suffer from the low conductivity due to the insulated connecting layer such as cysteamine [61]. Therefore, as shown above, vertically aligned CNTs directly grown from the surface of an electrode can effectively obviate this limitation attributed to the direct contact between CNTs and the substrate, which is highly preferred in electrochemical sensing applications. However, the tiny tip of CNT in nanoscale still cannot guarantee a reliable electrical contact between CNTs and the conductive substrate. Attempting to circumvent this problem, Fisher et al. developed a metallization technique for the effective wiring of CNTs produced by template-directed growth [112]. As shown in **Figure 11.11**, SWCNTs were directly synthesized from Fe by microwave plasma-enhanced chemical vapor deposition within the porous anodic alumina which could be produced by well-established method. Pd nanowire back contact for each SWCNT was electrodeposited from the underlying conductive layer (Ti) and extended to the initiation sites of SWCNTs within each pore, which ascertained the intimate contact between SWCNT and Ti substrate. Similarly, Pd nanoclusters also formed on SWCNTs extending out of the PAA surface during prolonged deposition. Two-terminal (bottom Ti-to-top Pd) electrical characteristics exhibited a linear *I*–*V* relationship, indicating the ohmic contact within the device.

Figure 11.11 Process flow and products for electrically contacted vertically oriented SWCNT arrays. Schematic of (a) film stack prior to anodization, (b) SWCNTs emerging from PAA pores after undergoing anodization and PECVD synthesis, and (c) electrodeposited Pd nanowires contacting the bottom of SWCNTs and Pd nanoclusters forming on SWCNTs located on the top PAA surface. Reprinted with permission from M. R. Maschmann, A. D. Franklin, A. Scott, D. B. Janes, T. D. Sands, T. S. Fisher. *Nano Lett.* 2006, 6, 2712. Copyright 2006 American Chemical Society.

11.5 Carbon Nanotube-Based Electrochemical Biosensors

Attributed to their unique structure and favorable electrochemical properties, CNTs have attracted considerable interest in the construction of biosensors with the current development of nanotechnology. Since the first paper reported by Santhanam et al. in 1996 [95], demonstrating the improved oxidative behavior of dopamine at CNT electrode, a great amount of CNT-based electrochemical biosensors have been developed toward the detection of hundreds of biological analytes. Various biological recognition elements, including antibodies, cofactors, DNA, enzymes, microorganisms, and proteins, have been used in the fabrication of CNT-based biosensors. Among all the biological elements, enzymes and redox proteins are recognized as the most widely used sensing elements in the construction of biosensors due to their excellent specificity and broad substrates of interest for clinical diagnostics and environmental monitoring [113].

In the past decade, many review articles have been published to summarize the state-of-the-art applications of CNTs in biosensors [30, 42, 103, 114, 113–124]. In this chapter, we will focus on the construction and application of CNT–enzyme and CNT–redox protein-based biosensors. Up to date, a variety of enzymes and redox proteins have been applied in the construction of CNT-based electrochemical biosensors, such as glucose oxidase (GOx), horse radish peroxidase (HRP), laccase, tyrosinase, ascorbate oxidase, choline oxidase (ChO), glucose dehydrogenase (GDH), alcohol dehydrogenase (ADH), lactate dehydrogenase (LDH), glutamat dehydrogenase (GLDH), organophosphorus hydrolase (OPH), acetylcholinesterase (AChE), hemoglobin (Hb), myoglobin (Mb), cytochrome c (Cyt c), and so on. Therefore, according to the nature of the biomolecule incorporated in the system CNT-based electrochemical biosensors can be categorized as follows: CNT–oxidase-based biosensors, CNT–dehydrogenase-based biosensors, CNT–hydrolase-based biosensors, CNT-bienzyme–based biosensors, and CNT–redox protein-based biosensors.

11.5.1 CNT–Oxidase-Based Biosensors

11.5.1.1 CNT–GOx-based biosensors

Glucose biosensors account for about 85% of the entire biosensors and have a market share over $5 billion because of the large amount of diabetic patients constituting about 5% of world's population [125]. Most glucose biosensors are based on the amperometric detection of glucose catalyzed by GOx. For more information about the fundamentals of the electrochemistry of glucose and the perspectives on the evolution of electrochemical glucose biosensors, please refer to the state-of-the-art review articles by Heller et al. [126] and Wang [127].

GOx catalyzes the oxidation of glucose to generate H_2O_2 in the presence of oxygen **(Eq. 11.1)**.

$$\text{Glucose} + O_2 \xrightarrow{\text{GOx}} \text{Gluconolactone} + H_2O_2 \qquad (11.1)$$

H_2O_2 can be further oxidized or reduced on the working electrode and the current response is correlated with the concentration of glucose.

CNTs possess the remarkable electrocatalytic activity towards H_2O_2 [15, 97, 98], and thus have been widely applied in the fabrication of GOx-based glucose biosensors. Wang et al. have demonstrated the classic design of GOx–CNT glucose sensors using CNT-Teflon composite electrode [97] and screen-printed CNT electrode[128]. The dramatically improved performances of the electrodes are shown in **Figure 11.12**. The pronounced current response obtained at such a favorable potential (0.1 V vs. Ag/AgCl) led to a high sensitivity toward glucose and good selectivity against common interference compounds (acetaminophen and uric acid). GOx was also immobilized on the MWCNT paste electrode with a satisfactory linear range and detection limit [129]. Furthermore, Lin et al. developed a CNT NEE-based biosensor with covalently attached GOx, which could quantify glucose up to 30 mM with a detection limit as low as 0.08 mM and excellent selectivity toward ascorbic acid, uric acid, and acetaminophen [53]. This work exemplified the prototype of the miniaturized biosensing probe.

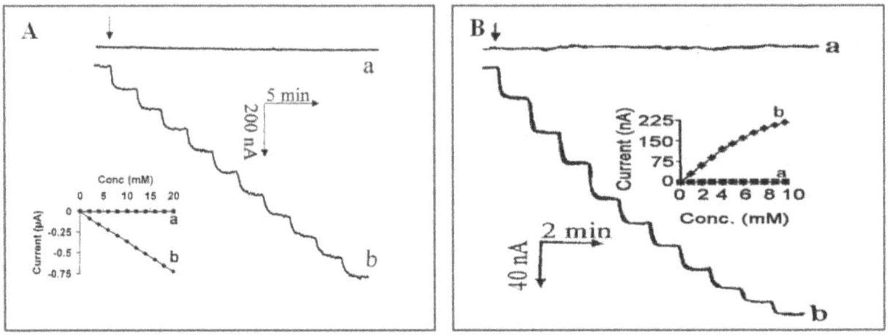

Figure 11.12 (A) Current–time recordings for successive additions of 2 mM glucose at the graphite-Teflon-GOx (a) and the MWCNT-Teflon-GOx (b) electrodes measured at +0.1 V (vs. Ag/AgCl). Reprinted with permission from J. Wang, M. Musameh. *Analytical Chemistry* 2003, 75, 2075. Copyright 2003 American Chemical Society. (B) Current–time recordings obtained upon the successive additions of 1 mM glucose at the graphite-GOx (a) and CNT-GOx (b) SP electrodes measured at +0.1 V (vs. Ag/AgCl). Reprinted with permission from J. Wang, M. Musameh. *Analyst* 2004, 129, 1. Copyright 2004 The Royal Society of Chemistry.

Since then, different methods have been developed to modify the traditional GOx–CNTs system. Lin et al. described a flow injection amperometric glucose biosensor based on alternatively assembled polyelectrolyte, GOx, and CNTs (PDDA–GOx–PDDA–CNT) [94]. The PDDA-GOx-PDDA sandwich-like structure provided a favorable microenvironment to keep the activity and to prevent the leakage of GOx without the compromise of the excellent electrocatalytic properties of CNTs in the inner layer. Operated at −0.1 V (vs. Ag/AgCl), this glucose biosensor exhibited a wide linear range of 15 μM to 6 mM with a detection limit of 7 μM. Schmidtke et al. developed an innovative method to construct a LBL glucose biosensor using redox polymer [93]. Sodium cholate suspension-dialysis was first applied to produce high-quality dispersion of GOx–SWCNTs composite. Then multilayer films of (PVP-Os)–GOx–SWCNTs were created by repeated, alternating exposure to the redox polymer (PVP-Os) and enzyme (GOx–SWCNTs) solutions. A very high sensitivity of 56 $\mu A \cdot mM^{-1} \cdot cm^{-2}$ was achieved ascribed to the combined effect of the LBL composite on electrocatalysis, biocompatibility, and enzyme loading. Compton et al. demonstrated the application of room temperature ionic liquid composite in the CNT-based glucose biosensor [98]. The resultant

GOx–OPFP–MWCNTs composite electrode for the detection of glucose was tested at 0.3 V (vs. SCE) and displayed good sensitivity (2 μM/mM) with the linearity up to 6 mM (without Nafion) or 12 mM (with Nafion).

CNTs have also been decorated by metal nanoparticles (NPs) to increase the catalytic performance of both components. Recently, an LBL assembly composed of MWCNTs, Au NPs, and GOx was designed for the specific detection of glucose and exhibited an excellent performance with a wide linear range (0.1–10 mM), good sensitivity (2.53 μA/mM), and a low detection limit (6.7 μM) [84]. Moreover, taking advantage of the synergistic effect of MWCNTs, Au NPs, IL, and CHIT, Lee et al. fabricated an amperometric glucose biosensor integrating these four components on an ITO electrode [130]. The association of Au NPs in the composite electrode greatly improved the sensitivity of the biosensor and negligible interference from uric acid and ascorbic acid was observed. Luong et al. reported the modification of Nafion dissolved SWCNTs with Pt NPs (diameter 2–3 nm) synthesized by wet chemistry [89]. The SWCNT--Pt NPs–GOx composite was deposited on the surface of GCE for the detection of glucose and displayed linearity up to 5 mM, sensitivity of 2.11 μA/mM, and a detection limit of 0.5 μM, which is much lower than that of SWCNTs–GOx–GCE (1 mM) and Pt NPs–GOx–GCE (400 μM). In another research, Xu et al. reported an amperometric glucose biosensor based on Pt NPs electrodeposited on MWCNTs and GOx immobilized in biocompatible CHIT–SiO_2 sol-gel [131]. A wide linear range of 1 μM to 23 mM, a short response time within 5 s, and good sensitivity as high as 58.9 $\mu A \cdot mM^{-1} \cdot cm^{-2}$ were achieved. Besides Au and Pt, Pd was also applied in the construction of glucose biosensor [132]. Electrodeposition was carried out to co-deposit GOx and Pd NPs onto Nafion-solubilized MWCNTs on GCE. The Pd/MWCNT/GOx greatly enhanced the storage stability and selectivity of the biosensor with linearity up to 12 mM and a detection limit of 0.15 mM.

Conducting polymers (CPs) are well known for their capability to increase active surface area and provide good electrical contact between the sensing materials and the electrode [133]. With regard to the application of CPs in CNT-based glucose biosensors, Wang et al. first described the facile preparation route of amperometric enzyme electrode on the basis of the incorporation of MWCNTs and GOx into an electropolymerized polypyrrole (PPy) film [91]. CNTs, herein, serving as the dopant for maintaining the electrical neutrality during the growth of PPy, retained its electrocatalytic activity and imparted good performance to the glucose biosensor with linearity up to 50 mM and a detection limit of 0.2 mM. Another glucose biosensor was fabricated by loading GOx into polyaniline (PANi)–MWCNTs and Nafion–silica nanocomposite [134]. Cyclic voltammograms of $Fe(CN)_6^{3-/4-}$ with the as-prepared nanocomposite-modified ITO electrode and a variety of other control electrodes revealed the improved electrocatalytic activity of PANi–MWCNTs. In addition, it has been demonstrated that the highly porous, biocompatible Nafion–silica not only provided a large surface area for the enzyme loading but also maintained 93% of enzyme activity over 20 days. In a recent study, 3,4-ethylenedioxythiophene (EDOT) was electrodeposited on the MWCNT/baked Prussian blue film to effectively entrap GOx [135]. The resultant PEDOT–GOx-baked PB–MWCNTs SPCE showed a linear response to glucose from 1 mM to 10 mM (2.67 $\mu A \cdot mM^{-1} \cdot cm^{-2}$) in a flow injection analysis mode at –0.1 V (vs. Ag/AgCl) with highly resolved and reproducible signals (R.S.D.=2.54%).

Because of their excellent mechanical strength, CNTs can also be processed to form CNT-based network which can also be applied in the fabrication of biosensors. Xu et al. reported the electrospinning of poly(acrylonitrile-*co*-acrylic acid) (PANCAA) filled with MWCNTs and covalent binding of GOx through the activation of carboxyl groups on the surface of PANCAA [136]. The nanofibrous membrane was directly deposited on PTE for the amperometric detection of glucose. The water-insoluble polymer (PANCAA) enabled the biosensor to be reused. Furthermore, due to the intrinsic conductivity of CNTs, this type of network can also be used as a flexible, free-standing electrode in the sensing application [137]. A homogeneous SWCNT film was first produced by filtering SWCNT solution through an anodic aluminum oxide membrane and then transfer onto a transparent poly(ethylene terephthalate) support after dissolving Al_2O_3. The glucose biosensor was realized by encapsulating GOx into the SWCNT film with the aid of Nafion. A linear range of 0.25–3 mM glucose with a detection limit of 97 μM was observed.

Last but not least, Fisher et al. developed a Au/Pd nanocube–SWCNT NEE glucose biosensor [138]. GOx was covalently immobilized on Au nanocubes. As mentioned in previous section, the Pd layer provided a seamless connect between SWCNTs and the conductive substrate, as well as a low-resistance contact between SWCNTs and the Au interface. Meanwhile, Au nanocubes offered the electrode surface ideal enzyme docking ports with excellent biocompatibility. Furthermore, the nanoelectrode environment produced a favorable mass transfer pathway for the access of glucose and the diffusion of H_2O_2. The resultant biosensor exhibited a wide linear range spanning from 10 μM to 50 mM with a detection limit of 1.3 μM.

GOx has also been extensively investigated in DET-based CNT biosensors. As a typical flavoprotein, holo-GOx consists of an apo-GOx and a FAD prosthetic group. The direct electrochemistry of GOx, in fact, is the DET between FAD active center and the electrode surface facilitated by CNTs. The redox reactions of FAD with its natural electron acceptor, oxygen, are shown in the following equations:

$$\mathrm{FAD} + 2e^- + 2\mathrm{H}^+ \leftrightarrow \mathrm{FADH_2} \quad (11.2) \quad \rightarrow \mathrm{DET}$$

$$\mathrm{FADH_2} + \mathrm{O_2} \rightarrow \mathrm{FAD} + \mathrm{H_2O_2} \quad (11.3)$$

Since the electron turnover rate of FAD to electrode in the presence of CNTs is much higher than that of FAD to oxygen, the backward reaction of Eq. 11.2 on the electrode is favored and a well-defined redox peak of FAD can be observed. In a glucose biosensor, with the addition of the analyte, glucose is converted to gluconolactone through following reaction.

$$\mathrm{Glucose} + \mathrm{FAD} \leftrightarrow \mathrm{Gluconolactone} + \mathrm{FADH_2} \quad (11.4)$$

Therefore, the increased concentration of glucose can be correlated with the decreased reduction peak of FAD.

Guiseppi-Elie et al. first demonstrated the DET of GOx on SWCNTs [139]. GOx was immobilized on the annealed SWCNT paper and displayed quasi-reversible one-electron transfer process. Herein, SWCNTs were able to "pierce" the glycoprotein shell of GOx and gain access to FAD, which was unapproachable on the smooth surface of traditional electrodes. Afterward, many attempts have been made to achieve DET of GOx by CNTs. For example, the DET of GOx was observed by PSS–MWCNT/Au/IL-modified GCE [140]. A linear range up to 20 mM glucose with a detection limit of 25 μM was obtained. In another research, LBL assembly of GOx/PDDA/SWCNT-modified GCE also displayed a pair of well-behaved redox peaks of FAD and a wide linear range of 1–40 mM for the detection of glucose.

Recently, the DET of GOx based on boron-doped CNTs (BCNTs) was reported [141]. As shown in **Figure 11.13**, the enhanced redox peaks (already taking the larger background current into account) of BCNTs was due to increased number of defective sites on BCNTs, which was beneficial to enzyme loading and DET between FAD and electrode. This DET-based glucose sensor had a linear range of 0.05–0.3 mM and a limit of detection of 0.01 mM. The BCNT-modified electrode exhibited good selectivity against the interference from uric acid and ascorbic acid. A similar strategy has been applied to accentuate the DET of FAD using nitrogen-doped CNTs [142]. The resultant glucose biosensor showed a wider linear range (0.02–1.02 mM) with the same detection limit (0.01 mM). These promising results demonstrated the applicability of BCNTs and NCNTs in the fabrication of DET-based, mediator-free, third generation of glucose biosensors [127].

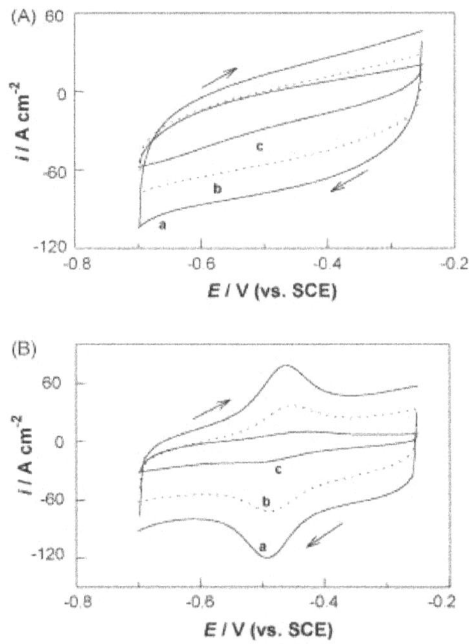

Figure 11.13 Cyclic voltammograms of different electrodes in 0.1 M deoxygenated phosphate buffer solution (pH 6.98) at the scan rate of 30 mV s^{-1}. (A) BCNTs/GCE (a), CNTs/GCE (b), and bare GCE (c); and (B) GOD/BCNTs/GCE (a), GOD/CNT/GCE (b), and GOD/GCE (c). Reprinted with permission from C. Deng, J. Chen, X. Chen, C. Xiao, L. Nie, S. Yao. *Biosens Bioelectron* 2008, 23, 1272. Copyright 2008 Elsevier B.V.

11.5.1.2 CNT–HRP-based biosensors

H_2O_2 is an electroactive compound generated as a by-product of oxidative metabolism in organisms. The detection of H_2O_2 by electrochemical method is of great importance and has been extensively applied in enzymatic sensors, immunosensors, and DNA sensors.

HRP, which contains a heme cofactor in the active site, can catalyze the reduction of H_2O_2 according to the following equations [143]:

$$\text{HRP(red)} + H_2O_2 \rightarrow \text{HRP(ox)} + H_2O \quad (11.5)$$

$$\text{HRP(ox)} + e^- \leftrightarrow \text{HRP(red)} \quad (11.6) \quad \rightarrow \text{DET}$$

The incorporation of CNTs with HRP can greatly bridge the electron transfer from the active site to the electrode. DET recreates the active center of HRP enabling H_2O_2 to be detected at -0.2 to 0 V, under which most interference would be effectively eliminated. Rusling et al. first reported the covalent attachment of HRP onto the ends of vertically oriented SWCNT forest arrays [108]. Quasi-reversible redox peaks of Fe^{III}/Fe^{II} at -0.25 V (vs. SCE) indicated excellent electron transfer of HRP conducted by "trees" in the SWCNT forest. Working at 0 V, the resultant biosensor demonstrated a sensitivity of 0.049 μA/μM and a detection limit of 50 nM towards H_2O_2. Xie et al. developed a H_2O_2 biosensor on the basis of the immobilization of HRP in a nanocomposite film of tetrathiafulvalene–tetracyanouinodimethane (TTF–TCNQ) and MWCNT modified gold electrode [144]. MWCNTs, along with TTF–TCNQ serving as the electron mediator, greatly promoted the reduction of H_2O_2 at 0 V (vs. SCE). A linear range from 0.005 to 1.05 mM and a detection limit of 500 nM for the amperometric detection of glucose was obtained. Recently, an ultrasensitive H_2O_2 biosensor was constructed using HRP covalent attached to the LBL-assembled, non-oriented MWCNT–PDDA-modified pyrolytic graphite electrode [145]. Under the optimized condition, this amperometric H_2O_2 biosensor gave a remarkable detection limit of 1.5 nM.

Peroxidase has also been widely used in the detection and remediation of a wide spectrum of aromatic pollutants [146]. The principle of the detection of phenolic compounds by HRP is shown in Eqs. 11.7 to 11.9.

$$\text{HRP(red)} + H_2O_2 \rightarrow \text{Compound(I)} + H_2O \quad (11.7)$$

$$\text{Compound(I)} + AH_2 \leftrightarrow \text{Compound(II)} + AH\cdot \quad (11.8)$$

$$\text{Compound(II)} + AH_2 \leftrightarrow \text{HRP(red)} + AH\cdot \quad (11.9)$$

In contrast to DET (Eq. 11.6), herein, the active site of HRP is restored by an external electron donor (AH_2, which denotes a phenolic compound), which is so-called mediated electron transfer [147]. The oxidized substrate (AH·) is highly electroactive and can be electrochemically reduced back to their initial form. Therefore, in a typical HRP-based phenolic compounds biosensor, certain amount of H_2O_2 exists in the working solution to initiate the oxidation reaction of phenolic compounds. Under this regime, a highly sensitive biosensor for the detection of phenolic compounds was developed [148]. LBL-assembled poly(allylamine hydrochloride) MWCNT/HRP was employed on the AUE. MWCNTs provided a suitable microenvironment to retain the activity of HRP and acted as a transducer to amplify the electrochemical signal. A linear response from 0.1 μM to 20.4 μM catechol with a detection limit of 0.06 μM was presented. This biosensor exhibited different analytical performance towards different phenolic compounds. It could be concluded that the sensitivities of different substituent phenolic compounds followed the trend: $NH_2 > OH > CH_3 > Cl > NO_2$, which was in good agreement with the electron donating ability of the substituent (Eqs. 11.8 and 11.9). A similar HRP/MWCNT was also constructed with the addition of methylene blue serving as the redox mediator [149].

11.5.1.3 CNT–tyrosinase-based biosensors

CNTs are also combined with tyrosinase for the biosensing of phenolic compounds which belong to United States Environmental Protection Agency priority pollutants [150]due to their high toxicity. Tyrosinase (also known as phenol oxidase, diphenol oxidase, polyphenol oxidase, etc.), which is a well-studied multicopper oxygenase, can efficiently catalyze the two-electron oxidation of phenolic compounds through the formation of *o*-diphenols and *o*-quinones [151–153], both of which are highly electroactive, enabling phenolic compounds to be facilely determined on the electrode. Tyrosinase was immobilized on SWCNT-modified GCE and showed good sensitivities toward phenol (155 μA/mM), catechol (355 μA/mM), *m*-cresol (267 μA/mM), and *p*-chlorophenol (343 μA/mM) at −0.1 V (vs. SCE) [154]. Another enzyme composite electrode based on tyrosinase/MWCNT/epoxy/graphite power was also described to measure catechol and displayed a detection range from 0.01 mM to 0.15 mM with a sensitivity of 294 $\mu A\cdot mM^{-1}\cdot cm^{-2}$ at an applied potential of −0.2 V (vs. Ag/AgCl) [155].

11.5.1.4 CNT–laccase-based biosensors

Laccase, a multicopper oxidase, is well-characterized in the catalysis of diphenol via a one-electron oxidation pathway [151–153]. Similar to the case with other oxidases, the highly electroactive intermediates can be easily oxidized on the surface of electrode, especially with the aid of CNTs. For example, several catechol biosensors have been developed on the basis of laccase–MWCNT system in acidic solutions (pH 3–4) [156–158] and the best detection limit as low as 20 nM catechol was reported [158]. Recently, a hydroquinone biosensor was described by casting the mixture of laccase, dopamine, and MWCNTs on the surface of GCE where laccase could induce the polymerization of dopamine to form a novel matrix [159]. The resultant laccase–polydopamine–MWCNTs–GCE composite electrode exhibited a linear range of 0.1–48 μM hydroquinone with sensitivity of 643 $\mu A\cdot mM^{-1}\cdot cm^{-2}$ and a detection limit of 20 nM at −0.05 V (vs. SCE).

Laccase is also well-known for their catalytic ability to oxygen reduction reaction (ORR) and has been widely used in enzymatic biofuel cells [160]. Dong et al. demonstrated the oxygen biosensor using laccase–CHIT–MWCNT-modified GCE with ABTS as the probe [156]. Working at 0.35 V (vs. Ag/AgCl), this oxygen biosensor could linearly quantify the dissolved oxygen up to 0.2 mM with sensitivity of 27.3 $\mu A\cdot mM^{-1}$ and a detection limit of 7.8 μM. In addition, laccase activity could be inhibited by cyanide, azide, and fluoride. Therefore, according to the decrease of the reduction current from ORR catalyzed by laccase, inhibition-type biosensors for these environmental pollutants could be constructed [157].

11.5.1.5 CNT–ascorbate oxidase-based biosensors

Ascorbate oxidase belongs to the class of multicopper oxidase and can catalyze the one-electron oxidation of ascorbic acid [151]. Mao et al. reported the integration of *in vivo* microdialysis sampling system with a SWCNT-modified GCE for the continuous and on-line monitoring of ascorbate depletion in the global ischemia induced rat striatum [161]. SWCNTs undergone thermal treatment was found to be able to enhance the electron transfer of ascorbate oxidation at a low working potential and to resist the surface fouling of the electrode. Operated at 0.03 V (vs. Ag/AgCl), this biosensor could

Table 11.1 Other CNT–oxidase-based biosensors

Enzyme	Target	Electrode	Working potential	Linear range (mM)	Sensitivity	Detection limit	Note	Reference
Alcohol oxidase	Ethanol	MWCNT/carbon film electrode	−0.3 V (vs. SCE)	up to 1.4	44.5 μA·mM^{-1}·cm^{-2}	86 μM	—	[162]
Amino acid oxidase	Phenylalanine	Pt NPs/MWCNT/GCE	0.7 V (vs. Ag/AgCl)	up to 0.05	25.4 ± 1.2 μA·mM^{-1}	0.25 μM	—	[163]
Cholesterol oxidase	Cholesterol	CHIT–IL/MWCNT(SH)–Au/ITO	−0.05 V (vs. Ag/AgCl)	0.5–5	0.2 μA·mM^{-1}	—	—	[92]
Cholesterol oxidase	Cholesterol	Pt NPs–MWCNTs–CHIT/AUE	0.1 V (vs. SCE)	0.01–3	3.82 μA·mM^{-1}·cm^{-2}	—	PSS (LBL)	[164]
Cholesterol oxidase	Cholesterol	MWCNT/AUE	0.7 V (vs. SCE)	up to 6	0.559 μA·mM^{-1}·cm^{-2}	200 μM	PDDA (LBL)	[165]
Choline oxidase	Choline	PANi/MWCNT/GCE	0.4 V (vs. SCE)	0.001–2	—	0.3 μM	—	[166]
Choline oxidase	Choline	N-doped CNT/GCE	0.3 V (vs. SCE)	0.1–2.9	5.4 μA·mM^{-1}	15 μM	Chronoamperometry	[33]

Choline oxidase + HRP	Choline	MWCNT/GCE	−0.2 V (vs. Ag/AgCl)	0.05–5	—	10 μM	FIA; PDDA (LBL)	[167]
Galactose oxidase	Galactose	CHIT–SWCNTs/GCE	−0.4 V (vs. Ag/AgCl)	up to 1	1.126 $\mu A \cdot mM^{-1}$	25 μM	FIA; ORR	[168]
Glucose oxidase + myrosinase	Glucosinolate	Au NPs/MWCNT/Teflon	0.5 V (vs. Ag/AgCl)	0.02–1	1.89 ± 0.04 $\mu A \cdot mM^{-1}$	5.9 μM	—	[169]
Glutamate oxidase	Glutamate	Pt NPs/MWCNT/GCE	0.7 V (vs. Ag/AgCl)	up to 0.1	29.9 ± 0.8 $\mu A \cdot mM^{-1}$	0.25 μM	—	[163]
Glutamate oxidase	Glutamate	Ppy/MWCNT/PTE	0.6 V (vs. Ag/AgCl)	up to 0.14	384 $\mu A \cdot mM^{-1} \cdot cm^{-2}$	0.3 μM	—	[170]
Lactate oxidase	Lactate	CHIT/PVI-Os/MWCNT/AUE	0.3 V (vs. Ag/AgCl)	up to 0.8	19.7 $\mu A \cdot mM^{-1} \cdot cm^{-2}$	5 μM	—	[171]
Lactate oxidase	Lactate	Pt NPs/MWCNT/GCE	0.7 V (vs. Ag/AgCl)	up to 0.1	30.1 ± 0.3 $\mu A \cdot mM^{-1}$	0.25 μM	—	[163]
Lactate oxidase	Lactate	MWCNT paste electrode	−0.1 V (vs. Ag/AgCl)	up to 7	14.4 ± 0.4 $\mu A \cdot M^{-1}$	300 μM	—	[172]
Pyranose oxidase + α-glucosidase	Maltose	CHIT–MWCNT/graphite rods	−0.7 V (vs. Ag/AgCl)	0.25–2	0.882 $\mu A \cdot mM^{-1} \cdot cm^{-2}$	—	ORR	[173]

linearly detect ascorbic acid in the range of 0.5–100 μM with a detection limit of 0.2 μM. The presence of ascorbate oxidase, which was supplied with artificial cerebrospinal fluid as the perfusion solution, would decrease the oxidation current on the electrode due to the consumption of ascorbic acid. Therefore, the presence of ascorbate oxidase could guarantee the elimination of other interference such as dopamine, uric acid, 5-hydroxytryptamine, and 3,4-dihydroxyphenylacetic acid.

11.5.1.6 Other CNT–oxidase-based biosensors

Similar to GOx, many other oxidases requiring a FAD prosthetic group, such as amino acid oxidase, cholesterol oxidase, choline oxidase, ethanol oxidase, galactose oxidase, glutamate oxidase, lactate oxidase, lysine oxidase, oxalate oxidase, and xanthine oxidase, can catalyze the oxidation of the corresponding substrate with the consumption of oxygen as well as the formation of H_2O_2 (Eq. 11.10).

$$\text{Substrate} + O_2 \xrightarrow{\text{oxidase}} \text{Product} + H_2O_2 \tag{11.10}$$

The concentration of the target analyte (substrate) can be correlated with either the consumption of oxygen or the production of H_2O_2. Owing to their favorable property for the immobilization of enzyme and their outstanding catalytic ability toward H_2O_2, CNTs have also been integrated with different kinds of FAD-based oxidases for various sensing purposes. Table 11.1 summarizes the recently published research articles in this area.

11.5.2 CNT–Dehydrogenase-Based Biosensors

Owing to the wide substrate range, dehydrogenases have been exploited as the recognition element for the specific determination of various analytes. In the category of these enzymes, NAD^+/NADH-dependent dehydrogenases are the most popular option in the fabrication of CNT-based biosensors due to the facile detection of NADH by electrochemical approach (Eqs. 11.11 and 11.12).

$$\text{Substrate} + NAD^+ + e^- \xrightarrow{\text{dehydrogenase}} \text{Product} + \text{NADH} \tag{11.11}$$

$$\text{NADH} \xrightarrow{\text{electrode}} NAD^+ + e^- \tag{11.12}$$

However, the high overpotential and the severe surface fouling are associated with the oxidation of NADH on conventional carbon electrodes, such as GCE and HOPG[113]. In this regard, CNTs emerge as an ideal electrocatalytic material for the anodic detection of NADH with considerable superiority ascribed to their excellent electrocatalytic and anti-fouling abilities. Wang et al. observed the substantially decreased overpotential and surface fouling in the oxidation of NADH using CNT-modified GCE at neutral pH [16]. The reaction started at −0.05 V (vs. Ag/AgCl) on the MWCNT/GCE, which was about 0.5 V lower than that on unmodified GCE **(Figure 11.14A)**. Meanwhile, a larger chronoamperometric response with a better stability was also obtained on the MWCNT/GCE **(Figure 11.14B)**. Compared to the ordinary CNTs, recently, BCNT-modified GCE was found to be able to further increase the current response and decrease the surface fouling at a lower working potential for the oxidation of NADH [32]. In addition, the improved electrocatalytic performances of CNT paste electrodes with different binders was also demonstrated toward the detection of NADH [96–98].

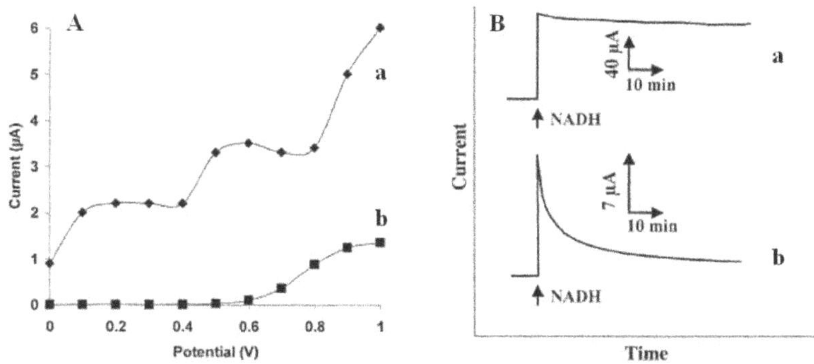

Figure 11.14 (A) Hydrodynamic voltammograms for 1×10^{-4} M NADH at unmodified (a) and MWCNT-coated (b) glassy carbon electrodes. (B) Stability of the response to 5×10^{-3} M NADH using unmodified (a) and MWCNT-modified (b) glassy carbon electrodes. Operating potential, +0.6 V (vs. Ag/AgCl); stirring rate, 500 rpm; electrolyte, phosphate buffer (0.05 M, pH 7.4). Reprinted with permission from M. Musame, J. Wang, A. Merkoci, Y. Lin. *Electrochem. Commun.* 2002, 4, 743. Copyright 2002 Elsevier B.V.

Many dehydrogenases, such as GDH, ADH, LDH, and GLDH, have been incorporated into CNT-based biosensing systems for the quantitative analysis of different targets on the basis of anodic detection of NADH.

11.5.2.1 CNT–GDH-based biosensors

Gorski et al. developed a CHIT–MWCNT biocompatible system and demonstrated its application in the electrochemical glucose sensing based on GDH, which was covalently immobilized on CHIT using glutaraldehyde [88]. In the pH 7.4 phosphate buffer solution with the addition of NAD^+, the GDH–CHIT–MWCNT-modified GCE could linearly quantitate glucose in the range of 5–300 μM with sensitivity of 80 $\mu A \cdot mM^{-1} \cdot cm^{-2}$ and a detection limit of 3 μM. Moreover, this GDH-based glucose biosensor displayed an impressive operational stability by recording the chronoamperometric response in a stirred solution containing 0.5 mM glucose and 0.1 mM NAD^+ at 0.4 V (vs. Ag/AgCl), which still kept constant over 24 hours. Compare to the GOx-based glucose biosensor, GDH-based glucose biosensor does not demand any oxygen, which, theoretically, can effectively eliminate the interference from the deviation of oxygen concentration in the solution, and thus provide a more accurate glucose measurement. However, in contrast to GOx, the cofactor of which (FAD) tightly binds within the apo-enzyme, GDH usually does not have a firmly bound cofactor (NAD^+); exogenous NAD^+ is required in the solution. Therefore, GDH cannot be easily incorporated into reagentless glucose sensor (no cofactor in the working solution). To overcome this limitation, Gorski et al. devised a novel GDH/MWCNT-based reagentless biosensor by covalently attaching GDH and NAD^+ on the polyamino-saccharide chains of CHIT with the aid of glutaraldehyde and EDC, respectively [174]. The addition of Nafion led to an enhanced selectivity and sensitivity. Under the optimized condition, this reagentless glucose biosensor exhibited sensitivity of 1.8 $\mu A \cdot mM^{-1} \cdot cm^{-2}$ in the linear range of 0.02–2 mM at 0.4 V (vs. Ag/AgCl). In another work, GDH/MWCNT-based reagentless biosensor was constructed by non-covalently binding NAD^+ on the surface of MWCNTs [175]. GDH was cross-linked on the surface of NAD^+-MWCNT-modified GCE by glutaraldehyde. The strong π–π stacking interaction between the adenine moiety in the NAD^+ and MWCNTs and the cross-linked GDH layer secured the stability of NAD^+ during the electrochemical experiment. This biosensor responded linearly to glucose in the range of 0.01–0.3 mM with a detection limit of 4.81 μM at 0.3 V (vs. Ag/AgCl). In another GDH-based reagentless biosensor, GDH, diaphorase, and NAD^+ were co-deposited on the surface of SWCNT paste electrode covered by osmium (Os) redox polymer [176]. $Os^{2+/3+}$, serving as an electron mediator, was reduced by converting NADH back to NAD^+ under the catalysis of diaphorase, and in turn, oxidized on the surface of the electrode. Working at 0.2 V (vs. Ag/AgCl), the proposed biosensor showed sensitivity of 13.4 $\mu A \cdot mM^{-1} \cdot cm^{-2}$ up to 0.3 mM with a detection limit of 10 μM, and has been successfully applied in the determination of glucose in two different sweet wines.

11.5.2.2 CNT–ADH-based biosensors

Wang et al. reported the prototype of ADH-based ethanol biosensor using CNT–Teflon composite electrode [97]. Another ethanol biosensor was designed by solubilizing MWCNTs and ADH in mesoporous titania–Nafion composite solution [177]. The sol-gel derived titania not only increased the pore size of Nafion but also improved the biocompatibility of the matrix, leading to a faster amperometric response as well as better enzyme stability of the biosensor. In a pH 7.0 buffer solution containing 3.0 mM NAD^+, the proposed biosensor displayed a linear range of 0.01–3 mM with sensitivity of 51.6 $\mu A \cdot mM^{-1} \cdot cm^{-2}$ at 0.5 V (vs. Ag/AgCl). Similarly, poly(vinyl alcohol) was applied to disperse MWCNTs and ADH towards the fabrication of ethanol biosensor [178]. Under the optimal condition, the current response changed linearly with the concentration of ethanol up to 1.5 mM with sensitivity and detection limit of 196 $nA \cdot mM^{-1}$ and 13 μM, respectively. In addition, the electron mediator can be applied to further decrease the working potential. Recently, SWCNTs were electrochemically modified by poly(brilliant cresyl blue) (PBCB) serving as the electron mediator in the application of ethanol biosensing [179]. ADH was first immobilized on carrageenan matrix and the enzymatic composite was then casted on the PBCB–SWCNT-modified GCE. Working at 0 V (vs. SCE) in the pH 7.5 buffer solution with 3.0 mM NAD^+, the biosensor could detect as low as 0.1 mM ethanol with a linear range of 0.4–2.4 mM.

11.5.2.3 CNT–GLDH-based biosensors

Glutamate is the major excitatory neurotransmitter in the central nervous system and has been used as an important indicator to various neurological disorders. Therefore, the sensitive and selective detection of glutamate is of great importance in neurological science. Glutamate is also a common additive in the food industry, which is widely used as a flavor enhancer. An amperometric glutamate biosensor was designed by casting GLDH on meldola's blue (MDB)–CHIT—MWCNT-modified GCE where MDB functioned as a redox mediator [180]. The composite electrode could efficiently facilitate the oxidation of NADH at –0.14 V (vs. Ag/AgCl) without any interference from ascorbic acid and uric acid. Glutamate was successfully detected by this biosensor at –0.1 V (vs. Ag/AgCl) in the presence of 4 mM NAD^+ with the detection limit of 2 μM. In an analogous research, glutamate biosensor was fabricated via the covalent immobilization of GLDH–BSA on thionine–SWCNT-modified GCE, where BSA served as the enzyme loading and binding matrix to stabilize GLDH, and thionine served as the redox mediator to further lower down the working potential [181]. The amperometric measurement was conducted in pH 8.3 phosphate buffer containing 2.5 mM NAD^+ at 0.19 V (vs. SHE). The excellent sensing performance was reflected by the wide linear range (0.5–400 μM), high sensitivity (137.3 $\mu A \cdot mM^{-1} \cdot cm^{-2}$), low detection limit (0.1 μM), rapid response (5 s), and impressive selectivity (against ascorbic acid, uric acid, and 4-acetaminophenol). As a comparison, a more complicated glutamate sensing system was developed with enhanced analytical parameters [182]. The positively charged, dendrimer encapsulated Pt nanoparticles (Pt-PAMAM) and negatively charged GLDH were alternately absorbed on MWCNTs by the electrostatic interaction, resulting in the multilayered $(GLDH–Pt–PAMAM)_n$–MWCNTs, which was then secured on the surface of GCE by the subsequently electropolymerized PPy. The proposed biosensor could quantify glutamate in the linear range of 0.1–60 μM with sensitivity of 51.48 $\mu A \cdot mM^{-1}$ and a detection limit of 10 nM at 0.2 V (vs. Ag/AgCl) in a physiological pH buffer solution containing 0.1 mM NAD^+. Furthermore, the biosensor exhibited outstanding selectivity with negligible interference from ascorbic acid, uric acid, glutathione, acetaminophen, L-cysteine at the same concentration as that of glutamate.

11.5.2.4 CNT–LDH-based biosensors

Lactate is an important analyte in clinical diagnosis, fermentation monitoring, and food quality control. A hybrid lactate biosensor was constructed by electrodepositing poly(3-methylthiophene) on the surface of GCE followed by the casting of MWCNTs and LDH [183]. Under the optimized condition, a linear range of 1–500 μM and a detection limit of 0.56 μM were obtained with good repeatability. The incorporation of Nafion could effectively alleviate the interference from ascorbic acid and uric acid without affecting the response of lactate. The applicability of the biosensor to the real sample was demonstrated by the determination of lactate in red wine and the results were consistent with the ones from HPLC. Electron mediator can also be integrated in the lactate sensing system. For example, meldola's blue (MDB) was mixed with SWCNTs in a biocompatible sol-gel silica matrix [184]. The resultant MDB–SWCNTs–SPE has shown dramatically increased electrocatalytic activity towards NADH at −0.05 V (vs. Ag/AgCl) compared to the electrode without mediator. With the addition of LDH, the composite biosensor could detect D-lactate as low as 16 μM with a linear range of 0.1–2.6 mM. Recently, Mao et al. developed a flow injection analysis system for on-line detecting physiological compounds, such as lactate and glucose, on the basis of SWCNTs and dehydrogenases [185]. SWCNTs were first modified by an electron mediator (methylene green), and casted on the split-disk plastic carbon film electrode; LDH or GDH was first dispersed in BSA solution and then cross-linked on the surface of working electrode by glutaraldehyde. The artificial cerebrospinal fluid containing NAD^+ was applied as the mobile phase. The dynamic linear range for lactate and glucose were both 0.05–1 mM with a detection limit of 1 μM and 0.5 μM, respectively , at 0 V (vs. Ag/AgCl). Moreover, the proposed system was successfully applied to continuously and simultaneously monitoring lactate and glucose in brain microdialysate of rat following global cerebral ischemia/reperfusion.

11.5.3 CNT–Hydrolase-Based Biosensors

11.5.3.1 CNT–OPH-based biosensors

CNT-based biosensors have also been exploited for environmental applications. Organophosphorus (OP) compounds which cause severe nerve and muscular disorders, have been widely applied in the production of pesticides, petroleum additives, and chemical warfare agents [186]. OPH hydrolyzes P–O bond or P–S bond of OP compounds to give an electroactive product such as 4-nitrophenol or a thiol moiety, respectively, which can further be oxidized or reduced on the electrode. Wang et al. reported an amperometric biosensor for OP pesticides on the basis of OPH, which hydrolyzed OP pesticides to produce 4-nitrophenol, and MWCNTs, which greatly improved the oxidation of 4-nitrophenol [187]. Under optimized conditions, the biosensor was able to detect as low as 0.15 μM paraoxon (P–O) with sensitivity of 25 nA/μM and 0.8 μM methyl parathion (P–S) with sensitivity of 6 nA/μM. Another OPH mutant hydrolyzing P–S bond was employed in the amperometric detection of V-type nerve agent [188]. The incorporation of MWCNTs dramatically increased the anodic current of enzymatically generated thiol compounds. Using Demeton-S as a nerve agent mimic, the developed biosensor exhibited a linear range up to 85 μM, sensitivity of 8 μA/mM, and a detection limit of 1 μM (258 ppb).

11.5.3.2 CNT–AChE-based biosensors

AChE, also known as acetylcholine hydrolase, has been applied in the fabrication of CNT-based biosensor to detect OP compounds based on the inhibition of AChE by the analyte. In this case, AChE catalyzed the hydrolysis of acetylthiocholine to produce thiocholine (Eq. 11.13), which could undergo the electrooxidation enhanced by MWCNTs at 0.2 V (vs. Ag/AgCl) [189].

$$\text{Acetylthiocholine} + H_2O \xrightarrow{\text{AChE}} \text{Thiocholine} + \text{Acetic acid} \quad (11.13)$$

The introduction of methyl parathion [190] or paraoxon [189] would irreversibly inhibit AChE activity, and thus dramatically decrease the current response. Therefore, detection limits of 50 nM for methyl parathion and 0.5 nM for paraoxon were obtained. It has been found that such naturally irreversible inhibition can become "reversible" with the aid of nucleophiles. As shown in **Figure 11.15A**, the OP-AChE can be reactivated by the nucleophilic substitution of AChE with 2-PAM. Based on this principle, Lin et al. described the construction of highly sensitive flow injection amperometric biosensor for OP compounds [191]. The entire sensing system is held in a compact case **(Figure 11.15B,C)** and the working electrode was a CNT-modified SCPE. The exposure to paraoxon with different concentrations for 30 min resulted in the linear decrease of AChE activity and the lowest detectable concentration of paraoxon was 0.1 nM **(Figure 11.15D)**.

Similarly, Other AChE–CNT-based biosensors for the flow injection analysis (FIA) of OP compounds were reported. In one research, GCE was modified by self-assembled PDDA–AChE–PDDA–MWCNT [192]. AChE was reactivated by pyridine 2-aldoxime methiodide (PAM). Under optimized conditions, the biosensor was able to distinguish 0.4 pM paraoxon with a 6-min inhibition time. In another work, GCE was casted by AChE–CHIT–MWCNTs [193]. AChE was reactivated by 2-pyridine aldoxime methiodide (2-PAM). Using Sulfotep as a model compound, the resultant biosensor gave a linear range of 1.5–80 nM with a detection limit of 1 nM at 10 % inhibition after a 14-min flow-stop incubation.

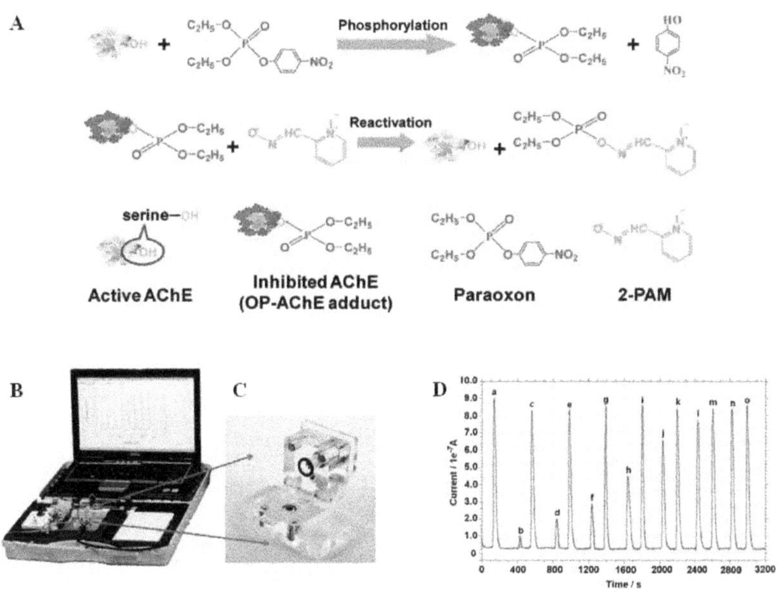

Figure 11.15 (A) Schematic illustration of OP-AChE Formation and the AChE Regeneration Process by the reactivator. (B) The entire analysis system and (C) the part of the sensing area. (D) Amperometric responses of 3 nM AChE in PBS containing 5 mM acetylthiocholine before paraoxon exposure (a) and after exposure to (b) 50, (d) 25, (f) 10, (h) 5, (j) 1, (l) 0.5, and (n) 0.1 nM paraoxon for 30 min and subsequent reactivation by 5 mM 2-PAM with incubation time of 15 min (peaks c, e, g, i, k, m, o). Reprinted with permission from D. Du, J. Wang, J. N. Smith, C. Timchalk, Y. Lin. *Anal. Chem.* 2009, 81, 9314. Copyright 2009 American Chemical Society.

11.5.4 CNT–Bienzyme System-Based Biosensors

11.5.4.1 CNT–GOx–HRP-based biosensors

As discussed before, GOx can oxidize glucose to produce H_2O_2, which can in turn be reduced by HRP to produce water. Therefore, a GOx + HRP bienzyme system can theoretically be applied in the glucose sensing. The key issues herein are the sufficient contact between the two enzymes to facilitate the diffusion of H_2O_2 from GOx to HRP and the electron transfer from HRP and electrode surface. In this regard, CNTs apparently can serve not only as docking sites to immobilize the enzymes adjacent to each other but also as the bridge to effectively promote the electrical communication between HRP and electrode surface. A mediator-free bienzyme amperometric glucose biosensor was developed on the basis on the mixture of GOx and HRP immobilized on MWCNT-modified GCE [194]. Operated at 0 V (vs. Ag/AgCl), this biosensor displayed a detection limit of 0.5 μM with sensitivity of 80.5 $\mu A \cdot mM^{-1} \cdot cm^{-2}$. In another research, GOx and HRP were entrapped on the surface of SWCNTs coated AUE by the electropolymerized PPy [195]. The linear range and sensitivity of the glucose biosensor were determined to be 0.03–2.43 mM and 7.01 $\mu A \cdot mM^{-1} \cdot cm^{-2}$, respectively. Prussian blue (PB) functions as an "artificial peroixidase" [196] and has been widely applied in H_2O_2 sensors [197]. Recently, an "bienzyme" glucose biosensor was constructed on the basis of GOx + PB on the matrix of MWCNT/PVP. The detection limit was determined as low as 2 μM at the working potential of 0.1 V (vs. Ag/AgCl) [85].

11.5.4.2 CNT–AChE–ChO-based biosensors

Another type of CNT–bienzyme system has also been developed by coupling AChE with ChO for the detection of OP compounds. In this bienzyme system, AChE hydrolyzed acetylcholine into acetate and choline (Eq. 11.14). ChO subsequently converted choline into H_2O_2 (Eq. 11.15), which could be easily detected on the MWCNT-modified electrode [190].

$$\text{Acetylcholine} + H_2O \xrightarrow{\text{AChE}} \text{Choline} + \text{Acetic acid} \qquad (11.14)$$

$$\text{Choline} + O_2 \xrightarrow{\text{Cho}} \text{Betaine aldehyde} + H_2O_2 \qquad (11.15)$$

11.5.5 CNT–Redox Protein-Based Biosensors

The DET of redox proteins–CNTs composites and their electrocatalytic abilities towards H_2O_2 have also been investigated. Ju et al. reported the DET of Hb immobilized in a grafted collagen–MWCNT matrix, where a pair of redox peaks centered at -0.32 and -0.40 mV (vs. SCE) and a surface-controlled electrode process were displayed [198]. Due to the biocompatibility and the porous structure leading to the high enzyme loading in the matrix, the Hb-grafted collagen–MWCNT biosensor exhibited a fast detection of H_2O_2 down to 130 nM and good operational stability over a month. Another H_2O_2 biosensor on the basis of the DET of Mb was described [199]. The fast DET of Mb and electrocatalysis of H_2O_2 was promoted by silver-doped CNT. A formal potential of -0.295 V (vs. Ag/AgCl) with the peak-to-peak separation of 0.16 V for the heme center and a linear range of 0.002–1.2 mM with a detection of 360 nM for the detection of H_2O_2 were obtained. The DET of Cyt c and its catalysis of H_2O_2 were realized using Au NPs–IL–MWCNT composite film [200]. A formal potential and peak-to-peak separation were determined to be -0.129 V (vs. Ag/AgCl) and 0.036 V, respectively. The response increased linearly to the H_2O_2 concentration in the range of 0.05–1.15 mM operating at -0.2 V.

References

1. Iijima, S. (1991). Helical microtubules of graphitic carbon, *Nature*, **354**, pp. 56–58.
2. Monthioux, M., and Kuznetsov, V. L. (2006). Who should be given the credit for the discovery of carbon nanotubes?, *Carbon*, **44**, pp. 1621–1623.
3. Ebbesen, T. W., and Ajayan, P. M. (1992). Large-scale synthesis of carbon nanotubes, *Nature*, **358**, pp. 220–222.
4. Iijima, S., and Ichihashi, T. (1993). Single-shell carbon nanotubes of 1-nm diameter, *Nature*, **363**, pp. 603–605.
5. Bethune, D. S., Klang, C. H., de Vries, M. S., Gorman, G., Savoy, R., Vazquez, J., and Beyers, R. (1993). Cobalt-catalysed growth of carbon nanotubes with single-atomic-layer walls, *Nature*, **363**, pp. 605–607.
6. Hamada, N., Sawada, S., and Oshiyama, A. (1992). New one-dimensional conductors: graphitic microtubules, *Phys. Rev. Lett.*, **68**, pp. 1579–1581.

7. Ajayan, P. M. (1999). Nanotubes from carbon, *Chem. Rev.*, **99**, pp. 1787–1799.
8. Wildoer, J. W. G., Venema, L. C., Rinzler, A. G., Smalley, R. E., and Dekker, C. (1998). Electronic structure of atomically resolved carbon nanotubes, *Nature*, **391**, pp. 59–62.
9. Amelinckx, S., Bernaerts, D., Zhang, X. B., Vantendeloo, G., and Vanlanduyt, J. (1995). A structure model and growth-mechanism for multishell carbon nanotubes, *Science*, **267**, pp. 1334–1338.
10. Thess, A., Lee, R., Nikolaev, P., Dai, H. J., Petit, P., Robert, J., Xu, C. H., Lee, Y. H., Kim, S. G., Rinzler, A. G., Colbert, D. T., Scuseria, G. E., Tomanek, D., Fischer, J. E., and Smalley, R. E. (1996). Crystalline ropes of metallic carbon nanotubes, *Science*, **273**, pp. 483–487.
11. Blase, X., Benedict, L. X., Shirley, E. L., and Louie, S. G. (1994). Hybridization effects and metallicity in small radius carbon nanotubes, *Phys. Rev. Lett.*, **72**, pp. 1878–1881.
12. Meyyappan, M. (2005) *Carbon Nanotubes: Science and Applications* (CRC Press, Boca Raton, Fla. ; London).
13. Niyogi, S., Hamon, M. A., Hu, H., Zhao, B., Bhowmik, P., Sen, R., Itkis, M. E., and Haddon, R. C. (2002). Chemistry of single-walled carbon nanotubes, *Acc. Chem. Res.*, **35**, pp. 1105–1113.
14. Haddon, R. C. (1993). Chemistry of the fullerenes: The manifestation of strain in a class of continuous aromatic molecules, *Science*, **261**, pp. 1545–1550.
15. Wang, J., Musameh, M., and Lin, Y. H. (2003). Solubilization of carbon nanotubes by Nafion toward the preparation of amperometric biosensors, *J. Am. Chem. Soc.*, **125**, pp. 2408–2409.
16. Musameh, M., Wang, J., Merkoci, A., and Lin, Y. H. (2002). Low-potential stable NADH detection at carbon-nanotube-modified glassy carbon electrodes, *Electrochem. Commun.*, **4**, pp. 743–746.
17. Banks, C. E., and Compton, R. G. (2006). New electrodes for old: From carbon nanotubes to edge plane pyrolytic graphite, *Analyst*, **131**, pp. 15–21.
18. Banks, C. E., Davies, T. J., Wildgoose, G. G., and Compton, R. G. (2005). Electrocatalysis at graphite and carbon nanotube modified electrodes: Edge-plane sites and tube ends are the reactive sites, *Chem. Commun.*, pp. 829–841.
19. Davies, T. J., Hyde, M. E., and Compton, R. G. (2005). Nanotrench arrays reveal insight into graphite electrochemistry, *Angew. Chem. Int. Ed.*, **44**, pp. 5121–5126.

20. Banks, C. E., Crossley, A., Salter, C., Wilkins, S. J., and Compton, R. G. (2006). Carbon nanotubes contain metal impurities which are responsible for the "electrocatalysis" seen at some nanotube-modified electrodes, *Angew. Chem. Int. Ed.*, **45**, pp. 2533–2537.

21. Banks, C. E., Moore, R. R., Davies, T. J., and Compton, R. G. (2004). Investigation of modified basal plane pyrolytic graphite electrodes: definitive evidence for the electrocatalytic properties of the ends of carbon nanotubes, *Chem. Commun.*, pp. 1804–1805.

22. Wantz, F., Banks, C. E., and Compton, R. G. (2005). Direct oxidation of ascorbic acid at an edge plane pyrolytic graphite electrode: A comparison of the electroanalytical response with other carbon electrodes, *Electroanalysis*, **17**, pp. 1529–1533.

23. Jurkschat, K., Wilkins, S. J., Salter, C. J., Leventis, H. C., Wildgoose, G. G., Jiang, L., Jones, T. G. J., Crossley, A., and Compton, R. G. (2006). Multiwalled carbon nanotubes with molybdenum dioxide nanoplugs — New chemical nanoarchitectures by electrochemical modification, *Small*, **2**, pp. 95–98.

24. Chou, A., Bocking, T., Singh, N. K., and Gooding, J. J. (2005). Demonstration of the importance of oxygenated species at the ends of carbon nanotubes for their favourable electrochemical properties, *Chem. Commun.*, pp. 842–844.

25. Heng, L. Y., Chou, A., Yu, J., Chen, Y., and Gooding, J. J. (2005). Demonstration of the advantages of using bamboo-like nanotubes for electrochemical biosensor applications compared with single walled carbon nanotubes, *Electrochem. Commun.*, **7**, pp. 1457–1462.

26. McCreery, R. L. (2008). Advanced carbon electrode materials for molecular electrochemistry, *Chem. Rev.*, **108**, pp. 2646–2687.

27. Banks, C. E., Ji, X. B., Crossley, A., and Compton, R. G. (2006). Understanding the electrochemical reactivity of bamboo multiwalled carbon nanotubes: the presence of oxygenated species at tube ends may not increase electron transfer kinetics, *Electroanalysis*, **18**, pp. 2137–2140.

28. Stephan, O., Ajayan, P. M., Colliex, C., Redlich, P., Lambert, J. M., Bernier, P., and Lefin, P. (1994). Doping graphitic and carbon nanotube structures with boron and nitrogen, *Science*, **266**, pp. 1683–1685.

29. Zhou, W. Y., Bai, X. D., Wang, E. G., and Xie, S. S. (2009). Synthesis, structure, and properties of single-walled carbon nanotubes, *Adv. Mater.*, **21**, pp. 4565–4583.

30. Merkoci, A., Pumera, M., Llopis, X., Perez, B., del Valle, M., and Alegret, S. (2005). New materials for electrochemical sensing VI: Carbon nanotubes, *Trends Anal. Chem.*, **24**, pp. 826–838.
31. Gong, K. P., Du, F., Xia, Z. H., Durstock, M., and Dai, L. M. (2009). Nitrogen-doped carbon nanotube arrays with high electrocatalytic activity for oxygen reduction, *Science*, **323**, pp. 760–764.
32. Deng, C. Y., Chen, J. H., Chen, X. L., Mao, C. H., Nie, Z., and Yao, S. Z. (2008). Boron-doped carbon nanotubes modified electrode for electroanalysis of NADH, *Electrochem. Commun.*, **10**, pp. 907–909.
33. Xu, X. A., Jiang, S. J., Hu, Z., and Liu, S. Q. (2010). Nitrogen-doped carbon nanotubes: High electrocatalytic activity toward the oxidation of hydrogen peroxide and Its application for biosensing, *ACS Nano*, **4**, pp. 4292–4298.
34. Pumera, M. (2009). The Electrochemistry of carbon nanotubes: Fundamentals and applications, *Chem. Eur. J.*, **15**, pp. 4970–4978.
35. Moore, R. R., Banks, C. E., and Compton, R. G. (2004). Basal plane pyrolytic graphite modified electrodes: Comparison of carbon nanotubes and graphite powder as electrocatalysts, *Anal. Chem.*, **76**, pp. 2677–2682.
36. Sljukic, B., Banks, C. E., and Compton, R. G. (2006). Iron oxide particles are the active sites for hydrogen peroxide sensing at multiwalled carbon nanotube modified electrodes, *Nano Lett.*, **6**, pp. 1556–1558.
37. Yang, W. R., Ratinac, K. R., Ringer, S. P., Thordarson, P., Gooding, J. J., and Braet, F. (2010). Carbon nanomaterials in biosensors: Should you use nanotubes or graphene?, *Angew. Chem. Int. Ed.*, **49**, pp. 2114–2138.
38. Gong, K. P., Chakrabarti, S., and Dai, L. M. (2008). Electrochemistry at carbon nanotube electrodes: Is the nanotube tip more active than the sidewall?, *Angew. Chem. Int. Ed.*, **47**, pp. 5446–5450.
39. Hirsch, A. (2002). Functionalization of single-walled carbon nanotubes, *Angew. Cheme. Int. Ed.*, **41**, pp. 1853–1859.
40. Banerjee, S., Kahn, M. G. C., and Wong, S. S. (2003). Rational chemical strategies for carbon nanotube functionalization, *Chem. Eur. J.*, **9**, pp. 1899–1908.
41. Tasis, D., Tagmatarchis, N., Georgakilas, V., and Prato, M. (2003). Soluble carbon nanotubes, *Chem. Eur. J.*, **9**, pp. 4001–4008.
42. Davis, J. J., Coleman, K. S., Azamian, B. R., Bagshaw, C. B., and Green, M. L. H. (2003). Chemical and biochemical sensing with modified single walled carbon nanotubes, *Chem. Eur. J.*, **9**, pp. 3732–3739.

43. Dyke, C. A., and Tour, J. M. (2004). Overcoming the insolubility of carbon nanotubes through high degrees of sidewall functionalization, *Chem. Eur. J.*, **10**, pp. 813–817.

44. Banerjee, S., Hemraj-Benny, T., and Wong, S. S. (2005). Covalent surface chemistry of single-walled carbon nanotubes, *Adv. Mater.*, **17**, pp. 17–29.

45. Balasubramanian, K., and Burghard, M. (2005). Chemically functionalized carbon nanotubes, *Small*, **1**, pp. 180–192.

46. Tasis, D., Tagmatarchis, N., Bianco, A., and Prato, M. (2006). Chemistry of carbon nanotubes, *Chem. Rev.*, **106**, pp. 1105–1136.

47. Peng, X. H., and Wong, S. S. (2009). Functional covalent chemistry of carbon nanotube surfaces, *Adv. Mater.*, **21**, pp. 625–642.

48. Karousis, N., Tagmatarchis, N., and Tasis, D. (2010). Current progress on the chemical modification of carbon nanotubes, *Chem. Rev.*, **110**, pp. 5366–5397.

49. Liu, J., Rinzler, A. G., Dai, H. J., Hafner, J. H., Bradley, R. K., Boul, P. J., Lu, A., Iverson, T., Shelimov, K., Huffman, C. B., Rodriguez-Macias, F., Shon, Y. S., Lee, T. R., Colbert, D. T., and Smalley, R. E. (1998). Fullerene pipes, *Science*, **280**, pp. 1253–1256.

50. Ziegler, K. J., Gu, Z. N., Peng, H. Q., Flor, E. L., Hauge, R. H., and Smalley, R. E. (2005). Controlled oxidative cutting of single-walled carbon nanotubes, *J. Am. Chem. Soc.*, **127**, pp. 1541–1547.

51. Aviles, F., Cauich-Rodriguez, J. V., Moo-Tah, L., May-Pat, A., and Vargas-Coronado, R. (2009). Evaluation of mild acid oxidation treatments for MWCNT functionalization, *Carbon*, **47**, pp. 2970–2975.

52. Huang, W. J., Taylor, S., Fu, K. F., Lin, Y., Zhang, D. H., Hanks, T. W., Rao, A. M., and Sun, Y. P. (2002). Attaching proteins to carbon nanotubes via diimide-activated amidation, *Nano Lett.*, **2**, pp. 311–314.

53. Lin, Y. H., Lu, F., Tu, Y., and Ren, Z. F. (2004). Glucose biosensors based on carbon nanotube nanoelectrode ensembles, *Nano Lett.*, **4**, pp. 191–195.

54. Patolsky, F., Weizmann, Y., and Willner, I. (2004). Long-range electrical contacting of redox enzymes by SWCNT connectors, *Angew. Chem. Int. Ed.*, **43**, pp. 2113–2117.

55. Nguyen, C. V., Delzeit, L., Cassell, A. M., Li, J., Han, J., and Meyyappan, M. (2002). Preparation of nucleic acid functionalized carbon nanotube Arrays, *Nano Lett.*, **2**, pp. 1079–1081.

56. Yu, X., Munge, B., Patel, V., Jensen, G., Bhirde, A., Gong, J. D., Kim, S. N., Gillespie, J., Gutkind, J. S., Papadimitrakopoulos, F., and Rusling, J. F.

(2006). Carbon nanotube amplification strategies for highly sensitive immunodetection of cancer biomarkers, *J. Am. Chem. Soc.*, **128**, pp. 11199–11205.

57. Huang, W. J., Lin, Y., Taylor, S., Gaillard, J., Rao, A. M., and Sun, Y. P. (2002). Sonication-assisted functionalization and solubilization of carbon nanotubes, *Nano Lett.*, **2**, pp. 231–234.
58. Lin, Y., Zhou, B., Fernando, K. A. S., Liu, P., Allard, L. F., and Sun, Y. P. (2003). Polymeric carbon nanocomposites from carbon nanotubes functionalized with matrix polymer, *Macromolecules*, **36**, pp. 7199–7204.
59. Zhang, Y. J., Li, J., Shen, Y. F., Wang, M. J., and Li, J. H. (2004). Poly-L-lysine functionalization of single-walled carbon nanotubes, *J. Phys. Chem. B*, **108**, pp. 15343–15346.
60. Alkire, R. C. (2009) *Chemically Modified Electrodes* (Wiley-VCH ; Chichester : John Wiley [distributor], Weinheim).
61. Merkoçi, A. (2009) *Biosensing Using Nanomaterials* (Wiley-Blackwell, Oxford).
62. Azamian, B. R., Davis, J. J., Coleman, K. S., Bagshaw, C. B., and Green, M. L. H. (2002). Bioelectrochemical single-walled carbon nanotubes, *J. Am. Chem. Soc.*, **124**, pp. 12664–12665.
63. Zhao, Y. L., and Stoddart, J. F. (2009). Noncovalent functionalization of single-walled Carbon nanotubes, *Acc. Chem. Res.*, **42**, pp. 1161–1171.
64. Tomonari, Y., Murakami, H., and Nakashima, N. (2006). Solubilization of single-walled carbon nanotubes by using polycyclic aromatic ammonium amphiphiles in water - Strategy for the design of high-performance solubilizers, *Chem. Eur. J.*, **12**, pp. 4027–4034.
65. Ogoshi, T., Takashima, Y., Yamaguchi, H., and Harada, A. (2007). Chemically-responsive sol-gel transition of supramolecular single-walled carbon nanotubes (SWNTs) hydrogel made by hybrids of SWNTs and cyclodextrins, *J. Am. Chem. Soc.*, **129**, pp. 4878– 4879.
66. Simmons, T. J., Bult, J., Hashim, D. P., Linhardt, R. J., and Ajayan, P. M. (2009). Noncovalent functionalization as an alternative to oxidative acid treatment of single wall carbon nanotubes with applications for polymer composites, *ACS Nano*, **3**, pp. 865–870.
67. Wu, P., Chen, X., Hu, N., Tam, U. C., Blixt, O., Zettl, A., and Bertozzi, C. R. (2008). Biocompatible carbon nanotubes generated by functionalization with glycodendrimers, *Angew. Chem. Int. Ed.*, **47**, pp. 5022–5025.

68. Holder, P. G., and Francis, M. B. (2007). Integration of a self-assembling protein scaffold with water-soluble single-walled carbon nanotubes, *Angew. Chem. Int. Ed.*, **46**, pp. 4370–4373.

69. Chen, R. J., Zhang, Y. G., Wang, D. W., and Dai, H. J. (2001). Noncovalent sidewall functionalization of single-walled carbon nanotubes for protein immobilization, *J. Am. Chem. Soc.*, **123**, pp. 3838–3839.

70. Besteman, K., Lee, J. O., Wiertz, F. G. M., Heering, H. A., and Dekker, C. (2003). Enzyme-coated carbon nanotubes as single-molecule biosensors, *Nano Lett.*, **3**, pp. 727–730.

71. Britz, D. A., and Khlobystov, A. N. (2006). Noncovalent interactions of molecules with single walled carbon nanotubes, *Chem. Soc. Rev.*, **35**, pp. 637–659.

72. Wang, J. Y., Chu, H. B., and Li, Y. (2008). Why single-walled carbon nanotubes can be dispersed in imidazolium-based ionic liquids, *ACS Nano*, **2**, pp. 2540–2546.

73. Rege, K., Raravikar, N. R., Kim, D. Y., Schadler, L. S., Ajayan, P. M., and Dordick, J. S. (2003). Enzyme-polymer-single walled carbon nanotube composites as biocatalytic films, *Nano Lett.*, **3**, pp. 829–832.

74. Shim, M., Kam, N. W. S., Chen, R. J., Li, Y. M., and Dai, H. J. (2002). Functionalization of carbon nanotubes for biocompatibility and biomolecular recognition, *Nano Lett.*, **2**, pp. 285–288.

75. Chen, R. J., Bangsaruntip, S., Drouvalakis, K. A., Kam, N. W. S., Shim, M., Li, Y. M., Kim, W., Utz, P. J., and Dai, H. J. (2003). Noncovalent functionalization of carbon nanotubes for highly specific electronic biosensors, *Proc. Nat. Acad. Sci. U. S. A.*, **100**, pp. 4984–4989.

76. Yang, H., Wang, S. C., Mercier, P., and Akins, D. L. (2006). Diameter-selective dispersion of single-walled carbon nanotubes using a water-soluble, biocompatible polymer, *Chem. Commun.*, pp. 1425–1427.

77. Star, A., Steuerman, D. W., Heath, J. R., and Stoddart, J. F. (2002). Starched carbon nanotubes, *Angew. Chem. Int. Ed.*, **41**, pp. 2508–2512.

78. Moulton, S. E., Maugey, M., Poulin, P., and Wallace, G. G. (2007). Liquid crystal behavior of single-walled carbon nanotubes dispersed in biological hyaluronic acid solutions, *J. Am. Chem. Soc.*, **129**, pp. 9452-9457.

79. Karajanagi, S. S., Yang, H. C., Asuri, P., Sellitto, E., Dordick, J. S., and Kane, R. S. (2006). Protein-assisted solubilization of single-walled carbon nanotubes, *Langmuir*, **22**, pp. 1392-1395.

80. Zorbas, V., Ortiz-Acevedo, A., Dalton, A. B., Yoshida, M. M., Dieckmann, G. R., Draper, R. K., Baughman, R. H., Jose-Yacaman, M., and Musselman,

I. H. (2004). Preparation and characterization of individual peptide-wrapped single-walled carbon nanotubes, *J. Am. Chem. Soc.*, **126**, pp. 7222–7227.

81. Zheng, M., Jagota, A., Semke, E. D., Diner, B. A., Mclean, R. S., Lustig, S. R., Richardson, R. E., and Tassi, N. G. (2003). DNA-assisted dispersion and separation of carbon nanotubes, *Nat. Mater.*, **2**, pp. 338–342.

82. Chen, X., Lee, G. S., Zettl, A., and Bertozzi, C. R. (2004). Biomimetic engineering of carbon nanotubes by using cell surface mucin mimics, *Angew. Chem. Int. Ed.*, **43**, pp. 6111–6116.

83. Chen, X., Tam, U. C., Czlapinski, J. L., Lee, G. S., Rabuka, D., Zettl, A., and Bertozzi, C. R. (2006). Interfacing carbon nanotubes with living cells, *J. Am. Chem. Soc.*, **128**, pp. 6292–6293.

84. Wu, B. Y., Hou, S. H., Yin, F., Zhao, Z. X., Wang, Y. Y., Wang, X. S., and Chen, Q. (2007). Amperometric glucose biosensor based on multilayer films via layer-by-layer self-assembly of multi-wall carbon nanotubes, gold nanoparticles and glucose oxidase on the Pt electrode, *Biosens. Bioelectron.*, **22**, pp. 2854–2860.

85. Li, J., Qiu, J. D., Xu, J. J., Chen, H. Y., and Xia, X. H. (2007). The synergistic effect of Prussian-Blue-grafted carbon nanotube/poly(4-vinylpyridine) composites for amperometric sensing, *Adv. Funct. Mater.*, **17**, pp. 1574–1580.

86. Komathi, S., Gopalan, A. I., and Lee, K. P. (2009). Fabrication of a novel layer-by-layer film based glucose biosensor with compact arrangement of multi-components and glucose oxidase, *Biosens. Bioelectron.*, **24**, pp. 3131–3134.

87. Wang, J., and Musameh, M. (2003). Enzyme-dispersed carbon-nanotube electrodes: A needle microsensor for monitoring glucose, *Analyst*, **128**, pp. 1382–1385.

88. Zhang, M. G., Smith, A., and Gorski, W. (2004). Carbon nanotube-chitosan system for electrochemical sensing based on dehydrogenase enzymes, *Anal. Chem.*, **76**, pp. 5045–5050.

89. Hrapovic, S., Liu, Y. L., Male, K. B., and Luong, J. H. T. (2004). Electrochemical biosensing platforms using platinum nanoparticles and carbon nanotubes, *Anal. Chem.*, **76**, pp. 1083–1088.

90. Karnicka, K., Miecznikowski, K., Kowalewska, B., Skunik, M., Opallo, M., Rogalski, J., Schuhmann, W., and Kulesza, P. J. (2008). ABTS-modified multiwalled carbon nanotubes as an effective mediating system for bioelectrocatalytic reduction of oxygen, *Anal. Chem.*, **80**, pp. 7643–7648.

91. Wang, J., and Musameh, M. (2005). Carbon-nanotubes doped polypyrrole glucose biosensor, *Anal. Chim. Acta*, **539**, pp. 209–213.

92. Gopalan, A. I., Lee, K. P., and Ragupathy, D. (2009). Development of a stable cholesterol biosensor based on multi-walled carbon nanotubes-gold nanoparticles composite covered with a layer of chitosan-room-temperature ionic liquid network, *Biosens. Bioelectron.*, **24**, pp. 2211–2217.

93. Tsai, T. W., Heckert, G., Neves, L. F., Tan, Y. Q., Kao, D. Y., Harrison, R. G., Resasco, D. E., and Schmidtke, D. W. (2009). Adsorption of Glucose Oxidase onto Single-walled carbon nanotubes and its application in layer-by-layer biosensors, *Anal. Chem.*, **81**, pp. 7917–7925.

94. Liu, G. D., and Lin, Y. H. (2006). Amperometric glucose biosensor based on self-assembling glucose oxidase on carbon nanotubes, *Electrochem. Commun.*, **8**, pp. 251–256.

95. Britto, P. J., Santhanam, K. S. V., and Ajayan, P. M. (1996). Carbon nanotube electrode for oxidation of dopamine, *Bioelectrochem. Bioenerg.*, **41**, pp. 121–125.

96. Valentini, F., Amine, A., Orlanducci, S., Terranova, M. L., and Palleschi, G. (2003). Carbon nanotube purification: Preparation and characterization of carbon nanotube paste electrodes, *Anal. Chem.*, **75**, pp. 5413–5421.

97. Wang, J., and Musameh, M. (2003). Carbon nanotube/teflon composite electrochemical sensors and biosensors, *Anal. Chem.*, **75**, pp. 2075–2079.

98. Kachoosangi, R. T., Musameh, M. M., Abu-Yousef, I., Yousef, J. M., Kanan, S. M., Xiao, L., Davies, S. G., Russell, A., and Compton, R. G. (2009). Carbon nanotube-ionic liquid composite sensors and Biosens, *Anal. Chem.*, **81**, pp. 435–442.

99. Gooding, J. J., Chou, A., Liu, J. Q., Losic, D., Shapter, J. G., and Hibbert, D. B. (2007). The effects of the lengths and orientations of single-walled carbon nanotubes on the electrochemistry of nanotube-modified electrodes, *Electrochem. Commun.*, **9**, pp. 1677–1683.

100. Diao, P., and Liu, Z. F. (2010). Vertically Aligned Single-Walled Carbon Nanotubes by Chemical Assembly - Methodology, Properties, and Applications, *Adv. Mater.*, **22**, pp. 1430–1449.

101. Gooding, J. J., Wibowo, R., Liu, J. Q., Yang, W. R., Losic, D., Orbons, S., Mearns, F. J., Shapter, J. G., and Hibbert, D. B. (2003). Protein electrochemistry using aligned carbon nanotube arrays, *J. Am. Chem. Soc.*, **125**, pp. 9006–9007.

102. Chattopadhyay, D., Galeska, I., and Papadimitrakopoulos, F. (2001). Metal-assisted organization of shortened carbon nanotubes in monolayer and multilayer forest assemblies, *J. Am. Chem. Soc.*, **123**, pp. 9451–9452.

103. Katz, E., and Willner, I. (2004). Biomolecule-functionalized carbon nanotubes: Applications in nanobioelectronics, *Chemphyschem*, **5**, pp. 1085–1104.

104. Knopf, G. K., and Bassi, A. S. (2007) *Smart Biosensor Technology* (CRC Press, Boca Raton).

105. Xiao, Y., Patolsky, F., Katz, E., Hainfeld, J. F., and Willner, I. (2003). "Plugging into enzymes": Nanowiring of redox enzymes by a gold nanoparticle, *Science*, **299**, pp. 1877–1881.

106. Contakes, S. M., Juda, G. A., Langley, D. B., Halpern-Manners, N. W., Duff, A. P., Dunn, A. R., Gray, H. B., Dooley, D. M., Guss, J. M., and Freeman, H. C. (2005). Reversible inhibition of copper amine oxidase activity by channel-blocking ruthenium(II) and rhenium(I) molecular wires, *Proc. Nat. Acad. Sci. U. S. A.*, **102**, pp. 13451–13456.

107. Leger, C., and Bertrand, P. (2008). Direct electrochemistry of redox enzymes as a tool for mechanistic studies, *Chem. Rev.*, **108**, pp. 2379–2438.

108. Yu, X., Chattopadhyay, D., Galeska, I., Papadimitrakopoulos, F., and Rusling, J. F. (2003). Peroxidase activity of enzymes bound to the ends of single-wall carbon nanotube forest electrodes, *Electrochem. Commun.*, **5**, pp. 408–411.

109. Davis, J. J., Coles, R. J., and Hill, H. A. O. (1997). Protein electrochemistry at carbon nanotube electrodes, *J. Electroanal. Chem.*, **440**, pp. 279–282.

110. Ren, Z. F., Huang, Z. P., Xu, J. W., Wang, J. H., Bush, P., Siegal, M. P., and Provencio, P. N. (1998). Synthesis of large arrays of well-aligned carbon nanotubes on glass, *Science*, **282**, pp. 1105–1107.

111. Tu, Y., Lin, Y. H., and Ren, Z. F. (2003). Nanoelectrode arrays based on low site density aligned carbon nanotubes, *Nano Lett.*, **3**, pp. 107–109.

112. Maschmann, M. R., Franklin, A. D., Scott, A., Janes, D. B., Sands, T. D., and Fisher, T. S. (2006). Lithography-free in situ Pd contacts to templated single-walled carbon nanotubes, *Nano Lett.*, **6**, pp. 2712–2717.

113. Wang, J. (2005). Carbon-nanotube based electrochemical biosensors: A review, *Electroanalysis*, **17**, pp. 7–14.

114. Yang, W. R., Thordarson, P., Gooding, J. J., Ringer, S. P., and Braet, F. (2007). Carbon nanotubes for biological and biomedical applications, *Nanotechnology*, **18**, pp. 412001.

115. Wang, Z. D., and Lu, Y. (2009). Functional DNA directed assembly of nanomaterials for biosensing, *J. Mater. Chem.*, **19**, pp. 1788–1798.

116. Trojanowicz, M. (2006). Analytical applications of carbon nanotubes: a review, *Trends Anal. Chem.*, **25**, pp. 480–489.

117. Wang, J., and Lin, Y. H. (2008). Functionalized carbon nanotubes and nanofibers for biosensing applications, *Trends. Anal. Chem.*, **27**, pp. 619–626.

118. Chen, H., Jiang, C. M., Yu, C., Zhang, S., Liu, B. H., and Kong, J. L. (2009). Protein chips and nanomaterials for application in tumor marker immunoassays, *Biosens. Bioelectron.*, **24**, pp. 3399–3411.

119. Jacobs, C. B., Peairs, M. J., and Venton, B. J. (2010). Review: Carbon nanotube based electrochemical sensors for biomolecules, *Anal. Chim. Acta*, **662**, pp. 105–127.

120. Zhao, Q., Gan, Z. H., and Zhuang, Q. K. (2002). Electrochemical sensors based on carbon nanotubes, *Electroanalysis*, **14**, pp. 1609–1613.

121. Yanez-Sedeno, P., Riu, J., Pingarron, J. M., and Rius, F. X. (2010). Electrochemical sensing based on carbon nanotubes, *Trends Anal. Chem.*, **29**, pp. 939–953.

122. Balasubramanian, K., and Burghard, M. (2006). Biosensors based on carbon nanotubes, *Anal. Bioanal. Chem.*, **385**, pp. 452–468.

123. Agui, L., Yanez-Sedeno, P., and Pingarron, J. M. (2008). Role of carbon nanotubes in electroanalytical chemistry — A review, *Anal. Chim. Acta*, **622**, pp. 11-47.

124. Kim, S. N., Rusling, J. F., and Papadimitrakopoulos, F. (2007). Carbon nanotubes for electronic and electrochemical detection of biomolecules, *Adv. Mater.*, **19**, pp. 3214–3228.

125. Newman, J. D., and Turner, A. P. F. (2005). Home blood glucose biosensors: A commercial perspective, *Biosens. Bioelectron.*, **20**, pp. 2435–2453.

126. Heller, A., and Feldman, B. (2008). Electrochemical glucose sensors and their applications in diabetes management, *Chem. Rev.*, **108**, pp. 2482–2505.

127. Wang, J. (2008). Electrochemical glucose biosensors, *Chem. Rev.*, **108**, pp. 814–825.

128. Wang, J., and Musameh, M. (2004). Carbon nanotube screen-printed electrochemical sensors, *Analyst*, **129**, pp. 1–2.

129. Rubianes, M. D., and Rivas, G. A. (2003). Carbon nanotubes paste electrode, *Electrochem. Commun.*, **5**, pp. 689–694.

130. Rappathy, D., Gopalan, A. Y., and Lee, K. P. (2009). Synergistic contributions of multiwall carbon nanotubes and gold nanoparticles in a chitosan-ionic liquid matrix towards improved performance for a glucose sensor, *Electrochem. Commun.*, **11**, pp. 397–401.

131. Zou, Y. J., Xiang, C. L., Sun, L. X., and Xu, F. (2008). Glucose biosensor based on electrodeposition of platinum nanoparticles onto carbon nanotubes and immobilizing enzyme with chitosan-SiO2 sol-gel, *Biosens. Bioelectron.*, **23**, pp. 1010–1016.

132. Lim, S. H., Wei, J., Lin, J. Y., Li, Q. T., and KuaYou, J. (2005). A glucose biosensor based on electrodeposition of palladium nanoparticles and glucose oxidase onto Nafion-solubilized carbon nanotube electrode, *Biosens. Bioelectron.*, **20**, pp. 2341–2346.

133. Hatchett, D. W., and Josowicz, M. (2008). Composites of intrinsically conducting polymers as sensing nanomaterials, *Chem. Rev.*, **108**, pp. 746–769.

134. Gopalan, A. I., Lee, K. P., Ragupathy, D., Lee, S. H., and Lee, J. W. (2009). An electrochemical glucose biosensor exploiting a polyaniline grafted multiwalled carbon nanotube/perfluorosulfonate ionomer-silica nanocomposite, *Biomaterials*, **30**, pp. 5999–6005.

135. Chiu, J. Y., Yu, C. M., Yen, M. J., and Chen, L. C. (2009). Glucose sensing electrodes based on a poly(3,4-ethylenedioxythiophene)/Prussian blue bilayer and multi-walled carbon nanotubes, *Biosens. Bioelectron.*, **24**, pp. 2015–2020.

136. Wang, Z. G., Wano, Y., Xu, H., Li, G., and Xu, Z. K. (2009). Carbon nanotube-filled nanofibrous membranes electrospun from Poly(acrylonitrile-co-acrylic acid) for glucose biosensor, *J. Phys. Chem. C*, **113**, pp. 2955–2960.

137. Pham, X. H., Bui, M. P. N., Li, C. A., Han, K. N., Kim, J. H., Won, H., and Seong, G. H. (2010). Electrochemical characterization of a single-walled carbon nanotube electrode for detection of glucose, *Anal. Chim. Acta*, **671**, pp. 36–40.

138. Claussen, J. C., Franklin, A. D., ul Haque, A., Porterfield, D. M., and Fisher, T. S. (2009). Electrochemical biosensor of nanocube-augmented carbon nanotube networks, *ACS Nano*, **3**, pp. 37–44.

139. Guiseppi-Elie, A., Lei, C. H., and Baughman, R. H. (2002). Direct electron transfer of glucose oxidase on carbon nanotubes, *Nanotechnology*, **13**, pp. 559–564.

140. Li, F. H., Wang, Z. H., Shan, C. S., Song, J. F., Han, D. X., and Niu, L. (2009). Preparation of gold nanoparticles/functionalized multiwalled carbon nanotube nanocomposites and its glucose biosensing application, *Biosens. Bioelectron.*, **24**, pp. 1765–1770.

141. Deng, C. Y., Chen, J. H., Chen, X. L., Mao, C. H., Nie, L. H., and Yao, S. Z. (2008). Direct electrochemistry of glucose oxidase and biosensing for glucose based on boron-doped carbon nanotubes modified electrode, *Biosens. Bioelectron.*, **23**, pp. 1272–1277.

142. Deng, S. Y., Jian, G. Q., Lei, J. P., Hu, Z., and Ju, H. X. (2009). A glucose biosensor based on direct electrochemistry of glucose oxidase immobilized on nitrogen-doped carbon nanotubes, *Biosens. Bioelectron.*, **25**, pp. 373–377.

143. Ruzgas, T., Csoregi, E., Emneus, J., Gorton, L., and MarkoVarga, G. (1996). Peroxidase-modified electrodes: Fundamentals and application, *Anal. Chim. Acta*, **330**, pp. 123–138.

144. Cao, Z. J., Jiang, X. Q., Xie, Q. J., and Yao, S. Z. (2008). A third-generation hydrogen peroxide biosensor based on horseradish peroxidase immobilized in a tetrathiafulvalene-tetracyanoquinodimethane/multiwalled carbon nanotubes film, *Biosens. Bioelectron.*, **24**, pp. 222–227.

145. Munge, B. S., Dowd, R. S., Krause, C. E., and Millord, L. N. (2009). Ultrasensitive hydrogen peroxide biosensor based on enzyme bound to layered nonoriented multiwall carbon nanotubes/polyelectrolyte electrodes, *Electroanalysis*, **21**, pp. 2241–2248.

146. Husain, Q., Husain, M., and Kulshrestha, Y. (2009). Remediation and treatment of organopollutants mediated by peroxidases: A review, *Crit. Rev. Biotechnol.*, **29**, pp. 94–119.

147. Ruzgas, T., Emneus, J., Gorton, L., and Markovarga, G. (1995). The Development of a peroxidase biosensor for monitoring phenol and related aromatic-compounds, *Anal. Chim. Acta*, **311**, pp. 245–253.

148. Liu, L. J., Zhang, F., Xi, F. N., and Lin, X. F. (2008). Highly sensitive biosensor based on bionanomultilayer with water-soluble multiwall carbon nanotubes for determination of phenolics, *Biosens. Bioelectron.*, **24**, pp. 306–312.

149. Santos, A. S., Pereira, A. C., Sotomayor, M. D. R. T., Tarley, C. R. T., Duran, N., and Kubota, L. T. (2007). Determination of phenolic compounds

based on co-immobilization of methylene blue and HRP on multi-wall carbon nanotubes, *Electroanalysis*, **19**, pp. 549–554.

150. http://water.epa.gov/scitech/swguidance/methods/pollutants.cfm.

151. Solomon, E. I., Sundaram, U. M., and Machonkin, T. E. (1996). Multicopper oxidases and oxygenases, *Chem. Rev.*, **96**, pp. 2563–2605.

152. Strack, D., and Schliemann, W. (2001). Bifunctional polyphenol oxidases: Novel functions in plant pigment biosynthesis, *Angew. Chem. Int. Ed.*, **40**, pp. 3791– 3794.

153. Duran, N., Rosa, M. A., D'Annibale, A., and Gianfreda, L. (2002). Applications of laccases and tyrosinases (phenoloxidases) immobilized on different supports: A review, *Enzyme Microb. Technol.*, **31**, pp. 907–931.

154. Zhao, Q., Guan, L. H., Gu, Z. N., and Zhuang, Q. K. (2005). Determination of phenolic compounds based on the tyrosinase-single walled carbon nanotubes sensor, *Electroanalysis*, **17**, pp. 85–88.

155. Lopez, B. P., and Merkoci, A. (2009). Improvement of the electrochemical detection of catechol by the use of a carbon nanotube based biosensor, *Analyst*, **134**, pp. 60–64.

156. Liu, Y., Qu, X. H., Guo, H. W., Chen, H. J., Liu, B. F., and Dong, S. J. (2006). Facile preparation of amperometric laccase biosensor with multifunction based on the matrix of carbon nanotubes-chitosan composite, *Biosens. Bioelectron.*, **21**, pp. 2195–2201.

157. Liu, Y., and Dong, S. J. (2008). Electrochemical characteristics of mediated laccase-catalysis and electrochemical detection of environmental pollutants, *Electroanalysis*, **20**, pp. 827–832.

158. Tan, Y. M., Deng, W. F., Ge, B., Xie, Q. J., Huang, J. H., and Yao, S. Z. (2009). Biofuel cell and phenolic biosensor based on acid-resistant laccase-glutaraldehyde functionalized chitosan-multiwalled carbon nanotubes nanocomposite film, *Biosens. Bioelectron.*, **24**, pp. 2225–2231.

159. Tan, Y. M., Deng, W. F., Li, Y. Y., Huang, Z., Meng, Y., Xie, Q. J., Ma, M., and Yao, S. Z. (2010). Polymeric Bionanocomposite Cast Thin Films with In Situ Laccase-Catalyzed Polymerization of Dopamine for Biosensing and Biofuel Cell Applications, *J. Phys. Chem. B*, **114**, pp. 5016–5024.

160. Barton, S. C., Gallaway, J., and Atanassov, P. (2004). Enzymatic biofuel cells for Implantable and microscale devices, *Chem. Rev.*, **104**, pp. 4867–4886.

161. Zhang, M. N., Liu, K., Gong, K. P., Su, L., Chen, Y., and Mao, L. Q. (2005). Continuous on-line monitoring of extracellular ascorbate depletion in the rat striatum induced by global ischemia with carbon nanotube-

modified glassy carbon electrode integrated into a thin-layer radial flow cell, *Anal. Chem.*, **77**, pp. 6234–6242.

162. Gouveia-Caridade, C., Pauliukaite, R., and Brett, C. M. A. (2008). Development of electrochemical oxidase biosensors based on carbon nanotube-modified carbon film electrodes for glucose and ethanol, *Electrochim. Acta*, **53**, pp. 6732–6739.

163. Male, K. B., Hrapovic, S., and Luong, J. H. T. (2007). Electrochemically-assisted deposition of oxidases on platinum nanoparticle/multi-walled carbon nanotube-modified electrodes, *Analyst*, **132**, pp. 1254–1261.

164. Yang, M. H., Yang, Y., Yang, H. F., Shen, G. L., and Yu, R. Q. (2006). Layer-by-layer self-assembled multilayer films of carbon nanotubes and platinum nanoparticles with polyelectrolyte for the fabrication of biosensors, *Biomaterials*, **27**, pp. 246–255.

165. Guo, M. L., Chen, J. H., Li, J., Nie, L. H., and Yao, S. Z. (2004). Carbon nanotubes-based amperometric cholesterol biosensor fabricated through layer-by-layer technique, *Electroanalysis*, **16**, pp. 1992–1998.

166. Qu, F. L., Yang, M. H., Jiang, J. H., Shen, G. L., and Yu, R. Q. (2005). Amperometric biosensor for choline based on layer-by-layer assembled functionalized carbon nanotube and polyaniline multilayer film, *Anal. Biochem.*, **344**, pp. 108–114.

167. Wang, J., Liu, G. D., and Lin, Y. H. (2006). Amperometric choline biosensor fabricated through electrostatic assembly of bienzyme/polyelectrolyte hybrid layers on carbon nanotubes, *Analyst*, **131**, pp. 477–483.

168. Tkac, J., Whittaker, J. W., and Ruzgas, T. (2007). The use of single walled carbon nanotubes dispersed in a chitosan matrix for preparation of a galactose biosensor, *Biosens. Bioelectron.*, **22**, pp. 1820–1824.

169. Serafin, V., Agui, L., Yanez-Sedeno, P., and Pingarron, J. M. (2009). Glucosinolate amperometric bienzyme biosensor based on carbon nanotubes-gold nanoparticles composite electrodes, *Electroanalysis*, **12**, pp. 1527–1532.

170. Ammam, M., and Fransaer, J. (2010). Highly sensitive and selective glutamate microbiosensor based on cast polyurethane/AC-electrophoresis deposited multiwalled carbon nanotubes and then glutamate oxidase/electrosynthesized polypyrrole/Pt electrode, *Biosens. Bioelectron.*, **25**, pp. 1597–1602.

171. Cui, X. Q., Li, C. M., Zang, J. F., and Yu, S. C. (2007). Highly sensitive lactate biosensor by engineering chitosan/PVI-Os/CNT/LOD network nanocomposite, *Biosens. Bioelectron.*, **22**, pp. 3288–3292.

172. Rubianes, M. D., and Rivas, G. A. (2005). Enzymatic biosensors based on carbon nanotubes paste electrodes, *Electroanalysis*, **17**, pp. 73–78.

173. Odaci, D., Telefoncu, A., and Timur, S. (2010). Maltose biosensing based on co-immobilization of alpha-glucosidase and pyranose oxidase, *Bioelectrochemistry*, **79**, pp. 108–113.

174. Zhang, M. G., Mullens, C., and Gorski, W. (2007). Coimmobilization of dehydrogenases and their cofactors in electrochemical biosensors, *Anal. Chem.*, **79**, pp. 2446–2450.

175. Zhou, H. J., Zhang, Z. P., Yu, P., Su, L., Ohsaka, T., and Mao, L. Q. (2010). Noncovalent attachment of NAD(+) cofactor onto carbon nanotubes for preparation of integrated dehydrogenase-based electrochemical biosensors, *Langmuir*, **26**, pp. 6028–6032.

176. Antiochia, R., and Gorton, L. (2007). Development of a carbon nanotube paste electrode osmium polymer-mediated biosensor for determination of glucose in alcoholic beverages, *Biosens. Bioelectron.*, **22**, pp. 2611–2617.

177. Choi, H. N., Lyu, Y. K., Han, J. H., and Lee, W. Y. (2007). Amperometric ethanol biosensor based on carbon nanotubes dispersed in sol-gel-derived titania — nafion composite film, *Electroanalysis*, **19**, pp. 1524–1530.

178. Tsai, Y. C., Huang, J. D., and Chiu, C. C. (2007). Amperometric ethanol biosensor based on poly(vinyl alcohol)-multiwalled carbon nanotube-alcohol dehydrogenase biocomposite, *Biosens. Bioelectron.*, **22**, pp. 3051–3056.

179. Yang, D. W., and Liu, H. H. (2009). Poly(brilliant cresyl blue)-carbonnanotube modified electrodes for determination of NADH and fabrication of ethanol dehydrogenase-based biosensor, *Biosens. Bioelectron.*, **25**, pp. 733–738.

180. Chakraborty, S., and Raj, C. R. (2007). Amperometric biosensing of glutamate using carbon nanotube based electrode, *Electrochem. Commun.*, **9**, pp. 1323–1330.

181. Meng, L., Wu, P., Chen, G. X., Cai, C. X., Sun, Y. M., and Yuan, Z. H. (2009). Low potential detection of glutamate based on the electrocatalytic oxidation of NADH at thionine/single-walled carbon nanotubes composite modified electrode, *Biosens. Bioelectron.*, **24**, pp. 1751–1756.

182. Tang, L. H., Zhu, Y. H., Yang, X. L., and Li, C. Z. (2007). An enhanced biosensor for glutamate based on self-assembled carbon nanotubes and dendrimer-encapsulated platinum nanobiocomposites-doped polypyrrole film, *Anal. Chim. Acta*, **597**, pp. 145–150.

183. Agui, L., Eguilaz, M., Pena-Farfal, C., Yanez-Sedeno, P., and Pingarron, J. M. (2009). Lactate dehydrogenase biosensor based on an hybrid carbon nanotube-conducting polymer modified electrode, *Electroanalysis*, **21**, pp. 386–391.

184. Arvinte, A., Sesay, A. M., Virtanen, V., and Bala, C. (2008). Evaluation of meldola blue-carbon nanotube-sol-gel composite for electrochemical NADH sensors and their application for lactate dehydrogenase-based biosensors, *Electroanalysis*, **20**, pp. 2355–2362.

185. Lin, Y. Q., Zhu, N. N., Yu, P., Su, L., and Mao, L. Q. (2009). Physiologically relevant online electrochemical method for continuous and simultaneous monitoring of striatum glucose and lactate following global cerebral ischemia/reperfusion, *Anal. Chem.*, **81**, pp. 2067–2074.

186. Singh, B. K. (2009). Organophosphorus-degrading bacteria: Ecology and industrial applications, *Nat. Rev. Microbiol.*, **7**, pp. 156–164.

187. Deo, R. P., Wang, J., Block, I., Mulchandani, A., Joshi, K. A., Trojanowicz, M., Scholz, F., Chen, W., and Lin, Y. H. (2005). Determination of organophosphate pesticides at a carbon nanotube/organophosphorus hydrolase electrochemical biosensor, *Anal. Chim. Acta*, **530**, pp. 185–189.

188. Joshi, K. A., Prouza, M., Kum, M., Wang, J., Tang, J., Haddon, R., Chen, W., and Mulchandani, A. (2006). V-type nerve agent detection using a carbon nanotube-based amperometric enzyme electrode, *Anal. Chem.*, **78**, pp. 331–336.

189. Joshi, K. A., Tang, J., Haddon, R., Wang, J., Chen, W., and Mulchandani, A. (2005). A disposable biosensor for organophosphorus nerve agents based on carbon nanotubes modified thick film strip electrode, *Electroanalysis*, **17**, pp. 54–58.

190. Lin, Y. H., Lu, F., and Wang, J. (2004). Disposable carbon nanotube modified screen-printed biosensor for amperometric detection of organophosphorus pesticides and nerve agents, *Electroanalysis*, **16**, pp. 145–149.

191. Du, D., Wang, J., Smith, J. N., Timchalk, C., and Lin, Y. H. (2009). Biomonitoring of organophosphorus agent exposure by reactivation of cholinesterase enzyme based on carbon nanotube-enhanced flow-injection amperometric detection, *Anal. Chem.*, **81**, pp. 9314–9320.

192. Liu, G. D., and Lin, Y. H. (2006). Biosensor based on self-assembling acetylcholinesterase on carbon nanotubes for flow injection/amperometric detection of organophosphate pesticides and nerve agents, *Anal. Chem.*, **78**, pp. 835–843.

193. Kandimalla, V. B., and Ju, H. X. (2006). Binding of acetylcholinesterase to a multiwall carbon nanotube-cross-linked chitosan composite for flow-injection amperometric detection of an organophosphorous insecticide, *Chem. Eur. J.*, **12**, pp. 1074–1080.

194. Yao, Y. L., and Shiu, K. K. (2008). A mediator-free bienzyme amperometric biosensor based on horseradish peroxidase and glucose oxidase immobilized on carbon nanotube modified electrode, *Electroanalysis*, **20**, pp. 2090–2095.

195. Zhu, L. D., Yang, R. L., Zhai, J. G., and Tian, C. Y. (2007). Bienzymatic glucose biosensor based on co-immobilization of peroxidase and glucose oxidase on a carbon nanotubes electrode, *Biosens. Bioelectron.*, **23**, pp. 528–535.

196. Karyakin, A. A. (2001). Prussian blue and its analogues: Electrochemistry and analytical applications, *Electroanalysis*, **13**, pp. 813–819.

197. Ricci, F., and Palleschi, G. (2005). Sensor and biosensor preparation, optimisation and applications of prussian blue modified electrodes, *Biosens. Bioelectron.*, **21**, pp. 389–407.

198. Zong, S. Z., Cao, Y., and Ju, H. X. (2007). Direct electron transfer of hemoglobin immobilized in multiwalled carbon nanotubes enhanced grafted collagen matrix for electrocatalytic detection of hydrogen peroxide, *Electroanalysis*, **19**, pp. 841–846.

199. Liu, C. Y., and Hu, J. M. (2009). Hydrogen peroxide biosensor based on the direct electrochemistry of myoglobin immobilized on silver nanoparticles doped carbon nanotubes film, *Biosens. Bioelectron.*, **24**, pp. 2149–2154.

200. Xiang, C. L., Zou, Y. J., Sun, L. X., and Xu, F. (2008). Direct electron transfer of cytochrome c and its biosensor based on gold nanoparticles/room temperature ionic liquid/carbon nanotubes composite film, *Electrochem. Commun.*, **10**, pp. 38–41.

Chapter 12

Carbon Nanotube-Based Gas Sensors: A State of the Art

P. Bondavalli,[a,b] L. Gorintin,[a] and P. Legagneux[a]
Nanocarb Lab,[a] UMR 137,[b] Physics Department, Thales Research and Technology, 128 Rt Departementale, 91767 Palaiseau, France
paolo.bondavalli@thalesgroup.com

In this chapter, we present the recent studies on carbon nanotube (CNTs)-based sensors with a major focus on CNT field effect transistor (CNTFETs). Although these devices have allowed to achieve sensors with an impressive sensitivity compared with existing technologies, the physical interpretation of the effect of interaction between the gas molecules and the CNTs has not been clarified yet. This is particularly evident for CNTFET-based sensors (doping or Schottky barrier effect?). In this contribution, we will try to find a coherence between the physical interpretations advanced by the different scientific teams and to answer unsolved questions. Moreover, considering that the selectivity is the main issue, we will talk about different routes that have been proposed to overcome this problem: functionalization using polymers, electrodes metal diversification, metal decoration of SWCNT mats, study of the desorption time of the different gases. For each technique, we will try to evaluate the advantages and the drawbacks.

Carbon Nanotubes and Their Applications
Edited by Qing Zhang

ISBN 978-981-4241-90-8 (Hardcover), 978-981-4303-18-7 (eBook)
www.panstanford.com

12.1 Introduction

Sensors are, in general terms, devices that convert chemical or physical quantities in a signal, containing information, suitable for an operator. For example some gas sensors translate the interaction of gaseous molecules to the variation of an electrical signal.

Gas sensors are employed in a large number of applications: aerospace, medicine, robotics, warfare, pollution control, security threats, safety, etc. Unfortunately, the commercially available gas sensors do not completely meet the expected specifications for most of applications especially concerning sensitivity, selectivity, compactness, versatility, lower power consumption and time response. In this context nanomaterials can allow people to obtain breakthrough performances and to constitute a real alternative to more traditional technologies (e.g., metal oxides [1]).

CNTs are one-dimensional molecular structures obtained by rolling up one graphene sheet (single-wall carbon nanotube, i.e., SWCNT) or more (multi-wall carbon nanotube, i.e., MWCNT). Since the first sensing measurements performed at Stanford in 2000 [2], many scientific teams have focused their interests on CNT-based chemical and biological sensors [3–5]. Indeed, different kinds of gas/chemical sensors, based on different working principles, have been fabricated thanks to CNTs: miniaturized ionizing gas sensors (also called "micro-gun" sensors) [6], CNT thin films with variable resistance as a function of the adsorbed gas properties [7–10] and finally carbon nanotube field effect transistor (CNTFET)-based sensors.

Concerning the so called "micro-gun sensors," briefly, these are sort of miniaturized gas chromatography systems exploiting the gas ionization produced by the nanotube tips. One of the main issues about these devices is the detection limit. Indeed they are able to detect only some percent of the gas in air (which corresponds to hundreds of ppm). The other concern is that the breakdown voltages are quite high (hundreds of volts) and so power consumption too.

CNT-based resistive sensors exploit the change of CNT mat after having interacted with the targeted gas molecules. In this case the main drawbacks are: the recovery time is too high (more than 30 min at least), the response time is quite long (these are larger devices than CNTFET-based sensors) and finally the physics of the phenomenon is not clearly understood; thus, it is difficult to improve their performances (the right parameters to improve are not well known).

Focusing on CNTFET-based sensors, these devices have the potential to be ultra compact, effective at room temperature, with low power consumption, with a very fast response, a short recovery time (a few seconds) and show a good versatility (they can be used for several gases). Moreover, we can achieve relatively low cost fabrication (using a CMOS-compatible technology). For this reason we have decided to deeply analyze the state of the art in the domain of this kind of gas/chemical sensors and to review the different physical arguments developed to explain the gas/device interaction, trying to find a coherence between them in order to answer the different unsolved questions.

A section is devoted to the different routes developed to improve selectivity. This is a very important issue considering that a high selectivity is the key factor to definitively strike huge markets such as security (e.g., detection of warfare gases such as nerve agents or of highly volatile explosive traces), healthcare (e.g., NO sensors for asthma monitoring or CO_2 for emergency airway management), safety (e.g., CO or BTEX sensors) and environmental monitoring (e.g., NO_2, benzene).

12.2 Gas Sensing Using CNTFET: Carbon Nanotubes Doping or Schottky Barrier Modulation?

As told previously, the first paper showing the great potentiality of CNTs for gas sensing applications was published in 2000 by Kong [2] *et al.* at the University of Stanford. In this paper the authors observed that a single SWCNT used as a transistor channel between two gold electrodes, fabricated on a SiO_2/Si substrate acting as a bottom gate, interacted with gas molecules, changing the CNTFET transfer characteristics (source/drain current as a function of the gate voltage).

The gases used for tests were NO_2 and NH_3 with concentrations between 2 and 10000 ppm. These gases were chosen because they exhibit two opposite electronic behaviors as a consequence of their chemical affinity: The first one is an "electron withdrawing" and the second one an "electron-donor" gas.

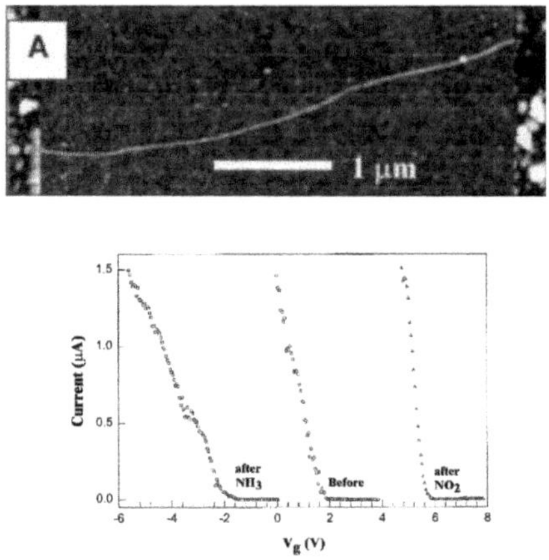

Figure 12.1 CNTFET transfer characteristic change after exposure to NO_2 and NH_3 [2].

The experimental results pointed out an increase and a reduction of the p-type CNTFET conductance for NO_2 and NH_3, respectively (Figure 12.1). Concerning NO_2, its interaction with the nanotube was interpreted as strictly connected to a bulk doping effect. Actually, NO_2 can be bound to a semiconducting nanotube with a subsequent electron charge transfer from the tube to the adsorbed molecules [11–14]: The hole carriers in the nanotube increase and thus does the conductance, with a consequent shift of the activation bias (V_{ON}) to a larger positive voltage. This effect has been assimilated to a sort of "molecular gating" on the CNT due to the gas molecules adsorption. On the contrary, NH_3 molecules, as it has been demonstrated, have no binding affinity with semiconducting SWCNTs [11–14]. In this case the transfer characteristic change as well as the V_{ON} shift to negative gate bias, can be caused by one of two following phenomena:

- The binding between NH_3 molecules and the hydroxyl groups on the SiO_2 substrate which leads to a reduction of the negative charges on the oxide.
- The interaction of NH_3 with oxygen species adsorbed on SWCNT (this will shown in the next pages).

In Kong's paper, the phenomenon interpretation for CNTFET-based sensing initially was related to a sort of doping of the body of the nanotube and not to the metal/SWCNT junctions. Authors stated that the gas molecules doped the nanotube changing its conductance.

However, several scientific teams have adopted a different point of view and have focused their studies on demonstrating that the metal/SWCNT junctions are the key players in the sensing mechanism.

Scientists from Georgestown University [15] in 2006 fabricated CNTFETs using two Pd electrodes on a SiO_2/Si substrate (bottom gate configuration). To evaluate the effect of metal/SWCNT contacts, they covered the electrodes with a 200 nm of polymethyl-methacrylate (PMMA) and a 2 μm layer of SU8 (Figure 12.2).

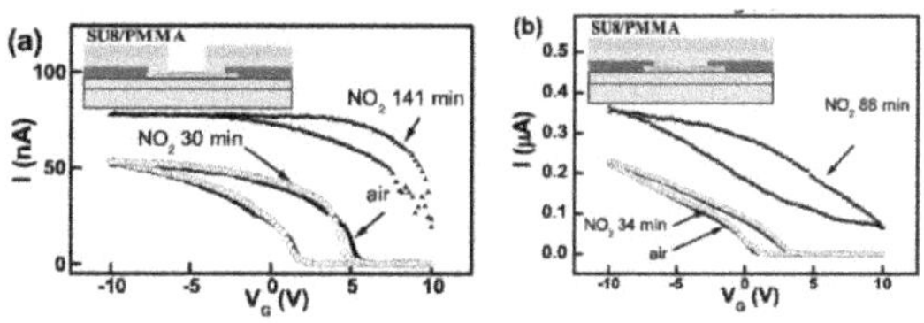

Figure 12.2 Results after exposure to different concentrations of NO_2 and NH_3 for the centre-covered configuration (left) and contact-covered configuration (right) [15].

After having exposed the CNTFET to a concentration raising up to 150 ppm of NO_2, they did not observe any change of the transfer characteristic. However, after 30 min of exposure to a concentration of 200 ppm, the transfer characteristics changed as a consequence of the diffusion of the gas molecules through the polymer layers. The same result was obtained for a whole covered device (Figure 12.2), showing that the NO_2 molecules had no effect on the channel doping of the CNTFET.

A similar experiment using the same kind of transistor-structure has been recently performed in 2009 by Peng *et al.* at the Nanyang Technological University (Singapore) in collaboration with MIT (USA). In this case, they passivated the body and the contact of a CNTFET using Si_3N_4. The results confirmed Georgestown results: The sensing is related to the interaction of gas with the metal/CNT junction and not to the body of the nanotube.

Peng *et al.* also demonstrated that the NH_3 sensing, in the contact covered configuration for temperatures higher than 150°C, was due to the presence of Oxygen near the SWCNT defects. The same measurement performed using N_2 as gas carrier at high temperatures showed no change of the transfer characteristics after exposure to NH_3 concentration up to 400 ppm [15B].

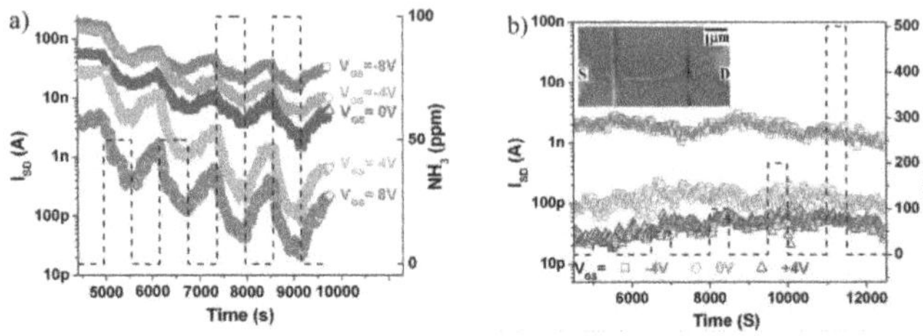

Figure 12.3 Effect of NH_3 on a CNTFET with, respectively, the center (a) and the metal/CNT junctions (b) passivated using Si_3N_4 [15B].

Another team that has been working for more 10 years on CNTFET-based physics, is the IBM team led by P. Avouris. Their work was not focused on sensor issues but on the mechanism of the effect of oxygen on the CNTFETs transfer characteristics. In fact their goal was to build logic circuit using CNTFETs (YES or NOT type circuits).

They underscored that CNTFETs operate as unconventional Schottky-barrier transistors and that the effect of adsorbed gases, i.e., O_2, can be explained by their influence on the metal electrode work-functions. IBM team demonstrated that if a CNTFET was annealed in vacuum at 600 K for several minutes, it showed an n-type electric behavior. They asserted that this phenomenon could not be explained by oxygen doping [16–18]. Indeed, assuming that the CNT was initially doped *p* by oxygen, they could not to explain why the annealing allowed to achieve a n-type SWCNT and not an intrinsic one. Moreover, if the SWCNT were doped by oxygen, the change of oxygen concentration, and so of doping, would have given rise to a shift of the V_{ON} of the CNTFET. Indeed, it was not the case: The off bias interval was unchanged (Figure 12.4).

IBM researchers formulated the hypothesis that the gas molecules adsorption changed the value of the Schottky barrier built-up between the SWCNT and metal. To validate this hypothesis, they simulated the effect on the CNTFET transfer characteristics caused by the change of the metal work function, as a consequence of the adsorption of oxygen. At the same time they simulated the effect of an "hypothetical" CNTFET oxygen doping. The results have been compared with the experimental ones and demonstrated unquestionably the IBM hypothesis (Figure 12.4).

In other studies issued by the same research team [19,20], it has been explained that the interaction of oxygen at the junction between metal electrodes and SWCNT, changes the metal work function and so the Fermi level alignment. The Fermi level at the contact is not pinned by "Metal–Induced-Gap states" (MIGs), as it happens for metal/semiconductor planar contacts, and is controlled by the metal work function. In the light of this analysis, they have concluded that oxygen rises the metal electrode (Au in this case) work function and so it permits the switching of the electrical behavior from n-type (in vacuum) to p-type.

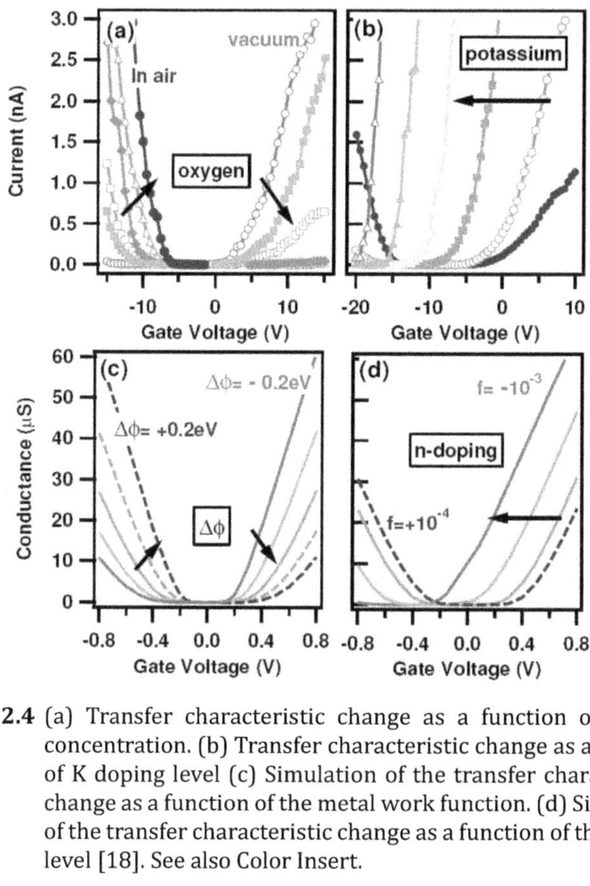

Figure 12.4 (a) Transfer characteristic change as a function of oxygen concentration. (b) Transfer characteristic change as a function of K doping level (c) Simulation of the transfer characteristic change as a function of the metal work function. (d) Simulation of the transfer characteristic change as a function of the doping level [18]. See also Color Insert.

The hypothesis of the gas effect on metal work function formulated by IBM researchers can be explained using an analogy with the model developed for organic compounds and metal junctions [21,22]. In this case the interface is characterized by the presence of interface dipoles which generate an electrostatic potential close to the electrodes. The presence of the interface dipoles at settled at the contact modifies the energy of the electronic levels in organic compounds and the work function of the metal surface. When the interface dipole points its negative pole toward the organic film and its positive pole toward the metal, the result is an increase of both the metal work function and the highest occupied molecular orbital (HOMO) energy of the organic film, by addition of an electrostatic energy. Therefore, the hole injection barrier is reduced. On the other hand, when the dipole direction is reversed, the metal work function is reduced and so is the electron injection barrier. In the case of a metal/CNT junction, similarly, the gas molecules introduce electric dipoles at the interfaces and modify the intensity of the dipole settled at the Metal/CNT junction. This change directly influences the metal work function and the Fermi level alignment. This consideration totally agrees with the simulations and experimental results in the case of oxygen exposition obtained by IBM: Changing the oxygen concentration, the dipole strength varies and the metal work function is increased, thus reducing the hole injection barrier (the hole current raises). The same interpretation can be made when the metal/SWCNT Schottky barrier is modulated using polar chemical compounds [22].

The IBM works have been analyzed and theoretically modeled by T. Yamada of NASA Ames Research Center [23,24]. Yamada has demonstrated that the simple Schottky model (a.k.a. "intimate model") does not adequately describe the effect of oxygen molecules on the modulation of the Schottky barrier value to obtain a p-type junction in air (Figure 12.5). In this model, it is considered a junction of an oxidized metallic electrode surface and a plain SWCNT or the opposite, and therefore we proceed in aligning bands at the contact. Yamada stated that this analysis was based on two wrong assumptions. First, the gold surface is inert and so it does not interact with oxygen. Secondly, if we merely align the Fermi level at the contact, considering that oxygen molecules are electronegative, they would increase the SWCNT electron affinity. Therefore, according to standard Schottky theory we should obtain the following:

$$\Phi_{Bh} = EG + \phi_s - \chi_m \tag{12.1}$$

$$\Phi_{Bh}' = EG + \phi_s' - \chi_m, \tag{12.2}$$

where Φ_{Bh}/Φ_{Bh}' and ϕ_s/ϕ_s' stand, respectively, for the Schottky barriers and the SWCNT electron affinities in vacuum and air. EG is the SWCNT band gap and χ_m, the metal work function.

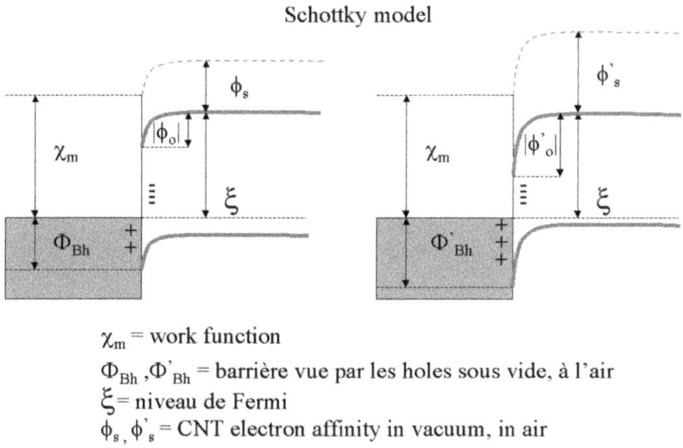

Figure 12.5 Band diagram at Metal/CNT junctions in vacuum (left) and exposed to air (right) using the standard Schottky model.

From these formulas, it can be deduced that an Au/SWCNT Schottky barrier should be lower in vacuum than if exposed to oxygen because these last are negatively charged. This is exactly the opposite of what that has been observed by IBM researchers. Yamada explains that the only way to coherently justify the effect of oxygen is to take into account a sort of transition region between the metal and the SWCNTs. This region is characterized by gold clusters on the electrode surface and charged oxygen molecules. The gold clusters are achieved during the metal deposition on the SWCNT extremities. In fact, the gold molecules are produced during deposition (to fabricated the CNTFET contacts) because Au molecules are attracted by the SWCNTs. Yamada affirms that in this zone the oxygen molecules can be chemisorbed at the clusters tips.

Considering that oxygen molecules are negatively charged, they induce a balancing positive charge in the metal and SWCNT. Consequently, a modification of the potential drop at the junction is observed (Figure 12.6). This potential drop will be larger for CNTFETs exposed to air than in vacuum. Therefore,

$$\Phi_{Bh} = EG + \phi_s - \chi_m - |\Delta U| \tag{12.3}$$

$$\Phi_{Bh}' = EG + \phi_s' - \chi_m - |\Delta U'|, \tag{12.4}$$

where $|\Delta U|$ and $|\Delta U'|$, respectively, stand for the potential drop in vacuum and in air.

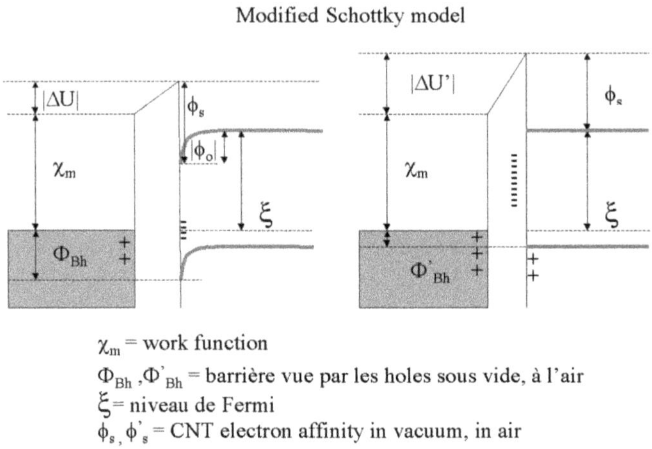

Figure 12.6 Band diagram at Metal/CNT junctions in vacuum (left) and exposed to air (right) using the modified Schottky model (a.k.a non-intimate model).

We can observe that in this case the Schottky barrier for holes in metal is lower if CNTFETs are exposed to air, according to the experimental results obtained by IBM researcher: Yamada's model seems to be satisfying and suitable to describe the effective interaction of gases on metal/SWCNT junctions.

12.3 CNTFET-Based Sensors Achieved Using SWCNT Networks

The previous paragraphs have dealt with CNTFET-based sensors fabricated using individual SWCNT. The main issue is that to effectively employ only one single SWCNT is extremely difficult. First, considering that no method up to now exists to fabricate only semiconducting SWCNTs, it is not possible to predict if a SWCNT is metallic or semiconductor. Secondly, it is quite laborious to identify the position of one single SWCNT using amplitude force microscope (AFM) technique. From an industrial point of view it is not a suitable solution for batch sensor fabrication.

For these reasons, many scientists have focused their work on CNTFET using SWCNT networks as channels. Indeed, as far as random networks are concerned, it has been shown that, through a percolation effect, an overall semiconductor behavior can be obtained for carefully controlled areal densities [24–27] Only two conditions must be fulfilled: The distance between the two electrodes must be larger than the SWCNT length (otherwise metallic nanotubes could cause a short-circuit) and the areal density of the SWCNT mat has to exceed slightly the percolation threshold. Actually, for too high densities, the conduction can reach an ohmic trend with no gating effect [28]. One of the first CNTFET-based sensors achieved using SWCNT mats, has been developed by Snow and al. of the Naval Research Laboratory [29,30] in Washington, DC.

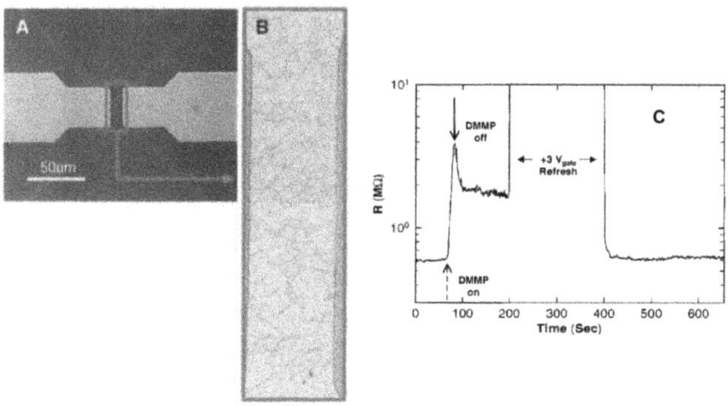

Figure 12.7 (A) CNTFET overview. (B) carbon nanotubes networks (CNTFET channel) [29]. (C) Effect of the bias gate inversion on the resistance change related to DMMP exposure [29].

They fabricated a sensor using lithographically patterned SWCNT networks between Ti drain and source contacts on a SiO_2 layer and Si substrate in a bottom gate configuration (Figure 12.7). These sensors were exposed to saturated vapor of dimethyl-methyl-phosphonate (DMMP) [31] which is a nerve agent stimulant. The researchers found a V_{ON} negative shift, compared with the results obtained in air, which was compatible with the "electron donating" character of DMMP. The phenomenon was explained as a consequence of the electron transfer from DMMP to the SWCNT. The observed recovery time was of several hours of merely exposure to air, but could be dramatically reduced simply applying an inverse bias for around 200 sec (Figure 12.7). Researchers thought that the interaction between the negative charges induced in the channel by the gate voltage (positive), removed the DMMP reducing the desorption barrier. The same method has been employed by researchers at Yonsei University in South Korea [32] for refreshing CNTFET-based sensors after exposure to 300 ppm of NO_2 and NH_3. In this case the, CNTFET gate voltage was inversed for 30 sec: a negative voltage to enhance desorption of NO_2 gas molecules and a positive gate voltage for NH_3 gas molecules. This technique is very interesting because it allows to not degrade the sensor (heating several times) and to achieve a room temperature refreshable sensor (low power consumption device).

An integrated sensor based on CNTFET transistors has been manufactured by researchers from the NASA Ames Research Centre [33]. They fabricated a sensor composed of interdigitated Pt electrodes (with finger distance ranging from 4 to 50 μm) with CNT chains, grown using an *in situ* chemical vapor deposition process (CVD) [34], as the transistor channel. This final device, composed of twelve CNTFETs, shows a very promising sensitivity for NO_2 and NH_3 concentrations, respectively, between 400 ppb and 10 ppm, 4 and 50 ppm in air.

One of the main novelties of this device resides in the fact that an embedded heating layer is positioned between the substrate and the dielectric layer (Si_3N_4). This one can be used to reduce the recovery time desorbing more rapidly the gas molecules (Figure 12.8). Indeed, in the paper this is not pointed out by the authors.

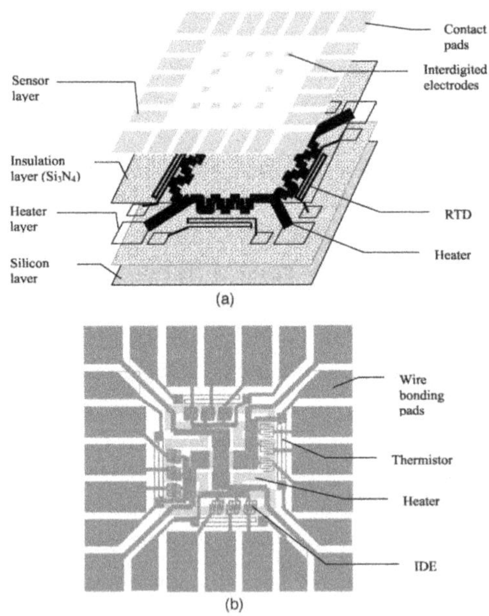

Figure 12.8 Schematic view of the different layers composing the integrated sensor [33].

Maybe the embedded heating is principally used to evaluate the desorption time of gases and so to improve selectivity (see the next paragraph). This example is particularly interesting because allows us to identify the principal issues to move from a lab prototype to a device for real life applications, introducing at the same time some solutions for the final device packaging. However, we have to point out that in the paper there is no clear information about the main principle or technique employed to achieve a selective sensing.

12.4 Routes to Improve the Selectivity

We have shown in the previous paragraphs that CNTFET-based sensors can be very sensitive to gas exposure (100 ppt for NO_2 [34]). However, the real concern is to demonstrate a suitable selectivity.

We can mention four different methods which have been proposed to achieve this goal:

- Functionalization of the channel with a specific polymer
- Functionalization of the channel with DNA
- Diversification of the metallic electrodes
- SWCNT decoration with diversified metallic nanoparticles.

The first team suggesting a method to clearly improve the CNTFET-based gas sensor selectivity was the team led by H. Dai at Stanford University [35]. Their idea was to functionalize the CNTFETs using specific polymers to enhance sensitivity to specific gases.

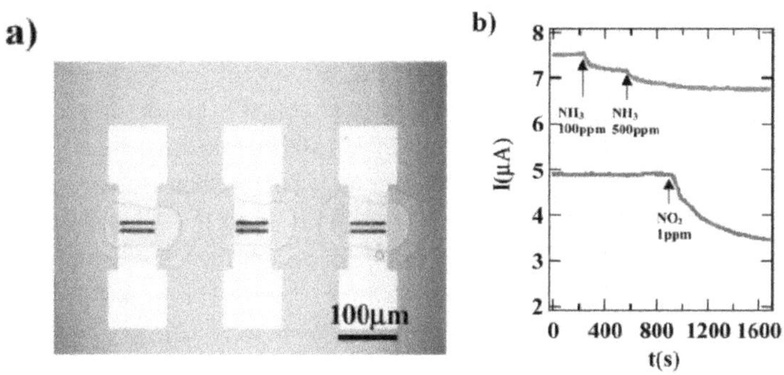

Figure 12.9 (a) CNTFET transistors functionalized using the micro-spotting technique. (b) Experimental results for CNTFETs functionalized with PEI (violet solid line) and with Nafion (red line) after exposure to NO_2 and NH_3 [35].

This technique echoes the functionalization technique used also for Sound Acoustic Wave (SAW) filter device for gas sensing [36,37] and has been used also for SWCNT-network-based sensors developed by Nanomix Inc. for CO_2 sensing [38,39]. The CNTFETs were fabricated bridging Mo electrodes with SWCNTs achieved by CVD in-situ growth on a SiO_2/Si substrate [40]. First the CNTFETs were functionalized using Polyethyleneimine (PEI) simply by immersion of the CNTFET platform. Following the PEI functionalization, a sudden change of CNTFET electrical character was observed: CNTFETs switched from p-type to n-type. This behavior was linked to the doping of high-density amine on PEI molecules [41] on the nanotubes.

Concerning the physical interpretation of the phenomenon, the effect of polymer on CNTFETs was explained by the doping on the nanotube and not by the modulation of the metal work function (Schottky barrier change). The results after exposure, for the PEI coated CNTFETs showed an extremely higher sensitivity to NO_2 compared to bare CNTFETs: The CNTFET sensors could detect a concentration of 100 ppt, and saturated at 3 ppb; 100 ppt constitutes up to now the lowest gas concentration detected by CNTFET gas sensors in the world.

This effect was explained by the reductions of electron number in the channel, and of the conductance, caused by NO_2 which is a strong "electron-withdrawing" gas. Concerning the exposure to interferent gases, the functionalized CNTFETs show no transfer characteristic change after exposure to NH_3 (up to 1000 ppm), CO, CO_2, CH_4, H_2, and O_2.

In the second part of their study, researchers functionalized a CNTFET array using Nafion, deposited by micro-spotting. The objective was to reduce the sensor sensitivity for "electron-donating" gas species. In this case, the sensors showed a good sensitivity for concentrations between 100 and 500 ppm but no response to 1 ppm of NO_2. No further results for higher NO_2 concentrations are given in the paper.

Stanford team employed a new approach for rapidly desorbing gas molecules: They refreshed the sensors using UV light illumination [42]. This approach has also been patented [43].

From our point of view, polymer functionalization is really attractive but some concerns could be raised. Polymers are highly sensitive to environment moisture and are the principal cause of the high false alarm rate observed for polymer-functionalized SAW-filter-based sensors. Moreover, the integrity of the polymer layer after several UV exposures must be checked.

Another example of utilization of CNTFET functionalization to improve sensitivity and selectivity has been shown by researchers from University of Pennsylvania [44,45]. They chose the "bio-functionalization" route using several DNA sequences. In more details, they functionalized *in situ* growth SWCNT, bridging two gold electrodes, in order to enhance the sensing sensitivity to specific gases. Subsequently, they exposed the CNTFET array to different vapors: methanol, propionic acid, trimetylamine (TMA), dinitrotoluene (DNT) and DMMP.

They found that the DNA decorated CNTFETs had an enhanced sensitivity (Table 12.1). However this enhancement depended on the DNA sequence used for a specific gas sensing. These results made researchers think that using a large array of CNTFET functionalized using "ad-hoc" DNA sequences, could permit to obtain a sort of electronic nose to identify univocally a large number of chemical and biological targets. However, up to now, the physical phenomenon at the basis of this dramatic sensitivity enhancement has not been adequately explained and no clear interpretation formulated.

Table 12.1 Source/drain current change after exposure to different analytes for a CNTFET functionalized using two different DNA sequences [69]

Odor	Vapor pressure (Torr)	Estimated concn (ppm)	% $\Delta I/I$ Bare swCN	swCN+ Seq 1	swCN+ Seq 2
water	17.5	700	0 ± 1	0 ± 1	0 ± 1
propionic acid	4	150	0 ± 1	17 ± 2	+8 ± 1
TMA	500	20000	-9 ± 2	−20 ± 2	−30 ± 2
Methanol	100	4000	0 ± 1	−12 ± 2	−20 ± 2
DMMP	0.6	25	0 ± 1	−14 ± 2	−7 ± 2
DNT	1	40	0 ± 1	−14 ± 4	−4 ± 2

Another approach to improve selectivity is to diversify the CNTFET metal electrodes. This concept has been developed by a joint-research team of Thales and Ecole Polytechnque in France (Nanocarb Lab.) [46–52]. More precisely, they exploited the mechanism described by IBM researchers (extended to all the gases and metal electrodes) the modulation of the metal work function operated by gas exposure, to achieve the fingerprinting of a specific gas using a network of CNTFETs. These last were fabricated using different metals for the contacts: The change of transfer characteristics of each transistor after gas exposure can identify a gas unambiguously. This idea was also introduced by G. Gu of Molecular Nanosystem Inc. in 2002 even if the concept of CNT network, for gas fingerprinting, had not been developed [53]. As prove-of-concept, Nanocarb team manufactured CNTFETs composed of four different metals as electrodes: Au, Pd, Pt and Ti. The work functions for these metals are, respectively, 5.1 eV for the first two, 5.65 eV and 4.33 eV for the last [54–58]. They employed these metals in order to verify if the gas interaction was strictly related only to the initial work function difference between the different metals or if it depended on the nature of gas and electrode interaction.

Authors performed several measurements using different electrodes design (20 transistors for each metal with different electrodes distance) and these results were in close agreement with the initial hypothesis: It seemed that the sensor sensitivity was not influenced by the number of nanotubes linking the electrodes. The results obtained after exposure of CNTFETs transistors to 1 ppm DMMP at ambient conditions are shown in Table 12.2. From these measurements, researchers formulated the hypothesis that different metals interact in a different way with the DMMP.

Table 12.2 A summary of the results and number of transistors tested for each different metal electrode

Current reduction after DMMP exposure	
Metals	I_{DS} current reduction % after exposure (1 ppm DMMP) Error ±5%
Ti/Pd	90
Ti/Au	65
Ti	50
Ti/Pt	20

They pointed out that the Pd and Au transfer characteristic change was very different. This confirmed that the effect of gas on CNTFET characteristics was not directly related to the initial metal work function but more probably to a specific chemical interaction between the metal and the gas.

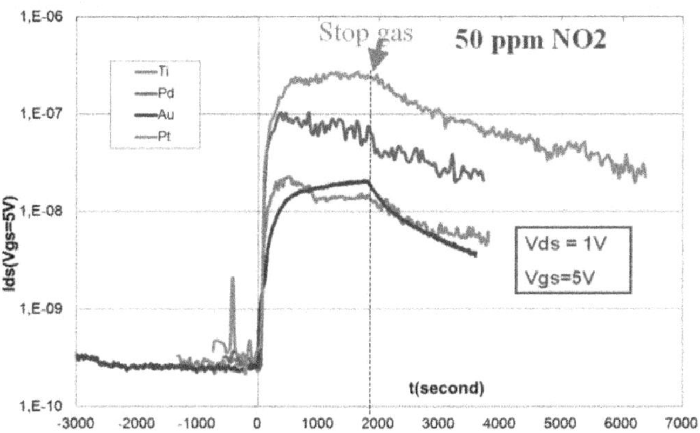

Figure 12.11 Change in I_{DS} currents of CNTFERT using different metals as electrodes under 50 ppm of NO_2.

Similar results were obtained for 3 other gases (NO_2, CO, NH_3) at the same concentration (50 ppm). And all of them show different behavior of the sensors which are summarized in Table 12.3 showing that each gas interact in different ways with each electrodes/ nanotubes systems.

Table 12.3 I_{DS} (in percent) change of CNTFET using different metals as electrodes under four types of analytes after 10 min exposure ±5%

Metals	CO (50 ppm)	NO_2 (50 ppm)	NH_3 (50 ppm)	DMMP (1 ppm)
Ti	10%	4400%	-50%	-50%
Pd	40%	38000%	-30%	-90%
Au	70%	6600%	-20%	-65%
Pt	~0%	84000%	-40%	-20%

These systems also show a very long desorption time because no special strategy was used to reinitialize sensors.

A very interesting approach has also been developed by researchers at Nanomix Inc. and Pittsburgh University. They exploited an idea first introduced in 2001 by Kong *et al.* of Stanford

University [59]: to decorate the SWCNT channel of a CNTFET with metal nanoparticles. They deposed nanoparticles of different metals on networks of SWCNTs connecting two Pd electrodes (Figure 12.12).

Therefore they fabricated an array of CNTFETs in which each one was characterized by a different metal "decoration" [60,61]. They exposed this array to several gases (NO [57], H_2, CO, CH_4, H_2S, NO_2, NH_3 [61]) and observed a specific change of the transfer characteristics of each transistor as a function of the nature of the nanoparticles and of the gas. These results made researchers think that a large array of "metal-decorated" CNTFETs could be used in order to recognize univocally the gases identifying their electronic fingerprinting.

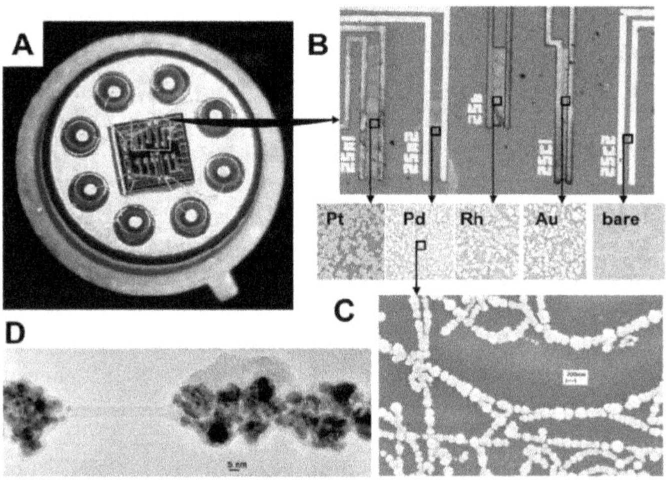

Figure 12.12 (a) A CNTFET-based sensor chip. (b) An array of CNTFETs decorated using different metals. (c) An SEM image of CNT with metal particles on their surfaces. (d) An HRTEM image of Pd decorated CNTs [81].

Regarding the physics of the phenomenon, researchers from Stanford, who had analyzed the electrical behavior of one CNTFET decorated with Pd nanoparticles after exposure to H_2, stated that H_2 dissociated on the Pd surface lowered the Pd work function. This reduction gave rise to an increased transfer of electron from Pd to SWCNT, naturally p-type, reducing the channel conductance. This hypothesis could explain the experimental results for H_2 but could not be applied for other gas species.

Kauffman *et al.* [61] failed to try an exhaustive explanation for the phenomenon but made a very interesting analysis. They underlined the main difference between the interactions of metal nanoparticles/SWCNTs and metal electrodes/SWCNTs. In the first case without any molecular interaction (exposure to gas), the nanoparticles acted exactly as inert hole scattering sites, reducing mobility. Only upon molecular interaction their behavior was exactly the same of the metal electrodes (Schottky barrier modulation caused by work function change). In this case the metal nanoparticle/SWCNTs junctions could be described using the Yamada model, introduced previously.

12.5 Conclusions

In this chapter, we exposed different recent scientific works dealing with CNTFET-based gas sensors. These devices show very promising performances for striking hugely the sensor market in the next years. A lot of advantages could be advanced: the extreme sensitivity (100 ppt for NO_2), the very short response and recovery time (some seconds), efficiency at room temperature, the low power consumption, the CMOS technology compatibility (relatively low cost), etc. One major issue remains: to demonstrate definitively a high selectivity compared with existing technology. Different routes to improve selectivity, showing interesting results, have been proposed. They are based on the electrodes metal diversification, polymer functionalization, metal particle deposition on the SWCNT mat, time desorption resolution for various testing gases. The common thing in all these approaches is the tendency to multiply the parameters (the metal electrodes nature, the functionalizing polymers, the metal particles on the mat, ...) in order to develop a sort of matrix that will allow people to identify accurately the gases. This seems to be a suitable route to achieve a sort of selectivity with this kind of devices. We could suggest that the utilization of these approaches jointly could improve dramatically the final device selectivity.

Finally, we can add that the theoretical efforts to understand deeply the interaction of gases with CNTFETs (obtained using a single SWCNT or a mat) must be enhanced. It is not completely clear, for example, the effect of gases on the polymer or DNA functionalized CNTFETs or on the metal electrodes. Only understanding these phenomena will allow us to improve the efficiency of these sensors. For this reason a joint effort of theoretical teams and engineering teams can help to overcome the technological and theoretical bottlenecks and boost the development of this new generation of ultra-miniaturized sensors.

References

1. Comini, E. (2006) Metal oxide nano-crystals for gas sensing, *Anal Chim Acta 568*, 28–40.
2. Kong, J., Franklin, N. R., Zhou, C., Chapline, M. G., Peng, S., Cho, K., and Dai, H. (2000) Nanotube molecular wires as chemical sensors, *Science 287*, 622–625.
3. Sinha, N., and Ma, J. (2006, Mars) Carbon nanotube-based sensors, *J Nanosci Nanotechnol.*
4. Chen, R. J., Choi, H. C., Bangsaruntip, S., Yenilmez, E., Tang, X., Wang, Q., Chang, Y., and Dai, H. (2004) An investigation of the mechanisms of electronic sensing of protein adsorption on carbon nanotube devices, *J Am Chem Soc 126*, 1563–1568.
5. Chen, R. J., Bangsaruntip, S., Drouvalakis, K. A., Wong Shi Kam, N., Shim, M., Li, Y., Kim, W., Utz, P. J., and Dai, H. (2003) Noncovalent functionalization of carbon nanotubes for highly specific electronic biosensors, *Proc Nat Acad Sci USA 100*, 4984–4989.
6. (WO/2004/059298) Miniaturized gas sensors featuring electrical breakdown in the vicinity of carbon nanotube tips.
7. Goldoni, A., Larciprete, R., Petaccia, L., and Lizzit, S. (2003) Single-wall carbon nanotube interaction with gases: sample contaminants and environmental monitoring, *J Am Chem Soc 125*, 11329–11333.
8. Ueda, T., Norimatsu, H., Bhuiyan, M. M. H., Ikegami, T., and Ebihara, K. (2006) NO sensing property of carbon nanotube based thin film gas sensors prepared by chemical vapor deposition techniques, *Jap J Appl Phys Part 1 Regular Papers Short Notes and Review Papers 45*, 8393.
9. Li, J., Lu, Y., Ye, Q., Cinke, M., Han, J., and Meyyappan, M. (2003) Carbon nanotube sensors for gas and organic vapor detection, *Nano Lett 3*, 929–933.
10. Valentini, L., Armentano, I., and Kenny, J. (2005) Electrically switchable carbon nanotubes hydrophobic surfaces, *Diamond Relat Mater 14*, 121–124.

11. Bauschlicher, C. W., and Ricca, A. (2004) Binding of NH_3 to graphite and to a (9,0) carbon nanotube, *Phys Rev B 70*, 115409.

12. Zhao, J., Buldum, A., Han, J., and Lu, J. P. (2002) Gas molecule adsorption in carbon nanotubes and nanotube bundles, *Nanotechnology 13*, 195.

13. Zhao, J., Buldum, A., Han, J., and Lu, J. P. (2001) Gas molecules adsorption on carbon nanotubes, in *Materials Research Society Symposium Proceedings*, p 13-13. Warrendale, Pa.; Materials Research Society; 1999.

14. Peng, S., and Cho, K. (2000) Chemical control of nanotube electronics, *Nanotechnology 11*, 57–60.

15. Zhang, J., Boyd, A., Tselev, A., Paranjape, M., and Barbara, P. (2006) Mechanism of NO2[2 to be sub] detection in carbon nanotube field effect transistor chemical sensors, *Appl Phys Lett 88*, 123112.

15B. Peng, N, Zhang Qing, Chow C.L., Tan O.K., and Marzari N. (2009) Sensing mechanisms for carbon nanotube based NH_3 gas detection, *Nano Lett 9*, 1626.

16. Derycke, V., Martel, R., Appenzeller, J., and Avouris, P. (2002) Controlling doping and carrier injection in carbon nanotube transistors, *Appl Phys Lett 80*, 2773–2775.

17. Heinze, S., Tersoff, J., Martel, R., Derycke, V., Appenzeller, J., and Avouris, P. (2002) Carbon nanotubes as schottky barrier transistors, *Phys Rev Lett 89*, 106801.

18. Heinze, S., Tersoff, J., and Avouris, P. (2003) Electrostatic engineering of nanotube transistors for improved performance, *Appl Phys Lett 83*, 5038–5040.

19. Cui, X., Freitag, M., Martel, R., Brus, L., and Avouris, P. (2003) Controlling energy-level alignments at carbon nanotube/Au contacts, *Nano Lett 3*, 783.

20. Léonard, F., and Tersoff, J. (2000) Role of fermi-level pinning in nanotube schottky diodes, *Phys Rev Lett 84*, 4693–4696.

21. Crispin, X., Geskin, V., Crispin, A., Cornil, J., Lazzaroni, R., Salaneck, W. R., and Brédas, J. L. (2002) Characterization of the interface dipole at organic/metal interfaces, *J Am Chem Soc 124*, 8131–8141.

22. Crispin, X. (2004) Interface dipole at organic/metal interfaces and organic solar cells, *Solar Energy Mater Solar Cells 83*, 147–168.

23. Yamada, T. (2004) Modeling of carbon nanotube Schottky barrier modulation under oxidizing conditions, *Phys Rev B Condensed Matter Mater Phys 69*, 125408–125408.

24. Yamada, T. (2006) Equivalent circuit model for carbon nanotube Schottky barrier: Influence of neutral polarized gas molecules, *Appl Phys Lett 88*, 083106.
25. Kumar, S., Blanchet, G. B., Hybertsen, M. S., Murthy, J. Y., and Alam, M. A. (2006) Performance of carbon nanotube-dispersed thin-film transistors, *Appl Phys Lett 89*, 143501.
26. Kumar, S., Murthy, J. Y., and Alam, M. A. (2005) Percolating conduction in finite nanotube networks, *Phys Rev Lett 95*, 066802.
27. Snow, E. S., Novak, J. P., Campbell, P. M., and Park, D. (2003) Random networks of carbon nanotubes as an electronic material, *Appl Phys Lett 82*, 2145.
28. Bekyarova, E., Itkis, M. E., Cabrera, N., Zhao, B., Yu, A., Gao, J., and Haddon, R. C. (2005) Electronic properties of single-walled carbon nanotube networks, *J Am Chem Soc 127*, 5990–5995.
29. Novak, J. P., Snow, E. S., Houser, E. J., Park, D., Stepnowski, J. L., and McGill, R. A. (2009) Nerve agent detection using networks of single-walled carbon nanotubes, *Appl Phys Lett 83*, 4026–4028.
30. Snow, E. S., Perkins, F. K., and Robinson, J. A. (2006) Chemical vapor detection using single-walled carbon nanotubes, *Chem Soc Rev 35*, 790–798.
31. Hopkins, A. R., and Lewis, N. S. (2001) Detection and classification characteristics of arrays of carbon black/organic polymer composite chemiresistive vapor detectors for the nerve agent simulants dimethylmethylphosphonate and diisopropylmethylphosponate, *Anal Chem 73*, 884–892.
32. Jeong, S., Kim, K., An, K., Hwang, H., Han, C., Yun, M., and Lee, Y. Fabrication of Gas Sensor Using Single-Walled Carbon Nanotubes Dispersed in Dichloroethane.
33. Li, J., Lu, Y., Ye, Q., Delzeit, L., and Meyyappan, M. (2005) A gas sensor array using carbon nanotubes and microfabrication technology, *Electrochem Solid State Lett 8*, H100.
34. Delzeit, L., Chen, B., Cassell, A., Stevens, R., Nguyen, C., and Meyyappan, M. (2001) Multilayered metal catalysts for controlling the density of single-walled carbon nanotube growth, *Chem Phys Lett 348*, 368–374.
35. Qi, P., Vermesh, O., Grecu, M., Javey, A., Wang, Q., Dai, H., Peng, S., and Cho, K. J. (2003) Toward large arrays of multiplex functionalized carbon nanotube sensors for highly sensitive and selective molecular detection, *Nano lett 3*, 347–352.
36. Nieuwenhuizen, M. S., and Harteveld, J. L. N. (1997) Studies on a surface

acoustic wave (SAW) dosimeter sensor for organophosphorous nerve agents, *Sens Actuators B Chem 40*, 167–173.

37. Williams, D., and Pappas, G. (1999) Rapid identification of nerve agents Sarin (GB) and Soman (GD) with the use of a field-portable GC/SAW vapor detector and liquid desorption front-end device, *Field Anal Chem Technol 3*, 45–53.
38. Star, A., Joshi, V., Thomas, D., Niemann, J., Gabriel, J. C. P., and Valcke, C. Nanoelectronic CO_2 breath sensors, *Tech Proc 2005 NSTI Nanotechnol* 8–12.
39. Star, A., Han, T. R., Joshi, V., Gabriel, J. C. P., and Grüner, G. (2004) Nanoelectronic carbon dioxide sensors, *Adv Mater 16*, 2049–2052.
40. Franklin, N. R., Wang, Q., Tombler, T. W., Javey, A., Shim, M., and Dai, H. (2002) Integration of suspended carbon nanotube arrays into electronic devices and electromechanical systems, *Appl Phys Lett 81*, 913.
41. Shim, M., Javey, A., Kam, N. W. S., and Dai, H. (2001) Polymer functionalization for air-stable n-type carbon nanotube field effect transistors, *J Am Chem Soc 123*, 11512–11513.
42. Chen, R. J., Franklin, N. R., Kong, J., Cao, J., Tombler, T. W., Zhang, Y., and Dai, H. (2001) Molecular photodesorption from single-walled carbon nanotubes, *Appl Phys Lett 79*, 2258.
43. Dai, H., and Chen, R. J. (2008) Photodesorption in carbon nanotubes. Patent WO 02/086480.
44. Staii, C., Johnson Jr, A. T., Chen, M., and Gelperin, A. (2005) DNA-decorated carbon nanotubes for chemical sensing, *Nano Lett 5*, 1774–1778.
45. Johnson, A. T., Staii, C., Chen, M., Khamis, S., Johnson, R., Klein, M. L., and Gelperin, A. (2006) DNA-decorated carbon nanotubes for chemical sensing, *Semiconductor Sci Technol 21*, S17.
46. Bondavalli, P., Legagneux, Pierre, Le Barny, Pierre, Pribat, Didier, and Nagle, Julien. (2006, Décembre 7) Patent WO/2006/128828, Array of FET transistors having a nanotube or nanowire semiconductor element and corresponding electronic device, for the detection of analytes.
47. Bondavalli, P., Legagneux, P., and Pribat, D. (2007) Capteurs de gaz a base de transistors a nanotubes de carbone: une nouvelle generation de capteurs nanometriques, *Revue De L Electricite ET DE l Electronique 4*, 54.
48. Bondavalli, P., Legagneux, P., Pribat, D., Balan, A., and Nazeer, S. (2008)

Gas fingerprinting using carbon nanotubes transistor arrays, *J Exp Nanosci 3*, 347–356.

49. Bondavalli, P., Legagneux, P., and Pribat, D. (2009) Carbon nanotubes based transistors as gas sensors: State of the art and critical review, *Sens Actuators B Chem 140*, 304–318.
50. Bondavalli, P., Gorintin, L., Legagneux, P., Pribat, D., Cailler, L., and Simonato, J. P. CNTFET Gas Sensors Using SWCNT Mats: Method for Low-cost Fabrication, Solution to Improve Selectivity, Experimental Results using Interfering Agents.
51. Bondavalli, P., Legagneux, P., Pribat, D., and Gorintin, L. highly selective sensing of DMMP and NH_3 using CNTFET array based gas sensors obtained using Spray-gun technique and fabricated with different metal electrodes to achieve gas electronic fingerprinting.
52. Gorintin, L., Bondavalli, P., Legagneux, P., and Pribat, D. (2009) High performances cntfets achieved using CNT networks for selective gas sensing, in *Proc SPIE*, p 739909.
53. Gu, G., and Pan, L. S. (2004, June 17) Nanotube chemical sensor based on work function of electrodes.
54. Morikawa, Y., Ishii, H., and Seki, K. (2004) Theoretical study of n-alkane adsorption on metal surfaces, *Phys Rev B 69*, 41403.
55. Mori, T., Kozawa, T., Ohwaki, T., Taga, Y., Nagai, S., Yamasaki, S., Asami, S., Shibata, N., and Koike, M. (2009) Schottky barriers and contact resistances on p type GaN, *Appl Phys Lett 69*, 3537–3539.
56. (1979) Solid surface physics /contributions by J. Holzl, F.K. Schulte, H. Wagner. Springer-Verlag, Berlin ; New York.
57. Yang, M. H., Teo, K. B. K., Milne, W. I., and Hasko, D. G. (2005) Carbon nanotube schottky diode and directionally dependent field-effect transistor using asymmetrical contacts, *Appl Phys Lett 87*, 253116.
58. Kim, K. K., Bae, D. J., Yang, C. M., An, K. H., Lee, J. Y., and Lee, Y. H. (2005) Nanodispersion of single-walled carbon nanotubes using dichloroethane, *J Nanosci Nanotechnol 5*, 1055–1059.
59. Kong, J., Chapline, M. G., and Dai, H. (2001) Functionalized carbon nanotubes for molecular hydrogen sensors, *Adv Mater 13*, 1384–1386.
60. Star, A., Joshi, V., Skarupo, S., Thomas, D., and Gabriel, J. P. (2006) Gas sensor array based on metal-decorated carbon nanotubes, *J Phys Chem B 110*, 21014–21020.

61. Kauffman, D. R., and Star, A. (2007) Chemically induced potential barriers at the carbon nanotube- metal nanoparticle interface, *Nano Lett* 7, 1863–1868.

Chapter 13

Carbon Nanotubes and Their Composites for Viscoelastic Applications

Edwin Hang Tong Teo,[a] Maziar Shakerzadeh,[b] and Beng Kang Tay[b]
[a]*Temasek Laboratories@NTU, 50 Nanyang Avenue, 639798 Singapore*
[b]*School of Electrical and Electronic Engineering, Nanyang Technological University, 50 Nanyang Avenue, 639798 Singapore*
htteo@ntu.edu.sg

Vibration-induced failure of devices has plagued the developers of mechanical systems since the development of early machinery. This problem has now manifested itself into design issues in small-scale mechanical devices. Since the first microelectromechanical device was reported in 1967, numerous issues with the longevity, energy lost, wear, instability and performance of these devices have been associated with parasitic vibration noises. This problem is especially acute now in the rapidly developing area of nanoelectromechanical systems because of the extreme surface-to-volume ratios these devices have. As such, numerous studies are currently devoted to improving the vibration damping in nano-devices. The traditional method for addressing these issues is to integrate dampers and low-friction materials together in a device. However, unlike macro environments, implementation of such protection on the sub-

Carbon Nanotubes and Their Applications
Edited by Qing Zhang

ISBN 978-981-4241-90-8 (Hardcover), 978-981-4303-18-7 (eBook)
www.panstanford.com

micrometer scale is either unfeasible or extremely costly because of the complex fabrication involved. A more reasonable solution is to use an intrinsic damping material built directly into the device for energy dissipation and load recovery. One promising solution is the use of carbon nanotubes and their composites. This is one of the main driving factors in the study of carbon nanotube films and carbon nanotube–polymer composites for their mechanical strength and viscoelastic properties. This chapter thus focuses on the strength and weaknesses associated with using carbon nanotubes and their composites in damping and viscoelastic applications.

13.1 Introduction

Carbon nanotubes' (CNTs) exceptional mechanical strength, low density, light weight and high elasticity make them ideal candidates as structural materials or for material reinforcements in many applications [1–6]. In many respects, CNTs are like ideal nanometer-sized strands of fiber or pillars. They are strong, resilient to repeated deformation and bending. Moreover, many of CNTs' properties can be tuned by adjusting their geometric dimensions as well as their type (single wall (SWCNT), double wall, multiwall (MWCNT), etc.). It is this possibility of tailoring the properties of CNTs that makes them so exciting for many applications.

Although individual CNTs are already remarkable in terms of mechanical strength, it is when they are group together that their true potential is realized [7]. Singly, CNTs have good prospects as nanowires or conducting filaments in nano-electronics but for current mechanical applications requiring high strain and high frequency loading-unloading cycles, a single strand of CNT is simply not suitable. However, by binding them together either as CNT-forest or CNT films, [8, 9] real practical mechanical applications can be realized. In fact, one of the most interesting mechanical properties of CNTs and their composites lies in their viscoelastic nature.

Individual CNTs have recently been reported to exhibit negative structural stiffness leading to enhance overall microstructure stiffness and damping [10, 11]. This effect of CNTs has been verified by both molecular dynamics and finite element simulations [5, 12]. More recently, Yap *et al.* have also experimentally demonstrated this through atomic force microscopy (AFM) cantilever loading on

MWCNTs [10, 11] (more on this work will be discussed later). In addition to this intrinsic effect of individual tubes, energy dissipation can be further enhanced when large amount of CNTs interact with each other or other elements creating frictional loses. This effect happens in CNT films as well as CNT-based composites. This enhancement is due to the CNTs moving or slipping against each other or their surrounding matrix material. This slippage causes energy losses due to frictional interactions, which in turn leads to energy losses, which are translated into enhanced damping behavior.

As there are two main types of CNT-based damping materials, this chapter is divided into two separate sections. The first part discusses pure CNT films while the next part focuses on CNT composites; mainly CNT/polymer composites.

13.2 Pure Carbon Nanotube Films

Current conventional damping treatments are based on several different techniques and materials, including elastomers, actuating (magnetic, piezoelectric, etc.) materials and fluids, and electronic feedback circuits, etc. [13–15]. These techniques have generally been reliable for large systems. However, they are usually heavy, non-compact and may not be suitable in high temperature environments. Moreover, for smaller devices, the complexity involve in miniaturizing these techniques make their implementation either not feasible or simply too costly. A more direct approach through the use of viscoelastic materials to damp vibration would be better. One of the more promising materials to be used in this form are CNT films.

To study the damping properties of CNT films, Koratkar *et al.* [16] used a piezo-silica sandwich beam with an embedded MWCNT film in a flat-wise bend testing experiment. The CNT film used in their setup was several tens of micrometers thick with tube diameters at ~30 nm. Chemical vapor deposition (CVD) using xylene–ferrocene precursors was used to synthesis their CNTs. [17, 18] Their sandwich films consisted of a top piezoelectric sheet with a bottom silica substrate encasing the MWCNT film between them. The $Pb(Zr,Ti)O_3$ (PZT) and silica sheets were bonded using a special cyanoacrylate adhesive. The test setup is schematically shown in Figure 13.1b. A reference sample with no CNT reinforcement was also tested to compare it with the CNT sandwiched film.

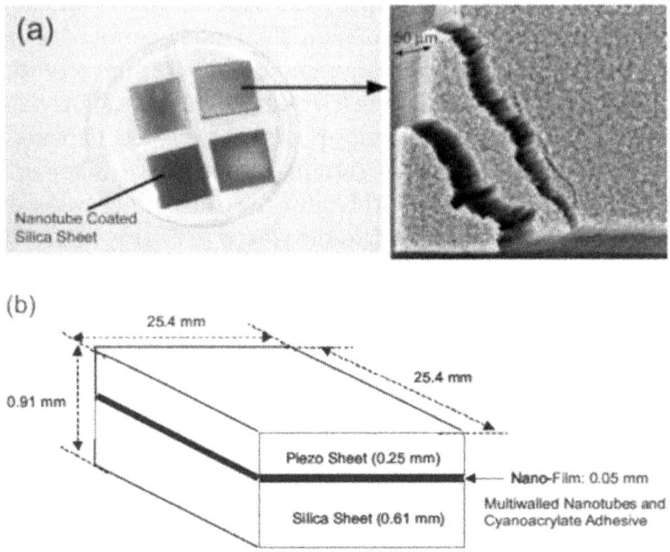

Figure 13.1 (a) CVD grown CNT films with thickness at a few tens of micrometers. Four prepared sheets are shown in the photograph (left). (b) Schematic of sandwich beam [16].

Figure 13.2 shows the results of a frequency-sweep test to compare the dynamic response between the CNT-based beam and the baseline beam. From the figure, it is obvious that the sandwich structure with CNT reinforcement shows a decrease in the dynamic amplification at resonance as well as an increase in the first bending frequency from 700 Hz to 840 Hz. The reduced amplification is an indication of increase damping while the frequency shift can be explained by an increase in stiffness of the overall structure. Importantly, both are the key qualities required in good damping materials. In fact, using a finite element beam model simulation to fit the experiment data, Koratkar *et al.* [16] also found that the CNT-reinforced sandwich beam has a damping ratio (Z_b) of 0.3 and a cross-sectional stiffness (EI) of 87.7×10^{-3} Nm^2. Comparatively, the baseline beam has a Z_b of 0.1 and EI of 65.1×10^{-3} Nm^2. Therefore, based on their calculations [16], the implementation CNT film would result in a 200% increase in structural damping and 30% increase in overall stiffness. This is quite remarkable considering that the addition of the CNT film adds almost no extra weight to the structure.

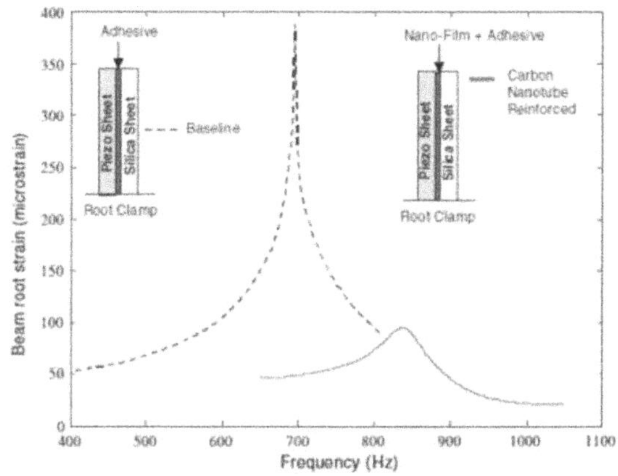

Figure 13.2 Dynamic frequency response of the test structures. Comparing the baseline beam and the nanotube reinforced sandwich beam for a 50 V rms frequency-sweep test, the nanotube reinforced structure shows a very significant increase both in damping and stiffness [16].

A more close-up examination of what exactly happens during loading cycles of CNT films was carried by Cao *et al.* [19] using vertically aligned MWCNTs film that were detached from their substrates. The lengths of the CNTs used in their case were ~860–1200 μm with diameter at ~20 nm. In their work, they examine the free standing aligned CNT films directly by compressive loading/unloading cycles. The entire CNT films were basically compressed and released along the film thickness direction at a constant strain, repeatedly for thousands of cycles. Throughout the entire test cycle range (lasted in excess of 10,000 cycles), the CNT films were never observed to fracture, tear or collapse under compression. In fact, initially even films compressed to only 15% of their original height were still observed to return to their original length upon unloading. However, it was observed that after a few hundred cycles, a noticeable (20% of total film length) shortening occurred. This is due to the slight buckling of the CNTs. The CNT film recovery was also reported to be extremely fast at >2000 μm/sec [19].

Scanning electron microscopy (SEM) images (Figure 13.3) shows the deformation of the CNT films after various load-unload cycles. From the images, wave-like folds appear across the entire length of the films. These clearly indicate how the CNTs in the films actually transform to absorb the loading force, which is through collective buckling in an ordered fashion. The reason for their apparent collective deformation is due to the density and compactness of the films. In these dense CNT films, it is very difficult for individual CNTs to buckle independently due to the proximity of the neighboring tubes. This type of cooperative buckling of the CNTs is thus the most space-efficient and energetically favorable configuration to adopt during compressive loading [19]. The deformation of the CNT film in this manner allows for maximum volume reduction (without extra space requirements) with the least compressive load.

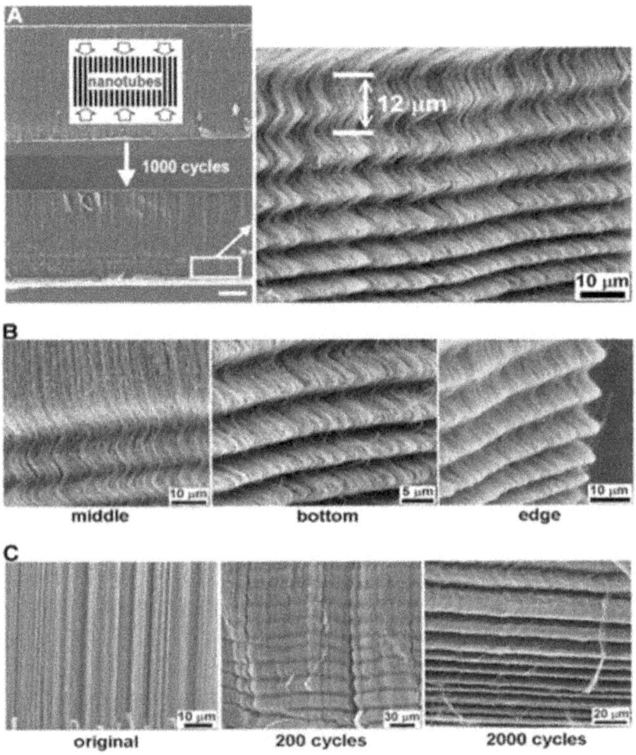

Figure 13.3 SEM micrographs of buckled CNTs under compression. (A) SEM of the original uncompressed freestanding film with thickness at 860 μm (left, top) and a compressed film (left, bottom), with a reduced thickness of 720 μm. Left scale bar, 200 μm. (Left, inset) (B) Buckling at different parts of the CNT film. (C) SEM image of a various 1.2 mm-thick films at different phases of the compression cycles [19].

More interesting is the stress–strain plot of the CNT films shown in Figure 13.4. The compressive stress of the film σ_{film} is taken as the total applied force divided by the film area whereas the strain ε, is the relative compressed length with respect to the initial length of the CNT. From the figure, three distinct phases were observed during loading and unloading. These phases could be classified as (i) an initial Hookean region with $\varepsilon < 22\%$ and elastic modulus at 50 MPa, followed by (ii) a phase of reduced modulus at 12 MPa where the ε is between 22% and 79% and finally (iii) when densification occurs and the stress raises rapidly due to the large volume reduction. These three phases are characteristic of reported open-cell foams, which show similar characteristics under loading [20–22]. More importantly, the large stress loops in both the curves shows that large amount of energy (64%) is absorbed during compression cycles, which are indicative of good damping properties.

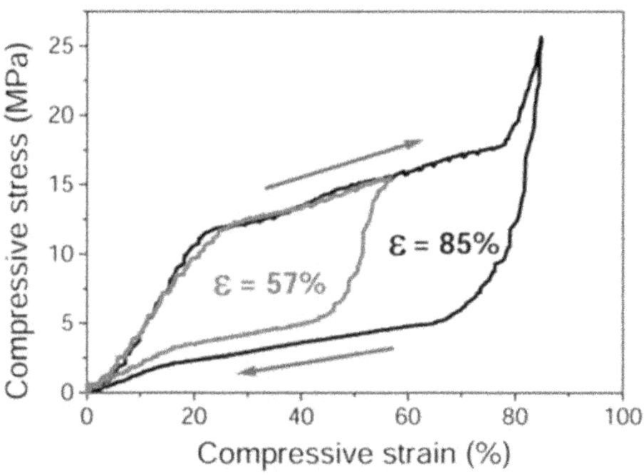

Figure 13.4 Compressive stress versus strain hysteresis of aligned CNT films. Stress-strain (σ_{film} versus ε) curves during the first cycle at set strains of ε = 85 (black curve) and 57% (red curve). Both curves show three distinct stages in their response to loading. [19]

Most of the energy dissipation in Cao *et al.*'s work was attributed to CNT to CNT interaction or the movement of air through the porous CNTs film [19]. However, recent work by Yap *et al.* [10, 11], indicated that at least some of the energy dissipation is due to the individual CNTs intrinsic viscoelastic properties. Through using AFM probes on

MWCNTs in compression experiments, Yap *et al.* have demonstrated that at different stages in the CNT bending regime (different stage of buckling of the CNTs), the force–displacement (FD) characteristic of the CNT actually changes drastically. Figure 13.5 shows the FD plot of a MWCNT with length of 1.2 μm and diameter of 30 nm. The initial linear increase in force at the beginning of the plot corresponds well with the initial Hookean characteristics of the CNT film shown in Figure 13.4. Beyond 170 nm displacement, a negative stiffness region is observed where there is a sudden force drop to a lower value, which is indicative of shell buckling. Their results also concurs with those of Cao *et al.* [19], where even after hundreds of load-unload cycles, no permanent deformation was observed.

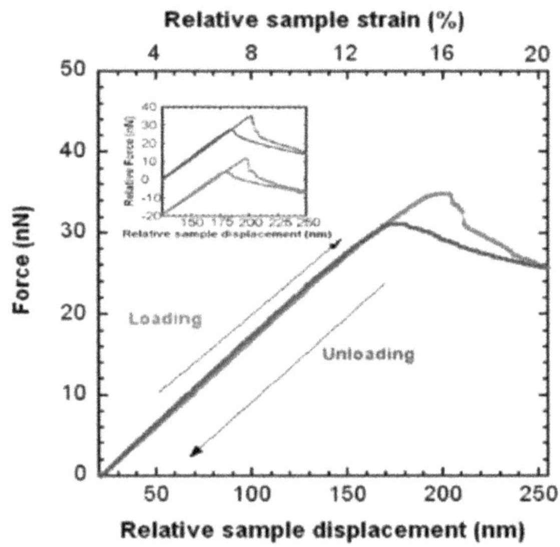

Figure 13.5 Curves of axial Force versus relative vertical displacement and strain on an AFM tip-mounted MWCNT [11].

From the FD, it can be deduced that most of the energy dissipation actually occurs only in the buckling phase of the bending regime. This is confirmed through their derivation of the tan δ and loss moduli per unit cross sectional area of the MWCNTs shown in Figure 13.6. From the figure, an increase in both the tan δ (~400–600% increase) and loss moduli (~600% increase) once the buckling starts clearly shows that the kinked phase of the bending regime contributes greatly to the overall damping property of the CNTs.

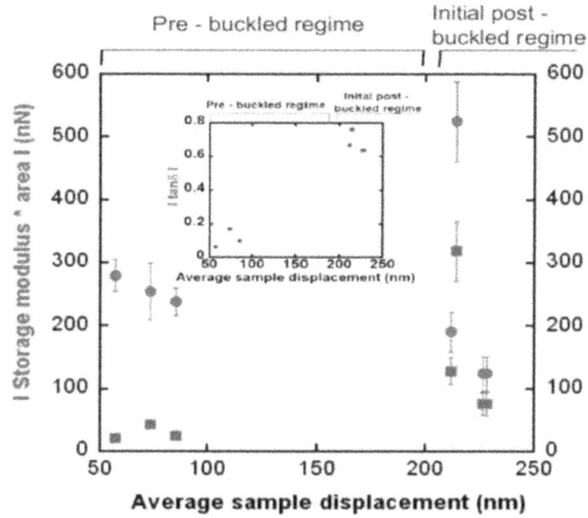

Figure 13.6 Absolute values of the storage and loss moduli (multiplied by the cross-sectional area) at different average sample displacement d_{av} in both prebuckled and initial postbuckled regimes. The inset shows the absolute value of tan δ at different average d_{av}. Marked increase in tan δ and the moduli are observed immediately after buckling [11].

One issue with CNT films is the large porosity due to the gaps between individual CNTs. These gaps reduce the overall density of the CNTs in the films, which then lowers the compression strength for energy-absorbing applications [23]. Moreover with higher porosity and thus larger distances between individual tubes, there is a weaker van der Waals force between tubes, which may affect the overall viscoelastic property of the film. As such, one challenge to the successfully application of CNT films for viscoelastic applications lies with the need to increase the overall CNT densities within the films. To reduce the porosity of CNT films, higher density CNT films have been produced by high temperature vacuum decomposition of 6H–SiC single crystals [24]. With at least 10 times higher densities (0.95 g cm^{-3}) then catalytic CVD growth nanotubes, these carbide-derived carbon (CDC) nanotube films are very promising in its use for energy absorption applications. [25] The films grown this way have CNTs with diameters at 1–3 nm and tubes density between 100,000–89,000 tubes μm^{-2}. Experimental work on these films have also shown for the first time negative stiffness in CNT films, which was previously only reported in individual CNTs [23].

Another problem with high porosity films is that these films are generally only held together by relative weak van der Waals forces and the underlying substrate, making their exposed surface prone to splitting. In essence this splitting of the CNTs means that not all the CNTs are bending uniaxially. This then effectively lowers the absolute compressive strength of the CNT film. To resolve this, Teo *et al.* attached a hard "roof" directly to the tips of the CNT film creating a sandwich structure trapping the CNTs in between the roof and the substrate [26]. The hard "roof" is actually a hard diamond like carbon film coated directly onto the aligned MWCNT film through an arc discharge plasma process. This 2-dimension-on-1-dimension approach essentially binds the CNTs on two ends and effectively eliminating splitting of the CNTs. The CNT film modified in this way is also shown to have interesting tunable viscoelastic characteristics.

13.3 Carbon Nanotube Composites

Because of their outstanding properties, polymers are of great interest for different biomedical [27] and microelectronic [28, 29] applications. One of the reasons for the introduction of polymers in these devices is to mitigate mechanical noises associated with their operations. These induced noises (or structural vibrations) could be self-inherent or more usually a result of the device's operating environment. Moreover, these excess mechanical vibrations are non-trivial impositions, which may affect the performance of the device as well as decreasing its life span. As such, polymers' roles in these devices as anti-vibration elements are extremely critical. For instance, it has been shown that the performance of a transistor is a function of the pentacene morphology, which itself is affected by the viscoelastic properties of the dielectric layer morphology of pentacene, which itself is affected by viscoelastic properties of the polymers [28]. Polymers used in this way are known as passive damping solutions.

Active treatments on the other hand use the signal acquired through a feedback control system to design the necessary control signal to rein the vibration. These techniques are generally cumbersome, overly sophisticated and simply not practical for small devices. How would one incorporate an individual sensor, actuator and feedback control unit into an already complicated nano-electro-mechanical-system (NEMS) without increasing the weight, size and design complexity? Traditional active dampers are generally reliable, robust and a main stay in large industrial systems. But for current exceedingly small devices, where every gram counts, the way to go will be passive systems, which use the intrinsic damping properties of the devices' constituents. In other words, the device uses its own components to reduce vibrations without adding any more weight and fabrication complexity. Different passive methods such as piezoelectric materials [30], incorporation of dampers [31, 32], electro-rheological/magneto-rheological fluids [33, 34] have been incorporated in practical structures to control the vibrations. However, issues such as reliability, weight and space limitations and thermal stability will be noticeable challenges once these structures are considered to be used practically [16]. Elastomers can also be subjoined to the devices to fulfill this requirement and control the vibrations.

Although the concept of using the passive approach to cutting down noises seems to be the most straight forward approach for small devices and polymers appear to be the ideal candidate for this role, other factors have to be taken into consideration. Polymers are light, easy to work with and cheap to produce, making them the material of choice. However, their viscoelastic properties may not be suitable for the entire range of vibrations the devices are subjected to. The need to enhance or vary the viscoelastic properties of these polymers are thus an important area of study. One of the most successful methods involves modifying the polymers through incorporation of nanofillers to change their viscoelastic properties.

The reason nanofillers could alter the viscoelastic property of polymers is due to the interaction between the interface of the polymers and nanofillers. This is because, in general, the interfacial slippage between the polymer matrix and the nanofillers results in considerable energy dissipation, which enhances their viscoelastic properties considerably. Obviously in this case, the choice of the type of nanofillers could play a very important role in the outcome of the polymers' viscoelastic performance. The most important aspect of the nanofillers to be considered is their surface area in contact with the polymer since the largest fractional losses (and thus energy absorption) could occur with the nanofillers having the most surface area. In order to have the largest interfacial contact area, the nanofillers should have a high surface-to-volume ratio. Moreover, in order to decrease the mass penalty (or the weight in general of the device), the nanofillers should have low mass densities. In addition, the load transfer between the polymer matrix and the nanofillers should be carefully studied. Due to these factors, there have been extensive studies of many different nanofillers such as nanospheres, nanorods and other forms of nanoparticles. In particular, the viscoelastic properties of CNTs as nanofillers have been getting increasing attention [35–38]. This is not surprising considering that nanotubes are light [39] and have a huge surface-to-volume aspect ratio. Furmaniak *et al.* [40] reported surface area of up to 650 m^2/g for double wall CNTs.

Suhr *et al.* [35] studied viscoelastic properties of nanotube/epoxy composite with (1–2%) weight fraction nanotube fillers and found that viscoelastic behavior was enhanced significantly. As shown in Figure 13.7, to disperse nanotubes in polymer matrix, both nanotubes and polycarbonate (PC) matrix were mixed in tetrahydrofuran (THF) Followed by 15 minutes sonication of the solution. The solution was then poured in methanol with THF to methanol ratio of 1:5. PC doesn't dissolve in methanol and hence immediately precipitates from it. It was the filtered and dried under

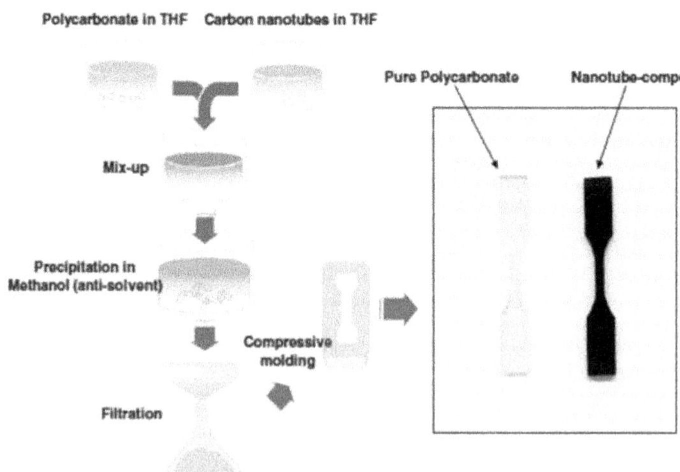

Figure 13.7 Schematics of polymer/CNT composite sample preparation [38].

vacuum conditions. In order to fabricate the dog-bone shaped tension samples (width = 3.2 mm, thickness = 3.2 mm and length = 63.25 mm) the dried precipitate was molded in a mold preheated at 205°C.

The loss and storage modulus of pure PC and PC/SWCNT composite with 1.5 wt% CNT is shown in Figure 13.8. As shown in Figure 13.8a, increasing the strain amplitude leads to a decrease in the elastic modulus of the composite remarkably, while the elastic modulus of the pure PC is almost unchanged. This implies that at large strain magnitudes, the reinforcement effect of the nanotube fillers is degraded. The loss modulus of the SWCNT reinforced polymer increased dramatically compare to that of pure PC sample.

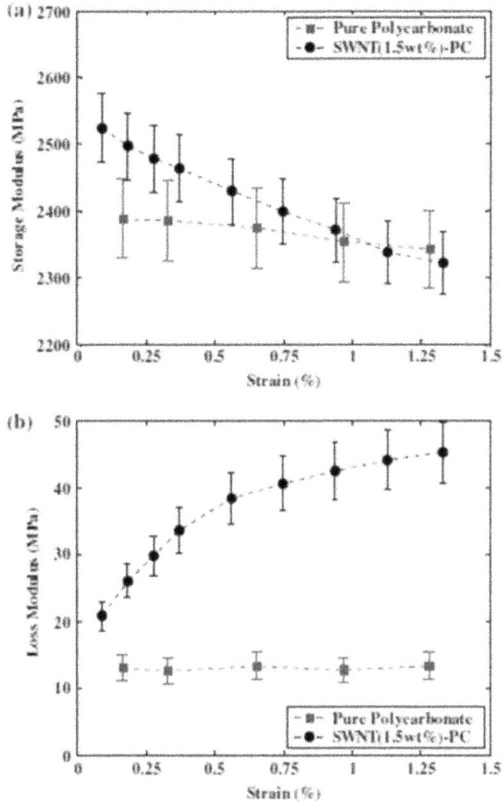

Figure 13.8 (a) Storage and (b) loss moduli vs. strain amplitude. The composite consists of 1.5 wt% SWCNT. Test frequency was kept at 10 Hz [38].

Moreover, increasing the strain amplitude in SWCNT-filled PC increases the loss modulus dramatically. Considering both the loss and storage modulus it can be deduced that at low strain amplitudes, the axial load is transferred to the CNT fillers and hence the storage modulus of the CNT-filled composite increases with respect to that of pure PC. However, increasing the strain rate results in slippage of CNTs on the polymer matrix and therefore lowers the efficiency of force transferred to the polymer matrix. As a result of this load transfer, the storage modulus decreases drastically. This slippage of the CNTs on the polymer matrix, results in considerable energy dissipation through the friction force between the nanotube fillers and the matrix. Increasing the strain amplitude activates the CNT/matrix slippage for the nanotubes that are not perpendicular to the force direction. Therefore, increasing the strain amplitude results in an increase in the loss modulus and a further decrease in the storage modulus. As it is shown, at the strain amplitude of 1.2%, a loss modulus of 45 MPa, which is 250% higher than that of PC baseline was achieved.

It should be noted that besides the CNT/polymer slippage, CNT/CNT slippage may also be responsible to frictional energy dissipation. Cumings and Zettl [41] studied the CNT/CNT friction behavior and found that 0.5 MPa shear stress is enough to activate the CNT/CNT slippage. The critical shear stress for the interfacial slippage at the nanotube/polymer was studied by Barber *et al.* [42] and it was found that the critical shear stress for CNT/polymer interfacial slippage is about 20 MPa. Since the stress level at which the loss modulus begins to increase was about 20 MPa, it can be deduced that the dominant mechanism for energy dissipation in CNT/polymer composite is the CNT/polymer slippage and not the CNT/CNT interfacial slip. This idea has also been examined by Suhr *et al.* through controlling the sonication time in the fabrication process. Increasing the sonication time results in a better dispersion of CNTs in the polymer matrix. Therefore, in the samples sonicated for short times, the CNTs will be present in the forms of bundles rather than single nanotubes. Therefore, by applying periodic tensile stress, there will be more CNT on CNT interaction whereas in longer time sonicated specimen, since the nanotubes are dispersed properly, the dominant slippage happens between the nanotubes and the polymer matrix.

As shown in Figure 13.9, increasing the sonication time and better dispersion of CNTs increases the loss modulus drastically. Hence, it can be deduced that polymer/CNT slippage is the main mechanism for the energy dispersion in CNT/polymer composites. It should be noted that since MWCNT can be dispersed to individual nanotubes, they can have more significant effect provided that a similar tube density is being used. However, since in any case it is only the outermost tube which interacts with the polymer matrix, with the same volume or mass ratio, SWCNTs tends toward better results compared with MWCNTs.

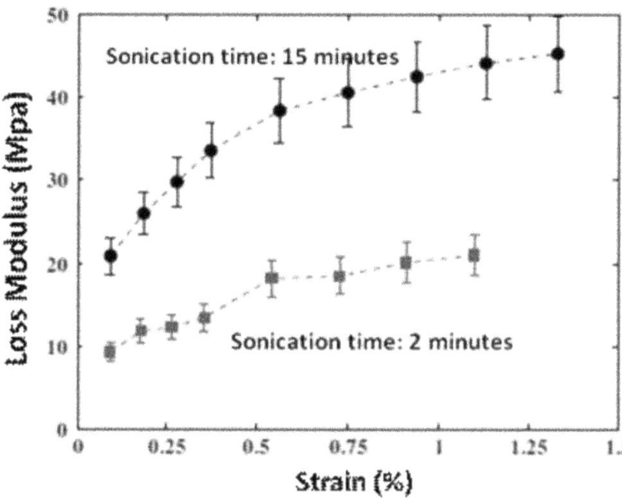

Figure 13.9 Loss modulus of polymer/CNT composite as a function of strain amplitude and sonication time [38].

As shown, dispersion of nanotubes in the polymer matrix plays an important role in the viscoelastic properties of the composite. However, the weak van der Waals force between the polymer and SWCNTs doesn't lead to proper dispersion of CNTs in the polymer matrix. In order to get a better dispersion of nanotubes, nanotubes can be oxidized by sonication in nitric acids. The dipole–dipole interactions resulted from the carboxylic groups result in better dispersion of the nanotubes and hence outstanding enhancement in loss modulus (Figure 13.10).

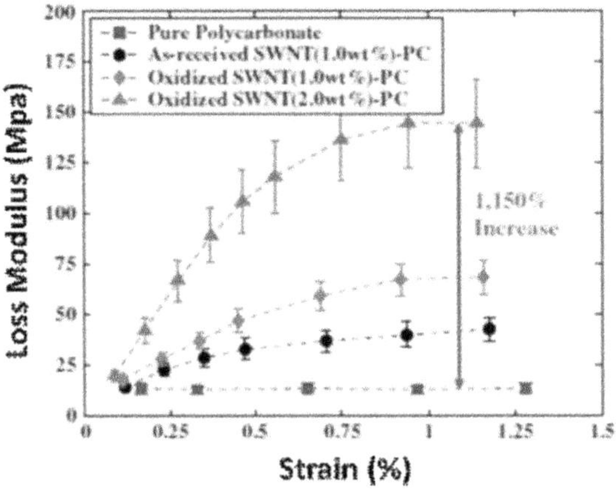

Figure 13.10 Loss moduli as a function of strain amplitude for composites with oxidized nanotubes [38].

As it is mentioned, the main mechanism for energy dissipation in CNT/polymer composites is the slippage of polymer matrix with respect to the CNT fillers. Therefore, the bonding status of the nanofillers and polymer matrix should play an important role in viscoelastic properties of the composite. The carboxylic acid groups on oxidized nanotubes, can be used to enable covalent bonding between the nanotubes and polymer matrix [43]. As shown in Figure 13.11, strong covalent bonding between the nanotubes and polymer degrades the loss modulus.

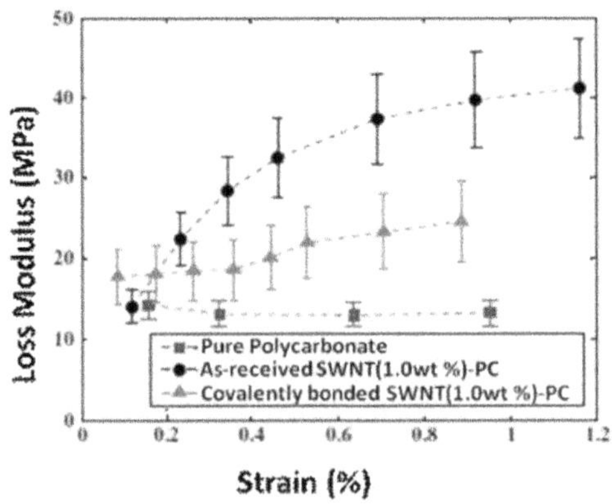

Figure 13.11 Loss moduli of PC, as received and covalently bonded CNTs as a function of strain [38].

The effect of temperature on viscoelastic properties of the nanocomposites has also been studied. As shown in Figure 13.12, loss modulus of CNT/polymer composite increases with increasing temperature. Increase in temperature results in enhancement of carbon chain mobility. This leads to easier slippage of polymer chain on nanotubes and hence increases the loss modulus.

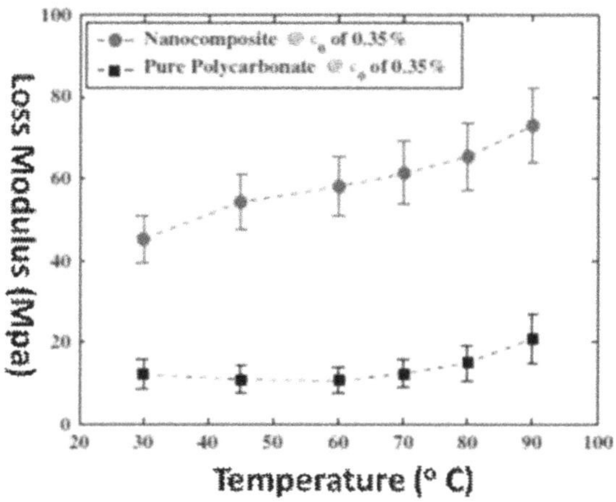

Figure 13.12 Loss modulus as a function of temperature of pure PC and PC-CNT composite [38].

Besides the experimental studies, the damping properties of CNT/polymer composites have been studied theoretically through finite element analysis [44], molecular dynamics (MD) [45], shear lag analysis [44] and beam vibration analysis. [16, 46] It should be noted that for such a one dimensional nanostructure with a very high aspect ratio, the interfacial stiffness is not uniform across the interface and hence should be studied statistically. Considering the spatially distributed nature of the CNT–polymer contact and based on the distributed-element friction model of Iwan, [47] Huang *et al.* [36] developed a new friction model to analyze the energy dissipation of CNT/polymer composites as a function of CNT parameters as well as a distribution of nanofillers in the polymer matrix. The system is modeled as different elasto-plastic elements of having different yields. As shown in Figure 13.13, CNTs are regarded as parallel-series distributed-element model of Jenkins elements. Considering the N as the total number of elements, each element is assumed as a spring with stiffness of K/N in series with a coulomb damper of a slip force f_i/N.

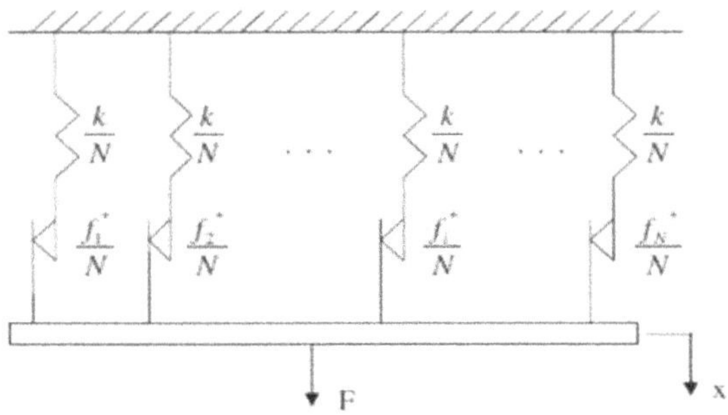

Figure 13.13 Distributed hysteresis model [36].

Upon the loading, the reaction of the system can be estimated by Equation 13.1 where n is the number of elements undergoing slippage while the others don't.

$$F = \sum_{i=1}^{n} f_i^* / N + kx(N-n)/N \tag{13.1}$$

In the case of large N, the net force can be written as an integral over individual forces using a distribution function (φ):

$$F = \int_0^{kx} f^* \varphi(f^*) df^* + kx \int_{kx}^{\infty} \varphi(f^*) df^* \tag{13.2}$$

At large displacements, the second term in Equation 13.2 vanishes and the slip force can be simplified as

$$F = \int_{kx}^{\infty} f^* \varphi(f^*) df^* \tag{13.3}$$

Distribution function, which itself is a function of contact compliance, coefficient of friction and the contact topography of the interface can be deduced experimentally. Besides, analytical models can also be used to estimate the distribution function [48]. The distribution function proposed by Iwan [47] (Equation 13.4) is widely used for this application:

$$\varphi(f^*) = \frac{1}{2\Delta f^*}; \quad F_y - \Delta f \le f^* \le F_y + \Delta f \tag{13.4}$$

Using Iwan's distribution function, the equation of motion of the ith CNT can be written as:

$$m_i v_i = K_c \Delta x_{i+1} - K_c \Delta x_i + F_i \tag{13.5}$$

where K_c and Δx_i are the axial stiffness and the length of the ith CNT element respectively. The energy dissipation can be then estimated by

$$E = \sum_{t=1}^{s} \int_{0}^{T} F_i(y_i - x_i)dt \tag{13.6}$$

Based on the model presented above, Huang *et al.* [36] studied the effect of many different parameters on the damping properties of polymer/CNT composites.

As shown in Figure 13.14, the largest energy dissipation happens at the natural frequency of the CNTs. Frequency ratio is the ratio of applied frequency to CNTs natural ratio $(\omega/(K_c/\mathrm{m})^{0.5})$. Axial stiffness of the CNTs is fixed. In order to study the effect of binding of the CNT ends to the matrix, simulations have been done at different values of end binding stiffness to axial stiffness ($\boldsymbol{e} = K_e/K_c$). Increasing $\boldsymbol{e}$ results in increase in higher frequencies, which is due to larger adhesion of the CNTs to the matrix. However, at the same frequency, increase in the CNT/matrix stiffness decreases the energy dissipation. Since the CNTs are bonded stiffer to the polymer, they have smaller sliding and hence the energy dissipation decreases (Figure 13.15).

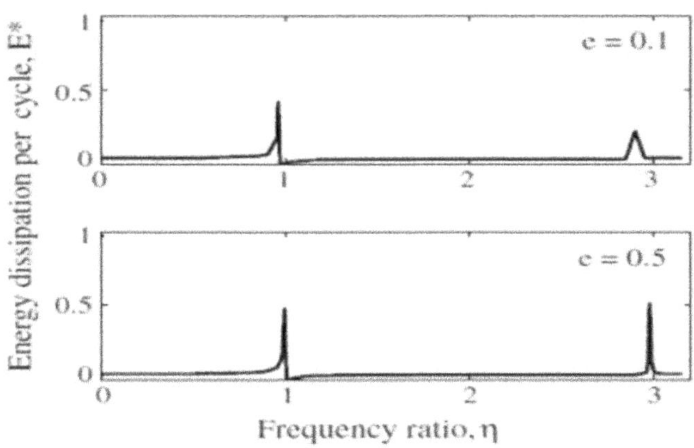

Figure 13.14 Energy dissipation vs. frequency ratio [36].

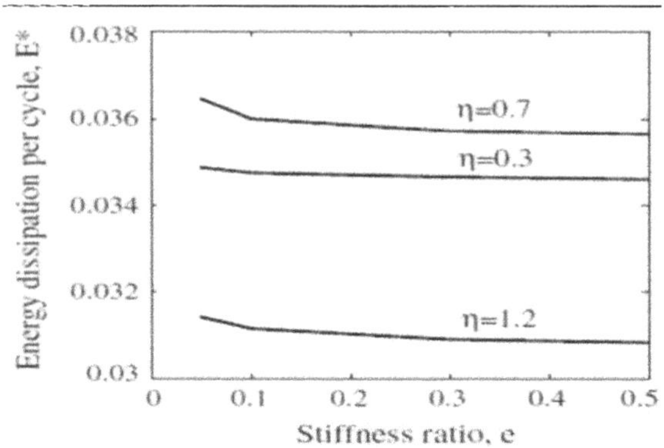

Figure 13.15 Effect of stiffness ratio on the energy dissipation at constant frequency [36].

The other parameter that affects the damping properties of the CNT/polymer composite structure is the excitation amplitude of the vibration. As shown in Figure 13.16, at low excitation frequencies, the energy dissipation behaves nonlinearly with increasing the amplitude. However, at higher amplitudes, the energy dissipation shows a linear behavior. At low amplitudes, a slip-stick combination is taking place. Increasing the amplitude increases the slip gradually and therefore at low amplitudes, energy dissipation shows a nonlinear behavior. On the other hand, at higher amplitudes, the system shows a full slip behavior and therefore, energy dissipation shows a linear behavior.

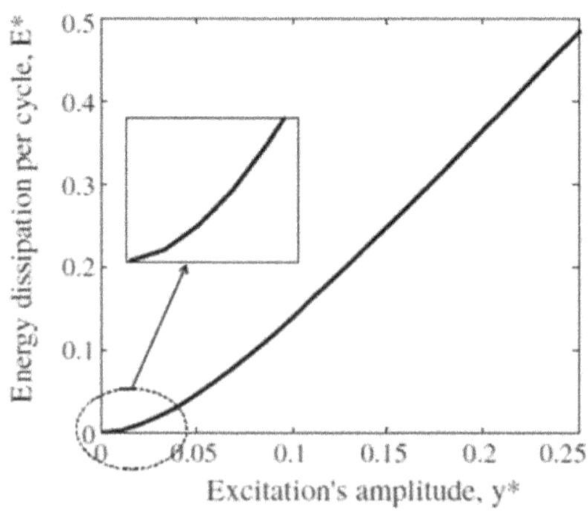

Figure 13.16 Energy dissipation vs. excitation amplitude [36].

Another factor that needs to be studied carefully is the effect of CNT concentration on damping properties of the composite structure. At the first glance it seems that since the main mechanism for the energy dissipation in these structures is the CNT/polymer slippage, increasing the CNT concentration enhances the overall damping properties. Increase in CNT concentration increases the total CNT/polymer interface and hence should lead to enhanced damping properties. However, experimental results [37, 49, 50] show different behavior from this expectation. It is found that there is an optimum CNT concentration at which the highest energy dissipation happens. It is found that the thickness of the polymer around the coating, dramatically affects the stiffness [51]. Thicker coating of the polymer on the CNT results in easier interfacial slippage. As shown in Figure 13.17, Huang *et al.* found that there is an optimum amount of CNT concentration (about 3%) at which the composite shows the highest damping properties.

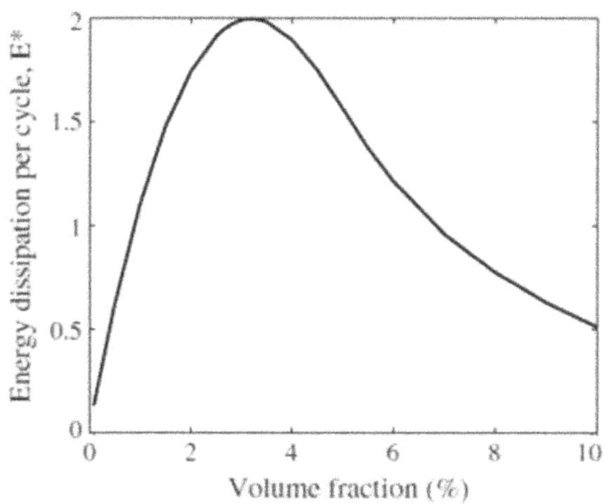

Figure 13.17 Effect of CNT concentration on damping properties of CNT/polymer composite [36].

13.4 Conclusions

In this chapter, various works on the viscoelastic properties of CNT and CNT-composites have been reviewed. In particular, the chapter focuses on two types of CNT-based films: pure CNT films and CNT–polymer composites. For pure CNT films, the main driving force behind the films' viscoelastic properties is due to the intrinsic mechanical property of the individual CNT involved as well as their frictional interaction with each other. During compression cycles, the region where the most damping would occur when the CNTs in the film starts to buckle and exhibit negative stiffness. In CNT–polymer composites, the CNTs' interaction with their surrounding polymer matrix is the main contributor to the viscoelastic properties observed. Importantly, it has been shown that CNT-based film systems have high resilience to plastic deformation, high temperature tolerance and remarkable viscoelastic characteristics. These combine with the fact that CNT is inherently light give CNT-based technology (either in the form of nanofillers or as a whole film) a key advantage in its use in many aspects of mechanical systems from the nano to the macro scale.

References

1. Baughman, R. H., Zakhidov, A. A., and Heer, W. A. (2002). Carbon nanotubes: The route toward applications, *Science,* **297**, pp. 787–792.
2. Treacy, M. M. J., Ebbesen, T. W., and Gibson, J. M. (1996). Exceptionally high young's modulus observed for individual carbon nanotubes, *Nature,* **381**, pp. 678–680.
3. Yu, M. F., Lourie, O., Dyer, M. J., Moloni, K., Kelly, T. F., and Ruoff, R. S. (2000). Strength and breaking mechanism of multiwalled carbon nanotubes under tensile load, *Science,* **287** pp. 637–640.
4. Wagner, H. D., Lourie, O., Feldman, Y., and Tenne, R. (1998). Stress-induced fragmentation of multiwall carbon nanotubes in a polymer matrix, *Appl. Phys. Lett.,* **72**, pp. 188–190.
5. Yakobson, B. I., Brabec, C. J., and Bernholc, J. (1996). Nanomechanics of carbon tubes: instabilities beyond linear response, *Phys. Rev. Lett.,* **76**, pp. 2511–2514.
6. Wong, E. W., Sheehan, P. E., and Lieber, C. M. (1997) Nanobeam mechanics: elasticity, strength, and toughness of nanorods and nanotubes, *Science,* **277**, pp. 1971–1975.
7. Tong, T., Zhao, Y., Delzeit, L., Kashani, A., Meyyappan, M., and Majumdar, A. (2008). Height independent compressive modulus of vertically aligned carbon nanotube arrays, *Nano Lett.,* **8**, pp. 511–515.

8. Waters, J. F., Riester, L., Jouzi, M., Guduru, P. R., and Xu, J. M. (2004). Buckling instabilities in multiwalled carbon nanotubes under uniaxial compression, *Appl. Phys. Lett.*, **85**, pp. 1787–1789.
9. Cao, A., Veedu, V. P., Li, X., Yao, Z., Ghasemi-Nejhad, M. N., and Ajayan, P. M. (2005). Multifunctional brushes made from carbon nanotubes, *Nat. Mater.*, **4**, pp. 540–545.
10. Yap, H. W., Lakes, R. S., and Carpick, R. W. (2007). Mechanical instabilities of individual multiwalled carbon nanotubes under cyclic axial compression. *Nano Lett.*, **7**, pp. 1149–1154.
11. Yap, H. W., Lakes, R. S. and Carpick, R. W. (2008). Negative stiffness and enhanced damping of individual multiwalled carbon nanotubes, *Phys. Rev. B*, **77**, p. 045423.
12. Pantano, A., Boyce, M. C., and Parks, D. M. (2004). Mechanics of axial compression of single and multi-wall carbon nanotubes, *J. Eng. Mater. Technol.*, **126**, pp. 279–284.
13. Liao, W. H., and Wang, K. W. (1997). On the analysis of viscoelastic materials for active constrained layer damping treatments, *J. Sound Vib.*, **207**, pp. 319–334.
14. Davis, C. L., and Lesieutre, G. A. (2000). An activity tuned solid-state vibration absorber using capacitive shunting of piezoelectric stiffness, *J. Sound Vib.*, **232**, pp. 601–617.
15. Kamath, G. M., Wereley, N. M., and Jolly, M. R. (1999). Characterization of magnetorheological helicopter lag dampers, *J. Am. Helicopter Soc.*, **44**, pp. 234–248.
16. Koratkar, N., Wei, B. Q., and Ajayan, P. M. (2002). Carbon nanotube films for damping applications, *Adv. Mater.*, **14**, pp. 997–1000.
17. Wei, B., Zhang, Z. J., Ramanath, G. and Ajayan, P. M. (2000). Lift-up growth of aligned carbon nanotube patterns, *Appl. Phys. Lett.*, **77**, pp. 2985–2987.
18. Zhang, Z. J., Wei, B. Q., Ramanath, G., and Ajayan, P. M. (2000). Substrate-site selective growth of aligned carbon nanotubes, *Appl. Phys. Lett.*, **77**, pp. 3764–3766.
19. Cao, A., Dickrell, P. L., Sawyer, W. G., Ghasemi-Nejhad, M. N., and Ajayan, P. M. (2005). Super-compressible foamlike carbon nanotube films, *Science*, **310**, pp. 1307–1310.
20. Kinney, J. H., Marshall, G. W., Marshall, S. J., and Haupt, D. L. (2001). Three-dimensional Imaging of large compressive deformations in elastomeric foams, *J. Appl. Polym. Sci.*, **80**, pp. 1746–1755.

21. Gibson, L. J. (2000). Mechanical behavior of metallic foams. *Annu. Rev. Mater. Sci.,* **30**, pp. 191–227.

22. Zhu, H. X., Mills, N. J., and Knott. J.F. (1997). Analysis of the high strain compression of open-cell foams, *J. Mech. Phys. Solids*, **45**, pp. 1875–1899.

23. Pathak, S., Cambaz, Z. G., Kalidindi, S. R., Swadener, J. G., and Gogotsi, Y. (2009). Viscoelasticity and high buckling stress of dense carbon nanotube brushes, *Carbon,* **47**, pp. 1969–1976.

24. Kusunoki, M., Rokkaku, M., and Suzuki, T. (1997). Epitaxial carbon nanotube film self-organized by sublimation decomposition of silicon carbide, *Appl. Phys. Lett.,* **71**, pp. 2620–2622.

25. Cambaz, Z. G., Yushin, G., Osswald, S., Mochalin, V., and Gogotsi, Y. (2008). Noncatalytic synthesis of carbon nanotubes, graphene and graphite on SiC, *Carbon,* **46**, pp. 841–849.

26. Teo, E. H. T., Yung, W. K. P., Chua, D. H. C., and Tay, B. K. (2007). A carbon nanomattress: A new nanosystem with intrinsic, tunable, damping properties, *Adv. Mater.,* **19**, pp. 2941–2945.

27. Krishnan, S., Weinman, C. J., and Ober, C. K. (2008). Advances in polymers for anti-biofouling surfaces, *J. Mater. Chem.*, **18**, pp. 3405–3413.

28. Choongik, K., Facchetti, A., and Marks, T. J. (2007). Polymer gate dielectric surface viscoelasticity modulates pentacene transistor performance, *Science,* **318**, pp. 76–80.

29. Ng, T. N., Wong, W.S., Chabinyc, M. L., Sambandan, S., and Street, R. A. (2008) Flexible image sensor array with bulk heterojunction organic photodiode, *Appl. Phys. Lett.,* **92**, p. 213303.

30. Collinger, J. C., Wickert, J. A., and Corr, L. R. (2009). Adaptive piezoelectric vibration control with synchronized switching, *J. Dyn. Syst.*, **131**, p. 041006.

31. Tangpong, X. W., Wickert, J. A., and Akay, A. (2008). Finite element model for hysteretic friction damping of traveling wave vibration in axisymmetric structures, *J. Vib. Acoust.*, **130**, pp. 1–7.

32. Tangpong, X. W., Wickert, J. A., and Akay, A. (2008). Distributed friction damping of travelling wave vibration in rods, *Philos. T. R. Soc. A*, **366**, pp. 811–827.

33. Farjoud, A., Cavey, R., Ahmadian, M., and Craft, M. (2009). Magneto-rheological fluid behavior in squeeze mode, *Smart Mater. Struct.*, **18**, p. 095001.

34. Farjoud, A., Vahdati, N., and Yap, F. F. (2008). MR-fluid yield surface determination in disc-type MR rotary brakes. *Smart Mater. Struct.*, **17**, p. 035021.

35. Suhr, J., Koratkar, N., Keblinski, P., and Ajayan, P. (2005). Viscoelasticity in carbon nanotube composites. *Nat. Mater.*, **4** pp. 134–7.

36. Huang, Y. and Tangpong, X. W. (2010). A distributed friction model for energy dissipation in carbon nanotube-based composites. *Commun. Nonlinear Sci.*, **15**, pp. 4171–4180.

37. Gibson, R. F., Ayorinde, E. O., and Wen, Y. F. (2007). Vibrations of carbon nanotubes and their composites: A review. *Compos. Sci. Technol.*, **67**, pp. 1–28.

38. Suhr, J., and Koratkar, N. A. (2008). Energy dissipation in carbon nanotube composites: A review, *J. Mater. Sci.*, **43**, p. 4370–4382.

39. Kim, S. H., Mulholland, G. W., and Zachariah, M. R. (2009). Density measurement of size selected multiwalled carbon nanotubes by mobility-mass characterization. *Carbon*, **47**, pp. 1297–302.

40. Furmaniak, S., Terzyk, A. P., Goembiewski, R., and Gauden, P.A. (2010). Surface area of closed carbon nanotubes determined from room temperature measurements of alcohols adsorption. *Chem. Phys. Lett.*, **499**, pp. 141–145.

41. Cumings, J., and Zettl, A. (2000). Low-friction nanoscale linear bearing realized from multiwall carbon nanotubes. *Science*, **289**, pp. 602–4.

42. Barber, A. H., Cohen, S. R., and Wagner, H. D. (2003). Measurement of carbon nanotube-polymer interfacial strength, *Appl. Phys. Lett.*, **82**, pp. 4140–4142.

43. Eitan, A., Kuiyang, J., Dukes, D., Andrews, R., and Schadler, L. S. (2003). Surface modification of multiwalled carbon nanotubes: toward the tailoring of the interface in polymer composites, *Chem. Mater.*, **15**, pp. 3198–3201.

44. Zhou, X., Shin, E., Wang, K. W., and Bakis, C. E. (2004) Interfacial damping characteristics of carbon nanotube-based composites, *Compos. Sci. Technol.*, **64**, pp. 2425–2437.

45. Suhr, J., Koratkar, N., and Ajayan, P. (2004). Damping characterization of carbon nanotube thin films, *Proc. SPIE*, **5386**, p. 153–161.

46. Koratkar, N. A., Bingqing, W., and Ajayan, P. M. (2003). Multifunctional structural reinforcement featuring carbon nanotube films, *Compos. Sci. Technol.*, **63**, p. 1525–1531.

47. Iwan, W. D. (1966). A distributed-element model for hysteresis and its steady-state dynamic response, *J. Appl. Mech.*, **33**, pp. 893–900.

48. Spanos, P. T. D. (1979). Hysteric structural vibrations under random load, *J. Acoust. Soc. Am.*, **65**, pp. 404–410.
49. Kireitseu, M., Hui, D., and Tomlinson, G. (2008). Advanced shock-resistant and vibration damping of nanoparticle-reinforced composite material, *Compos. Part B*, **39**, pp. 128–138.
50. Mahmoodi, S. N., Jalili, N., and Khadem, S. E. (2008). An experimental investigation of nonlinear vibration and frequency response analysis of cantilever viscoelastic beams, *J. Sound Vib.*, **311**, pp. 1409–1419.
51. Zhang, L. C., and Vondeitcharova, T. Bending and local buckling of a nanocomposite beam reinforced by a single-walled carbon nanotube, *Int. J. Solids Struc.*, **43**, pp. 3006–3024.

Chapter 14

Toward Multifunctional Carbon Nanotube Fibers

Xiaohua Zhang and Qingwen Li
Suzhou Institute of Nano-Tech and Nano-Bionics, Ruoshui Road 398, Suzhou 215123, China
xhzhang2009@sinano.ac.cn, qwli2007@sinano.ac.cn

Carbon nanotubes (CNTs) are the strongest one-dimensional materials ever discovered. For engineering purpose, CNTs should be assembled into continuous lightweight and high-performance fibers by various spinning methods. Recent efforts have been made to enhance the mechanical properties of CNT fibers, accompanied with different functionalities. This chapter provides a brief review of the synthesis methods, mechanical properties, and multifunctionalities of CNT fibers.

14.1 Introduction

Textiles composed of lightweight, high-strength and functional fibers are very attractive for daily life. The fabrication of such fiber depends severely on the properties of its constituents. CNTs are the strongest

Carbon Nanotubes and Their Applications
Edited by Qing Zhang

ISBN 978-981-4241-90-8 (Hardcover), 978-981-4303-18-7 (eBook)
www.panstanford.com

and stiffest materials ever discovered in terms of tensile strength and elastic modulus, up to 100 GPa and 1 TPa, respectively [1]. They are also excellent conductors for both electricity and heat. To exploit their excellent properties at a macroscopic level, it is desirable to assemble CNTs with length of tens to hundreds of micrometers into continuous fibers tens-of-meters long, in a head-to-tail manner. The method includes *direct spinning* during CNT growth [2–6], drawing and twisting (*dry spinning*) from aligned CNT arrays (forests) [7–12], coagulation-based *wet spinning* [13–21], and other assembling processes [22–30]. The spun fibers are composed of pure CNTs or have a large fraction of CNTs.

It has been proved that the mechanical properties of CNT fibers greatly correlate with the quality of CNTs and the spinning methods. Among the numerous processes, the dry-spin and direct-spin methods are more promising to make CNT fibers stronger than at least 2 GPa [4,5,12]; however, this is still far below the strength of their constituent nanotubes. The central factor in limiting their mechanical properties is the low shear stress of the interface between CNTs. Unlike the covalent bonds within an individual CNT, the bonding between CNTs is largely through van der Waals (vdW) forces, whose strength depends on the interfacial area and the packing density of CNTs.

In terms of their mechanical properties, CNT fibers have already shown their potential as new-type engineering materials. Compared with traditional carbon fibers, CNT fibers also exhibit other unique properties. For example, they are much more flexible than carbon fibers. By deposition of metal nanoparticles, the mechanically robust CNT fibers might have metal-like electrical conductivities [31]. The introduction of polydiacetylene (PDA) allows a dry-spun CNT fiber to change its color in response to an electrical current or mechanical stress [32]. The spun CNT fibers have also shown a better efficiency as a glucose biosensor than a traditional metal electrode [33].

In this chapter, we review advances in the synthesis of nanotube fibers. We first discuss the spinning methods in detail. In the second part, we describe a general dependence of the mechanical properties of the spun fibers on the nanotube structures and spinning processes. Finally, we present recent progresses in multifunctional CNT fibers.

14.2 Spinning of Carbon Nanotube Fibers

Individual CNTs and CNT bundles have been synthesized up to millimeters in length [34–36]. However, the way to assemble a centimeters-long fiber directly from them [22,24] has not been used widely. The arbitrarily long CNT fibers, with a large CNT volume or mass fraction (more than at least 50%), can be spun from the direct [2] and dry spinning [7,8]. For the wet spinning, the products are usually CNT/polymer composite fibers, whose properties strongly depend on the choice of solutions or gels [14,17,21,37,38]. These methods have demonstrated great success in making ultrastrong CNT fibers, many of them comparable or stronger than existing tough materials such as carbon fibers, poly(p-phenylene benzobisoxazole) (PBO) fibers, and aramid fibers [4].

14.2.1 Direct Spinning

The direct spinning during the chemical vapor deposition (CVD) synthesis of CNTs was first developed by Alan Windle's group at University of Cambridge [2]. In this method, a continuous fiber is spun from CNT aerogel formed in the CVD reaction zone, schematically shown in Figure 14.1a. The precursor is typically a liquid hydrocarbon feedstock with ferrocene [$Fe(C_5H_5)_2$] and thiophene (C_4H_4S). Ferrocene forms iron nanoparticles that act as nucleation sites for the growth of CNTs, and the thiophene is an established rate enhancer for vapor grown carbon fibers [39]. The solution is injected into a heated gas-flow reactor in flowing hydrogen. Figure 14.1b shows a snapshot of the formation process where a CNT fiber is drawn out of the aerogel at the furnace temperature of 1050–1200°C. The spun fibers are strong and flexible, and can be knotted without degradation of their tensile strengths [40] (see the scanning electron microscopy (SEM) image shown in Figure 14.1c).

The nanotubes in the fiber, multi-walled or single-walled (MWCNTs and SWCNTs), can be achieved by adjusting the reaction conditions. With ethanol (C_2H_5OH) feedstock containing 0.23–2.3 wt% of ferrocene, MWCNTs are formed with a thiophene concentration of 1.5–4.0 wt%, a hydrogen gas flow rate of 400–800 ml/min, and a synthesis temperature of 1100–1180°C, while SWCNTs are favorably formed at a lower thiophene concentration (~0.5 wt%), a higher gas flow rate (~1200 ml/min) and temperature

(>1200°C) [2]. To ensure a stable spinning, the Fe/S atomic ratio in the feedstock could be adjusted to ~0.7 [3]. If aromatic-hydrocarbons such as benzene (C_6H_6), hexane (C_6H_{14}), and mesitylene [$C_6H_3(CH_3)_3$] and mixed hydrocarbons such as petroleum are used, there could be deposition of carbon particles, thick fibers, or both, rather than a continuous spinning. Besides ethanol, diethyl ether [$(C_2H_5)_2O$], polyethylene glycol [$-(CH_2$-CH_2-$O)_9$], acetone [$(CH_3)_2CO$], and ethyl formate (CH_3CH_2COOH), all containing oxygen groups, can also be used for the continuous spinning.

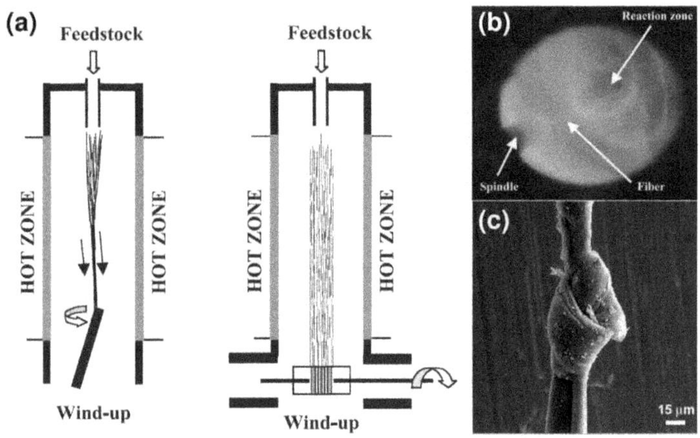

Figure 14.1 (a) Schematic of the direct spinning. The liquid feedstock is injected into the hot zone where an aerogel of CNTs forms. The aerogel is captured and continuously wound out of the furnace as a fiber or film. An offset rotating spindle is used for the wind-up (left), and can be operating at a lower temperature outside the furnace (right). (b) Snapshot showing the fiber formation. (c) SEM image of a knotted fiber. From refs 2,4.

A common characteristic of the spun fibers, comprising 10^5–10^6 individual CNTs, is their non-uniformity of cross sections, shown in Figure 14.2. Instead of the unit of GPa, Yuri is more often used to describe the tensile properties (strength and modulus) of such fibers, after Yuri Artsutanov who proposed the space elevator in 1960. Yuri is a unit by dividing the load by the linear mass density that is the mass (in grams) per kilometer fiber. One million Yuris (1 M Yuri) is somewhere between steel wire (0.5 M Yuri) and Spectra® 2000 fiber (3.5 M Yuri). It is also very common to use GPa cc/g or N/tex, both of which are equivalent to 1 M Yuri. The first reported direct-spun CNT fibers in 2004 had the tensile strengths ranging from 0.05 to 0.5 N/tex (equivalent to 0.10 and 1.0 GPa, assuming a density of 2.0 g/cc) [2]. The strength was improved to 0.7 N/tex [3], 2.2 N/tex [4], and a surprisingly high value of 9.8 N/tex measured with a gauge length of 1 mm [5] (Figure 14.3).

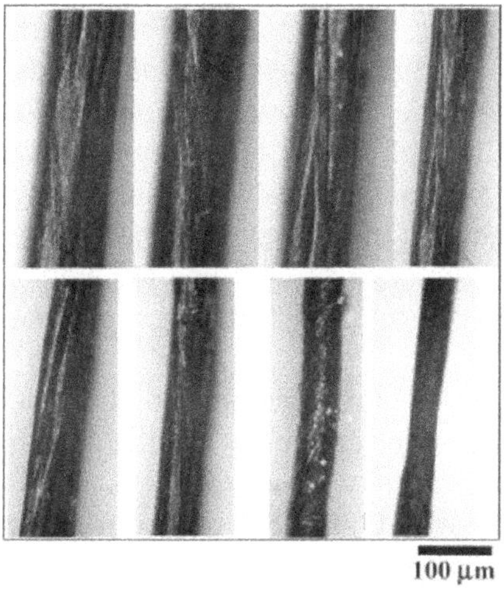

Figure 14.2 Optical microscopy images of different parts of a fiber showing variation in diameter. The diameters vary from 10 to ~100 μm. From ref 3.

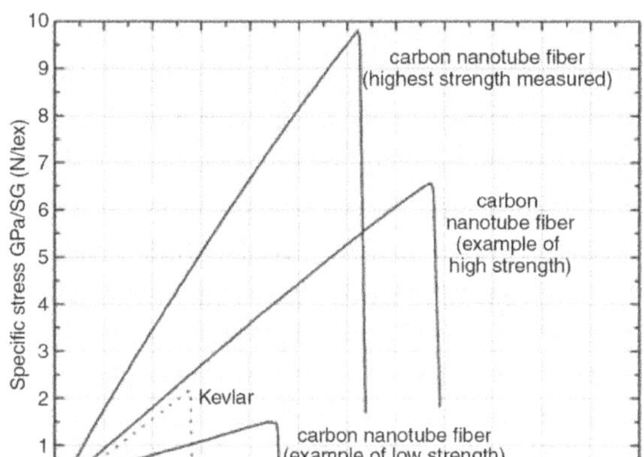

Figure 14.3 Stress–strain curves for 1-mm specimens of typical CNT fibers. A curve for Kevlar® 49 is included as a benchmark. From ref 5.

Another significant parameter for practical applications, probably more important than the strength, is the energy absorbed before the fracture, the toughness. For a direct spun fiber with a tensile strength of 0.7 N/tex, its toughness has been reported to be 40 J/g, comparable to Grafil® 34-700 carbon fibers (30 J/g), Kevlar® 49 aramid fibers (35 J/g), or Zylon® PBO fibers (45 J/g) [4]. The highest value reported so far for direct-spun fibers was 121 J/g [5], but still far below a remarkable record of 570 J/g for a poly(vinyl alcohol) (PVA)/SWCNT (~60 wt%) composite fiber with a tensile strength of 1.8 GPa (1.2 N/tex) [16]. The large value of the latter is due to the large strain at break of ~100%. Actually, rather than the fracture toughness, the energy absorbed at low strains is more crucial especially for applications such as body armor. Therefore, CNT fibers are very promising with their high modulus, see also Figure 14.3.

The direct-spun fibers usually contain 5–15 wt% of iron nanoparticles due to the continuous feeding of ferrocene during the growth, with a mass density of ~2 g/cc [2]. A recent study has shown a direct-spun fiber with CNT purity of 96.6 wt%, however, still with a large density of 1.67–2.0 g/cc [6]. The iron particles might play a crucial role, as the density can be decreased to smaller than 1 g/cc for dry-spun fibers where nearly no catalytic particles exist (to be discussed). Another problem hindering the further development of direct spinning is the defects associated with included carbonaceous particles [5]. Recently, by improving the spinning set-up, multi-layered CNT fibers have been synthesized [6].

It is important to note here that the direct spinning can be performed in a two-step way [27]. The first step is the growth of large-scale nonwoven CNTs, mostly single-walled, by using ferrocene and sulfur powder acting as the catalyst source and mixture of argon and methane (CH_4) as the carbon source [41]. The grown CNTs usually form a strong, highly conducting, and transparent film, which can be easily handled in the second step to make CNT fibers by post-synthesis treatments [27–30]. The process is quite similar to the direct spinning where the CNT film drawn out of the aerogel is collected directly into a fiber. By twisting the films into fibers, the constituent CNT bundles are pretightened and the inter-bundle junctions are strengthened. As a result, both the moduli and the strengths of fiber are promisingly enhanced by a factor of about 2, up to 9–15 GPa and 550–800 MPa, respectively [28].

Although the direct-spun fibers are strong, the method itself still has a large room to be improved in some aspects such as CNT alignment, purity, diameter uniformity, and strength consistency.

14.2.2 Dry Spinning

A breakthrough was made in 2002 by Shoushan Fan's group at Tsinghua University, who first discovered that vertically aligned CNT arrays could be spinnable, opening a new access to assemble CNTs into continuous fibers [7]. Following that, Ray H. Baughman's group at University of Texas at Dallas developed a draw-and-twist spinning method [8]. As most of the catalytic iron particles remain on the substrates [42], the mass percent of carbon is higher than that by the direct spinning, 96–98 wt% of carbon in the form of nanotubes and 2–4 wt% of Fe and amorphous carbon [8]. Some other groups, including a research team at Australian Commonwealth Scientific and Industrial Research Organisation (CSIRO), Lianxi Zheng's group at Nanyang Technological University, Yuntian T. Zhu's group at North Carolina State University, and our group at Suzhou Institute of Nano-Tech and Nano-Bionics, are improving the spin methods and the quality of CNT arrays in attempt to make strong and functional fibers.

Figure 14.4a shows a snapshot of the dry spinning, in which MWCNTs ~10 nm in diameter are simultaneously drawn from the array and twisted. The direction of drawing is orthogonal to the original growth direction of CNTs and parallel to the substrate. The initial angle between the draw direction and the substrate can be tuned from 0° to nearly 90°, as shown by the videos from refs 9,43. The CNT arrays used for fiber spinning are grown on an iron catalyst-coated substrate by CVD, based on the methods by Zhifeng Ren's [44] and Hongjie Dai's groups [45]. In these methods, the coated substrate is placed inside a quartz boat and then heated in a flowing argon gas to the growth temperature of 620–750°C [8,9,46]. Then the carbon source acetylene (C_2H_2) and H_2 are added to the Ar gas to start the growth of CNT arrays. The growth pressure can be atmospheric or as low as about 2 torr [9]. The gas flow for growing spinnable CNT arrays, however, differs greatly from each other. For example, a total flow rate of 580 sccm including 5 mol% C_2H_2 [8], a flow rate of 12 sccm C_2H_2 and 425 sccm Ar or 500 sccm C_2H_2 and 50 sccm H_2 [9], and an optimized condition of $Ar/H_2/C_2H_2$ = 200–300/100/30 sccm [46] have been used by different groups.

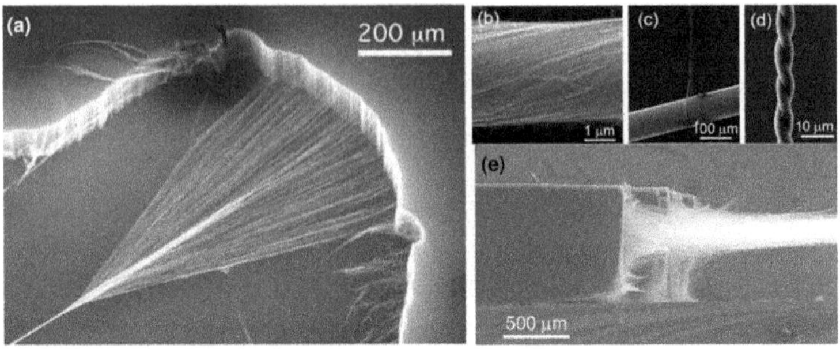

Figure 14.4 SEM images showing the draw-and-twist process (a), CNT fiber at high magnification (b), CNT fiber forming a loop around a Ni wire and then twisted (c), the corresponding twisted two-ply section (d), and side view of drawing a unidirectional CNT sheet. From refs 8,12,46.

As an alternative, the use of ethylene (C_2H_4) as the carbon source requires a slower total flow rate of 100-sccm C_2H_4 and 100-sccm mixture of H_2 and Ar [11]. Although the growth rate of ~60 μm/min by using C_2H_4 is much slower than that by C_2H_2, the grown CNTs usually have fewer walls (2–5) and are more super-aligned, resulting in higher performances of the spun fibers [47]. In the ethylene CVD, an Al_2O_3 buffer layer is always required to help the deposition of a more uniform Fe film [11,48].

Spinnable CNT arrays can be grown by other CVD methods [49,50]. For example, using iron(II) chloride ($FeCl_2$) as catalyst and C_2H_2 as the carbon source, spinnable CNT arrays with a height of ~2 mm can be produced at 8–10 Torr and a furnace temperature of 820°C. However, the CNTs are usually very large in diameter (17–30 nm) and have a wall number of 15 on the average.

In Figure 14.5, two different spinning method are schematically shown. The first way (Figure 14.5a), the simplest one, is the drawing and twisting of the CNT sheet directly pulled out from an array. As passing through a drop of polar solvent or polymer solution, the fibers' performance can be improved because stronger interfacial vdW interaction between CNTs in the fiber can be introduced either by the rapid solvent evaporation or by the infiltration of polymer molecules. With using the liquid, not only is there a shrink of diameter of the spun fibers, but also the capillary force and/or the introduction of polymer refine the structural morphology of the fibers.

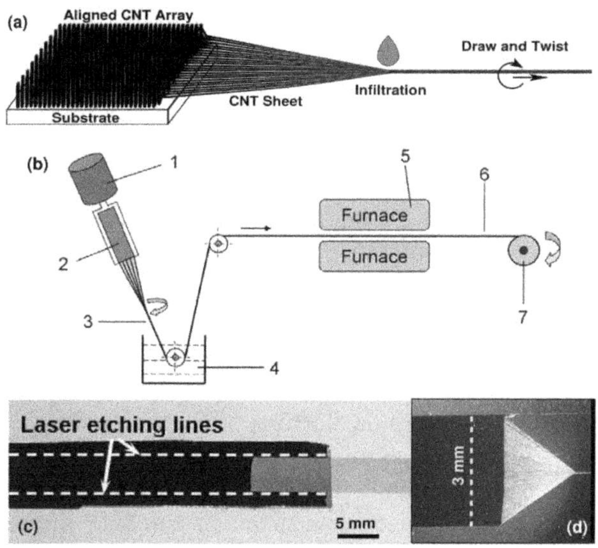

Figure 14.5 (a) Schematic view of CNT fiber spinning. The twist is applied directly with the drawing on the pulled-out CNT sheet from a CNT array. Infiltration can be performed by using a drop of polymer solution. (b) A separate rotation (1) is applied on an array (2) to introduce twist on the fiber (3). The fiber is immersed in a glass vessel containing solvent (4) immediately. After passing through a drying/heating zone (5), the fiber becomes shrunk (6) and is collected on a winder. (c,d) In order to get a fiber with a uniform diameter, the array width is fixed by laser etching or by growing CNT on a patterned catalytic zone. From refs 47,51,53.

In order to realize continuous spinning, a third motion of winding should be included. The three motions (draw, twist, and wind) can be performed separately [51] or simply integrated [52]. Figure 14.5b represents a set-up with separate twisting. Such separation is important for large-scale production of CNT fibers, as it makes that possible to manufacture multi-ply and twisted fibers directly from several CNT arrays.

In contrast to the direct spinning, the fibers from the dry method are cleaner and lighter, due to the fewer iron nanoparticles. The linear mass density of an 8 μm-diameter fiber from CNTs with a tube diameter of ~8 nm and ~6 walls, is 0.075 tex, corresponding to a mass per unit volume of 1.49 g/cc, both much smaller than the direct-spun fibers. For a fiber with a diameter of 7.7 μm from mainly double- and triple-walled CNTs, the linear mass density is only 0.035 tex, with a density of 0.75 g/cc.

The strongest dry-spun fiber had a tensile strength of 3.3 GPa [12]. However, the strongest repeatable and long fibers, at least longer than three meters, was 1.23 GPa and ~50 GPa in strength and modulus, respectively, by performing tensile tests every 10 cm along the fiber [47,53]. Simply after PVA infiltration, the strength can be improved by ~400 MPa. Very recently, Shoushan Fan's group reported a 2.0-GPa strength for a CNT/PVA composite fiber, by using the method shown in Figure 14.5b [54].

The dry spinning not only allows the spun fibers with high purity, strong strength and low density, but also shows some other advantages. With this method, the fiber diameter is a simple function of the ribbon width, which can be modulated by laser etching or by growing on a patterned catalytic zone, see Figure 14.5c,d. During the spinning, many other treatments such as stretching, impregnating, drying, and heating can be easily applied. Different from the direct spinning where the CNT aerogel has to be formed at temperature higher than 1000°C, with toxic thiophene and high hydrogen percentage as well, the growth of spinnable precursors, CNT arrays, is performed here at a lower temperature of 600–800°C. The dry spinning is obviously greener, safer, and more energy-saving.

14.2.3 Wet Spinning

Hitherto, CNTs are easy to be massively produced in forms of powders [55,56] and unspinnable arrays [57,58]. In order to create continuous fibers from such CNTs, coagulation spinning is often employed, which is a conventional method to manufacture engineering polymer fibers such as aramid, acrylic, and polyacrylonitrile (PAN) fibers. In this process, a polymer solution is extruded into a bath that contains a second liquid in which the solvent is soluble but the polymer is not. The polymer therefore phase separates and condenses to form a fiber. However, as CNTs are hydrophobic, chemically inert, and favorably bundled and entangled due to the strong intertube vdW interaction, they are difficult to be dissolved in a common aqueous, organic, or acidic media. An efficient way is the ultrasonic dispersion with the aid of surfactants such as sodium dodecyl sulfate (SDS) and cetyl trimethylammonium bromide (CTAB), or of polymer wrapping by DNA or polyvinylpyrrolidone (PVP). However, it is still a challenge to disperse CNTs at high concentration. Furthermore, the concentrations of SDS and CNT play important roles for the solution homogeneity with a single phase (well-dispersed) or the multi-phase of CNTs including the coexistence of dispersed tubes and CNT clusters.

The first wet spinning of CNT fibers was performed nearly at the same time by Andrew Rinzler's group at University of Florida [13] and Philippe Poulin's group at Université de Bordeaux [14,37]. In the process of the latter, for example, PVA molecules displaced the surfactant (SDS), causing the coagulation of CNT fibers. As shown schematically in Figure 14.6a, the SWCNT dispersion is slowly injected through a syringe needle or a thin glass capillary into a polymer solution that contains 5 wt% of PVA (molecular weight 70000). The addition of PVA to the suspensions initially stabilized by SDS induces aggregation of the CNTs, as in flowing systems. The amphiphilic PVA adsorbs onto the tubes and displaces some SDS molecules. Because the polymer solution is more viscous than the CNT dispersion, there is a shear contribution in the flow at the tip of the capillary, inducing alignment of CNTs. Therefore a stable CNT ribbon (mesh) is formed and can be drawn out of the PVA solution. After being washed with pure water, or acetone [16] to remove most of the surfactants and polymers, and then being dried and collated, it finally becomes a fiber, but with a worse CNT orientation and a lower volume fraction than the fibers derived from the previous two methods. Twisted fibers are spun if the spinning is performed in two dimensions (drawing and rotating).

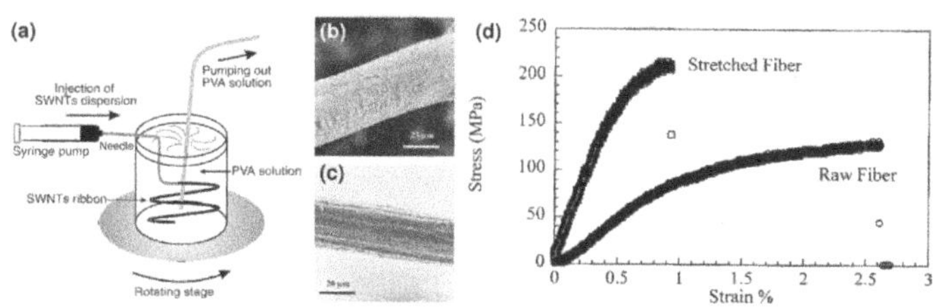

Figure 14.6 (a) Schematic of the wet spinning. SWCNT dispersion is injected through a syringe needle or a thin glass capillary into the PVA solution in a rotating container. The CNTs rapidly stack together as coming out of the capillary tube, due to the viscous PVA. Therefore a long CNT ribbon, with flow-induced alignment, is formed, and drawn out of the solution. (b,c) SEM images of a raw and a stretched CNT fiber. (d) Stress vs strain curve of a raw and a stretched fiber made from the same bath of CNTs. From refs 14,37.

The fiber diameter ranging from a few micrometers to 100 μm can be controlled by varying the parameters (injection rate, flow conditions, and dimensions of the capillary tube) that control the initial thickness of the ribbons. Experiments have shown there were no measurable effects on the CNT orientation in the resulting dried fibers for a wide range of these parameters [37]. It shows that the alignment should be mainly fixed during the drying stage, where the mesh is collated, not during the initial coagulation process. Therefore the fibers can be rewetted, swollen, and redried under a tensile stretching. The rewetted and swollen fibers can be stretched up to 160%, without disassembly, indicating that the CNTs and adsorbed polymers have formed cross-linked fiber structure. Figure 14.6b,c shows the SEM images of a raw and a stretched CNT fiber. The stretching process qualitatively modifies the aspect of the fiber. In agreement with the improvement of CNT orientation, the fiber's tensile strength and Young's modulus can be increased from 125 MPa and 10 GPa to 230 MPa and 40 GPa, respectively, by stretching up to 145% (see Figure 14.6d). If the stretching is performed in a flow of hot air (~180°C), a temperature above the PVA glass transition [19], the fiber strength can be increased up to 1.4–1.8 GPa, with a strain at break of ~11% and a toughness of ~55 J/g. The mass density is estimated to be 1.3–1.5 g/cc.

Coagulation spinning has also been done with systems other than PVA. For example, CNT biofibers can be spun by using biomolecules acted as both the dispersant and coagulant [20]. These fibers have strengths up to 183 MPa and moduli up to 6.9 MPa, depending on the type of biomolecule used [59]. Pure CNT fibers can also be produced by spinning from CNT/surfactant/water solutions followed by high-temperature treatment [18], and from ethanol/glycerol or ethanol/glycol mixture and washing with water and ethanol [60]. These polymer-free fibers are brittle and weak without the reinforcement of a polymer matrix. Their specific strength and modulus are 0.065 and 12 N/tex, respectively [18].

An alternative wet spinning was developed by Richard E. Smalley's group (now the Richard E. Smalley Institute for Nanoscale Science and Technology) at Rice University [17]. In this method, SWCNTs are dispersed with 8 wt% in 102% sulfuric acid (2 wt% excess SO_3). In superacids, SWCNTs form charge-transfer complexes of individual positively charged nanotubes surrounded by a finite number of sulfuric acid anions [61], as illustrated in Figure 14.7.

The mixture is extruded from a capillary tube and doped into the coagulation bath that contains diethyl ether, 5 wt% aqueous sulfuric acid, or water, to get CNT fibers. The fibers can be washed and dried to get purified and enhanced. The advantage of such fibers is the absence of supporting surfactant or polymer structure. However, they have a very limited tensile strength, only of 116 MPa, but a high Young's modulus of 120 GPa, and a mass density of <1.11 g/cc, larger than that of a dry-spun fiber from double- and triple-walled tubes (0.75 g/cc). The larger density arises from the fact that sulfuric acid molecules intercalate CNT bundles within the fiber and possibly fill the nanotubes.

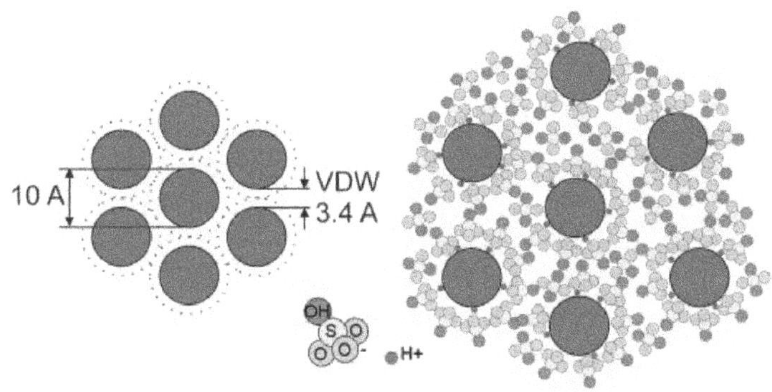

Figure 14.7 Illustration of the disperse of SWCNTs in superacids (100+% sulfuric acid). From ref 17. See also Color Insert.

Different from the surfactant-assisted dispersion, SWCNTs show isotropic, liquid-crystalline, crystal solvate, and solid phases in superacids [38]. Here the phase behavior refers to the various liquid microstructures that can form as a function of CNT and solvent parameters. Liquid-crystalline solutions of CNTs have also been achieved in water by introducing sidewall functional groups [62–64], and in ethylene glycol [21]. As the CNT phase is a function of the concentration, a new coagulation process is developed, which allows spinning from different types of nanotube such as pristine CNTs and nitrogen-doped MWCNTs (N-MWCNTs) [21]. These CNTs can be dispersed in ethylene glycol with the aid of ultrasonic dispersion, with concentration of 1–3 wt% to ensure a pure nematic CNT phase. By pumping the dispersions into a diethyl ether bath, the domains of the nematic CNT dispersions are shear oriented as passing through a syringe needle, forming a CNT fiber out of the needle. The spun fibers are flexible and have tensile strengths of 100–210 MPa. However, the N-MWCNT fiber has a much higher stiffness than the MWCNT fiber, indicating that the straight N-MWCNTs might increase the interaction between individual nanotubes.

Although the spun fibers are weak in terms of their tensile strengths, (there were only few fibers strong as ~2 GPa [16,19]), coagulation wet spinning provides a facile route for the fabrication of macroscopically well-aligned CNT fibers and CNT/polymer composite fibers, and can introduce multifunctionality by using various polymer molecules, for example, enhanced electrical conductivity [18,65,66], biofunctionality [66], and electrochemical behavior [67].

14.3 Mechanical Properties of CNT Fibers

In order to characterize CNT fibers, many issues have to be considered, for instance, pristine nanotube structure (tube diameter, wall thickness, defects, aspect ratio, etc.), assembly morphology (fiber diameter, twist angle, CNT alignment and waviness, CNT packing density, etc.), and mechanical properties (tensile strength, elastic modulus, strain at break, toughness, etc.). Before discussing the dependence of the mechanical properties on different parameters, we first provide an overall comparison between various CNT fibers. Then the structural and morphological dependences are presented.

14.3.1 Comparison Between Different Nanotube Fibers

Table 14.1 lists the fiber diameter (d_f), surface twist angle (θ_s), mass density (ρ), linear mass density (λ), CNT fraction, tensile strength (σ_f), elastic modulus (E_f), and toughness (at failure) for various CNT fibers. Many efforts have been devoted to improve their tensile strengths. On the average, the pure CNT fibers have been prepared with a strength higher than 2 GPa, either by the direct or dry spinning. By dividing by the mass density, the specific strength (up to 2.2–2.8 N/tex) is already comparable or larger than traditional carbon fibers and aramid fibers [4]. The toughness at failure (potentially larger than 100 J/g) is also encouraging for body armor.

As the mass density ρ differs greatly from fiber to fiber, ranging from 0.2 to 2.0 g/cc, there is also a large spread of the specific strength, from 0.02 to 9.8 N/tex. (Note here the linear mass density depends severely on the fiber diameter, and could not be used to compare the different fibers.) To improve their performance, CNT fibers should be tightly compacted and possibly infiltrated by polymer molecules, which are heavier than CNTs, and then have an increased mass density. Therefore, the constituent CNTs should have as few walls as possible to decrease the mass density. The decrease in wall number is positive to the mechanical properties. For example, the direct- [4] and un-infiltrated dry-spun fibers [47] from mainly double- and triple-walled CNTs have strengths of 2.2 and 1.6 N/tex, respectively, as their densities are both smaller than 1 g/cc. After infiltration with PVA or thermosetting polyimide (PI) prepolymer, poly(amic acid) (PAA), the dry-spun fibers are improved to be ~2.0 N/tex. The strength is further improved to 2.8 N/tex by curing the PAA molecules. However, the wet-spun fibers usually contain a large fraction of supporting surfactants or polymers. To date, the maximum specific strength is still within 1.5 N/tex [16,19].

Notice there were three fibers having a toughness of more than hundreds of J/g [12,16,19]. The 975 J/g of a dry-spun fiber might be overestimated as its density was only about 1/5 of the others. Generally speaking, the strain at break of direct- or dry-spun fibers is within 5–10%, and their toughness is nearly proportional to the modulus. A typical toughness at failure is about 50–100 J/g under today's spinning technique. For the wet-spun fibers, the strain at break can be much larger, even up to 100–400%, corresponding to the two highest values of toughness given in Table 14.1. The large strain arises of the bad alignment of polymer molecules along the constituent CNTs [19,37]. By stretching the fibers, the polymer chains become more compact with CNTs, forming a better CNT-polymer network with sufficiently small scattering of the orientation angle (within 4–6°, much smaller than that of ~27° before the stretching) [19]. After such post-spin treatment, the wet-spun fibers usually have the similar strain at break, modulus, and toughness with the direct- and dry-spun fibers.

Despite the multifunctionalities of the wet-spun fibers, it is hard to say which method is the best. However, considering the specific strength and modulus, the dry spinning shows superiority, as there are nearly no catalytic iron nanoparticles and molecules with a large mass density. Furthermore, the mechanical properties of a dry-spun fiber is very uniform along the whole fiber, and the fiber diameter and twist angle can be easily controlled by setting the ratio between the draw and twist speeds.

Table 14.1 Parameters summarized to evaluate CNT fibers

Method	d_f (μm)	θ_s (°)	ρ (g/cc)	λ (tex)	CNT fraction		σ_f		E_f		Toughness (J/g)
					(wt%)	(vol%)	(N/tex)	(GPa)	(N/tex)	(GPa)	
Direct [2]	—	—	2.0	—	85–95	—	0.05–0.5	0.1–1.0	—	—	—
Direct [3]	10–100	—	2.1	—	—	—	0.7	1.46	—	—	40
Direct [4]	2–20	—	1.0	0.05–0.5	—	—	2.2	2.2	—	—	46–60
Direct [5]	7–20	—	0.9	0.04	—	—	1.5–9.8	1.3–8.8	—	—	13–121
Direct [6]	—	—	1.67–2.0	0.11–0.92	96.6	—	0.2–0.75	0.4–1.25	—	—	—
Dry [8]	1–10	—	0.8	0.01	96–98	—	0.31–0.57	0.3–0.85	—	5–30	14–20
Dry [12]	—	—	0.2 [a]	—	—	—	—	1.35–3.3	—	100–263	975 [a]
Dry [10]	3–13	10–21	—	—	—	—	—	0.85–1.91	—	275–330	—
Dry [b]	2–15	10–40	0.75–1.49	0.035–0.075	—	—	0.57–1.60	0.86–1.23	30–80	16–55	30–45
Dry [c]	5–14	20	0.75	0.06	—	50–70	2.8	2.06	136	100	50
Dry [51]	5–40	10–40	0.38–0.64	—	—	—	1.56–1.71	0.6–1.1	—	48–56	—
Dry [d]	15–30	—	0.98	—	81	—	—	1.5–2.0	—	120–200	—

Dry [67]	5–60	0–70	0.8–1.7	—	—	—	1.0	0.8–1.04	—	—	—
Wet [14]	10–100	—	1.3–1.5	—	—	—	—	0.15	—	9–15	—
Wet [37]	35	—	—	—	—	—	—	0.23	—	40	—
Wet [16]	50	—	1.5	—	60	—	1.2	1.8	53	80	570 [e]
Wet [17]	50–80	—	0.87–1.11	—	—	—	—	0.12	—	120	—
Wet [19]	30	—	1.3–1.5	—	—	—	0.4	0.55	—	—	870 [f]
Wet [19]	30	—	1.3–1.5	—	—	—	1.05–1.35	1.4–1.8	26–34	35–45	40–60 [g]
Wet [18]	10–50	—	—	—	—	—	0.065–0.77	—	8.9–12	—	137

Note: d_f, θ_s, ρ, λ, σ_f, and E_f refer to the fiber diameter, surface twist angle, mass density, linear mass density, tensile strength, and elastic modulus, respectively. The values are collected from literature.

[a] The extremely large toughness and the small mass density were measured from a loosely packed fiber.

[b] From ref 47 and our unpublished work.

[c] From ref 69, where CNT/PI composite fibers were measured.

[d] From ref 54, where CNT/PVA composite fibers were measured.

[e] Obtained with a strain at break of 104%.

[f] Untreated CNT/PVA fibers with strain at break of 340%.

[g] Hot-stretched CNT/PVA fibers with strain at break of 11%.

14.3.2 Morphological and Structural Dependences

The mechanical properties of CNT fibers can be described as functions of different parameters. Figure 14.8a,b shows the tensile strength and modulus against the fiber diameter, for the dry-spun C_2H_2-I and C_2H_2-II fibers, defined according to the tube structure of their constituent CNTs [47]. Contrary to popular belief that a thinner fiber might be stronger, both σ_f and E_f increase *slowly* with d_f. The reason is that a thinner fiber is less compacted due to the smaller radial pressure inside the fiber. However, these values nearly reach a plateau for fibers with a diameter larger than 10 μm. When further increasing the fiber diameter, up to more than 20 μm, the strength and modulus might be maintained [54] (see Figure 14.8c), or decrease linearly [68]. The decrease is possibly due to the loose assembly structure inside the fiber, because the twisting/stretching treatment helps the alignment only of the CNTs on the outermost fiber surfaces, especially in the large-diameter fibers. By using infiltration or swelling-and-drying with the aid of polymer solution, the fiber can be overall densified, and then have a non-diameter-dependent tensile properties.

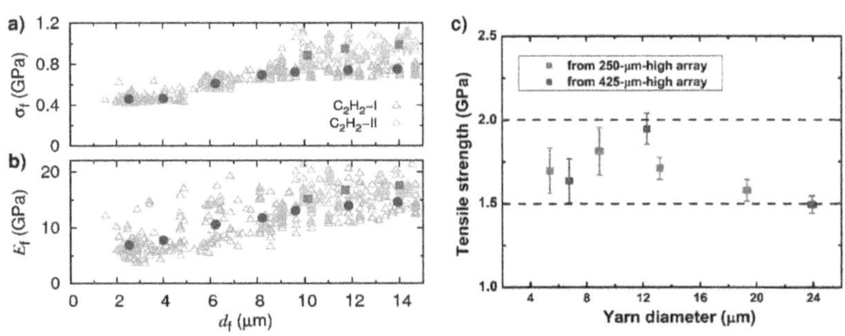

Figure 14.8 (a,b) Fiber strength (σ_f) and modulus (E_f) generally increase with fiber diameter (d_f). Blue triangles are the 40 strongest C_2H_2-I fibers for different diameter windows (0–3, 3–5, 5–7, 7–9, 9–11, 11–13, and 13–15 μm), averaged as the solid circles. Due to the fewer number of samples, the top 30 highest results of the largest three diameter windows are plotted and averaged, in red, for the C_2H_2-II fibers. (c) Tensile strengths of CNT/PVA fibers (yarns) with various diameters made from two arrays with different heights. From refs 47,54.

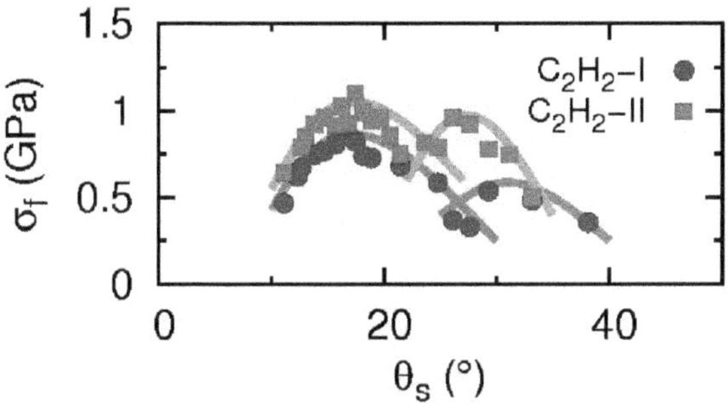

Figure 14.9 Fiber strength as a function of twist angle for dry-spun fibers with diameters larger than 10.5 μm. From ref 47.

However, the tensile properties strongly depend on the twist. Several studies have, even unconsciously, shown such twist dependence [10,47,51,68,70]. There is a strength peak at a twist angle of 15–20° [47,68], a normal peak that can be similarly observed in traditional fibers. For example, the optimal twist angle of cotton fibers is within the range of 20–23° [71,72]. Therefore the generic equation [8] or model [47] describing traditional fibers can be used for CNT fibers. With further increasing the angle, different from traditional ones whose strength drops greatly, CNT fibers show another peak around 27–30° (Figure 14.9), depending on the CNT structure: it appears at a smaller angle if the tubes have fewer walls. The new peak arises from the collapse transformation of the special hollow structure of CNTs, as one can find the evidence from Raman scattering (discussed in ref 47). The double-peak mechanical behavior can be explained by a continuum model which allows the introduction of the collapse. The two peaks are both important because the normal one corresponds to the optimal spin condition for high-performance CNT fibers, and the other can be used as a post-spin treatments (stretching and overtwisting) to enhance fibers.

The nanotube waviness and alignment also influence the quality of CNT fibers [73]. However it is still difficult to present a clear morphological dependence as the morphology is strongly affected by the spinning technique. For dry-spun fibers, the tube waviness might not show a strong influence on the fiber strength, see Figure 14.10. The three fibers are spun from highly spinnable CNT arrays, according to the definition of spinnability by Zheng et al. [73], and composed of CNTs with similar diameter and wall thickness but just different tube waviness. If the waviness is large, the CNT fiber shows a sticky tensile behavior that the tensile stress does not grow up rapidly until a certain large tensile strain. As decreasing the waviness, the modulus goes up and the strain at break decreases. However, their strengths are nearly the same, indicating the tube structure is playing a more important role.

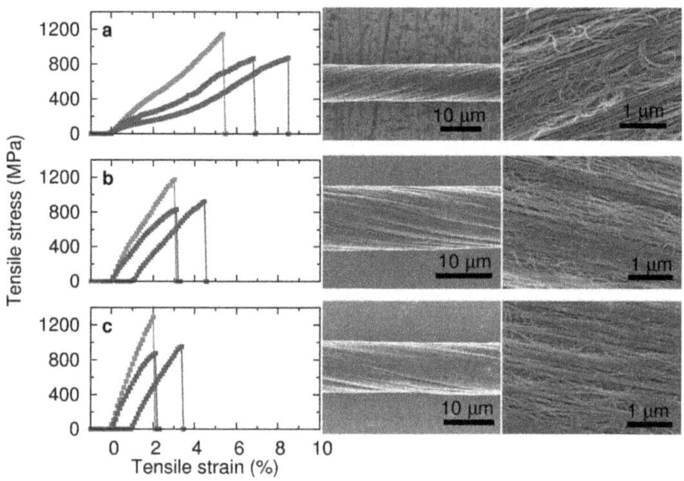

Figure 14.10 Tensile tests and fiber morphology of the C_2H_2 fibers with different tube waviness. In each case the blue curves are results of two as-spun fibers, and the red curve is of one PVA-infiltrated fiber. (a) The fiber with large tube waviness shows a sticky tensile behavior. (b,c) As decreasing the waviness, the modulus goes up and the strain at break decreases. As the CNT arrays are highly spinnable, the waviness does not show a strong influence on the fiber strength. From ref 53.

The number of walls (wall thickness) of the constituent CNTs is crucial to determine the mechanical properties. As already mentioned above, the double- and/or triple-walled CNT fibers have the highest tensile strength (up to 2.0 GPa) and modulus (up to 100 GPa). For MWCNT fibers, their strengths of 564 MPa (>10 walls) [9], 0.63–1.1 GPa (~9 walls) [51], 0.8–1.04 GPa (~9 walls) [68], and 0.29–1.3 GPa (2–10 walls) [73] are much smaller. The different contribution of the wall thickness mainly come from the different tube rigidity. As increasing the thickness, nanotubes become more rigid and difficult to deform the cross section [74]. When CNTs are assembled into a bundle or a fiber, they are surrounded by neighboring tubes. Only for the few-walled, especially the single- and double-walled tubes, they can change the cross section easily from circular to pentagonal to increase the intertube vdW attraction. For the few-walled fibers, the collapse of CNTs can be also easily introduced by twisting [47] or stretching [4], resulting in high tensile properties of the spun fibers.

There are many other structural and morphological issues playing roles in the enhancement of CNT fibers. For example, by increasing the tube length (aspect ratio), stronger fibers are spun [51,68]. Densified fibers are stronger too [51,52]. Modifying the interfacial structure and interaction also lead to higher mechanical properties. For example, PVA-infiltrated fibers are usually about 300–400 MPa stronger than the un-infiltrated fibers [8,47], and ~1 GPa stronger than un-shrunk pure CNT fibers [54]. The enhancement is due to the stronger interfacial vdW interaction between polymers and CNTs. The polymer-CNT interaction can be further increased by improving the polymer orientation along the tubes. Similarly, a certain defects might also improve the interfacial stress greatly, because the interfacial friction is indeed a function of the surface roughness. By spinning nitrogen-doped CNTs, high-modulus and high-toughness fibers can be achieved [21], where the modulus is increased from 69 GPa (from pristine CNTs) to 142 GPa. The situation becomes much more complicated for CNT/polymer composite fibers, especially from the wet spinning. The polymer type and structure, fraction, and orientation, and the post-spin treatments including stretching and heating, can change the fiber performance. However, in order to understand the underlying mechanism of fiber improvement, a well-designed measurement on the interfacial structure, interaction, and frictional characteristic is necessary.

14.4 Multifunctionalities of CNT Fibers

CNTs are nanomaterials with more functionalities than other known materials synthesized by mankind. For individual CNTs, functional groups can be easily formed on their surfaces in various ways [75]. When they are assembled into fiber, different polymer molecules can also be introduced around them. These methods enrich the spun fibers with new features. Therefore, besides their potential as a new generation of carbon fiber, CNT fibers can be used as a smart material with the ability to respond to electrical currents, mechanical stresses, temperature changes, and various environments.

CNT fibers are good conductors with a temperature-dependent conductivity of 450–600 S/cm [76]. Different from individual CNTs, the conductivity of a fiber is derived from two components: the resistance of the constituent CNTs and the contact resistance between them. The resistivity of individual tubes is usually two orders of magnitude lower than their assemblies [8,77], due to the large contact resistance between CNTs. Therefore, the conductivity of CNT fibers is still orders of magnitude smaller than a copper wire. Experiments have shown that the conduction is predominantly controlled by a three-dimensional hopping of electrons between CNTs [76,78], and that the temperature dependence of the fiber conductivity can be expressed as $\sigma \propto exp(-A/T^{1/4})$, where σ and T are the conductivity and temperature and A is a constant.

The incorporation of metal particles increases significantly the conductivity of CNT fibers. Galvanic deposition of gold nanoparticles on CNT fibers leads to a conductivity up to 1150 S/cm [76]. As the gold nanoparticles were found to be sparsely distributed on the fibers, Randeniya et al. improved the incorporation technique [31]. In their method, a CNT fiber is immersed in an aqueous solution of the salt of the metal to be deposited. The two fiber ends are connected to reducing metal anodes, which have a lower reduction potential than the metal in the solution. The uniform deposition of Cu, Au, Pd, and Pt increased the electric conductivity up to 3×10^5, 2×10^5, 2×10^4, and 5×10^3 S/cm, respectively. Due to their uniform face-centered cubic structure of Au and Cu in the fibers, and their strong vdW interaction with CNTs to encourage the initial deposition, the conductivities of the composite fibers were nearly 50% of the corresponding pure metals (copper 5.9×10^5 S/cm, gold 4.6×10^5 S/cm). However, the lower conductivity of other composite fibers was attributed to the presence of grain boundaries in the thin metal films. These composite fibers also showed a typical metal-like dependence of electrical resistivity on temperature. With introducing metal into the fiber, the mass density goes up largely. For example, a CNT/Cu fiber was about a factor of 2–3 lighter than a Cu wire, while the density of a pure CNT fiber could be only ~1/7 of that of Cu. There was also a drop in the tensile strength, from ~1 GPa of untreated fibers to

500–650 MPa of metal-coated fibers. Therefore, further research to reduce the loss of strength and to minimize the increase in weight is still necessary.

An exciting CNT based smart fiber has been reported recently by Peng et al. [32]. By soaking a dry-spun fiber in a diacetylenic precursor solution, followed by evaporation of solvent and polymerization under ultraviolet light, a CNT/PDA composite fiber is made. The fiber exhibits high electrical conductivities of 100–1000 S/cm, depending on the temperature. When an electric current is passing through the fiber, it changes rapidly (in less than 1 s) from blue to red, shown in Figure 14.11. After the turn-off of the current, the fiber reversely becomes blue. The reversible chromatic transition (color change) can continue for cycles when the current is lower than 30 mA. However, at higher current, the transition is irreversible. Such current-induced chromatism is not a temperature effect, but caused by the responsive feature of PDA backbones under the electric fields as electrons hopping from one CNT to another [76,78]. In the case of low currents, PDA conformation can return to the original state, corresponding to the reversible color change. However, higher currents might destroy the recovery capability, leading to irreversible chromatism. CNT/PDA fibers also respond chromatically to a wide variety of other environmental stimuli, for instance, mechanical abrasion and chemical/organic vapor such as tetrahydrofuran and *N,N*-dimethyl formamide. Nevertheless, it should be noted that abrasion-, chemical- and vapor-induced chromatism are not reversible.

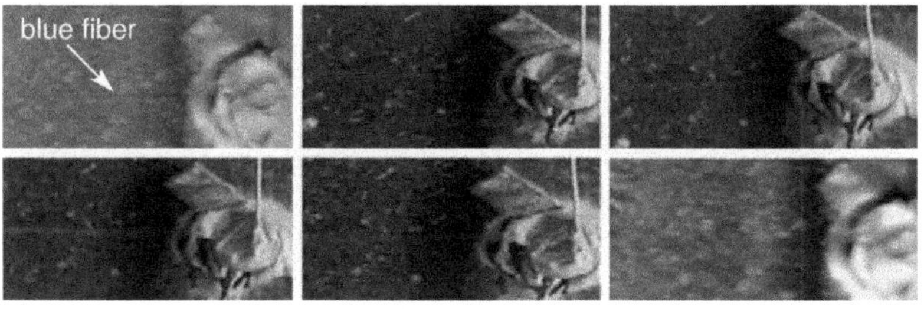

Figure 14.11 Passing a dc current of 10 mA through a CNT/PDA fiber causes a blue fiber (top left) to become red after 1 s (top middle). The current is turned off after 3 s (bottom left), and the fiber becomes blue again after a further 2 s (bottom right). From ref 32. See also Color Insert.

It is reported that CNTs are piezoresistive [79–83]. However, rather than using an individual CNT as a nanosensor, assembling these nanosensors into a microscopic one for easy integration into bulk structures is important for practical applications [84]. Besides the CNT/polymer composite sensors [85,86], pure CNT fiber can be utilized as a superior novel piezoresistive sensor with good repeatability and stability for structural health monitoring. The electrical resistance and the tensile stress of the fiber have a linear dependence on the strains within 1%. This resistance-strain behavior is repeatable, with no hysteresis, and stable at temperatures ranging from 77 to 373 K. Zhao et al. demonstrated that the sensors can be embedded into composite structures with minimal invasiveness and weight penalty, to monitor crack initiation and propagation [84].

CNT fiber as a glucose biosensor shows another success in potential applications [33]. Figure 14.12 shows the electrical response of a CNT fiber. The response current reaches a dynamic equilibrium within tens of seconds (response time) of each addition of glucose, generating a near steady-state current signal. It indicates a fast electron transfer between the redox center of the enzyme and the CNT fiber [87]. In order to achieve such sensibility, one end of the CNT fiber was quenched in liquid nitrogen, fractured, and manually unwound. Such end was brush-like with increased surface area for enzyme coating.

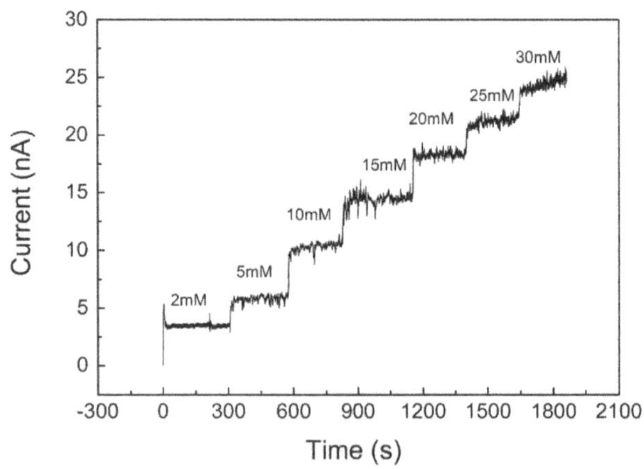

Figure 14.12 Amperometric response curve for a CNT fiber based glucose biosensor. Numbers in the chart represent the corresponding glucose concentration of the solution at different steps. The instantaneous rise in the response current corresponds to the rapid sensing of the fiber. From ref 33.

Moreover, CNT fiber based electrochemical devices such as supercapacitors or actuators can be made [20,88]. CNT fibers containing different biopolymers might have totally different response to humid atmosphere [59]. Clearly, these CNT-based multifunctional fibers are promising candidates for a wide range of applications.

14.5 Conclusion

Compared with the history of carbon fibers, which has spanned over 50 years and become an important engineering fiber, the development of CNT fibers is still in its infancy, but very rapid due to the progress in CNT growth. So far, different assembly methods have been proposed, and CNT fibers with different mechanical and electrical properties have been synthesized. Considering the repeatability and stability, the strongest fibers have the tensile strength of about 2 GPa or close to 3 N/tex, and the modulus of at least 100 GPa or 150 N/tex, both still far below those of individual CNTs. Although challenging, an efficient assembly of raw tubes into continuous fibers with desired geometry, size and much better mechanical properties is possible. Main issues concerned include the improvement of the tube aspect ratio, structure perfection, better connecting between CNTs, and the scale-up of spinning techniques. CNT fibers are also good to respond to electrical and thermal signals and sensitive to different environments. Several pioneering works, including the biosensor, strain sensor, and metallic and chromatic fibers, have been developed. As a new generation of carbon fiber, CNT fibers are lighter, stronger, stiffer, and more flexible, and essential to enable some advanced applications in future.

Acknowledgements

Funding from Hundred Talent Program for Q. Li and Knowledge Innovation Program (No. KJCX2.YW.M12) by Chinese Academy of Science, International Collaboration Project (No. 2009DFB50150) and National Basic Research Program (No. 2010CB934700) by Ministry of Science and Technology, National Natural Science Foundation of China (No. 10972152), and China Postdoctoral Science Foundation (No. 20100471394) is gratefully acknowledged.

References

1. R.S. Ruoff, D. Qian, and W.K. Liu, Mechanical Properties of Carbon Nanotubes: Theoretical Predictions and Experimental Measurements, *C. R. Phys.* **4**(9), 993–1008 (2003).
2. Y.-L. Li, I.A. Kinloch, and A.H. Windle, Direct Spinning of Carbon Nanotube Fibers from Chemical Vapor Deposition Synthesis, *Science* **304**(5668), 276–278 (2004).
3. M. Motta, Y.-L. Li, I. Kinloch, and A. Windle, Mechanical Properties of Continuously Spun Fibers of Carbon Nanotubes, *Nano Lett.* **5**(8), 1529–1533 (2005).
4. M. Motta, A. Moisala, I.A. Kinloch, and A.H. Windle, High Performance Fibres from 'Dog Bone' Carbon Nanotubes, *Adv. Mater.* **19**(21), 3721–3726 (2007).
5. K. Koziol, J. Vilatela, A. Moisala, M. Motta, P. Cunniff, M. Sennett, and A. Windle, High-Performance Carbon Nanotube Fiber, *Science* **318**(5858), 1892–1895 (2007).
6. X.-H. Zhong, Y.-L. Li, Y.-K. Liu, X.-H. Qiao, Y. Feng, J. Liang, J. Jin, L. Zhu, F. Hou, and J.-Y. Li, Continuous Multilayered Carbon Nanotube Yarns, *Adv. Mater.* **22**(6), 692–696 (2010).
7. K. Jiang, Q. Li, and S. Fan, Spinning continuous carbon nanotube yarns, *Nature* **419**(6909), 801–801 (2002).
8. M. Zhang, K.R. Atkinson, and R.H. Baughman, Multifunctional Carbon Nanotube Yarns by Downsizing an Ancient Technology, *Science* **306**(5700), 1358–1361 (2004).
9. X. Zhang, K. Jiang, C. Feng, P. Liu, L. Zhang, J. Kong, T. Zhang, Q. Li, and S. Fan, Spinning and Processing Continuous Yarns from 4-Inch Wafer Scale Super-Aligned Carbon Nanotube Arrays, *Adv. Mater.* **18**(12), 1505–1510 (2006).
10. X. Zhang, Q. Li, Y. Tu, Y. Li, J.Y. Coulter, L. Zheng, Y. Zhao, Q. Jia, D.E. Peterson, and Y. Zhu, Strong Carbon-Nanotube Fibers Spun from Long Carbon-Nanotube Arrays, *Small* **3**(2), 244–248 (2007).
11. Q. Li, X. Zhang, R.F. DePaula, L. Zheng, Y. Zhao, L. Stan, T.G. Holesinger, P.N. Arendt, D.E. Peterson, and Y.T. Zhu, Sustained Growth of Ultralong Carbon Nanotube Arrays for Fiber Spinning, *Adv. Mater.* **18**(23), 3160–3163 (2006).
12. X. Zhang, Q. Li, T.G. Holesinger, P.N. Arendt, J. Huang, P.D. Kirven, T.G. Clapp, R.F. DePaula, X. Liao, Y. Zhao, L. Zheng, D.E. Peterson, and Y. Zhu, Ultrastrong, Stiff, and Lightweight Carbon-Nanotube Fibers, *Adv. Mater.* **19**(23), 4198–4201 (2007).

13. H.H. Gommans, J.W. Alldredge, H. Tashiro, J. Park, J. Magnuson, and A.G. Rinzler, Fibers of aligned single-walled carbon nanotubes: Polarized Raman spectroscopy, *J. Appl. Phys.* **88**(5), 2509–2514 (2000).
14. B. Vigolo, A. Pénicaud, C. Coulon, C. Sauder, R. Pailler, C. Journet, P. Bernier, and P. Poulin, Macroscopic Fibers and Ribbons of Oriented Carbon Nanotubes, *Science* **290**(5495), 1331–1334 (2000).
15. A.V. Neimark, S. Ruetsch, K.G. Kornev, P.I. Ravikovitch, P. Poulin, S. Badaire, and M. Maugey, Hierarchical Pore Structure and Wetting Properties of Single-Wall Carbon Nanotube Fibers, *Nano Lett.* **3**(3), 419–423 (2003).
16. A.B. Dalton, S. Collins, E. Muňoz, J.M. Razal, V.H. Ebron, J.P. Ferraris, J.N. Coleman, B.G. Kim, and R.H. Baughman, Super-Tough Carbon-Nanotube Fibres, *Nature* **423**(6941), 703–703 (2003).
17. L.M. Ericson, H. Fan, H. Peng, V.A. Davis, W. Zhou, J. Sulpizio, Y. Wang, R. Booker, J. Vavro, C. Guthy, A.N.G. Parra-Vasquez, M.J. Kim, S. Ramesh, R.K. Saini, C. Kittrell, G. Lavin, H. Schmidt, W.W. Adams, W.E. Billups, M. Pasquali, W.-F. Hwang, R.H. Hauge, J.E. Fischer, and R.E. Smalley, Macroscopic, Neat, Single-Walled Carbon Nanotube Fibers, *Science* **305**(5689), 1447–1450 (2004).
18. M.E. Kozlov, R.C. Capps, W.M. Sampson, V.H. Ebron, J.P. Ferraris, and R.H. Baughman, Spinning Solid and Hollow Polymer-Free Carbon Nanotube Fibers, *Adv. Mater.* **17** (5), 614–617 (2005).
19. P. Miaudet, S. Badaire, M. Maugey, A. Derré, V. Pichot, P. Launois, P. Poulin, and C. Zakri, Hot-Drawing of Single and Multiwall Carbon Nanotube Fibers for High Toughness and Alignment, *Nano Lett.* **5**(11), 2212–2215 (2005).
20. C. Lynam, S.E. Moulton, and G.G. Wallace, Carbon-Nanotube Biofibers, *Adv. Mater.* **19**(9), 1244–1248 (2007).
21. S. Zhang, K.K.K. Koziol, I.A. Kinloch, and A.H. Windle, Macroscopic Fibers of Well-Aligned Carbon Nanotubes by Wet Spinning, *Small* **4**(8), 1217–1222 (2008).
22. C. Liu, H.-M. Cheng, H.T. Cong, F. Li, G. Su, B.L. Zhou, and M.S. Dresselhaus, Synthesis of Macroscopically Long Ropes of Well-Aligned Single-Walled Carbon Nanotubes, *Adv. Mater.* **12**(16), 1190–1192 (2000).
23. C. Liu, Y. Tong, H.-M. Cheng, D. Golberg, and Y. Bando, Field Emission Properties of Macroscopic Single-Walled Carbon Nanotube Strands, *Appl. Phys. Lett.* **86**(22), 223114, (2005).
24. H.W. Zhu, C.L. Xu, D.H. Wu, B.Q. Wei, R. Vajtai, and P.M. Ajayan, Direct Synthesis of Long Single-Walled Carbon Nanotube Strands, *Science* **296**(5569), 884–886 (2002).

25. B. Wei, R. Vajtai, Y.Y. Choi, P.M. Ajayan, H. Zhu, C. Xu, and D. Wu, Structural Characterizations of Long Single-Walled Carbon Nanotube Strands, *Nano Lett.* **2**(10), 1105–1107 (2002).
26. Q.H. Yang, S. Bai, T. Fournier, F. Li, G. Wang, H.M. Cheng, and J.B. Bai, Direct Growth of Macroscopic Fibers Composed of Large Diameter SWNTs by CVD, *Chem. Phys. Lett.* **370**(1–2), 274–279 (2003).
27. W. Ma, L. Song, R. Yang, T. Zhang, Y. Zhao, L. Sun, Y. Ren, D. Liu, L. Liu, J. Shen, Z. Zhang, Y. Xiang, W. Zhou, and S. Xie, Directly Synthesized Strong, Highly Conducting, Transparent Single-Walled Carbon Nanotube Films, *Nano Lett.* **7**(8), 2307–2311 (2007).
28. W. Ma, L. Liu, R. Yang, T. Zhang, Z. Zhang, L. Song, Y. Ren, J. Shen, Z. Niu, W. Zhou, and S. Xie, Monitoring a Micromechanical Process in Macroscale Carbon Nanotube Films and Fibers, *Adv. Mater.* **21**(5), 603–608 (2009).
29. W. Ma, L. Liu, Z. Zhang, R. Yang, G. Liu, T. Zhang, X. An, X. Yi, Y. Ren, Z. Niu, J. Li, H. Dong, W. Zhou, P.M. Ajayan, and S. Xie, High-Strength Composite Fibers: Realizing True Potential of Carbon Nanotubes in Polymer Matrix through Continuous Reticulate Architecture and Molecular Level Couplings, *Nano Lett.* 9(8), 2855–2861 (2009).
30. Y. Gao, J. Li, L. Liu, W. Ma, W. Zhou, S. Xie, and Z. Zhang, Axial Compression of Hierarchically Structured Carbon Nanotube Fiber Embedded in Epoxy, *Adv. Funct. Mater.* **20**(21), 3797–3803 (2010).
31. L.K. Randeniya, A. Bendavid, P.J. Martin, and C.-D. Tran, Composite Yarns of Multiwalled Carbon Nanotubes with Metallic Electrical Conductivity, *Small* **6**(16), 1806–1811 (2010).
32. H. Peng, X. Sun, F. Cai, X. Chen, Y. Zhu, G. Liao, D. Chen, Q. Li, Y. Lu, Y. Zhu, and Q. Jia, Electrochromatic Carbon Nanotube/Polydiacetylene Nanocomposite Fibres, *Nat. Nanotech.* **4**(11), 738–741 (2009).
33. Z. Zhu, W. Song, K. Burugapalli, F. Moussy, Y.-L. Li, and X.-H. Zhong, Nano-yarn Carbon Nanotube Fiber Based Enzymatic Glucose Biosensor, *Nanotechnology* **21**(16), 165501, (2010).
34. A. Cao, P.M. Ajayan, and G. Ramanath, Assembly of mm-scale macrobridges with carbon nanotube bundles, *Appl. Phys. Lett.* **83**(2), 356–358 (2003).
35. X. Wang, Q. Li, J. Xie, Z. Jin, J. Wang, Y. Li, K. Jiang, and S. Fan, Fabrication of Ultralong and Electrically Uniform Single-Walled Carbon Nanotubes on Clean Substrates, *Nano Lett.* **9**(9), 3137–3141 (2009).
36. Q. Wen, R. Zhang, W. Qian, Y. Wang, P. Tan, J. Nie, and F. Wei, Growing 20 cm Long DWNTs/TWNTs at a Rapid Growth Rate of 80–90 μm/s, *Chem. Mater.* **22**(4), 1294–1296 (2010).

37. B. Vigolo, P. Poulin, M. Lucas, P. Launois, and P. Bernier, Improved Structure and Properties of Single-Wall Carbon Nanotube Spun Fibers, *Appl. Phys. Lett.* **81**(7), 1210–1212 (2002).

38. V.A. Davis, A.N.G. Parra-Vasquez, M.J. Green, P.K. Rai, N. Behabtu, V. Prieto, R.D. Booker, J. Schmidt, E. Kesselman, W. Zhou, H. Fan, W.W. Adams, R.H. Hauge, J.E. Fischer, Y. Cohen, Y. Talmon, R.E. Smalley, and M. Pasquali, True Solutions of Single-Walled Carbon Nanotubes for Assembly into Macroscopic Materials, *Nat. Nanotech.* **4**(12), 830–834 (2009).

39. G.G. Tibbetts, C.A. Bernardo, D.W. Gorkiewicz, and R.L. Alig, Role of Sulfur in the Production of Carbon Fibers in the Vapor Phase, *Carbon* **32**(4), 569–576 (1994).

40. J.J. Vilatela, and A.H. Windle, Yarn-Like Carbon Nanotube Fibers, *Adv. Mater.* 22(44), 4959–4963 (2010).

41. L. Song, L. Ci, L. Lv, Z. Zhou, X. Yan, D. Liu, H. Yuan, Y. Gao, J. Wang, L. Liu, X. Zhao, Z. Zhang, X. Dou, W. Zhou, G. Wang, C. Wang, and S. Xie, Direct Synthesis of a Macroscale Single-Walled Carbon Nanotube Non-Woven Material, *Adv. Mater.* **16**(17), 1529–1534 (2004).

42. K. Liu, K. Jiang, C. Feng, Z. Chen, and S. Fan, A Growth Mark Method for Studying Growth Mechanism of Carbon Nanotube Arrays, *Carbon* **43**(14), 2850–2856 (2005).

43. K. Liu, Y. Sun, P. Liu, J. Wang, Q. Li, S. Fan, and K. Jiang, Periodically Striped Films Produced from Super-Aligned Carbon Nanotube Arrays, *Nanotechnology* **20**(33), 335705 (2009).

44. Z.F. Ren, Z.P. Huang, J.W. Xu, J.H. Wang, P. Bush, M.P. Siegal, and P.N. Provencio, Synthesis of Large Arrays of Well-Aligned Carbon Nanotubes on Glass, *Science* **282**(5391), 1105–1107 (1998).

45. S. Fan, M.G. Chapline, N.R. Franklin, T.W. Tombler, A.M. Cassell, and H. Dai, Self-Oriented Regular Arrays of Carbon Nanotubes and Their Field Emission Properties, *Science* **283**(5401), 512–514 (1999).

46. K. Liu, Y. Sun, L. Chen, C. Feng, X. Feng, K. Jiang, Y. Zhao, and S. Fan, Controlled Growth of Super-Aligned Carbon Nanotube Arrays for Spinning Continuous Unidirectional Sheets with Tunable Physical Properties, *Nano Lett.* **8**(2), 700–705 (2008).

47. J. Zhao, X. Zhang, J. Di, G. Xu, X. Yang, X. Liu, Z. Yong, M. Chen, and Q. Li, Double-Peak Mechanical Properties of Carbon Nanotube Fibers, *Small* **6**(22), 2612–2617 (2010).

48. Y. Zhang, G. Zou, S.K. Doorn, H. Htoon, L. Stan, M.E. Hawley, C.J. Sheehan, Y. Zhu, and Q. Jia, Tailoring the Morphology of Carbon Nanotube Arrays:

From Spinnable Forests to Undulating Foams, *ACS Nano* **3**(8), 2157–2162 (2009).

49. Y. Inoue, K. Kakihata, Y. Hirono, T. Horie, A. Ishida, and H. Mimura, One-Step Grown Aligned Bulk Carbon Nanotubes by Chloride Mediated Chemical Vapor Deposition, *Appl. Phys. Lett.* **92**(21), 213113 (2008).
50. Q. Zhang, D.-G. Wang, J.-Q. Huang, W.-P. Zhou, G.-H. Luo, W.-Z. Qian, and F. Wei, Dry Spinning Yarns from Vertically Aligned Carbon Nanotube Arrays Produced by an Improved Floating Catalyst Chemical Vapor Deposition Method, *Carbon* **48**(10), 2855–2861 (2010).
51. K. Liu, Y. Sun, R. Zhou, H. Zhu, J. Wang, L. Liu, S. Fan, and K. Jiang, Carbon Nanotube Yarns with High Tensile Strength Made by a Twisting and Shrinking Method, *Nanotechnology* **21**(4), 045708 (2010).
52. C.D. Tran, W. Humphries, S.M. Smith, C. Huynh, and S. Lucas, Improving the Tensile Strength of Carbon Nanotube Spun Yarns Using a Modified Spinning Process, *Carbon* **47**(11), 2662–2670 (2009).
53. J. Jia, J. Zhao, G. Xu, J. Di, Z. Yong, Y. Tao, C. Fang, Z. Zhang, X. Zhang, L. Zheng, Q. Li, *Carbon* **49**(4), 1333–1339 (2011).
54. K. Liu, Y. Sun, X. Lin, R. Zhou, J. Wang, S. Fan, and K. Jiang, Scratch-Resistant, Highly Conductive, and High-Strength Carbon Nanotube-Based Composite Yarns, *ACS Nano* **4**(10), 5827–5834 (2010).
55. P. Nikolaev, M.J. Bronikowski, R.K. Bradley, F. Rohmund, D.T. Colbert, K.A. Smith, and R.E. Smalley, Gas-Phase Catalytic Growth of Single-Walled Carbon Nanotubes from Carbon Monoxide, *Chem. Phys. Lett.* **313**(1–2), 91–97 (1999).
56. M.J. Bronikowski, P.A. Willis, D.T. Colbert, K.A. Smith, and R.E. Smalley, Gas-Phase Production of Carbon Single-Walled Nanotubes from Carbon Monoxide Via the HiPco Process: A Parametric Study, *J. Vac. Sci. Technol. A* **19**(4), 1800–1805 (2001).
57. R. Xiang, G. Luo, W. Qian, Q. Zhang, Y. Wang, F. Wei, Q. Li, and A. Cao, Encapsulation, Compensation, and Substitution of Catalyst Particles During Continuous Growth of Carbon Nanotubes, *Adv. Mater.* **19**(17), 2360–2363 (2007).
58. R. Xiang, G. Luo, W. Qian, Y. Wang, F. Wei, and Q. Li, Large Area Growth of Aligned CNT Arrays on Spheres: Towards Large Scale and Continuous Production, *Chem. Vap. Deposition* **13**(10), 533–536 (2007).
59. A.J. Granero, J.M. Razal, G.G. Wallace, and M. in het Panhuis, Spinning Carbon Nanotube-Gel Fibers Using Polyelectrolyte Complexation, *Adv. Funct. Mater.* **18**(23) 3759–3764 (2008).

60. J. Steinmetz, M. Glerup, M. Paillet, P. Bernier, and M. Holzinger, Production of Pure Nanotube Fibers Using a Modified Wet-Spinning Method, *Carbon* **43**(11), 2397–2400 (2005).

61. S. Ramesh, L.M. Ericson, V.A. Davis, R.K. Saini, C. Kittrell, M. Pasquali, W.E. Billups, W.W. Adams, R.H. Hauge, and R.E. Smalley, Dissolution of Pristine Single Walled Carbon Nanotubes in Superacids by Direct Protonation, *J. Phys. Chem. B* **108**(26), 8794–8798 (2004).

62. W. Song, I.A. Kinloch, and A.H. Windle, Nematic Liquid Crystallinity of Multiwall Carbon Nanotubes, *Science* **302**(5649), 1363–1363 (2003).

63. W. Song, and A.H. Windle, Isotropic-Nematic Phase Transition of Dispersions of Multiwall Carbon Nanotubes, *Macromolecules* **38**(14), 6181–6188 (2005).

64. S. Zhang, I.A. Kinloch, and A.H. Windle, Mesogenicity Drives Fractionation in Lyotropic Aqueous Suspensions of Multiwall Carbon Nanotubes, *Nano Lett.* **6**(3), 568–572 (2006).

65. E. Muñoz, D.-S. Suh, S. Collins, M. Selvidge, A.B. Dalton, B.G. Kim, J.M. Razal, G. Ussery, A.G. Rinzler, M.T. Martínez, and R.H. Baughman, Highly Conducting Carbon Nanotube/Polyethyleneimine Composite Fibers, *Adv. Mater.* **17**(8), 1064–1067 (2005).

66. J.M. Razal, K.J. Gilmore, and G.G. Wallace, Carbon Nanotube Biofiber Formation in a Polymer-Free Coagulation Bath, *Adv. Funct. Mater.* **18**(1), 61–66 (2008).

67. J.N. Barisci, M. Tahhan, G.G. Wallace, S. Badaire, T. Vaugien, M. Maugey, and P. Poulin, Properties of Carbon Nanotube Fibers Spun from DNA-Stabilized Dispersions, *Adv. Funct. Mater.* **14**(2), 133–138 (2004).

68. S. Fang, M. Zhang, A.A. Zakhidov, and R.H. Baughman, Structure and Process Dependent Properties of Solid-State Spun Carbon Nanotube Yarns, *J. Phys. Condens. Matter.* **22**(33), 334221 (2010).

69. C. Fang, J. Zhao, J. Jia, Z. Zhang, X. Zhang, and Q. Li, Enhanced Carbon Nanotube Fibers by Polyimide, *Appl. Phys. Lett.* **97**(18), 181906 (2010).

70. K. Sears, C. Skourtis, K. Atkinson, N. Finn, and W. Humphries, Focused Ion Beam Milling of Carbon Nanotube Yarns to Study the Relationship Between Structure and Strength, *Carbon* **48**(15), 4450–4456 (2010).

71. J.J. Hebert, R. Giardina, D. Mitcham, and M.L. Rollins, The Effect of Convolutions on Orientation Measurements in Cotton Fibers, *Text. Res. J.* **40**(2), 126–129 (1970).

72. N. Morosoff, and P. Ingram, The Spiral Angle of Cotton, *Text. Res. J.* **40**(3), 250–255, (1970).

73. L. Zheng, G. Sun, and Z. Zhan, Tuning Array Morphology for High-Strength Carbon- Nanotube Fibers, *Small* **6**(1), 132–137 (2010).

74. T. Hertel, R.E. Walkup, and P. Avouris, Deformation of Carbon Nanotubes by Surface Van Der Waals Forces, *Phys. Rev. B* **58**(20), 13870–13873 (1998).

75. D. Tasis, N. Tagmatarchis, A. Bianco, and M. Prato, Chemistry of Carbon Nanotubes, *Chem. Rev.* **106**(3), 1105–1136 (2006).

76. Q. Li, Y. Li, X. Zhang, S.B. Chikkannanavar, Y. Zhao, A.M. Dangelewicz, L. Zheng, S.K. Doorn, Q. Jia, D.E. Peterson, P.N. Arendt, and Y. Zhu, Structure-Dependent Electrical Properties of Carbon Nanotube Fibers, *Adv. Mater.* **19**(20), 3358–3363 (2007).

77. W. Zhou, J. Vavro, C. Guthy, K.I. Winey, J.E. Fischer, L.M. Ericson, S. Ramesh, R. Saini, V.A. Davis, C. Kittrell, M. Pasquali, R.H. Hauge, and R.E. Smalley, Single Wall Carbon Nanotube Fibers Extruded from Super-Acid Suspensions: Preferred Orientation, Electrical, and Thermal Transport, *J. Appl. Phys.* **95**(2), 649–655 (2004).

78. H. Peng, Aligned Carbon Nanotube/Polymer Composite Films with Robust Flexibility, High Transparency, and Excellent Conductivity, *J. Am. Chem. Soc.* **130**(1), 42–43 (2008).

79. T.W. Tombler, C. Zhou, L. Alexseyev, J. Kong, H. Dai, L. Liu, C.S. Jayanthi, M. Tang, and S.-Y. Wu, Reversible Electromechanical Characteristics of Carbon Nanotubes Under Local-Probe Manipulation, *Nature* **405**(6788), 769–772 (2000).

80. E.D. Minot, Y. Yaish, V. Sazonova, J.-Y. Park, M. Brink, and P.L. McEuen, Tuning Carbon Nanotube Band Gaps with Strain, *Phys. Rev. Lett.* **90**(15), 156401 (2003).

81. J. Cao, Q. Wang, and H. Dai, Electromechanical Properties of Metallic, Quasimetallic, and Semiconducting Carbon, *Phys. Rev. Lett.* **90**(15), 157601 (2003).

82. R.J. Grow, Q. Wang, J. Cao, D. Wang, and H. Dai, Piezoresistance of Carbon Nanotubes on Deformable Thin-Film Membranes, *Appl. Phys. Lett.* **86**(9), 093104 (2005).

83. T. Nishio, Y. Miyato, K. Kobayashi, K. Matsushige, and H. Yamada, Piezoresistive Properties of Carbon Nanotubes Under Radial Force Investigated by Atomic Force Microscopy, *Appl. Phys. Lett.* **92**(6), 063117 (2008).

84. H. Zhao, Y. Zhang, P.D. Bradford, Q. Zhou, Q. Jia, F.-G. Yuan, and Y. Zhu, Carbon Nanotube Yarn Strain Sensors, *Nanotechnology* **21**(30), 305502 (2010).

85. E.T. Thostenson, and T.-W. Chou, Carbon Nanotube Networks: Sensing of Distributed Strain and Damage for Life Prediction and Self Healing, *Adv. Mater.* **18**(21), 2837–2841 (2006).

86. E.T. Thostenson, and T.-W. Chou, Real-Time in Situ Sensing of Damage Evolution in Advanced Fiber Composites Using Carbon Nanotube Networks, *Nanotechnology* **19**(21), 215713 (2008).

87. C. Cai, and J. Chen, Direct Electron Transfer of Glucose Oxidase Promoted by Carbon Nanotubes, *Anal. Biochem.* **332**(1), 75–83 (2004).

88. E. Muňoz, A.B. Dalton, S. Collins, M. Kozlov, J. Razal, J.N. Coleman, B.G. Kim, V.H. Ebron, M. Selvidge, J.P. Ferraris, and R.H. Baughman, Multifunctional Carbon Nanotube Composite Fibers, *Adv. Eng. Mater.* **6**(10), 801–804 (2004).

Chapter 15

Mechanical Applications of Carbon Nanotubes

Yani Zhang, Gengzhi Sun, and Lianxi Zheng
School of Mechanical and Aerospace Engineering, Nanyang Technological University, 639798 Singapore
Lxzheng@ntu.edu.sg

Carbon nanotubes (CNTs) have excellent mechanical and electrical properties. Coupling these two properties could introduce new mechanical devices or improve existing mechanical system/ applications at different scales. At microscopic scale, CNTs hold great promise for nanoelectromechanical systems (NEMS), which provide unique motion capabilities and are useful for a range of sensing and detection applications, as well as for studying mechanical motion in quantum limit. At macro scale level, great progress has been made in CNT actuation. Large strain and high stress actuators have been studied based upon CNT electrostatic actuation, CNT electrochemical actuation, and CNT-polymer actuation. Eventually in structural applications, individual CNTs must be effectively aggregated into macroscopic products, so that their excellent properties would be retained at a larger scale and more practicable level. Recent

Carbon Nanotubes and Their Applications
Edited by Qing Zhang

ISBN 978-981-4241-90-8 (Hardcover), 978-981-4303-18-7 (eBook)
www.panstanford.com

progresses on CNT fibers have demonstrated such a possibility. In this chapter, we will discuss the current status, future research focuses and challenges of several major mechanical applications of CNTs, including CNT nanomechanics, CNT actuators, and high performance CNT fibers.

15.1 Introduction

CNT are the stiffest (Young's modulus) and strongest (yield strength) materials yet measured. Their Young's modulus is about 1 TPa, and the tensile strength is about 11–63 GPa for individual multiwall CNTs (MWCNTs) and 13–52 GPa for individual single-wall CNTs (SWCNTs) [1,2,3]. They are also good conductors of both electricity and heat [4,5,6,7]. These extraordinary properties make them attractive for advanced mechanical applications.

At microscopic scale, CNTs hold great promise for nanoelectromechanical systems (NEMS). Various nanomechanics, such as high-frequency oscillators [8], rotational actuators [9], nanometer tweezers [10] and nanometer cargoes [11], have been experimentally demonstrated using individual CNTs. These NEMS devices integrate electrical and mechanical functionalities at nanoscale, and thus possess unique motion capabilities that are useful for a range of sensing and detection applications. The ultraminiature sensors could provide spatial resolution at the atomic scale and vibrate at frequencies in the gigahertz range, making them possible to measure the forces between individual biomolecules [12], forces arising from magnetic resonance of single spin [13], and perturbations that arise from mass fluctuations involving single atoms and molecules [14]. In addition, high Young's modulus and low specific weight of SWCNTs qualify them as ultimate mechanical resonators for studying mechanical motion in quantum regime. Strong coupling between single-electron tunneling and nanomechanical motion in CNTs has been observed [15,16].

At macro scale level, great progresses have been made in CNT actuations by coupling CNT's excellent mechanical properties and electrical properties. Large strain and high stress actuators have been studied based upon CNT electrostatic actuation [17], CNT electrochemical actuation [18], and CNT-polymer actuation [19].

Although practical actuators are still under developing, the success of direct conversion of electrical energy into mechanical energy from these actuations offers the opportunities for a number of high-technology applications, including humanoid robots, artificial and damaged hearts, artificial limbs, medical prosthetic devices, bird- or insect-like air vehicles, etc. [20].

Ultimately, it is better to prepare CNTs as structural materials, in order to fully explore their superior mechanical properties. CNT-reinforcement seems to be a good approach toward this direction. However, the dimensions of CNTs, a few nanometers in diameter and up to millimeter in length [21], have confronted us with a great challenge: how to retain the great potentials of individual CNTs in macroscopic scale composites? To date, enormous efforts have been paid on "conventional" reinforcement approach, i.e., dispersing random CNTs into polymers to fabricate reinforced composite materials. This approach has faced at least two major problems [22,23,24]: poor dispersion and poor load transfer. Because of van der Waals attractive forces, CNTs, especially for long CNTs that are essential for effective load transfer, tend to aggregate into bundles, making uniform dispersion into a polymer extremely difficult. The non-reactive nature of CNT's outer wall yields relatively weak bonding at CNT-polymer interface; this in turn gives poor load transfer from the polymer matrix to CNTs. As a result, all currently reported experimental data are significantly below theoretical predictions. Even in the future when these two limiting factors are eliminated to some extent, low CNT volume fraction in composites will still limit the performance of the final CNT-polymer matrixes. Luckily, recent progresses [25,26,27,28] in neat CNT fibers demonstrate the possibility to retain CNT's excellent properties at larger scale and more practicable level. The CNT fibers have been reported to have tensile strength of 1~3 GPa, Young's modulus of 100~260 GPa, toughness of 100~900 J/g, and density of 0.2 g/cm^3. These progresses motivate further study of lightweight and high strength composites for possible structural applications.

In this chapter, we will discuss several major mechanical applications mentioned above. Particularly, the topics will include CNT nanomechanics, CNT actuators, and high-performance CNT fibers.

15.2 CNT Nanomechanics

15.2.1 CNT Resonators as Mass/Force Sensors

The nanometer-scale size, excellent mechanical and electrical properties of CNTs offer opportunities to couple the mechanical and electrical domains at nanoscale device level, making CNTs an ideal material for preparing NEMS. Among many CNT NEMS, mechanical oscillators hold bright future. Their wide range applications, including ultrasensitive mass/force detection, have trigged extensive study of CNTs' vibration properties.

Acoustoelectric effect [29] and thermal vibrations [30] have been first studied in suspended doubly clamped SWCNTs. Since these nanometer-sized resonators oscillate at very high frequencies but with very small vibration amplitudes, it is desirable to study mechanical vibration driven and detected by electrical signals. Static and dynamic mechanical deflections were then electrically activated and studied in cantilevered MWCNTs in an *in situ* transmission electron microscope (TEM) [31]. When a static potential is applied, CNTs become electrically charged and will be bent toward the counter electrode. Once a time-dependent voltage is applied, the CNTs receive a time-dependent force and are then resonantly excited at the fundamental frequency and higher harmonics. The deflection and resonation could be observed in TEM by monitoring CNT's deflected contours. This electrically excited mechanical resonance of CNTs opens up applications in nanoresonant devices such as sensors, oscillator circuits, and nanoballances [32].

It was also found that CNT's resonant frequency could be tuned electrically. In a field emission microscope, Purcell *et al.* [33] have observed that the resonant frequency varies linearly with applied voltages. Their experimental setup is shown in Figure 15.1. Anode voltage V_s was used to excite resonance, while voltage V_A was used to generate electron emission. When resonance is excited, the pattern on screen will become larger and emission current will become smaller. Using this setup, the resonant frequency can be exactly measured by monitoring the emission current. And more interestingly, the voltage V_A also creates a longitudinal tension on the CNTs, and thus the resonance could be tuned by changing the voltage V_A, similar to the tuning of a guitar string. In this study, the frequency could be tuned by over 10 times.

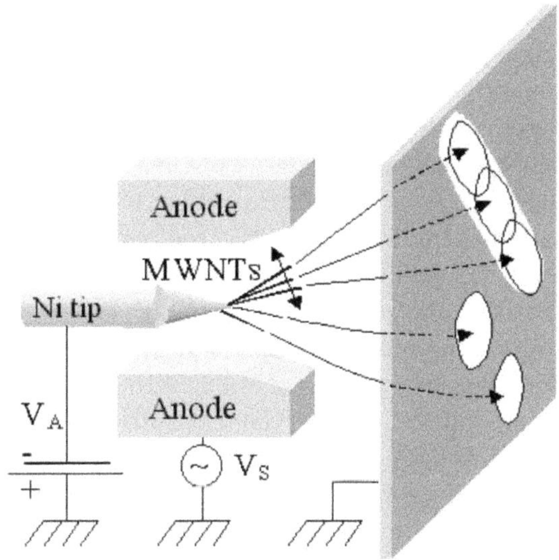

Figure 15.1 Scheme of the experimental field emission setup used to measure the resonances (distances: tip anode ~2 mm, tip screen ~3 cm, and screen size ~4 cm). (Reference 33).

Based upon the success in electrical excitation of CNT resonators, scientists further integrate a mechanical resonator with a field effect transistor (FET) to achieve electrical actuation and detection of the guitar-string-like oscillation modes of doubly clamped CNT oscillators [8]. In such a device structure, a CNT can act as a transistor and a resonator at the same time, and thus is able to sense its own motion. It is found that the frequency can be widely tuned and the devices can be used to detect very small forces. Figure 15.2 is a scanning electron microscopy (SEM) image of such a device. An SWCNT is suspended over a trench between two metal (Au/Cr) electrodes. The measurement is done in a vacuum chamber at pressures below 10^{-4} torr. CNT motion is actuated and detected using the electrostatic interaction with the gate electrode underneath the CNT. A gate voltage induces an additional charge on the CNT. The attraction between the charge and its opposite charge on the gate causes an electrostatic force downward on the CNT. The gate voltage has both a static (DC) component and a small time-varying (AC) component: the DC voltage at the gate produces a static force on the CNT that can be used to control its tension, while the AC voltage produces a periodic electric force, which sets the CNT into motion. As the driving frequency approaches the resonance frequency of the CNT, the displacement becomes large, and the current passing through the CNT shows distinctive feature — significant change in current — on top of a slowly changing background. The change of background current is due to the modulating gate voltage, while signal current is due to the resonant motion of the CNT. The response of the current to the frequency fits well to a Lorentzian function and could be used to identify resonant frequency. The DC voltage on the gate is found to be able to tune the tension in CNT and therefore the oscillation frequency. Resonant frequencies of 3–200 MHz at different samples and gate voltages have been observed. The smallest detected motion of such devices is about 0.5 nm, corresponding to a force sensitivity of 1 fN $Hz^{-1/2}$. Using a similar approach [34], CNT-based resonators with the fundamental mode frequency over 1.3 GHz have been obtained at room temperature in air and at atmospheric pressure. In addition to the distinctive current change near resonance, an abrupt transition in phase signal (~110° within 3 MHz) was also observed and attributed to the interference between the background and the resonance response. The sharp changes in the amplitude and phase could be used to detect frequency shift for sensing applications. Using these phenomena for mass detection, the author obtained a mass sensitivity of 10^{-18} g with $\delta\omega = 3$ MHz from phase change, and 10^{-18} g with $\delta\omega = 16$ MHz from amplitude change.

More recently, a room temperature CNT-based nanomechanical resonator with atomic mass resolution has been demonstrated [35]. Instead of using doubly clamped device geometry, the singly clamped device geometry was used in this study. Such a device structure gives a higher vibration range but lower clamping losses, and as a result the resonator has a very high quality factor (~1000), which has yielded a mass sensitivity of 1.3×10^{-25} kg $Hz^{-1/2}$, or 0.40 gold atoms $Hz^{-1/2}$. Using this extreme mass sensitivity, the atomic mass shot noise, analogous to the electronic shot noise, has been observed.

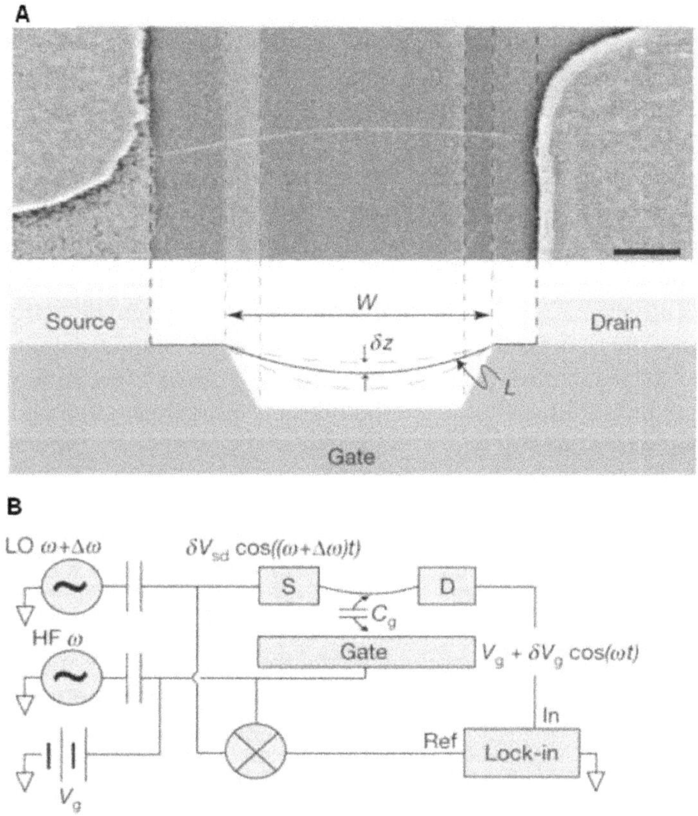

Figure 15.2 Device geometry and diagram of experimental set-up. (A) SEM image of a suspended device (top) and a schematic of device geometry (bottom). Scale bar, 300 nm. Metal electrodes (Au/Cr) are shown in yellow, and the silicon oxide surface in grey. (B) A diagram of the experimental set-up. A local oscillator (LO) voltage is applied to the source (S) electrode at a frequency offset from the high-frequency (HF) gate voltage signal by an intermediate frequency of 10 kHz. (Reference 8).

15.2.2 CNT Nanomechanics in Quantum Limit

Those high-frequency CNT resonators not only offer extreme mass and force sensitivity but also provide a unique way to observe the imprint of quantum phenomena directly [36], including uncertainty principle limits in position detection [37]. Particularly, as the continual progress in shrinking devices, these sensors could (or required to) offer a mass sensitivity close to zeptonewtons (10^{-21} N), a spatial resolution at the atomic scale, and vibration frequencies in the gigahertz range, which bring us to a new realm at which the intrinsic fluctuations (both quantum and thermal fluctuations) must be considered.

According to Heisenberg uncertainty principle [38], there exists a relationship between position x and momentum p with $\Delta x \bullet \Delta p \geq \hbar/2$ ($\hbar$ is Plank's constant divided by 2π) during continuous measurement, leading to a fundamental limit in position detection. Accordingly for a harmonic resonator used for position detection or in mass/force measurement, there is a so-called "standard quantum limit" for resonator's position. This limit is defined as the root mean square amplitude of quantum fluctuations of the resonator, and can be expressed as [39]

$$\Delta x_{\mathrm{SQL}} = \sqrt{\frac{\hbar}{2m\omega_0}}$$

At the same time, the energy of each vibration mode will be quantized as $E_N = \hbar\omega_0\,(N + 1/2)$, where $N = 0, 1, 2\ldots$is the occupation factor of the vibration mode with the frequency ω_0. The minimum energy ($\hbar\omega_0/2$) corresponds to the quantum ground state, or the vibration at an average position of $x = 0$ with a standard deviation of Δx_{SQL}.

The study of these quantum behaviors is of great practical and fundamental importance, because they set ultimate limits for sensitive force detection. The challenges for observing them in a nanoscale mechanical system are how to sense the small displacement, how to cool down the NEMS, and how to minimize back-action during motion detection. Since the absolute value of Δx_{SQL} is quite small (smaller than the size of an atom), how to read out the motion response (near quantum limit) into electrical signal is the first challenge confronting. Traditional techniques used in micro-scale mechanical systems, such as optical interferometry, cantilever reflection, piezoelectricity, and magnetomotive detection, cannot be applied to nanoscale mechanical systems due to the small size of nanosystems and energy dissipation. In addition, linear detector used previously in position detection will also impose back-action to the mechanical device, and then increase the total noise. For example, even if the minimum back-action in an optical measurement (at zero temperature) is considered, the back-action noise will be [37]:

$$\Delta x_{\text{QL}} = \sqrt{\frac{\hbar}{\ln 3 \bullet m\omega_0}} \approx 1.35 \bullet \Delta x_{\text{SQL}}$$

This is larger than standard quantum limit, adding more difficulties in observing the intrinsic quantum oscillation from mechanical device itself.

This problem was tackled by coupling nanomechanics to quantum systems. Mechanical resonators coupled with various quantum systems such as superconducting qubits, quantum dots, or single-electron transistors (SETs) have been studied. Among them, SET is the most promising device to enable quantum limit position detection. In such an integrated NEMS, as shown in Figure 15.3, the motion of the resonator induces a change of charge on the gate electrode of the SET, and the charge change can then be detected by directly monitoring the change of SET's conductance. In this way, the quantum system is capable to provide sufficient sensitivity and the lowest noise for quantum limit detection [36].

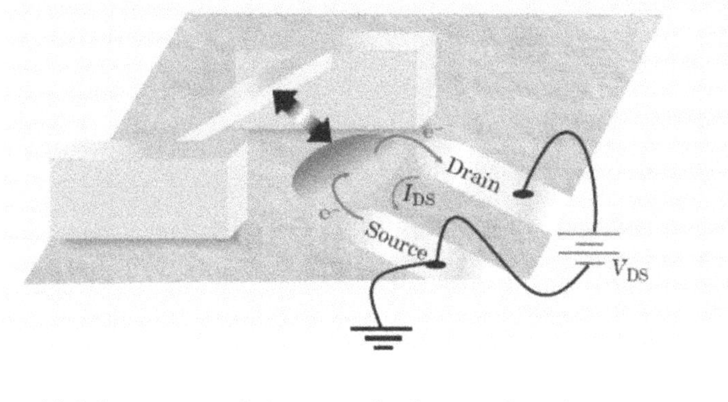

Figure 15.3 Structure of integrated device for detecting quantum fluctuation. (Reference 36).

Another prerequisite for observing quantum mechanical behaviors is a low temperature environment. To understand how close a mechanical device is approaching to the quantum limit, the average fluctuating energy of an individual mechanical mode coupled to a thermal bath is usually expressed as the function of so called thermal occupation factor N_{th} [36]:

$$\langle E \rangle = \hbar\omega N_{\text{th}} = \hbar\omega\left(\frac{1}{2} + \frac{1}{e^{\hbar\omega/k_B T} - 1}\right)$$

N_{th} follows the Bose-Einstein distribution. At high temperatures, each mode has the energy of $k_B T$, which belongs to the classical behavior. The quantum behavior can be observed only when $k_B T << \hbar\omega$, at which N_{th} is less than 1 and the mechanical mode becomes "frozen out". As shown in Figure 15.4, the "frozen out" temperature for a resonator's ground mode is a function of its vibration frequency and is in the mK level. This low temperature can be achieved by using standard dilution refrigerators. However, other active cooling schemes are also required, because at quantum regime the thermal conductance is also quantized, imposing a limit for thermal dispassion rate of the nanomechanical device [40]. Sideband cooling [41] and back-action cooling [42] were used to cool the mechanical device down below the temperature of its environment.

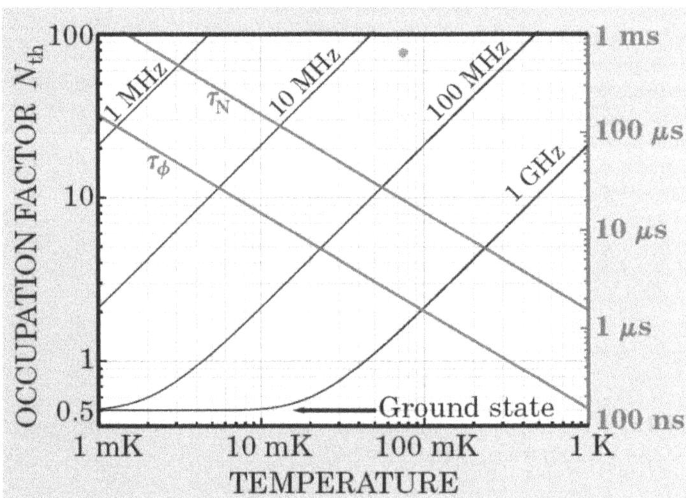

Figure 15.4 The occupation factor N_{th} (black curves) of various mechanical resonator frequencies changes with temperature T. (Reference 36).

Following these schemes, strong coupling between single-electron tunneling and nanomechanical motion has been observed in integrated CNT-based nanomechanical resonators [15,16]. In these studies, device consists of a CNT across a trench that makes electrical contact to two metal electrodes (similar device structure as shown in Figure 15.2). Electrons are confined in the CNT by Schottky barrier at the Pt metal contacts, forming a quantum dot in the suspended segment. When CNT resonator is actuated through a nearby antenna, the DC current through the CNT was measured to monitor CNT's mechanical motion. At low temperature of 20 mK, Coulomb peaks due to single-electron tunneling were observed in electron transport, shown in Figure 15.5A. Figure 15.5B shows the resonant frequency changing according to gate voltage. It is found that the resonant frequency changes linearly at Coulomb valleys but has big dips at Coulomb peaks. The sudden shift in frequency (~0.1 MHz), as indicated by the offset between dash lines and caused by one-electron tunneling, has been observed. This is consistent with previous predication that adding one electron will result in detectable frequency shift due to the change in static force [43]. The dips at Coulomb peaks indicate that single-electron tunneling process exerts the resonator a dynamic fluctuating force, in addition to static force and RF driving force.

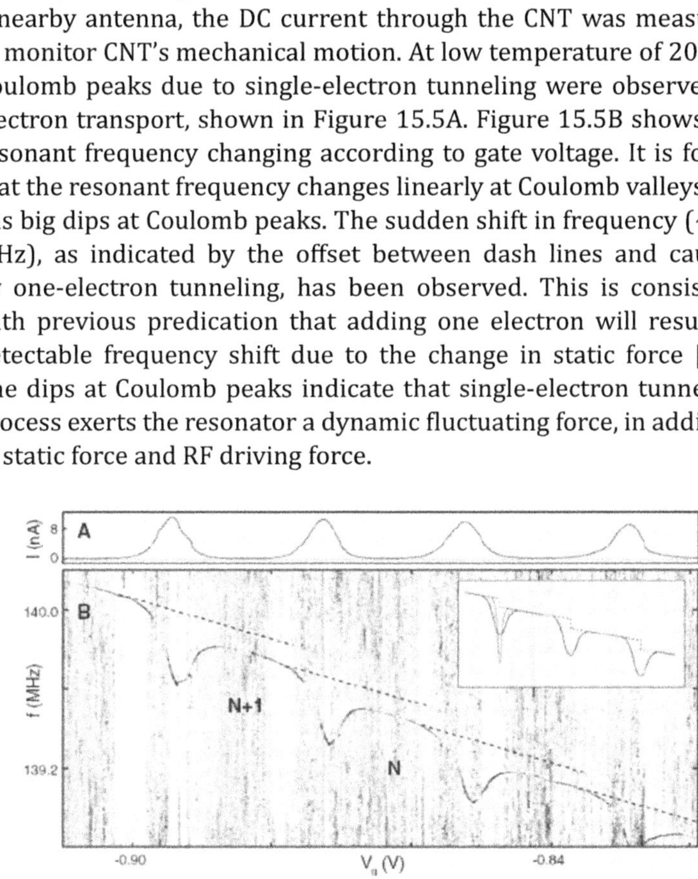

Figure 15.5 Single-electron tuning. (A) Nanotube current versus gate voltage showing single-electron tunneling at the peaks and Coulomb blockade in the valleys. (B) Normalized resonance signal $\Delta I/\Delta I_{\text{peak}}$ versus RF frequency and gate voltage (V_{sd} = 1.5 mV). The tuned mechanical resonance shows up as the darker curve with dips at the Coulomb peaks. N is the number of holes on the quantum dot. (Inset) The expected resonance behavior. (Reference 15).

According to Lassagne [16], the strong coupling in CNTs allows to cool the resonator toward ground state of motion to study its quantum behaviors. A thermal occupy factor of 3.8 was achieved in a SiN nanowire resonator by using active cooling strategy [41]. Considering CNT's high vibration frequency, small mass, and high stiffness, CNTs should be the best option for exploring quantum phenomena. Their dimensions not only make them susceptible to local forces, but also make it possible to integrate and tightly couple them to a variety of interesting electronic structures that exhibit quantum mechanical coherence. In addition, unlike fabricating resonators from other nanowire structures, preparing CNT resonators doesn't require any etching process. As a result, CNT resonators should have very low energy dissipation. Future study should focus on further improvement of quality factor of CNT resonators.

15.2.3 Other CNT Nanomechanics

Besides CNT resonators, other NEMS have also been demonstrated by taking advantages of CNT's small size and other unique properties. Here we briefly introduce nanometer tweezers, rotational actuators, and nanometer cargoes fabricated from individual CNTs or small CNT bundles.

Kim and Lieber [10] first proposed CNT nanotweezers based on static electrical attraction to manipulate nanostructures. As shown in Figure 15.6, freestanding electrical-independent electrodes were deposited onto tapered glass micropipettes, CNTs were then attached to each electrode. The fabricated CNT tweezers have two arms (CNT bundles) about 4 mm in length and 50 nm in diameter. When the voltages are applied to the electrodes, the tweezers open and close the free ends of the CNTs depending on the type of charges induced on CNTs. It is found that such nanotweezers are capable to grab and manipulate submicron clusters and nanowires. Other types of CNT nanotweezers have also been reported, including nanotweezers operated in an atomic force microscopy (AFM) [44], and nanotweezers closed by applied voltage but opened by attached macroactuators [45]. Because of their extreme small size, these nanotweezers could be used in the handling and manipulating small particles or living cells.

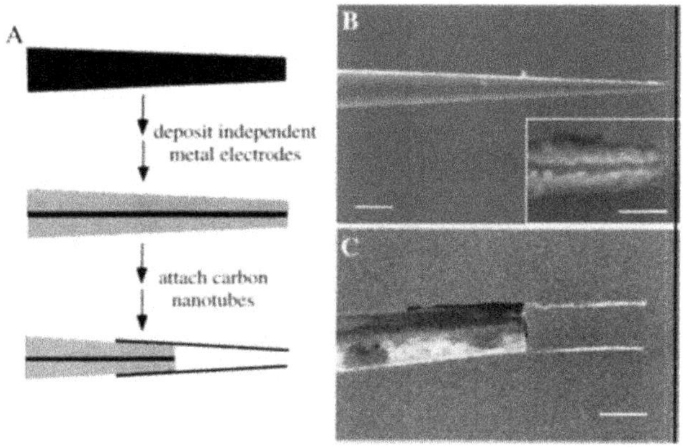

Figure 15.6 Overview of the fabrication of CNT nanotweezers. (A) Schematic illustrating the deposition of two independent metal electrodes and the subsequent attachment of CNTs. (B) SEM image of the end of a tapered glass structure (scale bar is 1 mm). The higher resolution inset shows clearly that the electrodes are separated (scale bar is 200 nm). (C) SEM image of nanotweezers after mounting two CNT bundles on each electrode (scale bar is 2 mm). (Reference 10).

Although nanotweezers can be operated at nanometer scale, the most interesting NEMS is still the electrically driven nanomotors along the development and miniaturization of mechanical systems. Following the success in CNT torsional springs [46,47,48] and CNT bearings [49], the CNT rotational actuators have been demonstrated by Fennimore and colleagues [9]. In their device, as shown in Figure 15.7, a metal plate is attached to a suspended MWCNT. The metal plate is the rotational element (R), and the CNT serves as the support shaft. The rotor plate assembly is surrounded by three stator electrodes (S1, S2, and S3), and the motion including the position, speed and direction of rotation of the rotor could then be controlled by applying voltage signals to the rotor and stators. More interestingly, they successfully shear the outer shells of the MWCNT using mechanical fatigue by applying very large stator voltages. As a result, the rotor can then freely rotate around the CNT without much friction due to the intrinsic low-friction bearing behavior of perfectly nested shells of MWCNTs [50,51,52]. By using a variety of suitably phased AC and DC voltage signals, this rotational actuator can perform at a finite frequency and no apparent wear or degradation in performance has been observed even after several thousands of rotational cycles.

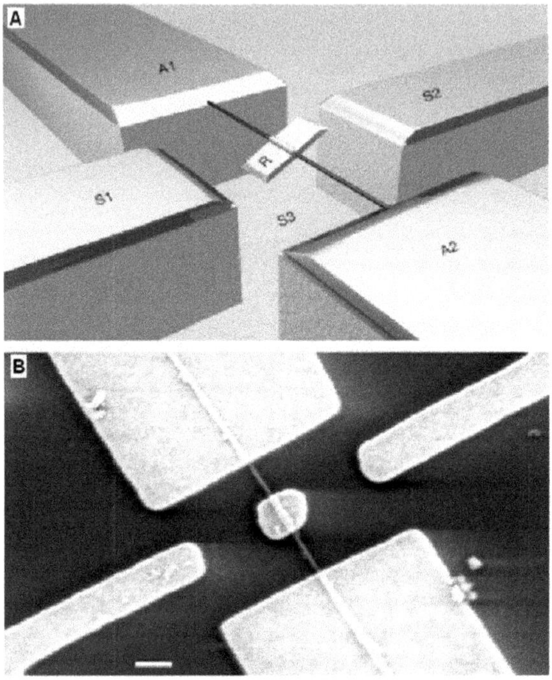

Figure 15.7 Integrated synthetic NEMS actuator. (A) Conceptual drawing of nanoactuator. A metal plate rotor (R) is attached to a MWCNT. Electrical contact to the rotor plate is made via the MWCNT and its anchor pads (A1, A2). Three stator electrodes, two on the SiO_2 surface (S1, S2) and one buried beneath the surface (S3), provide additional voltage control elements. (B) SEM image of nanoactuator (scale bar is 300 nm). (Reference 9).

Besides rotational motion, the translational motion was also realized using similar MWCNT approach. By coating an Au nanoparticle on a MWCNT and then selectively removing the outer shell of the MWCNT, Barreiro and colleagues [11] have demonstrated a CNT cargo that can perform subnanometer motion. The device structure is illustrated in Figure 15.8E. A so-called "electrical-breakdown" technique [53], i.e., passing a large current through the MWCNT, was used to selectively remove the outer shell of the CNT. Most of outer shell was removed due to the Joule heating, but the shell under the Au particle still remained due to the efficient heat absorption of the Au particle. As a result, it formed a structure in which a short shell is coaxially located over a long inner CNT, allowing free motion of Au particle along the CNT. As shown in Figure 15.8A–D, the cargo could be driven into translational motion when passing an electrical current through the CNT. However, it is found that the motion direction has no dependence on current's polarity. This finding points to thermal gradient induced motion [54]: In the presence of a thermal gradient, there will be a net current of phononic excitations traveling from the hot spot to the cold spot, which will collide with the mobile element and thus drive the cargo into translational motion. In the device shown in Figure 15.8E, the electrical current can induce heating effect to the device, but the generated heat mainly dissipates through the electrodes, so the hottest point could be expected at the middle of the CNT, and the nanoparticle would move along CNT toward the electrode sides.

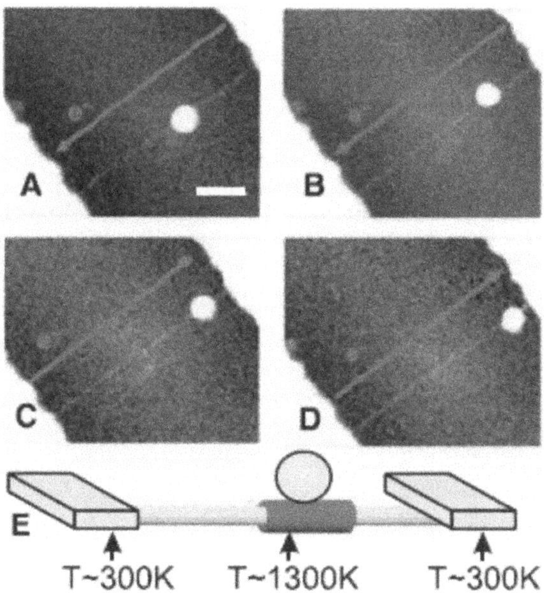

Figure 15.8 Translational motion. (A to D) Top-down SEM images where the gold cargo (sphere) is moving along the CNT. The direction of motion remains the same as the polarity of the current (red arrow) is reversed (scale bar is 400 nm). (E) Schematic illustration of the temperature along the CNT caused by Joule heating. (Reference 11).

CNT NEMS can not only realize a variety of motions for future practical applications, but also offer opportunity to study the fundamental motion phenomena at nanoscale, such as ratchet effect [55,56], vanishingly small friction [50], fraction resonances [57], and above-mentioned thermal gradient-induced motion. According to earlier theoretic studies, nanoparticles can move along a helical orbit around the CNT by taking advantages from atomic corrugation [58,59]. In the case of a double-walled CNT, it was found that the stability of a CNT depends on the diameter difference between inner and outer layers, but the potential barrier for the relative displacement of the inner and outer layers depends significantly on their chirality difference. For a variety of given chirality pairs, several unique energy surfaces could be obtained. The minimum energy could be along CNT axis (Figure 15.9B), or around the CNT (Figure 15.9D), or more generally in a helical pattern (Figure 15.9C). Accordingly, the motion could be either rotational or translational. In addition, in the energy surfaces shown in Figure 15.9B,D, one can find that the minimum energy shows periodic patterns, indicating the possibility of non-uniform motion. Experimentally, stepwise fashion of rotational movement was observed [11], strongly supporting the theoretical predication. These studies open a door for next generation CNT NEMS that could move in a new class of tracks defined by CNT's chiralities, and give the promise for NEMS in precise applications.

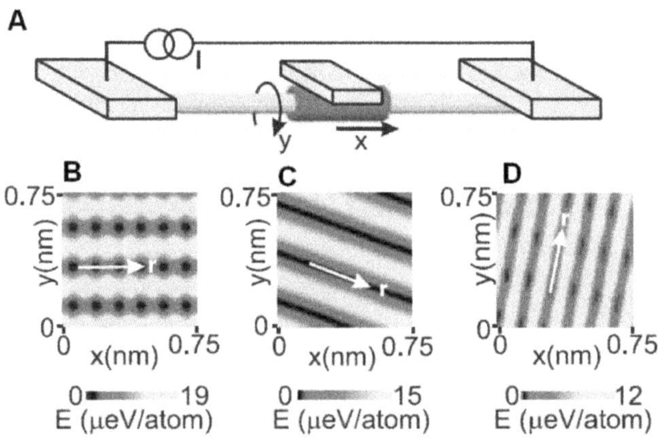

Figure 15.9 (A) Schematic of the CNT motor and its degrees of freedom. The outer (red) CNT moves with respect to the inner (yellow) CNT. (B to D) Shape of the energy barrier for the relative motion between two coaxial CNTs: (5,5)/(10,10), (29,9)/(38,8), and (27,12)/(32,17), respectively. The white arrows indicate the easy axis of motion (r). The motion is modulated by a series of small periodic barriers in (B) and (D), whereas vanishingly small friction is expected in (C). (Reference 11).

15.3 Macro-Scale CNT Actuators

Beyond individual CNT level, the direct conversion of electrical energy into mechanical energy was also realized through CNT films, CNT bundles/fibers, and CNT composites. Different actuation mechanisms including electrostatic actuation, electrochemical actuation, and shape-memory composite actuation, have been studied.

15.3.1 CNT Electrostatic Actuators

Electrostatic attraction and repulsion between individual CNTs were used for cantilever-based nanotweezers [10], and mechanical-based switches and logic elements [60]. This mechanism was also utilized to develop CNT electrostatic actuators from CNT aerogel sheets and CNT bundles.

Baughman's group [61] produced centimeter-wide, meter-long transparent CNT aerogel sheets, which are highly oriented, free-standing, and also have superior gravimetric strength compared with the sheet of high-strength steel. These CNT aerogel sheets can be easily stacked and can support liquid droplets that are 50,000 times more massive than the supporting sheet region in contact with the droplet. They used these CNT aerogel sheets to study the electrostatic actuation [17], and obtained giant stroke (220%) and high actuation speed (3.7×10^4%/s) [17]. As shown in Figure 15.10, when applying a voltage to the CNT sheet with respect to ground, the CNT sheet can show "ballooning" effect in the width direction due to electrostatic repulsive force. By tuning this voltage, the width of CNT sheet can change continuously from 0% (Figure 15.10A) to 220% (Figure 15.10B). Detail study showed the actuation in the width direction of CNT sheet is voltage dependent, with V^2 dependence at low voltages and $V^{2/3}$ dependence at high voltages. Large-stroke actuation by expansion in width (as well as in thickness) due to charge injection is also accompanied by the contraction in length. As a result, the actuator possesses large stress-generation capability and large work capability per cycle in length direction, with an isometric specific stress up to 4.0 MPa cm^3/g and the work per cycle about ~30 J/kg. More important, such electrostatic actuation can be realized at a wide temperature range, from room temperature to 1500 K (Figure 15.10C), and the stroke does not have significant change upon increasing temperature.

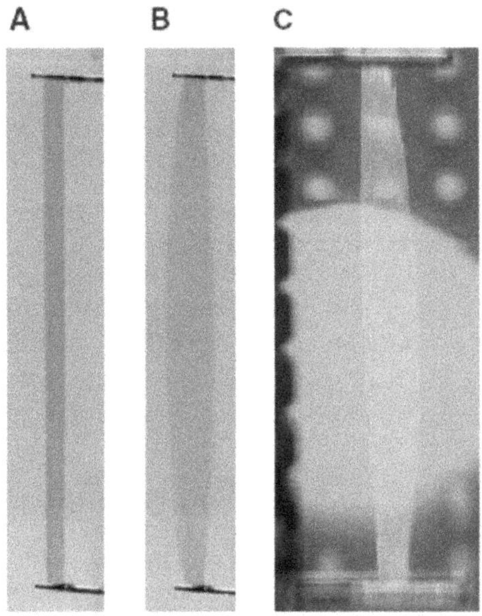

Figure 15.10 Photograph of a rigidly end-supported 50 mm-long and 2-mm-wide CNT sheet strip (A) and the same sheet strip expanded in width by applying 5 kV with respect to ground (B). (C) Photograph of a 25 mm-long CNT sheet strip actuated at 1500 K by applying 3 kV. (Reference 17).

These properties make CNT sheet actuator very promising for applications of electrically driven artificial muscle. Its stress-generation capacity is much superior to natural skeletal muscle, and its work per cycle is comparable with natural muscle [62]. In addition, it can work in a wide range of temperatures that is impossible to reach by natural muscle or actuators made from other materials such as ferroelectric and electrostrictive materials.

15.3.2 CNT Electrochemical Actuators

The CNT electrostatic actuators can generate large strain and stress, but they also require very high working voltage (more than 1000 V). One option to provide low-voltage actuators is to use electrochemical actuation, which is based on quantum-chemical effect and electrochemical double-layer charging effect. The charge injection in CNTs will induce dimensional changes in the length of the C–C covalent bonds. At low charge densities, earlier studies, both calculations [63,64] and experimental results [65,66], on charge transfer between graphite and conducting polymers showed that the strain in carbon material is an expansion for electron injection and a contraction for hole injection, indicating that quantum effect dominates. While in the case of a high-density charge injection of either sign, only expansion can be observed, suggesting that the static repulsive force dominates.

The electrochemical actuation property of CNTs was first studied by Baughman *et al.* [18]. When CNTs are immersed in electrolyte, an applied voltage on them will induce electronic charge injection into the CNTs, which will be compensated at the CNT/electrolyte interface by electrolyte ions, forming the so-called "double layer" (Figure 15.11A). When these CNTs are used as electrodes, for example using two "buckypaper" strips (macroscopic sheets made of randomly entangled bundles of SWCNTs) laminated together with an intermediate layer of double-sided Scotch tape as electrodes in an electrochemical cell (as shown in Figure 15.11B), electrochemical actuation can be realized.

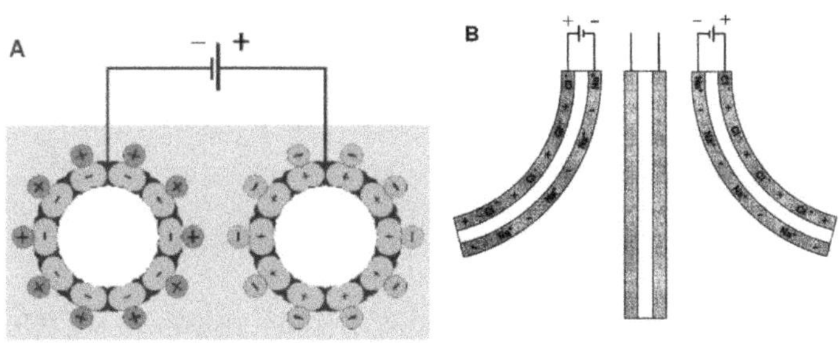

Figure 15.11 (A) Schematic illustration of charge injection and (B) electrochemical actuation in CNT electrochemical actuators. (Reference 18).

Based upon above-mentioned mechanism and earlier demonstrations, Baughman and colleagues proposed an actuator structure that could directly measure CNT's electrochemical response. In this structure, as shown in Figure 15.12A, one end of a CNT-paper strip was anchored to the bottom of the electrolyte bath, and the other end was attached to a cantilever, which was used to detect the length change of the CNT strip. Figure 15.12B shows the strain response of this actuator during cyclic voltammetry. The actuation is proportional to the applied potential, and the hysteresis increases with the potential as well. Further study shows that this device can be operated at a frequency of > 1 KHz, and only has ~33% decrease in actuator stroke after 140,000 cycles of scans (±0.5 V at 1 Hz). The maximum strain recorded on the actuators was 0.2%, and a stress of 0.75 MPa was generated during the isometric contraction, which is significantly higher than the peak capacity of human skeletal muscle [62]. The porous nature of the CNT films enables fast ion transport with response times of < 10 ms, with the maximum effective strain rate to date shown to be about 19%/s, and an effective power to mass ratio of 270 W/kg (half that of a high revving electric motor). Moreover, these characteristics were obtained at low voltages (≤ 2 V), and CNTs are chemically inert and stable over a broad range of temperatures, making this type of CNT actuators very promising for practical applications in various environments.

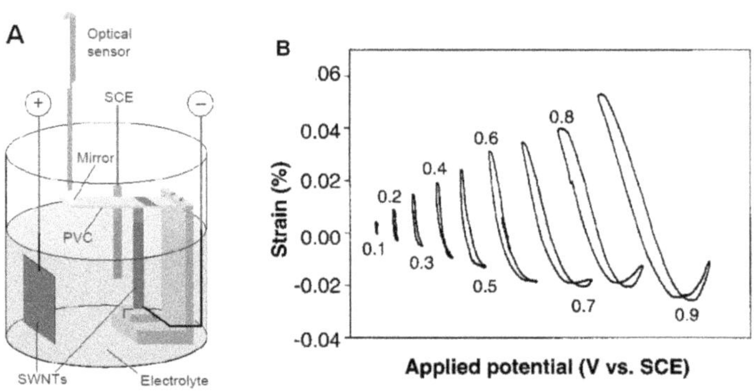

Figure 15.12 (A) The structure and (B) strain response of CNT-sheet electrochemical actuator. (Reference 18).

However, in applications that need to consume high-density energy such as in autonomous robots, an electrical-powered actuator must be wired to a stationary power source or be powered by a battery; either of these will limit the moving range or active time of the robots. Fuel-powered CNT actuators were then proposed and studied. In a recent study [67], catalyst-containing CNTs were used to simultaneously function as an actuator, a fuel-cell electrode, and a supercapacitor electrode. In such a fuel-cell actuator, the chemical energy from the fuel is converted into electrical energy by generating a potential on CNT electrode, and the electrical energy is then used to charge CNTs for actuation. When oxygen gas reaches the CNT electrode, chemical reaction will generate a potential of ~ 0.9 V versus a normal hydrogen electrode (NHE); while when hydrogen accesses CNT electrode, reduction reaction will generate a potential of ~ 0 V (versus a NHE). Through a well-designed device structure, the CNT electrode can be charged/discharged between 0 V and 0.9 V, and realize continuous actuation at the same time. Because a fuel cell and fuel source can deliver more than three times as much electrical energy as a battery of the same weight [20], CNT fuel-powered actuators are preferred for applications such as autonomous humanoid robots in hazardous environments.

Though the future is bright, there still exist several aspects that limit the efficiency of CNT-sheet actuators. First, a CNT sheet is composed of billions of individual actuator (each individual CNT) that are randomly oriented. This "dispersion" results in an actuation that is far from being optimal. Second, the natural agglomeration of tubes in ropes (under the effect of van der Waals forces) prevents the complete penetration of electrolyte ions inside the material. As a consequence, actuation phenomenon is limited to CNTs forming the periphery of the ropes. These limitations, arising from bundling and entanglement at the meso and macroscopic scale, prevent the direct extrapolation of intrinsic actuation properties of separated CNTs. Since the actuation strain in CNT sheets is a result of expansion in individual CNTs upon charging, these limitations could be partially overcome by using longer and more aligned CNTs, for example CNT fibers, to replace CNT sheets in actuators. Because CNT fibers usually have higher tensile strengths and moduli [68], high stress and high work densities could be achieved for CNT-fiber actuators.

15.3.3 CNT Composite Actuators

Shape-memory polymer is another type of important materials for actuation. They can exhibit large strain but usually associated with a low recovery stress, so it is interesting to utilize CNT reinforced shape-memory composites as potential actuating materials. Large actuation strain and large recovery stress could be expected if the flexibility of polymer and high strength and modulus of CNTs could be synergistically combined. Previous studies [19,69,70] indicate that the addition of CNTs in polymer will not only improve the actuation response of polymer materials, but also introduce new functions and new phenomena that either of material alone doesn't have. Here we introduce these CNT/polymer actuators according to the role of CNTs played in composite materials.

Firstly, addition of CNTs can enhance the mechanical response of polymer actuators. CNTs can be used as fillers in actuating polymers (shape-memory polymers). In those electroactive polymer-based actuator systems, CNTs serve primarily as a passive reinforcement to enhance the mechanical or/and electrical properties that contribute to the actuator performance. Most of the studies demonstrated that the reinforcement of CNTs had enabled higher actuation response. Levitsky *et al.* [71] investigated the electrochemical actuation behavior of layered Nafion/CNT composites by applying a thin coating of CNTs to the surfaces of the membrane as electrodes. The actuators were found to give a large mechanical response at low voltages (turn-on voltage of ~2.5 V) under open-air conditions, i.e., in the absence of a surrounding supporting electrolyte. Landi *et al.* [72] later found that the dispersed CNTs, which served as an embedded network of electrodes throughout the Nafion polymer matrix, could enhance the electro-osmotic effect, thus enhanced the actuation response. Lee *et al.* [73] studied CNT/Nafion actuators and concluded that the best performance in terms of maximum stress and stress rate occurred at uniform distribution of CNTs. Not only higher stress generated in the CNT/polymer composites, Tahhan *et al.* [74] found that the reinforcing influence of CNTs enabled higher strains to be maintained in electroactive polymer actuators and that the enhanced electrochemical efficiency was resulted from the increased conductivity. By incorporating CNTs in an electroactive polymer matrix, Zhang *et al.* [75] found that the actuation response of the composite was significantly enhanced over the pristine polymer.

As a good example for performance improvement, Koerner [19] demonstrated 50% increase in recovery stress by adding 1–5 vol.% CNTs in a thermoplastic elastomer (Morthane). As shown in Figure 15.13A, tremendous recovery of deformation, in excess of 300%, was realized, as indicated by the complete closing of a loose knot that was formed by room-temperature stretching (800%) of a composite containing 0.57 vol.% (1 wt%) CNTs. When exposure to near-infrared irradiation, the composite deformed to 300% and exerted ~19 J to lift 60 g weight more than 3 cm. The neat Morthane, on the other hand, remains unaffected (Figure 15.13C). Furthermore, the directionality of the recovery can be controlled (Figure 15.13B) — as the infrared absorption is very strong, the initial absorption, non-radiative energy decay and associated heating are initially confined to the near-surface region of the ribbon. The resulting non-homogeneous strain recovery within the ribbon causes it to "curl" toward the infrared source. Finally, Joule heating that arises from the finite resistivity of the composite enables current-induced actuation (Figure 15.13D). The enhancement of fixity and stress recovery behavior is also obtained for repeated cycling to 250% strain of a composite with 2.9 vol.% (5 wt%) CNTs. It is believed that the anisotropic CNTs increase the rubbery modulus and improve shape fixity by enhancing strain-induced crystallization. Non-irradiative decay of infrared photons absorbed by the CNTs raises the internal temperature, melts strain-induced polymer and remotely triggers the release of the stored strain energy.

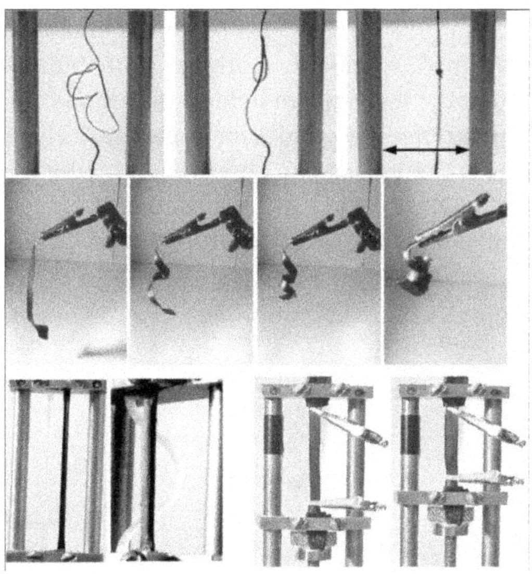

Figure 15.13 Optical images of the shape- and stress recovery of polymer-CNT composites. Left side: initial room-temperature deformed state; right side: recovered state. (A) Stretch Morthane ribbon containing CNTs shows that the knot closes on strain recovery. (B) Strain recovery toward and IR source. (C) Comparison of the stress recovery before (left) and after (right) actuation. Neat Morthane bend does not recover. (D) Electrically stimulated stress recovery. (Reference 19).

In addition to performance improvement, CNTs could add new functions to polymer actuators and bring some new phenomena associated with these new functions. Ahir, *et al.* [70] reported photo-induced mechanical actuation that was observed in a CNT/polydimethylsiloxane (PDMS) composite when exposed to infrared radiation. Significant photo-actuation response of the nanocomposites was obtained. The strength of photo-actuator response, at a given radiation intensity, is of the order of tens of kilopascals. Translated into the stroke, this corresponds to actuation strains of 2–4%. More interestingly, the composite tends to expand at small pre-strains, when stimulated by photons, by an amount that is orders of magnitude greater than the pristine PDMS, but it will contract at larger pre-strain even under identical infrared excitation. The photo-mechanical response changes sign at certain level of uniaxial pre-strain, as shown in Figure 15.14. This continuously reversible response of so large a magnitude makes it important for actuator applications. The authors claimed that this reversible response is due to CNT's orientation within a homogenous matrix, and they developed a model, using a probability distribution function, to study the mechanism behind. The model can reproduce the observed phenomenon, and seems conclusively indicate that CNT alignment is the reason for observed actuation phenomenon, but experiment evidence, such as IR response of individual CNTs, is still lacking, and will be the focus of future study.

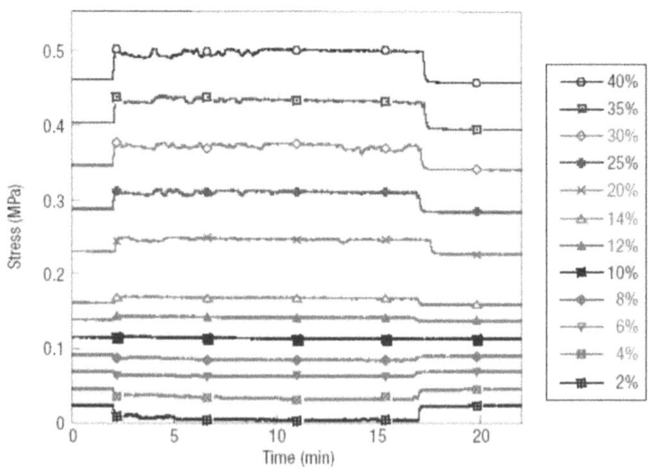

Figure 15.14 Response to IR radiation at different values of pre-strain. (Reference 70). See also Color Insert.

Poulin's group [69] reported the shape and temperature memory of nanocomposites that have an effect of broadened glass transition by using fibers that contain a large fraction of CNTs embedded in polyvinyl alcohol (PVA). Generally, in shape-memory polymer materials, large strain is usually associated with low stress recovery, and consequently the energy density is low. However, in this research, the CNT-PVA fibers can realize large strain of 800% plus large recovery stress (140 MPa). When the fibers are stretched at a deformation temperature of T_d and cooled down to room temperature under fixed strain, their length did not change when the load was released at room temperature, showing good "shape fixity" (Figure 15.15A). Particularly, these composite fibers can memorize the temperature at which they have been deformed. As shown in Figure 15.15B, when reheated at fixed strain the fibers can generate a strong stress with a peak value (1~2 orders of magnitude greater than the stress generated by conventional shape-memory polymer) at a well-defined temperature (T_s) that is roughly equal to T_d. This distinctive feature provides an opportunity to rationally control T_s without varying the chemical structure of the material. The mechanism behind is illustrated in Figure 15.15C. The presence of the CNTs substantially alters the thermo-mechanical properties of the composite fibers: the storage modulus is increased one order of magnitude and is much less temperature-dependent, and the transition point at T_g is no longer observed in composite. These observations suggest that a broadening of the glass transition induces observed temperature memory effect.

Although much efforts have been devoted and also significant improvement of actuation response has been obtained by adding dispersed CNTs or embedded CNTs in shape-memory polymers, there are still lots of potential issues/challenges should be further addressed on the way to greatly utilize the mechanical, electrical, and thermal characteristics of CNTs. The well-assembled CNT fibers can be an effective approach to assemble CNTs on a large scale along with a well-defined direction, and allow the optimization of properties of shape memory polymer fibers and electrochemical actuators. Future research will be valuable to achieve stiffer and stronger structures that will be capable of generating strong stresses combined with large strain deformations; thereby leading to novel and efficient technologies of actively moving materials.

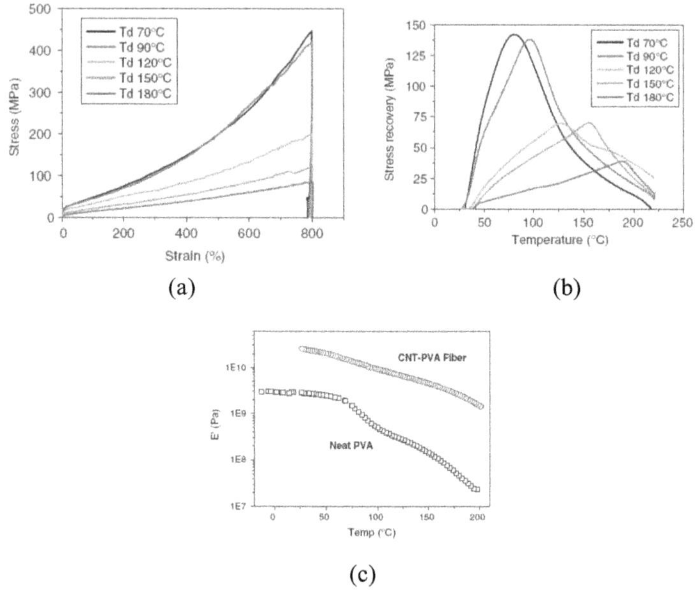

Figure 15.15 (A) Stress versus strain curves of CNT-composite fibers. (B) Stress generated by a nanocomposite fiber when it is reheated. (C) Storage modulus E' as a function of temperature for neat and dried PVA (squares) and for dried CNT-PVA fibers (circles). (Reference 69). See also Color Insert.

15.4 High-Performance CNT Fibers

As mentioned above, in future actuators, CNTs not only should be used as fillers, but also can be used as neat structures. It is then critical to order and assemble CNTs on the macroscopic scale to optimize material properties. The development of CNT fibers is an interesting approach toward this objective [76]. In addition to actuator application, high tensile strength and high modulus of CNTs make them very attractive for many structural applications, such as aerospace structures, firefighter clothing, bulletproof vests and sporting goods [77]. In these applications, CNTs must be assembled into macroscopic articles that retain enough of the properties of constituent CNTs. Fabricating continuous CNT yarns or fibers may be the first step toward high-performance composites for those applications.

Numerous methods have been developed to assemble such fibers [78]. Generally, these techniques may be divided into (1) solution-spinning methods [79,80,81,82,83], where CNTs need to be dispersed into a liquid first, and then spun into fibers, in a process similar to that used for polymeric fibers; (2) spinning from vertically aligned CNT arrays grown on a substrate [84,85]; (3) spinning from an aerogel of CNTs formed in chemical vapor deposition (CVD) reaction zone [86,87]; (4) other methods, like spinning from cotton-like CNT mats [88,89]. For simplicity, these techniques can also be grouped into wet spinning and dry spinning. The performances of CNT fibers are strongly dependent on processing methods.

15.4.1 Wet-Spun CNT Fibers

CNT fibers could be produced by using "solution-spinning" method, just like most synthetic fibers created from a concentrated, viscous liquid. These processes consist of dispersing the CNTs in solution and then recondensing the CNTs in a stream of another solution, which serves as coagulant. The first critical challenge in development of this method is the difficulty of processing CNTs in liquid state. CNTs are inert in pristine state, and tend to bundle together due to the strong van der Waals interactions, making them difficult to disperse uniformly in aqueous or any organic solvents. Some methods have been utilized to overcome this problem through oxidation and grafting with different functionalities [90,91,92,93], but these methods normally destroy CNT's intrinsic structures and properties. Thus, they are not favored for fiber spinning. Shaffer *et al.* [94] have previously suggested that CNTs could be viewed as analogous to high-aspect ratio, rigid-rod polymers. According to this analog, CNTs are supposed to be applicable to two types of spinning methods: coagulation spinning and liquid-crystal solution spinning.

15.4.1.1 Surfactant-based coagulation spinning

Generally, the principle of the "coagulation spinning" used for synthetic fiber processing could be depicted as: when a polymer solution is extruded through a thin capillary tube and injected into a bath that contains a second liquid in which the solvent is soluble but the polymer is not, the polymer will condense and form a fiber due to the phase separation. Employing this "coagulation spinning" method for CNT fiber fabrication, the CNTs need to be dispersed into a liquid solution at an almost molecule level so that they can be manipulated and aligned well. Surfactants are widely utilized for this purpose because of their ability to absorb/wrap at the surface of individual CNTs and prevent them from re-bundling. This spinning approach was initially adopted by Vigolo [79]. In their fabrication process, as shown schematically in Figure 15.16, arc-discharge-produced SWCNTs were firstly dispersed in an aqueous solution by using sodium dodecyl sulfate (SDS) as surfactant, then injected into a rotating bath of aqueous PVA solution, which serves as coagulant. During this process, PVA displaced the surfactant, causing CNTs collapse and forming ribbon-like elastomeric gel-fibers. These fragile fibers were pulled from the coagulation bath at a rate of about 1 cm min^{-1} in order to form solid fibers. Such fibers were washed by immersing in successive water container in order to remove excess PVA and surfactant residues, and then dried by pulling them out of water bath.

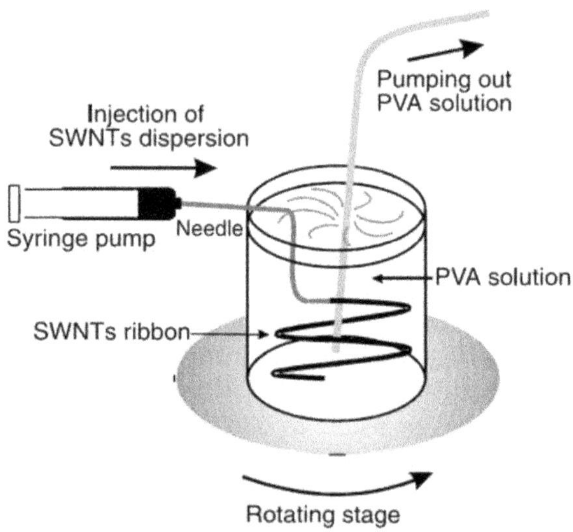

Figure 15.16 Schematic of the experimental setup used to make SWCNT ribbons. The capillary tip was orientated so that the SWCNT injection was tangential to the circular trajectory of the polymer solution. (Reference 79).

It is found that one critical parameter to obtain a good dispersion of CNTs is the amount of SDS. When the concentration of SDS is too low, large and dense clusters of the CNTs will be still found after sonication, which means that the amount of surfactant is too low to produce an efficient coating and induce enough electrostatic repulsion that could counterbalance van der Waals attractions. On the other hand, when the concentration of SDS is too high, the osmotic pressure of the excess micelles causes depletion-induced aggregation. They found that an optically homogeneous solution could be formed with 0.35 wt% CNTs and 1 wt% SDS for CNTs with particular diameter and length. Flow-induced alignment could lead to a preferential orientation of the CNTs in fibers, and has a close relationship to relative flow rate between injection solution and coagulant solution, as shown in Figure 15.17. The coagulant must flow faster than the gel-fiber in order to stretch the fiber along the axis direction and promote alignment of CNTs in the fiber. This could be accomplished by rotating the coagulant container [79].

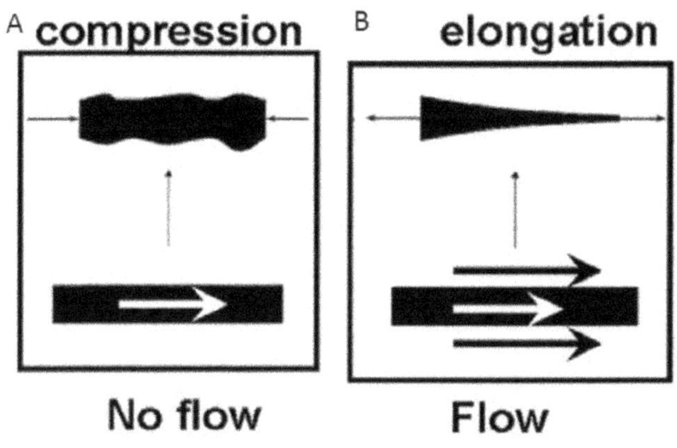

Figure 15.17 (A) When the coagulation bath is not flowed or slower than injecting rate, a net compressive force acts on the gel-like fiber, compromising alignment. (B) When the coagulant flows along with the extruded fiber and faster than the injecting rate, a net stretching force results and increases the alignment. (Reference 79).

This coagulation-based fiber spinning technique is exciting because of its simplicity and ability to produce fibers with very high CNT loadings (60 wt %). The final CNT/PVA composite fibers exhibited a tensile strength in the order of 0.1 GPa and a Young's modulus varying between 9 and 15 GPa. In contrast to most ordinary carbon fibers, CNT fibers (shown in Figure 15.18) can be heavily bent and even tightly tied without breaking. However, there are still some challenges existing in this method, including dispersion of SWCNTs at high concentrations, low processing rate, and the difficulty of handling.

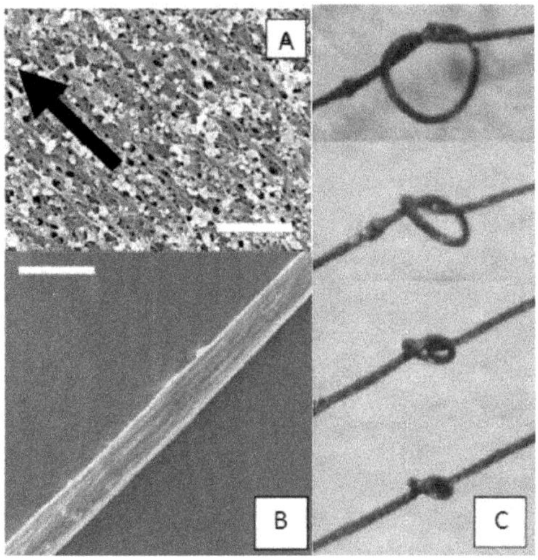

Figure 15.18 (A) A dry ribbon deposited on a glass substrate (The black arrow indicates the main axis of the ribbon, which corresponds to the direction of the initial fluid velocity). (B) A CNT fiber. (C) Knots reveal the high flexibility and resistance to torsion of the CNT microfibers. Scale bars: 500 nm and 25 μm for A and B. Reference 79).

In order to improve the mechanical performance of as-spun CNT fibers, various modified methods have been developed. By drying CNT fibers under load, improved mechanical properties were obtained with a tensile strength of 230 MPa and a Young's modulus to 45 GPa [95,96]. By hot drawing the fibers, Miaudet *et al.* [97] drew such PVA fibers at elevated temperatures, and the fibers yield a strength of 1.8 Gpa, a modulus of 45 Gpa, and a toughness of 55 Jg^{-1} at 11% strain. Dalton *et al.* [98] further advanced the spinning apparatus to spin fibers continuously by injecting CNT dispersion into a cylinder with the coagulant flowing in the same direction. They were able to spin a reel of CNT gel fibers and then converted it into 100 meters solid CNT-composite fibers, at a rate of more than 70 cm min^{-1}. The final fibers exhibit an increased mechanical performance with a tensile strength of 1.8 GPa and a Young's modulus of up to 80 GPa. Coagulation spinning has also been done with solutions other than

PVA. For example, Lynam *et al.* [99] produced CNT biofibers based on a wet-spinning process in which biomolecules acted as both the dispersant and coagulant. These fibers possessed strength of 0.17 GPa and modulus of 0.146 GPa.

Because the existence of second component polymer will add complicity of processing, and this second component is usually an insulator, which will compromise the conductive property of as-spun fibers, therefore, pure CNT fibers are favorable in some circumstances. Kozlov *et al.* [82] developed a polymer-free solution spinning method. Pure CNT fibers can be produced from CNT/surfactant/water solutions. However, the mechanical properties of the as-spun fibers are not impressive, showing a specific strength of 65 MPa g^{-1} cm^{-3}, specific modulus of 12 GPa $g^{-1}cm^{-3}$, and electrical conductivity of 140 S cm^{-1}.

15.4.1.2 Liquid crystal-based solution spinning

Spinning from lytropic liquid-crystalline solution of rigid-rod molecules is another important method used for fiber production. It can be seen in Figure 15.19 that CNTs could be viewed as high-aspect ratio, rigid-rod polymers and have been shown to exhibit liquid crystallinity by Song [100,101]. Ericson *et al.* [81] first successfully produced well-aligned macroscopic fibers composed solely of SWCNTs from lyotropic solutions in super acids. Fuming sulfuric acid charges SWCNTs and promotes them to order into an aligned phase with individual mobile CNTs surrounded by acid anions. This ordered dispersion was then extruded into a coagulant bath (diethyl ether, 5% sulfuric acid, or water) to form continuous macroscopic CNT fibers.

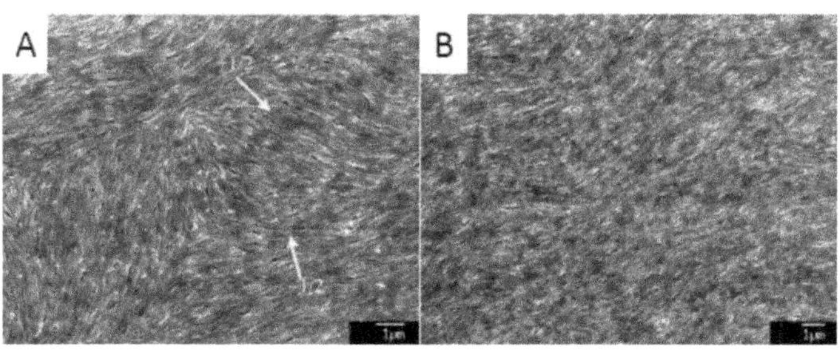

Figure 15.19 SEM images of a dried nematic MWCNT film. (A) The director fields around a pair of disclinations of topological strength +1/2 and −1/2 and (B) the region toward the edge of the film that is free of disclinations. (References 100,101).

The possible mechanism, that very high CNT concentrations in the spinning solutions are possible for this superacid spinning, is the repulsive interaction between CNTs generated in superacids (100+% sulfuric acid) due to formation of charge-transfer complexes: individual positively charged CNTs surrounded by a finite number of sulfuric acid anions. At very low concentration, such charged tube-anion complexes behave as Brownian rods. At higher concentration, as shown in Figure 15.20, a dense layer of sulfuric acid anions surrounds the individual CNTs, "swelling" CNT ropes and forming a charge-transfer complex. At very high concentration, the CNTs coalesce and form ordered domains, behaving similarly to nematic liquid crystalline.

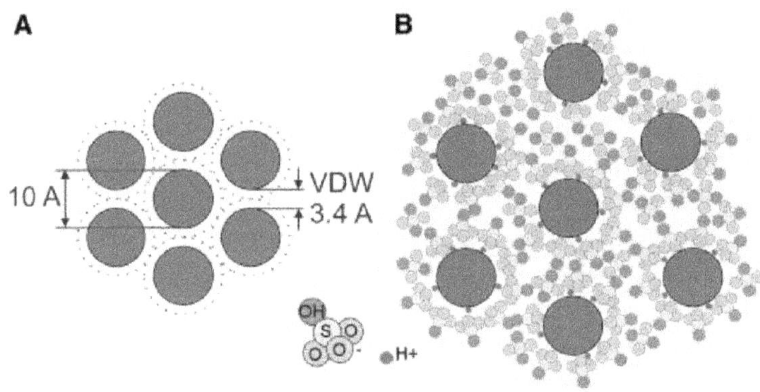

Figure 15.20 A model illustrating the swelling of SWCNT ropes in sulfuric acid. (A) A cartoon of SWCNTs in van der Waals contact within a neat fiber. (B) The same SWCNT fiber after re-exposure to sulfuric acid. (Reference 81). See also Color Insert.

Such spun fibers possess interesting structural and physical properties, including high orientation, good electrical and thermal conductivities, and reasonable mechanical properties. The alignment of CNTs within these fibers is within ±15.5°. The strength is 116 ± 10 MPa, and Young's modulus approaches 120 ± 10 GPa, similar to these of Kevlar fibers. However, some protonation of the material occurs because of prolonged contact with the sulfuric acid. The CNT/acid system is very sensitive to water; the introduction of even minimal moisture causes phase separation and precipitation of discrete needle-like crystal solvates. Superacid route is also found not effective for MWCNTs. To address the last problem, Zhang *et al.* [102] developed a new coagulation process, by which they spun MWCNTs from a liquid-crystalline ethylene glycol dispersion. The MWCNT fibers have a Young's modulus of 69 ± 41 GPa and a yield strain of 0.3%. Fracture occurred typically at strains below 3% and stresses of 0.15 ± 0.06 GPa.

15.4.2 Dry-Spun CNT Fibers

15.4.2.1 Spinning from vertical-aligned CNT arrays

In order to eliminate the dispersion problem existing in wet spinning methods, spinning CNT fibers directly from as-grown CNT materials seems to be a more convenient way. A breakthrough was made by Jiang [84] in 2002 by simply drawing a neat CNT yarn from a vertically super-aligned CNT array. They found that CNTs could be self-assembled into yarns of up to 30 cm in length. Following that, Zhang *et al.* [61] produced highly orientated, free-standing CNT transparent sheets using a similar method, and further assemble CNTs into fibers by using a draw-twisting spin method [85]. The typical SEM images of fiber spinning processes are shown in Figure 15.21.

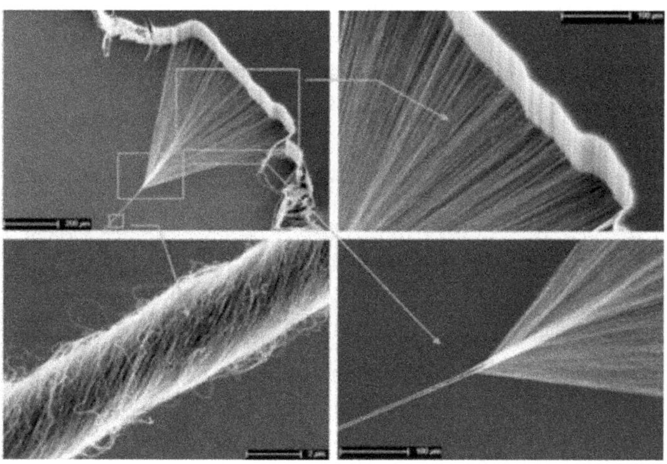

Figure 15.21 SEM images showing the structures formed during the draw-twist process. (Reference 85).

The mechanical performance of as-drawn CNT yarns and fibers are not satisfactory, although many applications of them have been proposed and demonstrated [103,104,105]. Different post-spinning methods were developed to improve their performances. Jiang [84] found that the strength and conductivity of their yarns could be improved after being heated at high temperatures. By introducing twist during spinning of CNTs from CNT forests to make multiply, torque-stabilized yarns, Zhang *et al.* [85] achieved yarn strength greater than 460 MPa. They emphasized that the load could be transferred effectively between CNTs because of the twist. In a twisted fiber, individual CNTs are inclined at an angle α with respect to the fiber axis, generating transverse forces that lock the fibers together as a coherent structure. They also found these twisted yarns deformed hysteretically over large strain ranges from 0.5% to 8%, providing up to 48% energy damping. These yarns could also retain their strength and flexibility even after being heated in air at 450°C for an hour or being immersed in liquid nitrogen. Instead of using post-spinning treatment, X. F Zhang *et al.* [25,26,27] found that mechanical properties could be significantly improved by using longer CNT arrays. The tensile strength and stiffness of their fibers spun from a 1-mm-long CNT array were measured in the range of 1.35 to 3.3 GPa and 100–263 GPa, respectively, which are many times stronger and stiffer per weight than the best existing engineering fibers and CNT fibers reported previously. It is obvious that the strength of CNT fibers increased with increasing CNT array length, which yields a much larger friction between CNTs. Moreover, longer CNTs will introduce fewer mechanical defects (like the ends of CNTs) per unit fiber length. Other factors like structure, purity, density, alignment and the straightness of CNTs [26,28] have all been investigated. For instance, in order to get dense packed CNT fibers, surface-tension-driven densification [106,107] was employed during fiber spinning. M. Zhang *et al.* found that after the CNT yarn was pulled through droplets of ethanol, the several centimeters wide yarn shrank into a tight fiber typically 20–30 μm in diameter and the strength of the CNT yarn was dramatically improved. L. X. Zheng *et al.* have observed a strong correlation between the array morphologies (the straightness of CNTs) and the fiber properties: well-aligned arrays yield high performance, while wavy arrays give poor performance. Figure 15.22 summarizes the influence of several parameters on fibers' mechanical performances.

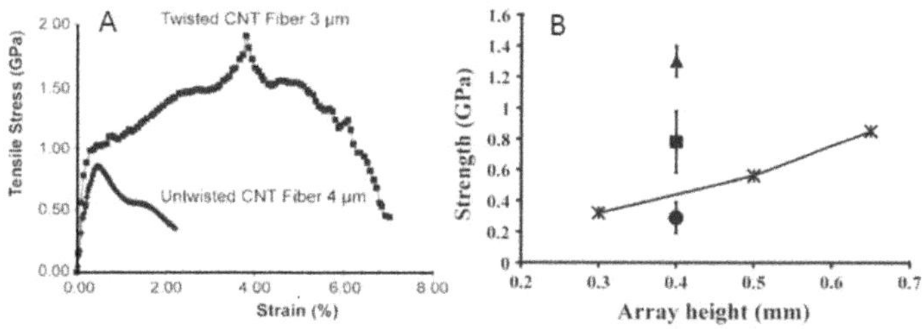

Figure 15.22 (A) As-spun and post-twisted small-diameter CNT fibers spun form a 650 μm array. (B) A comparison of fiber strengths at different array heights. The black line shows that the strength of CNT fibers increases with the array height, and marks shows the strength dependence on array morphology (the dot represents the data from oxygen-assisted growth, the square represents the data from the normal growth, and the triangle represents the data from the hydrogen-assisted growth). (References 26,28).

Since CNTs are nearly parallel aligned, the CNT yarn is intrinsically an anisotropic material and has a special axis along the drawing direction, which demonstrates many fascinating properties and applications. However, some key issues need to be solved in advance to realize their practical applications. Currently the growth of CNT arrays is easy, but not all CNT arrays could be spun into yarns or fibers. X. B. Zhang *et al.* [106] found that strong van der Waals interactions exist between individual CNTs within super-aligned arrays, and this van der Waals force makes the CNTs join end to end, thus forming a continuous yarn during pulling. Meanwhile, M. Zhang *et al.* [85] claimed that the formation of yarn was due to the disordered regions at the top and bottom of the CNT arrays, which entangled together forming a loop. Further investigation is needed to understand the underlying spinning mechanism.

15.4.2.2 Spinning from aerogel of CNTs

Zhu *et al.* [108] have first reported the formation of a 20 cm-long CNT thread after the pyrolysis of hexane, ferrocene, and thiophene. This work shows the possibility of fiber formation directly in a furnace. Based on this phenomenon, a totally different fiber spinning method was developed by Li *et al.* [86]. They were able to spin neat CNT fibers directly from an aerogel of CNTs formed in CVD reaction zone, as shown in Figure 15.23. The precursor materials include liquid hydrocarbon feedstock, ferrocene, which forms the iron nanoparticles that act as nucleation sites for the growth of CNTs, and thiophene, which is an established rate enhancer for vapor-grown carbon fibers [109]. The key requirements for continuous spinning are the formation of CNT aerogel and removal of the product from reaction zone. These were realized through the appropriate choice of reactants, control of the reaction conditions, and continuous withdrawal of the products with a rotating spindle used in various geometries.

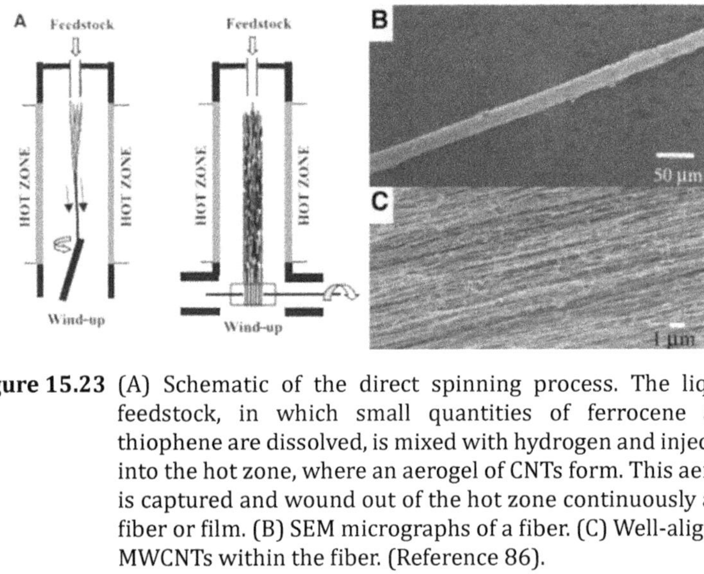

Figure 15.23 (A) Schematic of the direct spinning process. The liquid feedstock, in which small quantities of ferrocene and thiophene are dissolved, is mixed with hydrogen and injected into the hot zone, where an aerogel of CNTs form. This aergel is captured and wound out of the hot zone continuously as a fiber or film. (B) SEM micrographs of a fiber. (C) Well-aligned MWCNTs within the fiber. (Reference 86).

Recently, systematic studies of this method have been carried out [86,110,111]. From the view of reactants and growth conditions, the continuous spinning process is possible with a range of oxygen-containing carbon sources. Aromatic hydrocarbons lead to the deposition of carbon particles, thick fibers, or both, but cannot enable a continuous spinning process unless they are mixed with another oxygen containing source. Thiophene is found to be a necessary additive. It was used as an established rate enhancer for vapor grown carbon fibers [109], but its actual role played in CNT aerogel generation is still open to discussion. However, it is well accepted that sulfur, another additive, plays a major role in promoting carbon-hydrocarbon reactions, especially when associated with iron [109]. Through carefully controlling the growth conditions, the length and diameter of CNT in aerogel could be tuned, and it is found that lower concentrations of iron lead to a greater proportion of SWCNTs and DWCNTs, which are favorable for high-performance fibers. For example, it is found that the large-diameter double-wall CNTs may collapse within fibers, leading to an increase in friction between individual CNTs, which is beneficial for mechanical performance of CNT fibers [112]. Regarding the processing parameters, it is found that CNT alignment, the density and microstructures of fibers can be controlled by drawing/winding rate and post-processing methods. The degree of alignment could be manipulated by adjusting the winding rate as there is a tension introduced into this winding process, which supplies a force to align CNTs in the fiber. By introducing the wetting and evaporation of volatile organic liquids such as acetone, the condensation of the CNTs in fibers is greatly increased. Motta *et al.* [111] described improvement in mechanical strength and related the performance to unique aspects of fiber microstructure with "dog-bone" shape (shown in Figure 15.24). They have also shown that the mechanical properties of the fibers are directly related to the type of CNTs, which in turn, can be controlled by the careful adjustment of process parameters.

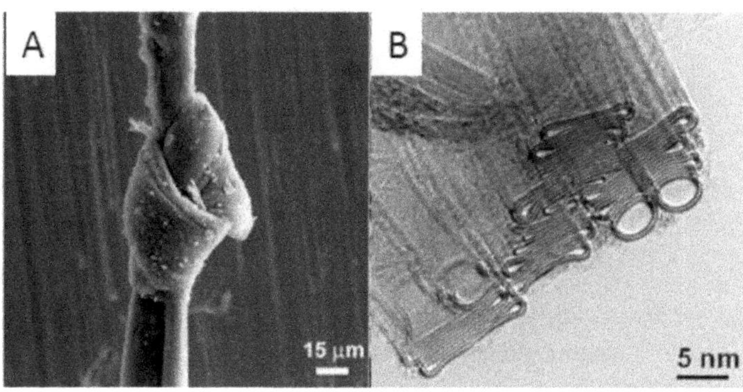

Figure 15.24 Structure of the fiber product. (A) SEM image of knotted fiber. (B) HR-TEM image of a bundle close to a fiber fracture revealing that the bundles consist, predominantly, of collapsed double-wall nanotubes greater than 5 nm. (Reference 111).

Through the optimization process, Koziol *et al.* [87] found that by drawing the aerogel at a winding rate of 20 m/min, the strength of the fiber, mainly containing DWCNTs, can reach around 10 GPa, which is the highest value reported so far. As can be seen in Figure 15.25, it shows the distribution of specific fiber strengths for a range of gauge lengths. From the figure, in the case of 20 mm gauge lengths, the strength of CNT fibers peaks at around 1 GPa. As the gauge length decreases, the strength distribution becomes bimodal with a second peak at 6.5 GPa, which is indicative of "weak point" along the fiber at intervals on the same order as the gauge length. These "weak points" exist inside the fibers without real interlock between individual CNTs, leading to a decrease in mechanical strength when the fiber is long.

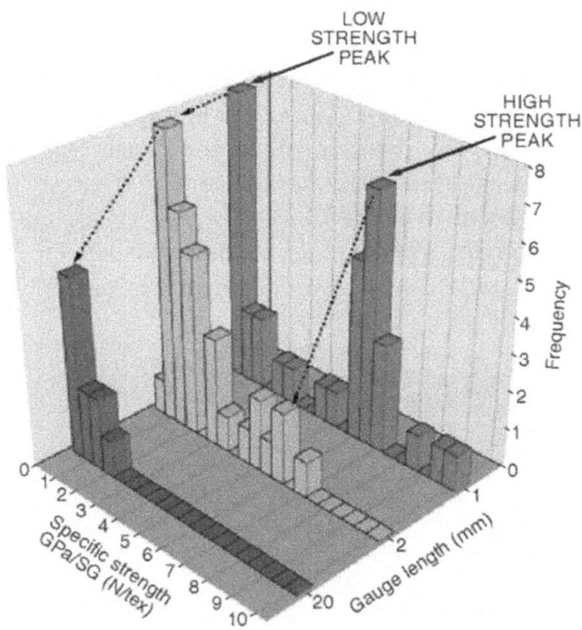

Figure 15.25 specific strength distribution of CNT fibers at different gauge lengths. (Reference 87).

15.5 Conclusions

In last few years, CNTs have been demonstrated great potentials in wide range mechanical applications, from nanoscale mechanical devices to structural composites. In this chapter, we have reviewed CNTs' applications in NEMS, actuators, and high-performance fibers. Various NEMS, including oscillators, rotational actuators, nanometer tweezers and nanometer cargoes, have been experimentally demonstrated. These NEMS devices can take advantages from atomic corrugation and then realize special moving functions. By coupling CNT motion with quantum systems, CNT resonators can be used to study mechanical motion in quantum regime, and provide exceptional capability on mass, force, and position detection. The success of CNT actuation, demonstrated by electrostatic actuation, electrochemical actuation, and CNT composite actuation, may offer an opportunity in artificial muscle applications. More excitingly, the progress on strong CNT fibers has shown a sign that superior properties of individual CNTs' could be retained at practical size level. These strong CNT fibers will eventually lead to high-performance composites for future structural applications.

References

1. Yu, M. F., Lourie, O., Dyer, M. J., Moloni, K., Kelly, T. F., and Rouff, R. S. (2000). Strength and breaking mechanism of multiwalled carbon nanotubes under tensile load, *Science,* **287**, pp. 637–640.
2. Yu, M. F., Files, B. S., Arepalli, S., and Ruoff, R. S. (2000). Tensile loading of ropes of single wall carbon nanotubes and their mechanical properties, *Phys. Rev. Lett.,* **84**, pp. 5552–5555.
3. Demczyk, B. G., Wang, Y. M., Cumings, J., Hetman, M., Han, W., Zettl, A., and Ritchie, R. O. (2002). Direct mechanical measurement of the tensile strength and elastic modulus of multiwalled carbon nanotubes, *Mater. Sci. Eng. A,* **334**, pp. 173–178.
4. Nikolaev, P., Dai, H. J., Petit, P., Robert, J., Xu, C. H., Lee, Y. H., Kim, S. G., Rinzler, A. G., Colbert, D. T., Scuseria, G., Tománek, D., Fischer, J. E., and Smalley, R. E. (1996). Crystalline ropes of metallic carbon nanotubes, *Science,* **273**, pp. 483–487.
5. Frank, S., Poncharal P., Wang, Z. L., and de Heer, W. A. (1998). Carbon nanotube quantum resistors, *Science,* **280**, pp. 1744–1746.
6. Hone, J., Whitney, M., and Zettle, A. (1999). Thermal conductivity of single-walled carbon nanotubes, *Synthetic Met.,* **103**, pp. 2498–2499.
7. Berber, S., Kwon, Y. K., and Tomànek, D. (2000). Unusually high thermal conductivity of carbon nanotubes, *Phys. Rev. Lett.,* **84**, pp. 4613–4616.

8. Sazonova, V., Yaish, Y., Ustunel, H., Roundy, D., Arias, T. A., and McEuen. P. L. (2004). A tunable carbon nanotube electromechanical oscillator, *Nature*, **431**, pp. 284–287.
9. Fennimore, A. M., Yuzvinsky, T. D., Han, W. Q., Fuhrer, M. S., Cumings, J., and Zettl, A. (2003). Rotational actuators based on carbon nanotubes, *Nature*, **424**, pp. 408–410.
10. Kim, P., and Lieber, C. M. (1999). Nanotube nanotweezers, *Science*, **286**, pp. 2148–2150.
11. Barreiro, A., Rurali, R., Hernandez, E. R., Moser, J., Pichler, T., Forro, L., and Bachtold, A. (2008). Subnanometer motion of cargoes driven by thermal gradients along carbon nanotubes, *Science* **320**, pp. 775–778.
12. Bustamante, C., Chemla, Y. R., Forde, N. R.,and Izhaky, D. (2004). *Annu. Rev. Biochem.*, **73**, pp. 705–748.
13. Rugar, D., Budakian, R., Mamin, H. J., and Chui, B. W. (2004). Single spin detection by magnetic resonance force microscopy, *Nature*, **430**, pp. 329–332.
14. Ekinci, K. L., Yang, Y. T., and Roukes, M. L. (2004). Ultimate limits to inertial mass sensing based upon nanoelectromechanical systems, *J. Appl. Phys.*, **95**, pp. 2682–2689.
15. Steele, G. A., Huttel, A. K., Witkamp, B., Poot, M., Meerwaldt, H. B., Kouwenhoven, L. P., and van der Zant, H. S. J. (2009). Strong coupling between single-electron tunneling and nanomechanical motion, *Science*, **325**, pp. 1103–1107.
16. Lassagne, B., Tarakanov, Y., Kinaret, J., Garcia-Sanchez, D., and Bachtold, A. (2009). Coupling mechanics to charge transport in carbon nanotube mechanical resonators, *Science*, **325**, pp. 1107–1110.
17. Aliev, A. E., Oh, J. Y., Kozlov, M. E., Kuznetsov, A. A., Fang, S. L., Fonseca, A. F., Ovalle, R., Lima, M. D., Haque, M. H., Gartstein, Y. N., Zhang, M. Zakhidov, A. A., and Baughman, R. H. (2009). Giant-stroke, superelastic carbon nanotube aerogel muscles, *Science*, **323**, pp. 1575–1578.
18. Baughman, R. H., Cui, C. X., Zakihidov, A. A., Iqbal, Z., Barisci, J. N., Spinks, G. M., Wallace, G. G., Mazzoldi, A., De Rossi, D., Rinzler, A. G., Jaschinski, O., Roth, S., and Kertzsz, M. (1999). Carbon nanotube actuators, *Science*, **284**, pp. 1340–1344.
19. Koerner, H., Price, G., Pearce, N. A., Alexander, M., and Vaia, R. A. (2004). Remotely actuated polymer nanocomposites – stress-recovery of carbon-nanotube-filled themoplastic elastomers, *Nat. Mater.*, **3**, pp. 115–120.

20. Baughman, R. H. (2005). Playing nature's game with artificial muscles, *Science*, **308**, pp. 63–65.

21. Zheng, L. X., O'Connell, M. J., Doorn, S. K., Hoffbauer, M. A., Roop, B. J., Jia, Q. X., Dye, R. C., Peterson, D. E., Huang, S. M., Liu, J., and Zhu, Y. T. (2004). Ultra-long single-wall carbon nanotubes, *Nat. Mater.*, **3**, pp. 673–676.

22. Szleifer, I., and Yerushalmi-Rozen, R. (2005). Polymers and carbon nanotubes — dimensionality, interactions and nanotechnology, *Polymer*, **46**, pp. 7803–7818.

23. Fiedler, B., Gojny, F. H., Wichmann, M. H. G., *et al.* (2006). Fundamental aspects of nano-reinforced composites, *Compos. Sci. Technol.*, **66**, pp. 3115–3125.

24. Wong, M., Paramsothy, M., Xu, X. J., *et al.* (2003). Physical interactions at carbon nanotube-polymer interface, *Polymer*, **44**, pp. 7757–7764.

25. Zhang, X. F., Li, Q. W., Holesinger, T. G., Arendt, P. N., Huang, J. Y., Kirven, P. D., Clapp, T. G., DePaula, R. F., Liao, X. Z., Zhao, Y. H., Zheng, L. X., Peterson, D. E., and Zhu, Y. T. (2007). Ultrastrong, stiff, and lightweight carbon-nanotube fibers, *Adv. Mater.*, **19**, pp. 4198–4201.

26. Zhang, X. F., Li, Q. W., Tu, Y., Li, Y., Coulter, Y., Zheng, L. X., Zhao, Y. H., Jia, Q. X., Peterson, D. E., and Zhu, Y. T. (2007). Strong carbon nanotube fiber spun from long CNT array, *Small*, **3**, pp. 244–248.

27. Li, Q. W., Zhang, X. F., DePaula, R. F., Zheng, L. X., Zhao, Y. H., Stan, L., Arendt, P. N., Peterson, D. E., and Zhu, Y. T. (2006). Sustained fast growth of long carbon nanotube arrays for fiber spinning, *Adv. Mater.*, **18**, pp. 3160–3163.

28. Zheng, L. X., Sun, G. Z., and Zhan, Z. Y. (2010). Tuning array morphology for high-strength nanotube fibers, *Small*, **6**, pp. 132–137.

29. Reulet, B., Kasumov, A. Y., Kociak, M., Deblock, R., Khodos, II., Gorbatov, Y. B., Volkov, V. T., Journet, C., and Bouchiat, H. (2000). Acoustoelectric effects in carbon nanotubes, *Phys. Rev. Lett.*, **85**, pp. 2829–2832.

30. Babic, B., Furer, J., Sahoo, S., Farhangfar, S. and Schonenberger, C. (2003). Intrinsic thermal vibrations of suspended doubly clamped single-wall carbon nanotubes, *Nano Lett.*, **3**, pp. 1577–1580.

31. Poncharal, P., Wang, Z. L., Ugarte, D., and deHeer, W. A. (1999). Electrostatic deflections and electromechanical resonances of carbon nanotubes, *Science*, **283**, pp. 1513–1516.

32. Gao, R. P., Wang, Z. L., Bai, Z. G., de Heer, W. A., Dai, L. M., and Gao, M. (2000). Nanomechanics of individual carbon nanotubes from pyrolytically grown arrays, *Phys. Rev. Lett.*, **85**, pp. 622–625.

33. Purcell, S. T., Vincent, P., Journet, C., and Binh, V. T. (2002). Tuning of nanotube mechanical resonances by electric field pulling, *Phys. Rev. Lett.*, **89**, pp. 276103(1–4).

34. Peng, H. B., Chang, C. W., Aloni, S., Yuzvinsky, T. D., and Zettl, A. (2006). Ultrahigh frequency nanotube resonators, *Phys. Rev. Lett.*, **97**, pp. 087203(1–4).

35. Jensen, K., Kim, K., and Zettl, A. (2008). An atomic-resolution nanomechanical mass sensor, *Nat. Nanotechnol.*, **3**, pp. 533–537.

36. Schwab, K. C., and Roukes, M. L. (2005). Putting mechanics into quantum mechanics, *Phys. Today*, **58**, pp. 36–42.

37. LaHaye, M. D., Buu, O., Camarota, B., and Schwab, K. C. (2004). Approaching the quantum limit of a nanomechanical resonator, *Science*, **304**, pp. 74–77.

38. Heisenberg, W. (1927). The physical principles of quantum theory, *Z. Phys.*, **43**, pp. 172–198.

39. Braginsky, V. B., and Khalili, F. Y. (1995). *Quantum Measurement* (Cambridge Univ. Press, Cambridge), pp. 12–15.

40. Schwab, K. C., Henriksen, E. A., Worlock, J. M., and Roukes, M. L. (2000). Measurement of the quantum of thermal conductance, *Nature*, **404**, pp. 974–977.

41. Rocheleau, T., Ndukum, T., Macklin, C., Hertzberg, J. B., Clerk, A. A., and Schwab, K. C. (2010). Preparation and detection of a mechanical resonator near the ground state of motion, *Nature*, **463**, pp. 72–75.

42. Naik, A., Buu, O., Lahaye, M. D., Armour, A. D., Clerk, A. A., Blencowe, M. P., and Schwab, K. C. (2006). Cooling a nanomechanical resonator with quantum back-action, *Nature*, **443**, pp. 193–196.

43. Sapmaz, S., Blanter, Y. M., Gurevich, L., and van der Zant, H. S. J. (2003). Carbon nanotubes as nanoelectromechanical systems, *Phys. Rev. B*, **67**, pp. 235414(1–7).

44. Akita, S., Nakayama, Y., Mizooka, S., Takano, Y., Okawa, T., Miyatake, Y., Yamanaka, S., Tsuji, M., and Nosaka, T. (2001). Nanotweezers consisting of carbon nanotubes operating in an atomic force microscope, *Appl. Phys. Lett.*, **79**, pp. 1691–1693.

45. Lee, J., and Kim, S. (2005). Manufacture of a nanotweezer using a length controlled CNT arm, *Sensor. Actuat. A Phys.*, **120**, pp. 193–198.

46. Jannik, C., Meyer, Paillet, M., and Roth, S. (2005). Single-molecular torsional pendulum, *Science*, **309**, pp. 1539–1541.

47. Williams, P. A., Papadakis, S. J., Patel, A. M., Falvo, M. R., Washburn, S., and Superfine, R. (2003). Fabrication of nanometer-scale mechanical

devices incorporating individual multiwalled carbon nanotubes as torsional springs, *Appl. Phys. Lett.,* **82**, pp. 805–807.

48. Papadakis, S. J., Hall, A. R., Williams, P. A., Vicci, L., Falvo, M. R., Superfine, R., and Washburn, S. (2004). Resonant oscillators with carbon-nanotube torsion springs, *Phys. Rev. Lett.,* **93**, pp. 146101(1–4).
49. Bourlon, B., Glattli, D. C., Miko, C., Forro, L., and Bachtold, A. (2004). Carbon nanotube based bearing for rotational motions, *Nano Lett.,* **4**, pp. 709–712.
50. Kolmogorov, A. N., and Crespi, V. H. (2000). Smoothest bearings: Interlayer sliding in multiwalled carbon nanotubes, *Phys. Rev. Lett.,* **85**, pp. 4727–4730.
51. Cumings, J., and Zettl, A. (2000). Low-friction nanoscale linear bearing realized from multi-walled carbon nanotubes, *Science,* **289**, pp. 602–604.
52. Yu, M. F., Yakobson, B. I., and Ruoff, R. S. (2000). Controlled sliding and pullout of nested shells in individual multiwalled carbon nanotubes, *J. Phys. Chem. B,* **104**, pp. 8764–8767.
53. Collins, P. G., Arnold, M. S., and Avouris, P. (2001). Engineering carbon nanotubes and nanotube circuits using electrical breakdown, *Science,* **292**, pp. 706–709.
54. Petta, J. R., Johnson, A. C., Taylor, J. M., Laird, E. A., Yacoby, A., Lukin, M. D., Marcus, C. M., Hanson, M. P., and Gossard, A. C.(2005). Coherent manipulation of coupled electron spins in semiconductor quantum dots, *Science,* **309**, pp. 2180–2184.
55. Tu, Z. C., and Hu, X. (2005). Molecular motor constructed from a double-walled carbon nanotube driven by axially varying voltage, *Phys. Rev. B,* **72**, pp. 033404(1–4).
56. Astumian, R. D., and Hanggi, P. (2002). Brownian motors, *Phys. Today,* **55**, pp. 33–39.
57. Tangney, P., Cohen, M. L., and Louie, S. G. (2006). Giant wave-drag enhancement of friction in sliding carbon nanotubes, *Phys. Rev. Lett.,* **97**, pp. 195901(1–4).
58. Saito, R., Matsuo, R., Kimura, T., Dresselhaus, G., and Dresselhaus, M. S. (2001). Anomalous potential barrier of double-wall carbon nanotube, *Chem. Phys. Lett.,* **348**, pp. 187–193.
59. Schoen, P. A. E., Walther, J. H., Arcidiacono, S., Poulikakos, D., and Koumoutsakos, P. (2006). Nanoparticle traffic on helical tracks: Thermophoretic mass transport through carbon nanotubes, *Nano Lett.,* **6**, pp. 1910–1917.

60. Kaul, A. B., Wong, E. W., Epp, L., *et al.* (2006). Electromechanical carbon nanotube switches for high-frequency applications, *Nano Lett.*, **6**, pp. 942–947.

61. Zhang, M., Fang, S. L., Zakhidov, A. A., Lee, S. B., Aliev, A. E., Williams C. D., Atkinson, K. R., and Baughman, R. H. (2005). Strong, transparent, multifunctional, carbon nanotube sheets, *Science*, **309**, pp. 1215–1219.

62. Madden, J. D. W., Vandesteeg, N. A., Anquetil, P. A., Madden, P. G. A., Takshi, A., Pytel, R. Z., Lafontaine, S. R., Wieringa, P. A., and Hunter, I. W. (2004). Artificial muscle technology: Physical principles and naval prospects, *IEEE J. Oceanic Eng.*, **29**, pp. 706–728.

63. Chan, C. T., Kamitakahara, W. A., Ho, K. M., and Eklund, P. C. (1987). Charge-transfer effects in graphite intercalates: Ab initio calculation and neutron-diffraction experiment, *Phys. Rev. Lett.*, **58**, pp. 1528–1531.

64. Baughman, R. H., Murthy, N. S., Eckhardt, H., and Kertesz, M. (1992). Charge oscillations and structure for alkali-metal-doped polyacetylene, *Phys. Rev. B*, **46**, pp. 10515–10539.

65. Baron, F., Flandrois, S., Hauw, C., and Gaultier, J. (1982). Charge-transfer and islands in metal-halides graphite-intercalation compounds - new evidence from x-ray-diffraction of intercalated $MnCl_2$, *Solid State Commun.*, **42**, pp. 759–762.

66. Kamitakahara, W. A., Zarestky, J. L., and Eklund, P. C. (1985). C-C bond distance and charge-transfer in D_2SO_4 - graphite compounds, *Synthetic Met.*, **12**, pp. 301–305.

67. Ebron, V. H., Yang, Z. W., Seyer, D. J., Kozlov, M. E., Oh, J. Y., Xie, H., Razal, J., Hall, L. J., Ferraris, J. P., MacDiarmid, A. G., and Baughman, R. H. (2006). Fuel-powered artificial muscles, *Science*, **311**, pp. 1580–1583.

68. Lucas, M., Vigolo, B., Badaire, S., Le Bolloc'h, D., Marucci, A., Durand, D., Hamilton, M., Zakri, C., Poulin, P., and Launois, P. (2002). Alignment of carbon nanotubes in macroscopic fibers, In *Structural and Electronic Properties of Molecular Nanostructures*, edited by H. Kuzmany, J. Fink, M. Mehring, and S. Roth, Vol. 633, pp. 579–582.

69. Miaudet, P., Derre, A., Maugey, M., Zakri, C., Piccione, P. M., Inoubli, R., and Poulin, P. (2007). Shape and temperature memory of nanocomposites with broadened glass transition, *Science*, **318**, pp. 1294–1296.

70. Ahir, S. V., and Terentjev, E. M. (2005). Photomechanical actuation in polymer-nanotube composites, *Nat. Mater.*, **4** , pp. 491–495.

71. Levitsky, L. A., Kanelos, P., and Euler, W. B. (2004). Electromechanical actuation of composite material from carbon nanotubes and ionomeric polymer, *J. Chem. Phys.*, **121**, pp. 1058–1065.

72. Landi, B. J., Raffaelle, R. P., Heben, M. J., Alleman, J. L., VanDerveer, W., and Gennett, T. (2005). Development and characterization of single wall carbon nanotube-Nafion composite actuators, *Mat. Sci. Eng. B-Solid State Mater. Adv. Technol.*, **116**, pp. 359–362.

73. Lee, D. Y., Park, I. S., Lee, M. H., Kim, K. J., and Heo, S. (2007). Ionic polymer-metal composite bending actuator loaded with multi-walled carbon nanotubes, *Sensor Actuat. A Phys.*, **133**, pp. 117–127.

74. Tahhan, M., Truong, V. T., Spinks, G. M., and Wallace, G. G. (2003). Carbon nanotube and polyaniline composite actuators, *Smart Mater. Struct.*, **12**, pp. 626–632.

75. Zhang, S. H., Zhang, N. Y., Huang, C., Ren, K. L., and Zhang, Q. M. (2005). Microstructure and electromechanical properties of carbon nanotube/ poly(vinylidene fluoride-trifluoroethylene-chlorofluoroethylene) composites, *Adv. Mater.*, **17**, pp. 1897–1901.

76. Viry, L., Derre, A., Garrigue, P., Sojic, N., Poulin, P., and Kuhn, A. (2007). Optimized carbon nanotube fiber microelectrodes as potential analytical tools, *Anal. Bioanal. Chem.*, **389**, pp. 499–505.

77. Baughman, R. H., Zakhidov, A. A., and de Heer, W. A. (2002). Carbon nanotubes — the route toward applications, *Science*, **297**, pp. 787–792.

78. Behabtu, N., Green, M. J., and Pasquali, M. (2008). Carbon nanotube-based neat fibers, *Nano Today*, **3**, pp. 24–34.

79. Vigolo, B., Penicaud, A., Coulon, C., Pailler, R., Journet, C., *et al.* (2000). Macroscopic fibers and ribbons of oriented carbon nanotubes, *Science*, **290**, pp. 1331–1334.

80. Dalton, A. B., Collins, S., Razal, J., Munoz, E., Ebron, V. H., *et al.* (2004). Continuous carbon nanotube composite fibers: Properties, potential applications, and problems, *J. Mater. Chem.*, **14**, pp. 1–3.

81. Ericson, L. M., Fan, H., Peng, H. Q., Davis, V. A., Zhou, W., *et al.* (2004). Macroscopic, neat, single-walled carbon nanotube fibers, *Science*, **305**, pp. 1447–1450.

82. Kozlov, M. E., Capps, R. C., Sampson, W. M., Ebron, V. H., Ferraris, J. P., *et al.* (2005). Spinning solid and hollow polymer-free carbon nanotube fibers, *Adv. Mater.*, **17**, pp. 614–617.

83. Neri, W., *et al.* (2006). Surfactant-free spinning of composite carbon nanotube fibers. *Macromolecular Rapid Commun.*, **27**, pp. 1035–1038.

84. Jiang, K. L., Li, Q. Q., and Fan, S. S. (2002). Nanotechnology: spinning continuous carbon nanotube yarns - carbon nanotubes weave their way into a range of imaginative macroscopic applications, *Nature*, **419**, pp. 801–801.

85. Zhang, M., Atkinson, K. R., and Baughman, R. H. (2004). Multifunctional carbon nanotube yarns by downsizing an ancient technology, *Science*, **306**, pp. 1358–1361.

86. Li, Y. L., Kinloch, I. A., and Windle, A. H. (2004). Direct spinning of carbon nanotube fibers from chemical vapor deposition synthesis, *Science*, **304**, pp. 276–278.

87. Koziol, K., Vilatela, J., Moisala, A., Motta, M., Cunniff, P., *et al.* (2007). High-performance carbon nanotube fiber, *Science*, **318**, pp. 1892–1895.

88. Zheng, L. X., Zhang, X. F., Li, Q. W., Chikkannanavar, S. B., Li, Y., *et al.* (2007). Carbon-nanotube cotton for large-scale fibers, *Adv. Mater.*, **19**, pp. 2567–2570.

89. Ci, L., Punbusayakul, N., Wei, J., Vajtai, R., Talapatra, S., and Ajayan, P. M. (2007). Multifunctional macroarchitectures of double-walled carbon nanotube fibers, *Adv. Mater.*, **19**, pp. 1719–1723.

90. Sun, G. Z., Liu, S. W., Hua, K. F., Lv, X. Y., Huang, L., and Wang, Y. J. (2007). Electrochemical chlorine sensor with multi-walled carbon nanotubes as electrocatalysts, *Electrochem. Commun.*, **9**, pp. 2436–2440.

91. Xu, J. B., Hua, K. F., Sun, G. Z., Lv, X. Y., and Wang, Y. J. (2006). Electrooxidation of methanol on carbon nanotubes supported Pt-Fe alloy electrode, *Electrochem. Commun.*, **8**, pp. 982–986.

92. Watts, P. C. P., Hsu, W. K., Kroto, H. W., and Walton, D. R. M. (2003). Are bulk defective carbon nanotubes less electrically conducting? *Nano Lett.*, **3**, pp. 549–553.

93. Star, A., Liu, Y., Grant, K., Ridvan, L., Stoddart, J. F., Steuerman, D. W., *et al.* (2003). Noncovalent side-wall functionalization of single-walled carbon nanotubes, *Macromolecules*, **36**, pp. 553–560.

94. Shaffer, M. S. P., and Windle, A. H. (1999). Analogies between polymer solutions and carbon nanotube dispersions, *Macromolecules*, **32**, pp. 6864–6866.

95. Vigolo, B., Poulin, P., Lucas, M., Launois, P., and Bernier, P. (2002). Improved structure and properties of single-wall carbon nanotube spun fibers, *Appl. Phys. Lett.*, **81**, pp. 1210–1212.

96. Munoz, E., Dalton, A. B., Collins, S., Kozlov, M., Razal, J., *et al.* (2004). Multifunctional carbon nanotube composite fibers, *Adv. Eng. Mater.*, **6**, pp. 801–804.

97. Miaudet, P., Badaire, S., Maugey, M., Derre, A., Pichot, V., *et al.* (2005). Hot-drawing of single and multiwall carbon nanotube fibers for high toughness and alignment, *Nano Lett.*, **5**, pp. 2212–2215.

98. Dalton, A. B., Collins, S., Munoz, E. Razal, J. M., Howard, V., *et al.*(2003). Super-tough carbon-nanotube fibres — These extraordinary composite fibres can be woven into electronic textiles, *Nature*, **423**, pp. 703–703.

99. Lynam, C., Moulton, S. E., and Wallace, G. G. (2007). Carbon-nanotube biofibers. *Adv. Mater.*, **19**, pp. 1244–1248.

100. Song, W. H., and Windle, A. H. (2005). Isotropic-nematic phase transition of dispersions of multiwall carbon nanotubes, *Macromolecules*, **38**, pp. 6181–6188.

101. Song, W. H., Kinloch, I. A., and Windle, A. H. (2003). Nematic liquid crystallinity of multiwall carbon nanotubes, *Science*, **302**, pp. 1363–1363.

102. Zhang, S. J., Koziol, K. K. K., Kinloch, I. A., and Windle, A. H. (2008). Macroscopic fibers of well-aligned carbon nanotubes by wet spinning, *Small*, **4**, pp. 1217–1222.

103. Xiao, L., Chen, Z., Feng, C., Liu, L., Bai, Z. Q., *et al.* (2008). Flexible, stretchable, transparent carbon nanotube thin film loudspeakers, *Nano Lett.*, **8**, pp. 4539–4545.

104. Feng, C., Liu, K., Wu, J. S., Liu, L., Cheng, J. S., *et al.* (2010). Flexible, stretchable, transparent conducting films made from superaligned carbon nanotubes, *Adv. Funct. Mater.*, **20**, pp. 885–891.

105. Sun, Y. H. Liu, K., Miao, J., Wang, Z. Y., Tian, B. Z., *et al.* (2010). Highly Sensitive surface-enhanced raman scattering substrate made from superaligned carbon nanotubes, *Nano Lett.*, **10**, pp. 1747–1753.

106. Zhang, X. B., Jiang, K. L., Feng, C., Liu, P., Zhang, L. N., *et al.* (2006). Spinning and processing continuous yarns from 4-inch wafer scale super-aligned carbon nanotube arrays, *Adv. Mater.*, **18**, pp. 1505–1510.

107. Wei, Y., Jiang, K. L., Feng, X. F., Liu, P., Liu, L., *et al.* (2007). Comparative studies of multiwalled carbon nanotube sheets before and after shrinking, *Phys. Rev. B*, **76**, pp. 045423 (7).

108. Zhu, H. W., Xu, C. L., Wu, D. H., Wei, B. Q., Vajtai, R., *et al.* (2002). Direct synthesis of long single-walled carbon nanotube strands, *Science*, **296**, pp. 884–886.

109. Tibbetts, G. G., Bernardo, C. A., Gorkiewicz, D. W., and Alig, R. L. (1994). Role of sulfur in the production of carbon-fibers in the vapor-phase, *Carbon*, **32**, pp. 569–576.

110. Motta, M., Li, Y. L., Kinloch, I., and Windle, A. (2005). Mechanical properties of continuously spun fibers of carbon nanotubes, *Nano Lett.*, **5**, pp. 1529–1533.

111. Motta, M., Moisala, A., Kinloch, I. A., and Windle, A. H. (2007). High performance fibres from 'Dog bone' carbon nanotubes, *Adv. Mater.*, **19**, pp. 3721–3726.

112. Zhang, X. H., and Li, Q. W. (2010). Enhancement of friction between carbon nanotubes: An efficient strategy to strengthen fibers, *ACS Nano*, **4**, pp. 312–316.

Chapter 16

Nonlinear Optical Properties of Carbon Nanotubes and Their Applications

Kin Kee Chow
School of Electrical and Electronic Engineering
Nanyang Technological University, 50 Nanyang Avenue, 639798 Singapore
kkchow@ntu.edu.sg

Single-wall carbon nanotube (SWCNT), due to its estimated ultra-high third-order nonlinearity, can be regarded as a potential new material for optical nonlinearity medium. The nonlinearity of carbon nanotubes (CNTs) is believed to be originated from the inter-band transitions of the π-electrons causing nonlinear polarization. In this respect CNTs are similar to other organic optical materials, which exhibit extremely high nonlinearity. CNT-based photonics devices offer several key advantages, including ultra-fast response, robustness, tunability of wavelength, and compatibility to fibers. This chapter describes various designs and fabrication techniques of CNT-based photonic devices. CNTs with suitable diameters, and thus suitable operation wavelengths, are deposited on different types of fibers or waveguides such as D-shaped fibers, taper fibers or planar waveguides to ensure effective CNT-light interaction.

Carbon Nanotubes and Their Applications
Edited by Qing Zhang

ISBN 978-981-4241-90-8 (Hardcover), 978-981-4303-18-7 (eBook)
www.panstanford.com

Optical nonlinear effects in particular four-wave mixing (FWM) are experimentally observed using the fabricated CNT-based devices. FWM-based wavelength conversion applications are further achieved with good performance.

16.1 Introduction

CNTs have recently drawn much research attention owing to their unique optical properties [1–3]. CNTs are mainly categorized as single-wall or multiwall. The structure of a SWCNT can be simply understood by wrapping a one-atom-thick layer of graphene into a cylinder. A multiwall CNT (MWCNT) is two or more CNTs nested within one another in a coaxial form. As the properties of MWCNTs are determined by contribution of all individual shells with different structures, they are usually more defective than SWCNTs. In the case of single-wall ones, due to their pure one-dimension properties (with typical diameter of ~1 nm and length of ~1 μm) as well as well-defined structures, they exhibit properties that are not shared by MWCNTs [1]. The structure of the SWCNTs is basically determined by a single parameter called chirality. Depending on the chirality, they exhibit two different electrical properties: metallic or semiconducting. Semiconducting CNTs have energy bandgaps like those in ordinary semiconductors, and photons having corresponding optical frequency can be absorbed by the CNTs. This absorption is exicitonic and saturable, and its recovery is very fast in order of femtoseconds. The bandgap energy in semiconducting CNTs can be controlled by the tube diameter. In particular, a tube diameter of around 1.2 nm gives absorption at a wavelength around the 1550 nm optical communication band [4,5]. However, it is not yet possible to synthesize only one selected chirality using any fabrication methods and further processing of sorting a narrow chirality distribution of CNTs is necessary. During recent years, much research effort had been focused on a number of photonics applications and studies of SWCNTs, including mode-locked lasers [4–6], ultra-fast optical response [7,8], and optical noise suppression [9]. In particular, such CNTs can be employed as an optical nonlinearity medium owing to the theoretically estimated ultrahigh third-order nonlinearity [10–12]. It is believed that the third-order nonlinearity of the CNTs is originated from the inter-band transitions of the π-electrons causing nonlinear polarization. In this sense CNTs can be regarded to have similar properties with other organic optical materials such as polyacetylene or polydiacetylenes, which exhibit extremely high third-order nonlinearity. Experimental investigation of such optical nonlinearities in CNTs has been initiated by a Kerr shutter-based optical switching using a few centimeters of CNT-deposited D-shaped fiber [13,14] as well as optical loop mirror incorporated with CNT-loaded planar waveguide [15], which indicated the possibilities of practical nonlinear CNT-based devices.

In this chapter, we describe the optical properties of SWCNTs and the considerations of CNT diameters in order to obtain suitable operation wavelength. The design and fabrication of CNT-based photonics devices with different platforms are introduced. Also, the experimental measurements of nonlinear effects generated from the fabricated CNT-based devices are discussed. In particular, the observation of four-wave mixing (FWM) in a CNT-deposited planar lightwave circuit (PLC) waveguide and its application in wavelength conversion are addressed. Such PLC waveguides can have much flexibility of integration and fabricating special waveguide structures. As a material for generating nonlinear effects, CNTs are deposited on the over-cladding removed PLC waveguide for CNT-light interaction. Tunable FWM-based wavelength conversion is obtained with the CNT device and a power penalty of 3 dB at 10^{-9} level is measured for 10 Gb/s non-return-to-zero (NRZ) wavelength converted signal in the bit-error-rate (BER) measurements.

16.2 Nonlinear Optical Properties of Carbon Nanotubes

Since its discovery, CNT can be synthesized with many techniques such as chemical vapor deposition, laser ablation, etc. Among different methods, SWCNTs can be made by a bulk production method called high-pressure CO conversion (HiPCO). Since the isolation of individual CNTs from each other is critical to obtain the maximum nonlinearity, the diameter and the diameter distribution of the CNTs are well controlled in the process. In order to further select the CNTs with suitable tube diameter, the fabricated CNT powder are dispersed in dimethylformamide (DMF), an effective solvent for separating and suspending individual CNTs. After several steps of centrifugal separations, only the homogeneous part of the CNT in DMF solution is adopted. Figure 16.1 shows the absorption spectrum of the CNTs measured by a spectrophotometer scanning from 400 nm to 2000 nm. By controlling the HiPCO process thus the nanotube diameters and the diameter distribution, the CNTs show a desirable absorption peak near 1550 nm. In the experiment, such properties of the CNTs with an absorption near 1550 nm are found to be highly suitable for generating nonlinear effects such as FWM at this wavelength range. The prepared CNT in DMF solution can be either directly sprayed on a waveguide surface with a spray gun or undergo optically-driven deposition to deposit onto an optical fiber. Figure 16.2 depicts the scanning electron microscope (SEM) image of some deposited CNTs on the surface of a waveguide, showing that the CNTs are successfully deposited on the surface.

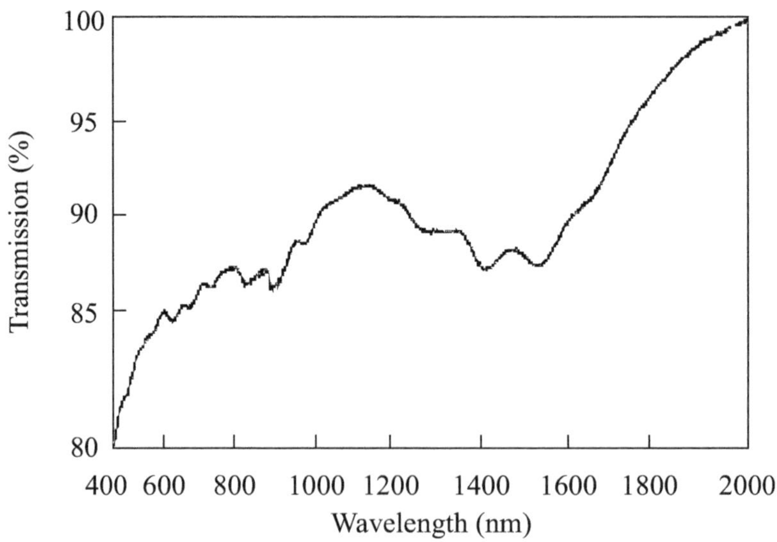

Figure 16.1 Absorption spectrum of single-wall carbon nanotubes measured by a spectrophotometer [17].

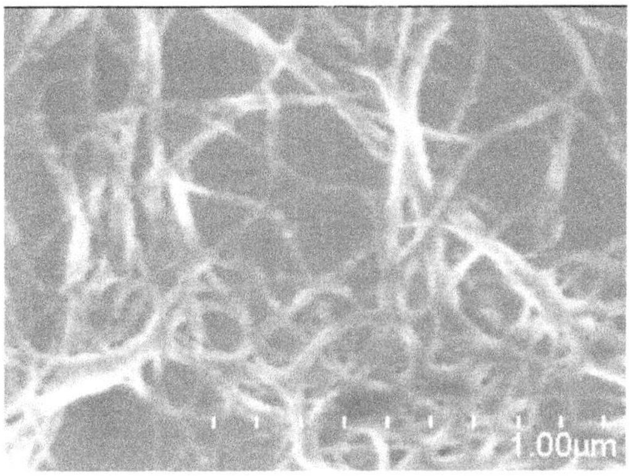

Figure 16.2 Scanning electron microscope image of the deposited CNTs on an optical waveguide [22].

16.3 Design and Fabrication of CNT-Based Optical Devices

One simple way to obtain a CNT-based fiber device is to directly spray the prepared CNT solution on a side-polished surface of a D-shaped fiber [13,14,16,17]. The working principle of the CNT-deposited D-shaped fiber is based on the interaction of CNTs with direct or evanescent field of the propagating light in the fiber. A D-shaped fiber can be prepared by polishing a segment of standard SMF held by a V-grooved block. The fiber together with the V-grooved block is polished with several steps in order to ensure the non-cracked and smooth surface of the D-shaped area, thus minimize the beam scattering through the polished face. The insertion loss is monitored during polishing therefore the amount of optical power leakage through the polished face can also be monitored. The CNTs are then deposited on the D-shaped surface by a spray method, and the fiber sample is finalized by adding a protection layer above the deposited CNTs. Figure 16.3a shows the schematic illustration of the D-shaped fiber and the corresponding SEM image of the deposited CNTs. Figure 16.3b is the photo of the finished device in a V-grooved block with the SMF pigtails. The overall insertion loss of the CNT-deposited D-shaped fiber adopted in this experiment is 12 dB with a CNT-light interaction length of around 5 cm. Note that the V-grooved block is a few cm longer as a buffer for protecting the junctions between the D-shaped region and the SMF pigtails. Since the CNTs are randomly sprayed on one polished side of the fiber, the finished fiber device is polarization sensitive with around 4-dB power variation to the polarization-dependent resonance of individual CNTs. It is worth noting that since the splicing loss of the device to sub-systems or laser cavities can be nearly neglected as the D-shaped fiber is made by standard SMF. However, it is difficult to obtain a fine polished surface for the CNT deposition in order to reduce propagation loss and light scattering, and a large volume of CNT solution is required for the device fabrication.

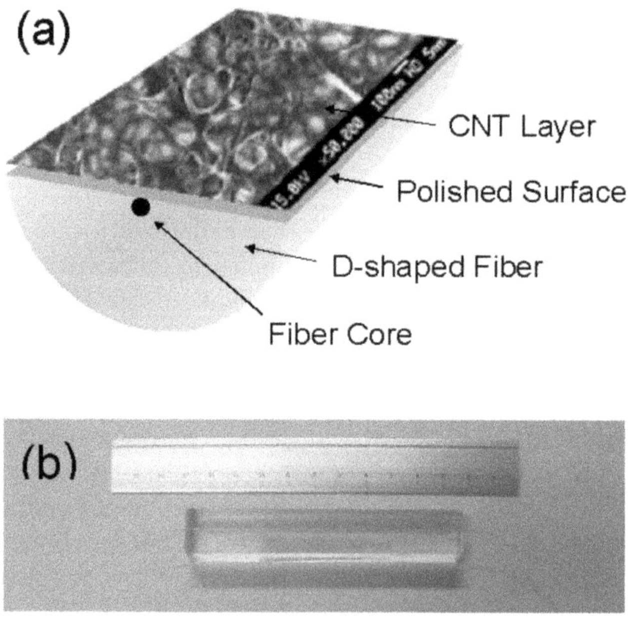

Figure 16.3 (a) Schematic illustration of D-shaped fiber with CNTs deposited on the polished surface; (b) photo of the finished CNT-deposited D-shaped fiber in a V-grooved block with single-mode fiber pigtails [17].

Besides directly spraying the CNT solution on a fiber, another effective method is optical deposition of CNTs on a fiber end or around a tapered fiber [18–20]. Such CNT-deposited tapered fiber works with the interaction between the deposited CNTs and the evanescent field of the propagating light in the fiber. Figure 16.4 shows the fabrication setup of the tapered fiber as well as for CNT deposition. The tapered fiber is fabricated using a meter of standard SMF patch cord with connectors in both ends. The coating of the fiber is first removed and fixed between two motor-driven transition stages. The propagation loss of the fiber is monitored by an optical power meter during the fabrication process. The fiber is then stretched with heat by a flame, where the stretching distance and the heating time are controlled in order to determine the fiber diameter of the tapered region. In this experiment, a tapered fiber with a diameter of around 10 μm is obtained with an extra insertion loss of 0.5 dB measured during the tapering process. A droplet of the prepared CNT in DMF solution is then applied and the tapered region of the fiber is immersed into the droplet. An amplified 1550 nm continuous wave (cw) light from an external cavity laser (ECL) is then launched into the fiber for optically-driven deposition of CNT around the tapered fiber. The deposition process is stimulated with a launched optical power of around +15 dBm and the propagation loss of the tapered fiber increases with the deposition time. Figure 16.5a,b show the microscopic image and the SEM image of the fabricated CNT-deposited tapered fiber, respectively. From the SEM image the tapered fiber is estimated to have a CNT-deposited length of around 100 μm. The overall propagation loss of the finished fiber device is measured to be 9 dB. With reference to the insertion loss of the tapered fiber before CNT deposition, approximately 86% of the light propagating through the tapered fiber is absorbed by the deposited CNTs. The fiber device is finished by drying up and put on another clean glass plate.

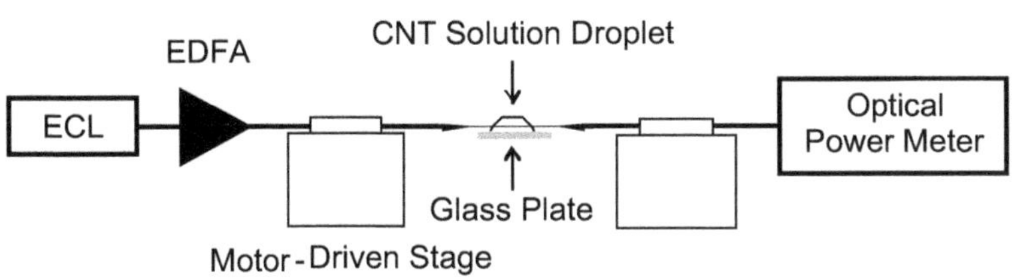

Figure 16.4 Fabrication setup of the CNT-deposited tapered fiber. ECL: external cavity laser; EDFA: erbium-doped fiber amplifier [20].

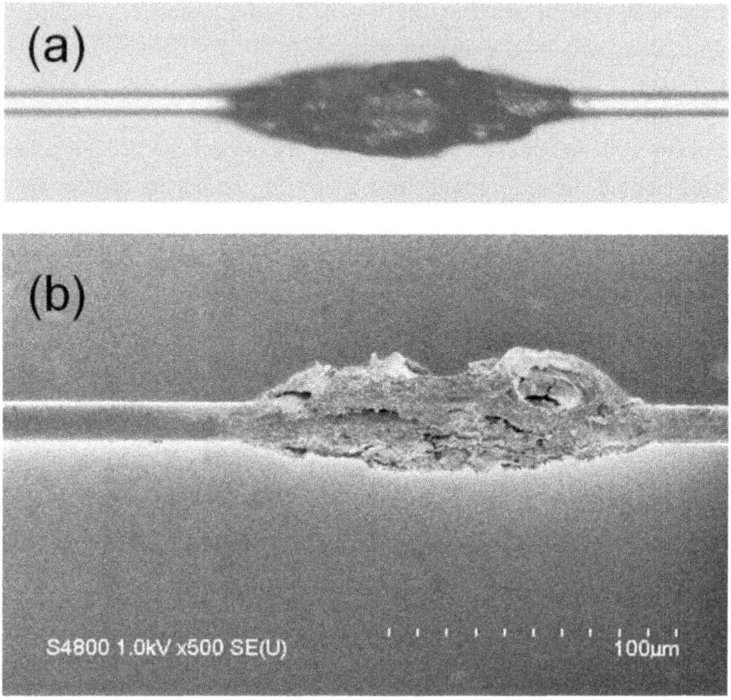

Figure 16.5 (a) Microscopic and (b) scanning electron microscope images of the CNT-deposited tapered fiber [20].

Since the fiber-based configurations are difficult for device integration or constructing different waveguide structures, a CNT-deposited planar lightwave circuit (PLC) becomes another suitable solution for making an effective device [15,21,22]. Such PLC waveguides can have much flexibility of integration and fabricating special waveguide structures. As a material for generating nonlinear effects, CNTs are deposited on the over-cladding removed PLC waveguide for CNT-light interaction. The CNT-deposited PLC waveguide works with the interaction between CNTs and the evanescent field of propagating light in the over-cladding-removed core of the waveguide. Figure 1.6a shows the schematic illustration of the adopted PLC waveguide. The waveguide is fabricated by the standard silicon-based PLC fabrication process and a number of straight waveguides can be simultaneously fabricated on the same PLC device. The over-claddings of the waveguides are intentionally removed for CNT deposition onto the core during the fabrication process. In the experiment, we have adopted a straight waveguide with 10 cm in length, 5 cm of over-cladding removed region, and 7 μm of waveguide width. The device is finished by evenly spraying the prepared CNT in DMF solution on the over-cladding removed region of the waveguide. Note that the absorption spectrum of the CNTs is measured before spraying to ensure that the properties are only from the CNTs. Figure 16.6b is the photo of the finished CNT-deposited PLC waveguide in the experimental setup. It is free-space coupled to other fiber components using cleaved dispersion-shifted fibers and index-matching gel between the waveguide and the fiber ends. The overall fiber-to-fiber insertion loss of the waveguide is 19 dB mainly due to the absorption of the deposited CNTs.

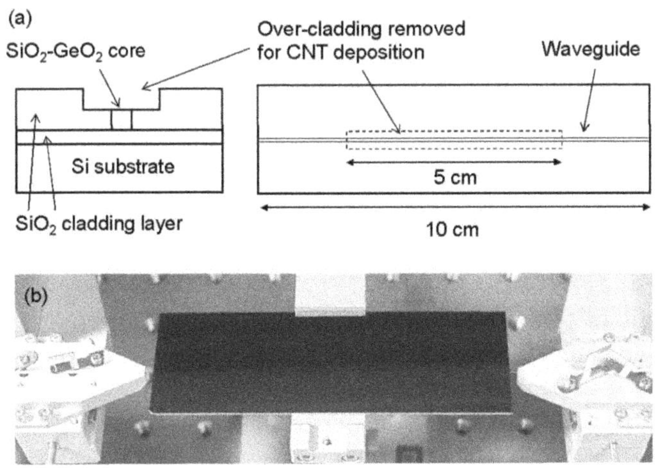

Figure 16.6 (a) Schematic illustration of planar lightwave circuit (PLC) waveguide with over-cladding removed for CNT deposition in cross-view (left) and top-view (right) and (b) the photo of CNT-deposited PLC waveguide with fiber coupling [22].

16.4 Nonlinear Applications of Carbon Nanotubes

This session focuses on the measurement of the optical nonlinear effects generated from the fabricated CNT devices. In particular, observation of four-wave mixing (FWM) generated in the CNT-deposited PLC waveguide and its application in wavelength conversion will be addressed. The experimental setup on FMW in CNT-deposited PLC waveguide is shown in Figure 16.7. A 2^{31}–1 bits pseudorandom NRZ signal (S) at 10 Gb/s is obtained by modulating the continuous-wave (cw) output of the wavelength tunable external cavity laser (ECL1). Another cw output (P) from the ECL2 is amplified with a high power erbium-doped fiber amplifier (EDFA) with suitable ASE filtering and combined with S through a 3 dB coupler. Then the combined light is launched into the CNT-deposited PLC waveguide through free-space coupling as shown in Figure 16.6b. The launched pump (P) power into the fiber is estimated to be +25 dBm. The pump (P) is fixed at 1550 nm and different converted wavelengths (C) are obtained by tuning the signal wavelength (S) in our experiment. Figure 16.8a shows the output FWM spectrum obtained after the CNT-deposited PLC waveguide and a significant idler (converted signal) at 1547.0 nm is observed. Figure 16.8b,c shows the close-up views of the converted signal and the input signal, respectively. From Figure 16.8b it is observed that the generated converted signal is spectrally broadened to have a 10 Gb/s modulation characteristics corresponding to the input signal, thus confirming the generation of the converted signal is the result of wave mixing between the input signal and the pump. The FWM conversion efficiency is defined as the ratio of the idler (C) power to the signal (S) power inside the CNT-deposited PLC waveguide. Assuming the pump wavelengths are close to each other and the propagation length is short, the conversion efficiency η can be approximately expressed as [23]

$$\eta(L) = (\gamma PL)^2, \tag{16.1}$$

where γ is the effective nonlinear coefficient, L is the light propagation distance, and P is the pump (P) power. From Eq. (16.1) the effective nonlinear coefficient of the CNT-deposited PLC waveguide in our experiment is calculated to be as high as 5.64×10^3 W^{-1} km^{-1}. As the CNTs are randomly sprayed on the PLC waveguide without any alignment, the derived γ is not as high as the corresponding predicted nonlinear refractive index n_2, which assuming all CNTs are well aligned to maximize the CNT-light interaction. Such limitation could be further improved by utilizing other deposition methods to obtain aligned CNTs [24,25]. Also, it is expected that further reduction of the waveguide dimension is possible if the γ of the deposited CNTs is maximized.

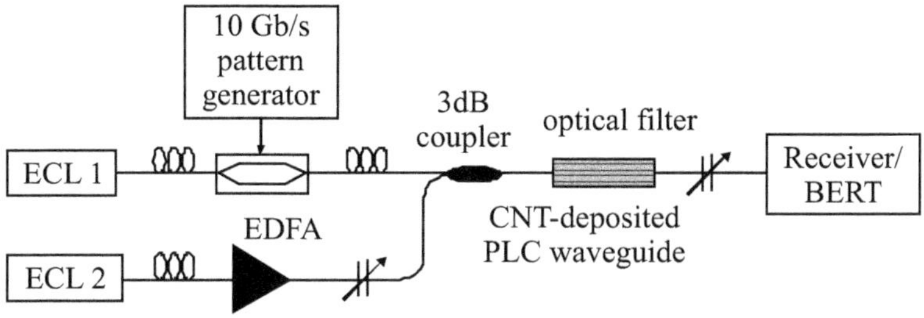

Figure 1.7 Experimental setup on four-wave mixing in a CNT-deposited PLC waveguide. ECL: external cavity laser; EDFA: erbium-doped fiber amplifier; OSA: optical spectrum analyzer; BERT: bit-error rate test set [22].

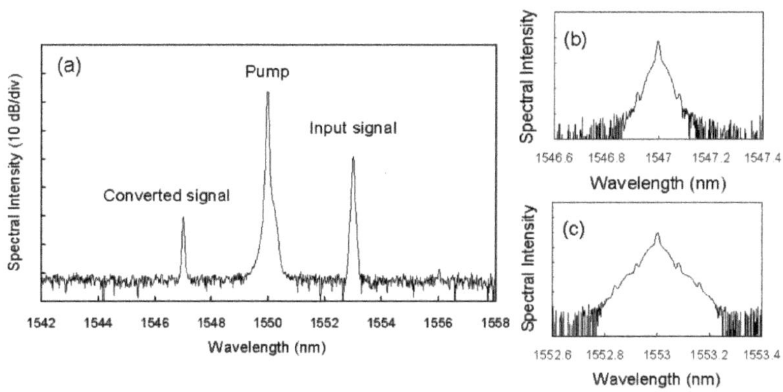

Figure 16.8 (a) Four-wave mixing spectrum obtained after the CNT-deposited PLC waveguide and the corresponding close-up views of (b) converted signal and (c) input signal [22].

Figure 16.9 plots the relation between the conversion efficiency and the S wavelength detuning against the fixed P. A 3 dB tuning range of around 8 nm is observed. One possible reason that limits the tuning range could be due to the response time of around 500 fs for the CNTs [4,5], in which the temporal pulse spreading Δt is corresponding to a bandwidth $\Delta\lambda$ of few nanometers. The performance of the FWM-based wavelength converter is further investigated by performing BER measurements. Figure 16.10 shows the output BER against the received optical power with the inset showing the 10 Gb/s eye diagrams of the input signal and the converted signal. In this measurement the wavelengths of S and C are the same with those in Figure 16.8. Note that the converted signal is obtained after the optical filter shown in Figure 16.7 followed by a low noise EDFA with suitable ASE filtering in the receiver. The figure shows the results of a 3 nm down-conversion and the power penalty is measured to be 3 dB at 10^{-9} BER level. It is worth noting that compared to other CNT devices described in Section 16.3, the power penalty is reduced as a relatively long length of CNT-light interaction can be obtained and the surface smoothness of the PLC waveguide can be well-controlled during the fabrication process. It is believed that if the CNTs are aligned with more evenly deposited on the waveguide to optimize the nonlinear interaction, further reduction of the power penalty can be obtained.

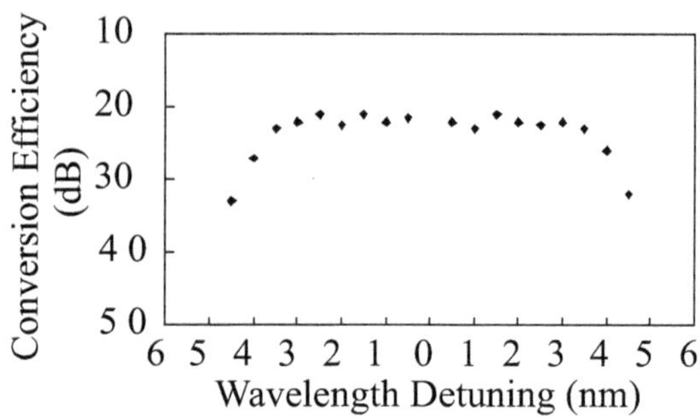

Figure 16.9 Plot of conversion efficiency against wavelength detuning [22].

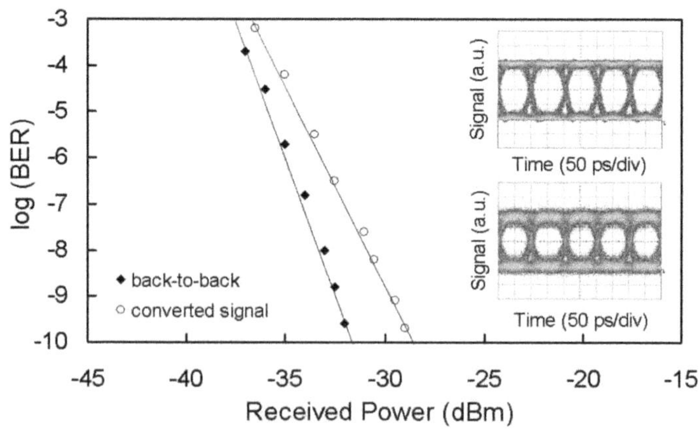

Figure 16.10 Plot of bit-error rate against received optical power; inset (upper) and (lower) show the 10 Gb/s eye-diagrams of the input and the converted signal, respectively [22].

16.5 Summary

Optical properties of SWCNTs, including absorption and third-order nonlinearities have been described in this chapter. Various CNT-based photonic devices using both fiber and waveguide platforms have been introduced. The nonlinear FWM-based wavelength conversion application using a CNT-deposited PLC waveguide has been further discussed. The nonlinear effects are generated with the interaction between CNTs and the evanescent field of propagating signal light in the over-cladding-removed core of the waveguide. A power penalty of 3 dB at 10^{-9} level in the 10 Gb/s BER measurements is obtained. Due to the design flexibility of the PLC waveguides, future work on CNT-deposited PLCs with different waveguide structures, including Mach–Zehnder interferometer and loop configurations, is foreseeable. Investigations of other optical nonlinear effects such as self-phase modulation and cross-phase modulation generated in CNTs are undergoing for different photonics applications. Since the intrinsic response time for CNT is in the order of less than 500 fs, this technology is expected to be developed into a practical and compact all-optical functional device platform, for applications in future ultra-high-speed applications such as all-optical signal processing and optical logic gates.

References

1. S. Iijima, and T. Ichihashi (1993) Single shell carbon nanotubes of one nanometer diameter, *Nature*, **363**, pp. 603–605.
2. A. Thess, R. Lee, P. Nikolaev, H. Dai, P. Petit, J. Robert, C. Xu, Y. H. Lee, S. G. Kim, D. T. Colbert, G. Scuseria, D. Tománek, J. E. Fischer, and R. E. Smalley (1996) Crystalline ropes of metallic carbon nanotubes, *Science*, **273**, pp. 483–487.
3. P. Avouris, M. Freitag, and V. Perebeinos (2008) Carbon nanotube optics and optoelectronics, *Nat. Photon.*, **2**, pp. 341–350.
4. S. Y. Set, H. Yaguchi, Y. Tanaka, and M. Jablonski (2004) Laser mode locking using a saturable absorber incorporating carbon nanotubes, *J. Lightwave Technol.*, **22**, pp. 51–56.
5. S. Y. Set, H. Yaguchi, Y. Tanaka, and M. Jablonski (2004) Ultrafast fiber pulsed lasers incorporating carbon nanotubes, *IEEE J. Select. Top. Quantum Electron.*, **10**, pp. 137–146.

6. S. Yamashita, Y. Inoue, S. Maruyama, Y. Murakami, H. Yaguchi, M. Jablonski, and S. Y. Set (2004) Saturable absorbers incorporating carbon nanotubes directly synthesized onto substrates/fibers and their applications to mode-locked fiber lasers, *Opt. Lett.*, **29**, pp. 1581–1583.
7. Y. C. Chen, N. R. Raravikar, L. S. Schadler, P. M. Ajayan, Y. P. Zhao, T. M. Lu, G. C. Wang, and X. C. Zhang (2002) Ultrafast optical switching properties of single-walled carbon nanotube polymer composites at 1.55 μm, *App. Phys. Lett.*, **81**, pp. 975–977.
8. S. Tatsuura, M. Furuki, Y. Sato, I. Iwasa, M. Tian, and H. Mitsu (2003) Semiconductor carbon nanotubes as ultrafast switching materials for optical telecommunications, *Adv. Mater.*, **15**, pp. 534–537.
9. S. Y. Set, H. Yaguchi, Y. Tanaka, and M. Jablonski (2003) A noise suppressing saturable absorber at 1550 nm based on carbon nanotube technology, *Proceedings of the Optical Fiber Communication Conference*, FL2.
10. V. A. Margulis, E. A. Gaiduk, and E. N. Zhidkin (1999) Third-order optical nonlinearity of semiconductor carbon nanotubes: Third harmonic generation, *Diamond Relat. Mater.*, **8**, pp. 1240–1245.
11. V. A. Margulis, E. A. Gaiduk, and E. N. Zhidkin (2001) Optical third-harmonic generation from an array of aligned carbon nanotubes with randomly distributed diameters, *Diamond Relat. Mater.*, **10**, pp. 27–32.
12. A. Maeda, S. Matsumoto, H. Kishida, T. Takenobu, Y. Iwasa, M. Shiraishi, M. Ata, and H. Okamoto (2005) Large optical nonlinearity of semiconducting single-walled carbon nanotubes under resonant excitations, *Phys. Rev. Lett.*, **94**, pp. 047404.
13. Y. W. Song, S. Y. Set, and S. Yamashita (2006) Novel Kerr shutter using carbon nanotubes deposited onto a 5-cm D-shaped fiber, *Proceedings of the Conference on Lasers and Electro-Optics*, CMA4.
14. K. K. Chow, S. Yamashita, and Y. W. Song (2009) A widely tunable wavelength converter based on nonlinear polarization rotation in a carbon-nanotube-deposited D-shaped fiber, *Opt. Express*, **17**, pp. 7664–7669.
15. K. Kashiwagi, S. Yamashita, H. Yaguchi, C. S. Goh, and S. Y. Set (2006) All optical switching using carbon nanotubes loaded planar waveguide, *Proceedings of the Conference on Lasers and Electro-Optics*, CMA5.
16. Y. W. Song, S. Yamashita, C. S. Goh, and S. Y. Set (2007) Carbon nanotube mode lockers with enhanced nonlinearity via evanescent field interaction in D-shaped fibers, *Opt. Lett.*, **32**, pp. 148–150.

17. K. K. Chow, and S. Yamashita (2009) Four-wave mixing in a single-walled carbon-nanotube-deposited D-shaped fiber and its application in tunable wavelength conversion, *Opt. Express*, **17**, pp. 15608–15613.
18. K. Kashiwagi, and S. Yamashita (2009) Deposition of carbon nanotubes around microfiber via evanascent light, *Opt. Express*, **17**, pp. 18364–18370.
19. J. W. Nicholson, R. S. Windeler, and D. J. DiGiovanni (2007) Optically driven deposition of single-walled carbon-nanotube saturable absorbers on optical fiber end-faces, *Opt. Express*, **15**, pp. 9176–9183.
20. K. K. Chow, M. Tsuji, and S. Yamashita (2010) Single-walled carbon-nanotube-deposited tapered fiber for four-wave mixing based wavelength conversion, *Appl. Phys. Lett.*, **96**, pp. 061104–061106.
21. K. Kashiwagi, S. Yamashita, Y. Nasu, H. Yaguchi, C. S. Goh, and S. Y. Set (2006) Planar waveguide-type saturable absorber based on carbon nanotubes, *Appl. Phys. Lett.*, **89**, pp. 081125–081127.
22. K. K. Chow, S. Yamashita, and S. Y. Set (2010) Four-wave mixing based wavelength conversion using a single-walled carbon-nanotube-deposited planar lightwave circuit waveguide, *Opt. Lett.*, **35**, pp. 2070–2072.
23. G. P. Agrawal (2001) *Nonlinear Fiber Optics*, 3rd Ed. (New York: Academic).
24. Y. Murakami, S. Chiashi, Y. Miyauchi, M. Hu, M. Ogura, T. Okubo, and S. Maruyama (2004) Growth of vertically aligned single-walled carbon nanotube films on quartz substrates and their optical anisotropy, *Chem. Phys. Lett.*, **385**, pp. 298–303.
25. Shinji Yamashita, Yusuke Inoue, Shigeo Maruyama, Youichi Murakami, Hiroshi Yaguchi, Tomoharu Kotake, and Sze Y. Set (2006) Mode-locked fiber lasers using adjustable saturable absorption in vertically aligned carbon nanotubes, *Jap. J. Appl. Phys.*, **45**, pp. L17–L19.

Index

Color Insert

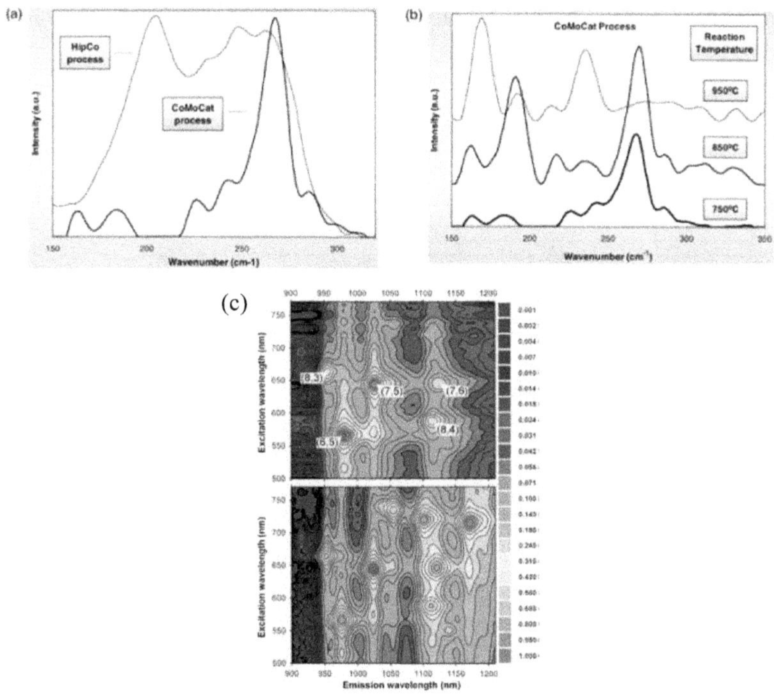

Figure 2.5

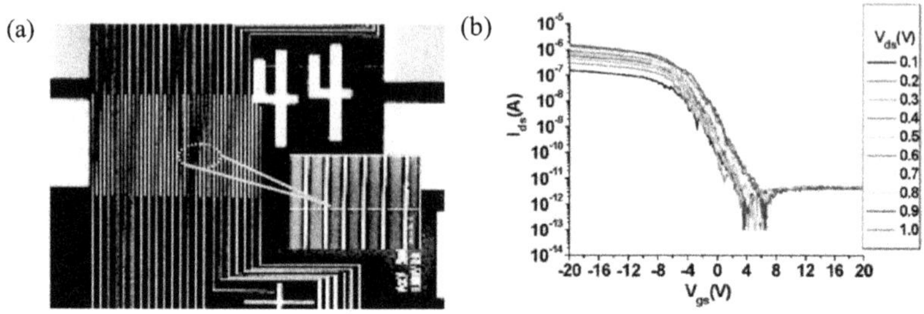

Figure 2.13

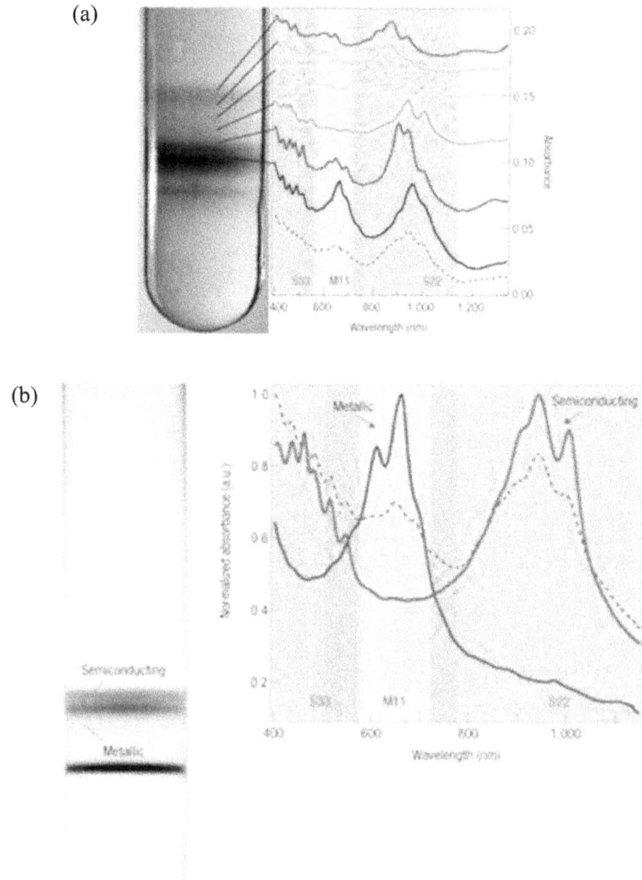

Figure 2.17

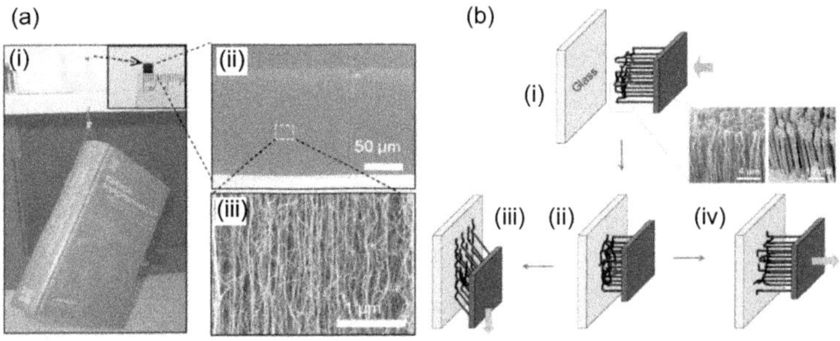

Figure 3.1

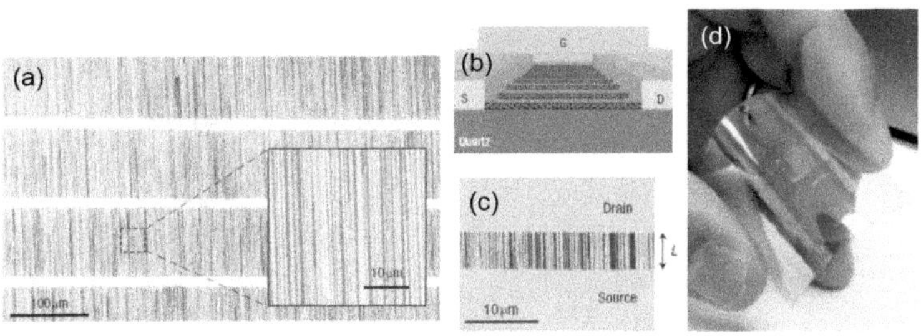

Figure 3.2

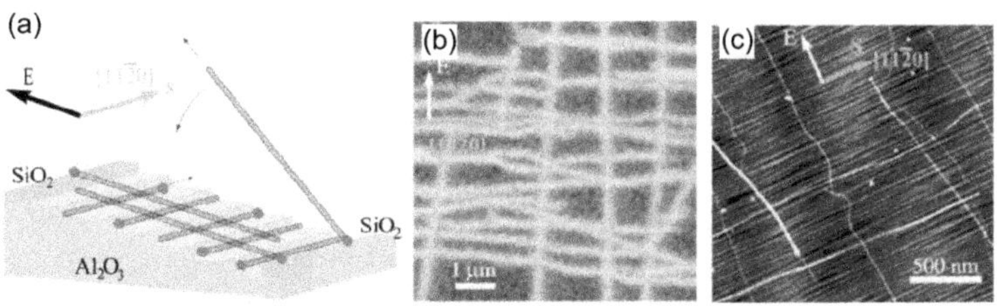

Figure 3.3

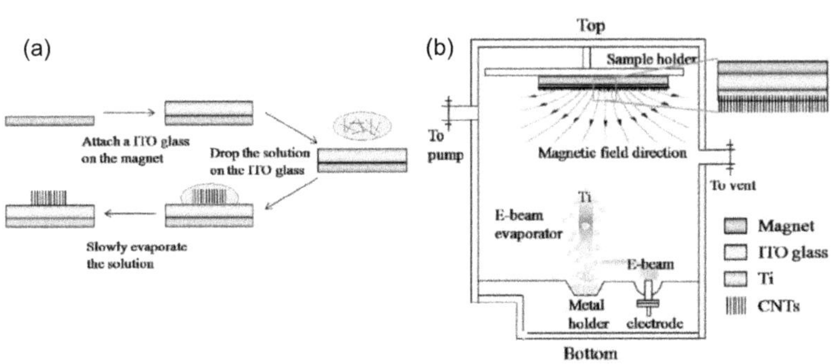

Figure 3.4

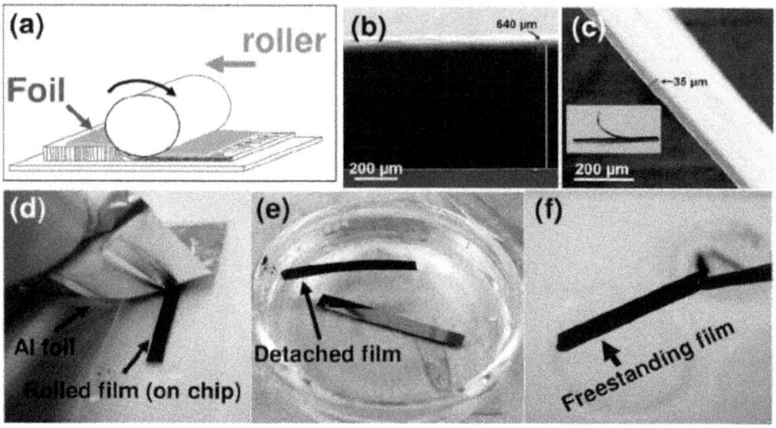

Figure 3.6

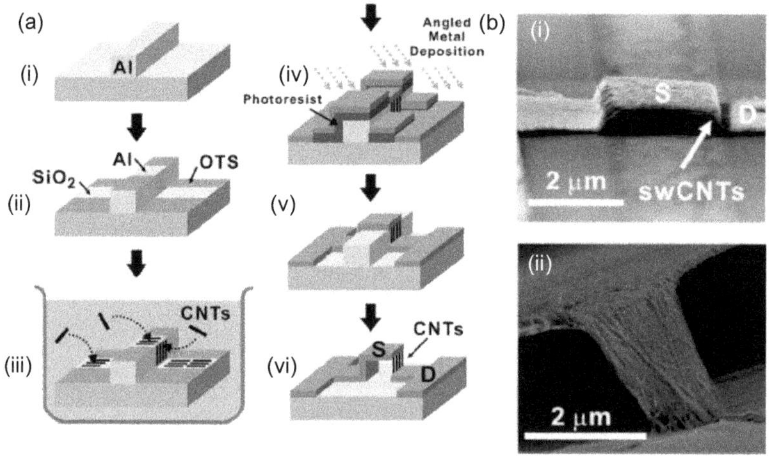

Figure 3.7

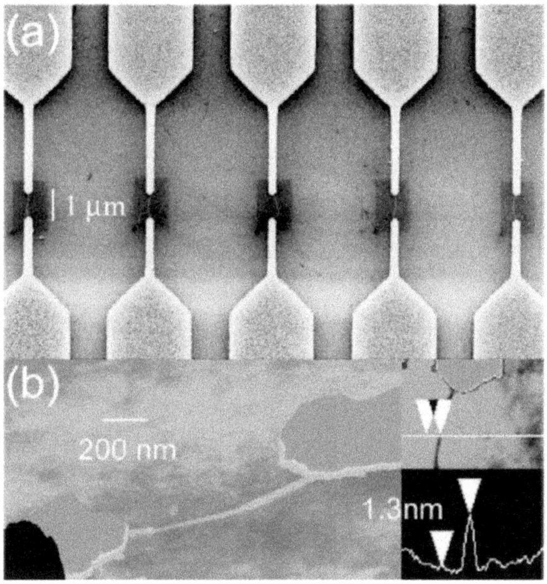

Figure 3.10

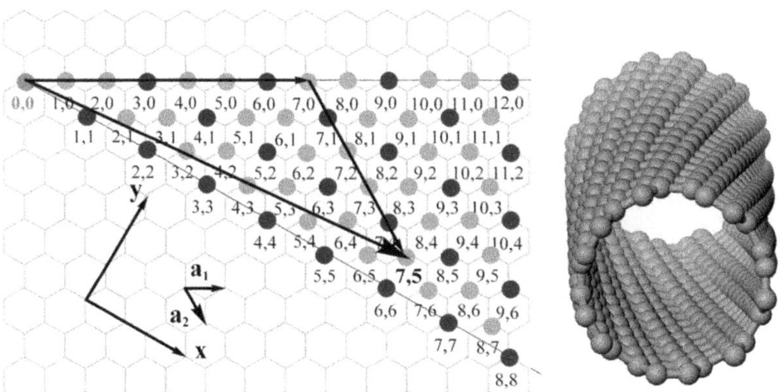

Figure 4.1

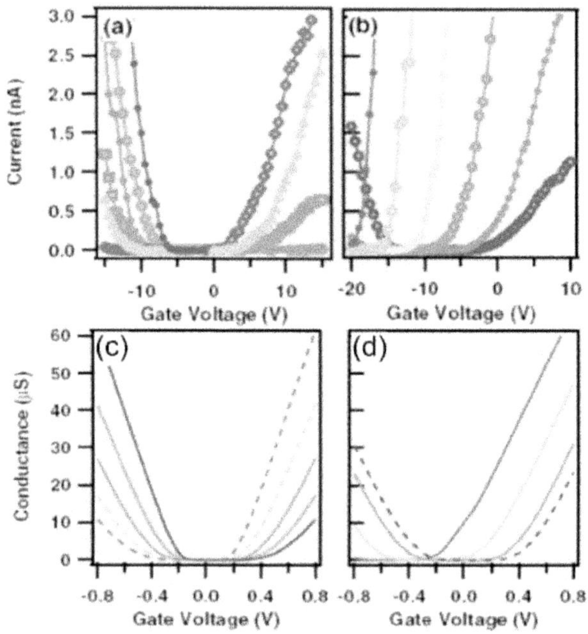

Figure 5.5

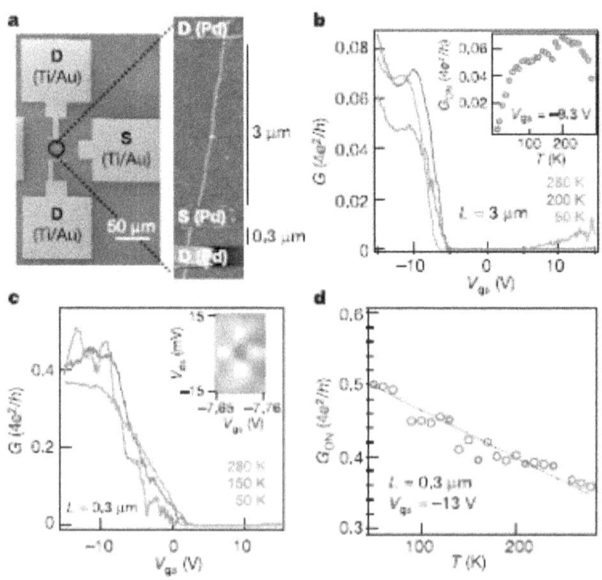

Figure 5.8

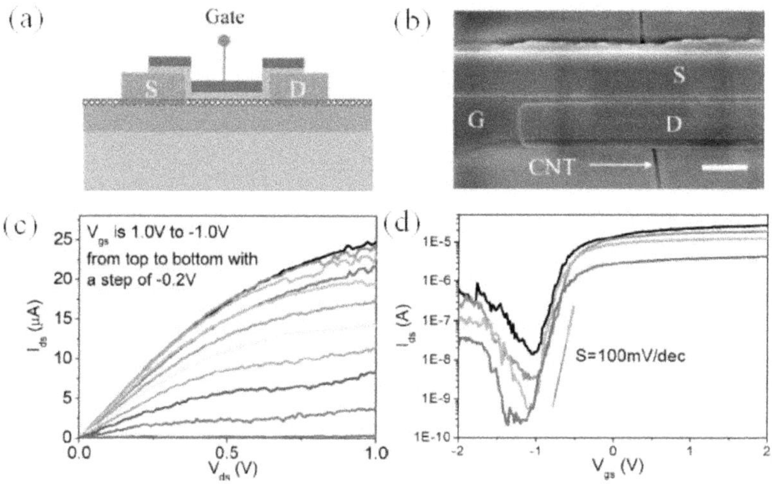

Figure 5.13

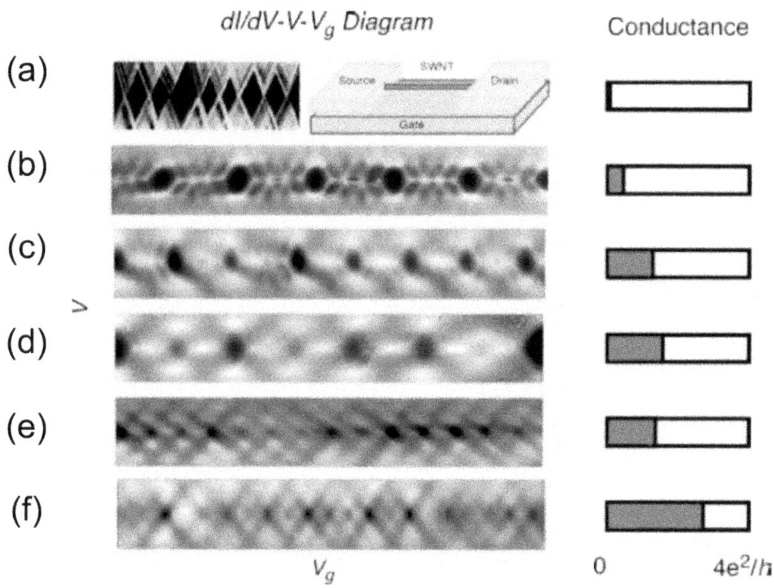

Figure 5.26

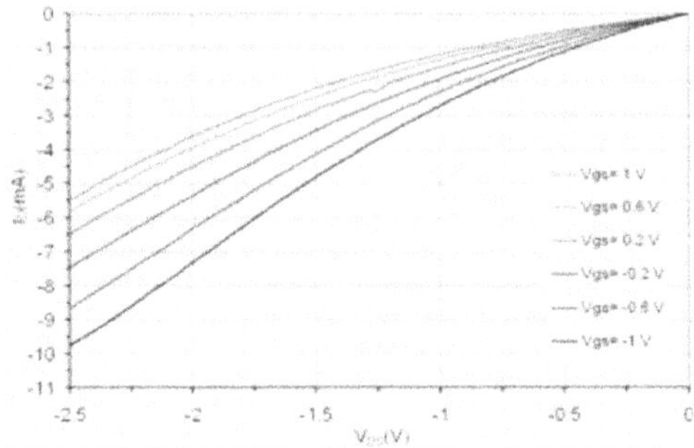

Figure 6.4

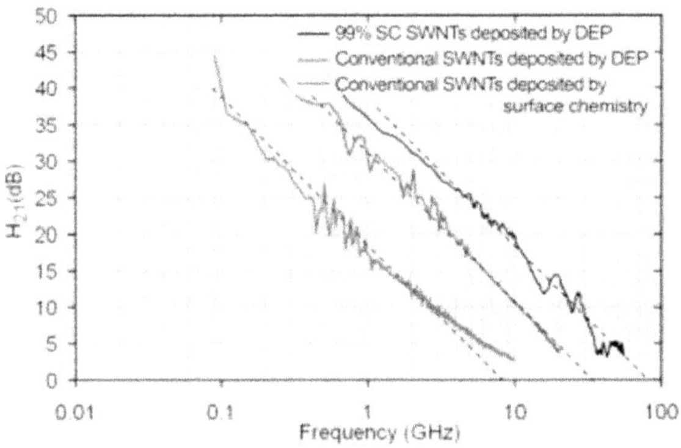

Figure 6.6

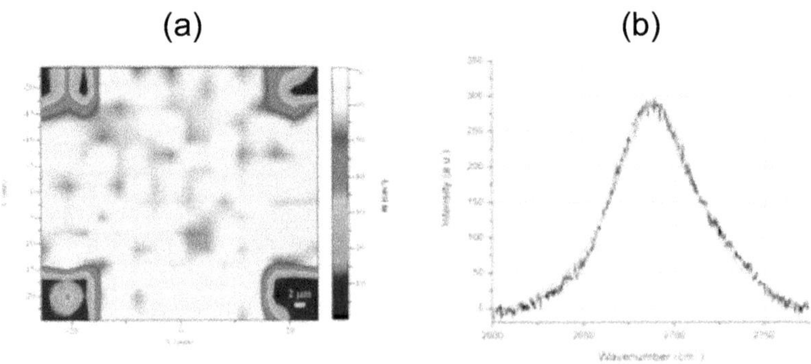

Figure 6.7

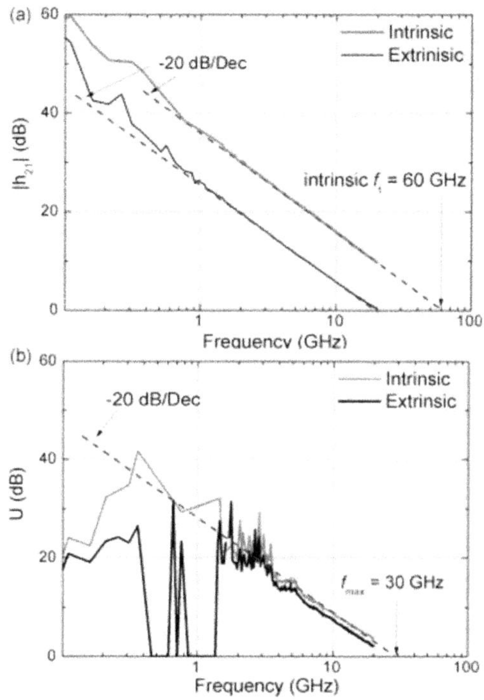

Figure 6.11

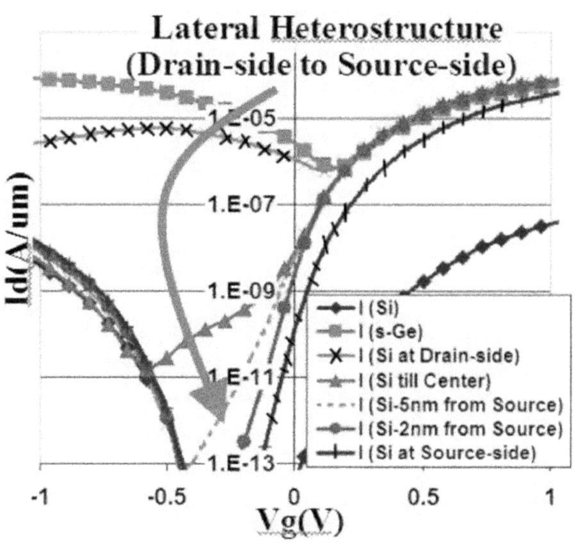

Figure 7.6

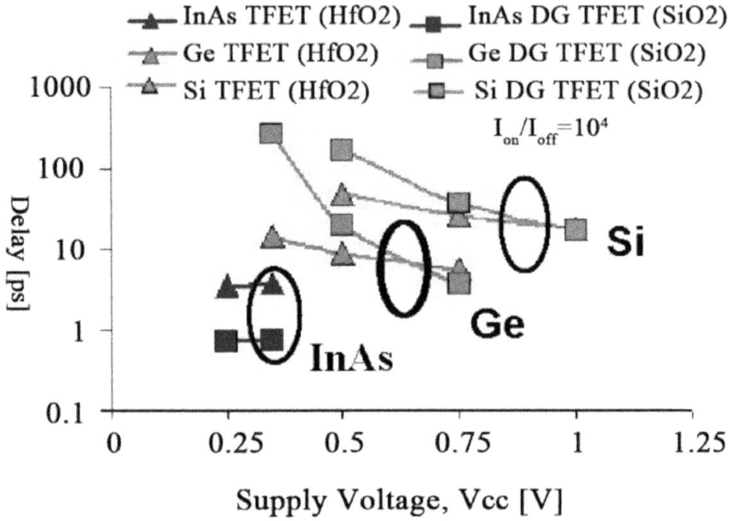

Figure 7.7

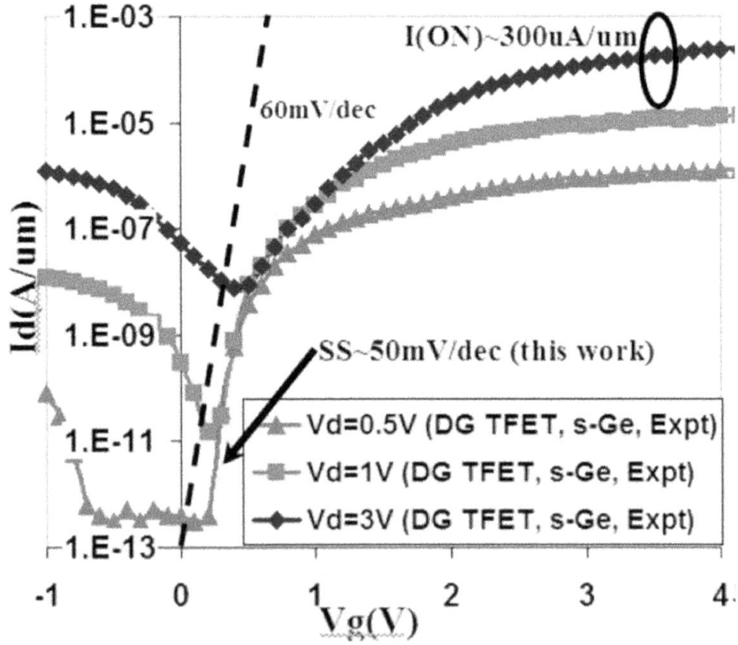

Figure 7.8

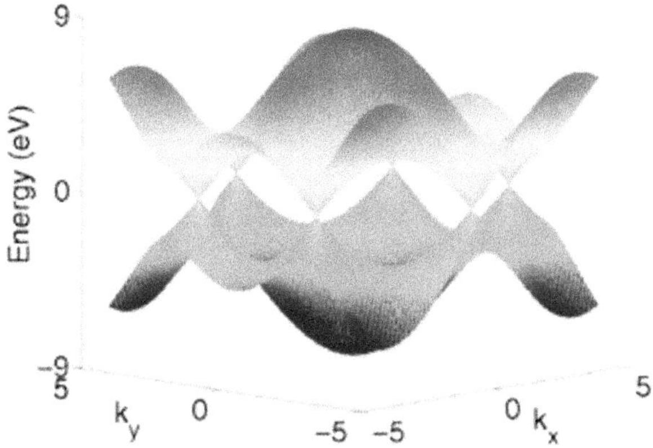

Figure 8.3

(a)

Chirality (n,m)	Sequences
(9,1)	$(TCC)_{10}$, $(TGA)_{10}$, $(CCA)_{10}$
(8,3)	$(TTA)_4TT$, $(TTA)_3TTGTT$, $(TTA)_5TT$
(6,5)	$(TAT)_4$, $(CGT)_3C$
(7,5)	$(ATT)_4$, $(ATT)_4AT$
(10,2)	$(TATT)_2TAT$
(8,4)	$(ATTT)_3$
(9,4)	$(GTC)_2GT$, $(CCG)_4$
(7,6)	$(GTT)_3G$, $(TGT)_4T$
(8,6)	$(GT)_6$, $(TATT)_3T$, $(TCG)_{10}$, $(GTC)_3$, $(TCG)_2TC$, $(TCG)_4TC$, $(GTC)_2$
(9,5)	$(TGTT)_2TGT$
(10,5)	$(TTTA)_3T$
(8,7)	$(CCG)_2CC$

(b)

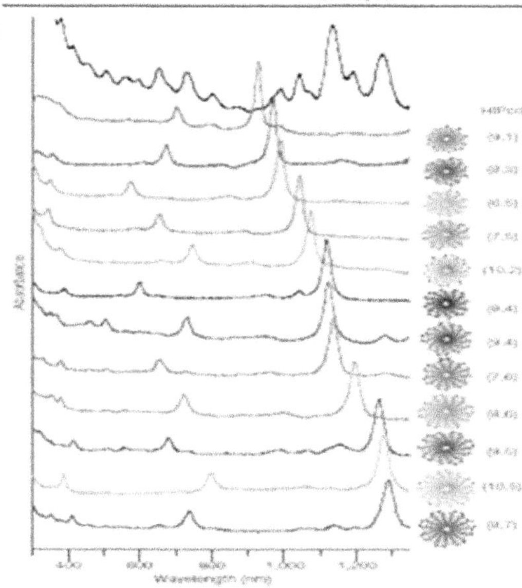

Figure 8.10

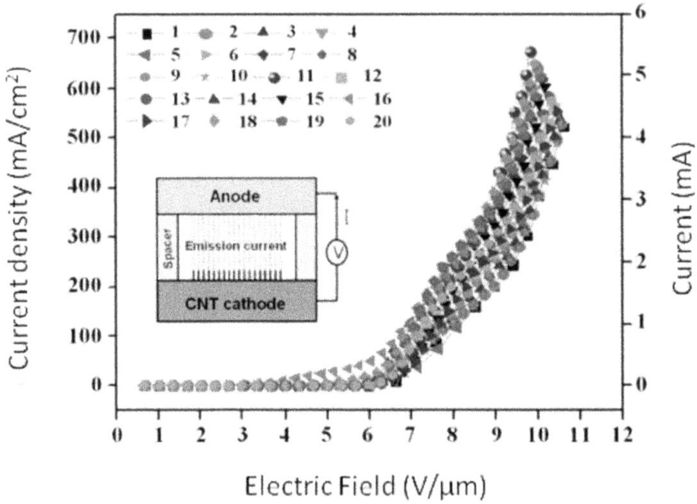

Figure 9.12

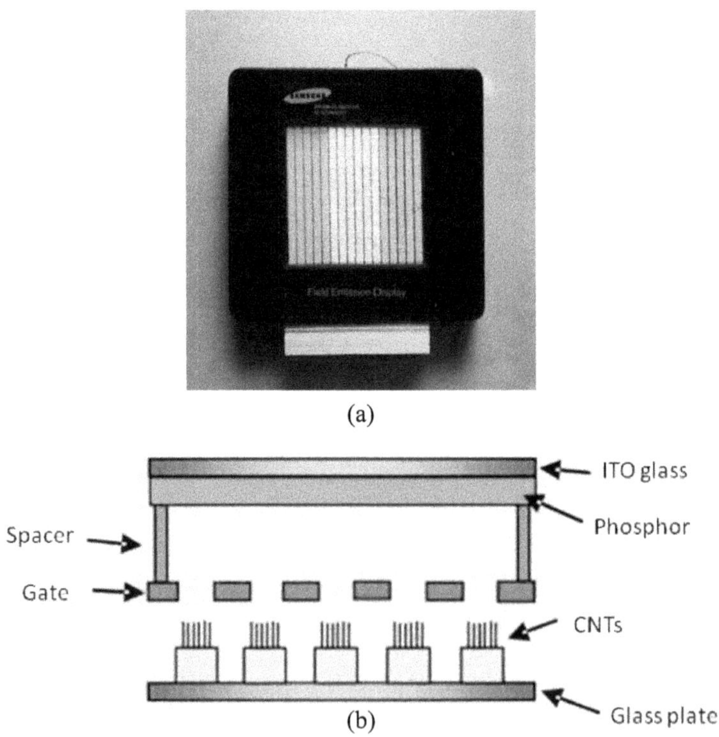

Figure 9.21

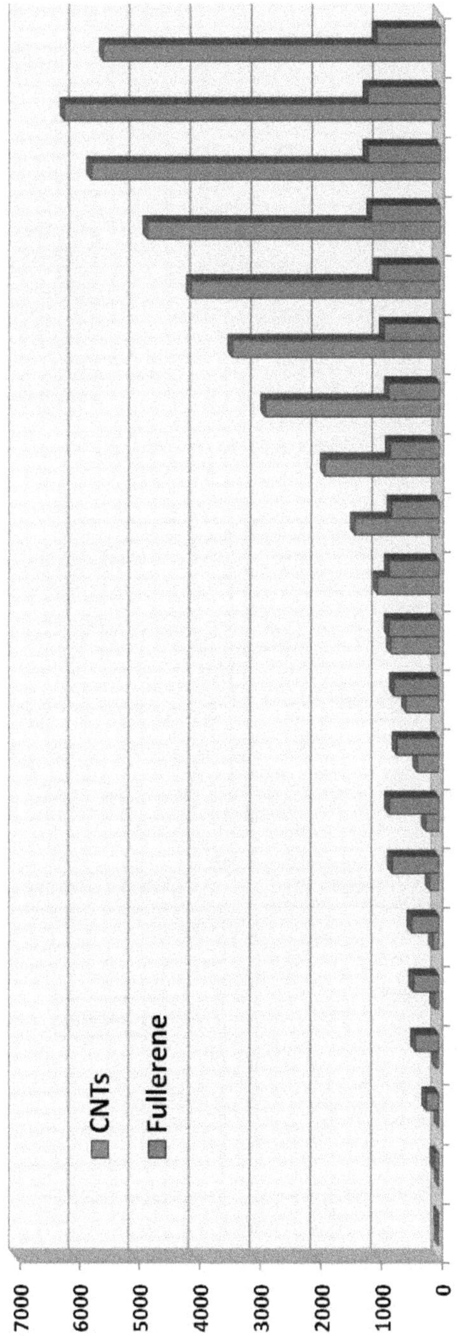

CNTs
Fullerene
0
1000
2000
3000
4000
5000
6000
7000
1990
1991
1992
1993
1994
1995
1996
1997
1998
1999
2000
2001
2002
2003
2004
2005
2006
2007
2008
2009
2010

Figure 11.1

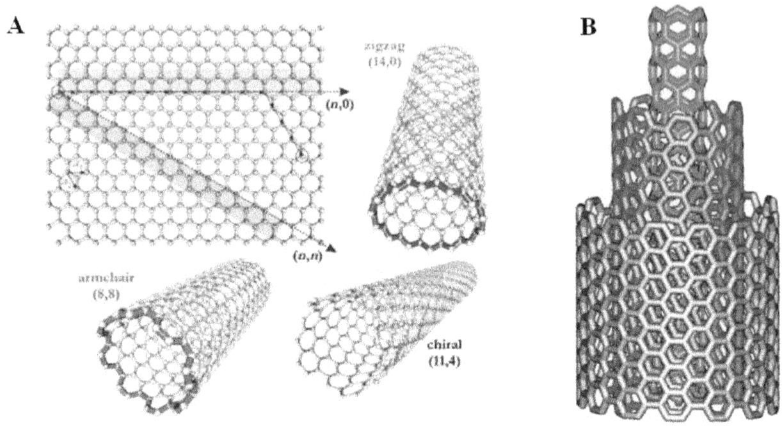

Figure 11.2

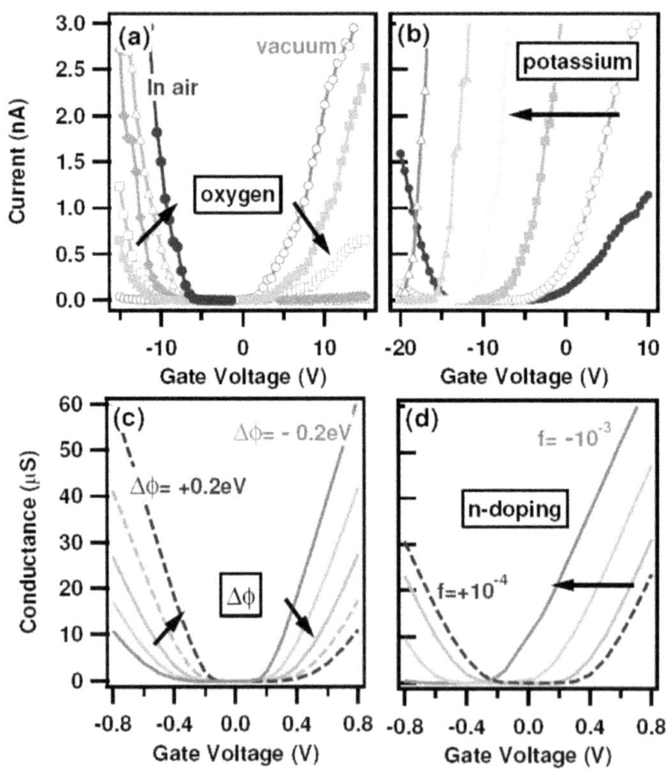

Figure 12.4

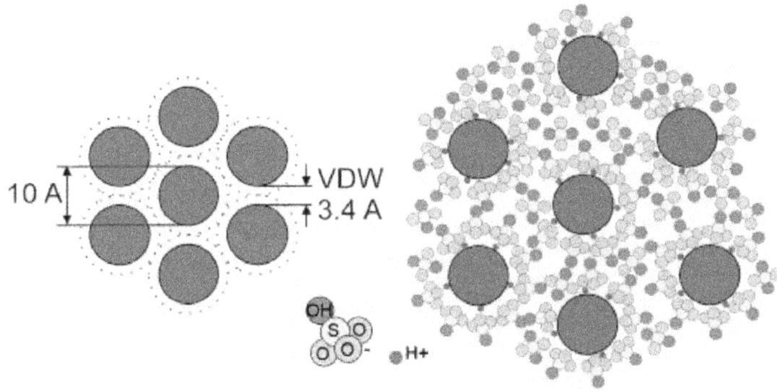

Figure 14.7

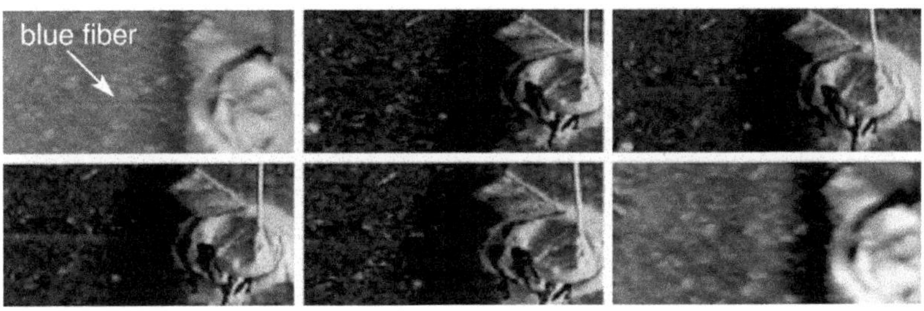

Figure 14.11

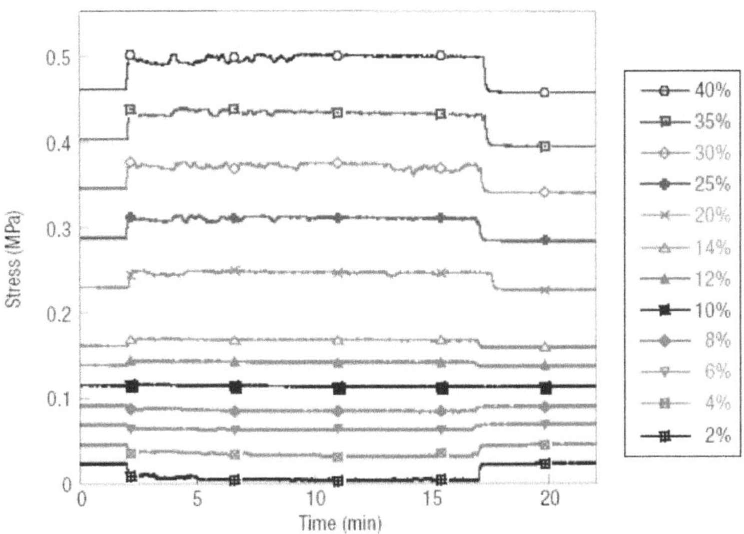

Figure 15.14

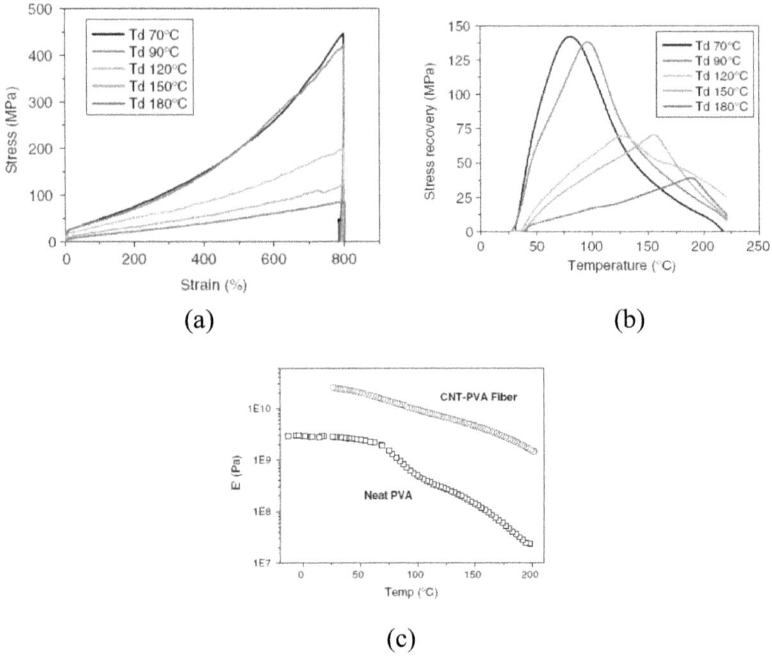

Figure 15.15

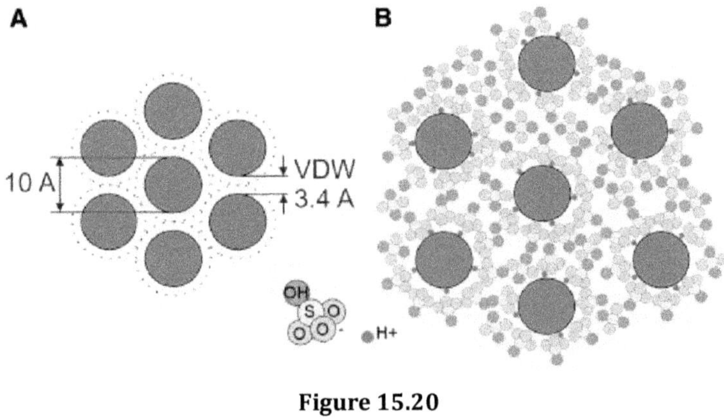

Figure 15.20

For Product Safety Concerns and Information please contact our EU
representative GPSR@taylorandfrancis.com
Taylor & Francis Verlag GmbH, Kaufingerstraße 24, 80331 München, Germany

www.ingramcontent.com/pod-product-compliance
Ingram Content Group UK Ltd.
Pitfield, Milton Keynes, MK11 3LW, UK
UKHW022208190926
475702UK00003B/262
* 9 7 8 9 8 1 4 2 4 1 9 0 8 *

The Power of Process

The Value of Due Process in Security Council Sanctions Decision-Making

THE POWER OF PROCESS

The Value of Due Process in Security Council Sanctions Decision-Making

DEVIKA HOVELL

OXFORD
UNIVERSITY PRESS

OXFORD
UNIVERSITY PRESS

Great Clarendon Street, Oxford, OX2 6DP,
United Kingdom

Oxford University Press is a department of the University of Oxford. It furthers the University's objective of excellence in research, scholarship, and education by publishing worldwide. Oxford is a registered trade mark of Oxford University Press in the UK and in certain other countries

First Edition published in 2016
Impression: 1

Published in the United States of America by Oxford University Press
198 Madison Avenue, New York, NY 10016, United States of America

British Library Cataloguing in Publication Data
Data available

Library of Congress Control Number: 2015944761

ISBN 978–0–19–871767–6

Printed and bound by
CPI Group (UK) Ltd, Croydon, CR0 4YY

For Mum and Dad, with love and thanks

Series Editors' Preface

The UN Security Council possesses a broad power pursuant to Article 41 of the UN Charter to impose economic sanctions against States and individuals. The Council's increasing use of 'targeted' sanctions against individuals has led to growing calls over the years to impose constraints on the exercise of its power. Dr Hovell's important work engages in a detailed enquiry into the theoretical and conceptual foundations of procedural fairness in Security Council decision-making, and in so doing it has been careful to take account of the considerable international and domestic judicial decision-making that has developed on the topic. The results of the enquiry are to be commended and provide no less than a basis for the development of constraints on the Security Council in the exercise of its power to sanction individuals.

CR
DS
ST

Acknowledgements

This book finds its source in my work as a DPhil student at the University of Oxford. The book itself has had a somewhat longer gestation period and I am grateful to Merel Alstein, Emma Endean, and Oxford University Press for their guidance and patience throughout the process. I wish to acknowledge that some of the chapters are based on other publications of mine. Parts of chapters 4, 5, 6 and 7 are to appear as 'Due Process in the United Nations' in the *American Journal of International Law*. Parts of chapters 3 and 6 appeared as 'A Dialogue Model: The Role of the Domestic Judge in Security Council Decision-Making' (2013) 25(3) *Leiden Journal of International Law* 579–597.

Primary thanks are due to my supervisors at the University of Oxford. Professor Dan Sarooshi was an expert guide in international law and I thank him for his kindness and support throughout the project. Of course, this project was not merely an apprenticeship in international law, but an application to an alternative guild of public law. I could not have asked for a greater guide in this respect than Professor Liz Fisher. In this book and all my future work, I hope to do justice to the academic generosity, rigour, ethics, and intellectual curiosity with which Liz instructed me. Ultimately, she taught me the importance of moving beyond instruction to find my own voice.

In the course of writing the book, I benefited from the support and influence of a number of intellectual environments and geographical backdrops. It is fitting that I completed this book at the beginning of my career at the London School of Economics where one of my main academic inspirations during the project, Professor Carol Harlow, spent the greater part of her career. I profited during the writing of the book from visiting fellowships at New York University, the European University Institute, and the University of Melbourne. I also wish to thank colleagues at the Australian National University, the European University Institute, the Hertie School of Governance, New York University, the University of Cambridge, the University of Glasgow, and the University of Melbourne who contributed generous comments at the seminars at which I presented my work over the years. Thank you to Vaughan Lowe, Dapo Akande, Guy Goodwin-Gill, and the range of astute (and fabulously argumentative) fellow DPhil students who took part in the international law research seminars at All Souls College, Oxford.

I am also very grateful to the friends, scholars, and colleagues who were willing to discuss my work more informally. There are too many to mention, but I wish to acknowledge the insights offered at different stages of the project, in conversations with Mark Aronson, Nehal Bhuta, Hilary Charlesworth, the wonderful Claire Charters, Simon Chesterman, Rob Cryer, David Dyzenhaus, Tom Franck, Conor Gearty, Carol Harlow, Gleider Hernández, Peter Johnston, Nico Krisch, Vaughan Lowe, Gerry Simpson, Angela Steininger, Antonios Tzanakopoulos, James Upcher, Guglielmo Verdirame, and George Williams. Particular thanks to my great friend Shelley Meagher who took the time to read through the entire manuscript and did her utmost to help sweep out the fustier corners.

For making my work day possible, I would like to thank Nananne, Suzu, Giovanna Nammur, Aleksandra Baran, Marissa Irwin, Monica Merryweather, and Katelyn Banks. As my best work is done with a fine coffee in hand, it would also be remiss of me not to thank Lori and the staff at Towpath café, Terry and the staff at the London Review of Books Cake Shop, and all the staff over the years at Caravan Exmouth Market for fuelling this book and hosting me for rather longer each day than the time it takes to drink a latte.

Finally, I wish to thank my family. This book is dedicated to Mum and Dad who, with unwavering support, encouragement, and good humour, provide my constant anchor. To Tom, Xavier and Lulu, thank you for adding the magic, sustaining my spirit, and enriching my life beyond measure.

DCH
London School of Economics
April 2015

Table of Contents

Table of Cases

INTERNATIONAL AND REGIONAL

United Kingdom

United States

Table of Resolutions, Treaties, and Legislation

RESOLUTIONS

TREATIES

LEGISLATION

National

Australia

Canada

China

India

Japan

South Africa

United Kingdom

1

Introduction

The Value of Due Process in Security Council Sanctions Decision-Making

The debate about due process in Security Council decision-making has played out for well over a decade. It is clear the critical scholarship has failed to galvanize the Security Council: the pace of procedural reform has been glacial and the Analytical Sanctions Support and Monitoring Team recently played down the impact of the scholarship, describing it as a 'small scholarly and generalist literature'.[1] The 'debate' increasingly resembles a conversation of the deaf. On the one hand, central to almost every reform proposal is an insistence on the inclusion of a judicial review mechanism in the sanctions decision-making process.[2] The UN Ombudsperson, a non-judicial office introduced by the

[1] 'Sixteenth Report of the Analytical Support and Sanctions Monitoring Team' (29 October 2014) UN Doc S/2014/770, para 43.

[2] Bardo Fassbender, *Targeted Sanctions and Due Process* (Council of Europe Committee of Legal Advisors on International Law 2006), 31, para 12.12; International Commission of Jurists, *Assessing Damage, Urging Action: Report of the Eminent Jurists Panel on Terrorism, Counter-Terrorism and Human Rights* (2009), 116–17; UN High Commissioner for Human Rights, 'Report on the Protection of Human Rights and Fundamental Freedoms while Countering Terrorism' (15 December 2010) UN Doc A/HRC/16/50, para 21; 'Report of the Special Rapporteur on the Promotion and Protection of Human Rights and Fundamental Freedoms while Countering Terrorism' (6 August 2010) UN Doc A/65/258, paras 56–7; Iain Cameron, 'UN Targeted Sanctions, Legal Safeguards and the European Convention on Human Rights' (2003) 72 *Nordic Journal of International Law* 159, 210; Larissa Van den Herik, 'The Security Council's Targeted Sanctions Regimes: In Need of Better Protection of the Individual' (2007) 20 *Leiden Journal of International Law* 797, 806–7; John Dugard, 'Judicial Review of Sanctions' in V Gowlland-Debbas (ed), *United Nations Sanctions and International Law* (Kluwer Law International 2001); Johannes Reich, 'Due Process and Sanctions Targeted Against Individuals Pursuant to Resolution 1267 (1999)' (2008) 33 *Yale Journal of International Law* 505, 510–11; Craig Forcese and Kent Roach, 'Limping into the Future: The UN 1267 Terrorism Listing Process at the Crossroads' (2010) 42 *George Washington International Law Review* 217, 265, 275; Jared Genser and Kate Barth, 'When Due Process Concerns Become Dangerous: The Security Council's 1267 Regime and the Need for Reform' (2010) 33 *Boston College International and Comparative Law Review* 1, 37; Erika de Wet, 'Human Rights Considerations and the Enforcement of Targeted Sanctions in Europe: The Emergence of Core Standards of Judicial Protection' in Bardo Fassbender (ed), *Securing Human*

Council to mitigate unfairness to individuals in sanctions decision-making in 2009, has been deemed inadequate by (among others) the Court of Justice of the European Union, the European Court of Human Rights, the UK Supreme Court, and the UN Special Rapporteur on Counter-terrorism and Human Rights on grounds that it is not a court.[3] On the other hand, the Security Council maintains its staunch opposition to any form of judicial review of its decision-making.[4] With most academics and critics focused on the need for a court-based process, the Council's failure to extend the Ombudsperson's mandate beyond the Al Qaeda sanctions regime (it currently applies to only one of sixteen existing sanctions regimes) has attracted little comment, let alone opposition.[5]

This book seeks to re-energize the debate, arguing that the intransigence on the part of both sides is less a product of ideology and more a question of (flawed) methodology. Over the last two decades, remarkably little attention has been given to the *value* of due process in the Security Council setting. Reform proposals have based themselves on a presumption that due process is essentially 'a good thing'. As such, the task of developing a procedural framework for the Security Council setting has been largely a-theoretical, with international lawyers resorting in large part to traditional international law source methodology to develop a procedural framework for sanctions

Rights?: Achievements and Challenges of the UN Security Council (Oxford University Press 2011), 169; Guglielmo Verdirame, *The UN and Human Rights: Who Guards the Guardians?* (Cambridge University Press 2011); Francesco Francioni, 'The Right of Access to Justice to Challenge the Security Council's Targeted Sanctions: After-thoughts on "Kadi"' in Ulrich Fastenrath, *From Bilateralism to Community Interest: Essays in Honour of Judge Bruno Simma* (Oxford University Press 2011); Grant L Willis, 'Security Council Targeted Sanctions, Due Process and the 1267 Ombudsperson' (2011) 42 *Georgetown Journal of International Law* 673, 737, 743–5; Lisa Ginsborg and Martin Scheinin, 'You Can't Always Get What You Want: The Kadi II Conundrum and the Security Council 1267 Terrorist Sanctions Regime' (2011) 8 *Essex Human Rights Review* 7, 11–12, 19; S Ghasem Zamani and Jamshid Mazaheri, 'The Need for International Judicial Review of UN Economic Sanctions' in Ali Z Marossi and Marisa R Bassett (eds), *Economic Sanctions Under International Law: Unilateralism, Multilateralism, Legitimacy, and Consequences* (TMC Asser Press 2015).

[3] *European Commission and Council v Kadi* [2013] ECR not yet reported, [133]; *Al-Dulimi and Montana v Switzerland* (App No 5809/08), 26 November 2013, [119]; *Nada v Switzerland* [2012] ECHR 1691, [209]–[14]; *Her Majesty's Treasury v Ahmed* [2010] UKSC 2, [145]–[156] per Lord Phillips; [184]–[185] per Lord Rodger; [246] per Lord Mance; Ben Emmerson QC, 'Report of UN Special Rapporteur on the Promotion and Protection of Human Rights while Countering Terrorism' (26 September 2012) UN Doc A/67/396, paras 14, 20–21.

[4] 'Thirteenth Report of the 1267 Monitoring Team' (31 December 2012) UN Doc S/2012/968, paras 15–16, 21; 'Eighth Report of the 1267 Monitoring Team' (14 May 2008) UN Doc S/2008/324, para 41. See also Security Council Report, 'Special Research Report: UN Sanctions' (25 November 2013), 14.

[5] Consider, for example, the repeated rejection of proposals by the Group of Likeminded States to extend the Ombudsperson's mandate: UN doc S/PV.6964 (10 May 2013); UN doc S/2012/805 (9 November 2012); UN doc S/PV.7285 (23 October 2014).

decision-making.[6] The aim has been to identify 'universally applicable' procedural standards under customary international law or general principles of civilized nations based on the recognition of such standards in 'x' number of treaties, or the representation of similar principles in 'y' number of states. The consequence of drawing on treaties and legal frameworks developed to apply to domestic settings has been to settle on a court-based adjudicatory model of due process for the sanctions context.

The problem with this approach is that the rationale has not been explained beyond the idea that such a model is consistent with principles that are 'binding' upon the Security Council or those implementing the Council's directives. It cannot be surprising that this formalist legal approach has failed to persuade the Security Council, a body that operates in a setting in which the relationship between law and politics is notoriously complex. Quite simply, the normative case in support of judicial review in the Security Council context has not yet been made. Why this fixation on courts? Are courts really necessary to instil fairness into sanctions decision-making about individuals? What are the due process values that court-based process promotes?

A second problem is that, when the issue is considered through a 'due process' rather than an international law lens, it becomes apparent that adoption of a 'one-size-fits-all' model of due process fundamentally misconceives the nature and role of due process. As any public lawyer will readily identify, while due process rights are recognized by most legal systems, this should not lead us into the error of thinking either that this makes the principles 'universal' or that they take the same shape or have the same scope in every legal system.[7] In the domestic literature on procedural fairness and due process, it is well established that due process is *contextual*: different legal contexts legitimately require different procedural standards and operate according to different principles and values.[8] The netting of similar principles from an

[6] By 'traditional international law source methodology', I refer to identification of international legal principles using the orthodox statement of sources in Article 38(1) of the Statute of the International Court of Justice (adopted 26 June 1945, entered into force 24 October 1945) 33 UNTS 993 ('ICJ Statute').

[7] Carol Harlow, 'Global Administrative Law: The Quest for Principles and Values' (2006) 17 *European Journal of International Law* 187, 204.

[8] One of the foremost comparative treatises on due process notes that 'there is no general blueprint to follow and the variety of approaches found in statutes is considerable': D J Galligan, *Due Process and Fair Procedures: A Study of Administrative Procedures* (Clarendon Press 1996), 315. For similar comments, see *Lloyd v McMahon* [1987] AC 625, per Lord Bridge; *R v Home Secretary ex p Doody* [1994] 1 AC 531, per Lord Mustill; *Morrissey v Brewer*, 408 US 471, 481 (1972); *Joint Anti-Fascist Refugee Committee v McGrath*, 341 US 123 (1951); Laurence Tribe, *American Constitutional Law* (2nd edn, Foundation Press 1988), 666.

array of different legal systems in accordance with traditional international law source methodology results in the artificial generation of a universally applicable set of norms that does not sit comfortably with the contextual nature of due process.

The aim in this book is to develop procedural principles for the international institutional context using a normatively rich rather than formalistic approach. In developing due process principles for the Security Council context, and indeed in tracing out the full implications of the assumption of 'governmental' powers of decision-making about individuals in the international sphere more generally, it is clear that a subtle shift in international law methodology is justified. Where the task is to construct a procedural framework for a new tier of governance, the focus should be less on formulating and systemizing rules reflected in existing state practice, and more on creating a new set of rules for an emerging governmental context. As Vaughan Lowe has recognized, the procedural principles that regulate international decision-making about individuals are more in the nature of 'interstitial' than primary norms, and should not be generated by the same processes as the traditional primary norms of international law.[9] Rather than working from practice to theory, the inverse is more appropriate, with the principal aim being to provide strong and enduring *theoretical* foundations to support institutional practice. To look at due process by identifying universally applicable standards is to look at the concept through the wrong end of the telescope. It is necessary first to consider which process value the procedure aims to advance, before determining the set of procedural standards that will enable its advancement. The metaphor of telescope is in itself misleading. Due process is more of a kaleidoscopic device that is able to advance a range of different process values through employment of a variety of procedural standards, which have the capacity to enhance the legitimacy of decision-making in different ways.

In Part I of the book, I undertake a review of the history of the targeted sanctions regime, identifying the key procedural reform proposals that have emerged in this context. I examine the formalist 'source-based' methodology underlying the development of these proposals, and the myths and misconceptions about procedural fairness that have arisen as a result of this methodology. In Part II, I lay out the foundations of a 'value-based' approach to the development of due process principles for the international legal order, with a focus on the Security Council sanctions context. Underlying the

[9] Vaughan Lowe, 'The Politics of Law-Making: Are the Method and Character of Norm Creation Changing?' in Michael Byers (ed), *The Role of Law in International Politics: Essays on International Relations and International Law* (Oxford University Press 2001), 219.

differences in form between procedural frameworks are different ideas as to the process values that a procedural framework can and should support in order to enhance the legitimacy of decision-making.[10] I begin with an analysis of the theoretical dimension of due process as a foundation for a better understanding in the international sphere as to the way or ways in which due process might enhance Security Council decision-making and indeed other forms of international decision-making. In doing so, I draw on the keystone Anglo-American jurisprudence of natural justice, procedural fairness and due process, from which the procedural law of many legal systems is derived.[11] My central finding is that the essence of due process is not to provide access to a court, but to provide access to 'dialogue'. Procedural standards aim to create a form of dialogue between decision-makers and the community to whom the decisions relate as a means of enhancing the legitimacy of the decision-making process. As might be expected, the choice of different procedural standards can affect the legitimacy of decision-making in different ways. In this part, I identify three value-based models of due process: (1) the instrumentalist model, which considers that the aim of procedures is to enhance the *accuracy* of decision-making outcomes; (2) the dignitarian model, which emphasizes the inherent value of procedures in preserving the *dignity* and *autonomy of the individual*; and (3) the public interest model, which regards due process as a mechanism to enhance *public accountability* in decision-making. Each of these models will be supported by different procedural frameworks. Moreover, closer analysis of each model reveals that they are based on different underlying theories about law and legal system and demonstrate a connection between due process and more deep-seated conceptions of community, law and values. It is clear there is an important connection between the debate about due process and the much bigger debate about structures for legitimation in the international legal order more generally. Development of a procedural framework for the Security Council sanctions regime is an exemplar of a construction site in global governance, with the capacity to entrench conceptions of law, community and values that could render the best-laid academic blueprints for the structure of the new world order obsolete.

Building from this theoretical analysis, it is interesting to assess how the various due process options would operate to ensure legitimacy of Security Council decision-making. As discussed, the primary option pushed by courts

[10] Jerry Mashaw, *Due Process in the Administrative State* (Yale University Press 1985), 42.

[11] Justice Frankfurter described due process as a principle that has evolved 'though centuries of Anglo-American constitutional history and civilization': *Joint Anti-Fascist Refugee Committee v McGrath*, 341 US 123 (1951).

and commentators is a court-based process or 'adjudicatory framework' for sanctions decision-making, under which a judicial organ would be vested with power to measure the legality of sanctions decision-making under international law ('internationalized judicial framework'). A second framework has emerged in a somewhat more organic fashion, with domestic and regional courts increasingly accepting challenges by individuals to sanctions decision-making and measuring the legality of sanctions decision-making under local law ('pluralist judicial framework'). In the meantime, the Security Council continues to experiment with a third non-court-based framework, with the latest initiative being the establishment of the Office of the UN Ombudsperson to protect the rights of individuals and entities listed under the sanctions regime ('Ombudsperson framework'). In Part III of the book, I compare and contrast these options. I draw conclusions about which model/framework is best equipped to enhance legitimacy in the Security Council context and in the international domain more generally, and resolve that it is not necessarily the court-based process favoured by most international lawyers.

PART I
PRACTICE

2

The Case for Procedural Reform

Due Process as Court Process

For over a decade, groups of states, scholars, committees, and courts have engaged in protracted criticism of the due process deficiencies in the implementation of targeted sanctions. The criticism has understandably resulted in a number of proposals for procedural reform. Early recognition of the need for procedural reform appeared in a series of multilateral reform initiatives,[1] which were followed up by other influential academic reports[2] and academic publications.[3] The sanctions regime has also been the subject

[1] [Interlaken process] Swiss Federal Office for Foreign Economic Affairs, 'Expert Seminar on Targeting UN Financial Sanctions' (Interlaken, 17–19 March 1998), 71; Swiss Federal Office for Federal Economic Affairs, 'Second UN Seminar on Targeting United Nations Financial Sanctions' (Interlaken, 29–31 March 1999), 211; [Bonn-Berlin process] Bonn International Centre for Conversion, 'Results of the Bonn-Berlin Process: Design and Implementation of Arms Embargoes and Travel and Aviation-Related Sanctions' (Bonn, 2001), 109; [Stockholm Process] Peter Wallensteen et al, 'Making Targeted Sanctions Effective: Results from the Stockholm Process on the Implementation of Targeted Sanctions' in *Results from the Stockholm Process on the Implementation of Targeted Sanctions* (Uppsala University, 2003), §43, §64.

[2] Thomas Biersteker and Sue Eckert, *Strengthening Targeted Sanctions through Fair and Clear Procedures* (Watson Institute Targeted Sanctions Project, Brown University 2006); Iain Cameron, *The European Convention on Human Rights, Due Process and United Nations Security Council Counter-terrorism Sanctions* (Committee of Legal Advisers on Public International Law 2006); Bardo Fassbender, *Targeted Sanctions and Due Process* (Council of Europe Committee of Legal Advisors on International Law 2006); International Commission of Jurists, *Assessing Damage, Urging Action: Report of the Eminent Jurists Panel on Terrorism, Counter-Terrorism and Human Rights* (2009); George Lopez et al, *Overdue Process: Protecting Human Rights while Sanctioning Alleged Terrorists* (Fourth Freedom Forum, Kroc Institute for International Peace Studies, University of Notre Dame 2009); Thomas Biersteker and Sue Eckert, *Addressing Challenges to Targeted Sanctions: An Update of the 'Watson Report'* (Watson Institute for International Studies 2009).

[3] Recent books and articles include: Ali Z Marossi and Marisa R Bassett (eds), *Economic Sanctions Under International Law: Unilateralism, Multilateralism, Legitimacy, and Consequences* (TMC Asser Press 2015); Bardo Fassbender (ed), *Securing Human Rights?: Achievements and Challenges of the UN Security Council* (Oxford University Press 2011); Guglielmo Verdirame, *The UN and Human Rights: Who Guards the Guardians?* (Cambridge University Press 2011); Lisa Ginsborg and Martin Scheinin, 'You Can't Always Get What You Want: The Kadi II Conundrum

of critical comment by the General Assembly,[4] the UN Secretary General,[5] the UN Special Rapporteur on the promotion and protection of human rights and fundamental freedoms while countering terrorism,[6] and the UN Human Rights Committee.[7] In addition, since the commencement of the sanctions regime, over 30 challenges have been lodged by individuals and entities in national and regional courts, including the European Court of Justice, the European Court of Human Rights, the Human Rights Committee, and domestic courts in Belgium, Canada, Italy, the Netherlands, Pakistan, Switzerland, Turkey, the United Kingdom, and the United States.[8] Certain courts have become increasingly sympathetic to claims regarding the procedural deficiencies of the Security Council sanctions regime, culminating in the European Court of Justice declaring the regulation implementing the relevant Security Council resolution into EU law to be invalid in the famous *Kadi* litigation.[9] Other courts have followed suit.[10]

The focus of the contemporary debate is on the difficulty of the individual to challenge his or her listing.[11] As the Watson Institute reported in 2009,

and the Security Council 1267 Terrorist Sanctions Regime' (2011) 8 *Essex Human Rights Review* 7; Grant L Willis, 'Security Council Targeted Sanctions, Due Process and the 1267 Ombudsperson' (2011) 42 *Georgetown Journal of International Law* 673; Jared Genser and Kate Barth, 'When Due Process Concerns Become Dangerous: The Security Council's 1267 Regime and the Need for Reform' (2010) 33 *Boston College International and Comparative Law Review* 1; Craig Forcese and Kent Roach, 'Limping into the Future: The UN 1267 Terrorism Listing Process at the Crossroads' (2010) 42 *George Washington International Law Review* 217.

[4] The World Summit Outcome document called upon the Security Council 'to improve its monitoring of the implementation and effects of sanctions, to ensure that sanctions are implemented in an accountable manner' and 'to ensure that fair and clear procedures exist for placing individuals and entities on sanctions lists and for removing them': UN General Assembly Resolution 60(1) (24 October 2005) UN Doc A/RES/60/1.

[5] 'Non-paper of the Secretary General' referred to in debate on 'Strengthening International Law: Rule of Law and Maintenance of Peace and Security' (22 June 2006) UN Doc S/PV.5474.

[6] Martin Scheinin, 'Report of the Special Rapporteur on the Promotion and Protection of Human Rights while Countering Terrorism' (16 August 2006) UN Doc A/61/267.

[7] UN Human Rights Committee, *Sayadi and Vinck v Belgium* (Communication No 1472/2006, 29 December 2008) UN Doc CCPR/C/94/D/1472/2006.

[8] 'Ninth report of the 1267 Monitoring Team' (13 May 2009) UN Doc S/2009/245, Annex 1; 'Eleventh Report of the 1267 Monitoring Team' (13 April 2011) UN Doc S/2011/245, Annex 1.

[9] Joined Cases C-402/05 P and C-415/05 P *Kadi v Council of the European Union* [2008] ECR I-0000.

[10] *Her Majesty's Treasury v Ahmed* [2010] UKSC 2 (UK Supreme Court); *Abousfian Abdelrazik v Minister of Foreign Affairs* 2009 FC 580 (Federal Court of Canada).

[11] The narrow focus of the debate about procedural deficiencies in sanctions decision-making is a worthy subject of further academic exploration, though this book is necessarily responsive to the current debate. For criticism of the narrowness of the debate, see Annalisa Ciampi, 'Security Council Targeted Sanctions and Human Rights' in Bardo Fassbender (ed), *Securing Human Rights?: Achievements and Challenges of the UN Security Council* (Oxford University Press 2011), 98–140; Larissa Van den Herik, 'Peripheral Hegemony in the Quest to Ensure Security Council

'[t]he rights to a fair trial and effective remedy lie at the heart of the debate on targeted sanctions violating individuals' human rights'.[12] In recent years, the debate has reached something of a stalemate between those pressing for reform to include a judicial (or quasi-judicial) review mechanism, and the Council's refusal to entertain such a prospect. Instead, the Security Council in 2009 created an Office of the UN Ombudsperson to hear individual challenges to sanctions listings. Faced with a stand-off between the Council and those asserting that the UN Ombudsperson does not meet due process standards, domestic courts have stepped into the breach, and have started assuming jurisdiction over challenges by individuals to sanctions decision-making. It is therefore possible to discern three structural frameworks around which debate about the structure of due process in sanctions decision-making currently revolves: (1) pressure for an international court-based mechanism; (2) the ad hoc practice of domestic court review; and (3) the Office of the UN Ombudsperson created by the Security Council. In this chapter, I provide a brief overview of the history of the evolution of a procedural framework for sanctions decision-making about individuals. I turn then to examine the focus of the contemporary debate on review mechanisms in sanctions decision-making, including the three emerging dominant structural frameworks for procedural reform.

A. Procedural History: Evolution of Listing, Notification, and De-listing Procedures

In the late 1990s, concern about the devastating humanitarian impact of blanket sanctions against states led to a strategic shift in policy by the Security Council.[13] Perhaps most infamously in the case of the sanctions regime against Iraq,[14] the impact of traditional sanctions had been felt—not by the targeted governments, but by their innocent and often oppressed civilian

Accountability for its Individualized UN Sanctions Regimes' (2014) 19(3) *Journal of Conflict and Security Law* 427, 435.

[12] Thomas Biersteker and Sue Eckert, *Addressing Challenges to Targeted Sanctions: An Update of the 'Watson Report'* (Watson Institute for International Studies 2009), 10.

[13] The shift in strategy was motivated by a consensus among states on the need to adopt a 'more prompt and effective response to present and future threats to international peace and security...designed so as to maximize the chance of inducing the target to comply with Security Council resolutions, while minimizing the negative effects of the sanctions on the civilian population': 'Report of the Secretary-General on the Work of the Organization' (30 August 2000) UN Doc A/55/1, 13 para 100.

[14] Save The Children, *Iraq Sanctions: Humanitarian Implications and Options for the Future* (2002), ch 2.

populations.[15] Beginning with the sanctions regime against the UNITA rebel movement in Angola, established in 1997, the Security Council decided to start targeting sanctions measures against relevant individuals, entities, and products, rather than in blanket fashion against states.[16] The traditional approach of imposing sanctions against states was replaced thereafter by the creation of a list (or perhaps more appropriately a blacklist) of individuals against which specific sanctions measures were imposed by Security Council resolutions. While there are currently sixteen targeted sanctions regimes in place,[17] I will focus in this overview on the listing, notification, and de-listing procedures developed under the regime that many consider to be the 'flagship' of the Council's sanctions practice (although it is in many ways atypical),[18] the Al Qaeda regime established by Security Council resolution 1267 (1999) ('the 1267 regime').[19]

The aim of the shift to targeted sanctions was to minimize harm against innocent civilian populations. However, perhaps because the transformation was motivated by a policy shift focused on outcomes, rather than a conscious institutional shift in decision-making focused on process, the transformation was not accompanied by any procedural reform to protect the rights of those included on sanctions blacklists. It is interesting to compare the assumption of direct decision-making authority over individuals in the sanctions context with the other major example in which this occurred, namely the Security Council's assumption of international criminal jurisdiction over individuals

[15] Press Release of UN Secretary General Kofi Annan, SG/SM/7625, 15 November 2000. See also Robin Geiss, 'Humanitarian Safeguards in Economic Sanctions Regimes: A Call for Automatic Suspension Clauses, Periodic Monitoring, and Follow-up Assessment of Long-Term Effects' (2005) 18 *Harvard Human Rights Journal* 167. See also Michael Reisman, 'Assessing the Lawfulness of Nonmilitary Enforcement: The Case of Economic Sanctions' (1995) 89 *American Society of International Law Proceedings* 350, 351.

[16] UN Security Council resolution 1127 (28 August 1997) UN Doc S/RES/1127; UN Security Council resolution 1173 (12 June 1998) UN Doc S/RES/1173. At the height of the UNITA sanctions regime, 157 individuals were subject to asset freezes and travel bans.

[17] The sixteen sanctions regimes impose sanctions against blacklisted individuals in relation to events in Somalia and Eritrea, Iraq, Liberia, the Democratic Republic of the Congo, Cote D'Ivoire, Sudan, Lebanon, North Korea, Iran, Libya, Guinea-Bissau, the Central African Republic, Yemen, South Sudan, and in the only two non-state-based regimes, against those associated with Al Qaeda and the Taliban.

[18] It is atypical because it is the only regime without a territorial connection directed predominantly at non-state actors and is presently the only sanctions regime to have access to the UN Ombudsperson.

[19] The 1267 regime was established by UN Security Council resolution 1267 (1999) and modified and strengthened by subsequent resolutions, chiefly 1333 (2000), 1390 (2002), 1526 (2004), 1617 (2006), 1735 (2006), 1822 (2008), and 1904 (2009). It was initially known as the Al Qaeda and Taliban sanctions regime, but in 2011 was divided into two separate sanctions regimes, dealing respectively with Al Qaeda and the Taliban: UN Security Council resolutions 1998 and 1999 (17 June 2011) UN Doc S/RES/1998 and UN Doc S/RES/1999.

with the establishment of the ad hoc international criminal tribunals for the former Yugoslavia and Rwanda. The contrast between the procedural initiatives accompanying the establishment of the sanctions regime, on the one hand, and the ad hoc tribunals, on the other, is striking. The UN Office of Legal Affairs and the UN International Law Commission, among other legal entities, were heavily involved in advising the Security Council over the legal and procedural framework for the Yugoslav and Rwanda Tribunals, and their establishment was accompanied by detailed rules of procedure and evidence.[20] By comparison, the targeted sanctions regime was established without reference to legal counsel at all.[21] When a member state suggested that the UN Office of Legal Counsel should be consulted, it was told that there were no legal issues involved in the listing or delisting of individuals on sanctions blacklists.[22] Though it is tempting to distinguish the international criminal tribunals in terms of their more burdensome impact on individuals, the consequence of being listed for an individual has been described as a form of 'civic death', and includes measures as serious as the comprehensive freezing of an individual's assets around the globe, and the denial of international travel, visas, and educational opportunities for the blacklisted individual and their families.[23]

Paradoxically, a shift in policy engineered to inject greater fairness into the sanctions regime has subsequently given rise to nearly two decades of debate about the lack of due process in sanctions decision-making.[24] In procedural

[20] Statute of the International Criminal Tribunal for the Former Yugoslavia, adopted in UN Security Council resolution 827 (25 May 1993) UN Doc S/RES/827; Statute of International Criminal Tribunal for Rwanda, adopted in UN Security Council resolution 955 (8 November 1994) UN Doc S/RES/955.

[21] Remarks by Hans Corell, 'Is the Security Council Bound by Human Rights Law' (103rd Annual Meeting of the American Society of International Law, Washington DC, 27 March 2009).

[22] Confidential communication, 24 January 2006; cited in Simon Chesterman, 'The Spy Who Came in from the Cold War: Intelligence and International Law' (2006) 27 *Michigan Journal of International Law* 1071, 1117.

[23] Individuals have spoken about being turned into 'international pariahs', with their reputations destroyed, and arguably placed in considerable danger because of the effect the listing process has had on immigration and asylum decisions: International Commission of Jurists, *Assessing Damage, Urging Action: Report of the Eminent Jurists Panel on Terrorism: Counter-terrorism and Human Rights* (2009), 117.

[24] Influential books and articles include Peter Gutherie, 'Security Council Sanctions and the Protection of Individual Rights' (2004) 60 *NYU Annual Survey of American Law* 491; Eric Rosand, 'The Security Council's Efforts to Monitor the Implementation of Al Qaeda/Taliban Sanctions' (2004) 98 *American Journal of International Law* 745; Iain Cameron, *The European Convention on Human Rights, Due Process and United Nations Security Council Counter-terrorism Sanctions* (Committee of Legal Advisers on Public International Law 2006); Andrea Bianchi, 'Assessing the Effectiveness of the UN Security Council's Anti-terrorism Measures: The Quest for Legitimacy and Cohesion' (2006) 17 *European Journal of International Law* 881; Larissa Van den Herik, 'The Security Council's Targeted Sanctions Regimes: In Need of Better Protection of the Individual' (2007) 20 *Leiden Journal of International Law* 797; Jeremy Farrall, *United Nations Sanctions and*

terms, the tale of the UN sanctions regime has been one of slow progress. For example, though the 1267 sanctions regime was established in 1999, it was not until 2002 that the 1267 Committee adopted guidelines for the conduct of its work. It was not until 2006 that the question as to how to address the fairness of the sanctions regime was said to have reached 'the top of the 1267 Committee's agenda'.[25] It was not until 2012 that the Monitoring Team was able to report, '[t]his is the first report of the Team in several years in which due process considerations have not taken centre stage'.[26] Though the Committee's Guidelines have been amended several times in response to continued pressure, there is still a sense that the procedural framework is woefully deficient, particularly in terms of listing, notification, reason-giving, and de-listing procedures.

(1) Listing Procedures

The Al Qaeda sanctions regime was established by UN Security Council resolution 1267 (1999) and subsequently modified and strengthened by resolutions 1333 (2000), 1390 (2002), 1526 (2004), 1617 (2006), 1735 (2006), 1822 (2008), 1904 (2009), 1989 (2011), 2083 (2012) and 2161 (2014). As described above, its primary focus is a list of individuals (the 'sanctions list') against which specific sanctions measures are imposed, including an assets freeze, travel ban, and arms embargo.[27] During the initial establishment of the 1267 regime, the creation of the list was self-confessedly based largely on 'political trust', with the 1267 Committee essentially rubber-stamping names submitted by member states.[28] Responsibility for listing lies clearly with members of the 1267 Committee; however, in practice there is little or no independent

the Rule of Law (Cambridge University Press 2007); Michael Bothe, 'Security Council's Targeted Sanctions against Presumed Terrorists: The Need to Comply with Human Rights Standards' (2008) 6 *Journal of International Criminal Justice* 541; Jared Genser and Kate Barth, 'When Due Process Concerns Become Dangerous: The Security Council's 1267 Regime and the Need for Reform' (2010) 33 *Boston College International and Comparative Law Review* 1; Craig Forcese and Kent Roach, 'Limping into the Future: The UN 1267 Terrorism Listing Process at the Crossroads' (2010) 42 *George Washington International Law Review* 217; Grant L Willis, 'Security Council Targeted Sanctions, Due Process and the 1267 Ombudsperson' (2011) 42 *Georgetown Journal of International Law* 673.

[25] Fifth Report of the Analytical Support and Sanctions Monitoring Team appointed pursuant to resolutions 1526 (2004) and 1617 (2005) (20 September 2006) UN Doc S/2006/750, para 2.

[26] Twelfth Report of the Analytical Support and Sanctions Monitoring Team, UN Doc S/2012/729, 1 October 2012, pp 4, 7.

[27] Since 2006, the assets freeze also applies to the provision of Internet hosting or related services: Security Council resolution 1735 (2006).

[28] Eric Rosand, 'The Security Council's Efforts to Monitor the Implementation of Al Qaeda/Taliban Sanctions' (2004) 98 *American Journal of International Law* 745.

evaluation of the evidence by the Committee.[29] Decisions are generally made on the papers according to a no-objection procedure, whereby a proposed name will be added to the list if no Committee Member objects to the listing within ten working days (it was initially restricted to five working days, though the period for consideration can still be shortened at the Committee Chair's discretion for emergency and time-sensitive listings).[30] While Committee members do not often object to the inclusion of names on the sanctions lists, states sometimes request for a hold to be placed on names, some of which have been maintained for years.[31] The United States has defended the limited reflection given to listing decisions by the Committee on the ground that it places every designation proposal through its domestic channels. However, this is self-confessedly not to check the adequacy of the evidentiary foundations for the listing, but rather to weigh intelligence equities into the balance to ensure that US security interests are not compromised by proposed listings.[32]

The arbitrariness of the listings stems largely from the fact that there is no real evidentiary burden that proposals must meet. For over five years, the only criteria for listing an individual, group, or entity was on the basis of an 'association with' Osama Bin Laden, Al Qaeda, or the Taliban. In the post-September 11 climate in 2001 and 2002, the US Office of Foreign Assets Control (the body responsible for the majority, if not all, early designations on the sanctions list)[33] is said to have been marked by 'chaos' and a 'rush to designate'.[34] An official in the US Treasury Department passed comment that 'we were so forward leaning we almost fell on our face', an express recognition that some of the evidentiary foundations for initial designations were extremely weak.[35] Faced with calls for more detailed criteria, the Chairman

[29] Iain Cameron, *The European Convention on Human Rights, Due Process and United Nations Security Council Counter-terrorism Sanctions* (Committee of Legal Advisers on Public International Law 2006), 5.

[30] Al Qaeda Committee, 'Guidelines of the Committee for the Conduct of its Work', available at <http://www.un.org/sc/committees/1267/pdf/1267_guidelines.pdf> ('1267 Committee Guidelines'), section 6(n).

[31] This situation has been ameliorated somewhat by sections 4(d) and (f) of the 1267 Committee Guidelines, which provide respectively that no matter may be left pending for longer than nine months, and that any holds placed by a Member will cease to have effect at the time its membership of the Committee ends (although this will not, of course, apply in relation to Permanent Members).

[32] Adam Szubin (US Office of Foreign Assets Control), 'The Terrorist Designation Process in the United States' (Countering Terrorism through Domestic and International Targeted Sanctions Conference, American University Washington College of Law, 15 September 2008).

[33] Eric Rosand, 'The Security Council's Efforts to Monitor the Implementation of Al Qaeda/Taliban Sanctions' (2004) 98 *American Journal of International Law* 745, 249.

[34] National Commission on Terrorist Attacks upon the United States, 'Staff Report to the Commission: Monograph on Terrorist Financing' (2004), 79, available at <http://www.9-11commission.gov/staff_statements/911_TerrFin_Ch5.pdf>.

[35] Ibid.

of the Sanctions Committee, Ambassador Alfonso Valdivieso of Colombia, responded, '[w]e are dealing with terrorism... It may be controversial, but all who have suggested alternatives are aware that we must not waste time on definitions, because the terrorists are acting'.[36]

The listing criteria were particularized more fully by the Security Council in 2005, but the effort to narrow the categories was undermined somewhat by the inclusion of a 'catch-all' provision extending the criteria to anyone 'otherwise supporting an affiliate' of Al Qaeda, Osama bin Laden, or the Taliban.[37] Designation requests have not as a rule been refused by the 1267 Committee on the basis of insufficient information. At the beginning of 2008, seventy-seven individuals lacked even the four basic identifiers considered necessary to allow effective implementation of the sanctions measures (that is, full name, date of birth, place of birth, and nationality).[38] The result of the overall approach has been to give 'an almost unlimited latitude of discretionary powers' to member states,[39] resulting in early listings being described by the Coordinator of the UN Al Qaeda and Taliban Monitoring Team as no more scientific than a 'back of an envelope thing'.[40] In 2012, the Monitoring Team recommended that the Committee provide further guidance on its criteria for listing beyond association, on the basis that this would not only be helpful guidance for submitting states, but would also help those listed understand what they had to do to be de-listed.[41]

The listing of individuals routinely takes place without providing those listed with the right to a hearing. States have complained that they have

[36] 'A Nation Challenged: Sanctions and Fallout', *New York Times*, 26 January 2002, A9.

[37] UN Security Council resolution 1617 (29 July 2005) UN Doc S/RES/1617; reaffirmed in UN Security Council resolution 1904 (17 December 2009) UN Doc S/RES/1904, paras 2 and 3 and UN Security Council resolution 2083 (2012), paras 2 and 3.

[38] During the review, only one of these names was removed for lack of sufficient information, while additional information was included in relation to thirty-eight entries: 'Report of Analytical Support and Sanctions Monitoring Team on Outcome of Comprehensive Review' (29 September 2010) UN Doc S/2010497, paras 10, 52.

[39] Francesco Francioni, 'The Right of Access to Justice to Challenge the Security Council's Targeted Sanctions: After-thoughts on "Kadi"' in Ulrich Fastenrath, *From Bilateralism to Community Interest: Essays in Honour of Judge Bruno Simma* (Oxford University Press 2011), 913.

[40] Richard Barrett, 'Fair and Clear Procedures in Sanctions Listing: An Insider Perspective', Countering Terrorism through Domestic and International Targeted Sanctions Conference, American University Washington College of Law, 15 September 2008. It is interesting that this 'rush to list' was repeated in the case of the Libyan sanctions regime established pursuant to resolution 1970 (2011) in relation to which 'Council members recall that there was little discussion of due process in deciding on the initial listings': Security Council Report, *Cross-cutting Report on the Rule of Law* (2011), 44.

[41] 'Thirteenth Report of the 1267 Monitoring Team' (31 December 2012) UN Doc S/2012/968, para 41.

limited influence over the decisions of the Committee, and states of residence and nationality have frequently complained that the Committee has added a name without consulting them, even though they are the ones most likely to know about the case.[42] Member states of the Security Council have defended the denial of the right to a hearing on grounds of the necessity of maintaining the confidentiality of the information upon which listings are based. They argue that the right of an individual to receive an oral hearing or to respond to the veracity and relevance of the underlying evidence must be denied on the basis that '[t]hose facts and that evidence, once classified as confidential or secret by the State which made the Sanctions Committee aware of them, are not, obviously, communicated to [the individual], any more than they are to the Member States of the United Nations to which the Security Council's resolutions are addressed'.[43]

(2) Notification Procedures

For almost a decade, there was no requirement for the Committee or member states to notify designated individuals or entities about their inclusion on the list, nor to provide them with reasons for the listing. The initial policy of non-notification was explained quite bluntly by one member of the Permanent Five: 'we don't notify terrorists; we kill them'.[44] Yet the policy has also been justified on the more defensible ground that it is necessary to maintain an element of surprise in listings in order to prevent the flight of individuals and assets, and the destruction of documents.[45] Provision has subsequently been made for retrospective notification of listings. In 2004, a voluntary request to notify individuals and entities of their listing was included in Security Council resolution 1526 (2004). Mandatory notification requirements were not introduced until 2008 with the passage of Security Council resolution 1822 (2008). This resolution provides that, in relation to future listings, the Secretariat must notify the state of nationality and residence of

[42] 'Report of Analytical Support and Sanctions Monitoring Team on Outcome of Comprehensive Review' (29 September 2010) UN Doc S/2010497, para 24.

[43] Joined Cases T-306/01 and T-315/01 *Kadi v Council and Commission* [2005] ECR II-0000, [276].

[44] Eric Rosand, 'Panel Discussion on UN Terrorist Designations and Sanctions: a Fair Process and Effective Regime?' (Center for Strategic and International Studies, Washington DC, 5 June 2008).

[45] In the early years of the targeted sanctions regime, there were instances when individuals and entities designated by sanctions committees reportedly found out about their listings from non-official sources: Thomas Biersteker and Sue Eckert, *Strengthening Targeted Sanctions through Fair and Clear Procedures* (Watson Institute Targeted Sanctions Project, Brown University 2006), 7.

listed individuals or entities within one week of the listing. Such member states must 'take all possible measures to notify or inform in a timely manner, the listed individual or entity of the designation and to include with this notification a copy of the publicly releasable portion of the statement of case'. The Committee itself only assumed a direct obligation to notify individuals in December 2009.[46]

(3) Reasons for Listing

States and other bodies continue to complain that listed parties have little opportunity to learn why they have been added to the list.[47] Denial of a right to receive a statement of reasons for placement on sanctions blacklists was initially justified by 'the need to protect secret intelligence material and sources, from which the names are usually derived, and by the desire to include suspected bad actors on the list as soon as possible'.[48] It was not until 2005 that the Security Council required states to provide a statement of case describing the basis of any listing proposal.[49] In 2006 and 2012, the Council provided further particulars, noting that the statement of case 'should provide as much detail as possible on the basis(es) for the listing, including: (i) specific findings demonstrating the association or activities alleged; (ii) the nature of the information [eg intelligence, law enforcement, judicial, media, admissions by subject, etc]; and (iii) supporting evidence or documents that can be supplied'. The failure to use mandatory language demonstrates that, though highly desirable, the information is not required for the purposes of listing.[50] In 2006, states were requested to identify what portions of the statement of case could be publicly released for the purpose of notifying the individual or entity subject to the listing, which strengthened into a requirement in 2008.[51] Notably, none of these obligations were retrospective. Indeed, by 2008, most of the entries on the list still did not state the reason for their inclusion, with

[46] UN Security Council resolution 1904 (17 December 2009) UN Doc S/RES/1904, Annex, para 15(b); 1267 Committee Guidelines, section 6(o).

[47] 'Ninth Report of the Analytical Support and Sanctions Monitoring Team Submitted Pursuant to Resolution 1822 (2008) Concerning Al-Qaida and the Taliban and Associated Individuals and Entities' (13 May 2009) UN Doc S/2009/245, para 47.

[48] Eric Rosand, 'The Security Council's Efforts to Monitor the Implementation of Al Qaeda/Taliban Sanctions' (2004) 98 *American Journal of International Law* 745, 749.

[49] UN Security Council resolution 1617 (29 July 2005) UN Doc S/RES/1617, para 4.

[50] Annalisa Ciampi, 'Security Council Targeted Sanctions and Human Rights' in Bardo Fassbender (ed), *Securing Human Rights?: Achievements and Challenges of the UN Security Council* (Oxford University Press 2011), 107.

[51] UN Security Council resolution 1735 (22 December 2006), UN Doc S/RES/1735, para 6; Security Council resolution 1822 (30 June 2008) UN Doc S/RES/1822, para 12.

over 400 names on the list lacking a full statement of case.[52] The list at that time was described by the Monitoring Team as 'dated and increasingly in need of a thorough overhaul'.[53] Sixteen individuals were known to be dead and, as described above, seventy-seven entries had fewer than four identifiers (that is, full name, date of birth, place of birth, and nationality), considered by the Team to be the minimum necessary to allow effective implementation of the sanctions measures.

In 2008, the Security Council called on the Sanctions Committee to carry out a 'Comprehensive Review' of the sanctions lists in order 'to ensure the Consolidated List is as updated and accurate as possible and to confirm the listing remains appropriate'.[54] This review, completed in July 2010, led to what was described as a 'much-needed paring down and updating of the sanctions list', including the de-listing of twenty-four individuals and twenty-one entities.[55] It proposed updates to 391 of the 488 entries then on the list, including changes to 217 narrative summaries of reasons for listing.[56] Narrative summaries of reasons have now been completed for all list entries, though these summaries have been criticized for containing less than obvious justification for an initial listing, let alone for its continuation.[57] Even after the Comprehensive Review in 2010, the Monitoring Team reflected with regret that many of the least revealing narrative summaries remained unchanged.[58]

Under Security Council resolutions 1822 (2008) and 2083 (2012), the Sanctions Committee is now mandated to conduct an annual review of those listings that have not been the subject of examination within a period of three or more years. The Sanctions Committee also conducts periodic reviews of the names of individuals reported to be deceased and of entities reported to be defunct. A similar review of entries lacking identifiers necessary to ensure effective implementation was mandated under Security Council resolutions 1904 (2009) and 2083 (2012). However, as the Monitoring Team noted, the outcome of all these reviews is highly dependent on the view of the designating state.[59] A recommendation by the Monitoring Team to implement

[52] 'Report of Analytical Support and Sanctions Monitoring Team on Outcome of Comprehensive Review' (29 September 2010) UN Doc S/2010497, para 19; Security Council Report, 'Counter-terrorism: Al Qaida and Taliban' (Update Report No 9, 26 June 2008).

[53] 'Report of Analytical Support and Sanctions Monitoring Team on Outcome of Comprehensive Review' (29 September 2010) UN Doc S/2010497, para 10.

[54] Security Council resolution 1822 (30 June 2008) UN Doc S/RES/1822, para 25.

[55] 'Report of Analytical Support and Sanctions Monitoring Team on Outcome of Comprehensive Review' (29 September 2010) UN Doc S/2010497, para 43.

[56] Ibid. [57] Ibid, para 57. [58] Ibid.

[59] 'Fourteenth Report of the Analytical Support and Sanctions Monitoring Team' (2 August 2013) UN Doc S/2013/467, para 24.

automatic de-listing where the designating state fails to argue for continued listing or provide detailed reasons for the listing was rejected by the Sanctions Committee.

(4) De-listing Procedure

The de-listing procedure has proved to be the most contentious issue affecting the UN sanctions regime, marked by a great deal of reluctance to implement reform on the part of the Permanent Five. Individuals and entities historically had no direct access to the Sanctions Committee to challenge their listings. It was initially the case that listed individuals and entities were dependent on the agreement of their state of nationality or residence to take up their case on their behalf, usually through diplomatic negotiations with the designating state. Sweden took up the case of three Swedish-Somali nationals listed in 2001, one of whom (Mr Abdirisak Aden) was an election candidate of the governing Social Democratic Party. The situation generated widespread publicity: prominent Swedes defied sanctions regulations by taking up a collection for their legal fees, and the editor of Sweden's premier business daily gave Mr Aden a job.[60] Sweden finally succeeded in negotiating their removal from the sanctions list in August 2002.[61] Those individuals unfortunate to come from countries who were either unwilling or unable to take up their case, or with whom Council members were unwilling to negotiate, had no access to the Sanctions Committee. It was said of the 1267 sanctions regime in 2008 that listed individuals 'stand even less chance of escaping this fate than they do of being released from Guantánamo Bay'.[62]

It is only belatedly that the Committee has introduced formal intermediaries through which individuals can petition the Committee. In December 2006, the Security Council established the 'focal point', an initiative that was intended to serve as a non-state-based forum to which designated individuals or entities could submit a request for de-listing.[63] This development was heralded by members of the Security Council as 'considerable progress in

[60] Serge Schmemann, 'A Nation Challenged: Sanctions and Fallout', *New York Times*, 26 January 2002, A9.

[61] 'US Drops Names of 2 Swedes from Al Qaeda List at UN', *New York Times*, 23 August 2002, A8.

[62] 'UN Law: Coming up Trumps', *The Economist* (31 January 2008).

[63] Some states such as France designated the focal point as the exclusive point of contact for their nationals and residents, declining to receive requests directly and requiring that all requests should go through the focal point.

terms of equity and transparency' (France); 'the best possible result at present' (Denmark); and a measure brought forward 'in the defence of human rights and in raising awareness of the need to fight terrorism from the point of view of respect for law and human rights' (Argentina).[64] However, in practice, the focal point turned out to be little more than a glorified postbox. Upon receiving a de-listing request from an individual, the focal point sent it to the designating state, the state of residence, and the state of nationality. It was only if one of those states supported the request that it was placed on the Committee's agenda.[65] Where a de-listing request was placed on the Committee's agenda, the focal point was entitled to attend deliberations, though played no role representing the individual's interests in the course of the decision-making. At the conclusion of the deliberations, the focal point would send a standard letter to the individual, setting out the result of their de-listing request, though not the reasons for the decision. The impact of the focal point on the due process rights of individuals was minimal. Between March 2007, when the focal point became operative, and 2009, when the Office of the UN Ombudsperson was established, the focal point received twenty-five de-listing requests relevant to the 1267 Committee (relating to eighteen individuals and twenty-two entities), leading to the removal of three individuals and seventeen entities.[66] Though the focal point was originally intended to be a full-time position, the initial holder of the post spent only one third of her time on focal point issues, and was subsequently appointed to an additional position serving two sanctions committees in the Department of Political Affairs in the UN Secretariat. Growing perceptions about the inadequacy of the focal point as a procedural safeguard for individuals led the Council to establish the Office of the UN Ombudsperson in 2009. The Ombudsperson has initially been established to serve only the 1267 regime. Calls to extend the Ombudsperson to the other sanctions regime have so far been rejected.[67] As the strongest

[64] 'Security Council Adopts Measures to Ensure Clear and Fair Procedures Exist for Delisting from Sanctions Committees' (19 December 2006) UN Press Release No SC/8913.

[65] Notably, in the event of conflicting opinions, such that one of these states supports and another objects to the request, the objection takes precedence. This has happened on at least three occasions. On the first occasion conflicting opinions arose, where the objecting state was a member of the Committee, the issue was subject to discussion within the Committee, and a precedent was set. The same result would likely follow even if the objecting state was not a member of the Committee.

[66] Statistics available at <http://www.un.org/sc/committees/dfp.shtml>.

[67] Consider, for example, the repeated rejection of proposals by the Group of Likeminded States and others to extend the Ombudsperson's mandate: Concept Paper on Security Council Working Methods (8 October 2014) UN Doc S/2014/725, discussed at the debate on Security Council Working Methods, 7285th meeting of Security Council (23 October 2014) UN Doc S/PV.7285; (10 May 2013) UN Doc S/PV.6964; (9 November 2012) UN Doc S/2012/805.

procedural measure introduced by the Security Council, it is examined in further detail later in this chapter.

B. The Contemporary Reform Agenda: A Court-Centred Conception of Procedural Justice

The evolution of listing, notification, and de-listing procedures evidences that important steps have been taken to improve the procedural framework accompanying Security Council decision-making. However, widespread dissatisfaction remains. There are continuing perceptions that the capacity of an individual to challenge his or her listing is inadequate. As discussed, the central focus of the contemporary debate has been on the choice of forum through which individuals might challenge sanctions decision-making and seek de-listing. The reform debate is currently weighted heavily in favour of a judicial or adjudicatory procedure for review of sanctions decision-making. While the structure of the body differs, the proposals are largely united in their agreement on the fact that such a tribunal would (1) have to be impartial and independent of the Security Council and (2) have the power to issue binding decisions. Those favouring an adjudicatory procedure can be divided loosely into two camps, with some favouring the adoption of an internationalized organ (reviewing sanctions decision-making under international law), and others advocating a decentralized arbitral option (under which domestic courts would assume jurisdiction to review sanctions decision-making under domestic law). Against the tide of the broader debate, the Security Council has established the Office of the UN Ombudsperson, a non-judicial alternative for the protection of the rights of individuals affected by sanctions decision-making. The contemporary debate can therefore be seen as divided between three competing procedural frameworks through which individuals might challenge sanctions decisions: (1) an internationalized judicial procedure; (2) a pluralist domestic-court-based judicial procedure; and (3) the non-adjudicatory Ombudsperson procedure.

(1) Internationalized Judicial Framework

The majority of reform proposals continue to push for a centralized arbitral body to be established within the UN with the power to compel de-listing through the issuing of binding decisions. Proponents of such a framework include academics,[68] courts,[69] and even one state legislature.[70]

Representative of the academic viewpoint is Bardo Fassbender's report, which concludes that an essential element of any procedural framework is 'the right of [a listed] person or entity to an effective remedy against an individual measure before an impartial institution or body previously established'.[71] While he does not specifically require the arbitral institution to be a judicial body, he states clearly his view that the remedy would not be 'effective' unless the relevant body had the power to issue binding decisions.[72] Both the UN High Commissioner for Human Rights[73] and the UN Special Rapporteur on

[68] Grant L Willis, 'Security Council Targeted Sanctions, Due Process and the 1267 Ombudsperson' (2011) 42 *Georgetown Journal of International Law* 673, 737, 743–5 ('effective due process relies on three principal concerns: the independence of the decision-maker, accessibility of the decision-maker to the individual, and the power of the decision-maker to grant effective relief... [T]he fact that listed individuals still do not have the right...to judicial review...leaves the regime not in compliance with commonly accepted international due process standards'); Jared Genser and Kate Barth, 'When Due Process Concerns Become Dangerous: The Security Council's 1267 Regime and the Need for Reform' (2010) 33 *Boston College International and Comparative Law Review* 1, 37 ('[a]n independent tribunal composed of security-savvy judges and selected by the UNSC, which has the power to hear target complaints and issue binding delisting decisions on the Sanctions Committee, constitutes the mechanism best-suited to the task'); Craig Forcese and Kent Roach, 'Limping into the Future: The UN 1267 Terrorism Listing Process at the Crossroads' (2010) 42 *George Washington International Law Review* 217, 265, 275 (criticizing the failure on the part of the UN 'to provide a judicial remedy for those who have been wrongly listed'); Johannes Reich, 'Due Process and Sanctions Targeted Against Individuals Pursuant to Resolution 1267 (1999)' (2008) 33 *Yale Journal of International Law* 505, 510–11; Larissa Van den Herik, 'The Security Council's Targeted Sanctions Regimes: In Need of Better Protection of the Individual' (2007) 20 *Leiden Journal of International Law* 797, 806–7; Iain Cameron, 'UN Targeted Sanctions, Legal Safeguards and the European Convention on Human Rights' (2003) 72 *Nordic Journal of International Law* 159, 210 ('to satisfy the human rights objections, the decision of the arbitral body would have to be binding on the sanctions committee'); John Dugard, 'Judicial Review of Sanctions' in V Gowlland-Debbas (ed), *United Nations Sanctions and International Law* (Kluwer Law International 2001).

[69] Case T-85/09 *Kadi v European Commission* [2010] EUECJ, [128]; *Her Majesty's Treasury v Ahmed* [2010] UKSC 2 (UK Supreme Court).

[70] Rapport de la Commission de Politique Extérieure du Parlement Suisse, 'Les Fondements de Notre Ordre Juridique Court-Circuités par l'ONU' (2010), translated in Council of Europe Parliamentary Assembly, Committee of Legal Affairs and Human Rights, 'Compatibility of UN Security Council and EU [terrorist] Black Lists with European Convention on Human Rights requirements' (7 December 2010).

[71] Bardo Fassbender, *Targeted Sanctions and Due Process* (Council of Europe Committee of Legal Advisors on International Law 2006), 30.

[72] Ibid, 31, para 12.12.

[73] 'Report of the United Nations High Commissioner for Human Rights on the Protection of Human Rights and Fundamental Freedoms while Countering Terrorism' (15 December 2010) UN Doc A/HRC/16/50, para 21.

human rights and counter-terrorism[74] have determined that the 1267 sanctions regime falls short of internationally recognized standards of due process because there is no recourse to independent judicial or quasi-judicial review either of a decision to list or denial of a request to de-list. The International Commission of Jurists has also advocated the 'introduction of a genuinely independent judicial or quasi-judicial complaint mechanism' into the UN sanctions regime.[75]

Those domestic and regional courts that have upheld challenges to the sanctions regime have largely done so on account of the absence of a judicial review mechanism. In *Kadi II*, the CJEU held that 'the improvements' (of which the Ombudsperson procedure was the most notable) do not provide the listed person with 'the guarantee of effective judicial procedure',[76] affirming views expressed by the European Court of Human Rights and the EU General Court that the Ombudsperson 'cannot be equated with the provision of an effective judicial procedure for review of decisions of the Sanctions Committee'.[77] The UK Supreme Court also explicitly considered whether the establishment of the Ombudsperson's office remedied due process concerns, but held that '[w]hile these improvements are to be welcomed, the fact remains that there was not when the designations were made, and still is not, any effective judicial remedy'.[78] Some domestic and regional courts have themselves purported to provide a form of judicial review, informally instating themselves as 'agents' of the international community and assessing the legality of sanctions decision-making under international law.[79] For example, in *Kadi*, the Court of First Instance assumed a power to review the validity of the Council's decision to list Mr Kadi under international law. This case

[74] 'Report of the Special Rapporteur on the Promotion and Protection of Human Rights and Fundamental Freedoms while Countering Terrorism' (2 August 2010) UN Doc A/65/258, paras 56–7.

[75] International Commission of Jurists, *Assessing Damage, Urging Action: Report of the Eminent Jurists Panel on Terrorism, Counter-Terrorism and Human Rights* (2009).

[76] *European Commission v Kadi* [2013] ECR not yet reported, [133].

[77] Case T-85/09 *Kadi v European Commission* [2010] EUECJ, [28]. See also *Nada v Switzerland* [2012] ECHR 169, [209]–[214]; *Al-Dulimi and Montana v Switzerland* (App No 5809/08), ECHR, 26 November 2013, [119]; Case T-85/09 *Kadi v European Commission* [2010] EUECJ, [128].

[78] *Her Majesty's Treasury v Ahmed* [2010] UKSC 2, [413]–[414].

[79] *Her Majesty's Treasury v Ahmed* [2010] UKSC 2 per Lord Philips and Lord Rodger; *Abousfian Abdelrazik v Minister of Foreign Affairs* 2009 FC 580; Joined Cases T-306/01 and T-315/01 *Kadi v Council and Commission* [2005] ECR II-0000; *Nada v SECO* 133 BGE II 450 (2007). Such a role has been envisaged more generally for national courts by, among others, Hersch Lauterpacht, 'Decisions of Municipal Courts as a Source of International Law' (1929) 10 *British Yearbook of International Law* 65, 67, 93; Richard Falk, *The Role of Domestic Courts in the International Legal Order* (Syracuse University Press 1964), 4, xii; Antonio Cassese, 'Remarks on Scelle's Theory of "Role-Splitting" (déboulement fonctionnel) in International Law' (1990) 1 *European Journal of International Law* 210, 228, 230–1.

is an interesting concrete example by which to assess the feasibility of an international court-based solution, and will be examined in further details in future chapters.

For the sake of completeness, it should be noted that courts have not been the only domestic organs to emphasize the need for a court-based process in sanctions decision-making. In March 2010, the Swiss Parliament adopted a motion declaring that it intended to refuse to apply sanctions to any targeted individual who had not been given judicial access after three years, was unable to appeal his or her listing before a judicial body, and had not had any further accusations made against him or her.[80]

(2) Pluralist Judicial Framework

The alternative proposal is for a decentralized court-based procedure under which domestic courts would exercise judicial review over sanctions decision-making under domestic law. A number of academics have made such a proposal.[81] However, support for such a framework has also emerged more organically, with a number of domestic courts agreeing to hear challenges from individuals caught up in the sanctions net. In the face of continued reluctance on the part of the Security Council to institute what is considered 'adequate' reform, these courts have become increasingly sympathetic to claims from individuals regarding the procedural deficiencies of the Security Council sanctions regime.

Two examples will be used to illustrate this trend. The European Court of Justice (ECJ) took a different approach from the Court of First Instance in the *Kadi* litigation discussed above, assuming jurisdiction to engage in full review of the legality of Community acts *under European*

[80] Rapport de la Commission de Politique Extérieure du Parlement Suisse, 'Les Fondements de Notre Ordre Juridique Court-Circuités par l'ONU' (2010), translated in Council of Europe Parliamentary Assembly, Committee of Legal Affairs and Human Rights, 'Compatibility of UN Security Council and EU [terrorist] Black Lists with European Convention on Human Rights requirements' (7 December 2010), 6.

[81] David Cortright and Erika de Wet, *Human Rights Standards for Targeted Sanctions* (Sanctions and Security Research Program 2010), 8; Jean L Cohen, 'A Global State of Emergency or the Further Constitutionalization of International Law: A Pluralist Approach' (2008) 15 *Constellations* 456, 474; Jessica Almqvist, 'A Human Rights Critique of European Judicial Review: Counter-Terrorism Sanctions' (2008) 57 *International and Comparative Law Quarterly* 303; Peter Gutherie, 'Security Council Sanctions and the Protection of Individual Rights' (2004) 60 *NYU Annual Survey of American Law* 491, 535. See also European Parliamentary Committee on Legal Affairs and Human Rights, *UN Security Council and European Union Blacklists* (2007), add, in which the Special Rapporteur on Legal Affairs and Human Rights for the Parliamentary Assembly of the Council of Europe expresses his hope that 'national courts could actually compel the UN authorities to improve their procedures and so help to increase the legitimacy of these lists'.

law, regardless of the origin of those acts in a Security Council resolution. The ECJ in *Kadi* thereby recognized its power to review the regulation implementing UN sanctions into European law for consistency with the fundamental rights of the European legal order. I refer to the European Courts as 'domestic' courts in this book because they position themselves in the *Kadi* litigation as courts of a 'quasi-domestic' legal order autonomous from the international legal order.[82] The approach of the ECJ was subsequently followed by the General Court (formerly the Court of First Instance) and the CJEU in *Kadi II*.[83] In *HM Treasury v Ahmed*, a majority of the UK Supreme Court similarly recognized that UK courts were entitled to engage in judicial scrutiny of Security Council actions as to whether they were incompatible with the basic premises of the UK legal order.[84] These cases shall be returned to in subsequent chapters as valuable test cases for the capacity of domestic courts to enhance the legitimacy of Security Council decision-making by engaging in review of Council decisions against domestic legal standards.

(3) International Non-Judicial Framework

The call to incorporate judicial review into UN sanctions decision-making has so far gone unheeded by the Security Council. The five Permanent Members of the Security Council remain staunchly opposed to any form of judicial or independent oversight on the basis that their own submissions have been carefully checked and reviewed by their respective governments.[85] The underlying issue is the reluctance on the part of these governments to allow foreign nationals, however well qualified, access to their secret information and confidential sources, particularly if this entails examination of the likely veracity of these sources.[86] One US official has indicated that the

[82] André Nollkaemper, 'The European Courts and the Security Council: Between Dédoublement Fonctionnel and Balancing of Values: Three Replies to Pasquale De Sena and Maria Chiara Vitucci' (2009) 20 *European Journal of International Law* 853, 863.

[83] Case T-85/09 *Kadi v European Commission* [2010] EUECJ.

[84] *Her Majesty's Treasury v Ahmed* [2010] UKSC 2, per Lord Hope, Lord Phillips, Lord Rodger, and Lord Mance.

[85] Security Council Report, 'Counter-terrorism: Al Qaida and Taliban' (Update Report No 9, 26 June 2008), 2.

[86] 'Tenth Report of the Analytical Support and Sanctions Monitoring Team Appointed Pursuant to Security Council Resolutions 1526 (2004) and 1617 (2005) Concerning Al-Qaida and the Taliban and Associated Individuals and Entities' (2 October 2009) UN Doc S/2009/502. The Ombudsperson has herself acknowledged that one of her biggest challenges will be 'access to classified or confidential information' possessed by states: 'Report of the Office of the Ombudsperson Pursuant to Security Council Resolution 1904' (24 January 2011) UN Doc S/2011/29, paras 33–5.

United States would stop submitting names to the Council in the event that full review was instituted.[87]

While the Council has remained impervious to calls for a judicial or quasi-judicial procedure, it continues nevertheless to respond to criticism of the UN sanctions regime. In December 2009, the Security Council adopted resolution 1904 (2009), establishing the Office of the Ombudsperson. The first office-holder, Judge Kimberly Prost, assumed the position on 3 June 2010.[88] Given the reluctance of the Security Council to institute procedural reform, the creation of the office was described as a 'miracle', though was also recognized as 'crucial to the survival' of the targeted sanctions regime.[89] The development was self-confessedly a response to the 'challenges, both legal and otherwise, to the measures implemented by Member States'.[90] Resolution 1904 and its annexes are directed at 'improving the gathering of relevant information pertaining to listings, expanding the flow of information between the Sanctions Committee and listed persons and entities, and ensuring that requests for delisting are more fully considered by the Sanctions Committee'.[91]

The Ombudsperson is not an adjudicatory process, but instead fulfils something of a fact-finding or inquisitorial role. Upon receiving a de-listing request, the Ombudsperson engages in a four-month period of information gathering (extendable for a further two months if necessary), during which she transmits the delisting request to relevant parties including the Committee, the designating state, and the state of nationality, seeking any additional information relevant to the request. In addition, the request is forwarded to the Monitoring Team, which is required to assemble a portfolio of relevant information drawn from court decisions, news reports, and fact-based assessments of the information provided by the petitioner. The Ombudsperson then engages in a two-month period of engagement, which may include a dialogue with the petitioner, to gain any additional information that may help the Committee with its decision. At the end of this process, the

[87] Adam Szubin (US Office of Foreign Assets Control), 'The Terrorist Designation Process in the United States' (Countering Terrorism through Domestic and International Targeted Sanctions Conference, American University Washington College of Law, 15 September 2008). See also 'Eighth Report of the Analytical Support and Sanctions Monitoring Team' (14 May 2008) UN Doc S/2008/324, 17 ('It is difficult to imagine that the Security Council could accept any review panel that appeared to erode its absolute authority to take action on matters affecting international peace and security, as enshrined in the Charter').

[88] UN Security Council resolution 1904 (17 December 2009) UN Doc S/RES/1904.

[89] US Embassy Confidential Cable, 'Commission Terrorism Sanctions Legal Challenges', 10USEUBRUSSELS 212 (24 February 2010).

[90] UN Security Council resolution 1904 (17 December 2009) UN Doc S/RES/1904, Preamble.

[91] David Cortright and Erika de Wet, *Human Rights Standards for Targeted Sanctions* (Sanctions and Security Research Program 2010), para 10.

Ombudsperson prepares a comprehensive report outlining the information she has obtained, and laying out the principal arguments concerning the delisting request. The Committee has thirty days to consider this report. The report is to be presented in person to the Committee, during which the Ombudsperson can answer any questions the Committee members may have regarding the request. If the Ombudsperson recommends de-listing, the individual or entity will be removed from the list unless, within sixty days, the Committee decides by consensus to maintain the listing. Under procedures introduced in resolutions 2083 (2012) and 2161 (2014), the Committee is now required to provide reasons for rejecting the request and provide an updated narrative summary of reasons for listing for the Ombudsperson to transmit to the petitioner within sixty days. The Ombudsperson must send the petitioner a letter within fifteen days communicating the Committee's decision, the reasons for its decision, and describing to the extent possible the process and publicly releasable factual information gathered by the Ombudsperson.

Under the first office-holder, the Ombudsperson procedure has produced results. The Ombudsperson has received fifty-five cases (of which forty-six have been concluded), leading to the de-listing of thirty-four individuals and twenty-seven entities. Three individuals were de-listed prior to the conclusion of the Ombudsperson process, though resort to the Ombudsperson process was a trigger for their de-listing. During this time, the de-listing of six individuals has been denied, though Ms Prost believes that in each of these cases, the individual received a fair process.[92] Though initially appointed for an eighteen-month period, Ms Prost's term has been renewed twice, most recently in January 2013 for a period of thirty months. Her term of office concludes in July 2015.

While the Ombudsperson procedure has received strands of support in the reform debate,[93] the assessment of the Office from a due process perspective has been mainly critical.[94] The main concern is that the Ombudsperson is

[92] Presentation by Kimberly Prost, 'Debating Kadi II', LSE/Matrix Seminar Series, 13 February 2014 (London School of Economics).

[93] Opinion of Advocate General Bot, *European Commission v Kadi* (19 March 2013), [81]–[87].

[94] [Cases] *European Commission and Council v Kadi* [2013] ECR not yet reported, [133]; *Al-Dulimi and Montana v Switzerland* (App No 5809/08), 26 November 2013, [119]; *Nada v Switzerland* [2012] ECHR 1691, [209]–[214]; *Her Majesty's Treasury v Ahmed* [2010] UKSC 2, [145]–[156] per Lord Phillips; [184]–[185] per Lord Rodger; [246] per Lord Mance; [Reports] Martin Scheinin, 'Report of the Special Rapporteur on the Promotion and Protection of Human Rights and Fundamental Freedoms while Countering Terrorism' (6 August 2010) UN Doc A/65/258, paras 56–7; David Cortright and Erika de Wet, *Human Rights Standards for Targeted Sanctions* (Sanctions and Security Research Program 2010), 10; Ben Emmerson QC, 'Report of UN Special Rapporteur on the Promotion and Protection of Human Rights while Countering Terrorism' (26 September 2012) UN Doc A/67/396, paras 14, 201; [Articles] Vanessa Arslanian, 'Great Accountability Should Accompany Great Power: The ECJ and the UN Security Council in Kadi I

'not a court', which seems to be short-hand for the dual criticism that the Ombudsperson is not sufficiently independent of the Security Council, and that it has no power to bind the Sanctions Committee. The Chair of the 1267 Committee acknowledged in August 2010 that de-listing decisions remain 'a political decision based on a political process'.[95] The CJEU held in *Kadi II* that 'the improvements' (of which the Ombudsperson procedure was the most notable) do not provide the listed person with 'the guarantee of effective judicial procedure',[96] affirming views expressed by the European Court of Human Rights and the EU General Court that the Ombudsperson 'cannot be equated with the provision of an effective judicial procedure for review of decisions of the Sanctions Committee'.[97] The UK Supreme Court in *Ahmed* found that, while the establishment of the Office of the Ombudsperson was to be welcomed, 'the fact remains that there... still is not any effective judicial remedy'.[98] In September 2012, Ben Emmerson, the Special Rapporteur on the protection of fundamental rights while countering terrorism, condemned the targeted sanctions regime as a 'ready means by which individual States can make executive decisions with far-reaching consequences, apparently unconstrained by domestic judicial review, or the international human rights treaties by which they are bound'.[99] It is clear that the perception in many quarters is that the procedural framework for sanctions decision-making remains deficient because of its non-judicial nature.

and II' (2012) 35 *Boston College International and Comparative Law Review* 1, 11, 15; Fremuth, 'Ein Prozess...: Zum Ausgleich zwischen der effektiven Bekämpfung des Terrorismus und der Beachtung der Menschenrechte in der Sanktionspraxis des Sicherheitsrates' [2012] *Die öffentliche Verwaltung* 81, 87 ff; Lisa Ginsborg and Martin Scheinin, 'You Can't Always Get What You Want: The Kadi II Conundrum and the Security Council 1267 Terrorist Sanctions Regime' (2011) 8 *Essex Human Rights Review* 7, 11–12, 19; Grant L Willis, 'Security Council Targeted Sanctions, Due Process and the 1267 Ombudsperson' (2011) 42 *Georgetown Journal of International Law* 673, 745; Craig Forcese and Kent Roach, 'Limping into the Future: The UN 1267 Terrorism Listing Process at the Crossroads' (2010) 42 *George Washington International Law Review* 217, 219, 264–5, 275; Jared Genser and Kate Barth, 'When Due Process Concerns Become Dangerous: The Security Council's 1267 Regime and the Need for Reform' (2010) 33 *Boston College International and Comparative Law Review* 1, 26, 41; Adele Kirschner, 'Security Council Resolution 1904 (2009): A Significant Step in the Evolution of the Al Qaida and Taliban Sanctions Regime?' (2010) 70 ZaöRV 585, 602, 604–5.

[95] UN Department of Public Information, 'Press Conference on Security Council Al-Qaida and Taliban Sanctions Committee' (2 August 2010).

[96] *European Commission v Kadi* [2013] ECR not yet reported, [133].

[97] Case T-85/09 *Kadi v European Commission* [2010] EUECJ, [28]. See also *Nada v Switzerland* [2012] ECHR 169, [209]–[214]; *Al-Dulimi and Montana v Switzerland* (App No 5809/08), 26 November 2013, [119].

[98] *Her Majesty's Treasury v Ahmed* [2010] UKSC 2, [145]–[156] per Lord Phillips; [184]–[185] per Lord Rodger (Lady Hale agreeing); [246] per Lord Mance.

[99] Ben Emmerson QC, 'Report of UN Special Rapporteur on the Promotion and Protection of Human Rights while Countering Terrorism' (26 September 2012) UN Doc A/67/396, paras 14, 20–1.

C. Conclusion

Over a decade after the 1267 regime was established, the Security Council is still mired in a debate about the need to provide greater procedural fairness to individuals in sanctions decision-making. There is presently something of a deadlock between those who consider that a court-based process is an essential pre-requisite to the legality of Security Council decision-making, and the Security Council, which remains impervious to calls for court-based reform. The Analytical Support and Sanctions Monitoring Team identified the problem in its 2008 Report:

> The [Sanctions] Committee has made a series of incremental improvements to its procedures which have addressed many of the concerns expressed about the fairness of the sanctions but one major issue remains: the suggestion that listing decisions by the Committee be subject to review by an independent panel. It is difficult to imagine that the Security Council could accept any review panel that appeared to erode its absolute authority to take action on matters affecting international peace and security, as enshrined in the Charter.[100]

My book takes the starting point that the problem is predominantly one of methodology. Reform proposals have relied chiefly on international law source methodology in defence of their insistence that procedural fairness requires a system of judicial or quasi-judicial review. Not only has this methodology failed to persuade the Security Council of the need for reform, it has also resulted in a distorted understanding of the concept of procedural fairness as it applies to the Security Council sanctions context. In the next chapter, I consider the methodology underlying the relevant reform proposals, and the myths and misconceptions about procedural fairness that have emerged in the international setting on account of this methodology.

[100] 'Eighth Report of the Analytical Support and Sanctions Monitoring Team' (14 May 2008) UN Doc S/2008/324, 17.

3

A 'Source-Based' Approach to Procedural Reform

Methodology, Myths, and Misconceptions

It is a natural tendency to fill voids with what we know.[1] Faced with inadequate procedural safeguards in the Security Council sanctions setting, it is unsurprising that international lawyers have drawn predominantly on existing procedural safeguards developed for domestic settings. Significantly, the transplant of such safeguards into the international sphere has been justified on the basis of legal methodology (in terms of the formal 'bindingness' of standards), rather than on the basis of suitability (in terms of the substantive contribution of such standards). Indeed, to date, the content and scope of procedural safeguards have largely been determined using classical international law source methodology, developing applicable principles in accordance with the pedigree sources of international law listed in Article 38 of the Statute of the International Court of Justice.[2] There have been two predominant approaches. The first is a 'top-down' approach, under which applicable due process standards have been derived from international treaties or customary international human rights law. The second is a 'bottom-up' approach, under which 'general principles' of procedural fairness have been derived from domestic legal regimes.

The problem with the heavy reliance on classical international law source methodology is that it threatens to lead to a distorted conception of procedural fairness in the international sphere. This chapter seeks to highlight a number of myths and misconceptions that have arisen as a

[1] Nico Krisch, *Beyond Constitutionalism: The Pluralist Structure of Postnational Law* (Martin Loughlin, John McCormick, and Neil Walker eds) (Oxford University Press 2010), 27.

[2] The classic exposition of the formal sources of international law is contained in Article 38 of the ICJ Statute. The primary sources recognized under Article 38 are: (1) international conventions, establishing rules expressly recognized by the contesting states; (2) international custom, as evidence of a general practice accepted as law; and (3) the general principles of law recognized by civilized nations.

direct result of the somewhat a-theoretical quest for universally binding principles promoted by international law source methodology. In the reform debate, it has given rise to a misconception that an 'adjudicatory' or court-based model is the only acceptable model of procedural fairness. It has also led to the incorporation of ill-fitting structures and classifications into the sanctions setting. For example, adoption of the human rights model of due process introduces a civil/criminal dichotomy that is an awkward fit in the sanctions context. Adoption of an administrative law model of procedural fairness has introduced an 'outdated' notion that procedural safeguards apply only to powers of a 'judicial' nature. Moreover, neither the 'top-down' nor 'bottom-up' approaches incorporate sufficient attention to the unique characteristics of Security Council decision-making on sanctions, in particular, the need for considered and creative mechanisms shaped to accommodate issues of international security and emergency. The intention in exposing distortions in the international legal order's understanding of procedural fairness is to lay the groundwork for a different methodological approach to procedural fairness in the Security Council setting.

A. Source-Based Methodology: Top-Down and Bottom-Up Approaches

The major approach taken to date in the reform literature has been a top-down approach to the development of procedural standards, which proceeds from the argument that the current procedural deficiencies violate international human rights, and that it is necessary for the sanctions regime to comply with a human rights model of due process. Academic literature, reform proposals, and judicial opinions have maintained that the current sanctions regime contravenes due process guarantees under international and regional human rights sources, such as the European Convention on Human Rights,[3]

[3] [Academic articles] Piet Eeckhout, 'Community Terrorism Listings, Fundamental Rights and UN Security Council Resolutions. In Search of the Right Fit' (2007) 3 *European Constitutional Law Review* 183, 206; Julia Hoffmann, 'Terrorism Blacklisting: Putting European Human Rights Guarantees to the Test' (2008) 15 *Constellations* 543, 549; Peter Gutherie, 'Security Council Sanctions and the Protection of Individual Rights' (2004) 60 *NYU Annual Survey of American Law* 491, 500, 502; [Case law] Joined Cases C-402/05 P and C-415/05 P *Kadi v Council of the European Union* [2008] ECR I-0000, [283], [326]; [Reports] David Cortright and Erika de Wet, *Human Rights Standards for Targeted Sanctions* (Sanctions and Security Research Program 2010), 1; Thomas Biersteker and Sue Eckert, *Strengthening Targeted Sanctions through Fair and Clear Procedures* (Watson Institute Targeted Sanctions Project 2006), 10–16.

international human rights treaties,[4] human rights under customary international law,[5] or 'general principles' of human rights law.[6] The general consensus

[4] [Academic articles] Craig Forcese and Kent Roach, 'Limping into the Future: The UN 1267 Terrorism Listing Process at the Crossroads' (2010) 42 *George Washington International Law Review* 217, 237; Machiko Kanetake, 'Enhancing Community Accountability of the Security Council through Pluralistic Structure: The Case of the 1267 Committee' (2008) 12 *Max Planck United Nations Yearbook* 113, 118–19; Andrew Hudson, 'Not a Great Asset: the UN Security Council's Counter-Terrorism Regime: Violating Human Rights' (2007) 25 *Berkeley Journal of International Law* 203, 215; Erika de Wet, 'The Security Council as a Law Maker: The Adoption of (Quasi-) Judicial Decisions' in Rüdiger Wolfrum and Volker Röben (eds), *Developments of International Law in Treaty-Making* (Springer 2005); Erika de Wet, 'The Role of Human Rights in Limiting the Enforcement Power of the Security Council: A Principled View' in Erika de Wet, André Nollkaemper, and Petra Dijkstra (eds), *Review of the Security Council by Member States* (Intersentia 2003); Peter Gutherie, 'Security Council Sanctions and the Protection of Individual Rights' (2004) 60 *NYU Annual Survey of American Law* 491, 499–503; [Case law] UN Human Rights Committee, *Sayadi and Vinck v Belgium* (Communication No 1472/2006, 29 December 2008) UN Doc CCPR/C/94/D/1472/2006; *A and others v The Netherlands* ILDC 1463 (NL 2010); [Reports] David Cortright and Erika de Wet, *Human Rights Standards for Targeted Sanctions* (2010), 1; Thomas Biersteker and Sue Eckert, *Strengthening Targeted Sanctions through Fair and Clear Procedures* (2006), 10–16; Report of the UN High Commissioner for Human Rights, 'Protection of Human Rights and Fundamental Freedoms while Countering Terrorism' (9 March 2007) UN Doc A/HRC/4/88, paras 25–6.

[5] [Academic articles] Daniel Halberstam and Eric Stein, 'The United Nations, the European Union and the King of Sweden: Economic Sanctions and Individual Rights in a Plural World Order' (2009) 46 *Common Market Law Review* 13, 21 ('the UN is bound by customary international law as well as general principles of law at least to the extent that the UN Charter does not provide otherwise'); Michael Bothe, 'Security Council's Targeted Sanctions against Presumed Terrorists: The Need to Comply with Human Rights Standards' (2008) 6 *Journal of International Criminal Justice* 541, 543, 550, 551–2 ('As the rights enshrined in the Universal Declaration [of Human Rights] are nowadays considered to constitute a core of customary law guarantees, these norms are particularly relevant for international organizations which are not bound by any of the human rights treaties, but "only" by customary law'); Andrea Bianchi, 'Security Council Anti-Terror Resolutions and their Implementation by Member States' (2006) 4 *Journal of International Criminal Justice* 1044, 1062; Iain Cameron, 'UN Targeted Sanctions, Legal Safeguards and the European Convention on Human Rights' (2003) 72 *Nordic Journal of International Law* 159, 167 ('the Security Council as such is only bound by those rights that have passed into general international law'); August Reinisch, 'Developing Human Rights and International Humanitarian Law Accountability of the Security Council for the Imposition of Economic Sanctions' (2001) 95 *American Journal of International Law* 851, 855; [Reports] Iain Cameron, *The European Convention on Human Rights, Due Process and United Nations Security Council Counter-terrorism Sanctions* (2006), 22; [Case Law] Joined Cases T-306/01 and T-315/01 *Kadi v Council and Commission* [2005] ECR II-0000, [231]; *Nada v SECO* 133 BGE II 450 (2007), [3].

[6] [Academic Articles] Daniel Halberstam and Eric Stein, 'The United Nations, the European Union and the King of Sweden: Economic Sanctions and Individual Rights in a Plural World Order' (2009) 46 *Common Market Law Review* 13, 21; Larissa Van den Herik, 'The Security Council's Targeted Sanctions Regimes: In Need of Better Protection of the Individual' (2007) 20 *Leiden Journal of International Law* 797, 806–7 ('Even if it is argued that the Security Council cannot be bound by human rights in precisely the same manner as are states, the core of the right to an effective remedy is a general and fundamental principle of the international order that the Security Council cannot just ignore'); August Reinisch, 'Developing

is that the current sanctions regime violates a number of due process rights recognized by the international human rights framework, most significantly, the right to a fair hearing,[7] the right to a remedy,[8] and the prohibition against retrospective laws.[9]

The second dominant source of due process norms applicable to Security Council sanctions decision-making has been the quest for 'general principles' of administrative law. The fecundity of administrative law as a source of procedural safeguards has been noted by many seeking reform in the sanctions context, including the authors of several of the most influential reform proposals. Bardo Fassbender's report, commissioned by the UN Office of Legal Affairs, asserted that certain standards of due process are recognized in the domestic law of a great number of states, and have thereby achieved the status of general principles of international law. These principles are said to bind international organizations when they exercise 'governmental' authority over individuals.[10] In the concluding section of its report on 'Strengthening Targeted Sanctions through Fair and Clear Procedures', the Watson Institute invoked the need for procedural fairness in the sanctions context, a term it defined by 'drawing on general standards of administrative law'.[11] A recommendation with slightly less clarity with regard to source was the recommendation from the Stockholm Process to 'introduce the possibility of administrative or judicial processes (eg regular reviews of the names on the lists) that fulfil ordinary expectations of due process to address mistakes that occur in listing'.[12] An equally vague though potentially related proposal in terms of source was the

Human Rights and International Humanitarian Law Accountability of the Security Council for the Imposition of Economic Sanctions' (2001) 95 *American Journal of International Law* 851, 855; [Reports] Bardo Fassbender, *Targeted Sanctions and Due Process* (2006), para 7 ('The due process rights of individuals recognized as general principles of law are... applicable to international organizations as subjects of international law when they exercise "governmental" authority over individuals').

[7] Universal Declaration of Human Rights (adopted 10 December 1948) UN General Assembly Resolution 217 A(III) ('UDHR'), Article 10; International Covenant on Civil and Political Rights (adopted 16 December 1966, entered into force 23 March 1976) 999 UNTS 171 ('ICCPR'), Article 14(1); European Convention for the Protection of Human Rights and Fundamental Freedoms (adopted 4 November 1950, entered into force 3 September 1953) 213 UNTS 221, ETS 5 ('ECHR'), Article 6.

[8] UDHR, Article 8; ICCPR, Article 2(3); ECHR, Article 13.

[9] UDHR, Article 29(2).

[10] Bardo Fassbender, *Targeted Sanctions and Due Process* (2006), 19–21.

[11] Thomas Biersteker and Sue Eckert, *Strengthening Targeted Sanctions through Fair and Clear Procedures* (2006), 49. See also Thomas Biersteker and Sue Eckert, *Addressing Challenges to Targeted Sanctions: An Update of the 'Watson Report'* (2009), 22.

[12] Peter Wallensteen, Carina Staibano, and Mikael Eriksson, *Making Targeted Sanctions Effective: Guidelines for the Implementation of UN Policy Options* (Results from the Stockholm Process on the Implementation of Targeted Sanctions 2003), 97, para 284.

proposal by a group of likeminded states that the Security Council sanctions regime was bound to comply with 'general principles of international law concerning fair procedure'.[13]

Both top-down and bottom-up approaches rely on pedigree sources of international law as a foundation for the identification of universally binding principles, therefore drawing largely on procedural safeguards developed for domestic contexts. In terms of international law orthodoxy, it is a methodological approach that is understandable and even commendable. As Boyle and Chinkin have noted in their book on *The Making of International Law*, legal positivism remains the *lingua franca* of international lawyers.[14] However, considered from the perspective of procedural fairness, such an approach gives rise to a distorted conception of this legal concept in the international sphere. In the following section, I outline three damaging myths and misconceptions about procedural fairness that follow from such an approach.

B. Myth No 1: Due Process is a 'One-Size-Fits-All' Adjudicatory Model

The central problem with international law source methodology is that it leads to a quest for a 'one-size-fits-all' or 'universally applicable' model of procedural fairness. As a consequence, the majority of reform proposals have fixated on a particular model of procedural fairness commonly found in domestic settings. I have characterized this as an 'adjudicatory' framework for procedural fairness, that is, a system structured around rules, adjudicatory trial-type hearings, and judicial review. As discussed in detail in Chapter 2, an aspect of almost every reform proposal under both approaches (including judicial findings against the sanctions regime) has been an insistence on the inclusion of a judicial mechanism in the sanctions decision-making process, described variously in terms of the 'right to judicial protection',[15] right to

[13] 'Annex: Elements for draft supplementary guidelines for the examination of Sanctions Committees' listing decisions' (2 July 2008) UN Doc A/62/891-S/2008/428. The like-minded states included Denmark, Germany, Liechtenstein, the Netherlands, Switzerland, and Sweden. Belgium and Costa Rica associated with the group in 2009.

[14] Alan Boyle and Christine Chinkin, *The Making of International Law* (Oxford University Press 2007).

[15] Joined Cases C-402/05 P and C-415/05 P *Kadi v Council of the European Union* [2008] ECR I-0000, [352]; Francesco Francioni, 'The Right of Access to Justice to Challenge the Security Council's Targeted Sanctions: After-thoughts on *Kadi*', in Ulrich Fastenrath et al, *From Bilateralism to Community Interest: Essays in Honour of Judge Bruno Simma* (Oxford University Press 2011), 917; Torsten Stein, 'Too "Smart" for Legal Protection?: UN Security Council's Targeted Sanctions and

'access' to a court,[16] 'right to judicial review',[17] 'right to a [binding] effective remedy before an impartial tribunal',[18] right to a 'judicial remedy',[19] 'right to an independent judicial mechanism',[20] or 'right to an effective judicial procedure'.[21] However, resort to this one-size-fits-all adjudicatory model in the existing literature gives rise to two problems: first, identification of a 'universally applicable' model is inconsistent with the contextual nature of procedural fairness; and, second, the adjudicatory model proposed is ill-suited to the Security Council setting.

a Pladoyer for another UN Tribunal' in Holger Hestermeyer et al, *Coexistence, Cooperation and Solidarity: Liber Amicorum Rüdiger Wolfrum* (Martinus Nijhoff 2012), 1539.

[16] *Her Majesty's Treasury v Ahmed* [2010] UKSC 2, [145]–[156] per Lord Phillips; [184]–[185] per Lord Rodger; [246] per Lord Mance; Rapport de la Commission de Politique Extérieure du Parlement Suisse, 'Les Fondements de Notre Ordre Juridique Court-Circuités par l'ONU' (2010), translated in Council of Europe Parliamentary Assembly, Committee of Legal Affairs and Human Rights, 'Compatibility of UN Security Council and EU [terrorist] Black Lists with European Convention on Human Rights Requirements' (7 December 2010), 6.

[17] [Academics] S Ghasem Zamani and Jamshid Mazaheri, 'The Need for International Judicial Review of UN Economic Sanctions' in Ali Z Marossi and Marisa R Bassett (eds), *Economic Sanctions Under International Law: Unilateralism, Multilateralism, Legitimacy, and Consequences* (TMC Asser Press 2015); Grant L Willis, 'Security Council Targeted Sanctions, Due Process and the 1267 Ombudsperson' (2011) 42 *Georgetown Journal of International Law* 673, 737, 743–5; Erika de Wet, 'Human Rights Considerations and the Enforcement of Targeted Sanctions in Europe: The Emergence of Core Standards of Judicial Protection' in Bardo Fassbender (ed), *Securing Human Rights?: Achievements and Challenges of the UN Security Council* (2011), 155, 169; John Dugard, 'Judicial Review of Sanctions' in V Gowlland-Debbas (ed), *United Nations Sanctions and International Law* (Kluwer Law International 2001); [Reports] 'Report of the United Nations High Commissioner for Human Rights on the Protection of Human Rights and Fundamental Freedoms while Countering Terrorism' (15 December 2010) UN Doc A/HRC/16/50, para 21; Martin Scheinin, 'Report of the Special Rapporteur on the Promotion and Protection of Human Rights and Fundamental Freedoms while Countering Terrorism' (6 August 2010) UN Doc A/65/258, paras 56–7; Peter Fromuth, 'The European Court of Justice Kadi Decision and the Future of UN Counterterrorism Sanctions, *ASIL Insight* (30 October 2009), available at <http://www.asil.org/insights/volume/13/issue/20/european-court-justice-kadi-decision-and-future-un-counterterrorism>; [Case law] *Her Majesty's Treasury v Ahmed* [2010] UKSC 2, [81] per Lord Hope.

[18] Bardo Fassbender, *Targeted Sanctions and Due Process* (2006), para 15; Case T-318/01 *Othman v Council and Commission*, 11 June 2009, [83]. See also Iain Cameron, 'UN Targeted Sanctions, Legal Safeguards and the European Convention on Human Rights' (2003) 72 *Nordic Journal of International Law* 159, 210; Larissa Van den Herik, 'The Security Council's Targeted Sanctions Regimes: In Need of Better Protection of the Individual' (2007) 20 *Leiden Journal of International Law* 797, 806–7; Johannes Reich, 'Due Process and Sanctions Targeted Against Individuals Pursuant to Resolution 1267 (1999)' (2008) 33 *Yale Journal of International Law* 505, 510–11.

[19] *Her Majesty's Treasury v Ahmed* [2010] UKSC 2 (UK Supreme Court) [413]–[414]; Craig Forcese and Kent Roach, 'Limping into the Future: The UN 1267 Terrorism Listing Process at the Crossroads' (2010) 42 *George Washington International Law Review* 217, 265, 275.

[20] International Commission of Jurists, *Assessing Damage, Urging Action: Report of the Eminent Jurists Panel on Terrorism, Counter-Terrorism and Human Rights* (2009).

[21] Case T-85/09 *Kadi v European Commission* [2010] EUECJ, [128].

(1) The Contextual Nature of Procedural Fairness

As any public lawyer will readily identify, procedural fairness is not a universally applicable concept, but is intended to be contextual. Procedural fairness or due process rights are accepted by all common law systems, and are recognized by many others.[22] Yet, as Carol Harlow warns, this should not lead us into the error of thinking either that this makes the principles 'universal' or that they take the same shape or have the same scope in every legal system.[23] In the domestic literature on procedural fairness and due process, it is well established that different legal contexts legitimately require different procedural standards and operate according to different principles and values.[24] One of the foremost comparative treatises on due process notes that 'there is no general blueprint to follow and the variety of approaches found in statutes is considerable'.[25] In the United Kingdom, the leading cases on procedural fairness recognize that the standards of fairness are not 'immutable' nor 'engraved on tablets of stone', but differ depending on 'the character of the decision-making body, the kind of decision it has to make and the statutory or other framework in which it operates'.[26] The US Supreme Court has also repeatedly insisted that the requirements of due process are flexible and must be tailored to particular circumstances.[27] Justice Frankfurter stated that due process 'is not a technical conception with a fixed content unrelated to time, place and circumstances' and 'cannot be imprisoned within the treacherous limits of any formula'.[28] Professor Laurence Tribe has observed that procedural fairness

[22] For a global survey, see S Guinchard et al, *Droit processuel, droit commun et droit comparé du procès* (3rd edn, 2005), cited in Carol Harlow, 'Global Administrative Law: The Quest for Principles and Values' (2006) 17 *European Journal of International Law* 187, 204.

[23] Carol Harlow, 'Global Administrative Law: The Quest for Principles and Values' (2006) 17 *European Journal of International Law* 187, 204.

[24] Lucia Zedner, 'Seeking Security by Eroding Rights: The Side-Stepping of Due Process' in Benjamin J Goold and Liora Lazarus (eds), *Security and Human Rights* (Hart Publishing 2007), 265.

[25] D J Galligan, *Due Process and Fair Procedures: A Study of Administrative Procedures* (Clarendon Press 1996), 315.

[26] *Lloyd v McMahon* [1987] AC 625 per Lord Bridge. In similar terms, it has been prescribed that '[w]hat fairness demands is dependent on the context of the decision, and this is to be taken into account in all its aspects': *R v Home Secretary ex p Doody* [1994] 1 AC 531 per Lord Mustill.

[27] See, for example, *Morrissey v Brewer* 408 US 471 (1972), 481 ('It has been said so often by this Court and others as not to require citation of authority that due process is flexible and calls for such procedural protections as the particular situation demands... Its flexibility is in its scope once it has been determined that some process is due; it is a recognition that not all situations calling for procedural safeguards call for the same kind of procedure').

[28] *Joint Anti-Fascist Refugee Committee v McGrath* 341 US 123 (1951). See also *Rochin v California* 342 US 432 (1957) ('These standards of justice are not authoritatively formulated anywhere as though they were specifics').

models are to be shaped by 'alternative conceptions of the primary purpose of procedural due process and by competing visions of how that purpose may best be achieved'.[29]

Consequently, different legal contexts will reflect different procedural frameworks. For example, the English common law tradition of *audi alteram partem* (right to a fair hearing) and the French tradition of *respect des droits de la défense* (respect for the rights of the defence) differ in terms of the procedural standards they recognize as essential to procedural fairness. As the differing phraseology of the principles imply, while the common law system privileges the right of individuals to a personal (and generally oral) hearing, in France, litigation before *juridictions administratives* takes a predominantly written form and there is no general principle of law or procedure providing an individual the right to be personally heard.[30] There is certainly no right in the latter system to cross-examine witnesses. The duty to provide reasons is another example of a principle upon which legal systems differ. Under British common law, there is no general duty to give reasons,[31] though such a requirement is recognized in American and Australian federal law,[32] and is deeply rooted in the jurisprudence of both French and Italian administrative courts.[33]

However, the aim of this book is to examine the contextual nature of procedural fairness in normatively rich rather than formulaic terms. Underlying the differences in form between procedural fairness frameworks are different ideas as to the process values that a procedural framework can and should support in order to enhance the legitimacy of decision-making.[34] For the purposes of this chapter, it is sufficient to note that procedural fairness is not intended to be a one-size-fits-all model, but is *contextual*. The netting of similar principles from an array of different legal systems in

[29] Laurence Tribe, *American Constitutional Law* (2nd edn, Foundation Press 1988), 666.

[30] Aubert Lefas, 'A Comparison of the Concept of Natural Justice in English Administrative Law with the Corresponding General Principles of Law and Rules of Procedures in French Administrative Law' (1978) 4 *Queen's Law Journal* 197.

[31] *R v Home Secretary ex p Doody* [1994] 1 AC 531, 564E. Note that there are multiple exceptions to this general principle to the extent that Wade and Forsyth recommend that '[t]he time has now surely come for the court to acknowledge that there is a general rule that reasons should be given for decisions, based on the principle of fairness which permeates administrative law, subject only to specific exceptions to be identified as cases arise': Wade and Forsyth, *Administrative Law* (Oxford University Press 2014), 443.

[32] Administrative Procedure Act of 1946, section 8(b); Administrative Decisions (Judicial Review) Act 1977, section 13.

[33] Giacinto della Cananea, 'Administrative Law in the European Legal Area: A Historic and Comparative Overview' in Armin von Bogdandy, Sabino Cassese, and Peter Huber (eds), *Handbuch Ius Publicum Europaeum, Band IV: Verwaltungsrecht in Europa* (CF Müller 2011).

[34] Jerry Mashaw, *Due Process in the Administrative State* (Yale University Press 1985), 42.

accordance with traditional international law source methodology, resulting in the artificial generation of a universally applicable set of norms, does not sit comfortably with the contextual nature of procedural fairness. Instead, I argue that a procedural framework for the Security Council sanctions regime should be developed against the backdrop of an understanding of the purpose or purposes that procedural fairness is able to serve in that setting.

(2) Unsuitability of the Adjudicatory Framework

The second problem arising from reliance on a 'one-size-fits-all' model of procedural fairness is that it results in the identification of an adjudicatory framework that is potentially ill-suited to the Security Council setting. This is a term I have already employed in previous chapters. By 'adjudicatory framework', I refer to a framework based on a logically coherent and complete system of principles and rules, where decisions are reached through the presentation of proofs in trial-type hearings and are amenable to authoritative review by an independent arbiter.[35] Such a framework was developed as a mechanism to deal with disputes that arise out of the distribution and power and the patterns of motivation typically found in domestic legal systems. Thomas Scanlon has recognized that this form of 'due process' is not readily exportable to regimes outside the domestic context on the ground that it depends on a minimal commitment to certain institutional tenets. These include, for example, the importance of belief in 'the possibility of an independent judiciary' and 'the need to defend a charge publicly and with reference to a known law'.[36] These tenets do not form part of the Security Council framework, and their introduction would require a fairly radical change of mindset on the part of the Permanent Members of the Security Council.

The potential inconsistency between the Security Council setting and an adjudicatory framework is an issue that will be examined in further detail in subsequent chapters. The relevant issue at this point is to recognize that the adjudicatory model, revolving around the necessity of judicial review, is not the only model of procedural fairness. Michael Bayles recognizes that '[t]he process benefits of adversary adjudication are often overestimated' and there are 'a number of more appropriate alternative decision-making

[35] See, for example, Lon Fuller, 'The Forms and Limits of Adjudication' (1978) 92 *Harvard Law Review* 353.

[36] T M Scanlon, 'Human Rights as a Neutral Concern' in TM Scanlon (ed), *The Difficulty of Tolerance: Essays in Political Philosophy* (Cambridge University Press 2003), 116.

models, particularly in the context of decisions to impose burdens on individuals'.[37] It is not so much judicial review as the right to 'recourse' that is a fundamental aspect of procedural fairness, meaning the right to have the primary decision re-examined 'to determine whether it is wrong, whether a mistake has been made, or whether it is simply not as good, wise or fair as it might have been'.[38] This need not take the form of an appeal to a judicial body. As Galligan cautions, 'The form and extent of recourse [as an aspect of due process] will vary according to the nature and quality of the primary decision process; different forms may be effective in different contexts, and imaginative thinking about new forms should not be discouraged by obstinate hankering for any set pattern.'[39] Against a better conceptual understanding of due process, it may be that alternative models of recourse are available that fit the Security Council setting better than that of adjudicatory decision-making and judicial review. This will be explored in further detail in subsequent chapters.

(3) Relationship to Global Administrative Law

It is worth making brief reference to the connection between the methodology underlying development of a procedural fairness framework for Security Council sanctions decision-making and the global administrative law project, a compelling project based out of New York University that is concerned with extending administrative law concepts (such as procedural fairness) to the international legal realm. It is arguable that the quest for a 'one-size-fits-all' model based on a set of so-called universally applicable due process rights, whether drawn from human rights law or domestic principles of administrative law, is inspired by, or at the very least echoes, proponents of the global administrative law project. However, such an assumption, it is submitted, misconceives the aims of the originators of the project. In their groundbreaking piece written in 2005, Benedict Kingsbury, Nico Krisch, and Richard Stewart describe the project of global administrative law as 'embryonic' and talk about the 'development of new mechanisms of administrative law at the global level'.[40] One of the authors describes the

[37] M Bayles, *Procedural Justice* (Kluwer 1990), 188.

[38] D J Galligan, *Due Process and Fair Procedures: A Study of Administrative Procedures* (Clarendon Press 1996), 392–3.

[39] Ibid, 402. See also M Bayles, *Procedural Justice* (Kluwer 1990), ch 8 (see in particular, 168–89).

[40] Benedict Kingsbury, Nico Krisch, and Richard B. Stewart, 'The Emergence of Global Administrative Law' (2005) 68 *Law and Contemporary Problems* 15, 15, 16.

global administrative law project as 'a self-consciously "modest" project'.[41] He recognizes that:[42]

> in [global administrative law], this move does not imply the prescriptive assumption that the tools of domestic administrative law ought to be transferred into the institutions of global governance, e.g. by establishing judicial review mechanisms wherever individuals are directly affected or by instituting public participation whenever global administrative bodies are engaged in rule-making.... Turning to administrative law for inspiration is mainly an attempt to expand the intellectual and practical resources for thinking about global governance, for bringing out similarities and differences, rather than for particular, transferable prescriptions.

As described, the use of the term 'global administrative law' is not diagnostic and corroborated, but is intended instead to be prognostic and exploratory. The danger of the project is that the caution deliberately built into the project has not been heeded in the way that it might. Therefore we find reform proposals in the Security Council sanctions context jumping immediately to the conclusion that 'standards of due process... recognized in the domestic law of a great number of states... bind international organizations when they [exercise] "governmental" authority over individuals'[43] and calls to extend procedural fairness to Security Council sanctions 'drawing on general standards of administrative law'.[44] Such assertions are made with a limited attempt to engage with the disparity in systems of administrative law around the globe or take into account the characteristics that set the Security Council framework apart from domestic administrative processes. The aim of this book is to reconfigure the debate so that it is more in line with the intention behind the global administrative law project, using administrative law as an inspiration for principles applicable to the international sphere, rather than as a body of principles to be directly transplanted.

[41] Nico Krisch, 'Global Administrative Law and the Constitutional Ambition', LSE Law, Society and Economy Working Papers 10/2009, 18. See also Benedict Kingsbury and Lorenzo Casini, 'Global Administrative Law Dimensions of International Organizations Law' (2009) 6 *International Organizations Law Review* 319, 358; Daniel Esty, 'Good Governance at the Supranational Scale: Globalizing Administrative Law' (2006) 115 *Yale Law Journal* 1490, 1561; Dan Sarooshi, 'The Role of Domestic Public Law Analogies in the Law of International Organisations' (2008) 5 *International Organizations Law Review* 237.

[42] Nico Krisch, 'Global Administrative Law and the Constitutional Ambition', LSE Law, Society and Economy Working Papers 10/2009, 12–13. See also Benedict Kingsbury, Nico Krisch, and Richard B. Stewart, 'The Emergence of Global Administrative Law' (2005) 68 *Law and Contemporary Problems* 15, 37–56.

[43] Bardo Fassbender, *Targeted Sanctions and Due Process* (Council of Europe Committee of Legal Advisors on International Law 2006), 19–21.

[44] Thomas Biersteker and Sue Eckert, *Strengthening Targeted Sanctions through Fair and Clear Procedures* (Watson Institute Targeted Sanctions Project, Brown University 2006), 49.

C. Myth No 2: Due Process and Inappropriate Domestic Labels

A second set of misconceptions arising from the application of domestic standards to the Security Council context is the perceived need to shoehorn Security Council decision-making into ill-fitting domestic classifications. One example is the debate as to whether sanctions should be classified as civil or criminal in nature, a line of inquiry following from the civil/criminal dichotomy that is the hallmark of most domestic legal systems. A second example emerging from reliance on administrative law under the bottom-up approach is the (ultimately misconceived and outdated) presumption that procedural fairness applies only to the exercise of 'judicial' power, leading to various attempts to characterize Security Council authority as 'judicial', 'quasi-judicial', 'administrative', or 'executive'.

(1) Procedural Fairness and the Civil/Criminal Divide

In the international debate about procedural fairness in sanctions decision-making, significant attention has been given to the question as to whether sanctions are civil or criminal in nature.[45] Several academics have conceptualized sanctions as 'criminal' measures.[46] Erika De Wet and André Nollkaemper consider that the assets freeze by the Security Council constitutes a sanction belonging to the criminal sphere in light of its punitive nature, severity, as well as the stigmatization resulting from it.[47] Andrea Bianchi considers that inclusion of an individual on a sanctions blacklist and the ensuing financial sanctions could reasonably be characterized as a criminal charge under the law of the European Convention on Human Rights.[48]

[45] See, for example, Melissa van den Broek, Monique Hazelhorst, and Wouter de Zanger, 'Asset-freezing: Smart Sanction or Criminal Charge' (2010) 27 *Utrecht Journal of International and European Law* 18.

[46] In addition to those cited below, see Grant L Willis, 'Security Council Targeted Sanctions, Due Process and the 1267 Ombudsperson' (2011) 42 *Georgetown Journal of International Law* 673, 696; Machiko Kanetake, 'Enhancing Community Accountability of the Security Council through Pluralistic Structure: The Case of the 1267 Committee' (2008) 12 *Max Planck United Nations Yearbook* 113, 118–19; George Lopez et al, *Overdue Process: Protecting Human Rights while Sanctioning Alleged Terrorists* (Fourth Freedom Forum, Kroc Institute for International Peace Studies, University of Notre Dame 2009), 3.

[47] Erika de Wet and André Nollkaemper, 'Review of Security Council Decisions by National Courts' (2002) 45 *German Yearbook of International Law* 166, 177.

[48] Andrea Bianchi, 'Security Council Anti-Terror Resolutions and their Implementation by Member States' (2006) 4 *Journal of International Criminal Justice* 1044, 1066. Another scholar

Ian Cameron labels the measures as 'quasi-criminal'.[49] Though Larissa Van den Herik does not come to a definitive conclusion, she considers that the link to the criminal offence of terrorism, and the profound impact that long-term freezing of assets has on an individual's life, support an argument that the 1267 sanctions should be characterized as a criminal law measure.[50] The High Commissioner for Human Rights has also declared that '[t]he longer an individual is on a list, the more punitive the effect will be'.[51]

The prevailing view is that sanctions are not criminal, but civil or administrative in nature. Perhaps most significantly, the UN Human Rights Committee in *Sayadi* determined that, although the sanctions regime has serious consequences for the individuals concerned, it does not concern a 'criminal charge' in the meaning of Article 14(1), and therefore discloses no violation of the relevant due process rights.[52] The Security Council has characterized targeted sanctions as being 'preventative in nature and... not reliant upon criminal standards set out under national law'.[53] This view is corroborated by the 1267 Committee Guidelines and in statements by the Al Qaeda and Taliban Sanctions Monitoring Team, which describe the sanctions as 'preventive' rather than punitive, including 'administrative measures such as freezing assets, prohibiting international travel and precluding arms sales'.[54] While a Belgian court came to the conclusion that sanctions could not reasonably be construed as an administrative and temporary measure,[55]

agreeing with this assessment concludes that the imposition of asset freezing should be considered for inclusion in the jurisdiction of the International Criminal Court: Jorge Godinho, 'When Worlds Collide: Enforcing United Nations Security Council Asset Freezes in the EU Legal Order' (2010) 16 *European Law Journal* 67, 90, 93.

[49] Iain Cameron, 'UN Targeted Sanctions, Legal Safeguards and the European Convention on Human Rights' (2003) 72 *Nordic Journal of International Law* 159, 192.

[50] Larissa Van den Herik, 'The Security Council's Targeted Sanctions Regimes: In Need of Better Protection of the Individual' (2007) 20 *Leiden Journal of International Law* 797, 806.

[51] High Commissioner for Human Rights, 'Report of the United Nations High Commissioner for Human Rights on the Protection of Human Rights and Fundamental Freedoms while Countering Terrorism' (2 June 2008) UN Doc A/HRC/8/13, paras 47, 50.

[52] UN Human Rights Committee, *Sayadi and Vinck v Belgium* (Communication No 1472/2006, 29 December 2008) UN Doc CCPR/C/94/D/1472/2006, [10.11]. See also Individual opinion (partly dissenting) by Sir Nigel Rodley, Ivan Shearer, and Iulia Antoanella Motoc, 27 ('Nor do we understand on what basis [the Committee] believes that articles 14 and 15 could be relevant to actions that the State party quite rightly maintains are administrative, not criminal').

[53] UN Security Council resolution 2083 (17 December 2012) UN Doc S/RES/2083, Preamble; UN Security Council resolution 1735 (22 December 2006) UN Doc S/RES/1735, para 14.

[54] Al Qaeda Committee, 'Guidelines of the Committee for the Conduct of its Work', available at <http://www.un.org/sc/committees/1267/pdf/1267_guidelines.pdf>, para 6; 'Third report of the Analytical Support and Sanctions Monitoring Team appointed pursuant to resolution 1526 (2004) concerning Al-Qaida and the Taliban and associated individuals and entities' (9 September 2005) UN Doc S/2005/572, paras 39–44.

[55] Decision of the Tribunal de Première Instance de Bruxelles, *Sayadi & Vinck v l'Etat Belge*, 18 February 2005.

the EU Court of First Instance rejected the classification of sanctions as criminal measures. Using the example of assets freezes, the Court held that 'the freezing of the applicant's funds is not a criminal conviction for the purposes of that case-law, since there has not been any classification as a criminal offence, since the measure in question concerns only a specific group of persons and since the severity of the measure is not sufficient for such purpose'.[56] The standard of review of sanctions measures was decided by the Court of Justice of the European Union in *Kadi II* '[h]aving regard to the preventive nature of the restrictive measures at issue'.[57] Academics such as Peter Gutherie and Bardo Fassbender consider sanctions to be more in the nature of civil measures as they are intended to be more 'preventive' or 'persuasive' than punitive.[58]

Efforts to position Security Council sanctions within the civil/criminal dichotomy derive from the use of this distinction in due process guarantees in the majority of human rights instruments.[59] For example, Article 6 of the European Convention on Human Rights distinguishes between civil and criminal trials, and identifies at least ten procedural rights applicable to those charged with criminal offences, though not to civil cases.[60] This two-track procedural system in human rights law derives in turn from the fact that the civil/criminal dichotomy is the hallmark of most domestic legal systems to which the relevant human rights instruments were developed to apply.[61]

The problem is that, while it may be an element of most domestic legal systems, the civil/criminal divide is not a feature of Security Council decision-making. The ongoing attempt to force sanctions decision-making into either a civil or criminal mould has been a distracting sideshow to the

[56] Case T-47/03 *Jose Maria Sison v Council of the European Union* [2007] ECR II-73, [101], [129].

[57] *European Commission v Kadi* [2013] ECR not yet reported, [130].

[58] Bardo Fassbender, *Targeted Sanctions and Due Process* (2006), para 12.6; Peter Gutherie, 'Security Council Sanctions and the Protection of Individual Rights' (2004) 60 *NYU Annual Survey of American Law* 491, 503–6. Gutherie notes, however, that there is a very plausible argument that Security Council sanctions closely resemble criminal sanctions '[g]iven the all-encompassing scope of the freeze, the international notoriety that accompanies inclusion on the sanctioning list, and the indeterminate length of the sanctions': 505.

[59] See, for example, ICCPR, Article 14; ECHR, Article 6; American Convention on Human Rights (adopted 22 November 1969, entered into force 18 July 1978) OAS Treaty Series No 36, 1144 UNTS 123 ('ACHR'), Article 8.

[60] These include: (1) the right to be presumed innocent; (2) the privilege against self-incrimination; (3) the right of silence; (4) the right to legal aid and assistance; (5) the right to be brought promptly before a court; (6) the right to release pending trial; (7) the right to disclosure of documents; (8) the right to confront witnesses; (9) the right to be tried on evidence not obtained by violation of fundamental rights; (10) the right not to be placed in double jeopardy: Andrew Ashworth, *Human Rights, Serious Crime and Criminal Procedure* (Sweet & Maxwell 2002).

[61] Paul Robinson, 'The Criminal-Civil Distinction and the Utility of Desert' (1996) 76 *Boston University Law Review* 201, 201.

main debate. It also potentially distorts the debate in a dangerous way as it opens up the possibility that procedural fairness might be deemed inappropriate or unnecessary in the sanctions context because sanctions cannot be construed as 'criminal' in nature (as was the case in *Sayadi*). As it happens, this bifurcation is in any event losing significance at the domestic level and becoming increasingly outdated.[62] The line between criminal penalties and civil sanctions is dissolving, and there is increasing fuzziness at the edges of the civil/criminal distinction. More recently, a number of hybrid orders have been introduced into the domestic setting, occupying a position midway between the civil and criminal paradigms. Good examples are innovations of the British government, including the anti-social behaviour order (or 'ASBO'),[63] and the terrorism control order.[64] ASBOs have been directed against a range of offenders such as prostitutes, beggars, and young people, and also less obvious categories such as noisy trampolinists, roaming pigs, and feeders of seagulls. As is clear from this list of targets, the behaviour against which an ASBO can be issued is not necessarily criminal, and is defined by reference simply to its likelihood of causing 'harassment, alarm or distress' to persons in the community. Similarly, though terrorism control orders are more clearly connected with criminal activity, a criminal charge is not a precondition to their imposition, and control orders may be imposed where there are 'reasonable grounds for suspecting that the individual is or has been involved in terrorism-related activity'. Both ASBOs and terrorism control orders are civil and preventive in their orientation, issued in civil proceedings. Yet both orders effectively create individual penal obligations, and the full might of the criminal law is brought to bear if there is a breach of the civil order, including in both cases a penalty of up to five years' imprisonment. The relevant issue for our purposes is the procedural safeguards that attach to the issue of such orders. Significantly, UK courts have come to the conclusion that, even though the orders are civil, the seriousness of their potential consequences for an individual should increase the measure of procedural protection afforded to him or her.[65] Judges in the United Kingdom increasingly regard

[62] Andrew Ashworth, *Human Rights, Serious Crime and Criminal Procedure* (Sweet & Maxwell 2002); Carole S Steiker, 'Civil and Criminal Divide' in Joshua Dressler (ed), *Encyclopedia of Crime and Justice*, vol 1 (Macmillan Reference USA 2001); 'Symposium: The Civil-Criminal Distinction' (1996) 7 *Journal of Contemporary Legal Issues* i-269; 'Symposium: The Intersection of Tort and Criminal Law' (1996) 76 *Boston University Law Review* 1-373.

[63] Introduced by the Blair government in the Crime and Disorder Act 1998 (UK), and expanded by Anti-social Behaviour Act 2003 (UK).

[64] Introduced by the Prevention of Terrorism Act 2005 (UK).

[65] *Secretary of State for the Home Department v MB* [2007] UKHL 46, [2008] 1 AC 440, [24] per Lord Bingham; *R (McCann) v Crown Court at Manchester* [2002] UKHL 39, [2003] 1 AC 787, [37] per Lord Steyn, [82]–[83] per Lord Hope of Craighead. See also Lucia Zedner, 'Preventative Justice or Pre-Punishment? The Case of Control Orders' (2007) 60 *Current Legal Problems* 174;

the classification of proceedings as criminal or civil as less important than the question of what protections are required for a fair trial,[66] and have held that the gravity and complexity of the charges and of the defence will impact on what fairness requires.[67]

The diminished significance of the civil/criminal dichotomy is also reflected more broadly. Since the 1980s, there has been an 'explosion' of litigation about whether civil fines and forfeitures or new forms of incapacitative incarceration are subject to any or all of the special criminal procedural protections.[68] In order to prevent states simply 're-labeling' proceedings to circumvent the safeguards attaching to criminal proceedings, courts are interrogating the motives lying behind the categorization of an offence as civil and criminal, and developing criteria by which to determine which context is appropriate. As Lucia Zedner has recognized, '[i]f the immediate or likely subsequent effect of a civil order is inherently punitive, then all the labels in the world will not render it justifiably civil, and the question of appropriate jurisprudential considerations must remain open to challenge'.[69] The European Court of Human Rights has insisted that the term 'charged with a criminal offence' should be given an autonomous meaning that looks to the substance rather than the form.[70] In a long line of decisions since the leading case of *Engel v Netherlands*,[71] a set of criteria has been developed and applied to determine whether criminal procedural safeguards apply. The criteria are threefold: (1) the characterization of the offence by the state in question; (2) the nature of the offence; and (3) the degree of the severity of the restrictions or deprivations flowing from the order that the court can make. The emphasis is firmly on the third criterion.[72] Essentially, in determining whether criminal procedural safeguards should be applied, the increasing trend is to look at the impact of the measure on the rights of the individual, rather than on the classification of the offence as civil or criminal.

A P Simester and Andrew Von Hirsch, 'Regulating Offensive Conduct through Two-Step Prohibitions' in A P Simester and Andrew Von Hirsch (eds), *Incivilities* (Hart Publishing 2006).

[66] *International Transport Roth GmbH v Secretary of State for the Home Department* [2002] EWCA Civ 158, [2003] QB 728, [33], [148].

[67] *R v Securities and Futures Authority Ltd ex p Fleurose* [2001] EWCA Civ 2015, [2002] IRLR 297, [14].

[68] Carole S Steiker, 'Civil and Criminal Divide' in Joshua Dressler (ed), *Encyclopedia of Crime and Justice*, vol 1 (Macmillan Reference USA 2001), 161.

[69] Lucia Zedner, 'Seeking Security by Eroding Rights: The Side-Stepping of Due Process' in Benjamin J Goold and Liora Lazarus (eds), *Security and Human Rights* (Hart Publishing 2007), 266.

[70] *Öztürk v Germany* (1984) 6 EHRR 409, [53]; *Lauko v Slovakia* (1998) 33 EHRR 994, [58].

[71] (1976) 1 EHRR 647.

[72] *Garyfallou AEBE v Greece* (1999) 28 EHRR 344, [33]; *Benham v UK* (1996) 22 EHRR 293; *Ezeh and Connors v UK* (2002) 35 EHRR 691.

In the Security Council context, a preoccupation with the issue as to whether sanctions should be characterized as 'civil' or 'criminal' in nature is misconceived and misleading. Instead of attempting to derive procedural protections by reference to inappropriate and increasingly outdated domestic labels, it is important to have regard to the sanctions measures themselves, and develop protections appropriate to the gravity of the potential consequences for the affected individual.

(2) Procedural Fairness and 'Judicial' Power

The bottom-up approach is similarly afflicted by reliance on outdated conceptions of due process under domestic law. The debate about the application of procedural fairness to Security Council decision-making has at times concerned itself with the question as to whether Security Council decision-making can be characterized as 'judicial', 'quasi-judicial', 'executive', 'administrative', or 'legislative' out of a sense that a functional characterization of Security Council decision-making authority in these terms is relevant to the question as to whether procedural safeguards apply. The impasse between the Security Council and member states on the need for procedural safeguards has been described as stemming in part from the Council's classification of the listing enterprise as 'an administrative political act, beyond judicial review, aimed at constraining potential terrorist actions'.[73] Security Council members are reported to have argued on several occasions, both publicly and privately, that the Council was not a judicial body and thus did not need to adopt 'judicial or semi-judicial rules'.[74] In his influential report on *Targeted Sanctions and Due Process*, Bardo Fassbender characterizes sanctions resolutions as 'legislative' acts that cannot be challenged by an individual, while classifying the listing of an individual name as an 'administrative' measure that can be challenged (though even this latter argument is somewhat weakened by his earlier acknowledgement that in many states administrative acts are exempt from judicial review).[75] Erika de Wet argues that the Security Council is bound to respect basic principles of procedural justice such as independence, impartiality, and even-handedness 'when creating a subsidiary body which will exercise a judicial function'.[76] For other authors, such direct links are not made,

[73] George Lopez et al, *Overdue Process: Protecting Human Rights while Sanctioning Alleged Terrorists* (Fourth Freedom Forum, Kroc Institute for International Peace Studies, University of Notre Dame 2009), 2.

[74] Security Council Report, *Cross-cutting Report on the Rule of Law* (2011), 38.

[75] Bardo Fassbender, *Targeted Sanctions and Due Process* (Council of Europe Committee of Legal Advisors on International Law 2006), paras 12.5, 1.10.

[76] Erika de Wet, 'The Security Council as a Law Maker: The Adoption of (Quasi-) Judicial Decisions' in Rüdiger Wolfrum and Volker Röben (eds), *Developments of International Law in Treaty-Making* (Springer 2005), 185.

though they reflect an impulse to characterize sanctions decision-making along domestic constitutional lines with the clear implication that this has some relevance for the applicable legal framework. Both Ian Cameron and Peter Gutherie describe the Security Council as a hybrid body, with Cameron characterizing the Council as 'an executive body... imposing a legal penalty',[77] and Gutherie describing it as 'an embryonic bureaucracy with executive power', which (like the ad hoc international criminal tribunals) is 'vested with power to make judgments concerning an individual's involvement in activities condemned by the international community'.[78] Jeremy Farrall characterizes the application of sanctions as 'quasi-legislative' and describes the legal determinations it makes incidentally to the sanctions regime as 'quasi-judicial'.[79]

The assumption made by international lawyers is that the characterization of sanctions decision-making in functional terms such as 'judicial' or 'administrative' has implications for the nature of procedural safeguards that the Security Council must respect. Yet this characterization stems from a misconception about the nature of due process. If international lawyers looked at due process in a conceptual rather than a derivative way, they would observe that there has been a marked shift in thinking about due process away from a conceptualization of decision-making along functional lines. Upon examining the historical foundations of the connection between procedure and judicial power, it becomes clear that the division of due process along constitutional-functional lines relies on an outdated conception of the source of procedural guarantees. In the common law world, procedural fairness was at one time considered to stem from a broader 'duty to act judicially'. Historically, magistrates had a range of functions, some of which were judicial (for example, the consideration of criminal charges such as theft) and some of which were administrative (for example, decisions related to the granting of licences for markets, or fixing the price of corn). An early attempt of the courts to decide the proper ambit of the rules of procedural fairness (or natural justice as it was known) led to the notion that a decision-maker ought to comply with the rules of natural justice (and so behave in a more or less adversarial fashion) if he or she was under *a duty to act judicially*. In the early twentieth century, a period of extreme discomfort with the emergence of the continental phenomenon of 'administrative law' led to a revival of this distinction.[80]

[77] Iain Cameron, 'UN Targeted Sanctions, Legal Safeguards and the European Convention on Human Rights' (2003) 72 *Nordic Journal of International Law* 159, 183.

[78] Peter Gutherie, 'Security Council Sanctions and the Protection of Individual Rights' (2004) 60 *NYU Annual Survey of American Law* 491, 492, 496.

[79] Jeremy Farrall, *United Nations Sanctions and the Rule of Law* (Cambridge University Press 2007), 16.

[80] *Nakkuda Ali v Jayaratne* [1951] AC 66, 75–7; *R v Metropolitan Police Commissioner ex p Parker* [1953] 1 WLR 1150, 1153; *R v St Lawrence's Hospital Statutory Visitors ex p Pritchard* [1953] 1 WLR

This 'discomfort' was epitomized by the writing of Lord Hewart on *The New Despotism* characterizing administrative law as 'not really a system at all, but... simply an exercise of arbitrary power in relation to certain matters',[81] and led to a highly formalist jurisprudence in which judges (encouraged by the findings of the Donoughmore Committee)[82] attempted to distinguish the 'judicial', 'quasi-judicial', and 'administrative' functions of government in order to determine whether the principles of natural justice applied.[83]

However, this approach to procedural fairness has long been overturned. In the landmark case of *Ridge v Baldwin*,[84] decided by the House of Lords in 1964, the so-called 'heresy' of the separation along functional lines was scotched.[85] The House of Lords held that the relevant characteristic attracting the application of natural justice was not the nature of the decision-making power, but the question whether the decision affected a person's rights. There was no additional requirement that the decision-making authority had to be acting judicially.[86] The effect of the case was to reinterpret the duty to act 'judicially' as a duty to act 'fairly', not tied to the classification of governmental functions. The principle of natural justice was thereby unshackled from its functional constraints, and applied to a broad spectrum of interests and a wide variety of decision-makers.[87] International lawyers have overlooked this landmark case, described by one administrative lawyer as the 'Magna Carta of natural justice'.[88] This is notwithstanding the fact that this development has particular advantages in the Security Council context in the sense that it precludes the need to shoehorn Security Council decision-making

1158; *R v Legislative Committee of the Church Assembly ex p Haynes-Smith* [1928] 1 KB 411, 415, interpreting *R v Electricity Commissioners ex p London Electricity Joint Committees Co (1920) Ltd* [1924] 1 KB 171, 205.

[81] Hewart, *The New Despotism* (Ernest Benn Limited 1929), 46.

[82] The Donoughmore Committee, a committee of inquiry set up in 1932, which recommended that a distinction be drawn along procedural lines between judicial functions (that is, the finding of facts and application of law to facts) and administrative functions (the use of discretion by administrators in applying policy rather than law). The effect was to insulate administrative functions from the common law doctrine of natural justice.

[83] See *Local Government Board v Arlidge* [1915] AC 120; *Errington v Minister of Health* [1935] 1 KB 249; *Franklin v Minister of Town and Country Planning* [1948] AC 87, cited in Carol Harlow and Richard Rawlings, *Law and Administration* (3rd edn, Cambridge University Press 2009), 622.

[84] *Ridge v Baldwin* [1964] AC 40.

[85] *R v Gaming Board for Great Britain ex p Benaim and Khaida* [1970] 2 All ER 528 per Lord Denning. See also S de Smith, Lord Woolf, and J Jowell, *Judicial Review of Administrative Action* (Sweet & Maxwell, London 1995), para 1-001.

[86] *Ridge v Baldwin* [1964] AC 40, 72–8 per Lord Reid.

[87] Carol Harlow and Richard Rawlings, *Law and Administration* (3rd edn, Cambridge University Press 2009), ch 14. See *Schmidt v Secretary of State for the Home Department* [1969] 2 WLR 337; *R v Gaming Board for Great Britain ex p Benaim and Khaida* [1970] 2 All ER 528.

[88] C K Allen, *Law and Orders* (3rd edn, Stevens 1965), 242.

into ill-suited domestic characterizations such as whether a decision-making power is legislative, judicial, or administrative. As recognized above, international lawyers should recognize that due process is a contextual concept, and the germane issue is not whether Security Council power can be characterized as 'judicial', but rather the extent to which any particular decision impacts upon individual rights.

D. Myth No 3: Procedural Safeguards Cannot Bend to the Emergency Context

A third problem with the resort to procedural frameworks developed for domestic settings, and in particular the human rights model under the top-down approach, is that they fail adequately to account for the emergency context within which the Security Council necessarily operates. As a consequence, polarized perspectives have arisen as to the scope of procedural safeguards applicable to the Security Council setting, from the Security Council's initial determination that no legal issues were involved in the listing or de-listing of individuals on sanctions blacklists,[89] to the determination by European courts that the full gamut of procedural safeguards apply.[90] Most human rights regimes recognize that governments may take measures derogating from rights including the right to a fair hearing in times of public emergency;[91] however, this 'right to derogate' does not help to define the core due process rights to which individuals are entitled in situations of emergency, nor to elucidate how these might be imagined in the Security Council sanctions context.[92]

[89] Confidential communication, 24 January 2006, cited in Simon Chesterman, 'The Spy Who Came in From the Cold War: Intelligence and International Law' (2006) 27 *Michigan Journal of International Law* 1071, 1117.

[90] *European Commission and Council v Yassin Abdullah Kadi* [2013] ECR not yet reported, [97]–[98], [125]; Case T-85/09 *Kadi v European Commission* [2010] EUECJ, [126]; Joined Cases C-402/05 P and C-415/05 P *Kadi v Council of the European Union* [2008] ECR I-0000, [317], [326].

[91] ICCPR, Article 4; ECHR, Article 15; ACHR, Article 27 (though interestingly this provision does not allow suspension of 'judicial guarantees essential for the protection of such rights').

[92] International humanitarian law and the 'right to a fair trial in times of war' under Common Article 3(1)(d) of the Geneva Conventions and Article 75(4) of Additional Protocol 1 is no greater help in this respect. First, it is unlikely that individuals on sanctions blacklists are entitled to such a right (see, for example, UN Human Rights Committee 'General Comment 32: Right to equality before courts and tribunals and to a fair trial' (23 August 2007) UN Doc CCPR/C/GC/32); and, second, such provisions do not deal in any specificity with the unique problems faced in the Security Council sanctions setting.

It is clear that greater balance needs to be achieved between these polarized positions. The difficulty is that, while the level of secrecy engaged in by the Security Council may at times have been disproportionate,[93] it is also clearly necessary to preserve the effectiveness of the sanctions regime so as not to give those placed on sanctions blacklists an opportunity to undermine the system, and to protect information relating to national security and defence interests necessarily implicated in sanctions decision-making. At the same time, the solution is not to abandon procedural protections altogether. Instead, imaginative institutional responses may be necessary in order to navigate an appropriate balance between the individual rights of those placed on sanctions blacklists, and the interests of international peace and security. In developing a suitably contextual procedural framework, it may be helpful to take inspiration from creative solutions developed in the course of the substantial domestic and UN history of intelligence sharing.[94] For example, in lieu of recognized rights such as the right of an individual to notice of the charge against them, to knowledge of the adverse evidence against them, to public judgment, and to access to information, it may be more appropriate to adopt techniques of delayed notice,[95] a limited right for an individual to know the 'gist' of the case against them,[96] the use of special advocates,[97]

[93] In Paul Conlon's analysis of the inner workings of the Iraq Sanctions Committee, he concluded that discussion of sensitive confidential matters accounted for a mere 2.5% of the committee's meeting time, and also that the closed meeting format in fact elicited little candour and frankness from committee members, and aided Iraq and sanctions evaders as it shielded them from the public spotlight: P Conlon, *United Nations Sanctions Management: A Case Study of the Iraq Sanctions Committee* (Transnational Publishers 2000) 34, 151.

[94] Simon Chesterman, *Shared Secrets: Intelligence and Collective Security* (Lowy Institute 2006), 70–71; Simon Chesterman, 'The Spy Who Came in From the Cold War: Intelligence and International Law' (2006) 27 *Michigan Journal of International Law* 1071.

[95] See 8 USC § 1189 (2004) and Executive Order No 13,224, 66 Fed Reg 49,079 (23 September 2001); upheld in *National Council of Resistance of Iran v Dept of State* 251 F 3d 192 (DC Cir 2001), 196, 208; *Global Relief Foundation v O'Neill* 315 F3d 748 (7th Cir 2002), cert denied, 124 S Ct 531 (2003). See further Eric Broxmeyer, 'The Problems of Security and Freedom: Procedural Due Process and the Designation of Foreign Terrorist Organizations Under the Anti-Terrorism and Effective Death Penalty Act' (2004) 22 *Berkeley Journal of International Law* 439.

[96] [Canada] *Chiarelli v Canada (Minister of Employment and Immigration)* [1992] 1 SCR 711; [United Kingdom] *R v Home Secretary ex p Fayed* [1997] 1 All ER 228 (CA); *R (Roberts) v Parole Board* [2005] 2 AC 738, 779; *Secretary of State for the Home Department v AF* [2009] UKHL 28; [United States] Classified Information Procedure Acts 18 USC app3 §4; *Hamdi v Rumsfeld* 542 US 507, 533 (2004); *Boumediene v Bush* 553 US 723 (2008). See further Serrin Turner and Stephen J Schulhofer, *The Secrecy Problem in Terrorism Trials* (Liberty and National Security Project 2005), 61.

[97] [United Kingdom] Special Immigration Appeals Commission Act 1997, sections 5(3)(c), 6; Special Immigration Appeals Commission (Procedure) Rules 2003, rule 37(3); Prevention of Terrorism Act 2005, Schedule, para 7; Civil Procedure (Amendment No 2) Rules 2005 (SI 2005/656), rules 76.23–76.25, 76.28, 76.29; [Canada] Immigration and Refugee Protection Amendment Act 2008, sections 85.1–85.5, passed following the Canadian Supreme Court's

independent management of evidence by a neutral fact-finder[98] and a limited right of public access to information.[99] The employment of alternative techniques that accommodate both the rights of individuals and the imperative interests of international peace and security should be explored in the Security Council setting, with due regard to the specific issues faced in the context of the sanctions regime.

E. Conclusion

A formalist approach to the development of a procedural framework for Security Council sanctions decision-making has proved limiting. It risks relegating international lawyers to little more than magical (but ultimately misguided) bean-counters, declaring universally applicable principles to apply to the Security Council based on the recognition of such rights in 'x' number of treaties, or the representation of similar principles in 'y' number of states. More damagingly, the transplantation of procedural principles developed for the domestic setting into the Security Council context raises the 'danger of distortion' famously identified by Jennings and Watts.[100] A number

decision in *Charkaoui v Canada* [2007] SCC 9 (now Immigration and Refugee Protection Act 2001, c 27, §83 (1.2), §85); [New Zealand] *Zaoui v Attorney General* [2004] NZLR 339. See further, John Ip, 'The Rise and Spread of the Special Advocate' (2008) *Public Law* 717; Daphne Barak-Erez and Matthew C Waxman, 'Secret Evidence and the Due Process of Terrorist Detentions' (2009) 48 *Columbia Journal of Transnational Law* 3.

[98] [United Kingdom] Special Immigration Appeals Commission Act 1997; [Israel] *A v State of Israel* Crim A.6659/06 (2008), [43]; *Khadri v IDF Commander in Judea and Samaria* [2004] HCJ 110006/04, [6]; *Agbar v IDF Commander in Judea and Samaria* [2007] HCJ 9441/07, [8]; *Federman v Commander of the Central Command* [2005] HCJ 5555/05 Israel SC 59(2) 865, 869 per Justice Rubinstein; [Special Tribunal for Lebanon] Lukasz Korecki, 'Procedural Tools for Ensuring Cooperation of States with the Special Tribunal for Lebanon' (2009) 7 *Journal of International Criminal Justice* 927.

[99] The law relating to access to information in several states incorporates a proportionality test requiring authorities: (1) to strike a balance between public interest in disclosure and the possible harm that disclosure will cause to protected interests, and/or (2) to take the least restrictive measures available: see, for example: [China] People's Republic of China Ordinance on Openness of Government Information 2007, entered into effect 1 May 2008, article 14; [India] Right to Information Act 2005, section 8(2); [Japan] Law Concerning Access to Information Held by Administrative Organs 1999, entered into effect April 2001, articles 6, 7; [South Africa] Promotion of Access to Information Act 2000, section 46; [UK] Freedom of Information Act 2000, entered into effect January 2005, section 2(2)(b).

[100] 'Unless there is some sufficient counterpart to [national laws and procedures] in the international sphere, or sufficient allowance is made for them in abstracting the principles from the various municipal rules, the operation of the principles as a source of particular rules of international law will be distorted': Robert Jennings and Arthur Watts, *Oppenheim's International Law*, vol I (9th edn, Longman 1996), 37. See also Judge Sir Arnold McNair's much-quoted statement in the *Advisory Opinion on the International Status of South West Africa* [1950] ICJ Rep 128, 132: 'The way

of myths and misconceptions about procedural fairness are already beginning to entrench themselves, including the notion that procedural fairness is a 'one-size-fits-all' adjudicatory framework built around a civil/criminal dichotomy that applies only to 'judicial' decision-making, and does not readily apply in emergency contexts. This fundamentally misconceives the contextual nature of procedural fairness, and the need to adapt its precepts to particular legal settings.

The aim of this book is to reconfigure the approach to procedural fairness in sanctions decision-making from a formalist source-based methodology to a more substantive 'value-based' approach based on a thicker normative understanding of the concept of procedural fairness. Development of a procedural fairness framework for the Security Council requires a more context-sensitive approach, focused not on the formal question as to which procedural safeguards can be said to be legally binding on the Council, but on the process values underlying such a concept, and the way or ways they can actually enhance Security Council decision-making. As the UN Ombudsperson stated in a Council debate, '[f]air process is supportive of, and in fact essential to, the *effective* implementation of sanctions measures'.[101] A value-based appreciation of the contribution of due process will not only increase the persuasiveness of the case for procedural reform, but more adequately fulfils the responsibility of international lawyers to develop international legal principles that respond to and improve the international legal system.

in which international law borrows from [municipal law] is not by means of importing private law institutions "lock, stock and barrel", ready made and fully equipped with a set of rules…In my opinion the true view of the duty of international tribunals in this matter is to regard any features or terminology which are reminiscent of the rules and institutions of private law as an indication of policies and principles rather than as directly importing these rules and institutions'. See also Brunée and Toope's more recent observation that 'dutiful incantations of caution notwithstanding, international lawyers and commentators on international law…frequently draw on domestic law in their assessments of or prescriptions for its international relative. The resulting picture of international law is rarely good, often ugly, and always distorted': Jutta Brunée and Stephen J Toope, *Legitimacy and Legality of International Law* (Cambridge University Press 2008).

[101] Kimberly Prost, Debate on Security Council Working Methods, 7285th meeting of Security Council (23 October 2014) UN Doc S/PV.7285.

PART II

THEORY

4

A 'Value-Based' Approach

Three Models of Due Process

Even within domestic legal systems, it was felt historically that insufficient attention had been paid to the theoretical aspects of procedures, reflecting a sentiment that procedural rules relate 'only to the nuts and bolts of legal machinery, whereas central theoretical issues lie elsewhere'.[1] Yet the vast literature on the subject in the last few decades reflects that the theoretical dimension of due process is extraordinarily rich.[2] This literature has been virtually untapped in the debate about the application of due process to the Security Council setting. Instead, it has been common in the international debate about sanctions decision-making to discuss due process in terms of a set of procedural rights, including (1) the right to notice; (2) the right to a hearing; (3) the right to reasons; (4) the right of appeal to an independent tribunal; and (5) the right of public access to information.[3] In proposing applicable procedural standards, little consideration has been given to the centralizing theoretical foundations that bind the constituent standards of due process to the broader concept, in particular the purpose or purposes underlying this set of rights. Yet, as this chapter seeks to show, due process is far more than the sum of its parts. An important aim of this value-based analysis is to demonstrate that due process is more than a set of discrete legal

[1] Gerry Maher, 'Natural Justice as Fairness' in Neil MacCormick and Peter Birks (eds), *The Legal Mind: Essays for Tony Honoré* (Clarendon Press 1986), 104.

[2] See, for example, J Roland Pennock and John W Chapman, *Due Process* (New York University Press 1977); Jerry Mashaw, *Due Process in the Administrative State* (Yale University Press 1985); M Bayles, *Procedural Justice* (Kluwer 1990); D J Galligan, *Due Process and Fair Procedures: A Study of Administrative Procedures* (Clarendon Press 1996), 10; Carol Harlow and Richard Rawlings, *Law and Administration* (3rd edn, Cambridge University Press 2009), ch 14.

[3] See, for example, Bardo Fassbender, *Targeted Sanctions and Due Process* (Council of Europe Committee of Legal Advisors on International Law 2006), 8, 28; 'Non-paper of the Secretary General' referred to in the debate on 'Strengthening international law: rule of law and maintenance of peace and security' (22 June 2006) UN Doc S/PV.5474, 5; International Commission of Jurists, *Assessing Damage, Urging Action: Report of the Eminent Jurists Panel on Terrorism: Counter-terrorism and Human Rights* (2009), 122.

standards, but is a touchstone for the way in which a particular legal order conceives of far larger issues.

This chapter will make three key points about due process. The first is that procedural fairness is ultimately about legitimacy in the allocation of decision-making power. An historical analysis of the concept reveals that due process has traditionally served as a mechanism for moderating the conflict between the exercise of public decision-making authority and the demands of the legal community in respect of which the authority is exercised. Second, it does this by enabling a form of 'dialogue' between decision-makers and the community to whom the decisions relate. Third, and most importantly, the purpose of this dialogue will differ between legal contexts. A survey of the literature reveals that it is possible to identify three dominant 'process values' underlying the dialogue enabled by due process. These different process values will serve to enhance the legitimacy of decision-making in different ways, and are connected to *meta* theories including legal conceptions of community, law, and the representation of values.

In writing about the theoretical aspects of due process, I do not expect that my analysis will contain anything newsworthy for public lawyers. My hope is that the conclusions reached will have public lawyers either nodding their heads in agreement, or condemning their obviousness. My intended audience in writing this chapter is the range of practitioners and academics of public international law involved in the important task of developing a procedural fairness framework for the Security Council context. My aim is to contribute a broader understanding of the concept of due process to the important debate about the adoption of procedural safeguards within the Security Council sanctions regime. It is only against the backdrop of this broader understanding that we can develop an appropriate procedural framework for the Security Council setting, with full knowledge of the broader implications of the choice of framework for the legitimacy of Security Council decision-making and the international legal order more generally.

A. Due Process: History, Power, and Legitimacy

Due process has played a very important historical role in shaping a society's legal, political, and social decision-making structures. Shifts in governmental decision-making power have often been matched by widespread procedural reform. During the eighteenth, nineteenth, and twentieth centuries, there was a shift in the centre of gravity of governmental authority from the legislative and judicial to the administrative branches of government across a range of countries, including the United States, England, and France.[4] In many cases, this shift in the locus of governmental authority was accompanied by crises of legitimacy, generally stemming from concerns about the concentration of decision-making authority in the hands of unelected officials, and attempts to insulate this authority from interference by the judiciary.[5] It was

[4] See, for example, Freedman's comment that 'as the role of the federal government in national life has expanded, the center of gravity of the powers it exercises has gradually shifted, from the legislature in the first half of the nineteenth century, to the judiciary in the second half of the nineteenth century, to the executive and administration in the twentieth century': James O Freedman, *Crisis and Legitimacy: The Administrative Process and American Government* (Cambridge University Press 1978), 3. In the United Kingdom, Professor Robson remarked that it is 'true to say that the centre of gravity in English government has shifted from legislation to administration during the past half-century, and the hegemony of the executive, whether we like it or not, is an accomplished fact': William A Robson, *Justice and Administrative Law: A Study of the British Constitution* (2nd edn, Stevens & Sons 1947), 33.

[5] In the United States, nearly every President since the period of the New Deal has been sufficiently disturbed by the functioning of administrative power to order at least one major study seeking recommendations for improvement to the functioning of administrative power, with the 1938 Special Committee on Administrative Law reporting that the 'fundamental evils' of administrative power included the combination of judicial with executive or legislative functions, and the lack of effective independent review or judicial control of administrative decisions': (1938) 61 ABAR 720. In the United Kingdom, Lord Hewart (then Lord Chief Justice in the United Kingdom) launched a scathing attack against the increasing assumption of unchecked rule-making, enforcement, and adjudicative power by executive offices and administrative agencies, asserting that 'organised administrative lawlessness' was 'certain in the long run to become despotism and there is danger if the . . . lawlessness is greatly extended, of the loss of those hardly won liberties which it has taken centuries to establish': Lord Hewart, *The New Despotism* (Ernest Benn 1929), 52. See also A V Dicey, *Introduction to the Study of Law of the Constitution* (E C S Wade ed) (9th edn, Macmillan 1939); C K Allen, *Bureacracy Triumphant* (Oxford University Press 1931); J A R Marriott, *The Mechanism of the Modern State* (Clarendon Press 1927). In France, the establishment of a separate administration insulated from the ordinary courts was said to be at 'the very heart of the gravest political and constitutional debates' of the *ancien regime* (Michel Antoine, *Le Conseil du roi sous le règne de Louis XV* [Droz 1970]) and sparked 'a fairly continuous and often too intense war between . . . the jurisdictional and the ministerial': Antoine-Jean-Baptiste Auget de Montyon (1733–1820), *Des agents de l'administration* (Archives de l'Assistance publique de Paris, Fonds Montyon), section titled 'Des intendants de province', both cited in Peter Lindseth, 'Always Embedded Administrations: The Historical Evolution of Administrative Justice as an Aspect of Modern Governance' in Christian Joerges, Bo Stråth, and Peter Wagner (eds), *The Economy as a Polity: The Political Constitution of Contemporary Capitalism* (Routledge Cavendish 2005), 121. See also Charles Louis de Secondat baron de Montesquieu, *De l'esprit des lois* (Flammarion 1979), Livre XI, ch VI, 'De la Constitution d'Angleterre', 1:294–5 and Livre VI, ch VI, 'Que, dans la monarchie, les ministres ne doivent pas juger', 1:207–8; Keith Michael Baker, 'Science and Politics at the End of the Old Regime' in K M Baker (ed), *Inventing the French Revolution: Essays on the French Political Culture in the 18th Century* (Cambridge University Press 1990), 153–166. This pressure for public accountability is said to have been in direct reaction to the changes in the structures of public governance in France that took place over the century stretching from the 1640s to the 1740s: Peter Lindseth, 'Always Embedded Administrations: The Historical Evolution of Administrative Justice as an Aspect of Modern Governance' in Christian Joerges, Bo Stråth, and Peter Wagner (eds), *The Economy as a Polity: The Political Constitution of Contemporary Capitalism* (Routledge Cavendish 2005), 117–35.

largely through the reformation of procedural fairness and due process law that these concerns were redressed. In the American context, the elaboration of due process principles has been described as the 'primary mechanism' for the redefinition of the modern administrative state.[6] Napoleon's 'rejudicialisation' of French administrative justice, through the establishment of the *Conseil d'Etat* and, later, the investiture of the *Conseil d'Etat* with the power of review over administrative decision-making authority, was described by Napoleon as necessary or 'the government will fall into scorn'.[7] In the United Kingdom, the constitutional significance of the House of Lords' decision in *Ridge v Baldwin*,[8] which extended procedural fairness to the exercise of all administrative functions, is expressed in its description by C K Allen as the 'Magna Carta of natural justice'.[9]

The idea that due process has served historically to enhance or restore legitimacy in the wake of shifts in decision-making authority is an interesting one in the present context of Security Council sanctions decision-making. The assumption of decision-making authority over individuals by the Security Council might be viewed as the next important shift in decision-making authority, this time from the domestic to the international sphere.[10] This shift in the locus of decision-making authority has

[6] Jerry Mashaw, *Due Process in the Administrative State* (Yale University Press 1985), 1. See also Stewart's description of the extension of procedural protections to decision-making by governmental agencies as a 'due process revolution' that was part of the 'struggle to adapt legal controls to expanded governmental authority': Richard B Stewart, 'The Reformation of American Administrative Law' (1975) 88 *Harvard Law Review* 1667, 1717–22. See also Charles A Reich, 'The New Property' (1964) 73 *Yale Law Journal* 733.

[7] Pelet de la Lozère, *Opinions de Napoléon* (F Didot 1833), 191; cited in Peter Lindseth, 'Always Embedded Administrations: The Historical Evolution of Administrative Justice as an Aspect of Modern Governance' in Christian Joerges, Bo Stråth, and Peter Wagner (eds), *The Economy as a Polity: The Political Constitution of Contemporary Capitalism* (Routledge Cavendish 2005), 117. See also the *Cadot* decision of December 1889, in which the Conseil declared that it had original jurisdiction to hear any administrative dispute from the moment it arose, which has been described as the ultimate realization on the part of the members of the Conseil d'Etat that there was still a place in the republican order for independent legal control of executive power, and that hierarchical political control by politicians or government ministers was not enough: Lindseth (above), 123.

[8] [1964] AC 40.

[9] C K Allen, *Law and Orders* (3rd edn, Stevens 1965), 242. See also *Wiseman v Borneman* [1971] AC 297; *Bushell v Secretary of State for the Environment* [1981] AC 75 per Lord Diplock; *Lloyd v McMahon* [1987] AC 628; *R v Home Secretary ex p Doody* [1993] 3 WLR 821, 878 per Lord Mustill.

[10] It has been described by one scholar as a 'rehearsal for world governance': Andrea Bianchi, 'Assessing the Effectiveness of the UN Security Council's Anti-terrorism Measures: The Quest for Legitimacy and Cohesion' (2006) 17 *European Journal of International Law* 881, 891. See also Dan Sarooshi, *International Organizations and their Exercise of Sovereign Powers* (Oxford University Press 2005); Bruce Cronin and Ian Hurd, *The UN Security Council and the Politics of International Authority* (Routledge 2008), 5; Joseph Weiler, 'The Geology of International Law—Governance, Democracy and Legitimacy' (2004) 64 *Zeitschrift für ausländisches öffentliches Recht und Völkerrecht* 547, 550.

certainly sparked similar concerns to those emerging during the rise of the modern administrative state, namely fears about the exercise of power over individuals by an unaccountable body and the absence of judicial review. Perceived against the backdrop of other historical shifts in governmental decision-making authority, the failure of the Security Council to respond to pressure to establish adequate procedural fairness safeguards regulating its assumption of decision-making authority over individuals has been something of an historical anomaly.

B. Due Process as 'Dialogue'

Though due process has proved a valuable tool in comparable historical contexts, there has been a failure in the international sphere to appreciate the precise way or ways in which procedural law can positively contribute to Security Council sanctions decision-making. It is often said, and history confirms, that the essential aim of procedural fairness is to enhance the legitimacy of decision-making.[11] However, legitimacy is a wide lens. If the case for procedural reform is to be persuasive, it is necessary to bring the essential contribution of due process into sharper focus.

I argue that due process serves to establish a 'peculiar form of dialogue' in decision-making.[12] Safeguards associated with procedural fairness, including the right to notice, hearing, reasons, review, and access to information, aim collectively to open up a structured dialogue between decision-making authority and those affected by decisions. Broadly, the aim of this dialogue is to enhance legitimacy. I am not so concerned in this book with examining the general link between dialogue and legitimacy.[13] My aim is not to demonstrate the superiority of dialogue theory over other established bases for the acceptance of governing authority, though the role of dialogue is particularly relevant in the international sphere where it is able to provide a substitute for

[11] Jerry Mashaw, *Greed, Chaos and Governance* (Yale University Press 1997), 108; T M Scanlon, 'Due Process' in J Roland Pennock and John W Chapman (eds), *Due Process* (New York University Press 1977), 94; Carol Harlow and Richard Rawlings, *Law and Administration* (3rd edn, Cambridge University Press 2009), 621.

[12] This expression is borrowed from Fuller's discussion of the related concept of adjudication: Lon Fuller, 'The Forms and Limits of Adjudication' (1978) 92 *Harvard Law Review* 353, 364.

[13] On this, see Jürgen Habermas, *The Theory of Communicative Action: Reason and the Rationalization of Society*, vol 1 (Thomas McCarthy tr) (Beacon 1984); Jürgen Habermas, *Between Facts and Norms: Contribution to a Discourse Theory of Law and Democracy* (William Rehg tr) (MIT Press 1996); Jens Steffek, 'The Legitimation of International Governance: A Discourse Approach' (2003) 9 *European Journal of International Relations* 249; Tom Tyler, *Why People Obey the Law* (Princeton University Press 2006).

other sources of democratic legitimacy.[14] I am more interested in the 'peculiar' form of dialogue enabled by procedural fairness, and the way or ways in which this procedural dialogue is able to contribute to greater legitimacy.

This does not escape the need to start with a general understanding of what I mean by the omnibus term of legitimacy. Legitimacy is the capacity of authority to pull toward consensual compliance.[15] Whether 'legitimacy' is described by reference to a Habermas-ian 'worthiness to be recognized',[16] a Franck-ian 'connection to a community's evolving standards',[17] or Beetham's requirement that decision-making authority should find its foundation in 'beliefs shared by both dominant and subordinate',[18] the concept of legitimacy envisages a connection between authority and community values sufficient to ground acceptance of that authority in the relevant community. Procedural fairness acts in the service of legitimacy by shoring up the connection that acts as legitimacy's source, providing legal standards that serve to establish a dialogue between decision-makers and the community affected by decisions to ensure decision-making takes place in accordance with relevant community values. Setting up this general connection with legitimacy is important because it reveals that due process is necessarily intricately bound up with conceptions of community, values, and the role of law in achieving these values. The contextual nature of due process arises from the fact that conceptions of community, law, and values will differ between legal contexts.

In developing a normatively rich understanding of procedural fairness, it is possible to identify a number of distinct schools of thought as to the precise way in which procedural fairness is thought to enhance the legitimacy of decision-making. These are commonly discussed in terms of process values, with scholars differing over the central process value or values advanced by procedural frameworks. While these process values are not mutually exclusive and may coexist, it is helpful to examine them in terms of distinct conceptual

[14] Daniel Esty, 'Good Governance at the Supranational Scale: Globalizing Administrative Law' (2006) 115 *Yale Law Journal* 1490, 1496. See also Ian Johnstone, 'Legislation and Adjudication in the UN Security Council: Bringing Down the Deliberative Deficit' (2008) 102 *American Journal of International Law* 275; Allen Buchanan and Robert O Keohane, 'The Legitimacy of Global Governance Institutions' (2006) 20 *Ethics and International Affairs* 405; Erika de Wet, 'The International Constitutional Order' (2006) 55 *International and Comparative Law Quarterly* 51, 74; Mattias Kumm, 'The Legitimacy of International Law: A Constitutionalist Framework of Analysis' (2004) 15 *European Journal of International Law* 907, 924–7.

[15] Thomas M Franck, 'The Power of Legitimacy and the Legitimacy of Power: International Law in an Age of Power Disequilibrium' (2006) 100 *American Journal of International Law* 88, 93.

[16] J Habermas, 'Legitimation Problems in the Modern State' in *Communication and the Evolution of Society* (T McCarthy tr) (Beacon Press 1979), 178–9.

[17] Thomas M Franck, *Fairness in International Law and Institutions* (Clarendon Press 1995), 26.

[18] David Beetham, *The Legitimation of Power* (Peter Jones and Albert Weale eds) (Macmillan 1991).

models. If we examine each model individually, we can better appreciate the different ways in which each model purports to contribute to the achievement of legitimacy in decision-making. These models can be described as: (1) the instrumentalist model, which considers that the aim of procedures is to enhance the *accuracy* of decision-making outcomes; (2) the dignitarian model, which emphasizes the inherent value of procedures in providing a form of *interest representation* for those affected by decision-making; and (3) the public interest model, which regards procedural fairness as a mechanism to enhance *public accountability* in decision-making. As we shall see in the following analysis, underlying these different models is in fact a difference of opinion about theories of community, law, and values, which impact upon the work done by due process to enhance legitimacy in each case.

C. The Instrumentalist Approach: A Model Based on Accuracy

The classical model of procedural fairness contributes to the legitimacy of decision-making by seeking to minimize the incidence of error. This model ascribes an instrumental role to procedural fairness, regarding it above all as a mechanism by which to achieve greater *accuracy* in the application of the substantive law to the facts. The preceding sentence is deliberately qualified: the focus of the instrumentalist model is not on achieving accuracy or truth in decision-making per se, but on the somewhat narrower aim of ensuring that decision-makers accurately apply the substantive law in their decision-making.

This model is associated with Jeremy Bentham, the original and most influential proponent of an instrumental approach to procedural justice. Unsurprisingly given this connection, it has been interpreted to form part of a broader utilitarian approach to decision-making, accepting that the aim is not merely accuracy, but a level of accuracy that will contribute to greater overall efficiency or effectiveness in decision-making. As Michael Bayles remarked, 'to aim at the truth, the whole truth and nothing but the truth' would be too expensive, committing society to an intolerable level of expenditure on decision-making procedures.[19] Under an 'economic' interpretation of

[19] M Bayles, *Procedural Justice* (Kluwer 1990), 117. See also Robert G Bone, 'Agreeing to Fair Process: The Problem with Contractarian Theories of Procedural Fairness' (2003) 83 *Boston University Law Review* 485, 493; J Bowring (ed), *Works of Jeremy Bentham*, vol 7 (William Tait 1838–1843), 335, cited in Gerald J Postema, *Bentham and the Common Law Tradition* (Clarendon Press 1986), 346.

the instrumentalist model, advocated in different formulations by Michael Bayles,[20] Ronald Dworkin,[21] Dennis Galligan,[22] and the US Supreme Court,[23] it is necessary to balance the cost of error[24] with the direct and indirect costs of procedural safeguards to the decision-making process, including financial cost,[25] delay, and complication.[26]

(1) Link between the Instrumentalist Model and Legal Positivism

The reason for examining these different models of procedural fairness is to understand the way in which they are able to contribute to the legitimacy of decision-making. Deeper analysis of the connection between the instrumentalist model and legitimacy in decision-making reveals that the causal relationship is more complex than might first appear. The legitimizing influence of the instrumentalist model has been described in terms of the certainty and stability that result from accurate and consistent application of the law.[27] Yet this argument is a circular one: legitimacy collapses into formal validity, while formal validity is said to guarantee legitimacy.[28] It is well recognized that fidelity to law cannot be achieved by mere formalism. There is no legitimacy in the consistent application of laws that do not command the broader support or acceptance of the community to which they are applied.

The source of legitimacy underlying the model can be better understood through recognition of the link between an instrumentalist model of procedural fairness and legal positivism. The instrumentalist model is essentially

[20] M Bayles, *Procedural Justice* (Kluwer 1990), 117–18.

[21] R Dworkin, *A Matter of Principle* (Clarendon Press 1986), ch 3.

[22] D J Galligan, *Due Process and Fair Procedures: A Study of Administrative Procedures* (1996), 201–4.

[23] *Mathews v Eldridge* 424 US 35 (1976). See criticism of this approach in Note, 'Specifying the Procedure Required by Due Process: Towards Limits on the Use of Interest Balancing' (1975) 88 *Harvard Law Review* 1510.

[24] M Bayles, *Procedural Justice* (Kluwer 1990), 118. Note that Dworkin includes in this both the 'bare harm' of being denied an advantage or being subjected to an imposition, and the 'moral harm' or injustice factor which occurs when a person is treated unjustly: R Dworkin, *A Matter of Principle* (Clarendon Press 1986), 80.

[25] Bayles explains that these may include 'the public costs for officials' salaries, hearings, and so on, and the private costs of parties in hiring lawyers, obtaining expert testimony, and so forth': M Bayles, *Procedural Justice* (Kluwer 1990), 119.

[26] Jerry Mashaw, *Due Process in the Administrative State* (Yale University Press 1985), 117.

[27] D J Galligan, *Due Process and Fair Procedures: A Study of Administrative Procedures* (Clarendon Press 1996), 10; D J Galligan, *Discretionary Powers* (Clarendon Press 1986), 68.

[28] James C Ketchen, 'Revisiting Fuller's Critique of Hart—Managerial Control and the Pathology of Legal Systems: The Hart-Weber Nexus' (2003) 53 *University of Toronto Law Journal* 1, 29.

the handmaiden of a positivist conception of the rule of law.[29] Bentham, an archetypal positivist, described procedural law as 'adjective law', which acts in the service of the ends of the substantive law in much the same way that the adjective is dependent upon and subservient to the noun that it modifies.[30] The nature of this subservience is captured in the description of the instrumentalist model as a 'transmission belt' for implementing legislative directives.[31] Basically, the only but necessary role of procedural law under this model is to make sure that the rules are applied as enacted, thus fulfilling the positivist's aim of achieving the accurate and objective implementation of the substantive law. Recognition of the link between a positivist theory of law and instrumentalism has three very important implications when we turn to consider the application of this model to the Security Council sanctions setting.

(a) Community: The Role of a Legitimately Representative Lawmaker

The first implication of the link between an instrumental model and positivism is that it presupposes the existence of a 'legitimately representative lawmaker' in the legal system within which the model applies. The accurate application of the substantive law only enhances legitimacy where the substantive law reflects the will of the broader legal community. Under a positivist theory of the rule of law, legitimacy does not emerge merely from the accurate application of law but from a deeper internal acceptance of the authority of the lawmaker to enact the law.[32] In a domestic context, the instrumentalist model contributes legitimacy to a legal system by ensuring that unrepresentative bureaucratic decision-makers accurately apply the commands of a legitimate source of authority, usually the legislature.[33] The duty to submit to the rules derives from the legitimacy of the whole compromise rather than from that of the particular command.[34] The existence

[29] H L A Hart described procedural law as the guarantor of impartiality and objectivity in the application of legal rules: H L A Hart, *The Concept of Law* (Clarendon Press 1961), 156.

[30] Gerald J Postema, 'The Principle of Utility and the Law of Procedure: Bentham's Theory of Adjudication' (1977) 11 *Georgia Law Review* 1393; citing Jeremy Bentham, 'Constitutional Code' in J Bowring (ed), *Works of Jeremy Bentham*, vol 9 (William Tait 1838–1843), 25 and Jeremy Bentham, 'The Principles of Judicial Procedure' in ibid, 6, 8.

[31] Richard B Stewart, 'The Reformation of American Administrative Law' (1975) 88 *Harvard Law Review* 1667, 1672.

[32] H L A Hart, *The Concept of Law* (Clarendon Press 1961); Duncan Kennedy, 'Legal Formality' (1973) 2 *Journal of Legal Studies* 351, 358–9.

[33] Richard B Stewart, 'The Reformation of American Administrative Law' (1975) 88 *Harvard Law Review* 1667.

[34] Duncan Kennedy, 'Legal Formality' (1973) 2 *Journal of Legal Studies* 351, 369. As Professor Michelman remarks, in such circumstances, we are 'imagined to have consented in advance to what appear as constraints or harms imposed upon us by others': Frank I Michelman, 'Formal and

of a 'legitimately representative lawmaker' is therefore of central importance to the capacity of the instrumentalist model to enhance the legitimacy of decision-making.

(b) Law: An Adjudicatory Legal Framework

Second, the instrumentalist model manifests a preference for a particular form of legal authority, best described as an adjudicatory legal framework. Fuller described such a framework as one that 'gives formal and institutional expression' to the influence of legal reason in human affairs.[35] As discussed above, it refers to a framework based on a logically coherent and complete system of principles and rules, where decisions are reached through the presentation of proofs in trial-type hearings and are amenable to authoritative review by an independent arbiter.[36] A number of implications follow in terms of applicable procedures: (a) clear and determinate legal standards; (b) the right of an affected party to a legal hearing, consisting in the opportunity to present proofs and reasoned arguments based on these standards; and (c) the giving of reasons for the decision. The lynchpin of such a model is (d) an independent judicial arbiter capable of authoritatively testing the accuracy of the application of these legal standards in the decision-making process. The fairness and effectiveness of adjudication will be promoted by (e) the giving of reasons demonstrating how a judge has applied the standards in the relevant decision.[37]

The adjudicatory framework is well placed to achieve the instrumentalist model's goal of accuracy in the implementation of the substantive law. In a logically coherent and complete system of principles and rules, decision-makers are provided with legal guidance in approaching individual cases, assisted by proofs and reasoned arguments from affected individuals as to how these legal standards should be applied. The idea is that, in a system based on clear and determinate standards, decision-makers are more likely to apply the law, instead of their personal notions of fairness, justice, or appropriateness. In turn, decision-making is amenable to review

Associational Aims in Procedural Due Process' in J Roland Pennock and John W Chapman (eds), *Due Process* (New York University Press 1977), 129.

[35] Lon Fuller, *The Morality of Law* (Rev edn, Yale University Press 1969).

[36] See, for example, Lon Fuller, 'The Forms and Limits of Adjudication' (1978) 92 *Harvard Law Review* 353.

[37] Ibid, 369, 373, 382, 388, 365. See also M Bayles, *Procedural Justice* (Kluwer 1990), 14, who distinguishes adjudication from other forms of decision-making on the following bases: '(1) There is a particular dispute. (2) Particular parties are involved;...(3) A third party is involved. (4) A hearing is held in which information concerning the dispute is presented. (5) The third party "settles" the controversy by rendering a decision. (6) This decision is based on principles and rules—those of the substantive law—with reference to the information presented at the hearing.'

in accordance with these standards to determine whether decisions can be counted as correct or incorrect.[38] Bentham's instrumentalist description of procedural fairness as 'adjective law', discussed above, emphasizes the co-ordinative role of procedural law under this model, with the key aim being to achieve the accurate application of the facts to the substantive law. Stewart's description of procedural law in terms of a 'transmission belt' for legislative directives captures the same idea, underlining the role of procedure in delivering an impartial application of agreed standards to particular cases.

(c) *Values: Absence of Substantive System of Values*

In terms of values, the instrumentalist model lacks any substantive normative dimension. It emphasizes the accurate fulfilment of the substantive law and does not differentiate between acceptable and unacceptable aspects of that law. Broader functions such as accountability, the balancing of interests, and representation of community values need not be met by procedural protections because they are assumed already to be reflected in the substantive law. This impacts upon the nature and scope of procedural safeguards required under the model. Under the instrumentalist model, the affected individual is regarded almost purely as a source of information, and their right to participate is limited to the extent required in the aid of accurate and efficient problem-solving.[39] The chief purpose of reason-giving under the instrumentalist model is to provide a mechanism by which the review body can assess whether there have been any errors of fact or law impinging upon the decision. The role of review is 'to scrutinize a decision or action to determine whether it has been properly made or performed'.[40] A decision will be made 'properly' according to the instrumentalist model where the law has been accurately and properly applied and authoritative standards are honoured. The role of process under the instrumentalist model is essentially to augment the information available to both decision-maker and subject, enabling the latter to point out errors of fact or reasoning and perhaps persuade the decision-maker to change his or her mind.[41]

[38] Frank I Michelman, 'Formal and Associational Aims in Procedural Due Process' in J Roland Pennock and John W Chapman (eds), *Due Process* (New York University Press 1977), 130.

[39] Jenny Steele, 'Participation and Deliberation in Environmental Law: Exploring a Problem-solving Approach' (2001) 21 *Oxford Journal of Legal Studies* 415.

[40] D J Galligan, *Due Process and Fair Procedures: A Study of Administrative Procedures* (Clarendon Press 1996), 395.

[41] *R v Secretary of State for the Home Department ex p Doody* [1994] AC 531, 551 (Lord Mustill).

(d) Conclusion

In conclusion, underlying the instrumentalist model is a set of much larger issues related to law, community, and values. The model is based on a positivist conception of the rule of law. Its operation depends upon the existence of a legitimately representative lawmaker, such that the law the subject of accurate application can be said to reflect the will of the broader community. It is ideally suited to an adjudicatory legal system that operates according to clear and determinate standards by which the 'accuracy' of decision-making can be measured. Moreover, it does not contribute a set of values by which to measure decision-making beyond those already contained in the substantive law. These conceptual foundations of the instrumentalist model will need to be factored in when considering the appropriateness of its application to the Security Council sanctions setting.

D. The Dignitarian Approach: A Model Based on Interest Representation

The quest for accuracy reflected in the instrumentalist approach to procedural fairness has been overshadowed in more recent times by a competing approach that recognizes individual dignity as one of the primary contemporary foundations for procedural fairness. This model emerged at the domestic level from an increasing sense that the accurate application of legislative standards does not define the 'core ideal' of procedural fairness.[42] The growing concentration of public and private power in bureaucratic hands over the last half-century across many legal traditions gradually withered faith in the ability of the legislature adequately to control agency discretion. It was not enough to ensure that decision-making accurately fulfilled the commands of the legislature, but rather it was seen as necessary to redress enhanced concerns about the diminished opportunity for expressions of human individuality.[43] Consequently, through the writings of a number of academics, due process and procedural fairness have come to be associated with values including 'basic notions of dignity',[44] process values including 'humaneness

[42] Gerry Maher, 'Natural Justice as Fairness' in Neil MacCormick and Peter Birks (eds), *The Legal Mind: Essays for Tony Honoré* (Clarendon Press 1986).

[43] Gerald E Frug, 'The Ideology of Bureaucracy in American Law' (1984) 97 *Harvard Law Review* 1276.

[44] Richard B Saphire, 'Specifying Due Process Values: Toward a More Responsive Approach to Procedural Protection' (1978) 127 *University of Pennsylvania Law Review* 111, 119.

and respect for individual dignity',[45] 'the value of autonomy, of controlling the events that affect you... and the respect and self-respect that come with it',[46] and as 'rights to interchange [expressing] the idea that to be a person, rather than a thing, is at least to be consulted about what is done with one'.[47] This approach is often discussed in terms of a remedial solution to the Kafkaesque situation facing Josef K in *The Trial*. According to this model, procedural fairness is a guarantee against an individual being manipulated by a faceless bureaucracy, and is an essential pre-requisite to the satisfaction of the innate need to be treated as a responsible, independent, and autonomous human entity.[48]

Of course, the problem with such 'intuitive' values or concerns as dignity, self-respect, and autonomy is that they are ultimately amorphous and indeterminate, and therefore difficult to translate into a workable procedural framework. As Professor Galligan remarks, 'any idea which starts with a concept as fundamental, but at the same time as abstract, as human dignity is liable to suffer from misunderstanding and misinterpretation'.[49] There is a sense that, for some scholars, it is tempting to regard dignity as a universal norm, respect for which is essential to any legal system, such that procedural fairness becomes some sort of 'concrete specification of the principles of natural law'.[50] However, as Professor McCrudden has recognized in his detailed exegesis of the concept of dignity, there is no consensus conception of its legal use, such that dignity can only become 'other than impossibly vague when it is tethered to a coherent community of interpretation'.[51] In other words, dignity will be subject to different interpretations across different community settings. Other scholars of procedural fairness have acknowledged the inadequacy of intuition as a normative foundation, and have eschewed the value of a natural rights jurisprudence 'justly excoriated for its intuitive assertions of fundamental fairness'.[52] It is felt that some theoretical defence must be

[45] Robert Summers, 'Evaluating and Improving Legal Processes—a Plea for "Process Values"' (1974) 60 *Cornell Law Review* 1, 23.

[46] E Dauer and T Gilhool, 'The Economics of Constitutionalized Repossession' (1973) 47 *Southern California Law Review* 116, 148–9.

[47] Laurence Tribe, *American Constitutional Law* (2nd edn, Foundation Press 1988), 666.

[48] Richard B Saphire, 'Specifying Due Process Values: Toward a More Responsive Approach to Procedural Protection' (1978) 127 *University of Pennsylvania Law Review* 111, 152.

[49] D J Galligan, *Due Process and Fair Procedures: A Study of Administrative Procedures* (Clarendon Press 1996), 75.

[50] Paolo G Carozza, 'My Friend is a Stranger: The Death Penalty and the Global Ius Commune of Human Rights' (2003) 81 *Texas Law Review* 1031, 1081.

[51] Christopher McCrudden, 'Human Dignity and Judicial Interpretation of Human Rights' (2008) 19 *European Journal of International Law* 655, 723.

[52] Jerry Mashaw, *Due Process in the Administrative State* (Yale University Press 1985), 47. '[I]f we remain at the level of intuition, these values can only be asserted as desirable and attractive,

offered to translate intuitive claims into rights. Accordingly, there have been several attempts to connect the dignitarian concept of process to a broader legal or constitutional tradition, and to situate it within a coherent account of individual rights.

(1) The Dignitarian Model and Liberal-Democratic Theory?

The obvious candidate is to anchor the dignitarian model in the liberal-democratic intellectual tradition, which rests on a bottom-up view of politics as a response to the demands of the individual and societal groups.[53] The Kantian strand of the liberal tradition in particular sets out the core value of the dignitarian model in its second reformulation of the categorical imperative: that each person be treated as an end, never merely as a means.[54] In this sense, procedural fairness has been rebranded in terms of a 'Kantian injunction'.[55] It has been called into the service of 'a preferred conception of social and political life, in which self-respect is recognized as the fundamental human good which social life affects', and in which an individual can steadfastly demand to 'be treated as an end and not solely as a means'.[56] For example, the right to a hearing is justified in terms that it serves to fill 'a potentially destructive gap in the individual's conception of himself... to have played a part in, to have made one's apt contribution to, decisions which are about oneself may be counted important even though the decision, as it turns out, is the most unfavourable one imaginable and one's efforts have not proved influential'.[57] The right to explanatory reasons is explained in

eclectically developed in contingent contexts, and balanced against each other (and perhaps other values) in a continuingly intuitive fashion': ibid, 182.

[53] Andrew Moravcsik, 'Taking Preferences Seriously: A Liberal Theory of International Politics' (1997) 51 *International Organization* 513.

[54] 'What sets us apart from other natural beings is our quality as a "person", that is, a subject of a moral practical reason... above and beyond any price, because... he is not merely a means to others... ends but is to be estimated as an end in himself, ie. he has *dignity* (an absolute intrinsic value) whereby he compels all other rational beings in this world to give him *respect*, and can measure himself among all the others of his kind and estimate himself as equal': Immanuel Kant, *Foundation of the Metaphysics of Morals* (Lewis White Beck tr) (Michie Company 1959), 60, 116.

[55] Edmund Pincoffs, 'Due Process, Fraternity and a Kantian Injunction' in J Roland Pennock and John W Chapman (eds), *Due Process* (New York University Press 1977), 172. See also Jerry Mashaw, *Due Process in the Administrative State* (Yale University Press 1985), 199.

[56] Frank I Michelman, 'Formal and Associational Aims in Procedural Due Process' in J Roland Pennock and John W Chapman (eds), *Due Process* (New York University Press 1977), 131.

[57] Ibid, 127. For similar reasons, the dignitarian model would incorporate a strong preference for an oral rather than a written hearing on grounds that 'the personal interaction accompanying a face-to-face meeting is more likely to convey the impression that the individual is not being treated as a mere means to the achievement of a utilitarian end': Richard B Saphire, 'Specifying Due Process Values: Toward a More Responsive Approach to Procedural Protection' (1978) 127 *University of Pennsylvania Law Review* 111, 163–4.

terms that it 'confirms the participant, even in the face of substantive disappointment, as engaging in an ongoing process of rational and self-regarding action'[58] and serves to fill 'a potentially destructive gap in the individual's conception of himself'.[59]

However, Professor Mashaw's illuminating scholarship on due process has demonstrated the limited determinate value of a dignitarian model of procedural fairness derived from liberal theory. Ultimately, Mashaw finds that liberal theory fails to provide a robust vision of process values.[60] Where procedural fairness is grounded in liberal foundations, it becomes either too strong or too weak in its capacity to safeguard the individual. Where the Kantian injunction is taken to its logical conclusion, procedural protections become too restrictive in that an individual would be entitled to demand any process that he or she deems necessary to the pursuit of his or her purposes in life. Yet clearly individual autonomy cannot be an absolute value: other fundamental values such as the preservation of society and the fundamental importance of social cohesion and solidarity beget necessary limits on individual autonomy.[61] Kant himself acknowledged that the categorical imperative was never intended to be applied to the phenomenal world to generate rules of conduct.[62] The problem in structuring procedural fairness around an 'atomistic' notion of dignity and autonomy, imagining the individual as self-contained and always complete within his or herself, is that it is ultimately 'at war with man's basic sociability'.[63] Situating the dignitarian model within too robust a vision of liberal theory fails to capture the essence of the model, in which decision-making is regarded as a social process, and procedures serve what has been described as 'relational', 'communal', 'fraternal', or 'interpersonal' aims.[64] While certain rights might be classified as 'the rights of...egotistic man, man separated from other men and the community', the right to participate in decision-making is more aptly characterized in the category of rights of the citizen which, far from

[58] Jerry Mashaw, *Due Process in the Administrative State* (Yale University Press 1985), 176.

[59] Frank I Michelman, 'Formal and Associational Aims in Procedural Due Process' in J Roland Pennock and John W Chapman (eds), *Due Process* (New York University Press 1977), 127.

[60] Jerry Mashaw, *Due Process in the Administrative State* (Yale University Press 1985), 216.

[61] Martin Loughlin, *Public Law and Political Theory* (Clarendon Press 1992), 110–11.

[62] Jerry Mashaw, *Due Process in the Administrative State* (Yale University Press 1985), 194. See Immanuel Kant, *Critique of Practical Reason* (Lewis White Beck tr) (Macmillan Publishing Co 1956), 50–2, 60–4.

[63] Jerry Mashaw, *Due Process in the Administrative State* (Yale University Press 1985), 216.

[64] Frank I Michelman, 'Formal and Associational Aims in Procedural Due Process' in J Roland Pennock and John W Chapman (eds), *Due Process* (New York University Press 1977), 128; D J Galligan, *Due Process and Fair Procedures: A Study of Administrative Procedures* (Clarendon Press 1996), 239.

being atomistic, are 'political rights that are only exercised in community with other men'.[65]

Yet even if we attempt to situate procedural fairness within a more social interpretation of liberal theory, there is no satisfactory theoretical stopping point when trading the claims of individual liberty for the demands of social necessity.[66] For example, Rawls' theory of social justice renders Kant's vision more determinate against a societal backdrop, describing his original position as 'procedural interpretation of Kant's conception of autonomy and the categorical imperative'.[67] Self-respect plays a critical role in Kant's *A Theory of Justice.* Though we must be cautious not to make more of the demand for self-respect in procedural frameworks than Rawls himself makes,[68] a fundamental demand for rational processes of social decision-making is a fair implication from Rawlsian theory. Yet Mashaw laments that it would be enough for a procedural framework to establish that decision-making is *comprehensible* to the rational individual in order to avoid a liberal critique, and confirm the participant as engaging in an ongoing process of rational and self-regarding action.[69] Such a position renders procedural safeguards inadequate. The aim of the dignitarian model is about more than merely providing individuals with a reason or statement to internalize, but is about involving individuals in a 'conversation', in which they gain a sense of being engaged in an ongoing dialogue in a community of equals.[70]

Of course, creating distance between the dignitarian model and liberal theory is not without benefits as we consider the transfer of procedural fairness to the international legal setting, and more particularly the Security Council context. A liberal theory of international law has not gained

[65] Jeremy Waldron, *Law and Disagreement* (Oxford University Press 1999), 232; citing Karl Marx, 'On the Jewish Question' in Jeremy Waldron (ed), *Nonsense Upon Stilts: Bentham, Burke and Marx on the Rights of Man* (Methuen 1987), 147.

[66] Jerry Mashaw, *Due Process in the Administrative State* (Yale University Press 1985), 185.

[67] John Rawls, *A Theory of Justice* (Harvard University Press 1971), 256.

[68] Rawls has little to say about actual decision procedures, and what he does say is more consistent with an instrumentalist model of procedural fairness. He states that 'the rule of law requires some form of due process: that is, a process reasonably designed to ascertain the truth, in ways consistent with the other ends of the legal system' and that '[t]he precepts of natural justice are to insure that the legal order will be impartially and regularly maintained': ibid, 239. See further D J Galligan, *Due Process and Fair Procedures: A Study of Administrative Procedures* (Clarendon Press 1996), 133–4, 147–8.

[69] Jerry Mashaw, *Due Process in the Administrative State* (Yale University Press 1985), 175–6. Mashaw notes that this does not even entail a requirement of rationality: '[t]hat the agency or official is mistaken or foolish in some—even many—cases may thwart our desires, but it need not impair our capacity to understand what has happened and to receive whatever instruction is appropriate to the occasion': 199.

[70] Ibid, 216–17.

widespread support, and any theory of procedural fairness grounded in liberal theory would be subject to critiques of liberal triumphalism, shrinking the domain of international law at the expense of international cooperation.[71]

(2) The Dignitarian Model and Pluralism

Taking into account Mashaw's criticism, especially as it relates to the international legal setting where the role of the individual is in particular danger of being subsumed, a more satisfactory theoretical framework is to situate the dignitarian approach within a pluralist conception of legal and political culture. Such an interpretation of the dignitarian model maintains the connection between procedural fairness and autonomy, but adopts a broader, more socially embedded interpretation of autonomy as a concept associated with self-government. Autonomy relates essentially to a form of freedom of association on the part of individuals to organize themselves voluntarily into groups as a means through which to impress their private preferences on decision-makers. In a pluralist order, where there is recognized divergence in subjective beliefs, values, and ends, it is felt that decision-making will only be legitimate where it takes account of all perspectives, and becomes something more like a process of free negotiation or uninhibited bargaining among the various participants.[72] The idea is that the numbers and intensities of preferences will ultimately be reflected in decision-making outcomes.[73] Within this framework, procedural fairness becomes a mechanism by which to assure 'interest representation' of individuals and groups within the broader community, with the aim of procedure being to involve in the process of decision-making all of the persons or groups affected by any decision-making action.[74]

Recognition of the link between the dignitarian approach and pluralism provides a more robust foundation for process values, but also provides a more appropriate basis than liberal-democratic theory upon which to consider its

[71] José E Alvarez, 'Interliberal Law: Comment' (2000) 94 *American Society of International Law Proceedings* 249; José E Alvarez, 'Do Liberal States Behave Better? A Critique of Slaughter's Liberal Theory' (2001) 12 *European Journal of International Law* 183.

[72] Cass R Sunstein, 'Interest Groups in American Public Law' (1985–1986) 38 *Stanford Law Review* 29, 32.

[73] Ibid, 32–3; Cass R Sunstein, 'Beyond the Republican Revival' (1988) 97 *Yale Law Journal* 1539, 1542–3.

[74] Richard B Stewart, 'The Reformation of American Administrative Law' (1975) 88 *Harvard Law Review* 1667, 1759. See also Edmund Pincoffs, 'Due Process, Fraternity and a Kantian Injunction' in J Roland Pennock and John W Chapman (eds), *Due Process* (New York University Press 1977), 179; Laurence Tribe, *American Constitutional Law* (2nd edn, Foundation Press 1988), 666.

application to the international legal sphere. Yet, as with the link between instrumentalism and positivism, recognition of the connection between the dignitarian model and pluralism will have several broader implications. In particular, it promotes a more 'open architecture' in relation to conceptions of community, law, and values beyond those traditionally respected within legal systems.[75]

(a) Community: Striking the Balance between Representation and Elitism

In terms of legal community, the idea of the dignitarian model is to open up decision-making to a range of diverse interests and preferences, instilling 'flexible diversity' in decision-making so that it is able to remain responsive to the multiple interests at play. Normative conflict among multiple overlapping systems is regarded as unavoidable, and perhaps even desirable both as a source of innovation and a site for discourse among multiple community affiliations.[76] A criticism of this model is that it may lead to the privileging of the interests and preferences of well-organized and powerful bodies and groups.[77] One of the key problems Professor Sunstein identified as stemming from the pluralist approach in terms of community representation was the problem of 'unequal power', transforming decision-making into a 'series of accommodations among competing elites'.[78] This issue becomes relevant when we consider the application of the model to the international sphere and in particular the Security Council setting.

(b) Law: A Pluralist Vision of the Rule of Law (or Politics)

Under a dignitarian model, the main aim of procedural safeguards is to provide individuals with a modicum of dignity, autonomy, and self-respect, such as might be expected in a pluralist society, enabling them to assert their interests and preferences, and placing an obligation on decision-makers to take these interests into account in decision-making.[79] In concrete terms,

[75] This term is borrowed from Nico Krisch, 'The Open Architecture of European Human Rights Law' (2008) 71(2) *Modern Law Review* 183.

[76] Paul Berman, 'Global Legal Pluralism' (2007) 80 *Southern California Law Review* 1155, 1155.

[77] Richard B Stewart, 'The Reformation of American Administrative Law' (1975) 88 *Harvard Law Review* 1667, 1684–6, 1712.

[78] Cass R Sunstein, 'Interest Groups in American Public Law' (1985–1986) 38 *Stanford Law Review* 29, 29; Cass R Sunstein, 'Beyond the Republican Revival' (1988) 97 *Yale Law Journal* 1539, 1543–4.

[79] Richard B Stewart, 'The Reformation of American Administrative Law' (1975) 88 *Harvard Law Review* 1667, 1759. See also Edmund Pincoffs, 'Due Process, Fraternity and a Kantian Injunction' in J Roland Pennock and John W Chapman (eds), *Due Process* (New York University

the focus of the dignitarian model in the Security Council context would be on procedural safeguards such as: (1) providing individuals with the right to notice of decisions affecting them; (2) providing individuals with the right to a hearing to present their version of events; (3) providing individuals with the right to receive reasons for the decision sufficient to explain the individual's treatment, and to demonstrate that the individual's arguments against his or her treatment at the hands of decision-makers have been taken into account; and (4) right of access to a court to provide individuals with the opportunity to make an effective challenge to the reasons for their treatment. The aim of the dignitarian model is to forge a role for affected individuals in decision-making, reinforcing their dignity and worth by providing a mechanism through which their interests can be considered and taken into account.

Unlike the other normative models, the aim of the dignitarian model is not to work within a legal hierarchy to ensure decision-making accords with a formal set of laws or standards (as in the case of the instrumentalist model), or the solution that accords best with a set of universally agreed norms or principles (as in the case of the public interest model discussed in the next section). True to its pluralist underpinnings, the dignitarian model rejects a hierarchical conception of law in favour of a 'heterarchical' order, that is to say a horizontal rather than a vertical structure. Pluralism provides a 'juris-generative' rather than a harmonizing or centralizing legal framework, encouraging legitimate diversity in decision-making rather than providing a structure for its avoidance. The aim is to promote a continuous 'agonistic process of negotiation' between alternative visions on the basis of consent and mutual respect.[80] Procedural law becomes a device to enable affected interests to engage in a form of institutionalised 'signalling' (a term I have borrowed from Nico Krisch) of the weight of the various interests and values affected by the decision-making. The aim of decision-making then becomes to bring 'the competing visions into an informal balance'.[81]

Press 1977), 179; Laurence Tribe, *American Constitutional Law* (2nd edn, Foundation Press 1988), 666.

[80] Neil Walker, 'The Idea of Constitutional Pluralism' (2002) 65 *Modern Law Review* 317, 354. See also James Tully, *Constitutionalism in an Age of Diversity* (Cambridge University Press 1995).

[81] Nico Krisch, *Beyond Constitutionalism: The Pluralist Structure of Postnational Law* (Martin Loughlin, Mohn McCormick, and Neil Walker eds) (Oxford University Press 2010), 183. See also Paul Berman, 'Global Legal Pluralism' (2007) 80 *Southern California Law Review* 1155, 1166; Janet Levit, 'Bottom-Up International Lawmaking: Reflections on the New Haven School of International Law' (2007) 32 *Yale Journal of International Law* 393, 417.

(c) Values: Focus on Interests and Preferences

It could be argued that the focus of the dignitarian model is on providing an opportunity for the assertion of interests and preferences, rather than to contribute to the realization of shared values.[82] The pluralist system has been described as 'indifferent among preferences', with its chief concern being the aggregation of private preferences and the achievement of equilibrium among contending social forces.[83] This is to be distinguished from a system in which decisions have to be supported by argument and reasons and justified through an appeal to a broader public good. Under pluralist assumptions, there is no sense of a 'common good' or 'universal' set of shared values, which is regarded as alternatively mystical and tyrannical.[84] Under the dignitarian model, as I have termed it, there is no requirement for reasoned argument, and deliberation is directed toward ensuring that decision-making 'takes account of' all perspectives. The emphasis is on providing affected parties with an opportunity to assert their interests and preferences, rather than encouraging participants to phrase their objections as if they were thinking through what is best for all participants. It encourages a form of self-interested participation, concerned with providing parties with an opportunity to speak, rather than to contribute toward the development of shared values.

(3) Conclusion

In conclusion, like the instrumentalist model, underlying the dignitarian model are broader conceptions about community, law, and values. It aims to open up decision-making to the range of affected interests. It does not really cater to a 'rule of law' framework, such that the aim is not to ensure that decision-making accords with a set of legal rules. Rather, the emphasis is on providing affected parties with an opportunity to assert their interests and preferences. Decision-making is conceived as the process of weighing up the various interests and preferences affected, and arriving at an appropriate accommodation.

[82] Jenny Steele draws this helpful distinction between 'interests' and 'values': J Steele, 'Participation and Deliberation in Environmental Law: Exploring a Problem-solving Approach' (2001) 21(3) *Oxford Journal of Legal Studies* 415.

[83] Cass R Sunstein, 'Beyond the Republican Revival' (1988) 97 *Yale Law Journal* 1539, 1548.

[84] That is, 'tyrannical because pluralists see the change of preferences, or the subordination of private interests to the public good, as inevitably coercive and rarely the product of reasoned argument; mystical because pluralists take private preferences as exogenous variables': Cass R Sunstein, 'Interest Groups in American Public Law' (1985–1986) 38 *Stanford Law Review* 29, 83; Cass R Sunstein, 'Beyond the Republican Revival' (1988) 97 *Yale Law Journal* 1539, 1554.

E. The Public Interest Approach: A Model Based on Public Accountability

By contrast to the dignitarian model, which encourages a focus on self-interest, the public interest model regards procedural fairness as a mechanism through which to enhance representation of the *public interest* through decision-making. Proponents of this approach, including TRS Allan and Frank Michelman, view due process as a sort of analogue of an individual's participation in the political process.[85] This model is sometimes also discussed under the rubric of digitarianism; however, I consider that it is more appropriately distinguished from the dignitarian model in that it shifts the theoretical foundations away from individual autonomy, as defined in a Kantian or even a pluralist sense, and towards something more representative of social solidarity. It also takes a broader perspective on those who should be involved in decision-making, focusing less on individuals as discrete entities and more on individuals in the context of their membership of a broader legal community. From a theoretical perspective, it is underpinned by a 'communitarian' mentality, in which members of a community see their lives as bound up with the good of the communities that 'constitute' their identity.[86] The aim of the model is to infuse decision-making with a sense of what is good for all, rather than encouraging self-interested claims.

(1) Link between the Public Interest Model and the Public Interest

The public interest model of due process contributes to the legitimacy of decision-making by enhancing the connection between decision-making

[85] T R S Allan, 'Procedural Fairness and the Duty of Respect' (1998) 18 *Oxford Journal of Legal Studies* 497, 502, 509; Frank I Michelman, 'Formal and Associational Aims in Procedural Due Process' in J Roland Pennock and John W Chapman (eds), *Due Process* (New York University Press 1977), 136. See also Christopher Peters' observation that '[l]itigant-driven adjudication and democratic government both stem from the premise of participation—the idea that government decisions are more politically legitimate to the extent they reflect the participation of the governed in the process of making them': Christopher J Peters, 'Participation, Representation, and Principled Adjudication' (2002) 8 *Legal Theory* 185, 190.

[86] '[T]o say that members of a society are bound by a sense of community is not limply to say that a great many of them profess communitarian sentiments and pursue communitarian aims, but rather that they conceive their identity—the subject and not just the object of their feelings and aspirations—as defined to some extent by the community of which they are a part': M J Sandel, *Liberalism and the Limits of Justice* (Cambridge University Press 1982), 150.

and the public interest.[87] In advancing this aim, procedural fairness serves essentially as a process by which to enable community participation. It becomes a mechanism for effective interaction between decision-makers and individuals in a way that ensures that individuals are not merely the subjects of decisions, but have the opportunity to participate in the decision-making process, thereby rendering decision-making more representative. Increased opportunities for participation in decision-making serves to feed into discourse about the development of a shared body of values, with which decision-making should in turn accord. This enhances the legitimacy of decision-making by strengthening the bonds of rational consent between individuals and decisions. The idea is that the decision-maker's claim to authority and legitimacy is enhanced by an acknowledgment that 'the law or policy in question has been fairly adopted by procedures which enable *all citizens* to exert an influence, however limited, in any particular case'.[88]

(a) Community: A 'Communitarian' Conception

As with the other models, underlying the public interest model are broader theories about community, law, and values. The public interest model is based on a 'responsive' conception of community. It posits an 'ideal-involving' conception of community, in which the identities of the community and its members are integrally oriented in part by an acknowledgement of shared values.[89] Rather than supporting the notion that an individual should be treated as an end, the idea is that individuals should collectively be involved in the selection of ends.[90] The legitimacy of the system depends on a process of will-formation by the broader legal community leading to justified decisions about the pursuit of collective goals and the normative regulation of life in common.[91] It therefore does not seek to concentrate authority, as in the case of the instrumentalist model, or to insulate decision-making from

[87] T R S Allan, 'Procedural Fairness and the Duty of Respect' (1998) 18 *Oxford Journal of Legal Studies* 497, 504. See also RA Duff, 'Legal Obligations and the Moral Nature of Law' (1980) 25 *Juridical Review* 61, 82.

[88] T R S Allan, 'Procedural Fairness and the Duty of Respect' (1998) 18 *Oxford Journal of Legal Studies* 497, 508 (emphasis added).

[89] Charles Taylor, *Sources of the Self: The Making of the Modern Identity* (Cambridge University Press 1989), 99; Michael Walzer, 'The Communitarian Critique of Liberalism' (1990) 18 *Political Theory* 6, 9–10. See also John Tasioulas, 'In Defence of Relative Normativity: Communitarian Values and the Nicaragua Case' (1996) 16 *Oxford Journal of Legal Studies* 95, 117.

[90] Cass R Sunstein, 'Beyond the Republican Revival' (1988) 97 *Yale Law Journal* 1539, 1548, 1557.

[91] Jürgen Habermas, *Between Facts and Norms: Contribution to a Discourse Theory of Law and Democracy* (William Rehg tr) (MIT Press 1996), 158.

the political order, but rather encourages a *wider* sharing of legal authority.[92] The public interest model is based on the idea that popular consent and the development of shared beliefs can only arise from the public deliberation of the broad membership of the community. Participation is accordingly not restricted to those who are directly affected by decision-making authority, as in the case of the dignitarian model, but represents a 'guarantee of the opportunity for all to play their part in the political process'.[93]

(b) Law: A Responsive Approach to Law

In terms of the theory of law underlying the model, the public interest model encourages a 'responsive' approach to law and legal system. Scholars such as Nonet and Selznick depict a public interest model of the rule of law, drawing in turn upon the writings of Lon Fuller.[94] Underpinning this model is the idea that no legal regime can endure without a foundation in consent. Responsive law borrows much from the experience of democracy, but considers that 'gross legitimation' through majority rule provides only crude accountability.[95] As a model of law, it sets itself against more positivist accounts, such as that underlying the instrumentalist model. The argument is that accountability to determinate legal standards has the potential to breed 'formalism and retreatism', rendering institutions rigid and incapable of coping with new contingencies.[96] Instead of relying on the substantive law as blanket certification of the source of the decision-making power, the public interest model advocates continuing participation of individuals in decision-making in order to contribute to 'sustained justification for its use'.[97] Under a public interest model of law, respect for decision-making authority is negotiated, not won by subordination to formal rules.

Responsive law underscores the centrality of purpose in the legal enterprise.[98] The idea is that decision-making powers are conferred upon institutions for specific purposes, which are concrete enough to provide guidance to decision-makers in making decisions. Purposive regulation presumes a far wider and inclusive conception of the legal process, less concerned with

[92] Philippe Nonet and Philip Selznick, *Law and Society in Transition: Toward Responsive Law* (Octagon Books 1978), 95.

[93] T R S Allan, 'Procedural Fairness and the Duty of Respect' (1998) 18 *Oxford Journal of Legal Studies* 497, 509.

[94] Philippe Nonet and Philip Selznick, *Law and Society in Transition: Toward Responsive Law* (Octagon Books 1978). See also Lon Fuller, *The Morality of Law* (Rev edn, Yale University Press 1969); D J Galligan, *Discretionary Powers* (Clarendon Press 1986).

[95] Philippe Nonet and Philip Selznick, *Law and Society in Transition: Toward Responsive Law* (Octagon Books 1978), 56.

[96] Ibid, 76. [97] Ibid, 56.

[98] Lon Fuller, *The Morality of Law* (Rev edn, Yale University Press 1969), 194–5.

values such as consistency, predictability, and regularity, and focused instead on law as a rational, problem-solving, facilitative enterprise.[99] It is purpose and not fixed legal rules that should set the standards for criticizing established practice, thereby opening ways to change. Where purposes are left unclear, they may be extrapolated by reflecting upon the nature and terms of power in the light of community values.[100]

This reliance on the regulating potential of purpose emphasizes the importance of a union of legal authority and political will. The achievements of the responsive model depend on the will and resources of the political community.[101] Herein lies the connection between responsive law and a public interest model of due process. The hallmark of a responsive model of law is participatory decision-making. Participation within such a legal framework acts as 'a source of knowledge, a vehicle of communication, and a foundation for consent'.[102] Social pressures act as opportunities for self-correction. In terms described by Lon Fuller, the idea is that decision-makers should be seen as more than inert mirrors reflecting legal principles, but active participants in the enterprise of articulating the implications of shared purposes informed by contributions from affected individuals and the broader public.[103]

(c) Values: Relationship with Shared Value System

In terms of the representation of values, the public interest model is heavily reliant on the existence of a shared body of values existing alongside (or, more accurately, hierarchically superior to) the substantive law, according to which the law should be interpreted through decision-making. Not only does the public interest model provide a mechanism by which both affected individuals and the broader public can contribute to the gradual and modest generation of shared beliefs, but the model is thought to enhance legitimacy by encouraging greater acceptance of decision-making by the affected individual and the broader public as a consequence of the greater connection between decision-making and this higher body of shared values. T R S Allan acknowledges the importance of understanding or comprehension of decision-making by the individuals affected, noting that 'the giving of reasons by officials, when valid and pertinent, enables people to understand their treatment, as a prelude to acceptance'.[104] This

[99] Philippe Nonet and Philip Selznick, *Law and Society in Transition: Toward Responsive Law* (Octagon Books 1978), 98–9, 110.

[100] D J Galligan, *Discretionary Powers* (Clarendon Press 1986), 100.

[101] Philippe Nonet and Philip Selznick, *Law and Society in Transition: Toward Responsive Law* (Octagon Books 1978), 113.

[102] Ibid, 100.

[103] Lon Fuller, 'The Forms and Limits of Adjudication' (1978) 92 *Harvard Law Review* 353, 378.

[104] T R S Allan, 'Procedural Fairness and the Duty of Respect' (1998) 18 *Oxford Journal of Legal Studies* 497, 498.

understanding requires more than mere explanation, but an explanation that enables an individual to accept that his or her treatment amounts to a reasonable sacrifice for the public good.[105]

(2) Conclusion

In conclusion, the public interest model of procedural fairness is based on different underlying conceptions of community, law, and values than the instrumentalist or dignitarian approaches discussed previously. Its aim is to extend participation in decision-making not just to affected interests, but to the broader public. The rule of law is not seen as fixed, but as subject to influence though the process of communication enabled by procedural safeguards. Increased opportunities for participation assist to infuse decision-making with an understanding of the shared interests and values, with which decision-making should in turn apply.

F. Conclusion

Professor Joseph Sax once said of the redemptive quality of procedural reform that it is 'about nine parts myth and one part coconut oil'.[106] The Security Council's failure to implement serious procedural reform in the Security Council sanctions setting indicates that a similar viewpoint may exist among influential members of the international legal order. There is a sense that due process is regarded by the Council as a 'restriction' inflicted upon it, rather than a set of legal principles that have value as a legitimizing mechanism.[107] This chapter seeks to expose the myopia afflicting this viewpoint, drawing on the significant literature on the theoretical and conceptual foundations of procedural fairness to show the range of beneficial process values promoted by procedural fairness.

In this chapter I have highlighted an integral association between procedural fairness, legitimacy, and dialogue. As will be developed in subsequent chapters, the capacity of procedural fairness to enhance legitimacy through increased opportunities for dialogue provides a persuasive focus as to why the adoption of a procedural framework would be beneficial in the international legal order, a sphere regularly criticized for its under-inclusiveness and lack

[105] Ibid, 515.

[106] Joseph Sax, 'The Unhappy Truth about NEPA' (1973) 26 *Oklohoma Law Review* 239, 239.

[107] See more broadly on this theme of the political advantages of limitations on political authority: Stephen Holmes, *Passions and Constraint* (University of Chicago Press 1995).

Table 4.1 Models of Due Process

	Key Process Value	Key Community Participants	Theory of Law	Values
Instrumentalist	Accuracy	Lawmakers	Positivist	Value-Neutral
Dignitarian	Interest Representation	Stakeholders	Pluralist	(Powerful) Interests
Public Interest	Public Interest	Broad Community	Responsive	Universal Values

of representative opportunities. However, it is equally important to recognize that the capacity of dialogue to contribute to greater legitimacy differs between legal contexts dependent in part on different conceptions of community, law, and values in different settings. In this chapter, I have identified three predominant models of procedural fairness. Each model differs in terms of the way it purports to enhance the legitimacy of decision-making, advancing different process values, and supporting different theories of community, law, and representation of values. I summarize the differences in Table 4.1 above.

Applied to the Security Council sanctions setting, it is possible to see a connection between the normative models discussed in this chapter and the competing structural frameworks emerging in the sanctions context discussed in Chapter 2. The analysis in the following chapters demonstrates how the three competing structural frameworks (international court-based; domestic court-based; and ombudsperson) would advance the process values of the three normative models discussed above. This approach provides a more productive terrain for analysis and comparison between procedural frameworks than the formalist underpinnings of the contemporary legal debate, which assesses the relative worth of frameworks in terms of their 'bindingness'. Using value-based methodology, it becomes possible to appreciate the potential impact of these different frameworks on the legitimacy of decision-making, both in the Security Council sanctions context, and in the international legal order more generally.

PART III

APPLICATION

5

A Model Based on Accuracy

The Instrumentalist Model in the UN Sanctions Context

We begin with discussion of the adoption of an instrumentalist model of due process in the Security Council sanctions setting. As discussed above, this model is supported and advanced by a court-based or adjudicatory procedural framework.[1] This is of course the framework around which critics seeking reform in the Security Council setting have converged. It is also the model that a number of courts and tribunals have adopted, with the prominent examples being the Court of First Instance in the original *Kadi* decision,[2] certain judges of the UK House of Lords in the *Ahmed* decision,[3] and certain members of the UN Human Rights Committee in *Sayadi*.[4] These are examples of courts and tribunals that, faced with challenges to the Security Council sanctions regime, have recast themselves as agents of the international community or 'internationalized' courts, assessing the legality of Security Council decision-making based on standards of international law.[5]

Recalling the focus of the instrumentalist model on greater accuracy, the model appears well positioned to contribute to the legitimacy of Security Council sanctions decision-making. The incidence of error has undoubtedly weakened the effectiveness and credibility of the sanctions regime.[6] The

[1] See Chapter 4, section C.(1)(b).

[2] *Kadi v Council and Commission* [2005] ECR II-0000.

[3] *Her Majesty's Treasury v Ahmed* [2010] UKSC 2.

[4] UN Human Rights Committee, *Sayadi and Vinck v Belgium* (Communication No 1472/2006, 29 December 2008) UN Doc CCPR/C/94/D/1472/2006.

[5] I distinguish here domestic courts that assess Security Council decision-making under domestic law, discussed in more detail below as examples of the dignitarian model in action: see Chapter 6.

[6] 'Eleventh Report of the 1267 Monitoring Team' (13 April 2011) UN Doc S/2011/245, para 26; 'Eighth Report of the 1267 Monitoring Team' (14 May 2008) UN Doc S/2008/324, paras 2, 23, 25; 'Seventh Report of the 1267 Monitoring Team' (29 November 2007) UN Doc S/2007/677, para 3; 'Sixth Report of the 1267 Monitoring Team' (8 March 2007) UN Doc S/2007/132, para 14; 'Fifth Report of the 1267 Monitoring Team' (20 September 2006) UN Doc S/2006/750, para 22.

Analytical Support and Sanctions Monitoring Team for the 1267 Al Qaeda and Taliban Sanctions Committee has acknowledged, 'if states do not recognize the list as accurate and relevant, their implementation of the measures will be half-hearted and will not bring about the intended results',[7] concluding that 'the key to successful implementation of the sanctions regime is the amount of accurate detail on the consolidated list'.[8] The 1267 sanctions blacklist has been heavily criticized for errors such as listing based on insufficient or inaccurate information,[9] the continued listing of individuals reported to be deceased,[10] listing of individuals for an improper purpose (for example, to stifle political opposition),[11] and erroneous failure to list guilty targets.[12]

[7] 'Seventh Report of the Analytical Support and Sanctions Monitoring Team appointed pursuant to Security Council resolutions 1617 (2005) and 1735 (2006) concerning Al-Qaida and the Taliban and associated individuals and entities' (29 November 2007) UN Doc S/2007/677, para 3. See also similar statements in 'Eighth Report of the 1267 Monitoring Team' (14 May 2008) UN Doc S/2008/324, paras 2, 23, 25; 'Sixth Report of the 1267 Monitoring Team' (8 March 2007) UN Doc S/2007/132, para 14; 'Fifth Report of the 1267 Monitoring Team' (20 September 2006) UN Doc S/2006/750, para 22.

[8] 'Seventh Report of the Analytical Support and Sanctions Monitoring Team appointed pursuant to Security Council resolutions 1617 (2005) and 1735 (2006) concerning Al-Qaida and the Taliban and associated individuals and entities' (29 November 2007) UN Doc S/2007/677, para 90. In its Eleventh Report, the Monitoring Team emphasized that '[t]he Security Council has no interest in subjecting people to sanctions who do not meet the criteria for listing: to do so would not only be unfair, but would waste resources and undermine the credibility of the regime': 'Eleventh Report of the 1267 Monitoring Team' (13 April 2011) UN Doc S/2011/245, para 26.

[9] At the end of March 2008, fifty-seven entries for individuals on the Al Qaeda and Taliban blacklist did not contain a full name and date of birth, a further five entries had a full name and date of birth, but nothing else, while another twenty-six entries had a name, date, and place of birth, but no other identifiers such as nationality, address, or country of residence. In May 2011, the 1267 Monitoring Team reported that '[f]inancial institutions continue to express frustration with List entries that lack identifiers and therefore yield false positive matches in screening processes': 'Eleventh Report of the 1267 Monitoring Team' (13 April 2011) UN Doc S/2011/245, para 56.

[10] As at May 2011, forty-eight individuals on the list were reported to be deceased: 'Briefing by Ambassador Wittig, Chairman of the Security Council Committee established pursuant to resolution 1267' (16 May 2011) UN Doc S/PV.6536, 5.

[11] Rosand gives the example of Russian officials seeking to include Chechen rebel groups on the sanctions list as part of an effort to show the world that these groups are 'international terrorists as opposed to rebels fighting for a separatist cause' (Steve Gutterman, 'Russian Security Chief Blames Chechen Warlord Basayev for Suicide Bombings', *Associated Press*, 19 December 2003) and US support for the listing of an East Turkestan Islamic Movement, a Chinese separatist group as part of an effort to lessen Chinese opposition to a possible attack on Iraq (Eckholm, 'US Labeling of Group in China as Terrorist is Criticized' *New York Times*, 13 September 2006, A6): Eric Rosand, 'The Security Council's Efforts to Monitor the Implementation of Al Qaeda/Taliban Sanctions' (2004) 98 *American Journal of International Law* 745, 761, 752.

[12] 'Sixth Report of the 1267 Monitoring Team' (8 March 2007) UN Doc S/2007/132, para 18. Inattention to the development of the list might be implied from the fact that, between 2001 and 2009, the Committee only added one name to the Taliban section of the list: 'Eleventh Report of the 1267 Monitoring Team' (13 April 2011) UN Doc S/2011/245, para 11.

However, before jumping to conclusions about the appropriate normative model for the Security Council setting, it is worth revisiting one of the core arguments of this book, namely the idea that procedural fairness is contextual. While factual errors such as those described above require redress, it must be remembered that the instrumentalist model and adjudicatory framework that supports it were developed to deal chiefly with legal errors. As importantly, this model and the adjudicatory framework supporting it were developed to respond to typical domestic governmental context, that is, a form of 'stovepipe governance' defined by strong command chains, vertical hierarchies, representative institutions, and an independent judiciary. The question of contextualism of due process is not a one-way street. It is not only a question of adapting a particular model of procedural fairness to suit the Security Council sanctions setting. It also involves the question as to whether the Security Council setting satisfies certain institutional preconditions that underpin the capacity of the various normative models to deliver legitimacy. As we shall see, when it comes to applying the instrumentalist model to the Security Council context, the unique features of the Security Council setting create problems in terms of the model's capacity to deliver legitimacy. Ultimately, a value-based approach to the question of due process reveals that an international adjudicatory framework may have a less beneficial impact in normative terms on the legitimacy of Security Council decision-making than formalist approaches might suggest.

In the following chapter, I argue that there are a number of institutional features of the Security Council that would severely compromise the capacity of the instrumentalist model to deliver enhanced legitimacy in Security Council sanctions decision-making. The impact would not be limited to the Security Council sanctions setting. The connection between normative models of procedural fairness and conceptions of community, law, and values reveals that the choice of procedural framework for the sanctions setting has the capacity to impact at a broader level. I argue that adoption of an instrumentalist model would serve further to solidify a *classical* conception of the international legal order. In particular, adoption of an instrumentalist model and associated adjudicatory framework in the Security Council context would: (a) entrench a narrow conception of international community as confined to nation states; (b) give judges an inflated role in international law-making; and (c) have a stagnating effect on the international value system.

A. The Instrumentalist Model and International Community: The Problem of the Unrepresentative Lawmaker

Adoption of a procedural framework for the Security Council setting offers a rare chance to achieve a subtle adjustment in the scope of participation granted to individuals in what is commonly perceived to be an under-inclusive legal domain. Yet the scope provided by the instrumentalist model to achieve such an adjustment is very narrow indeed. The instrumentalist model adopts a narrow conception of community, limited to 'lawmakers', which, in the Security Council setting, can be reduced to the fifteen members of the Security Council and arguably narrowed further to the will of the Permanent Five. Through its complicity with this narrow and unrepresentative conception of international community, I show how the instrumentalist model becomes unhinged from its capacity to enhance the legitimacy of decision-making and risks, instead becoming highly authoritarian for the majority of international society.

(1) A Narrow Conception of International Community

The predominant effect of an instrumentalist model would be to police the traditional gateway of the international community, giving privileged access to state interests, and providing only a narrow role for non-state actors. The traditional conception of international community, developed from the Treaty of Westphalia and reinforced at the conclusion of the Second World War in the UN Charter, regards international law as the province of nation states.[13] In Prosper Weil's reductive slogan, 'the international community means states'.[14] According to classical conceptions of international law, the acceptance that individuals were to have no independent personality was 'a matter of moral principle'[15] and all

[13] As recognized in the first edition of Oppenheim, international law is 'a law between states only and exclusively': L Oppenheim, *International Law: A Treatise* (Longmans 1905), 341.

[14] Prosper Weil, 'Towards Relative Normativity in International Law?' (1983) 77 *American Journal of International Law* 413, 426.

[15] E Lauterpacht (ed), *International Law, being the Collected Papers of Hersch Lauterpacht* (Cambridge University Press 1970), vol I, 141. See also George Manner, 'The Object Theory of the Individual in International Law' (1952) 46 *American Journal of International Law* 428, 428–9.

representation of non-state actors was to occur through the state. As Philip Alston has noted, even the terminology of 'non-state actors' reinforces 'the assumption that the state is not only the central actor; but also the indispensable and pivotal one around which all other entities revolve'.[16]

Rather than working to transform the traditional conception of international community, the adoption of an instrumentalist model of procedural fairness would serve instead to further entrench it. As discussed in Chapter 4, the focus of the instrumentalist model is on advancing the will of *lawmakers*, with procedural law acting as a mechanism to ensure the accurate application of the substantive law. When applied to the Security Council setting, the scope of participant interests advanced by the model becomes very narrow. The 'law' that is the subject of the accurate application under the model would encompass for the most part Security Council resolutions and guidelines. The model would therefore above all be concerned with enforcing the will of the fifteen member states responsible for drafting the relevant resolutions. In reality, the participants can be reduced even further if it is accepted that Security Council resolutions represent more accurately the will of the Permanent Five. While the instrumentalist model would provide some role in the decision-making process to non-state actors, including affected individuals, the affected individual is regarded almost purely as a source of information under this model, and their right to participate is limited to the extent required in the aid of accurate application of the will of lawmakers.[17] Courts that have taken an instrumentalist approach have been forced to concede that '[r]esolutions adopted by the Security Council . . . are . . . binding on all the Member States of the Community' and it follows that Member States 'must leave unapplied any provision of . . . law [including human rights provisions] . . . that raises any impediment to the proper performance of their obligations under the Charter of the United Nations'.[18]

[16] Philip Alston, 'The "Not a Cat" Syndrome: Can the International Human Rights Regime Accommodate Non-State Actors and Human Rights?' in Philip Alston (ed), *Non-State Actors and Human Rights* (Oxford University Press 2005), 3.

[17] J Steele, 'Participation and Deliberation in Environmental Law: Exploring a Problem-solving Approach' (2001) 21(3) *Oxford Journal of Legal Studies* 415, 417–18.

[18] *Kadi v Council and Commission* [2005] ECR II-0000, [189]–[190]. Along the same lines, see UN Human Rights Committee, *Sayadi and Vinck v Belgium* (Communication No 1472/2006, 29 December 2008) UN Doc CCPR/C/94/D/1472/2006, Individual opinion (partly dissenting) by Sir Nigel Rodley, Ivan Shearer, and Iulia Antoanella Motoc, 27; Individual (dissenting) opinion of Ruth Wedgewood, 30.

(2) The Under-representativeness of Narrow Conception of International Community

This state-centric conception of international community supported by the instrumentalist model has resisted attempts at revision and remains the cornerstone of the positivistic analysis of international law. Yet it is also a conception that belies the reality of the ever-increasing complexity and hybridity of contemporary international society. There has been something of a 'participant revolution' in international affairs, transforming the international sphere into a political and legal order that encompasses a multiplicity of constituencies beyond nation states.[19] Bruno Simma's 1994 Hague Academy lectures on the shift 'From Bilateralism to Community Interest' recognized that the international domain is increasingly 'a community that encompasses not only states, but in the last instance all human beings'.[20] As the sanctions regime itself reflects, individuals are no longer merely objects who benefit from international law, but also subjects who can be bound by it. There is accordingly a widespread sense that the 'traditional subjects doctrine' is no longer able to provide a satisfactory account of the social realities underlying international law.[21] Writing as a scholar, Judge Higgins disputes this traditional doctrine on the basis that, by construing the reality of international law in terms of 'subjects and objects', distinguishing between state and non-state actors, international lawyers have 'erected an intellectual prison of their own choosing, and then declared it to be an unalterable construct'.[22] The solipsistic vision of states as the only actors in the international community must give way to the contemporary recognition of a wider community whose members have 'values, identities and roles distinct from the geographic limitations of states'.[23]

In contemporary society, it is a widely held view that the Security Council is under-representative of the majority of international society.[24] On any

[19] José E Alvarez, *International Organizations as Law-makers* (Oxford University Press 2005), 156, 608.

[20] Bruno Simma, 'From Bilateralism to Community Interest in International Law' (1994) 250 *Recueil des Cours de l'Académie de droit international de La Haye* 217, 231–4.

[21] Rosalyn Higgins, 'Conceptual Thinking about the Individual in International Law' (1978) 24 *New York Law School Law Review* 11; LB Sohn, 'The New International Law: Protection of the Rights of Individuals Rather than States' (1982) 32 *American University Law Review* 1.

[22] Rosalyn Higgins, *Problems and Process: International Law and How We Use It* (Clarendon Press 2004), 49–50.

[23] Robert McCorquodale, 'Beyond State Sovereignty: The International Legal System and Non-State Participants' [2006] *Revista Colombiana de Derecho Internacional* 103, 149.

[24] See, for example, Gráinne de Búrca, 'Developing Democracy Beyond the State' (2008) 46 *Columbia Journal of Transnational Law* 221; Robert A Dahl, 'Can International Organizations Be Democratic? A Skeptic's View' in Ian Shapiro and Casiano Hacker-Cordón (eds), *Democracy's Edges* (Cambridge University Press 1999), 19; Andrew Moravcsik, 'Is There a "Democratic Deficit" in

conception of representation, the range of those affected by Security Council decision-making with a valid claim to participate goes beyond the fifteen member states vested with the power to make Security Council resolutions under the UN Charter. Writing in 1948, Morgenthau noted the contrast between pretence and constitutional actuality in the Council's structure, 'between the democratic expectations roused by the words of the Charter and the autocratic performance envisaged by the actual distribution of functions which characterizes the constitutional provisions of the United Nations'.[25] This perception of the gap between the actual distribution of power in international society and the distribution of the Council's permanent seats has only increased since that time.[26] There are numerous continuing calls for expansion of the Council, though no reform proposal has as yet succeeded due to continuing failure to reach agreement on which states should be granted the additional prized seats.[27]

The formal state-centric approach to the concept of international community is especially difficult to reconcile with a sanctions regime that generates intrusive rules intended to produce severe legal restrictions for individuals. Moreover, it would go against the contemporary trend that has been towards providing ever more access and participation rights to 'international civil society' in its various forms.[28] In *Kadi*, the Advocate General recognized the potentially detrimental impact of marginalizing the role of individuals in Security Council decision-making, noting that 'when the risks to public security are believed to be extraordinarily high, the pressure is particularly strong to take measures that disregard individual rights, *especially in respect of individuals who have had little or no access to the political process*'.[29] Given the increasing impact of Security Council decision-making on individuals

World Politics? A Framework for Analysis' (2004) 39 *Government and Opposition* 336; Joseph Nye, 'Globalization's Democratic Deficit: How to Make International Institutions More Accountable' (2001) 80 *Foreign Affairs* 2; Eric Stein, 'International Integration and Democracy: No Love at First Sight' (2001) 95 *American Journal of International Law* 489.

[25] Hans Morgenthau, *Politics Among Nations: The Struggle for Power and Peace* (McGraw Hill 1993 (1948)), 321.

[26] Bruce Cronin and Ian Hurd, *The UN Security Council and the Politics of International Authority* (Routledge 2008), 209.

[27] United Nations Secretary-General High-Level Panel on Threats, Challenges and Change, *A More Secure World: Our Shared Responsibility* (United Nations 2004), paras 250–6.

[28] S Charnowitz, 'Non-governmental Organizations and International Law' (2006) 100 *American Journal of International Law* 348; D Otto, 'Nongovernmental Organizations in the United Nations System: The Emerging Role of International Civil Society' (1996) 18 *Human Rights Quarterly* 107; E Suy, 'The Status of Observers in International Organizations' (1978) 160 *Recueil Des Cours de l'Académie de droit international de La Haye* 74.

[29] Case C-402/05 P *Kadi v Council of the European Union* [2008] ECR I-0000, [35], Opinion of Advocate General Maduro (emphasis added).

and domestic actors, a procedural framework that focuses predominantly on advancing the will of the member states of the Security Council is unlikely to do much to enhance legitimacy. Instead, there is the danger that it will serve to heighten concern with the undemocratic aspects of this regime, which only directly represents an extremely narrow segment of interests in the international community.

(3) The Absence of a Legitimately Representative Lawmaker

The central problem with the instrumentalist model in the Security Council setting is that, through its complicity in a narrow and unrepresentative conception of the international community, the model becomes unhinged from the true source of its legitimacy. As discussed in Chapter 4, the capacity of the instrumentalist model to enhance legitimacy rests on the assumption that the law-makers embody the consent or acceptance of the broader community. A key element of the legitimacy framework underlying this model is accordingly the existence of a 'legitimately representative lawmaker'.[30]

The problem in applying the instrumentalist model to the Security Council context is that the Council does not operate within a system in which the applicable law can be said to have been formulated by a law-maker that is broadly representative of the international community. As discussed above, an instrumentalist model would reinforce a classical and still dominant positivist account of international law, which reflects that international law is no more or less than the rules to which *states* have expressed their express consent or acquiescence.[31] According to a traditional conception of international law, state consent replaces representativeness as the prevailing source of international law's bindingness. Yet this classical account of international law fails to accommodate the increasing complexity of the international legal order. Much of its structure and content remains in a form bequeathed by the bilateral and voluntarist model of international law developed out of the Treaty of Westphalia, and arguably unsuited for the making of international law that genuinely serves the interests of contemporary international society as it has expanded over the decades.[32] The notion that the representation of non-state actors within state populations can be assured through state consent is recognized as increasingly illusory.[33]

[30] See text accompanying nn 34–6 in ch 4. [31] ICJ Statute, Article 38(1).

[32] Benedict Kingsbury and Megan Donaldson, 'From Bilateralism to Publicness in International Law' in Ulrich Fastenrath et al (eds), *From Bilateralism to Community Interest: Essays in Honour of Bruno Simma* (Oxford University Press 2011), 81.

[33] Nico Krisch, *Beyond Constitutionalism: The Pluralist Structure of Postnational Law* (Martin Loughlin, Mohn McCormick, and Neil Walker eds) (Oxford University Press 2010), 18–19;

This pertains especially to law-making by international institutions such as the Security Council. The element of state consent to the powers of the international legal order and its institutions has been diluted by the expanding membership of the international community, such that states are able to maintain only a fraction of control.[34] Moreover, the delegation of powers to international institutions is generally thin, and the founding treaties of international institutions typically contain only vague guidance as regards the scope of the powers.[35] The UN Security Council in particular was never intended to be a representative body.[36] It was accepted that any pretension of majorities in the UN to 'governance' would lack political as well as legal authority and be doomed to failure.[37] Instead, the structure of Security Council decision-making is deliberately hegemonic. A 'concert of the great powers' was established on the basis that it was only through the harmony of the powerful that the UN could ensure peace and security.[38] The veto power places grossly disproportionate power over Security Council 'law-making' in the hands of the five Permanent Members, China, France, Russia, the United Kingdom, and the United States.[39] The other ten seats are shared between the other member states of the United Nations, who sit for two-year terms, meaning stints on the body are few and far between.[40] States that do not sit on the Council have a limited voice in the Council's decision-making,[41] and certain of those who do are open to the critique that they are themselves undemocratic and therefore hardly sit as representatives

Eyal Benvenisti, 'Exit and Voice in the Age of Globalization' (1999) 98 *Michigan Law Review* 167, 200–1.

[34] Nico Krisch, *Beyond Constitutionalism: The Pluralist Structure of Postnational Law* (Martin Loughlin, Mohn McCormick, and Neil Walker eds) (Oxford University Press 2010), 19.

[35] Ibid, 18.

[36] Bruce Cronin and Ian Hurd, *The UN Security Council and the Politics of International Authority* (Routledge 2008), 207; David Caron, 'The Legitimacy of the Collective Authority of the Security Council' (1993) 87 *American Journal of International Law* 552, 565.

[37] Oscar Schachter, 'The Crisis of Legitimation in the United Nations' (1981) 50 *Nordisk Tidsskrift Int'l Ret* 3, 3.

[38] Ibid. See also Gerry Simpson, *Great Powers and Outlaw States* (Cambridge University Press 2004).

[39] David Caron, 'The Legitimacy of the Collective Authority of the Security Council' (1993) 87 *American Journal of International Law* 552, 565–6. Caron cites Childers' statement that '[n]o longer is it morally possible for democratic people to accept that 5 out of 165 member governments can have such deadly power in the name of a peace-dedicated world organization': E Childers, 'Gulf Crisis Lessons for the UN' (1992) 23 *Bulletin of Peace Proposals* 129, 133.

[40] Jeremy Farrall, 'Does the UN Security Council Compound the Democratic Deficit?' (2009) 46 *Alberta Law Review* 913, 918.

[41] José E Alvarez, 'Hegemonic International Law Revisited' (2003) 97 *American Journal of International Law* 873; Charlotte Ku and Harold K Jacobson (eds), *Democratic Accountability and the Use of Force in International Law* (Cambridge University Press 2003), 358–9.

of their citizens.[42] Non-state actors have a very limited scope for input into Security Council decision-making, and the Council has remained largely aloof from NGO demands for access.[43] As is well appreciated, decision-making in the Security Council is carried out 'without the benefit of lawmakers representative of the demos these rules purport to affect'.[44]

A procedural framework that is focused on ensuring the accurate application of the resolutions drafted by the Security Council will not contribute to greater representation of the broader international community. In its application to the Security Council context, the main achievement of the model will be to emphasize formal application of the law developed by the fifteen members of the Security Council, which will do little to enhance the legitimacy of Council decision-making. A constituency in which 'voice and vote are reserved exclusively for governments' has been declared to be manifestly unfair.[45] There is an increasing sense that individuals possess a right to participation in international decision-making—'the right to speak out and raise one's *voice*'.[46] An instrumentalist approach to procedural law would do little to enhance the involvement, participation, or representation of the interests of non-state actors, but would focus on advancing state will, and more particularly the will of the fifteen members of the Security Council. The effect would be to further circumvent the capacity for individuals or non-state groupings to participate in the development of international law. Rather than enhancing legitimacy, adoption of an instrumentalist model is just as likely to perpetuate the idea that there is a 'democratic deficit' in decision-making by international organizations such as the Security Council, carried out 'without the benefit of lawmakers representative of the demos these rules purport to affect'.[47]

[42] Allen Buchanan and Robert O Keohane, 'The Legitimacy of Global Governance Institutions' (2006) 20 *Ethics and International Affairs* 405, 413.

[43] Alan Boyle and Christine Chinkin, *The Making of International Law* (Oxford University Press 2007), 78–9. An exception is the Arria Formula, devised in 1993, which enables a member of the Council to invite other Council members to an informal meeting, held outside Council chambers, and chaired by the inviting member. The meeting is called for the purpose of a briefing given by one or more persons, considered as experts in a matter of concern to the Council: 'Report of the Secretary-General in Response to the Report of the Panel of Eminent Persons on United Nations-Civil Society Relations' (13 September 2004) UN Doc A/59/354, para 13.

[44] José E Alvarez, *International Organizations as Law-makers* (Oxford University Press 2005), 630.

[45] Thomas M Franck, *Fairness in International Law and Institutions* (Clarendon Press 1995), 480.

[46] J E Nijman, *The Concept of International Legal Personality: An Inquiry into the History and Theory of International Law* (TMC Asser Press 2004).

[47] José E Alvarez, *International Organizations as Law-makers* (Oxford University Press 2005), 630. For further discussion of the undemocratic nature of an international legal system based on state consent, see Joseph Weiler, 'The Geology of International Law—Governance, Democracy and Legitimacy' (2004) 64 *Zeitschrift für ausländisches öffentliches Recht und Völkerrecht* 547.

B. The Instrumentalist Model and International Law: The Limitations of Bindingness and Judicial Supremacy

A second set of problems in transplanting the instrumentalist model to the Security Council sanctions context stems from the nature and role of law in the Security Council setting. As discussed in Chapter 4, the instrumentalist model relies on a conception of law as a logically coherent and complete system of principles and rules, where consistency and predictability are the most important values, and where there is a concern to minimize the opportunity for arbitrary or unpredictable exercises of power. The Security Council does not operate in such a legal environment. The Security Council is by contrast a deliberately hegemonic institution in which consistency counts for little and power is always unpredictable. In such a context, the instrumentalist model struggles to deliver legitimacy.

The problem stems from the foundations of the instrumentalist model and the adjudicatory framework supporting it in a *domestic* concept of the rule of law. As discussed in Chapter 4, the motivation in granting judicial organs the power of judicial review is to authorize them to act as 'transmission belts' for the existing law, with their main task being to ensure governmental decision-makers stay within the boundaries set by the democratic branches of government.[48] This model rests on two central underlying tenets that prove problematic in the Security Council context: a criterion of *bindingness*, reducing the reviewing court's toolkit to binding law, and the use of *supremacy* as a means to determine whether judicial or political organs are authoritative in a particular context. This reduces pivotal questions of law and authority to binary choices: law enforcement or law creation, binding or non-binding principles, judicial or political supremacy, international or domestic law. Transferred to the Security Council setting, where the parameters of law and authority remain in question, such clear-line dichotomies are difficult to draw. In the following section, I explore three essential problems with the translation of the instrumentalist model to the Security Council sanctions setting, namely: (1) the absence of formal binding rules; (2) the role of discretion in Security Council decision-making; and (3) the inappropriateness of judicial supremacy in the Security Council sanctions context.

[48] Richard B Stewart, 'The Reformation of American Administrative Law' (1975) 88 *Harvard Law Review* 1667, 1672. See also John Laws, 'Law and Democracy' [1995] *Public Law* 72.

(1) The Problem of Bindingness in the Absence of Formal Binding Rules

Under the instrumentalist model, domestic courts are heavily reliant on a criterion of bindingness, which recognizes that a court's toolkit for the resolution of disputes is typically limited to existing binding law. This is largely a consequence of the separation of powers typically built into democratic legal systems, where law creation is recognized as the province of the representative political organs while courts are restricted to the enforcement of this law. In exercising their power of review, the role of courts is conventionally circumscribed by recognition of a separation between 'judicial' and 'elective' power, which translates into a separation between the power of 'law enforcement' vested in courts and the power of 'law creation', generally vested in the democratic legislature.[49] In simplified terms, the traditional model of the judicial function casts courts, not as creators of new law, but as 'transmission belts' for the existing law, with their main task being to ensure governmental decision-makers stay within the boundaries set by the democratic branches of government.[50]

Applied to the Security Council sanctions setting, the criterion of 'bindingness' is an overly limiting concept. Though the Security Council has held a number of thematic debates on the rule of law,[51] and there is a growing consensus building around the view that the Security Council's authority depends on respect for the rule of law,[52] a significant degree of

[49] See, for example, Charles Louis de Secondat baron de Montesquieu, *De l'esprit des lois* (Flammarion 1979); Alexander Hamilton, James Madison, and John Jay, *The Federalist: A Commentary on the Constitution of the United States* (Tudor 1947), Nos 47, 48, 51; M J C Vile, *Constitutionalism and the Separation of Powers* (Oxford University Press 1967); E Barendt, 'Separation of Powers and Constitutional Government' [1995] *Public Law* 599; Lord Bingham, 'The Rule of Law' [2007] *Cambridge Law Journal* 67.

[50] Richard B Stewart, 'The Reformation of American Administrative Law' (1975) 88 *Harvard Law Review* 1667, 1672. See also John Laws, 'Law and democracy' [1995] *Public Law* 72. Even within common law systems, it is recognized that judges must develop the law within the established normative system, ensuring historical continuity with existing legal principles, and that this body of law can be overruled by legislative enactment. R Dworkin, *Law's Empire* (Harvard University Press 1986), 229; Roger J Traynor, 'The Limits of Judicial Creativity' (1978) 29 *Hastings Law Journal* 1025.

[51] 'Statement by President of the Security Council' (24 September 2003) UN Doc S/PRST/2003/15; 'Statement by President of the Security Council' (26 January 2004) UN Doc S/PRST/2004/2; 'Statement by President of the Security Council' (10 September 2004) UN Doc S/PRST/2004/32; 'Statement by President of the Security Council' (12 July 2005) UN Doc S/PRST/2005/30; 'Statement by President of the Security Council' (22 June 2006) UN Doc S/PRST/2006/28; 'Statement by President of the Security Council' (29 June 2010) UN Doc S/PRST/2010/11.

[52] 'World Summit Outcome Document', UN General Assembly Resolution 60(1) (24 October 2005) UN Doc A/RES/60/1.

uncertainty continues to surround the meaning of the 'rule of law' in the Security Council context.[53] There is a school of thought propagated by Hans Kelsen that casts the Security Council as unbound by law, particularly where it acts under Chapter VII.[54] Though this is not a view that finds much contemporary favour, there is nevertheless significant ambiguity surrounding the scope of legal rules that do bind the Security Council. This is problematic for the instrumentalist model as it complicates the identification of legal principles and rules by which the 'accuracy' of decision-making can be measured. As will be discussed below, whether the substantive law is considered in terms of the textual constraint under the UN Charter, general international law, or the higher body of norms known as *jus cogens*, the set of legal rules binding the Security Council in its decision-making about sanctions are limited, vague, and, to an extent, undecided.

(a) Textual Constraint under UN Charter

Textual constraint on the Council's decision-making capacity under the UN Charter is practically non-existent. The Charter gives the Security Council broad discretion to 'determine the existence of any threat to the peace, breach of the peace or act of aggression' and to 'decide what measures shall be taken' in response.[55] As the International Criminal Tribunal for the former Yugoslavia observed, '[o]nce the Security Council determines that a particular situation poses a threat to the peace or that there exists breach of the peace or an act of aggression, it enjoys a wide margin of discretion in choosing the course of action'.[56] Any body called upon to review Security Council sanctions decision-making in terms of its accurate fulfilment of the substantive law would be working with concepts as indeterminate as a 'threat to the peace' and the 'purposes and principles' of the United Nations.[57]

[53] Simon Chesterman, '"I'll Take Manhattan": The International Rule of Law and the United Nations Security Council' (2009) 1 *Hague Journal on the Rule of Law* 67.

[54] Kelsen stated that, when acting under Chapter VII, the Security Council is not bound by existing law, but may create 'new law for the concrete case': Hans Kelsen, *The Law of the United Nations: A Critical Analysis of its Fundamental Problems* (E A Praeger 1950), 295. See also G Oosthuizen, 'Playing the Devil's Advocate: The United Nations Security Council is Unbound by Law' (1999) 12 *Leiden Journal of International Law* 549.

[55] Charter of the United Nations (adopted 26 June 1945, entered into force 24 October 1945) 1 UNTS XVI ('UN Charter'), Article 39.

[56] *Prosecutor v Tadic (Jurisdiction)* (1995) 105 ILR 453, 467.

[57] Martti Koskenniemi, 'The Police in the Temple: Order, Justice and the UN—A Dialectical View' (1995) 6 *European Journal of International Law* 325, 327.

(b) General International Law

Beyond the broad and loose textual restraints of the UN Charter, there is also considerable uncertainty as to the general body of international law by which the Security Council is bound. It is widely acknowledged that the United Nations is not a party to the vast majority of treaties, including international human rights treaties, and is therefore not directly bound by them.[58] There is an argument that the Security Council is bound by the rules of customary international law, though certain authorities refute this. For example, the Commentary to the UN Charter rejects the contention that humanitarian law and human rights customary norms establish precise limits to Chapter VII powers, construing them instead as 'guidelines in the exercise of those powers', determining that 'it is up to the [Security Council] to strike the concrete balance between humanitarian and human rights concerns and the goal of maintaining peace'.[59] In order to confirm the application of international humanitarian law to the Council's activities, the Secretary General felt compelled to issue a bulletin expressing an intention to commit the UN to these obligations.[60] There is also the complicating factor of Article 103, which provides that, in the event of a conflict between Charter obligations and a state's obligations under any other international agreement, the Charter obligations prevail. There is an argument that the combined reading of Articles 25 and 103 is that a state's obligation to comply with Security Council resolutions will prevail over its obligations to comply with those under customary international law.[61] If review of Council decision-making was undertaken in

[58] Daniel Halberstam and Eric Stein, 'The United Nations, the European Union and the King of Sweden: Economic Sanctions and Individual Rights in a Plural World Order' (2009) 46 *Common Market Law Review* 13, 8; Michael Bothe, 'Security Council's Targeted Sanctions against Presumed Terrorists: The Need to Comply with Human Rights Standards' (2008) 6 *Journal of International Criminal Justice* 541, 552; Andrea Bianchi, 'Security Council Anti-Terror Resolutions and their Implementation by Member States' (2006) 4 *Journal of International Criminal Justice* 1044, 1061; Iain Cameron, 'UN Targeted Sanctions, Legal Safeguards and the European Convention on Human Rights' (2003) 72 *Nordic Journal of International Law* 159, 167; August Reinisch, 'Developing Human Rights and International Humanitarian Law Accountability of the Security Council for the Imposition of Economic Sanctions' (2001) 95 *American Journal of International Law* 851, 854. Though see August Reinisch, 'Securing the Accountability of International Organizations' (2001) 7 *Global Governance* 131, 141–3, arguing that the UN is bound 'transitively' by international human rights standards as a result and to the extent that its members are bound ('functional treaty succession by international organizations to the position of their member states').

[59] J Frowein and N Krisch, 'Introduction to Ch VII' in Bruno Simma (ed), *The Charter of the United Nations: A Commentary*, vol II (2nd edn, Oxford University Press 2002), 711.

[60] Secretary General Kofi Annan, 'UN Secretary General's Bulletin on Applicability of IHL to UN Forces' (6 August 1999) UN Doc ST/SGB/1999/13.

[61] Michael Wood, 'First Lecture: The Legal Framework of the Security Council' (Lauterpacht Lectures, University of Cambridge, 7 November 2006), para 55. For dissenting views on Article 103's applicability to customary norms, see Derek Bowett, 'The Impact of Security Council

accordance with an instrumentalist model of procedural fairness, there is a high probability that reviewing bodies will take differing interpretations to the scope of law by which the Council is bound, which in itself contributes uncertainty and lack of generality of the application of law in this context.

(c) *Jus Cogens Norms*

A position commanding somewhat more support is that the Security Council is bound by *jus cogens* norms. However, the scope of recognized *jus cogens* norms is very narrow indeed. Former President of the ICJ, Rosalyn Higgins, stated that '[t]he examples [of such norms] are likely to be very, very few in number'.[62] Ian Brownlie famously described the concept of *jus cogens* as the Bentley that never leaves the garage.[63] There is a notorious level of disagreement as to the scope of these norms, and significant indeterminacy as to their content at such a high level of abstraction.[64] According to the International Law Commission's Commentary to the Articles on State Responsibility, the list of norms clearly accepted and recognized as peremptory extend to the prohibition of aggression, genocide, slavery, racial discrimination, apartheid, crimes against humanity, torture, and the right to self-determination.[65] As a set of standards by which to measure the accuracy of Security Council conduct, it is undoubtedly a narrow set of norms, but also one beset with continuing problems of uncertainty.

The problem with vesting a judicial organ with the capacity to enforce *jus cogens* norms against the Security Council was exhibited by the Court of First Instance in *Kadi*. In its 2005 judgment, the CFI effectively re-styled itself as an agent of the international community providing a clear example of the instrumentalist model in action. In a statement that many international lawyers would agree with, the CFI recognized that there was one limit

Decisions on Dispute Settlement Procedures' (1994) 5 *European Journal of International Law* 89, 92; Rain Lilvoja, 'The Scope of the Supremacy Clause of the United Nations Charter' (2008) 57 *International and Comparative Law Quarterly* 583, 602 ff.

[62] Rosalyn Higgins, 'A Babel of Judicial Voices? Ruminations from the Bench' (2006) 55 *International and Comparative Law Quarterly* 791, 801. For an expansive view on the scope of *jus cogens*, see Alexander Orakhelashvili, *Peremptory Norms in International Law* (Oxford University Press 2006), 53–60.

[63] Ian Brownlie, 'Comment' in Joseph Weiler and Antonio Cassese (eds), *Change and Stability in International Law-Making* (Berlin 1988), 108, 110.

[64] Antonios Tzanakopoulos, 'Human Rights and United Nations Security Council Measures' in Erika de Wet and Jure Vidmar (eds), *Hierarchy in International Law: The Place of Human Rights* (Oxford University Press 2012), 54. See also Marko Milanović, 'Norm Conflict in International Law: Whither International Law?' (2009) 20 *Duke Journal of Comparative and International Law* 69.

[65] International Law Commission, 'Commentary to Articles on State Responsibility' in *Yearbook of the International Law Commission* (2001), vol II, Pt II, 85, para 5 and 112–13, paras 4–5.

to the binding effect of Security Council resolutions, namely that the Council 'must observe fundamental peremptory provisions of *jus cogens*'.[66] The next step was a little more controversial, with the Court claiming that it was 'empowered to check, indirectly, the lawfulness of the resolutions of the Security Council in question with regard to *jus cogens*, understood as the body of higher rules of public international law binding on all subjects of international law, including the bodies of the United Nations, and from which no derogation is possible'.[67] Yet the Court's next steps were more controversial still. Contrary to the accepted view described above that *jus cogens* norms are 'very very few in number' and citing minimal authority, the CFI took an expansive reading of *jus cogens*, essentially putting all human rights within that category, including the right to a fair hearing, the right not to be arbitrarily deprived of property, and the right to an effective remedy.[68] Perhaps unseated by the enormity of this first step, the CFI then took a second extraordinary step, holding that there had been no breach of these norms in the case in question, despite the fact that it had previously been widely accepted that the UN sanctions regime fell foul of these due process norms.

It is perhaps no surprise that the CFI judgment was roundly criticized.[69] In terms of its approach to the interpretation of applicable international legal and human rights norms, it was criticized on the one hand for artificially inflating the set of *jus cogens* norms by which the Security Council could be held to account. In doing so, the Court's approach became vaguely reminiscent of that derided in the title to Anthony D'Amato's 2010 article, 'It's a bird, it's a plane, it's jus cogens!' On the other hand, the CFI was criticized for its anaemic approach to human rights protections, initially in declining

[66] *Kadi v Council and Commission* [2005] ECR II-0000, [229].

[67] Ibid, [226]. [68] Ibid, [226]–[229].

[69] Marko Milanović, 'Norm Conflict in International Law: Whither International Law?' (2009) 20 *Duke Journal of Comparative and International Law* 69, 91, 93; J Klabbers, 'Setting the Scene' in J Klabbers, A Peters, and G Ulfstein, *The Constitutionalization of International Law* (Oxford University Press 2009), 1; C Eckes, 'Judicial Review of European Anti-Terrorism Measures—the *Yusuf* and Kadi Judgments of the Court of First Instance' (2008) 14 *European Law Journal* 74; Piet Eeckhout, 'Community Terrorism Listings, Fundamental Rights and UN Security Council Resolutions: In Search of the Right Fit' (2007) 3 *European Constitutional Law Review* 183, 195; Martin Nettesheim, 'UN Sanctions against Individuals—A Challenge to the Architecture of European Union Governance' (2007) 44 *Common Market Law Review* 567; Mielle Bulterman, 'Fundamental Rights and the United Nations Financial Sanctions Regime: The Kadi and Yusuf Judgments of the Court of First Instance of the European Communities' (2006) 19 *Leiden Journal of International Law* 753; Alexandre Vandepoorter, 'L'application communautaire des decisions du Conseil de Sécurité' (2006) 52 *Annuaire francais de droit international* 102, 122; Enzo Cannizzaro, 'A Machiavellian Moment? The UN Security Council and the Rule of Law' (2006) 3 *International Organizations Law Review* 189, 202; Benedetto Conforti, 'Decisioni del Consiglio disicurezza e diritti fondamentali in una bizarre sentenze del Tribunale communitario di primo grado' (2006) 11 *Diritto dell'Unione europea* 333, 342.

jurisdiction over individuals caught up in the UN sanctions net and, then, by narrowly interpreting the application of international human rights principles to the relevant facts. The Court's approach to the relationship between legal orders can be characterized as both monist and hierarchical. The judgment rests on the claim that the relevant law derives its validity and authority from its source in a Security Council resolution, and that obligations under Security Council resolutions prevail over all other legal obligations according to the terms of the EU treaty framework and UN Charter. The mooted *jus cogens* qualification is discarded as a fig-leaf with the CFI's rather unfortunate interpretation of this category both as unusually broad yet seemingly impossible to violate. The influence of the decision by the CFI in this respect has been minimal, and was ultimately overturned by the European Court of Justice. The judgment demonstrates, not only that the scope of applicable norms by which Security Council conduct can be judged is very narrow, but also that there is a strong danger in courts taking different interpretations as to the content of these norms. This is unlikely to contribute to greater overall legitimacy in Council decision-making.

(2) The Problem of Bindingness and the Role of Discretion

While the above discussion I hope highlights some of the problems with identifying legal rules that operate in the Security Council context, it is ultimately something of an artificial exercise to walk through formal legal rules that may or may not bind the Security Council as if 'bindingness' operates in the usual way in the Security Council setting. The ambiguity of the legal rules applicable to the Security Council setting is not ultimately a symptom of juridical ineptitude, but rather the by-product of a deliberate institutional choice to vest broad discretion in the Security Council.[70] The Security Council exercises a hybrid of political and legal authority, and the relationship between law and politics in the Security Council setting is a notoriously complex one.[71] According to the leading commentary to the UN Charter,

[70] As the drafting history of the UN Charter shows, in order to reach the primordial goal of peace maintenance, states were ready to submit to the Security Council in a manner unprecedented in the international order: R B Russell and J E Muther, *A History of the United Nations Charter* (Brookings Institution 1958), 646. For discussion see J Frowein and N Krisch, 'Introduction to Ch VII' in Bruno Simma (ed), *The Charter of the United Nations: A Commentary*, vol II (2nd edn, Oxford University Press 2002), 711; Gerry Simpson, *Great Powers and Outlaw States* (Cambridge University Press 2004), 184–188.

[71] Rosalyn Higgins, 'The Place of International Law in the Settlement of Disputes by the Security Council' (1970) 64 *American Journal of International Law* 1; Martti Koskenniemi, 'The Place of Law in Collective Security' (1996) 17 *Michigan Journal of International Law* 455; S Chesterman, '"I'll Take Manhattan": The International Rule of Law and the United Nations Security Council' (2009) 1 *Hague Journal on the Rule of Law* 67.

the Council's decision-making procedure is very different from that of courts such as the ICJ because, while '[t]he ICJ has to decide exclusively on the basis of international law (Art 38 of the ICJ Statute)...the S[ecurity] C[ouncil] has to decide primarily according to political criteria'.[72] Though it may be a difficult concept for lawyers to grapple with, under the United Nations framework established by the Charter, the interests of political 'order' and legal 'justice' are not necessarily aligned.[73]

As a consequence of the broad discretion vested in the Council, Security Council resolutions are often vague, and are not drafted with the level of attention to detail that might be expected in the case of domestic legislation.[74] The demands of compromise in the passage of Security Council resolutions, and the need to take account of political considerations such as reaching consensus or compromise among Security Council members in order to achieve the safe passage of resolutions means that the decisions of the Security Council 'while striving for a principled application based on UN Charter requirements are subject to "the achievement of the possible" '.[75] Deliberate flexibility is built into the drafting of resolutions to retain the capacity to respond to unforeseen, complex or urgent situations, and to change and adjust decisions that have continuing effect against a backdrop of shifting international emergency. Moreover, the implementation of Security Council decisions is commonly delegated, and Council resolutions often deliberately leave a margin of appreciation (or discretion) to states in interpreting the resolutions.

The broad level of discretion in Security Council decision-making, coupled with the narrow (and uncertain) scope of applicable legal principles discussed in the previous section, raises further questions about the appropriateness of an instrumentalist model of procedural fairness in this setting. Essentially, the instrumentalist model depends for its successful operation upon the existence of clear and determinate standards, rather than flexible

[72] J Delbrock, 'Functions and Powers: Article 24' in B Simma et al (eds), *The Charter of the United Nations: A Commentary*, vol 1 (Oxford University Press 2002), 446. See also Rosalyn Higgins, 'The Place of International Law in the Settlement of Disputes by the Security Council' (1970) 64 *American Journal of International Law* 1, 18.

[73] Martti Koskenniemi, 'The Police in the Temple: Order, Justice and the UN—A Dialectical View' (1995) 6 *European Journal of International Law* 325.

[74] See also Michael Byers, 'Agreeing to Disagree: Security Council Resolution 1441 and Intentional Ambiguity' (2004) 10 *Global Governance* 165; Richard Builder, 'The Office of the Legal Adviser: The State Department Lawyer and Foreign Affairs' (1962) 56 *American Journal of International Law* 654. For similar arguments in relation to the drafting of international conventions, see Francis Boyle, 'International Law in Time of Crisis: From the Entebbe Raid to the Hostages Convention' (1980) 75 *Northwestern University Law Review* 831.

[75] Rosalyn Higgins, 'The ICJ, the United Nations System and the Rule of Law' (Speech at the London School of Economics, 13 November 2006), 2.

guiding principles. The idea is that, in a system based on clear and determinate standards, decision-makers are more likely to apply the law, instead of their personal notions of fairness, justice, or appropriateness. As Professor Michelman notes (albeit in his critique of a purely positivistic conception of procedural fairness), 'unless there are objective standards in terms of which decisions can be counted correct or incorrect, it is hard to see in what sense we can say that a decision serves to secure to an individual that which is rightfully his'.[76]

The problem is that, in the Security Council context, discretion rather than clear and determinate standards is the gold standard. Ultimately, in the Security Council setting, 'accuracy' is a slippery term that may not be an appropriate or entirely realizable goal. When asked to interpret whether the power granted to states under Security Council resolution 1730 to place a 90-day hold on a de-listing could be renewed indefinitely, the UN Office of Legal Affairs responded that a sanctions resolution means 'whatever the sanctions committee wants it to'. In normative terms, the wide discretion vested in the Council is arguably inconsistent with an instrumentalist model of procedural fairness. It is well recognized in procedural fairness theory that the scope of discretion granted to decision-makers is a relevant factor in the determination of an appropriate procedural framework. Discretion is recognized as existing on a spectrum, from rules-based authority to that guided more loosely by broad policy goals. It is generally appreciated that, as decision-making moves away from rule-based decision-making and becomes more clearly discretionary, the adjudicatory procedural framework (focusing on the settlement of dispute between parties according to settled standards) may need to give way to more broadly consultative processes (focused on the representation of pluralist interests, or the realization of certain goals in the public interest).[77] In the case of Security Council decision-making, the needle swings sharply toward the discretionary end of the spectrum, arguing strongly in favour of consultative processes over adjudication in this context.

There is an argument that the Security Council should work harder to reduce ambiguity in Security Council decision-making as it creates problems for states seeking to implement sanctions decisions.[78] However,

[76] Frank I Michelman, 'Formal and Associational Aims in Procedural Due Process' in J Roland Pennock and John W Chapman (eds), *Due Process* (New York University Press 1977), 130.

[77] Nicola Lacey, 'The Jurisprudence of Discretion: Escaping the Legal Paradigm' in Keith Hawkins (ed), *The Uses of Discretion* (Clarendon Press 1992); D J Galligan, *Discretionary Powers* (Clarendon Press 1986), 88, 343; Philippe Nonet and Philip Selznick, *Law and Society in Transition: Toward Responsive Law* (Octagon Books 1978); G Ganz, *Administrative Procedures* (Sweet & Maxwell 1974), 1.

[78] States have encountered inconsistent responses from the Sanctions Committee on particular points of interpretation. For example, it has been known for the Sanctions Committee to respond

ultimately, it is an issue that arguably only the Security Council can rectify. Under an instrumentalist model, the role of plugging the gaps left by the Council, and resolving uncertainties in the surrounding law, would be vested in courts. As will be explored in the next section, this does not redeem the model, but arguably provides another argument for its rejection in the Security Council setting.

(3) The Problem of Supremacy: Judicial Abdication or the Prospect of Judicial Overreach

Even if the legal standards binding the Council could be identified with greater certainty, recognition that the Security Council acts within legal limits is very different from the proposition that these legal limits may be enforced against the Council by a judicial tribunal.[79] The brief history of the sanctions regime has shown that, where courts assume the role of reviewing Security Council decision-making under international law, they become strung between the poles of inert deference or over-representative defiance. The scope of discretion granted to the Council under the Charter forces courts into something of an 'all-or-nothing' approach under which they either abdicate any effective judicial role in reviewing the Council's exercise of discretion, or assume an inflated role that entitles them to subsume the Council's role in exercising that discretion.

Certain courts have already taken the former approach, determining that an 'accurate' interpretation of the applicable law (chiefly the combined effect of Articles 25 and 103 of the Charter) leaves the Security Council essentially free to take whichever measures it chooses in response to any threat to international peace and security.[80] In *Kadi*, the CFI's approach was based

affirmatively to a request for clarification by a state as to whether a certain product was sanctioned, in circumstances where members of the Sanctions Committee had not themselves outlawed that product under their domestic sanctions regimes.

79 According to Judge Schwebel, the subjection of the acts of the Security Council to the law 'relies not upon judicial review, but on self-censorship': *Questions of Interpretation and Application of the 1971 Montreal Convention Arising from the Aerial Incident at Lockerbie (Libyan Arab Jamahiriya v United Kingdom) (Preliminary Objections)* [1998] ICJ Rep 9, 76 (Dissenting Opinion of Judge Schwebel). See in similar terms Ian Brownlie, 'The Decisions of Political Organs of the United Nations and the Rule of Law' in Ronald St John MacDonald (ed), *Essays in Honour of Wang Tieya* (Martinus Nijhoff 1993), 91, 100; Simon Chesterman, '"I'll Take Manhattan": The International Rule of Law and the United Nations Security Council' (2009) 1 *Hague Journal on the Rule of Law* 67, 73.

80 See, for example, the decision of the Administrative Appeals Board of the Turkish Council of State in *Al-Qadi v The State* (TK 2007) ILDC 311; UN Human Rights Committee, *Sayadi and Vinck v Belgium* (Communication No 1472/2006, 29 December 2008) UN Doc CCPR/C/94/D/1472/2006, Individual opinion (partly dissenting) by Sir Nigel Rodley, Ivan Shearer, and

on an acknowledgement of the Security Council's position in the hierarchy. Its judgment rests on the claim that the relevant EU regulation derives its validity and authority from its source in a Security Council resolution, and that obligations under Security Council resolutions prevail over all other legal obligations according to the terms of the EU treaty framework and UN Charter. Whether deliberately or not, the CFI adopts a command theory of law that is, at best, misplaced and, at worst, an abdication of judicial responsibility against the backdrop of the broader negotiation of the architecture for global governance of which the judgment forms part. The critical reaction to the judgment, coupled with its lack of influence, indicates that continuing attachment to the power of Security Council 'command' and traditional notions of hierarchy is regarded as increasingly misplaced. The only institution that stands to benefit from this attachment is the Security Council itself. However, even this can be questioned. Despite the dearth of formal legal limitations on the Council, it has become clear that the Council's lawmaking capacity is not self-sufficient. The Council must share power with others and accept limits, if only to secure obedience to its commands. There is a growing sense that 'pillars' of Security Council supremacy such as Articles 25 and 103 are no longer adequate (if they ever were) to secure the effective implementation of Security Council resolutions.

The issue is not merely one of legitimacy, but also (linked to this) of effectiveness. The Security Council can no longer afford to rely on formal hierarchy, but must ensure that the decision-making takes account of widely held interests and values if it wishes its decisions to achieve widespread acceptance and compliance. In the sanctions setting, the Security Council is becoming increasingly reliant on non-state actors for co-operation, implementation, and enforcement of its decision-making. If these actors feel sidelined or are otherwise dissatisfied with Security Council decision-making, they may seek to undermine the Council's decisions.[81] Representatives of private industry, including banks, airlines, and other representatives of private industry, have the capacity to undermine the effectiveness of the sanctions regime if they facilitate offers of employment, educational, or travel opportunities, decline to freeze funds, or actively contribute funds to those on sanctions blacklists. Indeed, experience has shown that, even where states have formally complied with sanctions measures, elements of civil society have engaged in deliberate violations of sanctions decision-making where they do not

Iulia Antoanella Motoc, 27; Individual (dissenting) opinion of Ruth Wedgewood, 30; Individual opinion of Ivan Shearer, 32.

[81] M Zürn, 'Global Governance and Legitimacy Problems' (2004) 39 *Government and Opposition* 260, 283–4.

agree with the Council's decision to list certain individuals, as in the case of Mr Aden discussed above in Chapter 2(A)(4).[82] There is a wide perception that sanctions decision-making is unfair and that there is a need to recognize a broader system of legal principles by which to check their conduct.[83]

On the other hand, as has also been the case in the sanctions context, courts may be led under the instrumentalist model to assume a far greater role in international lawmaking than is normatively justifiable. Though the legal limits on the Security Council are narrow, they are also decidedly nebulous. The indeterminate limitations on the Security Council under the Charter and international law, coupled with the ambiguity of many Security Council resolutions, could end up giving courts disproportionate power to determine international policy. While in the domestic setting, courts are accustomed to filling gaps left by the legislature, in the Security Council context, these gaps become chasms. Courts vested with the power of review would inevitably be required to make difficult and largely unguided substantive choices among competing values, and indeed among inevitably controverted political, social, and moral conceptions of appropriate responses to threats to international peace and security. In *Ahmed*, certain judges of the UK Supreme Court who adopted what might be termed an 'instrumentalist' approach to their review of Security Council decision-making, inserted a requirement into the resolution that the asset freeze regime applied only to 'criminals' or those subject to a 'criminal charge'.[84] This interpretation is expressly refuted by Sanctions Committee Guidelines, which provide that neither a criminal charge nor a conviction is a precondition to being proposed or listed,[85] and was criticized by another judge on the same court on the grounds that it 'would emasculate the international system that the Security Council wishes to create'.[86]

The instrumentalist model positions courts as the central procedural actor. Yet, where courts interpret this as an invitation to judicial supremacy, this

[82] See Per Cramér, 'Recent Swedish Experiences with Targeted UN Sanctions: The Erosion of Trust in the Security Council' in Erika de Wet and André Nollkaemper (eds), *Review of the Security Council by Member States* (Intersentia 2003) 88; Serge Schmemann, 'A Nation Challenged: Sanctions and Fallout', *New York Times*, 26 January 2002; 'A Nation Challenged: In Bush's Words: Attack on "2 terrorist-supporting financial networks"', *New York Times*, 8 November 2001; P Koring, 'Federal lawyers argue they have no obligation to bring Abdelrazik home', *Globe and Mail*, 8 May 2009.

[83] José E Alvarez, *International Organizations as Law-makers* (Oxford University Press 2005), 585.

[84] *Her Majesty's Treasury v Ahmed* [2010] UKSC 2, [136]–[137] per Lord Phillips, and [203] Lord Brown.

[85] Al Qaeda Committee, 'Guidelines of the Committee for the Conduct of its Work', available at <http://www.un.org/sc/committees/1267/pdf/1267_guidelines.pdf>, para 6(c).

[86] *Her Majesty's Treasury v Ahmed* [2010] UKSC 2, [170] per Lord Rodger.

entails a fairly radical transformation of the balance between law and politics built into the Security Council setting. It is not the function of courts to create new aims for international society, or to impose on society new basic directives.[87] As Gerry Simpson has cautioned, '[a]las... notions of community [values] and common sense are contentious enough in homogenous national societies. Any attempt to incorporate judicial decision-making into the international legal order is likely to result in influence of even more egregious biases and ideological preferences than in domestic systems'.[88] While courts are recognized under Article 38 of the Statute of the International Court of Justice as having a role in the development of international law, this role is necessarily recognized as 'subsidiary', and not one of 'supremacy'.[89] Judicial supremacy was expressly rejected at the San Francisco Conference at which the Charter was drafted.[90] Even today, significant controversy attaches to the issue as to whether the ICJ has the power to review Security Council decision-making.[91] The ICJ has no power to bind the Security Council

[87] This was recognized by Lord Hoffmann in *Jones v Saudi Arabia* [2006] UKHL 26, [63]: 'It is not for a national court to "develop" international law by unilaterally adopting a version of that law which, however desirable, forward-looking and reflective of values it may be, is simply not accepted by other states.' See also L Fuller, 'The Forms and Limits of Adjudication' (1978) 92 *Harvard Law Review* 353, 392; John Hart Ely, *Democracy and Distrust* (Harvard University Press 1980).

[88] Gerry Simpson, 'Is International Law Fair?' (1996) 17 *Michigan Journal of International Law* 615, 625.

[89] UN Charter, Article 38(1)(d).

[90] A proposal advanced by the Belgian delegate to the San Francisco Conference, ultimately rejected by the other delegates, stated that 'Any state, party to a dispute brought before the Security Council, shall have the right to ask the Permanent Court of International Justice whether a recommendation or a decision made by the Council or proposed in it infringes on its essential rights. If the Court considers that such rights have been disregarded or are threatened, it is for the Council either to reconsider the question or to refer the dispute to the Assembly for decision': 'Dumbarton Oaks Proposals Concerning the Establishment of a General International Organisation: Amendments Submitted by the Belgian Delegation' (4 May 1945) UN Doc 2, G/7(k)(1), 3 UNCIO Docs 335, 336.

[91] The ICJ has clearly stated that it has no power to review Security Council resolutions: *Advisory Opinion on Legal Consequences for States of the Continued Presence of South Africa in Namibia (South West Africa) notwithstanding Security Council resolution 276 (1970)* [1971] ICJ Rep 45. For further debate on the issue, see, for example, Thomas M Franck, 'The "Powers of Appreciation": Who Is the Ultimate Guardian of UN Legality?' (1992) 86 *American Journal of International Law* 519; Michael Reisman, 'The Constitutional Crisis in the United Nations' (1993) 87 *American Journal of International Law* 83; GR Watson, 'Constitutionalism, Judicial Review, and the World Court' (1993) 34 *Harvard International Law Journal* 1; K Roberts, 'Second-Guessing the Security Council: The International Court of Justice and Its Powers of Judicial Review' (1995) 7 *Pace International Law Review* 281; Dapo Akande, 'The International Court of Justice and the Security Council: Is There Room for Judicial Control of the Decisions of the Political Organs of the United Nations?' (1997) 46 *International and Comparative Law Quarterly* 309; Kathleen Cronin-Furman, 'The International Court of Justice and the United Nations Security Council: Rethinking a Complicated Relationship' (2006) 106 *Columbia Law Review* 435.

through its judgments.[92] Instead, the drafters of the Charter endowed the Security Council with overwhelming power to protect international order, placing it at the apex of decision-making about international peace and security and vesting it with 'primary responsibility for the maintenance of international peace and security'.[93] Judicial activism by a court purporting to act as an unrepresentative and largely uninvited 'guardian' of the international legal order would arguably undermine rather than enhance the legitimacy and effectiveness of Security Council decision-making.

C. The Instrumentalist Model and International Values: The Stagnating Effect on the International Value System

The third overarching problem with the instrumentalist model in the Security Council context is that the model lacks any substantive normative dimension. Its concern is to ensure the accurate fulfilment of the substantive law, while failing to provide a basis to differentiate between acceptable and unacceptable aspects of that law. For the most part, adoption of such a model has resulted in reviewing courts adopting a position of deference to Security Council decision-making, as was exhibited by the CFI in *Kadi* and the Swiss Federal Tribunal in *Nada*.[94] Even worse, it may serve to have a 'chill-effect' on the development of international law, in effect confirming the legality of Security Council resolutions, and potentially forestalling positive developments.

In terms of the development of a procedural model to govern the future conduct of international institutions such as the Security Council, a value-free model would do little to enhance legitimacy. The view that international institutions are manifestly a good thing that can do no wrong, and whose functions could contribute only to the 'salvation of mankind', is the

[92] José E Alvarez, 'Judging the Security Council' (1996) 90 *American Journal of International Law* 1, 6.

[93] UN Charter, Article 24. For discussion of the drafters' intentions, see R Russell and J Muther, *A History of the United Nations Charter* (Brookings Institution 1958), 646; J Frowein and N Krisch, 'Introduction to Ch VII' in Bruno Simma (ed), *The Charter of the United Nations: A Commentary*, vol II (2nd edn, Oxford University Press 2002), 711; Gerry Simpson, *Great Powers and Outlaw States* (Cambridge University Press 2004), 184–8.

[94] Joined Cases T-306/01 and T-315/01 *Kadi v Council and Commission* [2005] ECR II-0000, [276]; *Nada v SECO* 133 BGE II 450 (2007). See also the decision of the Administrative Appeals Board of the Turkish Council of State in *Al-Qadi v The State* (TK 2007) ILDC 311, discussed in Antonios Tzanakopoulos, 'Domestic Court Reactions to UN Security Council Sanctions' in August Reinisch (ed), *Challenging Acts of International Organizations Before National Courts* (Oxford University Press 2010), 64.

perspective of a bygone era.[95] There was certainly a time in which international institutions were quite simply the agents of states—'loose associations for occasional specific joint action'—whose legitimacy was ultimately measured according to whether they were satisfactorily executing, and operating within the scope of, the functions they were established to perform with state consent.[96] However, faith is withering in this 'functionalist' approach to the conduct of international institutions, and some scholars have argued that it should be discarded.[97] It is increasingly recognized that international institutions are capable of all manner of missteps, omissions, and sins, including, in some cases, human rights violations,[98] and that there is a need to recognize a broader system of values and interests by which to check their conduct. It is accepted in contemporary international society that it is necessary to place more substantive limits on the conduct of international institutions. Professor Alvarez has described international organizations as akin to Mary Shelley's Frankenstein, established to fulfil certain functions, though gradually developing a powerful and uncontrollable force beyond the vision of their architects.[99] In such circumstances, the legitimacy of decision-making may not be sufficiently protected by a model that reins in international organizations merely to the extent that they diverge from compliance with the formal rules by which they were originally established.

The case law has shown already that review based solely on the process value of 'accuracy' is not sufficient to ensure the legitimacy of sanctions decision-making. It is indicative that in both *Ahmed* and *Sayadi*, those judges and Committee members that took such an approach were forced to reserve their most damning indictments of the deficiencies in the 1267 sanctions regime for non-binding comments made *obiter dicta* (or, in the case of the Committee, comments subsidiary to the main decision). In *Ahmed*, Lord Brown made comments *obiter* that '[t]he draconian nature of the regime imposed under these asset-freezing orders can hardly be over-stated'[100] and

[95] Nagendra Singh, *Termination of Membership of International Organisations* (Praeger 1958), vii.

[96] David Mitrany, 'The Functional Approach to World Organization' (1948) 24 *International Affairs* 350, 351.

[97] Jan Klabbers, 'Kadi Justice at the Security Council' (2007) 4 *International Organizations Law Review* 293; Jan Klabbers, *An Introduction to Institutional Law* (Cambridge University Press 2001); M Singer, 'Jurisdictional Immunity of International Organizations: Human Rights and Functional Necessity Concerns' (1995) 36 *Virginia Journal of International Law* 53.

[98] Frédéric Mégret and Florian Hoffmann, 'The UN as a Human Rights Violator? Some Reflections on the United Nations Changing Human Rights Responsibilities' (2003) 25 *Human Rights Quarterly* 314.

[99] José E Alvarez, *International Organizations as Law-makers* (Oxford University Press 2005), 585.

[100] *Her Majesty's Treasury v Ahmed* [2010] UKSC 2, [192].

that the regime maintained by the latter Order in Council was 'contrary to fundamental principles of human rights'.[101] This was despite the fact that the limitations of the instrumentalist model on which he relied ultimately led him to confirm the validity of one of the Orders in Council implementing the regime into UK law. In *Sayadi*, certain Committee members who criticized the Committee's decision to review the Security Council's decision-making on human rights grounds on the basis the Council was immune from review preceded this finding with the powerful statement:[102]

> We acknowledge that the authors may have been unjustly harmed by operation of the extravagant powers the Security Council has arrogated to itself, including the obstacles it has created to the correction of error. It is more than a little disturbing that the executive branches of 15 member states appear to claim a power, with none of the consultation or checks and balances that would be applicable at the national level, to simply discard centuries of States' constitutional traditions of providing bulwarks against exorbitant and oppressive executive action.

These judgments reinforce the perspective that it may be necessary to develop a procedural model that advances values apart from accuracy. Procedural law has the capacity to encourage dialogue on the operative norms and values that should govern international decision-making. An approach that restricts review of decision-making to accuracy in the implementation of the substantive law may result in courts stalling the development of international law by refusing to recognize emerging norms.[103] Courts that view themselves exclusively or primarily as impartial law enforcers could effectively freeze or crystallize international norms at a particular moment in time, thereby entrenching old understandings and providing a drag-effect on new developments.[104] The interests of legitimacy of the Security Council, and the international legal order more generally, arguably require more of a procedural model in contemporary international society.

D. Conclusion

The quest for accuracy is undoubtedly a worthy aim, which should by no means be rejected as one of the underlying values of a procedural fairness

[101] Ibid, [203].

[102] UN Human Rights Committee, *Sayadi and Vinck v Belgium* (Communication No 1472/2006, 29 December 2008) UN Doc CCPR/C/94/D/1472/2006, Individual opinion (partly dissenting) by Sir Nigel Rodley, Ivan Shearer, and Iulia Antoanella Motoc, 27.

[103] Anthea Roberts, 'Comparative International Law? The Role of National Courts in Creating and Enforcing International Law' (2011) 60 *International and Comparative Law Quarterly* 57.

[104] Ibid, 73.

framework for the Security Council. As observed in the introduction to this section, error in the implementation of the Security Council sanctions regime has served to undermine the credibility and effectiveness of decision-making. By enhancing the information available to decision-makers, requiring decision-makers to provide reasons for decisions, and expanding the opportunities for review, the accuracy of decision-making will undoubtedly be enhanced.

However, a procedural framework that advances accuracy as its sole or predominant aim will have a number of limitations when applied to the Security Council sanctions setting. As discussed in Chapter 4, the overarching aim of the instrumentalist model is to enhance legitimacy by ensuring the substantive law is accurately applied. The model is designed to operate in a political and legal framework where the substantive law the subject of 'accurate' application is created by a legitimately representative lawmaker, and forms part of a broader logically coherent and complete system of rules and principles that work together to ensure the values of the broader community are achieved. The Security Council is not such a setting. First, the Security Council is widely perceived to be unrepresentative in its mode of decision-making. Second, there is no certainty as to the legal principles and rules binding the Council by which the 'accuracy' of decision-making can be measured. Third, Council decision-making is not based on the application of clear and determinate legal standards, but more commonly involves significant and intentional levels of discretion and political compromise, and a margin of appreciation in their implementation. Fourth, and consequently, Security Council decision-making is not straightforwardly amenable to judicial review by an independent arbiter in accordance with ascertainable legal standards. The capacity of the instrumentalist model to enhance legitimacy in sanctions decision-making is limited by virtue of these fundamental characteristics of the Security Council decision-making context. Adoption of an instrumentalist model would perpetuate respect for a classical model of international law and institutions, adopting a laissez faire approach to the Council's decision-making, and abdicating responsibility for the standards, principles, and values guiding decision-making almost exclusively to the Security Council. For all these reasons, it may be necessary to look beyond an instrumentalist approach to procedural fairness to enhance the legitimacy of decision-making in the Security Council context.

6

A Model Based on Interest Representation

The Dignitarian Model in the UN Sanctions Context

Given the slow pace of reform in the Security Council sanctions context, and continuing refusal by the Council to respond to calls for an 'internationalized' judicial de-listing procedure, an alternative de-centralized judicial framework has emerged more organically, with a number of domestic and regional courts agreeing to hear challenges from individuals caught up in the sanctions net. Domestic courts have a long tradition of acting as guarantors of individual rights against erosion by governmental authority. In the absence of adequate protection for individuals at the Security Council level, it is therefore unsurprising that domestic courts have taken it upon themselves to act as guardians of individual rights. Responding to increasingly urgent challenges by individuals to the UN sanctions regime, certain courts have recognized a role for themselves in protecting and upholding fundamental individual rights, eschewing a traditional international law analysis of the relationship between the Security Council and domestic legal systems,[1] and enforcing *domestic or regional* interpretations of individual rights even where inconsistent with Security Council resolutions.

It is undoubtedly the case that this domestic and regional case law has been a most influential (and long overdue) source of individual rights protection in the sanctions context. The neglected voice of individuals placed on sanctions blacklists has been one of the major sources of criticism against the sanctions regime. The regime has been described as one that renders individuals 'effectively prisoners of state'[2] that 'does not begin to achieve fairness for

[1] For examples of the traditional approach to Security Council resolutions in other contexts, see *Behrami and Behrami v France; Saramati v France* (2007) 45 EHRR SE10; *Beric and others v Bosnia and Herzegovina* (2007) 46 EHRR SE6; *Al Jedda v Secretary of State for Defence* [2007] UKHL 58.

[2] *Her Majesty's Treasury v Ahmed* [2010] UKSC 2, [60] per Lord Hope (quoting Sedley LJ); cited with approval in Case T-85/09 *Kadi v European Commission* [2010] EUECJ, [149].

the person who is listed'.[3] The situation of an individual listed under the 1267 regime is not uncommonly compared to 'that of Josef K in Kafka's *The Trial*, who awakens one morning and for reasons never revealed to him or the reader, is arrested and prosecuted for an unspecified crime'.[4] Perceptions of the unfairness of the sanctions regime for individuals have been found by the Analytical Support and Sanctions Monitoring Team for the 1267 Al Qaeda and Taliban Sanctions Committee to have detracted from the credibility and effectiveness of the regime among member states.[5] The influence of domestic and regional courts has served to vault the individual into a more central position in Security Council decision-making, providing a route in the Council context to greater legitimacy focusing on enhancing representation of individual interests in sanctions decision-making. Prominent examples of such case law in the sanctions context include the decision of the Grand Chamber of the European Court of Justice in *Kadi*,[6] the decision of the General Court and Court of Justice of the European Union in *Kadi II*,[7] the majority's decision in the *HM Treasury v Ahmed* before the UK Supreme Court,[8] and the decision of the European Court of Human Rights in *Nada*.[9]

However, expansion of this de-centralized adjudicatory framework has impact beyond more robust protection of individual rights. As other scholars have noted, judicial review of Security Council decision-making in accordance with domestic and regional standards advances a pluralist vision of the international legal order.[10] In line with this scholarship, I argue that this

[3] *A, K, M, Q & G v HM Treasury* [2008] EWHC 869 (Admin), [18]. See also the General Court's description of the regime as 'draconian' in Case T-85/09 *Kadi v European Commission* [2010] EUECJ, [149].

[4] *Abousfian Abdelrazik v Minister of Foreign Affairs* 2009 FC 580, [53].

[5] 'Thirteenth report of the 1267 Monitoring Team' (31 December 2012) UN Doc S/2012/968, para 17; 'Ninth report of the 1267 Monitoring Team' (13 May 2009) UN Doc S/2009/245, para 16; 'Seventh Report of the 1267 Monitoring Team' (29 November 2007) UN Doc S/2007/677, para 26; 'Sixth report of the 1267 Monitoring Team' (8 March 2007) UN Doc S/2007/132, para 16; 'Fourth Report of the 1267 Monitoring Team' (10 March 2006) UN Doc S/2006/154, para 57; 'Second Report of the 1267 Monitoring Team' (15 February 2005) UN Doc S/2005/83, para 54. See also Security Council Report, 'Special Research Report: UN Sanctions' (25 November 2013), 14; C Whitlock, 'Terrorism Financing Blacklist at Risk', *Washington Post*, 2 November 2008.

[6] *Kadi v Council of the European Union* [2008] ECR I-0000.

[7] Case T-85/09 *Kadi v European Commission* [2010] EUECJ.

[8] *Her Majesty's Treasury v Ahmed* [2010] UKSC 2.

[9] *Nada v Switzerland* [2012] ECHR 1691.

[10] This has been recognized by scholars such as Nico Krisch, *Beyond Constitutionalism: The Pluralist Structure of Postnational Law* (Martin Loughlin, John McCormick, and Neil Walker eds) (Oxford University Press 2010); Gráinne de Búrca, 'The European Court of Justice and the International Legal Order after *Kadi*' (2010) 51 *Harvard Journal of International Law* 1; Daniel Halberstam and Eric Stein, 'The United Nations, the European Union and the King of Sweden: Economic Sanctions and Individual Rights in a Plural World Order' (2009) 46 *Common Market Law Review* 13; Samantha Besson, 'European Legal Pluralism after *Kadi*' (2009) 5 *European Constitutional Law Review* 237.

intervention by domestic courts, or 'pluralist judicial framework', can be analysed in theoretical terms as an example of the dignitarian model in action. Recognition of this connection between the pluralist judicial framework and dignitarian model enables us to undertake a normatively rich analysis of this framework in terms of its potential impact on conceptions of international community, international law, and the international value system in the Security Council and the international legal order more broadly.

A. The Dignitarian Model and International Community: Trading International Community for a 'Multiplicity of Publics'

Under a dignitarian model, the main aim of procedural safeguards is to provide individuals with a modicum of dignity, autonomy, and self-respect, such as might be expected in a pluralist society, enabling them to assert their interests and preferences, and placing an obligation on decision-makers to take these interests into account in decision-making.[11] In contrast to the instrumentalist approach discussed in the previous chapter, the focus is not on the accuracy of decision-making, but rather on the opportunity for the individual affected to state his or her case, and have that case taken into account. That is not to say that individuals must in all cases directly represent their own interests before the decision-making body in question. The structure for interest representation under the dignitarian model is potentially more complex. As discussed in Chapter 4, the sturdiest theoretical foundations of the dignitarian model rest on a pluralist conception of legal and political culture. Connected with the pluralist notion of autonomy underpinning the dignitarian model is the idea that individuals are self-legislating equals who are able to choose freely where their interests might be represented. The model is therefore not focused solely on individual rights, but on the individual's capacity to choose the forum within which to achieve optimal interpretation and representation of these rights.

The use of domestic and regional courts under the dignitarian model is, in these terms, a manifestation of the right of individuals to choose freely the manner of their representation in the international domain. It builds upon a pluralist notion of public autonomy and the right of individuals to

[11] Richard B Stewart, 'The Reformation of American Administrative Law' (1975) 88 *Harvard Law Review* 1667, 1684–6, 1712. See also Edmund Pincoffs, 'Due Process, Fraternity and a Kantian Injunction' in J Roland Pennock and John W Chapman (eds), *Due Process* (New York University Press 1977), 179.

determine which polity they want to be governed in and by. However, in opening up decision-making in this way, the dignitarian model would not serve to strengthen the bonds of 'international community'. Under a dignitarian model, individuals caught up in sanctions decision-making are less likely to characterize themselves as members of an international community, and more as victims of it. The impact of the dignitarian model would not be to broaden the scope of the 'international community' as such. Individuals do not claim rights under the dignitarian model as members of the international community or as citizens of a particular state, but as individuals exercising their right to choose between a multiplicity of overlapping and conflicting identities and loyalties depending upon the situation or issue at hand.[12] This is consistent with a pluralist conception of the international legal order, which rejects the idea of a singular constituency known as the 'international community' in favour of recognition of a 'multiplicity of publics'.[13]

Though this would update a classical conception of international community as confined to nation states, there is a downside to adoption of the dignitarian model in the Security Council context, a consequence that needs to be weighed into the overall assessment of the capacity of the model to enhance legitimacy. Adoption of a dignitarian model raises the prospect that the Security Council will increasingly be cast in the role of a 'broker of interests' rather than a guardian of the purposes and principles of the UN Charter and international community. Arguably, the international legal order should not operate in accordance with principles of market-ordering so as to advance the interests and preferences of the most powerful and organized elites, but was created to ensure respect for certain fundamental values in international relations, which include substantive principles of equality and non-discrimination.

The situation is magnified in the Security Council setting on account of power and wealth differentials, language and culture barriers, and the lack of identification with a common project, which render meaningful communication and deliberation beyond a narrow elite very difficult.[14] It is certainly

[12] Nico Krisch, *Beyond Constitutionalism: The Pluralist Structure of Postnational Law* (Martin Loughlin, John McCormick, and Neil Walker eds) (Oxford University Press 2010), 98.

[13] Nico Krisch, 'The Pluralism of Global Administrative Law' (2006) 17 *European Journal of International Law* 247; John Dryzek, *Deliberative Global Politics: Discourse and Democracy in a Divided World* (Polity Press 2006); James Bohman, *Democracy Across Borders* (MIT Press 2007); D Kennedy, 'One, Two, Three, Many Legal Orders: Legal Pluralism and the Cosmopolitan Dream' (2007) 31 *NYU Review of Law and Social Change* 642.

[14] Nico Krisch, *Beyond Constitutionalism: The Pluralist Structure of Postnational Law* (Martin Loughlin, John McCormick, and Neil Walker eds) (Oxford University Press 2010), 56. See also Eyal Benvenisti and G W Downs, 'The Empire's New Clothes: Political Economy and the Fragmentation of International Law' (2007) 60 *Stanford Law Review* 595.

the case that a geographical and socio-economic bias is evident already in the impact of the emerging pluralist adjudicatory framework in the sanctions setting.[15]Most judicial challenges to the sanctions regime have come from Europe and other advantaged nations, such as the United States and Canada. Even the informal debate about reform is marked by a systemic bias in favour of the most powerful voices in the international community. Over a decade's existence of the targeted sanctions regime, a raft of organizations, academics, and smaller states have agitated for reform, though many have reported that it has been impossible even to get a conversation started on the issue. By contrast, the decision of the European Court of Justice in *Kadi*, which confronted the Council with the prospect of non-compliance by a region as influential as the European Union, yielded an immediate response from members of the Permanent Five in terms of procedural reform with the establishment of the Office of the Ombudsperson.[16]

In terms of its impact on legitimacy, the dignitarian model's recognition of a voice for non-state actors will serve to modernize the classical conception of international organizations as entities created by states, composed of states, with the task of governing the relations between states. Recognition of the voice of non-state actors is not merely justified in the interests of fairness for those upon whom sanctions decision-making directly impacts, but also recognizes the imperative of enhancing representation in international decision-making of a significantly broader scope of actors than nation states.[17] However, in opening up decision-making in this way, the dignitarian model would not serve to strengthen the bonds of 'international community'. Rather, the use of domestic courts to vindicate the rights of individuals caught up in the sanctions regime may serve instead to alienate individuals from any notion of international community. For all intents and purposes, unless the Security Council itself

[15] Craig Forcese and Kent Roach, 'Limping into the Future: The UN 1267 Terrorism Listing Process at the Crossroads' (2010) 42 *George Washington International Law Review* 217, 217, 274.

[16] The Office of the Ombudsperson was established as a response to 'challenges, both legal and otherwise, to the measures implemented by Member States': UN Security Council resolution 1904 (17 December 2009) UN Doc S/RES/1904. See also 'Statement of French representative on the adoption of SC Resolutions 1988 and 1989' (2011) UN Doc S/PV.6557.

[17] Joseph Weiler, 'The Geology of International Law—Governance, Democracy and Legitimacy' (2004) 64 *Zeitschrift für ausländisches öffentliches Recht und Völkerrecht* 547; Robert O Keohane, 'Global Governance and Democratic Accountability' in David Held and M Koenig-Archibugi (eds), *Taming Globalization: Frontiers of Governance* (Polity 2003), 145; I M Young, *Inclusion and Democracy* (Princeton University Press 2000), 265–71; E Benvenisti, 'Exit and Voice in the Age of Globalization' (1999) 98 *Michigan Law Review* 167; Thomas M Franck, *Fairness in International Law and Institutions* (Clarendon Press 1995); David Held, *Democracy and the Global Order: From Modern State to Cosmopolitan Governance* (Polity Press 1995), 226–38; Thomas W Pogge, 'Cosmopolitanism and Sovereignty' (1992) 103 *Ethics* 48, 57–69.

develops a mechanism through which individual rights can be adequately protected, individuals caught up in sanctions decision-making will not identify themselves as constituents of the international community, but instead as wronged targets of it.

B. The Dignitarian Model and International Law: A Pluralist Conception of International Law

The adoption of a dignitarian model of due process in the Security Council context would have serious implications for the 'rule of law' in that setting. First, judicial review of sanctions measures by a range of domestic and regional courts introduces further risk of fragmentation in the international legal order, threatening to devalue qualities that might be thought to be fundamental to any framework asserting itself as a legal system, such as certainty, predictability, consistency, and generality of application, and raises the risk of 'fragmentation' in the international legal order.[18] Second, the risk that certain courts will invalidate sanctions measures threatens to lead to a breakdown in traditional notions of hierarchy, including those presently codified in Articles 25 and 103 of the UN Charter. That is not to say that these would necessarily be negative developments. Indeed, recognition of the strengths of the dignitarian model in this context might do much to expose the misguidedness of certain 'rule of law' debates related to the Security Council and introduce important qualifications to the use of the term in that setting.[19]

(1) The Problem of Uneven Justice

Concern has been expressed that judicial review by domestic and regional courts will open a 'Pandora's box', resulting in the fragmentation of Security Council resolutions along the borders of national and supra-national jurisdictions.[20] The danger of cases like *Kadi* and *Ahmed* is that they will set informal precedents for the invalidation of domestic mechanisms

[18] J Baquero Cruz, 'The Legacy of the Maastricht Judgment and the Pluralist Movement' (2008) 14 *European Law Journal* 389, 391–403. See also Pavlos Eleftheriadis, 'Pluralism and Integrity' [2010] *Ratio Juris* 365.

[19] For a summary of the thematic debates held on the rule of law in the Security Council since 2003, see Security Council Report, 'The Rule of Law: The Security Council and Accountability' (18 January 2013).

[20] Johannes Reich, 'Due Process and Sanctions Targeted Against Individuals Pursuant to Resolution 1267 (1999)' (2008) 33 *Yale Journal of International Law* 505, 510.

implementing the sanctions regime, opening up dangerous holes in the sanctions net.[21] The consequences for the Security Council sanction regime are not insignificant. The sanctions regime is uniquely vulnerable to failure by individual states to comply. Sanctions measures such as travel bans and asset-freezing rely heavily on universal compliance, or else individuals and funds will be channelled into the breach. A steady stream of domestic case law along the lines of *Kadi* and *Ahmed* could see the Security Council sanctions framework buckle under the weight of opposition by domestic courts.[22]

Yet it is possible to adopt a more positive take on this potentially damaging implication of the dignitarian model. The idea that all law must originate in a single power source is a distinctly domestic idea. While stability, uniformity, and certainty may be appropriate in more representative legal systems, a pluralist perspective cautions against exaggerating the importance of such qualities, particularly in an unrepresentative system such as the international legal order. Indeed, 'certainty' in the application of the law in a markedly unrepresentative system may in fact be a source of friction and instability where it clashes with the strong preferences of excluded actors.

In such a setting, legitimacy is more appropriately advanced by a model of law that preserves space for contestation and local variation. Within such a setting, normative conflict can become a source of innovation. A decision by a domestic court to invalidate the domestic mechanism implementing a Security Council resolution has the capacity to act as a spur to the Security Council to respond to the concern underlying the judgment. Where the concern is one felt more broadly among members of the international community, such that the judgment is likely to form a persuasive precedent for other domestic courts, the Security Council will (appropriately) feel pressured to implement reform. Litigation around sanctions has been influential in motivating reform because it was part of 'a broader movement for change'.[23] On the other hand, where a domestic judgment does not resonate more broadly, it will be marginalized and seen as exceptional.[24] In such circumstances, it

[21] Indeed, *Kadi* has already been followed in Joined Cases C-399/06 P and C-403/06 P *Hassan v Council* [2009] ECR I-11393; Case T-318/01 *Othman v Council* [2009] ECR II-01627; Joined Cases T-135/06 to 138/06 *Al-Faquih v Council* (General Court, 29 September 2010).

[22] 'Twelfth Report of the 1267 Monitoring Team' (1 October 2012) UN Doc S/2012/729, para 33; R Barrett, 'Al-Qaeda and Taliban Sanctions Threatened', *Policy Watch*, 6 October 2008, available at <http://www.washingtoninstitute.org/templateC05.php?CID=2935>; 'Statement by Representative of Switzerland' (13 November 2009) UN Doc S/PV.6217, 28.

[23] Nico Krisch, *Beyond Constitutionalism: The Pluralist Structure of Postnational Law* (Martin Loughlin, John McCormick, and Neil Walker eds) (Oxford University Press 2010), 184.

[24] Alan Boyle and Christine Chinkin, *The Making of International Law* (Oxford University Press 2007), 301.

would arguably be contingent upon the Security Council (and potentially others in the international community) to continue the dialogue, explaining why the decision is inappropriate to apply in the sanctions setting, and placing pressure on the violating state to bring its domestic machinery into conformity with the sanctions regime. Despite the potentially dire consequences for the sanctions regime, it may be that occasional non-compliance by states is appropriate if it ultimately renders the Council more responsive to the fundamental values of domestic legal orders, and thereby enhances the legitimacy of its decision-making.

(2) The Prospect of Hierarchical Breakdown

The second potentially problematic implication is for notions of hierarchy in international law. Adoption of a dignitarian model of due process will almost inevitably lead to a breakdown in traditional notions of hierarchy codified in Articles 25 and 103 of the UN Charter, which recognize the supremacy of Security Council decisions over other legal principles and legal orders. As the General Court recognized in *Kadi II*, the necessary consequence of the decision of the European Court of Justice (ECJ) in *Kadi* has been 'to render that primacy [of Security Council resolutions] ineffective in the Community legal order'.[25]

However, as with the concerns about fragmentation, it could be argued that contemporary attachment to traditional notions of hierarchy is misplaced. The positivist hierarchical account, which continues to be the dominant approach to international law, has been criticized as concealing the contested nature of fundamental issues in the international legal order, and as legitimizing the dominance of particular states in international society over others.[26] There is a sense that 'pillars' of Security Council supremacy such as Articles 25 and 103 might jeopardize rather than facilitate the effective implementation of Security Council resolutions on the basis that they downplay the significance of diversity in the global polity, and the importance of a normative role for non-state actors.[27]

[25] *Kadi v European Commission* [2010] EUECJ, [118].

[26] Benedict Kingsbury and Megan Donaldson, 'From Bilateralism to Publicness in International Law' in U Fastenrath et al (eds), *From Bilateralism to Community Interest: Essays in Honour of Bruno Simma* (Oxford University Press 2011), 81; Nico Krisch, 'Global Administrative Law and the Constitutional Ambition', LSE Law, Society and Economy Working Papers 10/2009, 10; James Tully, *Constitutionalism in an Age of Diversity* (Cambridge University Press 1995).

[27] See related discussion above in Chapter 5(B)(3).

As such, adoption of a dignitarian model of due process, providing a mechanism for stakeholders to assert their interests, does not so much represent a breakdown of existing principles of hierarchy and authority as recognize the position that these principles do not work in practice. The normative force of the dignitarian model, and the pluralist adjudicatory framework supporting it, does not stem from coercive notions of hierarchy (particularly fragile in the international setting where the capacity for enforcement is thin), but from the persuasiveness of competing visions, and the extent to which they are shared by the broader international community. Ideally, judicial intervention under this model will not seek to determine the outcome for the entire international community, but represents part of a collaborative attempt to achieve (limited) convergence on the observance of certain standards. Such an approach would advance a pluralist vision of the international legal order in which a multiplicity of distinct autonomous and authoritative legal orders relate through a continuous 'agonistic process of negotiation' and 'pragmatic adjustment' between alternative orders on the basis of consent and mutual respect.[28] The accommodation between interests achieved through pluralist dialogue is not arbitrary, but is dependent upon the weight of interests or values asserted.

(3) Solution: The 'Dialogue' Model of Judicial Function

Ultimately, then, it may be justice itself, not a departure from justice, for a domestic court to dis-apply domestic mechanisms implementing Security Council resolutions where it considers that the Security Council has struck an incorrect balance between the interests of international peace and security and fundamental domestic rights. However, for the dignitarian model to work effectively, domestic and regional courts may need to reflect on the nature and scope of their role in the broader global setting. As domestic courts begin to take on a larger role within the broader international legal system, assuming the capacity to review decision-making by international institutions, it may be necessary to generate a model of the judicial function appropriate to the broader system which a domestic court is operating.[29] The

[28] Neil Walker, 'The Idea of Constitutional Pluralism' (2002) 65 *Modern Law Review* 317, 354.

[29] On this point, see the discussion of 'systemic integration' in the International Law Commission's Report, 'Fragmentation of International Law: problems caused by the diversification and expansion of international law' (13 April 2006) UN Doc A/CN4/L.682, paras 37–43.

judiciary regularly exhibits a capacity to adapt to society's changing needs.[30] This capacity for change extends to the judicial function itself, which has been recognized as a function of place and time, influenced by its environment and the constantly changing factual and social reality.[31] Lon Fuller recognized the need for the judicial function to be constantly cognizant of the problem of *system*, in the sense that judges must always consider the coherence of the system in which they operate, and the powers and limitations of the institution of the judiciary as defined within that system.[32]

The dignitarian model of due process offers the potential to reconceptualize the judicial function where courts are engaging in review of international decision-making. As discussed in the previous chapter, the instrumentalist model of the judicial function based on traditional notions of bindingness and supremacy leads to a polarization between law creation and law enforcement, binding and non-binding law, judicial and political supremacy, and international and domestic legal orders. I argue that the future of domestic court engagement should not be in the perpetuation of these polarizing dichotomies, but in the fertile *dialogic* middle ground between these extremes.

In the following section, I suggest that the role of domestic and regional courts may need to be recalibrated in two important ways when they engage in review of international decision-making: first, domestic courts are entitled to claim *enhanced authority* of 'sources' rather than merely 'enforcers' of international law; and, second, domestic courts should recognize *qualified authority* in the sense that they are merely persuasive but not determinative authorities on the legal position. Taking these features into account, it is possible to develop a conception of the judicial function that positions domestic and regional courts in a more theoretically supportable way within the broader framework of the international legal system.

(a) Enhanced Power: Domestic Courts as a Source of Law

The domestic model of the judicial function in democratic systems is based in part on recognition of a necessary separation between the judicial power of law enforcement and the political power of law creation. It is the representative source of the law being applied by courts that lends legitimacy to the

[30] It has been said that a thousand years of common law are a thousand years of changes in the law in order to adapt it to the needs of a changing reality: Julius Stone, *Legal System and Lawyers' Reasoning* (Stanford University Press 1964), 209–98.

[31] A Barak, 'A Judge on Judging: The Role of a Supreme Court in a Democracy' (2002) 116 *Harvard Law Review* 19, 25.

[32] Lon Fuller, *Anatomy of the Law* (Frederick A Praeger 1968), 94.

judicial role. However, where courts are reviewing the decisions of international decision-makers such as the Security Council, the normative basis for this separation is lost. When domestic courts seek to review decisions made by international institutions, domestic courts are no longer acting as legitimate guardians and enforcers of domestic law. Instead, they are engaging with a body of law that finds its source in the broader international legal system.

In the international legal order, domestic courts are not mere enforcers, but are also recognized as sources of international law. First, decisions of domestic courts are evidence of state practice, relevant to the three primary sources of international law listed in Article 38(1) of the Statute of the International Court of Justice (ICJ), including interpretation of treaties, the development of international custom, and elaboration of general principles of law.[33] Second, judicial decisions are referred to expressly in Article 38(1) as a 'subsidiary means for the determination of rules of law'.[34] It is accepted that the reference to 'judicial decisions' in this context includes decisions of domestic courts.[35] In practice, it has been noted that domestic court decisions play an extremely important role in the identification and formation of international law, and are routinely cited as evidence of the meaning of international law.[36]

In such circumstances, a polarization between law enforcement and law creation in elaborating the role of domestic courts is 'a simplistic and often inaccurate approach'.[37] Instead, domestic courts must recognize their capacity to contribute to the development of international law. The authority provided to domestic courts in Article 38 is based in recognition of the special legal expertise of judges, and their capacity to develop and fill gaps in the law utilizing powers of legal reasoning.[38] As Judge Higgins has

[33] P Moremen, 'National Court Decisions as State Practice: A Transjudicial Dialogue?' (2006) 32 *North Carolina Journal of International Law and Commercial Regulation* 259.

[34] Statute of the International Court of Justice, Article 38(1)(d).

[35] Rosalyn Higgins, *Problems and Process* (Clarendon Press 1994), 218; Robert Jennings and Arthur Watts, *Oppenheim's International Law* (9th edn, Longman 1996), 41–2.

[36] A Nollkaemper, 'The Independence of the Domestic Judiciary in International Law' (2006) XVII *Finnish Yearbook of International Law* 1, 12–13; A Nollkaemper, 'Decisions of National Courts as Sources of International Law: An Analysis of the Practice of the ICTY' in G Boas and WA Schabas (eds), *International Law Developments in the Case Law of the ICTY* (Martinus Nijhoff 2003), 277–96; Anne-Marie Slaughter, 'A Typology of Transjudicial Communication' (1994) 29 *University of Richmond Law Review* 99.

[37] Anthea Roberts, 'Comparative International Law? The Role of National Courts in Creating and Enforcing International Law' (2011) 60 *International and Comparative Law Quarterly* 57, 60.

[38] Lon Fuller, *The Morality of Law* (Rev edn, Yale University Press 1969); R Dworkin, *Law's Empire* (Harvard University Press 1986); T R S Allan, 'The Rule of Law as the Rule of Reason: Consent and Constitutionalism' (1999) 115 *Law Quarterly Review* 221.

recognized, international law would never develop beyond its rudimentary state if courts felt that the distinction between *lex lata* and *lex ferenda* forever prevented it from applying the law in a progressive manner. Indeed, part of the problem of the traditional 'bindingness' approach is that it may have a 'chill effect' on the law, freezing or crystallizing international norms at a particular moment in time, and stalling important developments.[39] Instead, domestic courts should take the opportunity to exercise a more creative function, using legal reasoning to build on existing legal principles, and crafting applicable legal principles for new contexts. When operating in relation to the international legal sphere, the judicial function must include 'developing and applying international law to hitherto untested situations in order to obtain socially desirable and enlightened results'.[40]

(b) Qualified Authority: The Persuasive Power of Domestic Courts

Though domestic courts have a role in law creation when acting in relation to the international legal order, it is important to recognize that this role is a qualified one. Under Article 38(1), domestic courts are not empowered to make a final or authoritative determination of the applicable law.[41] Rather, their role in the determination of law is 'subsidiary'. They provide mere evidence of the primary sources of international law.[42] Ultimately, '[t]he authoritativeness of [their decisions] will…be founded upon the constitutional function, perceived role and reputation of the court'.[43] If international actors accept the relevant court's decision, its law-making impact will be substantial. If not, the decision 'may become marginalized, seen as exceptional and have limited law-making effect'.[44] The picture is therefore one of graded authority, leaving behind the binary scheme of binding/not binding and political/judicial supremacy, and instead associating norms with levels of persuasiveness tied to broader conceptions of an institution's legal reasoning and reputation.[45]

[39] Anthea Roberts, 'Comparative International Law? The Role of National Courts in Creating and Enforcing International Law' (2011) 60 *International and Comparative Law Quarterly* 57, 73.

[40] R Higgins, 'Aspects of the Case Concerning the Barcelona Traction, Light and Power Company, Ltd' (1971) 11 *Virginia Journal of International Law* 327, 341.

[41] See *Jones v Saudi Arabia* [2006] UKHL 26, [63] per Lord Hoffmann.

[42] I Brownlie, *Principles of Public International Law* (6th edn, Oxford University Press 2003), 6.

[43] M Shaw, 'A Practical Look at the International Court of Justice' in M Evans (ed), *Remedies in International Law: The Institutional Dilemma* (Oxford University Press 1998), 11, 27.

[44] Alan Boyle and Christine Chinkin, *The Making of International Law* (Oxford University Press 2007), 301.

[45] Nico Krisch, *Beyond Constitutionalism: The Pluralist Structure of Postnational Law* (Martin Loughlin, John McCormick, and Neil Walker eds) (Oxford University Press 2010), 12;

(c) The Dialogue Model of Judicial Function

The challenge then is to recalibrate the judicial function to take account of the enhanced power of domestic courts as a source of international law, while recognizing the qualified authority of domestic courts in the international legal order. My conclusion is that a 'dialogue model' of the judicial function positions courts in the most theoretically supportable way in the broader international legal system. This model does not purport to cast domestic courts in an entirely new and foreign role, but instead draws on accepted models for domestic court engagement with the domestic human rights. In the human rights context, domestic legal systems have developed a number of ways to reallocate power between judicial and political organs in a way that brings them into greater balance than under the traditional lopsided models of either constitutional supremacy (where courts are supreme) and parliamentary sovereignty (where parliament is supreme). While it has taken a number of forms in Canada, the United Kingdom, Australia, and New Zealand, the 'dialogue model' refers to a relationship in which judicial decisions about the compatibility of legislation with human rights are open to reversal by the legislature.[46]

The relevant aspect for the purposes of the present discussion is that the authority of domestic courts is construed, not in terms of supremacy, but in terms of a spectrum. Translated to the Security Council sanctions context, domestic courts could be seen as having the choice: (a) to *invalidate* Security Council resolutions; (b) to *interpret* resolutions in a way that is consistent with human rights; (c) to *declare* resolutions incompatible with human rights; or (d) to *abstain* from review. As discussed in Chapter 5, the first option potentially oversteps the permissible and practicable boundaries of the judicial function in the international legal system, while the final option would open domestic courts to the criticism that they had abdicated their responsibility to uphold the fundamental principles of their domestic legal systems. It is in the two central choices that we find the practically and theoretically supportable options for domestic courts. Each of these options can be seen as promoting dialogue in that they contribute to development of

B Kingsbury, 'Weighing Regulatory Rules and Decisions in National Courts' [2009] *Acta Juridica* 90; Mayo Moran, 'Shifting Boundaries: The Authority of International Law' in J Nijman and A Nollkaemper (eds), *New Perspectives on the Divide between National and International Law* (Oxford University Press 2007), 163.

[46] For relevant literature, see for example Tom Hickman, 'Constitutional Dialogue, Constitutional Theories and the Human Rights Act' [2005] *Public Law* 303; Kent Roach, 'Dialogic Judicial Review and Its Critics' (2004) 23 *Supreme Court Law Review* (2d) 49.

the law, yet are open to reversal by the Security Council should the Council fundamentally disagree with the court's decision. Regardless of which option is taken, the idea is for a legal culture to develop in which Council and courts come to see themselves as involved in a dialectical partnership or dialogue in which they are both working toward an appropriate balance between international, regional, and domestic values.[47] As we will see in the next section, courts have taken a range of interpretations as to how this balance should be struck.

C. The Dignitarian Model and International Values: Subjugating International Values to Parochial Interests

A key advantage of the dignitarian model is that it has greatly strengthened individual rights protection in the sanctions context. As recognized in Chapter 5, domestic courts adopting an 'instrumentalist' approach, acting as international agents and measuring the legality of Security Council decision-making under international law, have not provided much by way of human rights protection for individuals caught up in the sanctions net. By contrast, the domestic case law discussed in this chapter has been the most influential source to date of individual rights protection under the UN sanctions regime.[48] As is made clear in the relevant judgments, such courts are not driven by an 'instrumentalist' focus on the formal implementation on the will of the lawmaker, but by a 'dignitarian' focus on the protection of individual rights.[49]

The danger is that, taken too far, the dignitarian model can lead to an overemphasis on elite interests and parochial conceptions of rights and values. I describe the dignitarian model in Chapter 4 as one concerned above all with 'interest representation' enabling representation of pluralist rights and interests. The danger of this pluralist mentality is that procedural safeguards will ultimately lead to an overemphasis on diverse 'interests', paying little

[47] For a depiction of the 'dialectical' nature of the dialogue between courts and legislatures at a domestic level, see Grégoire Webber, 'The Unfulfilled Potential of the Court and Legislature Dialogue' (2009) 42 *Canadian Journal of Political Science* 443.

[48] Dire Tladi and Gillian Taylor, 'On the Al Qaida/Taliban Sanctions Regime: Due Process and Sunsetting' (2011) 10 *Chinese Journal of International Law* 771, 776, 789.

[49] *European Commission and Council v Kadi* [2013] ECR not yet reported, [111], [119], [121], [134]; *Kadi v Council of the European Union* [2008] ECR I-0000, [368]; *Her Majesty's Treasury v Ahmed* [2010] UKSC 2, [6] per Lord Hope (Lord Walker and Baroness Hale agreeing).

heed to the overarching need to contribute productive input into the development of a bedrock of shared fundamental 'values' that some regard as an essential complement to emerging structures of global governance.[50] Scholars such as Cass Sunstein and Jenny Steele have criticized pluralist approaches by invoking this contrast between 'interests' and 'values'.[51] In the course of distinguishing his republican alternative from a pluralist system, Professor Sunstein describes a pluralist system as 'indifferent among preferences', with its chief concern being the aggregation of private preferences and the achievement of equilibrium among contending social forces.[52] He distinguishes this from a system in which decisions have to be supported by argument and reasons and justified through an appeal to a broader public good. Jenny Steele illustrates the implications of such an approach for conceptions of deliberation. She describes the form of deliberation that would be encouraged under the dignitarian model as concerned with ensuring that decision-making 'takes account of' all perspectives, without building in any requirement for reasoned argument. In essence, the emphasis is on a form of narrow-minded and self-interested participation, concerned with providing parties with an opportunity to speak, rather than providing a process for communication and argument under which participants might be encouraged to phrase their objections as if they were thinking through which is best for all participants.

The domestic and regional case law to date reflects a range of interpretations of an emerging pluralist relationship between the Council and courts. In normative terms, it is helpful to differentiate between radical and moderate pluralist approaches to the dignitarian model.[53] Under the radical approach, domestic courts are almost entirely internally or domestically focused, and pay little regard to the overarching framework within which the decision under review was taken. Under a moderate approach, courts assign

[50] Stephen Gardbaum, 'Human Rights and International Constitutionalism' in Jeffrey Dunoff and Joel Trachtman (eds), *Ruling the World: Constitutionalism, International Law and Global Governance* (Cambridge University Press 2009), 249–51, 255–6; Andrew Hurrell, *On Global Order: Power, Values and the Constitution of the International Society* (Oxford University Press 2007), chs 3, 4; Erika de Wet, 'The International Constitutional Order' (2006) 55 *International and Comparative Law Quarterly* 51, 57–61

[51] Cass R Sunstein, 'Beyond the Republican Revival' (1988) 97 *Yale Law Journal* 1539; Jenny Steele, 'Participation and Deliberation in Environmental Law: Exploring a Problem-solving Approach' (2001) 21(3) *Oxford Journal of Legal Studies* 415.

[52] Cass R Sunstein, 'Beyond the Republican Revival' (1988) 97 *Yale Law Journal* 1539, 1544, 1548.

[53] This 'pluralism of pluralisms' draws on literature developed within the context of the EU as a means of theorizing the impact of the conflicting supremacy claims of the national and European levels in the EU, as adapted in Nico Krisch, *Beyond Constitutionalism: The Pluralist Structure of Postnational Law* (Martin Loughlin, John McCormick, and Neil Walker eds) (Oxford University Press 2010).

equal positions to different legal systems, though see the potential for their co-ordination having regard to a common point of reference or framework. Though a range of cases could be discussed, the latter phases of the *Kadi* litigation offer an interesting case study on the different ways in which courts have interpreted a 'dignitarian' approach to due process. We have already discussed in Chapter 4 how, in the first decision in the *Kadi* litigation, the Court of First Instance (CFI) acted in effect as an 'agent' of the international community, adopting an instrumentalist model of due process. In the later phases of the litigation, three versions of the dignitarian model emerge. In this section, I assess the different approaches taken at different phases of the *Kadi* litigation, including the politically radical 'subversive' approach of the ECJ in 2008 (ECJ *Kadi*),[54] the normatively radical 'subsidiarity' decision of the Court of Justice in the European Union (CJEU) in 2013 (CJEU *Kadi*),[55] and the normatively moderate 'Solange equivalence' approach taken by the Advocates General in *Kadi*.[56] My aim in doing so is to determine which is the best interpretation of the dignitarian model in the Security Council context in terms of its capacity to contribute to the overall legitimacy of Security Council decision-making.

(1) Politically Radical Pluralism: *Kadi* in the ECJ

In 2008, the ECJ handed down its decision on appeal from the CFI in the *Kadi* case. As is well known, the ECJ took a directly contrasting approach to the CFI, declining to defer to the Security Council and ultimately invalidating the regulation giving effect to the relevant Security Council resolution on the basis that it violated fundamental rights of the European legal order. At the heart of the decision is the Court's determination that the EC Treaty establishes an 'autonomous legal system which is not to be prejudiced by an international agreement'.[57] In contrast to the CFI's deference to the source of the initial decision-making power, namely the Security Council, the Security Council attempts to separate the regulation from its source and imagines

[54] *Kadi v Council of the European Union* [2008] ECR I-0000. In 2010, the General Court handed down a decision following the 2008 decision of the ECJ: Case T-85/09 *Kadi v European Commission* [2010] EUECJ. Detailed reference will not be made to this decision.

[55] *European Commission and Council v Kadi* [2013] ECR not yet reported.

[56] Opinion of Advocate General Maduro, *Kadi v Council of the European Union*, 16 January 2008; Opinion of Advocate General Bot, *European Commission and Council v Kadi*, 19 March 2013.

[57] ECJ *Kadi*, [316]. See also at [282]: 'an international agreement cannot affect the allocation of powers fixed by the [European] Treaties or, consequently, the autonomy of the Community legal system'.

that it is deciding the dispute 'in the context of the internal and autonomous legal order of the Community, within whose ambit the contested regulation falls'.[58] The Court underlines its duty to review EU regulations against the EU's basic constitutional charter 'regardless of the origin of those acts' in a Security Council resolution.[59] By amputating the international roots of the regulation, the ECJ emphasizes that it is not engaging in the review of the lawfulness of 'a resolution adopted by an international body', but rather of 'the *Community act* intended to give effect to that resolution'.[60] It then takes the seemingly straightforward step of measuring the regulation against EU standards and annulling the regulation on the basis that it violates Mr Kadi's rights of defence (including his right to be heard and to effective judicial protection) and his right to respect for property.

In terms of the dignitarian model, the ECJ takes what might be described as a 'radically' pluralist approach. In her seminal and influential article, Gráinne de Búrca was critical of the ECJ's approach to the relationship between legal orders, interpreting the judgment as radically 'dualist' in nature, inserting itself in the tradition of nationalism,[61] taking the path of 'European particularism'.[62] De Búrca criticized the Court's approach on the basis that it '[eschews] engagement in the kind of international dialogue that has generally been presented as one of the EU's strengths as a global actor' and determines the relationship between the EU and the international order 'in accordance with its own internal values and priorities rather than in accordance with any common principles or norms of international law'.[63] Kunoy and Dawes invoke criticism of the ECJ perspective as 'solipsistic',

[58] ECJ *Kadi*, [317].

[59] ECJ *Kadi*, [317], [326].

[60] ECJ *Kadi*, [278] (emphasis added).

[61] B Fassbender, 'Triepel in Luxemburg, Die dualistische Sicht des Verhältnisses zwischen Europa- und Völkerrecht in der "Kadi-Rechtsprechung" des EuGH als Problem des Selbstverständnisses der Europäischen Union' [2010] *Die Öffentliche Verwaltung* 333, 340; quoted in J Kokott and C Sobotta, 'The *Kadi* Case—Constitutional Core Values and International Law—Finding the Balance?' (2012) 23(4) *European Journal of International Law* 1015, fn 3. See also C Tomuschat, 'The *Kadi* Case: What Relationship is there between the Universal Legal Order under the Auspices of the United Nations and the EU Legal Order' [2009] *Yearbook of European Law* 654, 658 ff; E Cannizzaro, 'Security Council Resolutions and EC Fundamental Rights: Some Remarks on the ECJ Decision in the *Kadi* Case' [2009] *Yearbook of European Law* 593, 597 ff.

[62] Daniel Halberstam and Eric Stein, 'The United Nations, the European Union and the King of Sweden: Economic Sanctions and Individual Rights in a Plural World Order' (2009) 46 *Common Market Law Review* 13, 72.

[63] Gráinne de Búrca, 'The ECJ and the International Legal Order after *Kadi*' (2009) 51 *Harvard Journal of International Law*.

'imperialistic', and smacking of 'European self-centredness'.[64] The decision has been condemned as a disquieting gesture of bad citizenship against the international legal order, criticized by some scholars as a form of 'Texasization' of the EU's relationship with international law (invoking an analogy to the infamous *Medellin* case of the US Supreme Court in which the Supreme Court held that an order by the ICJ not to execute certain individuals was not binding on Texas under US domestic law).[65] It has also been argued that the decision will have a grave impact on the effectiveness of Security Council measures. Though the Council itself has not publicly addressed the issue, the Security Council's Monitoring Team has expressed concern that decisions such as *Kadi* have placed the sanctions regime 'at a crossroads' and could have a 'significant impact on the regime' jeopardizing the implementation of sanctions by member states with the 'potential to damage the regime or to distract it from looking forward'.[66]

However, my own view is that the ECJ should not be seen as proposing a normative approach to the relationship between EU and international legal orders. Indeed, there is a sense in the judgment that the ECJ itself was aware that it was pedalling in fiction by drawing this separation between international and EU legal orders. Though it insists that it is merely reviewing the implementation of Security Council measures by EU institutions, it does engage in a brief review of the structures and procedures at the UN level to ascertain whether they satisfy EU standards. The ECJ finds that, with respect to the 'system of restrictive measures set up by the United Nations with regard both to entry in the summary list and to removal from it', namely the focal point, 'the fact remains that the procedure before [the Sanctions Committee] is still in essence diplomatic and inter-governmental, the persons or entities concerned having no real opportunity of asserting their rights and that committee

[64] Bjørn Kunoy and Anthony Dawes, 'Plate Tectonics in Luxembourg: the ménage à trois between EC Law, International Law and the European Convention on Human Rights Following the UN Sanctions Cases' (2009) 46 *Common Market Law Review* 73; citing M Koskenniemi, 'International Law: Constitutionalism, Managerialism and the Ethos of Legal Education' (2007) 1 *European Journal of Legal Studies* 1.

[65] T M Franck, 'NYU Panel Discussion on the *Kadi* judgment', 17 September 2008, available at <http://globaladminlaw.blogspot.co.uk/2008/09/nyu-panel-discussion-on-kadi-judgment.html>. See also Gráinne de Búrca, 'The European Courts and the Security Council: Between Dédoublement Fonctionnel and Balancing of Values: Three Replies to Pasquale De Sena and Maria Chiara Vitucci' (2009) 20(3) *European Journal of International Law* 853, 862.

[66] See also Tzanakopoulos' description of the decision as 'hugely de-stabilizing': Antonios Tzanakopoulos, 'The Solange Argument as a Justification for Disobeying the Security Council in the *Kadi* Judgments' in M Avbelj, F Fontanelli, and G Martinico (eds), *Kadi on Trial: A Multifaceted Analysis of the Kadi Judgment* (Routledge 2014).

taking its decisions by consensus, each of its members having a right of veto'.[67] Moreover, the ECJ acknowledges that any judgment it hands down will have effect beyond the EU legal order. Somewhat undermining its insistence that review of the regulation would not entail any challenge to the primacy of the Security Council resolution, the ECJ acknowledges that the annulment will have the effect of 'seriously and irreversibly prejudicing the effectiveness of the restrictive measures imposed by the regulation'. As such, the ECJ builds in a three-month suspension of the effects of the judgment to give the relevant European institutions time to remedy the infringements.[68]

In my view, it is a mistake to interpret the 2008 *Kadi* judgment as a precedent advocating disassociation of domestic implementing measures from international measures in a re-affirmation of traditional dualist logic. Instead, we need to appreciate the judgment in its *political* context as more in the nature of an act of open judicial revolt against years of fruitless political dialogue. Türküler Isiksel invites us to see the ECJ's judgment in 2008 *Kadi* as 'an act of civil disobedience' rendered necessary by the UN Security Council's misapplication of foundational principles of the international order.[69] The judgment undoubtedly encourages violation of international legal norms in favour of regional norms. However, Türküler Isiksel argues (and I agree) that the ECJ's evasiveness towards international law in *Kadi* should not be regarded as lawless unilateralism.[70] Instead, she argues that the act of civil disobedience was justifiable because it was not self-serving nor did it threaten the structure of the international legal order. On the contrary, the decision ultimately protected norms that are at the foundation of the edifice of both Community law and international law from being casually jettisoned by another international body. Certainly the act of rebellion was effective. It was undoubtedly the single most important factor in persuading the Security Council finally to undertake meaningful procedural reform, with the establishment of the Office of the UN Ombudsperson.[71] Considering the 2008 *Kadi* decision in the fairest (and politically and normatively appropriate) light, the ECJ was acknowledging and embracing a slightly different role, providing legal counsel, though in a wider political forum.

[67] ECJ *Kadi*, [320], [323]. [68] ECJ *Kadi*, [373]–[376].

[69] N Türküler Isiksel, 'Fundamental Rights in the EU after *Kadi and Al Barakaat*' (2010) 16(5) *European Law Journal* 551.

[70] Ibid, 552.

[71] UN Security Council resolution 1904 (17 December 2009) UN Doc S/RES/1904. See also 'Statement of French representative on the adoption of SC Resolutions 1988 and 1989' (2011) UN Doc S/PV.6557.

(2) Normatively Radical Pluralism: *Kadi* in the CJEU

The problem is that, in *Kadi II*, the CJEU made the mistake of transforming the exception, a justifiable act of rebellion, into the rule. In *Kadi II*, the CJEU heard an appeal from a decision of the CFI (by then known as the General Court) in which the General Court applied the reasoning of the 2008 ECJ decision to annul the regulation enacted to replace the regulation annulled in ECJ *Kadi*.[72] It (somewhat reluctantly) accepted the task the ECJ had spelled out for it in its 2008 decision, namely to ensure 'in principle the full review of the lawfulness of the contested regulation in the light of fundamental rights, without affording the regulation any immunity from jurisdiction on the ground that it gives effect to resolutions adopted by the Security Council under Chapter VII'.[73]

On appeal, the European Council, Commission, and the United Kingdom challenged the dualist logic underlying the 2008 ECJ decision, arguing that to review the legality of Security Council resolutions in light of European law 'wholly ignores the fact that it is the Security Council which has primary responsibility for determining the measures necessary for the maintenance of international peace and security and ignores the primacy of obligations under the UN Charter over those arising under any other international agreement' (at [61]) (arguments based on Articles 24, 25, and 103 of the Charter respectively). This ground of appeal received short shrift. The CJEU declined to accept the European Council's invitation to restore respect for the supremacy of Security Council decision-making. Instead, it reinforced the separation between legal orders that was the hallmark of the ECJ 2008 decision. The CJEU determined that 'without the primacy of a Security Council resolution at the international level thereby being called into question', the Court was required to implement the 'constitutional guarantee' of judicial review of the lawfulness of all EU measures, including those which, as in the present case, implement an international law measure, in light of the fundamental rights guaranteed by the European Union.[74]

[72] The General Court admitted to finding 'foundation' in the criticisms from the EU institutions and intervening governments of the serious difficulties they found with ECJ *Kadi* judgment: *Kadi v European Commission* [2010] EUECJ, [121], [123].

[73] *Kadi v European Commission* [2010] EUECJ, [126]. It is interesting that, in the following paragraph, the General Court expressly implants the *Solange* qualification that was not express in the ECJ judgment, adding that this review was appropriate 'at the very least so long as the re-examination procedure operated by the Sanctions Committee clearly fails to offer the guarantees of effective judicial protection', citing the decision of Advocate General Maduro.

[74] CJEU *Kadi*, [67].

The key issue on appeal was the 'level of intensity' of judicial review by the European courts. Here, the European Commission and Council, supported by all the Member States intervening in the appeals, cautioned the CJEU against taking the 'excessively interventionist' approach followed by the General Court. Yet the CJEU was not for turning. In a move that is difficult to reconcile with the disassociation between legal orders at the heart of its response to the first ground of appeal, the CJEU assumed a power, not merely to review, but to second-guess Security Council decision-making. Where an individual challenges the lawfulness of a listing decision in the courts of the European Union, the Court held that the courts had an obligation to review whether 'rules as to procedure and rules as to competence, including whether or not the legal basis is adequate', had been observed. As interpreted by the CJEU, this meant that European courts were not merely entitled but obliged to enter into a review of the initial decision to list by the Security Council Sanctions Committee 'to ensure that the decision to list the individual was taken on a sufficiently solid factual basis'.

In such circumstances, the CJEU held that the relevant EU authority was under an obligation 'to produce information or evidence, confidential or not, relevant to such examination'. The Court here gives little quarter to concerns about international security or confidentiality, stating that 'the secrecy or confidentiality of that information or evidence is no valid objection' before the courts of the European Union. The CJEU does recognize that courts have the task to apply 'techniques which accommodate, on the one hand, legitimate security considerations about the nature and sources of information taken into account in the adoption of the act concerned and, on the other, the need sufficiently to guarantee to an individual respect for his procedural rights, such as the right to be heard and the requirement for an adversarial process'. However, in circumstances where the relevant authorities declined to disclose information, it was for the courts to determine 'whether the reasons relied on by that authority as grounds to preclude that disclosure are well founded'. If it is impossible for the Court to find that the reasons are well founded, including presumably where it cannot get its hands on the relevant information, those reasons cannot be relied upon as the basis for the contested listing decision. Ultimately, the CJEU found that none of the allegations presented against Mr Kadi in the narrative summary provided by the Sanctions Committee justified the adoption of sanctions measures. In those circumstances, it confirmed the annulment of the regulation.

In *Kadi II*, the CJEU made the mistake of transforming the justifiable act of rebellion by the ECJ in 2008 *Kadi* into an enduring normative approach. The problem is not so much with the idea that Security Council decisions should be subject to judicial review as with the *parochialism* of this move. At [131], the CJEU suggests that it is seeking to ensure 'a fair balance between the maintenance of international peace and security and the protection of fundamental rights and freedoms of the person concerned…, those being shared values of the UN and the European Union'. Yet, though the CJEU purports to apply fundamental rights, it is clear that it makes no attempt to understand those rights in the global context. The values it applies are not 'shared values', but 'fundamental rights guaranteed by the European Union'.

In *Kadi II*, the CJEU opts for a radically pluralist approach akin to a principle of subsidiarity, with the presumption that regional constitutional values prevail over international ones, even in the context of international decision-making. The effect is not to dissolve, but to invert the hierarchy established in the heavily disputed 2005 CFI *Kadi* decision. Between 2005 and 2013, the arc swung 180 degrees from a recognition of Security Council supremacy by the CFI to a claim to judicial supremacy by the CJEU. There is no attempt to reconcile the two orders. The CJEU held that, while Security Council resolutions had primacy at the international level, they became subject to the primacy of constitutional guarantees of the European Union when implemented at the European level. As such, the 'right to effective judicial protection' is interpreted as a 'declaration from a court, by means of a judgment ordering annulment whereby the contested measure is retroactively erased from the legal order'.[75] Here the CJEU applies, seemingly without reflection, a peculiarly domestic interpretation of due process. There is no attempt by the CJEU to evaluate the Ombudsperson procedure established as a review process in the Security Council sanctions context. The Office of the Ombudsperson is not even mentioned in the Court's findings. Instead, the Court merely alludes to the 'improvements added', with a fairly abrupt conclusion that 'they do not provide to the person… listed… the guarantee of judicial protection'. Paradoxically, application of the right to 'effective judicial protection' in the Security Council sanctions context is rendered inappropriate precisely because it is *ineffective*. The consequence of the decision is to implement a system of judicial supremacy mandating judicial review of Security Council decision-making. In doing so, the CJEU arrogates to itself the duty to perform a role that it cannot effectively exercise. States, in particular members of the Permanent Five, will be extremely reluctant to give up information to a foreign court. Yet the CJEU acknowledges that, if it cannot

[75] CJEU *Kadi*, [134].

get its hands on any information that supports the listing, it will be forced to annul the relevant sanctions measures.

The problem with the CJEU approach is that it provides no attempt to unify the different legal orders. Instead, the choice of frame, domestic, regional, or international, determines the decision, where the danger is that each legal order will seek to make its law govern the whole and to transform its preference into the general preference. Law and courts become part of the problem, not of the solution. Even under a pluralist approach, one could argue that the focus of courts should be on mechanisms by which to achieve the accommodation between conflicting values and interests in international society, and not on the triumph of one set of institutions or norms over another.

(3) Normatively Moderate Pluralism: Opinions of the Advocates General

In this respect, it is worth contrasting the decisions of the European courts with the opinions of the Advocates General delivered to assist the judges in this litigation.[76] Writing to assist the ECJ with its 2008 judgment, Advocate General Maduro notes the autonomy of the European Community legal order,[77] but immediately qualifies this with the statement that 'this does not mean ... that the Community's municipal legal order and the international legal order pass by each other like ships in the night'.[78] The core of his approach is to encourage a form of continuing dialogue between legal orders, stating that '[i]n an increasingly interdependent world, different legal orders will have to endeavour to accommodate each other's jurisdictional claims'.[79] He was firm that 'courts ought not to be institutionally blind', could not always claim a monopoly on determining how certain fundamental interests should be reconciled, and should, where possible, recognize the authority of institutions, such as the Security Council, that were sometimes better placed to weigh those fundamental interests.[80] It was Advocate General Maduro, rather than the ECJ itself, that proposed the application of an equivalence principle such as that applied in the famous *Solange* case,[81] promising respect for the primacy of Security Council resolutions on the condition that fundamental rights benefited from an equivalent degree of protection at the

[76] Opinion of Advocate General Maduro, *Kadi v Council of the European Union*, 16 January 2008; Opinion of Advocate General Bot, *European Commission and Council v Kadi*, 19 March 2013.

[77] Opinion of Advocate General Maduro, *Kadi v Council of the European Union*, 16 January 2008, [21], [37].

[78] Ibid, [22]. [79] Ibid, [44]. [80] Ibid, [44].

[81] *Solange I* (1974) BVerfGE 37; *Solange II* (1986) BVerfGE 73.

international level.[82] The flip-side was that, where the test of equivalence was not met, the Advocate General maintained the right to invalidate EU mechanisms giving effect to Security Council resolutions, and indeed resolved in his Opinion that the relevant regulation should be annulled in the case before the Court.[83]

Advocate General Bot picked up where Advocate General Maduro left off in his opinion delivered prior to the CJEU decision. In his opinion in *Kadi II* on the extent and intensity of review, Advocate General Bot emphasized the need to take account of the origin and context of the EU act it is reviewing. In particular, he noted the need to recognize 'the fact that inclusion on the list is decided on the basis of a centralised, universal procedure at the level of the United Nations or that such a decision is based on a summary of reasons drawn up by the Sanctions Committee on the basis of information or evidence which is provided to it by the State(s) which made the listing request, in most cases in confidence, and which is not intended to be made available to the EU institutions'.[84] Advocate General Bot paid express attention to the improvements introduced following the ECJ's 2008 judgment in *Kadi*, in particular the establishment of the Office of the Ombudsperson. He recognized that '[t]his process reflects a realization within the United Nations that, despite confidentiality requirements, the listing and delisting procedures must now be implemented on the basis of a sufficient level of information, that the communication of that information to the person concerned must be encouraged, and that the statement of reasons must be adequately substantiated'.[85] He held that the Ombudsperson helped to guarantee that listings are based on sufficiently serious evidence and recognized that 'excessively high regional or national requirements could, in truth, prove to be counterproductive' if states were 'less inclined in future to transmit confidential information to the Sanctions Committee'.[86] While emphasizing that the solution was not to give *carte blanche* to the Security Council, he advocated a position of 'mutual confidence and effective collaboration', arguing that 'an effective global fight against terrorism requires confidence and collaboration between the participating international, regional and national institutions, rather than mistrust'.[87]

In considering the form of the dignitarian model going into the future, I argue that the moderate form is the preferable approach. Ultimately, the purpose of the dialogue promoted by the dignitarian model is to motivate the

[82] Opinion of Advocate General Maduro, *Kadi v Council of the European Union*, 16 January 2008, [44], [54].
[83] Ibid, [56]. [84] Ibid, [75]. [85] Ibid, [82].
[86] Ibid, [84]. [87] Ibid, [85].

Council to adopt a contextually appropriate set of rights that will strengthen the effectiveness of the Council *as well as* the protection of fundamental values.[88] This is unlikely to be achieved by a radical approach that proceeds according to a 'blind' application of domestic principles, regardless of their bindingness or appropriateness in the Security Council context. Instead, a more flexible moderate approach, seeking 'equivalence' of protection, is arguably better equipped to lead ultimately to the adoption of a mutually satisfactory version of the rights that should govern at the UN level. Karen Knop has recognized that domestic courts play an important role in the 'translation' of international norms to the domestic setting, and that the particularization of international law will have a legitimating effect on both domestic legal systems and international law more generally.[89] There is no reason why this situation should not be reversed. Rather than acting as conveyor belts for local law, domestic and regional courts can play an important role under the dignitarian model in translating fundamental domestic or regional values to the international context.[90] These hybrid national/international norms will serve to feed into a broader dialogue about the development of applicable law. This in turn will serve to lend greater legitimacy to international law both from an international and domestic perspective, through a process of 'particularization and justification'.[91]

D. Conclusion

It is possible to conclude overall that a dignitarian model, and the pluralist adjudicatory framework supporting it, has the capacity to enhance the legitimacy of Security Council sanctions decision-making. It is clear that this model has already led to more robust protection of individual rights in

[88] Daniel Halberstam and Eric Stein, 'The United Nations, the European Union and the King of Sweden: Economic Sanctions and Individual Rights in a Plural World Order' (2009) 46 *Common Market Law Review* 13, 24, 27.

[89] Karen Knop, 'Here and There: International Law in Domestic Courts' (2000) 32 *NYU Journal of International Law and Politics* 501, 533, 535.

[90] Interestingly, in his extra-judicial writing, Maduro has suggested a 'reinforced' teleological approach that would 'force courts to articulate the normative preferences they attribute to particular rules and to relate them to the normative preferences of the overall legal order': Miguel Polares Maduro, 'Courts and Pluralism: Essay on a Theory of Judicial Adjudication in the Context of Legal and Constitutional Pluralism' in Jeffrey Dunoff and Joel Trachtman, *Ruling the World? Constitutionalism, International Law and Global Governance* (Cambridge University Press 2009), 368.

[91] Karen Knop, 'Here and There: International Law in Domestic Courts' (2000) 32 *NYU Journal of International Law and Politics* 501, 535.

Security Council sanctions decision-making. At the same time, it preserves important scope for Security Council discretion, owing largely to the more limited normative effect of domestic court judgments under this model and its more qualified approach to questions of supremacy. To the extent that the dignitarian model comes with the additional risk of opening up dangerous holes in the sanctions net, this would be less a symptom of the dignitarian model itself than of deeper problems with the positivist hierarchical account of the relationship between international and domestic law. In a diverse and contested space such as the international legal order, the pluralism promoted by the dignitarian model is a preferable alternative to a Procrustean positivist system that freezes the political order in an artificial consensus.[92] A far greater problem with this model is its vulnerability to differential measures of power and influence in the international legal order. There is a risk that it may end up reflecting rather than counter-balancing the existing power imbalance in the international domain, meaning that certain voices with legitimate concerns never penetrate the legal discourse, while the viewpoints of the most influential members of the international community will invariably be promoted. Moreover, if courts take a normatively radical instead of a normatively moderate approach, there is a risk that such a model will encourage an overly narrow and self-interested approach to problems affecting the entire international community, and offers little by way of contribution to the development of a comprehensive and coherent international legal system, underwritten by a set of common values. As an alternative, in the next chapter, I turn to discuss the public interest model of due process, under which due process becomes a mechanism through which to generate discourse about the development of a shared body of values, and to infuse decision-making with a sense of what is good for all, rather than encouraging self-interested claims.

[92] Nico Krisch, 'Global Administrative Law and the Constitutional Ambition' LSE Law, Society and Economy Working Papers 10/2009, 10.

7

A Model Based on Public Accountability

The Public Interest Model in the UN Sanctions Context

As has been shown so far, certain normative models of procedural fairness are compromised in their capacity to enhance the legitimacy of Security Council decision-making. The instrumentalist model would promote accuracy in the application of a narrow set of rules that already arguably lack legitimacy on the basis that they are unrepresentative of the majority of international society. The dignitarian model fares better, but still risks sidelining all but a set of powerful elite interests, and (at least in its radical form) encourages participation based on self-interest rather than the promotion of the common interest. The final model for discussion, the public interest model, is underpinned by a more solidaristic mindset than the positivist foundations of the instrumentalist model or the pluralist mentality underlying the dignitarian model. As discussed in Chapter 4, the public interest model has been described as a kind of democratic 'analogue', and provides a 'guarantee of the opportunity for all to play their part in the political process'.[1] It therefore strikes at the heart of one of the central critiques of the Security Council and the international legal order more generally: the idea that there is a 'democratic deficit' in international governance.

The Security Council sanctions regime invites a distilled version of the criticism made of the international legal order more generally that it increasingly represents 'Governance without Government',[2] with the added sting that the Security Council has greater capacity than the rest of the international order to enforce its (undemocratic) diktats. 'Democratically-based' critique about representation and accountability deficiencies are connected integrally

[1] T R S Allan, 'Procedural Fairness and the Duty of Respect' (1998) 18 *Oxford Journal of Legal Studies* 497, 502, 509; Frank I Michelman, 'Formal and Associational Aims in Procedural Due Process' in J Roland Pennock and John W Chapman (eds), *Due Process* (New York University Press 1977), 136.

[2] Joseph Weiler, 'The Geology of International Law—Governance, Democracy and Legitimacy' (2004) 64 *Zeitschrift für ausländisches öffentliches Recht und Völkerrecht* 547, 559–60.

to unease about the lack of transparency in the sanctions regime.[3] Concerns about information-sharing and the lack of accountability in the sanctions regime were present during the three multilateral reform processes that contributed to the development of targeted sanctions,[4] were expressly emphasized in the September 2005 World Summit Outcome Document,[5] and have not dissipated.[6] The overall accusation is that, as the Security Council assumes increasingly intrusive and coercive decision-making authority over individuals, it is—at the same time—becoming significantly less transparent, less representative of, and less accountable to those individuals.

Of course, some might argue that this concern is misplaced. It is not that the Security Council was ever intended to behave as a democracy. Certainly, a vision of *global* democracy is unrealistic and theoretically unsustainable. As Weiler has noted in emphatic terms, '[a] simplistic application of the majoritarian principle in world arenas would be normatively ludicrous'.[7] Indeed, democratic decision-making is expressly

[3] In using the term 'transparency', I rely on Elizabeth Fisher's definition of transparency as 'the decision to make visible, or provide access to, the resources on which an exercise of public... power may be based': Elizabeth Fisher, 'Transparency and Administrative Law: A Critical Evaluation' (2010) 63 *Current Legal Problems* 272, 274.

[4] [Interlaken process] Swiss Federal Office for Foreign Economic Affairs, 'Expert Seminar on Targeting UN Financial Sanctions' (Interlaken, 17–19 March 1998), 71; Swiss Federal Office for Federal Economic Affairs, 'Second UN Seminar on Targeting United Nations Financial Sanctions' (Interlaken, 29–31 March 1999), 211; [Bonn-Berlin process] Bonn International Centre for Conversion, 'Results of the Bonn-Berlin Process: Design and Implementation of Arms Embargoes and Travel and Aviation-Related Sanctions' (Bonn, 2001), 109; [Stockholm Process] Peter Wallensteen et al, 'Making Targeted Sanctions Effective: Results from the Stockholm Process on the Implementation of Targeted Sanctions' (2003), §43, §64.

[5] 'World Summit Outcome Document', adopted in General Assembly resolution 60/1 (24 October 2005) UN Doc A/RES/60/1.

[6] See, for example, Security Council Report, 'Special Research Report: UN Sanctions' (25 November 2013), 13–14; 'Twelfth Report of the 1267 Monitoring Committee' (1 October 2012) UN doc S/2012/729, para 35; Ben Emmerson QC, 'Report of UN Special Rapporteur on the Promotion and Protection of Human Rights while Countering Terrorism' (26 September 2012) UN Doc A/67/396, paras 43, 50–51; Bardo Fassbender, 'The Role for Human Rights in the Decision-making Process of the Security Council' in Bardo Fassbender (ed), *Securing Human Rights? Achievements and Challenges of the UN Security Council* (Oxford University Press 2011), 83–8; Gráinne de Búrca, 'The European Court of Justice and the International Legal Order after *Kadi*' (2010) 51 *Harvard Journal of International Law* 1; Ian Johnstone, 'Legislation and Adjudication in the UN Security Council: Bringing Down the Deliberative Deficit' (2008) 102 *American Journal of International Law* 275, 298; Thomas Biersteker and Sue Eckert, *Strengthening Targeted Sanctions through Fair and Clear Procedures* (Watson Institute Targeted Sanctions Project, Brown University 2006), 3.

[7] Joseph Weiler, 'The Geology of International Law—Governance, Democracy and Legitimacy' (2004) 64 *Zeitschrift für ausländisches öffentliches Recht und Völkerrecht* 547, 561. See also Oscar Schachter, 'The Crisis of Legitimation in the United Nations' (1981) 50 *Nordisk Tidsskrift Int'l Ret* 3, 18.

and deliberately prevented by the Council's structure.[8] The value of the public interest model lies in its potential to provide an alternative, and arguably more appropriate, analogue of an individual's participation in a democratic political process. The model builds upon theories of deliberative democracy that emphasize access to information and deliberation as obvious ways by which to enhance the representativeness and therefore legitimacy of decision-making. A number of scholars have argued for the adaptation of deliberative democracy theory to the Security Council context, identifying procedural safeguards as the key means through which to enhance the representativeness and therefore legitimacy of decision-making by international institutions.[9] Ian Johnstone goes so far as to describe the representative deficit in the Security Council as 'largely procedural in nature'.[10]

In this chapter, I argue that the Ombudsperson is a superior framework to internationalized or pluralist judicial frameworks in terms of its capacity to achieve the goals of the public interest model. The Office of the Ombudsperson was established by Security Council resolution 1904 in December 2009 and the first office-holder, Kimberly Prost, assumed the position on 3 June 2010.[11] Given the reluctance of the Security Council to institute procedural reform, the creation of the office was described as a 'miracle', though was also recognized as 'crucial to the survival' of the targeted sanctions regime.[12] Moreover, as discussed above in Chapter 2(B)(3), during the term of the first office-holder, the Ombudsperson procedure has produced results.

[8] Ian Hurd, *After Anarchy: Legitimacy and Power in the United Nations Security Council* (Princeton University Press 2007), 191.

[9] Benedict Kingsbury, 'The Concept of "Law" in Global Administrative Law' (2009) 20 *European Journal of International Law* 23, 48–50; Daniel Esty, 'Good Governance at the Supranational Scale: Globalizing Administrative Law' (2006) 115 *Yale Law Journal* 1490, 1490, 1520, 1522; Erika de Wet, 'The International Constitutional Order' (2006) 55 *International and Comparative Law Quarterly* 51, 74; Allen Buchanan and Robert O Keohane, 'The Legitimacy of Global Governance Institutions' (2006) 20 *Ethics and International Affairs* 405, 432, 434; Mattias Kumm, 'The Legitimacy of International Law: A Constitutionalist Framework of Analysis' (2004) 15 *European Journal of International Law* 907, 924–7; David Caron, 'The Legitimacy of the Collective Authority of the Security Council' (1983) 87 *American Journal of International Law* 552, 561, 576.

[10] Ian Johnstone, 'Legislation and Adjudication in the UN Security Council: Bringing Down the Deliberative Deficit' (2008) 102 *American Journal of International Law* 275, 276. See also Ian Johnstone, 'Security Council Deliberations: The Power of Better Argument' (2003) 14 *European Journal of International Law* 437; Ian Johnstone, 'The Security Council as Legislature' in Bruce Cronin and Ian Hurd (eds), *The UN Security Council and the Politics of International Authority* (Routledge 2008).

[11] UN Security Council resolution 1904 (17 December 2009) UN Doc S/RES/1904.

[12] US Embassy Confidential Cable, 'Commission Terrorism Sanctions Legal Challenges', 10USEUBRUSSELS 212 (24 February 2010).

However, while the Ombudsperson procedure has received strands of support in the reform debate,[13] the assessment of the Office from a due process perspective has been mainly critical.[14] The main concern is that the Ombudsperson is 'not a court'. The CJEU held in *Kadi II* that 'the improvements' (of which the Ombudsperson procedure was the most notable) do not provide the listed person with 'the guarantee of effective judicial procedure',[15] affirming views expressed by the European Court of Human Rights and the EU General Court that the Ombudsperson 'cannot be equated with the provision of an effective judicial procedure for review of decisions of the Sanctions Committee'.[16] The UK Supreme Court in *Ahmed* found that, while the establishment of the Office of the Ombudsperson was to be welcomed, 'the fact remains that there... still is not any effective judicial remedy'.[17] In September 2012, Ben Emmerson, the Special Rapporteur on the protection of fundamental rights while countering terrorism, condemned the targeted

[13] Opinion of Advocate General Bot, *European Commission v Kadi* (19 March 2013), [81]–[87].

[14] [Cases] *European Commission and Council v Kadi* [2013] ECR not yet reported, [133]; *Al-Dulimi and Montana v Switzerland* (App No 5809/08), 26 November 2013, [119]; *Nada v Switzerland* [2012] ECHR 1691, [209]–[214]; *Her Majesty's Treasury v Ahmed* [2010] UKSC 2, [145]–[156] per Lord Phillips; [184]–[185] per Lord Rodger; [246] per Lord Mance; [Reports] Martin Scheinin, 'Report of the Special Rapporteur on the promotion and protection of human rights and fundamental freedoms while countering terrorism' (6 August 2010) UN Doc A/65/258, paras 56–7; David Cortright and Erika de Wet, *Human Rights Standards for Targeted Sanctions* (Sanctions and Security Research Program 2010), 10; Ben Emmerson QC, 'Report of UN Special Rapporteur on the Promotion and Protection of Human Rights while Countering Terrorism' (26 September 2012) UN Doc A/67/396, paras 14, 20–1; [Articles] Vanessa Arslanian, 'Great Accountability Should Accompany Great Power: The ECJ and the UN Security Council in *Kadi I* and *II*' (2012) 35 *Boston College International and Comparative Law Review* 1, 11, 15; Fremuth, 'Ein Prozess...: Zum Ausgleich zwischen der effektiven Bekämpfung des Terrorismus und der Beachtung der Menschenrechte in der Sanktionspraxis des Sicherheitsrates' [2012] *Die öffentliche Verwaltung* 81, 87 ff; Lisa Ginsborg and Martin Scheinin, 'You Can't Always Get What You Want: The *Kadi II* Conundrum and the Security Council 1267 Terrorist Sanctions Regime' (2011) 8 *Essex Human Rights Review* 7, 11–12, 19; Grant L Willis, 'Security Council Targeted Sanctions, Due Process and the 1267 Ombudsperson' (2011) 42 *Georgetown Journal of International Law* 673, 745; Craig Forcese and Kent Roach, 'Limping into the Future: The UN 1267 Terrorism Listing Process at the Crossroads' (2010) 42 *George Washington International Law Review* 217, 219, 264–5, 275; Jared Genser and Kate Barth, 'When Due Process Concerns Become Dangerous: The Security Council's 1267 Regime and the Need for Reform' (2010) 33 *Boston College International and Comparative Law Review* 1, 26, 41; Adele Kirschner, 'Security Council Resolution 1904 (2009): A Significant Step in the Evolution of the Al Qaida and Taliban Sanctions Regime?' (2010) 70 *ZaöRV* 585, 602, 604–5.

[15] *European Commission v Kadi* [2013] ECR not yet reported, [133].

[16] Case T-85/09 *Kadi v European Commission* [2010] EUECJ, [28]. See also *Nada v Switzerland* [2012] ECHR 169, [209]–[214]; *Al-Dulimi and Montana v Switzerland* (App No 5809/08), 26 November 2013, [119].

[17] *Her Majesty's Treasury v Ahmed* [2010] UKSC 2, [145]–[156] per Lord Phillips; [184]–[185] per Lord Rodger (Lady Hale agreeing); [246] per Lord Mance.

sanctions regime as a 'ready means by which individual States can make executive decisions with far-reaching consequences, apparently unconstrained by domestic judicial review, or the international human rights treaties by which they are bound'.[18] It is clear that the perception in many quarters is that the procedural framework for sanctions decision-making remains deficient because of its non-judicial nature.

Based on the value-based approach outlined in this article, I reach a different conclusion. My conclusion is that the Ombudsperson is in fact superior to a court process as it offers the most appropriate response to legitimacy gaps in Security Council sanctions decision-making. As discussed in Chapter 4, the aim of the public interest model is not to measure the Council's decision-making against binding international legal norms (as is the case under the instrumentalist model), nor to reject the relevance of binding international legal norms in favour of negotiation between pluralist interests (as is the case under the dignitarian model), but to engage the community in debate that will ultimately contribute to the development of a more representative set of decisions and applicable norms. A non-judicial framework is preferable to a judicial framework in working toward this goal. Indeed, the central aim of inviting and responding to public opinion does not sit comfortably with the judicial function. As Richard Stewart noted in his enduring critique of a judicially implemented system of interest representation, judicial review was traditionally an instrument for checking governmental power and does not touch on the 'affirmative side' of government, which has to do with the representation of individuals and interests.[19] While courts tend to respect a strict separation between legal and political authority, the Ombudsperson is able to take account of the interests of the petitioner and the Security Council, while remaining responsive to the concerns of the broader international community. This is not to say that the Ombudsperson framework cannot be improved upon. However, by undertaking a normatively rich analysis of the framework through an understanding of the public interest model of due process which it supports, it is possible to better understand the way in which the framework is able to enhance the legitimacy of Security Council decision-making and the international legal order more broadly. In particular, I argue that the Ombudsperson framework offers the

[18] Ben Emmerson QC, 'Report of UN Special Rapporteur on the Promotion and Protection of Human Rights while Countering Terrorism' (26 September 2012) UN Doc A/67/396, paras 14, 20–1.

[19] Richard B Stewart, 'The Reformation of American Administrative Law' (1975) 88 *Harvard Law Review* 1667, 1687.

capacity to promote a subtle shift in classical conceptions of community, law, and values consistent with the emerging 'cosmopolitan constitutionalist' school of thought. Far from dismissing the Ombudsperson framework as inadequate, those involved in the reform debate, including practitioners, courts, and scholars, should encourage further strengthening and extension of the Ombudsperson's mandate so as to promote the key goals of the public interest model within the 1267 sanctions regime as well as other sanctions regimes.

A. The Public Interest Model and International Community: Access and Representation under the Ombudsperson Process

The public interest model of due process does not reject the classical conception of the international legal order, but seeks to regenerate it, situating it within a broader and recognizedly more heterogeneous framework of other legal orders and non-state entities. The Ombudsperson framework has undoubtedly opened up decision-making to individuals directly affected by it. The Ombudsperson is far more accessible than courts, and has the capacity to travel to the petitioner (rather than the reverse) for face-to-face interviews (or alternatively through email and telephone discussions) to ensure the petitioner's side of the case is heard. This was emphasized in a recent case where the Ombudsperson had to rely on the 'diligent and extraordinary efforts of officials' in the UN and other states to gain access to the petitioner, resources that one could imagine would not be available to domestic or regional courts.[20] The cost and delay of the Ombudsperson process is certainly less than judicial equivalents. While the absence of compulsory legal representation has been criticized,[21] the Ombudsperson argues that the fact that petitioners are not required to be represented by legal counsel in fact serves to make the process more accessible for petitioners.[22] The *Kadi* case, already discussed in some detail, is clear evidence of the expeditiousness of the Ombudsperson process in comparison to courts. The CJEU reached

[20] 'Eighth Report of the Office of the Ombudsperson' (31 July 2014) UN Doc S/2014/73, para 37.

[21] Ben Emmerson QC, 'Report of UN Special Rapporteur on the Promotion and Protection of Human Rights while Countering Terrorism' (26 September 2012) UN Doc A/67/396, para 52.

[22] In fact, petitioners were assisted by legal counsel in twenty-six of the fifty-five cases processed by the Ombudsperson to date: 'Eighth Report of the Office of the Ombudsperson' (31 July 2014) UN Doc S/2014/73, para 7.

its final 'non-decision' in Mr Kadi's case (in terms of its practical effect on Mr Kadi's listing) after twelve years and untold legal costs. By contrast, the report of the Ombudsperson that led to the de-listing of Mr Kadi took nine months to achieve and was praised by Mr Kadi and his lawyers as a 'proper hearing', 'formal and probing', that made 'an enormous difference to the person involved in the process'.[23]

The public interest model does not merely seek to improve access for the petitioner, but also works to ensure decision-making is representative and inclusive of the international community more generally. While Ian Johnstone describes the Security Council as a setting in which 'legal discourse looms large', discourse and debate do not generally extend beyond the Council Chamber.[24] By and large, Security Council deliberations are not celebrated for their inclusiveness. They operate to a large extent behind closed doors, without a public record being taken.[25] Though early official records of the Council's formal meetings are more detailed,[26] at some point it seems that Security Council members began to feel constrained by the responsibility of negotiating in the public eye, and increasingly substantive negotiations began to take place in private. Eventually, the practice of 'informal consultations' was institutionalized, and in 1978 a purpose-built room was constructed to host such consultations.[27] In time, the Council was conducting the bulk of its business in the consultations room rather than the Security Council chamber.[28] There are no official records of these consultations, and attendance is tightly controlled. Draft resolutions are rarely

[23] 'UN Sanctions, Human Rights and the Ombudsperson' (17 May 2013, Chatham House, London).

[24] Ian Johnstone, 'Legislation and Adjudication in the UN Security Council: Bringing Down the Deliberative Deficit' (2008) 102 *American Journal of International Law* 275, 276.

[25] From the early 1990s the Security Council has tended to conduct a substantial portion of its business in closed meetings. For example, in 2006 the Security Council held 272 formal meetings and 193 informal consultations: Security Council Report, *Security Council Transparency, Legitimacy and Effectiveness* (2007).

[26] For example, see *Repertoire of the Practice of the Security Council*, Supplement for 1946–51, Ch VIII.

[27] S D Bailey and S Daws, *The Procedure of the UN Security Council* (3rd edn, Clarendon Press 1998), 62.

[28] At a debate on the Council's working methods and procedure on 16 December 1994, the French Permanent Representative, Ambassador Merimee, suggested that, 'informal consultations have become the Council's characteristic working method, while public meetings, originally the norm, are increasingly rare and increasingly devoid of content: everyone knows that when the Council goes into public meeting everything has been decided in advance... informal meetings are not even real Council meetings at all; they have no official existence and are assigned no number. Yet it is in these meetings that all the Council's work is carried out': UN Doc S/PV.3483.

tabled for discussion in formal meetings until members of the Council are prepared to vote on them, at which point the outcome of the vote is fairly pre-determined. The Council sometimes adopts decisions with no discussion at all, meaning that there is no public record of the negotiations leading up to the vote, the justification for the decision, or a clear picture of the resolution's objectives.

This reflects the practice in the sanctions context. Between the establishment of the Al Qaeda and Taliban Sanctions Committee in October 1999 and the end of 2014, the Committee held forty-five formal meetings and 367 informal consultations.[29] This 'culture of confidentiality' was initially a prized technique, with the Al Qaeda and Taliban Sanctions Committee paying tribute in its 2004–5 Annual Report to a Chairman who had 'wisely determined that much of the work should be performed at informal meetings of the Committee to allow for enough flexibility in convening them and the free exchange of views, without a record'.[30] However, the Committee's lack of transparency has been the subject of trenchant criticism, to the extent that it is now widely considered to have detracted from the effectiveness and legitimacy of the sanctions regime.

The establishment of the UN Ombudsperson has done more to open up sanctions decision-making than any previous reform. The Ombudsperson's role is structured to focus on information-gathering, consultation, and outreach, and not just with the petitioner. Upon receiving a de-listing request, the Ombudsperson engages in a four-month period of information gathering (extendable for a further two months if necessary), during which she transmits the de-listing request to relevant parties including the Committee, the designating state, and the state of nationality, seeking any additional information relevant to the request. In addition, the request is forwarded to the Monitoring Team, which is required to assemble a portfolio of relevant information drawn from court decisions, news reports, and fact-based assessments of the information provided by the petitioner. The Ombudsperson then undertakes a two-month period of engagement, which may include a dialogue with the petitioner, to gain any additional information that may help the Committee with its decision. At the end of this process, the Ombudsperson prepares a comprehensive report outlining the information she has obtained, and laying out the principal arguments

[29] As reported in the Annual Reports of the 1267 Sanctions Committee between 2000 and 2014.

[30] 'Report of the Security Council Committee established pursuant to resolution 1267 (1999) concerning Al-Qaida and the Taliban and associated individuals and entities Annual Report' (31 December 2004) UN Doc S/2004/1039, 10.

concerning the de-listing request. The Committee has thirty days to consider this report. The report is to be presented in person to the Committee during which the Ombudsperson can answer any questions the Committee members may have regarding the request. If the Ombudsperson recommends de-listing, the individual or entity will be removed from the list unless, within sixty days, the Committee decides by consensus to maintain the listing. Under procedures introduced in resolutions 2083 and 2161, the Committee is now required to provide reasons for rejecting the request and provide an updated narrative summary of reasons for listing for the Ombudsperson to transmit to the petitioner within sixty days. The Ombudsperson must send the petitioner a letter within fifteen days communicating the Committee's decision, the reasons for its decision, and describing to the extent possible the process and publicly releasable factual information gathered by the Ombudsperson.

It is notable that the discourse created through the Ombudsperson process is not confined to member states of the Security Council, or even member states of the United Nations. The Ombudsperson engages in significant outreach in the course of her work, meeting regularly with states, intergovernmental organizations, UN bodies, judges of national, regional, and international courts, prosecutors, private lawyers, academics, representatives of non-governmental organizations, and civil society.[31] As Security Council decision-making impacts increasingly on non-state actors, both in terms of enforcement and implementation, it is crucial that discourse about Security Council decision-making involves networks of citizens and entities beyond nation states. In opening up decision-making in this way, the Ombudsperson process makes important political space, recognizing the role of NGOs, individuals, corporations, and other levels of government beyond the executive.[32] More than court-based processes, the Ombudsperson process has the potential to function as both a forum and audience for democratization by increasing access to information and opening up deliberation to a wider cross-section of the international community.

The transparency, visibility, and publicity of the Ombudsperson framework is fundamental to its capacity to promote access, debate, and

[31] See, for example, 'Third Report of the UN Ombudsperson', paras 21–9; 'Seventh Report of the Office of the Ombudsperson' (31 January 2014) UN Doc S/2014/73, paras 14–20.

[32] Harold Hongju Koh, 'Why Do Nations Obey?' (1997) 106 *Yale Law Journal* 2599; Mary Ellen O'Connell, 'New International Legal Process' (1999) 93 *American Journal of International Law* 334.

deliberation in sanctions decision-making. This publicity works, not only to ensure listed persons and entities are aware of the Ombudsperson process, but also to generate understanding and acceptance of the sanctions decision-making process among the broader community. In light of these goals, one of the key reforms should be to enable the broader dissemination of the reasons for de-listing decisions, and in particular the Ombudsperson's comprehensive report. As things stand, the comprehensive report is not made available to interested states, the petitioner, or the public. As a consequence, the reasoning of the Ombudsperson for arriving at a recommendation is not generally available, with the sole exception of the Committee's statement of reasons conveyed to the petitioner. Provided there are protections in place for confidential material, it is difficult to rationalize why the petitioner who is directly affected by the sanctions measures and the state and non-state entities charged with implementing those measures should not have the opportunity to review and understand the findings and analysis of the Ombudsperson. To enhance the fairness and transparency of the process, the most effective remedy would be to provide for public disclosure of the reports with proper measures in place to ensure the protection of confidential material. As the Ombudsperson points out, the absence of a prescribed mechanism for disclosing reasons has ramifications not merely for the fairness of the process, but also for its effectiveness, particularly at a time when there is increasing parallel consideration of individual cases at the national, regional, and international levels.[33]

B. The Public Interest Model and International Law: A Contextual and Responsive Set of Legal Standards

The Ombudsperson framework is not a one-way street in terms of outreach with the broader international community. By opening up the information flow between decision-making and the broader community, the aim is not solely to enhance public awareness about the sanctions regime, but also to increase the Council's responsiveness to public concerns about decision-making, and the values that should underlie it. There are two concrete ways in which the Ombudsperson process has the potential to render

[33] 'Seventh Report of the Office of the Ombudsperson' (31 January 2014) UN Doc S/2014/73, paras 49–52.

Security Council decision-making more responsive to contemporary values of the broader international community: (1) the Ombudsperson engages in *de novo* review of Security Council decision-making according to consistent and contextually appropriate standards; and (2) the Ombudsperson process encourages compliance through a 'justificatory discourse' rather than the concept of 'bindingness'.

(1) *De novo* Review and Contextual Standards

One of the most important distinctive aspects of the Ombudsperson process is that it enables *de novo* review. The Ombudsperson process examines the circumstances as they stand at the time of the de-listing request to determine the appropriateness of a continued listing, rather than at the time the initial decision was made. Unlike judicial review, the Ombudsperson's review process is not frozen in time, but can be responsive to present circumstances. This exclusive focus on the question as to 'whether today the continued listing of the individual or entity is justified based on all of the information now available' has undoubtedly facilitated the removal of people from the list and ensures the sanctions regime continues to perform a preventative (as opposed to punitive) role.

The Ombudsperson process also has advantages over other models in terms of the standards of review it applies. Under an instrumentalist model, courts are confined to the toolkit of 'binding law' which, as discussed in Part II, has led courts to derive 'universally applicable' standards of due process using standards that were largely developed to apply to the domestic governmental context. By contrast, under the dignitarian model, domestic and regional courts have applied a myriad of different domestic or regional standards, with the consequence that 'justice' for individuals will differ depending upon which court they can access. Even the European Courts have found it difficult to settle on a standard of review, and the raft of decisions before European Courts in the sanctions context show that the judges and Advocates General are completely at odds over the applicable standard of review (where the choice does not always come down in favour of human rights).[34] The Ombudsperson, on the other hand, has been granted

[34] See, for example, the different views on intensity of review expressed in *People's Mojahedin Organization of Iran v Council* [2008] ECR II-3487, [74]–[75]; Case T-85/09 *Kadi v European Commission* [2010] EUECJ, [138]–[147]; Opinion of Advocate General Bot, *European Commission v Kadi* (19 March 2013), [89]. See also *Youssef v Secretary of State* [2013] EWCA Civ 1302 (29 September 2013).

discretion to develop a standard of review applicable specifically to the Security Council sanctions context. When asked what standard she uses to assess the applications, she has been quick to confess 'I made it up!', but also emphasizes that she has chosen a standard that is appropriate to the unique context of the Security Council operating in an international framework. The Ombudsperson was conscious that the benchmark used could not be premised on the precepts of one particular legal system. Instead, she looked to the national and regional law and jurisprudence to develop a standard that incorporated concepts generally accepted as fundamental across legal systems, though particularly in the context of asset freezing or other restrictions in counter-terrorism regimes. In developing the appropriate standard, she also considered the aim of the measures, which she characterizes as two-fold: (1) to hamper access to resources and (2) to encourage a change of conduct. She characterizes these aims as preventative rather than criminal in nature, while at the same time acknowledging the seriousness of the restrictions. In these circumstances, she concluded that it was inappropriate to use standards applicable to a determination of criminal guilt or innocence. Rather, the standard she settled on was 'whether there is sufficient information to provide a reasonable and credible basis for the listing'.[35] In her view, this standard recognized a lower threshold appropriate to preventative measures, but set a sufficient level of protection for the rights of individuals and entities in this context.

(2) A 'Qualified' Notion of Bindingness

Reference to the 'non-judicial' nature of the Ombudsperson in critical literature is usually shorthand for two main criticisms: (1) that the Ombudsperson's report is not binding on the Sanctions Committee; and (2) the Ombudsperson is therefore not independent from the Security Council as her report can be overturned by the Sanctions Committee. In fact, under procedures introduced in Security Council resolution 1989, where the Ombudsperson recommends de-listing, the individual or entity will be removed from the list unless, within sixty days, the 1267 committee decides unanimously that the individual or entity should remain subject to the sanctions. Resolution 1989 also builds in a 'fire-alarm' mechanism under which a Committee member can request referral to the Security Council for decision. However, the eventuality has not occurred over the first five-year operation of the Office and the Monitoring Team has

[35] Office of the Ombudsperson, 'Approach to and Standard for Analysis, Observations, Principal Arguments and Recommendation' (August 2011), available at <http://www.un.org/en/sc/ombudsperson/approach.shtml>.

expressed the view that it is 'extremely unlikely' that the Committee would reject the Ombudsperson's recommendation or refer it to the Security Council unless it was already evident that at least nine members of the Council agreed that the decision of the Ombudsperson was wrong and that no permanent member thought that it was right.[36]

Moreover, in substantive terms, the possibility of Security Council intervention does not distinguish the Ombudsperson from other *judicial* contexts in the international legal order. The literature on international courts abounds with articles recognizing the limits of judicial independence in the global context, with Helfer and Slaughter referring to 'constrained independence',[37] Ginsburg referring to 'bounded discretion',[38] and Steinberg referring to the 'strategic space' within which courts can operate.[39] It is not the exception but the norm that states tend to ensure 'fire-alarm' controls are built into instruments establishing international courts, which can be triggered if judges are perceived by states to exceed the bounds of their delegated authority. The Security Council can stop the International Criminal Court investigating a particular case.[40] Carla del Ponte talks in her memoirs about why she felt constrained as Chief Prosecutor at the Yugoslav Tribunal in her investigation of alleged offences by NATO troops: 'I understood that I had collided with the edge of the political universe in which the Tribunal was allowed to function.'[41] The only body that can enforce decisions of the International Court of Justice (ICJ) is the Security Council, a factor that led the US Supreme Court in *Medellin* to conclude that ICJ judgments were not meant to be enforceable in US courts.[42] Of course, it has never been disputed that the ICJ is a court that issues 'binding' decisions.

Under a public interest model, it is not the bindingness of the decision that is the most relevant consideration. The public interest model relies on participatory procedures and 'justificatory discourse' over formal respect for the binding rules as a means to strengthen the legitimacy and accountability

[36] 'Thirteenth Report of the 1267 Monitoring Team' (31 December 2012) UN Doc S/2012/968, para 12.

[37] Laurence Helfer and Anne-Marie Slaughter, 'Why States Create International Tribunals: A Response to Professors Posner and Yoo' (2005) 95 *California Law Review* 1, 44 ff.

[38] Tom Ginsburg, 'Bounded Judicial Discretion in International Law-making' (2005) 45 *Virginia Journal of International Law* 1.

[39] Richard Steinberg, 'Judicial Lawmaking at the WTO: Discursive Constitutional and Political Constraints' (2004) 98 *American Journal of International Law* 247. See also Y Shany, 'No Longer a Weak Department of Power? Reflections on the Emergence of a New International Judiciary' (2009) 20 *European Journal of International Law* 73.

[40] Rome Statute of the International Criminal Court, Article 16.

[41] Carla del Ponte and Chuck Sudetic, *Madame Prosecutor: Confrontations with Humanity's Worst Criminals and the Culture of Impunity* (Feltrinelli Editore 2009), 60.

[42] *Medellin v Texas* 552 US 491 (2008), 13.

of international decision-making. Through the information-gathering process, dialogue with the petitioner, and the production of a comprehensive report containing recommendations to the Sanctions Committee, the Ombudsperson feeds into a dialogue with the Council, the petitioner, and the broader public based on contextual standards and principles. The input of the Ombudsperson serves to remove the discussion a step away from a conversation defined by a clash of self-interests between Security Council members, and provide recommendations and justifications oriented toward the development of a set of common guiding principles. The Council is encouraged to enhance the legitimacy and effectiveness of its decision-making through an appeal not to power politics, but to public reason. These reasons are in turn assessable by the 'court of public opinion' within which decision-making authority may be deemed illegitimate with consequences for its effectiveness if it persists in failing to provide public justifications for its decisions and withholds other information critical to the evaluation of institutional performance.[43]

In the end, the issue as to whether the Ombudsperson's decision-making is binding or non-binding is less important than the capacity of review by the Ombudsperson to extract 'reputational costs'.[44] As Johnstone elaborates in the Security Council context, '[i]f the interpretive community of governmental and non-governmental actors casts a negative judgment, the credibility of the Security Council will be undermined and those who must carry out the decisions will be less likely to comply'.[45] This assessment finds reflection in the broader community response to the sanctions framework. Already there is a sense that international civil society is closely monitoring the Security Council sanctions process, with academics, non-governmental organizations, and the media regularly passing judgment on the fairness of sanctions decision-making, with a corresponding effect on the legitimacy and effectiveness of the sanctions regime. Once again, the reform focus should be on ensuring greater transparency in sanctions decision-making through the work of the Ombudsperson. The legitimacy of the sanctions decision-making regime will be enhanced if petitioners, interested states, and the international community more broadly are in a position to understand and assess the reasons for the Sanctions Committee's (and the Ombudsperson's) decisions.

[43] Allen Buchanan and Robert O Keohane, 'The Legitimacy of Global Governance Institutions' (2006) 20 *Ethics and International Affairs* 405, 429.

[44] Ian Johnstone, 'Legislation and Adjudication in the UN Security Council: Bringing Down the Deliberative Deficit' (2008) 102 *American Journal of International Law* 275, 307.

[45] Ibid, 307.

C. The Public Interest Model and International Values: Striking the Balance between International and Individual Values

In developing a procedural framework for Security Council decision-making, one of the most complex problems is undoubtedly how to reconcile due process rights of individuals with the competing imperative of protecting the confidentiality and sensitivity of intelligence information at issue in the sanctions context. In terms of a clash between values, this is undoubtedly the knife-edge along which the most problematic and intransigent divisions in the due process debate have arisen. The Security Council cites confidentiality concerns as the key reason to limit due process protections on the basis that it is too complicated to '[find] a way to keep such intelligence, and how it was gathered, confidential'.[46] On the other hand, in the recent *Kadi II* decision, the CJEU has claimed the power to engage in full review of sanctions decision-making, while insisting that 'the secrecy or confidentiality of that information is no valid objection before the courts of the EU'.[47]

This deadlock is clearly unsatisfactory. The solution should neither be to abandon procedural protections altogether, nor to impose unrealistic obligations upon the Council to release information to domestic or regional agencies where that obligation is highly unlikely to be complied with. While it is well understood that transparency is a desirable institutional value and a core attribute of good governance,[48] secrecy and non-disclosure can also have value in particular contexts.[49] The Council has been described as 'a body

[46] 'Press Conference by Security Council President Ambassador Gérard Araud' (2 February 2010), available at <http://www.un.org/News/briefings/docs/2010/100202_Araud.doc.htm>. See also 'Remarks of Sanctions Committee Chairman Thomas Mayr-Harting: Press Conference on the Work of Security Council, Al Qaida Taliban Sanctions Committee' (27 October 2009), available at <http://www.un.org/News/briefings/docs/2009/091027_1267.doc.htm>.

[47] *European Commission and Council v Kadi* [2013] ECR not yet reported, [125].

[48] Daniel Esty, 'Good Governance at the Supranational Scale: Globalizing Administrative Law' (2006) 115 *Yale Law Journal* 1490, 1530; Thomas N Hale and Anne-Marie Slaughter, 'Transparency: Possibilities and Limitations' (2006) 30 *The Fletcher Forum of World Affairs* 153; ILA, *Final Report on Accountability of International Organisations* (2004), 8–9; F Fukuyama, *State-Building: Governance and World Order in the 21st Century* (Cornell University Press 2004); Anne-Marie Slaughter, 'Agencies on the Loose? Holding Government Networks Accountable' in G A Bermann et al (eds), *Transatlantic Regulatory Cooperation* (Oxford University Press 2000), 521, 529.

[49] Liz Fisher has recognized that secrecy and non-disclosure have value in particular contexts: Elizabeth Fisher, 'Transparency and Administrative Law: A Critical Evaluation' (2010) 63 *Current Legal Problems* 272, 280. See also Anne-Marie Slaughter, *A New World Order* (Princeton University Press 2004), 220.

in which confidentiality and informality regarding the decision-taking process are part of the business'.[50] Its effectiveness has been attributed to a 'procedure of confidentiality', which provides a climate for free-ranging and uninhibited debates, and the achievement of consensus.[51] As one scholar has argued, 'the Council can tolerate only so much consultation, publicity and review before it becomes hopelessly inefficient'.[52] The need to operate behind closed doors has been advocated as a 'functional necessity',[53] justified by the need to protect the confidentiality and sensitivity of the issues under discussion, and the desire to diminish political 'grandstanding' to the detriment of genuine discussion and negotiation.[54] At the same time, it is widely considered that the extent to which information about sanctions decision-making has been shielded from the public and affected individuals has been disproportionate.[55] The challenge is to devise imaginative institutional measures that can achieve the most appropriate balance between the individual rights of those placed on sanctions blacklists, and the interests of international peace and security.

Though it is early days, I argue that the Office of the UN Ombudsperson offers the greatest potential for reconciliation of due process and confidentiality concerns. The Ombudsperson has described gaining access to classified or confidential information as 'one of the key challenges' she faces.[56] However, she is also aware that 'this question of access is a critical one for due process',[57] and—in tangent with the Security Council—she has negotiated

[50] 'Statement by Ambassador Ronaldo Mota Sardenberg, Permanent Representative of Brazil to the UN' (New York, 13 October 2003).

[51] 'Statement of Ambassador Marker of Pakistan' (16 December 1994) UN Doc S/PV.3483.

[52] Ian Johnstone, 'Legislation and Adjudication in the UN Security Council: Bringing Down the Deliberative Deficit' (2008) 102 *American Journal of International Law* 275, 303.

[53] Paul Conlon, 'Lessons from Iraq: The Functions of the Iraq Sanctions Committee as a Source of Sanctions Implementation Authority and Practice' (1995) 35 *Virginia Journal of International Law* 633, 655.

[54] Farrall, *United Nations Sanctions and the Rule of Law* (Cambridge University Press 2007), 213. For example, France has argued that the effectiveness of the Security Council 'requires an appropriate split of Security Council activity between public meetings and the private consultations that are necessary to the conduct of negotiations and to the development of compromises': 'Security Council, Debate on Working Methods' (27 August 2008) UN Doc S/PV.5968, 9.

[55] Paul Conlon's analysis of the inner workings of the Iraq Sanctions Committee concluded that discussion of sensitive confidential matters accounted for a mere 2.5% of the Committee's meeting time, and that the closed meeting format in fact elicited little candour and frankness from Committee members. Even more damning is Conlon's conclusion that the lack of transparency in the proceedings of the Iraq Sanctions Committee actually aided Iraq and sanctions evaders as it shielded them from the public spotlight: Paul Conlon, *United Nations Sanctions Management: A Case Study of the Iraq Sanctions Committee* (Transnational Publishers 2000), 74, 151.

[56] 'Second Report of the Office of the Ombudsperson pursuant to Security Council resolution 1904' (22 July 2011) UN Doc S/2011/447, para 26.

[57] 'First Report of the Office of the Ombudsperson pursuant to Security Council resolution 1904' (24 January 2011) UN Doc S/2011/29, para 33.

a set of procedures enhancing her capacity to access critical information. In this respect, the Ombudsperson has three key advantages over other institutional frameworks: (1) expertise; (2) potential to locate pressure points; and (3) capacity to negotiate specific arrangements.

(1) Expertise

As Elizabeth Fisher has noted, it is not enough to provide access to information. The system also needs to provide a process of interpretation of that information, which requires expertise.[58] This is particularly the case in the sanctions context, where the Ombudsperson is required to deal with 'intelligence', from which inferences are drawn, more than 'evidence'. It is clear that the Ombudsperson is in a far better position than courts to assess the weight of this information. The Ombudsperson has acknowledged the importance of experience and 'institutional memory' in enabling her to assess the key questions and issues of concern for the Sanctions Committee.[59] In particular, the Ombudsperson has emphasized the value of her personal access to petitioners in the dialogue phase of the process, during which she has the opportunity to ask petitioners to respond to and explain inferences that might be drawn from relevant intelligence while working classified material into the background.[60] The Monitoring Team is also available to provide operational support and assistance to the Ombudsperson and has provided expert advice and analysis of audio-visual material in several cases.[61] While there are legitimate questions as to whether it is ever satisfactory to rely on secret information to which the petitioner does not have access, the Ombudsperson has expressed her confidence that in all cases (with one exception) the petitioner has been informed of the case underlying the listing and has had the opportunity to respond and be heard by the decision-maker.[62] In the exceptional

[58] Elizabeth Fisher, 'Transparency and Administrative Law: A Critical Evaluation' (2010) 63 *Current Legal Problems* 272, 306–7.

[59] 'Third Report of the UN Ombudsperson' (20 January 2012) UN Doc S/2012/49, para 7; 'Seventh Report of the Office of the Ombudsperson' (31 January 2014) UN Doc S/2014/73 para 30.

[60] It should be noted that certain lawyers representing individuals in delisting proceedings have argued that this dialogue sheds little light on the nature of the secret allegations against their clients.

[61] 'Seventh Report of the Office of the Ombudsperson' (31 January 2014) UN Doc S/2014/73, para 17.

[62] 'Eighth Report of the Office of the Ombudsperson' (31 July 2014) UN Doc S/2014/73, para 34; 'Seventh Report of the Office of the Ombudsperson' (31 January 2014) UN Doc S/2014/73, para 32; 'Sixth Report of the Office of the Ombudsperson' (31 July 2013) UN Doc S/2013/452, paras 32–5; 'Fifth Report of the Office of the Ombudsperson' (31 January 2013) UN Doc S/2013/71, para 29.

case, the Ombudsperson acknowledges that the petitioner was prejudiced as the relevant information was obtained at such a late stage that it could not be disclosed to the petitioner before preparation of the comprehensive report and has invited comments from the petitioner with a view to deciding whether he meets the threshold for a new petition.[63]

(2) Pressure Points

Even more persuasively, while the Ombudsperson has no power to compel production of confidential information, the Ombudsperson is in a unique position to place pressure on states to provide such information. Certainly the limitations that exist for domestic courts where information is not held by their domestic authorities do not exist for the Ombudsperson. Moreover, there are several factors that combine to give strength to requests by the Ombudsperson to provide confidential information upon which decisions have been based. First, the Ombudsperson's request for information is mandated by a Chapter VII Security Council resolution. In resolution 1989, the Security Council strongly urges member states to provide all relevant information to the Ombudsperson. The Ombudsperson has confirmed that reliance on this resolution has proved useful in encouraging states to co-operate with the Ombudsperson.[64] Second, as the Ombudsperson has recognized in a recent report, 'any lack of detail does not work to the prejudice of the petitioner'.[65] The conclusion reached in the Ombudsperson's Comprehensive Report is based on 'whether there is sufficient information to provide a reasonable and credible basis for the listing'.[66] It follows that if a state declines to disclose information upon which the listing decision is based, it is likely that the Ombudsperson will recommend de-listing. The Ombudsperson's decision has a 'triggering effect', and can only be reversed if the Sanctions Committee decides by consensus to do so.[67] This means that a refusal by a state to provide relevant information will likely attract the consequence that it will need to persuade every Committee member to reverse the Ombudsperson's decision if it wishes to maintain the listing. Third, the Ombudsperson is

[63] 'Sixth Report of the Office of the Ombudsperson' (31 July 2013) UN Doc S/2013/452, paras 33–5.

[64] Security Council resolution 1989 (17 June 2011) UN Doc S/RES/1989, para 25. See also 'Third Report of the UN Ombudsperson' (20 January 2012) UN Doc S/2012/49, para 41.

[65] 'Third Report of the UN Ombudsperson' (20 January 2012) UN Doc S/2012/49, para 42.

[66] Office of the UN Ombudsperson, 'Approach to and Standard for Analysis, Observations, Principal Arguments and Recommendation', August 2011, available at <http://www.un.org/en/sc/ombudsperson/approach.shtml>.

[67] UN Security Council resolution 1989 (17 June 2011) UN Doc S/RES/1989, para 23.

directed to update the 1267 Committee as individual cases progress, specifying 'details regarding which States have supplied information'.[68] Fourth, the Ombudsperson reports biannually directly to the Security Council, and in these reports typically discusses the level of state co-operation. These reports are also publicly available on the Ombudsperson's website. In her Third Report, the Ombudsperson noted that certain of the information lacked sufficient specificity to enable meaningful analysis, and also noted that certain state responses were received late, reducing the time available to the Ombudsperson to engage in dialogue with the petitioner and prepare her Comprehensive Report. Through such reports, both the Council and the broader international community are made aware of failures in state co-operation, increasing pressure on states to comply.

(3) Specific Arrangements

An additional innovation of the UN Ombudsperson is to instigate a practice of entering into specific arrangements with individual states to obtain access to confidential information. These arrangements are naturally based on a level of trust achieved between the state and the Ombudsperson. Theoretically, this level of trust should not be difficult to achieve given that the Ombudsperson is appointed by the Secretary-General in consultation with the Sanctions Committee, and must be 'an eminent individual of high moral character, impartiality and integrity with high qualifications and experience in relevant fields, such as legal, human rights, counter-terrorism and sanctions',[69] criteria more than fulfilled by the first office-holder, Kimberly Prost. To date, the Ombudsperson has entered into arrangements for access to classified or confidential information with fourteen states including the United Kingdom and France. The United States has expressed willingness, and demonstrated an ability, to share confidential information on an ad hoc basis. The Ombudsperson is said to be 'aggressively pursuing' the issue with a large number of states in an attempt to increase the number of agreements and arrangements, with a particular focus on those states that are most frequently involved in specific cases.[70]

In conclusion, the unique position of the Ombudsperson, who is both independent although connected to the Council, offers valuable potential for a process of independent review within the sanctions regime that includes

[68] Security Council resolution 1904 (17 December 2009) UN Doc S/RES/1904, Annex II, para 4.

[69] Ibid, para 20.

[70] 'Third Report of the UN Ombudsperson' (20 January 2012) UN Doc S/2012/49, para 19.

consideration of confidential or classified information. It is open to the same criticism as 'special advocates' or 'special counsel' in that the confidential information can generally not be shown to the petitioner, who is often the most important repository of information to contradict it.[71] However, the Ombudsperson is clearly a preferable option to domestic or international courts for those states wishing to maintain control over confidential information, while enabling it to be deployed in the context of the sanctions regime. While the review of confidential information by the Ombudsperson alone does not serve to increase public access to information, the procedure may nevertheless contribute to greater public acceptance of decision-making if the Ombudsperson continues to be regarded as an impartial and trusted entity who has engaged in an independent assessment that the available evidence justifies the listings.

D. Conclusion

The public interest model of due process takes a less radical route than the dignitarian model to the regeneration of the classical international legal order so as to respond to contemporary concerns about its legitimacy. It aims to correct the 'deliberative deficit' in Security Council sanctions decision-making by increasing public access to information and opportunities for deliberation about sanctions decision-making, with the ultimate goal of establishing an inclusive legal discourse though which decision-making is able to be justified and measured according to a dynamic conception of shared understandings.

The adoption of a public interest model in the UN sanctions regime would have a positive impact on overall legitimacy. It would provide a valuable and relatively un-intrusive means to enhance representativeness in the Security Council context, opening up decision-making to a range of actors who are affected by the Council's decisions and have traditionally been deprived of any voice. Perhaps the most interesting aspect of this model is that it enhances the legitimacy of Security Council decision-making without imposing a requirement of access to a court. Indeed, the non-judicial nature of the Ombudsperson's Office is potentially an advantage as it provides a mechanism by which to hold the Council to account, yet without relying on techniques of bindingness, coercion, or supremacy, which are inconsistent

[71] See *Secretary of State for the Home Department v AF* [2009] UKHL 28, [36]; citing Sedley LJ in *AM* [2008] EWCA Civ 1148, [2009] 2 WLR 423, [113]. See also Serrin Turner and Stephen J Schulhofer, *The Secrecy Problem in Terrorism Trials* (Liberty and National Security Project 2005), 12.

with fundamental pillars of the Council's make-up. The Ombudsperson's office also offers valuable techniques through which to reconcile the competing tensions of due process and the imperative of maintaining the confidentiality of certain of that information in the sanctions context.

At a broader level, the public interest model would serve to strengthen one of the competing visions for the future structure of the international legal order, the 'cosmopolitan constitutionalist' school of thought. The principal aim of the model is to encourage the Council to engage in reasoned argumentation, and the interpretive community to engage in principled deliberation, contributing to the development of a set of shared understandings by which the Council could be held to account. The result would be an altered conception of international community and law, opening up international law's development to influence from the broader interpretive community's principled interchange about the appropriate standards of international behaviour and the extent and limits of the Council's authority to regulate that behaviour.[72] For those in favour of steering the international legal order in the direction of a more 'constitutionalist' framework, the model not only has legitimating potential for the Security Council sanctions context, but also represents a valuable opportunity to shape the future structure of the international legal order.

[72] Ian Johnstone, 'Security Council Deliberations: The Power of Better Argument' (2003) 14 *European Journal of International Law* 437, 438.

8

Conclusion

This book invites attention to the issue of due process in sanctions decision-making as one of far greater significance than the prosaic terminology of 'process' might suggest. The establishment of the targeted sanctions regime, and the Security Council's assumption of direct decision-making authority over individuals, represents more than a shift in Security Council policy. Rather, it is an exemplar of an emerging system of international governance. The Council's initial assessment that there were 'no legal issues' involved in listing and de-listing of individuals is a stark example of the failure of international organizations adequately to assess the consequences of an increasing assumption of governmental authority.[1] At the domestic level, the activity of 'governing' is conditioned and sustained by 'an assemblage of rules, principles, canons, usages and manners' known in many systems as public law.[2] Failure by the Council to recognize even a narrow set of guiding principles threatened to do what Dan Sarooshi cautioned international organizations against, namely, 'to dispense with, by the stroke of a pen, the limitations on governmental tyranny that peoples have fought hard to win within their domestic polity'.[3]

It has become clear in the evolution of the targeted sanctions regime that the broader international community will not tolerate such a dispensation. In response to Council inertia on procedural issues, a dynamic and influential reform project has emerged, led by international lawyers and domestic courts, emphasizing the imperative of Security Council respect for procedural safeguards. The great majority of procedural reform proposals have drawn on legal safeguards developed for domestic legal settings. In more recent times,

[1] Confidential communication, 24 January 2006; cited in Simon Chesterman, 'The Spy Who Came In From the Cold War: Intelligence and International Law' (2006) 27 *Michigan Journal of International Law* 1071, 1117; Remarks by Hans Corell, 'Is the Security Council Bound by Human Rights Law?', 103rd Annual Meeting of the American Society of International Law, 27 March 2009.

[2] Martin Loughlin, *The Idea of Public Law* (Oxford University Press 2003), 30.

[3] Dan Sarooshi, *International Organizations and their Exercise of Sovereign Powers* (Oxford University Press 2005), 14.

the primary focus has been on the need for a court-based procedure. There is a widespread perception that if the regime fails to provide access to a judicial organ that can issue binding decisions against the Council, the regime will be inadequate. In the meantime, the Security Council remains staunchly opposed to any form of judicial oversight of sanctions decision-making. This deadlock has already had damaging implications for the effectiveness of the UN sanctions regime. Opposition to the Ombudsperson procedure on grounds it is 'not a court' has enabled the Council to avoid criticism for failing to extend the protections of the Office beyond the 1267 Sanctions Committee. If the recent trend continues, it will likely lead to further holes in the sanctions net as an increasing number of domestic courts succumb to the pressure and precedent of the international reform project and invalidate domestic implementing mechanisms on the basis that they fail to provide for access to a court.

While the Council's reluctance to institute adequate procedural safeguards is problematic, there are also problems with current methodology underlying the majority of procedural reform proposals. To date, international lawyers have predominantly adopted a formalist approach to the identification of applicable procedural safeguards, relying on the traditional sources of international law. As a consequence, the task of developing a procedural fairness framework has been under-theorized. The problem with this classical formalist approach to methodology is that it can tend to overemphasize the value of (descriptive) state practice to the detriment of (normative) theoretical appeal. The value of the descriptive route is that it focuses on what state practice has been, and ensures international legal principles correspond to the reality of state conduct.[4] However, the history of due process has largely unfolded in response to domestic legal contexts, practice that is arguably unsuited to the quite different legal and political context of the Security Council.

I argue in this book that a subtle shift in international law source methodology is justified. The solution to procedural deficiencies in sanctions decision-making is not simply to transplant procedural safeguards developed for domestic legal context to the Security Council setting. The Security Council setting is a unique legal context, quite distinct from domestic governmental systems. The gradual tracing out of the full implications of the assumption of 'governmental' powers in this setting is a major undertaking. To quote Alexander Hamilton from the Federalist Papers (and I do not

[4] Anthea Roberts notes that this 'descriptive' approach is the hallmark of traditional approaches to customary international law: Anthea Roberts, 'Traditional and Modern Approaches to Customary International Law: A Reconciliation' (2001) 95 *American Journal of International Law* 757, 761–4.

consider the analogy to the challenges faced in constructing a constitutional regime from a diverse substrata of states to be inapt):[5]

The erection of a new government, whatever care or wisdom may distinguish the work, cannot fail to originate questions of intricacy and nicety; and these may, in a particular manner be expected to flow from the establishment of a constitution founded upon the total or partial incorporation of a number of distinct sovereignties. 'Tis time only that can mature and perfect so compound a system, can liquidate the meaning of all the parts, and can adjust them to each other in a harmonious and consistent whole.

Where the task is to construct a procedural framework for a new tier of governance, the focus should be less on formulating and systemizing rules that already effectively exist in practice, and more on creating a new set of rules for an emerging governmental context. Rather than working from practice to theory, the inverse is more appropriate, with the principal aim being to provide strong and enduring theoretical foundations to support institutional practice. In a broader sense, it is more illuminating not to see theory and practice as competing sources of international law. Instead, international law should be regarded in the nature of a magnetic field within which theory and practice serve as different poles exerting influence. International legal rules result where there is balance between acceptable theory and acceptable practice.

In Chapter 4, I propose an alternative to the formalist practice-based methodology, and establish the foundations of a value-based approach to the construction of a procedural framework for sanctions decision-making. The essence of the theoretical inquiry into applicable procedural principles is a quest for legitimacy. Legitimacy is the capacity of authority to pull toward consensual compliance. If the Security Council wishes to exercise governmental authority over individuals, this authority will only be effective to the extent that it is broadly regarded as legitimate in the eyes of those over whom it is exercised. A focus on legitimacy is important in developing a procedural framework for the Security Council sanctions setting for two reasons. First, due process is all about enhancing the legitimacy of decision-making authority. It follows that any inquiry into an appropriate procedural framework needs to work from the starting point of which procedural principles will operate best to enhance legitimacy in the Security Council context. Due process works by way of a form of 'procedural dialogue' directed at strengthening the connection between decision-makers and the community to whom the decisions apply. This dialogue should aim to establish an 'equilibrium of

[5] Alexander Hamilton, James Madison, and John Jay, *The Federalist: A Commentary on the Constitution of the United States* (Tudor 1947), No 82, vol 2, 130.

involvement' in the particular context, or the point at which the connection between decision-makers and the relevant community is sufficient to engender acceptance of decisions within the community to which they relate.[6]

The second rationale for a focus on legitimacy in the development of procedural principles is that 'legitimacy' has far more persuasive appeal than 'law' in the Security Council context. Although in recent years, the Council has displayed a growing recognition of the need to incorporate the rule of law, this is as a matter of policy, rather than a matter of law.[7] To the Security Council, the question of due process is a political rather than a legal question. As Security Council debates demonstrate, the primary interest of Council members in due process is as a way of addressing the legitimacy of its sanctions regimes in order to enhance their implementation and therefore their effectiveness.[8] Accordingly, if Security Council members are persuaded that procedural principles will enhance the legitimacy of decision-making, they are far more likely to push for reform than they would be on the basis of a formalist argument that such principles are 'legally binding'.

It follows from the integral connection between due process and legitimacy that procedural fairness is not the 'one-size-fits-all' universally applicable concept that formalist approaches imply. Rather, due process is *contextual.* Given that the focus of my book is on the legitimating potential of due process, my analysis focuses on the identification of normatively rich conceptions of the principle. A survey of the extensive literature on the underlying normative foundations of due process reveals that there are a number of viewpoints about the central process value advanced by procedural frameworks. Different process values will be effective in enhancing legitimacy in different contexts, depending on underlying conceptions of community, law, and values in those contexts. The instrumentalist model focuses on ensuring the accuracy of the decision-maker's application of the substantive law. It works to enhance legitimacy by acting as a 'transmission belt' for the law adopted by the legitimately representative lawmaker, and in this way serves to ensure that decision-making respects the will of the broader polity. The dignitarian model, by contrast, operates to ensure a more direct means of interest representation. Its focus is on procedural mechanisms that will give voice to those affected by decision-making, and ensure decision-makers are aware of the rights and interests implicated in the decision-making process. Unlike the

[6] I gratefully acknowledge the source of this realization in a discussion with the late Professor T M Franck.

[7] Security Council Report, *Cross-Cutting Report on the Rule of Law* (2011), 41.

[8] See, for example, 'Debate on Security Council Working Methods', 7285th meeting of Security Council (23 October 2014) UN Doc S/PV.7285 (in particular contributions by Ms Kimberly Prost, Australia, Chile, Russia, and Nigeria).

instrumentalist model, it does not cater to a 'rule of law' framework in which arguments must appeal to legal reason, but is more concerned to ensure that decision-making 'takes account' of the diverse range of interests and preferences implicated in decision-making. The public interest model is different again. It does not focus on self-interested participation, but regards procedural fairness as a mechanism through which to enhance respect for the public interest in decision-making. Unlike the dignitarian model, it operates within a legal framework, though, unlike the instrumentalist model, it does not regard law as fixed. Instead, it encourages participation for the very purpose of ensuring operative legal principles remain responsive to shared purposes informed by contributions from affected individuals and the broader public.

Returning to the 'magnetic field' of international law, this theory only gains relevance in the Security Council setting against the backdrop of practice. The process of building an international system of law has not typically proceeded as a coherent legal enterprise, but has been more of a 'messy and contradictory and confused' endeavour in which institutional structures have been cobbled together as a political compromise.[9] The development of a due process framework for sanctions decision-making has been no exception. In practice, little theoretical attention has been given to the construction of a procedural framework for sanctions decision-making. Rather, procedural safeguards have developed largely organically. In Chapter 2, I described three competing structures: (1) an internationalized judicial procedure; (2) a pluralist judicial procedure; and (3) an international non-adjudicatory procedure. It is possible to connect the frameworks that have emerged in practice with a broader theoretical analysis. I argue that the internationalized judicial procedure would serve predominantly to advance an instrumentalist model of procedural fairness concerned to ensure the accuracy of decision-making. A pluralist judicial procedure would operate to promote dignitarian aims, and place greater emphasis on the dignity and autonomy of individuals in decision-making, and representation of their rights and interests. An international non-adjudicatory procedure would work to promote a public interest model of procedural fairness, with the chief goal being to enhance respect for the public interest in Security Council sanctions decision-making. Through these means, the theoretical implications of competing frameworks in the sanctions setting can be assessed, providing more fertile and fruitful terrain in which to debate the most appropriate framework.

The major contribution of this book is to demonstrate the differing extents to which different models (and therefore different frameworks) would enhance

[9] David Kennedy, 'The Disciplines of International Law and Policy' (1999) 12 *Leiden Journal of International Law* 9.

legitimacy in Security Council sanctions decision-making. It is of particular interest that the internationalized judicial framework emerges as the model least likely to enhance legitimacy in the Security Council sanctions context. An internationalized judicial authority would be restricted to measuring sanctions decision-making in terms of its accurate application of international law. The combined effect of Articles 25 and 103 of the UN Charter, which provides that the obligation on states to comply with Security Council resolutions prevails over all competing international obligations, means that the goal of procedural law would be limited to ensuring the Council does not contravene the clear terms of its own resolutions (a body of law that at best represents the will of fifteen member states) or those norms that can be said to fall within the stratospheric category of *jus cogens*. The possible detrimental consequences are twofold. On the one hand, such an approach threatens to narrow the scope of review of sanctions decision-making to an extent that could be described as an abdication of the judicial role to protect individual rights. On the other hand, it may result in courts expanding the judicial role beyond acceptable limits, assuming supremacy over the Council and overruling Security Council decisions based on an expansive interpretation of Council resolutions or *jus cogens* norms. The pluralist judicial solution fares better in terms of its legitimating potential. It would allow a more diverse set of viewpoints to penetrate the legal discourse about sanctions decision-making, provide more robust protection of individual rights, and place pressure on the Council to capitulate to domestic concerns that the sanctions regime was inconsistent with domestic legal standards. However, it also threatens to create a situation in which the Council is held hostage to the views of an elite group of states, such that the interests of international peace and security risk being subjugated to the rights of a smaller number of individuals.

It is the UN Ombudsperson that provides the greatest prospect of enhancing the legitimacy of Security Council decision-making. Though further reform is needed, the sanctions framework has the capacity to support a public interest model of procedural fairness that would enhance the legitimacy of decision-making through an enhanced procedural framework of more inclusive deliberations, greater access to information (including public notice, public reason-giving, and independent review), and greater responsiveness on the part of the Security Council to the concerns of the broader 'interpretive community'. Domestic courts could also have a role in this framework, though their most productive role would be less as an agent of enforcement, and more in the form of an expert contribution to a broader dialogue. The UN Ombudsperson and associated procedures provide a framework that enhances the representativeness and responsiveness of Security Council decision-making, yet without eroding important pillars of the Security

Council, chiefly its wide discretion and primary responsibility for the maintenance of international peace and security. The Ombudsperson framework would also provide the best potential for reconciliation of the due process obligation to provide access to information to those most affected by Security Council decision-making, without compromising important confidentiality concerns of state-based information forming the basis of sanctions decisions.

Finally, the development of a procedural framework is not only important in terms of its legitimating potential in the Security Council sanctions context. The choice of framework will plant a foothold in the broader debate about the *reformation of the international legal order*. The connection between procedural fairness and legitimacy relates procedural frameworks to broader conceptions of international community, international law, and the role (if any) of an international value system. An instrumentalist model, which acts as a 'transmission belt' for existing law, will serve to entrench a classical state-centric perspective on international law and community. By contrast, a dignitarian model would diminish the relevance of the rule of law in international decision-making, eschewing traditional conceptions of international law and hierarchy in favour of a pluralist model of the international legal order. A public interest model promotes similar conceptions of international law and community to the cosmopolitan constitutionalist school of thought, though potentially strengthens the foundations of this account of the international legal order by providing for a more representative contribution to the debate about the governing values of international society.

In the end, the development of a procedural fairness framework represents a far greater theoretical and practical challenge for the international legal order than has so far been acknowledged. I summarize in the Appendix the major implications of the various models. Though this book takes a position on the model that is best equipped to advance legitimacy in the international legal order, the significant difference of opinion on the future direction of the international legal order means that minds will legitimately differ over the most appropriate model. The aim of this book is to stimulate thinking, rather than foreclose debate. The generation of a framework for international governance is too important to leave to the 'messy and contradictory and confused' trajectory of international politics. Legal theory must play its part. To put a spin on the public law expression made famous by UK public lawyers Carol Harlow and Rick Rawlings, it could well be that, 'behind every theory of procedural fairness, there is a theory of international law'.[10]

[10] Harlow and Rawlings begin the latest edition of *Law and Administration* with the recognition that 'behind every theory of administrative law is a theory of government': Carol Harlow and Richard Rawlings, *Law and Administration* (3rd edn, Cambridge University Press 2009), 1.

Appendix

Table summarizing the impact of different normative models of procedural fairness.

	Key Process Value	Key Participants	Theory of International Law	Key International Value	Key Procedural Actor
Instrumentalist	Accuracy	Nation States	Classical Positivist	Sovereignty	Internationalized Court
Dignitarian	Autonomy	Stakeholders	Pluralist	Interest Representation	Domestic/ Regional Courts
Responsive	Public Interest	International Community	Cosmopolitan Constitutionalist	Accountability	UN Ombudsperson

Bibliography

Ackerman (1996). 'The Rise of World Constitutionalism' *Virginia Law Review* **83**: 771.

Akande, D (1997). 'The International Court of Justice and the Security Council: Is There Room for Judicial Control of the Decisions of the Political Organs of the United Nations?' *International and Comparative Law Quarterly* **46**: 309.

Allan, T R S (1998). 'Procedural Fairness and the Duty of Respect' *Oxford Journal of Legal Studies* **18**: 497.

Allen, C K (1931) *Bureacracy Triumphant*. Oxford, Oxford University Press.

Almqvist, J. (2008). 'A Human Rights Critique of European Judicial Review: Counter-Terrorism Sanctions' *International and Comparative Law Quarterly* **57**: 303.

Alston, P. (2005). 'The "Not a Cat" Syndrome: Can the International Human Rights Regime Accommodate Non-State Actors and Human Rights' in P Alston, *Non-State Actors and Human Rights*. Oxford, Oxford University Press: 3.

Alvarez, J E (1996). 'Judging the Security Council' *American Journal of International Law* **90**: 1.

Alvarez, J E (2000). 'Interliberal Law: Comment' *American Society of International Law Proceedings* **94**: 249.

Alvarez, J E (2001). 'Do Liberal States Behave Better? A Critique of Slaughter's Liberal Theory' *European Journal of International Law* **12**: 183.

Alvarez, J E (2003). 'Hegemonic International Law Revisited' *American Journal of International Law* **97**: 873.

Alvarez, J E (2005). *International Organizations as Law-makers*. New York, Oxford University Press.

Antoine, M (1970). *Le Conseil du roi sous le règne de Louis XV*. Geneva, Droz.

Ashworth, A (2002). *Human Rights, Serious Crime and Criminal Procedure*. London, Sweet & Maxwell.

Avbelj, M, F Fontanelli and G Martinico (eds) (2014). *Kadi on Trial: A Multifaceted Analysis of the Kadi Judgment*. New York, Routledge.

Bailey, S and S Daws (1998). *The Procedure of the UN Security Council*. Oxford, Clarendon Press.

Baker, K M (1990). 'Science and Politics at the End of the Old Regime' in K M Baker (ed.), *Inventing the French Revolution: Essays on the French Political Culture in the 18th Century*. Cambridge, Cambridge University Press: 153.

Baquero Cruz, J (2008). 'The Legacy of the Maastricht Judgment and the Pluralist Movement' *European Law Journal* **14**: 389.

Barak, A (2002). 'A Judge on Judging: The Role of a Supreme Court in a Democracy' *Harvard Law Review* **116**: 19.

Barak-Erez, D and M C Waxman (2009). 'Secret Evidence and the Due Process of Terrorist Detentions' *Columbia Journal of Transnational Law* **48**: 3.

Bayles, M (1990). *Procedural Justice.* Dordrecht, Kluwer.
Beetham, D (1991). *The Legitimation of Power.* London, Macmillan.
Benhabib, S (1996). 'Toward a Deliberative Model of Democratic Legitimacy' in S Benhabib, *Democracy and Difference: Contesting the Boundaries of the Political.* Princeton, Princeton University Press: 67.
Bentham, J (1838–1843). 'Constitutional Code' in J Bowring, *Works of Jeremy Bentham.* William Tate **9**: 8.
Bentham, J (1838–1843) 'The Principles of Judicial Procedure' in J Bowring, *Works of Jeremy Bentham.* William Tate: **2**: 1.
Benvenisti, E (1999). 'Exit and Voice in the Age of Globalization' *Michigan Law Review* **98**: 167.
Benvenisti, E and G W Downs (2007). 'The Empire's New Clothes: Political Economy and the Fragmentation of International law' *Stanford Law Review* **60**: 595.
Berman, P (2007). 'A Pluralist Approach to International Law' *Yale Journal of International Law* **32**: 301.
Berman, P (2007). 'Global Legal Pluralism' *Southern California Law Review* **80**: 1155.
Besson, S (2009). 'European Legal Pluralism after *Kadi*' *European Constitutional Law Review* **5**: 237.
Bianchi, A (2006) 'Assessing the Effectiveness of the UN Security Council's Anti-terrorism Measures: The Quest for Legitimacy and Cohesion' *European Journal of International Law* **17**(5): 881.
Bianchi, A (2006). 'Security Council Anti-Terror Resolutions and their Implementation by Member States' *Journal of International Criminal Justice* **4**: 1044.
Biersteker, T and S Eckert (2006). *Strengthening Targeted Sanctions through Fair and Clear Procedures.* Watson Institute Targeted Sanctions Project, Brown University.
Biersteker, T and S Eckert (2009). *Addressing Challenges to Targeted Sanctions: An Update of the 'Watson Report'.* Watson Institute for International Studies.
Bohman, J (1999). 'International Regimes and Democratic Governance: Political Equality and Influence in Global Institutions' *International Affairs* **75**: 499.
Bohman, J (2007). *Democracy Across Borders.* Cambridge, MIT Press.
Bone, R G (2003). 'Agreeing to Fair Process: The Problem with Contractarian Theories of Procedural Fairness' *Boston University Law Review* **83**: 485.
Bothe, M (2008). 'Security Council's Targeted Sanctions against Presumed Terrorists: The Need to Comply with Human Rights Standards' *Journal of International Criminal Justice* **6**: 541.
Bowett, D (1994). 'The Impact of Security Council Decisions on Dispute Settlement Procedures' *European Journal of International Law* **5**: 89.
Boyle, A and C Chinkin (2007). *The Making of International Law.* Oxford, Oxford University Press.
Boyle, F (1980). 'International Law in Time of Crisis: From the Entebbe Raid to the Hostages Convention' *Northwestern University Law Review* **75**: 831.
Brierly, J (1958). 'The Basis of Oligation in International Law' in H Lauterpacht, *The Basis of Obligation in International Law and other Papers of the Late James Brierly.* Oxford, Clarendon Press.

Brownlie, I (1993). 'The Decisions of Political Organs of the United Nations and the Rule of Law' in R S J MacDonald, *Essays in Honour of Wang Tieya*. Dordrecht, Martinus Nijhoff: 91.

Broxmeyer, E (2004). 'The Problems of Security and Freedom: Procedural Due Process and the Designation of Foreign Terrorist Organizations Under the Anti-Terrorism and Effective Death Penalty Act' *Berkeley Journal of International Law* **22**: 439.

Brunée, J and S J Toope (2008). *Legitimacy and Legality of International Law.* Cambridge, Cambridge University Press.

Buchanan, A and R O Keohane (2006). 'The Legitimacy of Global Governance Institutions' *Ethics and International Affairs* **20**: 405.

Builder, R (1962). 'The Office of the Legal Adviser: The State Department Lawyer and Foreign Affairs' *American Journal of International Law* **56**: 654.

Bull, H (1966). 'The Grotian Conception of International Society' in H Butterfield and M Wight, *Diplomatic Investigations*. Cambridge, Harvard University Press: 12.

Bull, H (1977). *The Anarchical Society: A Study of Order in World Politics*. London, MacMillan.

Bulterman, M (2006). 'Fundamental Rights and the United Nations Financial Sanctions Regime: The *Kadi* and *Yusuf* Judgments of the Court of First Instance of the European Communities' *Leiden Journal of International Law* **19**: 753.

Burke-White, W (2004). 'International Legal Pluralism' *Michigan Journal of International Law* **25**: 963.

Byers, M (2004). 'Agreeing to Disagree: Security Council Resolution 1441 and Intentional Ambiguity' *Global Governance* **10**: 165.

Cameron, I (2003). 'UN Targeted Sanctions, Legal Safeguards and the European Convention on Human Rights' *Nordic Journal of International Law* **72**: 159.

Cameron, I (2006). *The European Convention on Human Rights, Due Process and United Nations Security Council Counter-terrorism Sanctions*. Strasbourg, Committee of Legal Advisers on Public International Law.

Cannizzaro, E (2006). 'A Machiavellian Moment? The UN Security Council and the Rule of Law' *International Organizations Law Review* **3**: 189.

Cannizzaro, E (2009). 'Security Council Resolutions and EC Fundamental Rights: Some Remarks on the ECJ Decision in the *Kadi* Case' *Yearbook of European Law* 593.

Caron, D (1993). 'The Legitimacy of the Collective Authority of the Security Council' *American Journal of International Law* **87**: 552.

Carozza, P G (2003). 'My Friend is a Stranger: The Death Penalty and the Global Ius Commune of Human Rights' *Texas Law Review* **81**: 1031.

Cassese, A (1990). 'Remarks on Scelle's Theory of "Role-Splitting" (déboulement fonctionnel) in International Law' *European Journal of International Law* **1**: 210.

Cassese, A (2005). *International Law*. Oxford, Oxford University Press.

Chesterman, S (2006). 'The Spy Who Came In From The Cold War: Intelligence and International Law' *Michigan Journal of International Law* **27**: 1071.

Chesterman, S (2006). *Shared Secrets: Intelligence and Collective Security*. Sydney, Lowy Institute for International Policy.

Chesterman, S (2009). '"I'll Take Manhattan": The International Rule of Law and the United Nations Security Council' *Hague Journal on the Rule of Law* **1**: 67.

Choudhry, S (2006). 'Migration as the New Metaphor in Comparative Constitutional Law' in S Choudhry, *The Migration of Constitutional Ideas*. Cambridge, Cambridge University Press: 1.

Ciampi, A (2011). 'Security Council Targeted Sanctions and Human Rights' in B Fassbender (ed), *Securing Human Rights?: Achievements and Challenges of the UN Security Council*, Oxford, Oxford University Press: 98.

Cohen, J (1997). 'Deliberation and Democratic Legitimacy' in J Bohman and W Rehg, *Deliberative Democracy: Essays on Reason and Politics*. Cambridge, MIT Press: 67.

Cohen, J L (2008). 'A Global State of Emergency or the Further Constitutionalization of International Law: A Pluralist Approach' *Constellations* **15**(4): 456.

Conforti, B (2006). 'Decisioni del Consiglio disicurezza e diritti fondamentali in una bizarre sentenze del Tribunale communitario di primo grado' *Diritto dell'Unione europea* **11**: 333.

Conforti, B and F Francioni (eds) (1997). *Enforcing International Human Rights in Domestic Courts*, Martinus Nijhoff.

Conlon, P (1995). 'Lessons from Iraq: The Functions of the Iraq Sanctions Committee as a Source of Sanctions Implementation Authority and Practice' *Virginia Journal of International Law* **35**: 633.

Conlon, P (2000). *United Nations Sanctions Management: A Case Study of the Iraq Sanctions Committee*. New York, Transnational Publishers.

Cortright, D and E de Wet (2010). *Human Rights Standards for Targeted Sanctions*. Sanctions and Security Research Program.

Cossman, B (ed.) (2006). *Migrating Marriages and Comparative Constitutionalism. The Migration of Constitutional Ideas*. Cambridge, Cambridge University Press.

Cover, R (1983). 'Nomos and Narrative' *Harvard Law Review* **97**: 4.

Cramér, P (2003). 'Recent Swedish Experiences with Targeted UN Sanctions: The Erosion of Trust in the Security Council' in E de Wet and A Nollkaemper, *Review of the Security Council by Member States*. Intersentia: 88.

Cronin, B and I Hurd (2008). *The UN Security Council and the Politics of International Authority*. Abingdon, Routledge.

Cronin-Furman, K (2006). 'The International Court of Justice and the United Nations Security Council: Rethinking a Complicated Relationship' *Columbia Law Review* **106**: 435.

Dahl, R A (1999). 'Can International Organizations Be Democratic? A Skeptic's View' in I Shapiro and C Hacker-Cordón, *Democracy's Edges*. Cambridge, Cambridge University Press: 19.

Dauer, E and T Gilhool (1973). 'The Economics of Constitutionalized Repossession' *Southern California Law Review* **47**: 116.

de Búrca, G (2008). 'Developing Democracy Beyond the State' *Columbia Journal of Transnational Law* **46**: 221.

de Búrca, G (2009). 'The European Court of Justice and the International Legal Order after *Kadi*' *Harvard Journal of International Law* **51**: 1.

de Búrca, G (2009). 'The European Courts and the Security Council: Between Dédoublement Fonctionnel and Balancing of Values: Three Replies to Pasquale De Sena and Maria Chiara Vitucci' *European Journal of International Law* **20**(3): 853.

de Wet, E (2003). The Role of Human Rights in Limiting the Enforcement Power of the Security Council: A Principled View' in E de Wet, A Nollkaemper, and P Dijkstra, *Review of the Security Council by Member States*. Antwerp, Oxford, New York, Intersentia: 7.

de Wet, E (2005). The Security Council as a Law Maker: The Adoption of (Quasi-) Judicial Decisions' in R Wolfrum and V Röben, *Developments of International Law in Treaty-Making*. Berlin, Springer: 183.

de Wet, E (2006). 'The International Constitutional Order' *International and Comparative Law Quarterly* **55**: 51.

de Wet, E (2009). 'The Role of European Courts in the Development of a Hierarchy of Norms Within International Law: Evidence of Constitutionalisation' *European Constitutional Law Review* **5**: 284.

de Wet, E and A Nollkaemper (2002). 'Review of Security Council Decisions by National Courts' *German Yearbook of International Law* **45**: 166.

Delbrock, J (2002) 'Functions and Powers: Article 24' in B Simma et al (eds), *The Charter of the United Nations: A Commentary*, vol 1. Oxford, Oxford University Press: 446.

della Cananea, G (2003). 'Beyond the State: The Europeanization and Globalization of Procedural Administrative Law' *European Public Law* **9**: 563.

della Cananea, G (2010). 'Administrative Law in the European Legal Area: A Historic and Comparative Overview' in A von Bogdandy, S Cassese, and P Huber, *Handbuch Ius Publicum Europaeum, Band III: Verwaltungsrecht in Europa*. Heidelberg, CF Müller: 447.

del Ponte, C and Sudetic, C (2009). *Madame Prosecutor: Confrontations with Humanity's Worst Criminals and the Culture of Impunity*. Milan, Feltrinelli Editore.

Dicey, A V (1939). *Introduction to the Study of Law of the Constitution* (E C S Wade (ed.). 9th edn, Macmillan, London.

Dryzek, J (2000). *Deliberative Democracy and Beyond: Liberals, Critics, Contestations*.

Dryzek, J (2006). *Deliberative Global Politics: Discourse and Democracy in a Divided World*. Cambridge, Polity Press.

Duff, R (1980). 'Legal Obligations and the Moral Nature of Law' *Juridical Review* **25**: 61.

Dugard, J (2001). 'Judicial Review of Sanctions' in V Gowlland-Debbas (ed.), *United Nations Sanctions and International Law*. Kluwer Law International: 83.

Dunoff, J and J Trachtman (2009). *Ruling the World: Constitutionalism, International Law and Global Governance*. New York, Cambridge University Press.

Dworkin, R (1986). *A Matter of Principle*. Oxford, Clarendon Press.

Dyzenhaus, D (1996). 'Herman Heller and the Legitimacy of Legalit.' *Oxford Journal of Legal Studies* **16**(40): 641.

Eckes, C (2008). 'Judicial Review of European Anti-Terrorism Measures—the *Yusuf* and *Kadi* Judgments of the Court of First Instance' *European Law Journal* **14**: 74.

Eeckhout, P (2007). 'Community Terrorism Listings, Fundamental Rights and UN Security Council Resolutions. In Search of the Right Fit' *European Constitutional Law Review* **3**: 183.

Eleftheriadis, P (2010). 'Pluralism and Integrity' *Ratio Juris*: 365.

Ely, J H (1980). *Democracy and Distrust*. Cambridge, Harvard University Press.

Esty, D (2006). 'Good Governance at the Supranational Scale: Globalizing Administrative Law' *Yale Law Journal* **115**: 1490.

Falk, R (1964). *The Role of Domestic Courts in the International Legal Order*. New York, Syracuse University Press.

Farrall, J (2007). *United Nations Sanctions and the Rule of Law*. New York, Cambridge University Press.

Farrall, J (2009). 'Does the UN Security Council Compound the Democratic Deficit?' *Alberta Law Review* **46**(4): 913.

Fassbender, B (1998). 'The United Nations Charter as Constitution of the International Community' *Columbia Journal of Transnational Law* **36**: 529.

Fassbender, B (2006). *Targeted Sanctions and Due Process*. Strasbourg, Council of Europe Committee of Legal Advisors on International Law.

Fassbender, B (2010). 'Triepel in Luxemburg, Die dualistische Sicht des Verhältnisses zwischen Europa- und Völkerrecht in der "Kadi-Rechtsprechung" des EuGH als Problem des Selbstverständnisses der Europäischen Union'. *Die Öffentliche Verwaltung* 333.

Fassbender, B (ed.) (2011). *Securing Human Rights?: Achievements and Challenges of the UN Security Council*. Oxford, Oxford University Press.

Fassbender, B (2011). 'The Role for Human Rights in the Decision-making Process of the Security Council' in B Fassbender, *Securing Human Rights? Achievements and Challenges of the UN Security Council*. Oxford, Oxford University Press: 74.

Fisher, E (2010). 'Transparency and Administrative Law: A Critical Evaluation' *Current Legal Problems* **63**(1): 272.

Forcese, C and K Roach (2010). 'Limping into the Future: The UN 1267 Terrorism Listing Process at the Crossroads' *George Washington International Law Review* **42**: 217.

Francioni, F (2011). 'The Right of Access to Justice to Challenge the Security Council's Targeted Sanctions: After-thoughts on *Kadi*' in Ulrich Fastenrath et al, *From Bilateralism to Community Interest: Essays in Honour of Judge Bruno Simma*. Oxford, Oxford University Press: 908.

Franck, T M (1990). *The Power of Legitimacy Among Nations*. Oxford, Oxford University Press.

Franck, T M (1992). 'The "Powers of Appreciation": Who Is the Ultimate Guardian of UN Legality?' *American Journal of International Law* **86**: 519.

Franck, T M (1995). *Fairness in International Law and Institutions*. Oxford, Clarendon Press.

Franck, T M (1999). *The Empowered Self: Law and Society in the Age of Individualism*. Oxford, Oxford University Press.

Franck, T M (2006). 'The Power of Legitimacy and the Legitimacy of Power: International Law in an Age of Power Disequilibrium' *American Journal of International Law* **100**(1): 88.

Franck, T M and G H Fox (eds) (1996). *International Law Decisions in National Courts*. New York, Transnational Publishers.

Freedman, J O (1978). *Crisis and Legitimacy: The Administrative Process and American Government*. Cambridge, Cambridge University Press.

Frowein, J and Krisch, N (2002). 'Introduction to Ch VII' in Bruno Simma, *The Charter of the United Nations: A Commentary*, vol II. 2nd edn, Oxford, Oxford University Press: 711.

Frug, G E (1984). 'The Ideology of Bureaucracy in American Law' *Harvard Law Review* **97**: 1276.

Fukuyama, F (2004). *State-Building: Governance and World Order in the 21st Century*. New York, Cornell University Press.

Fuller, L (1969). *The Morality of Law*. New Haven, Yale University Press.

Fuller, L (1978). 'The Forms and Limits of Adjudication' *Harvard Law Review* **92**: 353.

Galligan, D J (1986). *Discretionary Powers*. Oxford, Clarendon Press.

Galligan, D J (1996). *Due Process and Fair Procedures: A Study of Administrative Procedures*. Oxford, Clarendon Press.

Ganz, G (1974). *Administrative Procedures*. London, Sweet & Maxwell.

Gardbaum, S (2009). 'Human Rights and International Constitutionalism' in J Dunoff and J Trachtman, *Ruling the World: Constitutionalism, International Law and Global Governance*. Cambridge, Cambridge University Press: 233.

Geiss, R (2005). 'Humanitarian Safeguards in Economic Sanctions Regimes: A Call for Automatic Suspension Clauses, Periodic Monitoring, and Follow-up Assessment of Long-Term Effects' *Harvard Human Rights Journal* **18**: 167.

Genser, J and K Barth (2010). 'When Due Process Concerns Become Dangerous: The Security Council's 1267 Regime and the Need for Reform' *Boston College International and Comparative Law Review* **33**: 1.

Ghasem Zamani, S and J Mazaheri (2015) 'The Need for International Judicial Review of UN Economic Sanctions' in A Z Marossi and M R Bassett, *Economic Sanctions Under International Law: Unilateralism, Multilateralism, Legitimacy, and Consequences*. The Hague, TMC Asser Press: 219.

Ginsborg, L and M Scheinin (2011). 'You Can't Always Get What You Want: The *Kadi II* Conundrum and the Security Council 1267 Terrorist Sanctions Regime' *Essex Human Rights Review* **8**: 7.

Godinho, J (2010). 'When Worlds Collide: Enforcing United Nations Security Council Asset Freezes in the EU Legal Order' *European Law Journal* **16**(1): 67.

Guinchard, S et al (2005). *Droit processuel, droit commun et droit comparé du procès*. Paris, Dalloz.

Gutherie, P (2004). 'Security Council Sanctions and the Protection of Individual Rights' *NYU Annual Survey of American Law* **60**: 491.

Habermas, J (1979). 'Legitimation Problems in the Modern State' in J Habermas, *Communication and the Evolution of Society*. Boston, Beacon Press: 178.

Habermas, J (1984). *The Theory of Communicative Action: Reason and the Rationalization of Society*. Boston, Beacon.

Habermas, J (1996). *Between Facts and Norms: Contributions to a Discourse Theory of Law and Democracy*. Cambridge, MIT Press.

Habermas, J (2001). 'Why Europe Needs a Constitution' *New Left Review* **11**: 8.

Haggenmacher, P (1990). 'Grotius and Gentili: A Reassessment of Thomas Holland's Inaugural Lecture' in H Bull, B Kingsbury, and A Roberts, *Hugo Grotius and International Relations*. Oxford, Clarendon Press: 133.

Halberstam, D and E Stein (2009). 'The United Nations, the European Union and the King of Sweden: Economic Sanctions and Individual Rights in a Plural World Order' *Common Market Law Review* **46**(1): 13.

Hale, T N and A-M Slaughter (2006). 'Transparency: Possibilities and Limitations' *The Fletcher Forum of World Affairs* **30**(1): 153.

Hamilton, A et al (1947). *The Federalist: A Commentary on the Constitution of the United States*. New York, Tudor.

Harlow, C (2006). 'Global Administrative Law: The Quest for Principles and Values' *European Journal of International Law* **17**(1): 187.

Harlow, C and R Rawlings (2009). *Law and Administration*. 3rd edn, Cambridge, Cambridge University Press.

Hart, H L A (1961). *The Concept of Law*. Oxford, Clarendon Press.

Held, D (1995). *Democracy and the Global Order: From Modern State to Cosmopolitan Governance*. Cambridge, Polity Press.

Heupel, M (2009). 'Multilateral Sanctions against Terror Suspects and the Violation of Due Process Standards' *International Affairs* **85**: 209.

Hewart (1929) *The New Despotism*. London, Ernest Benn Limited.

Higgins, R (1970). 'The Place of International Law in the Settlement of Disputes by the Security Council' *American Journal of International Law* **64**: 1.

Higgins, R (1978). 'Conceptual Thinking about the Individual in International Law' *New York Law School Law Review* **24**: 11.

Higgins, R (2004). *Problems and Process: International Law and How We Use It*. Cambridge, Clarendon Press.

Higgins, R (2006). 'A Babel of Judicial Voices? Ruminations from the Bench' *International and Comparative Law Quarterly* **55**: 791.

Hoexter, C (2000). 'The Future of Judicial Review in South African Administrative Law' *South African Law Journal* **17**: 484.

Hoffmann, J (2008). 'Terrorism Blacklisting: Putting European Human Rights Guarantees to the Test' *Constellations* **15**(4): 543.

Holmes, S (1995), *Passions and Constraint*. Chicago, University of Chicago Press.

Hong, J-H (2000). 'Administrative Law in the Institutionalized Administrative State in D-K Yoon, *Recent Transformations in Korean Law and Society*.

Huber, K T and A Rodiles (2012). 'An Ombudsperson in the United Nations Security Council: A Paradigm Shift?' *Anuario Mexicano de Derecho Internacional*: 107.

Hudson, A (2007). 'Not a Great Asset: the UN Security Council's Counter-Terrorism Regime: Violating Human Rights' *Berkeley Journal of International Law* **25**: 203.

Hurd, I (2007). *After Anarchy: Legitimacy and Power in the United Nations Security Council.* Princeton, Princeton University Press.

Hurrell, A (2001). 'Global Inequality and International Institutions' in T W Pogge, *Global Justice.* Oxford, Blackwell: 32.

Hurrell, A (2007). *On Global Order: Power, Values and the Constitution of the International Society.* Oxford, Oxford University Press.

International Commission of Jurists (2009). *Assessing Damage, Urging Action: Report of the Eminent Jurists Panel on Terrorism, Counter-Terrorism and Human Rights.* Geneva.

Ip, J (2008). 'The Rise and Spread of the Special Advocate' *Public Law*: 717.

Jain, M P (1997). 'Administrative Law of Malaysia and Singapore', *Malayan Law Journal.*

Jennings, R and A Watts (1996). *Oppenheim's International Law.* London and New York, Longman.

Johnstone, I (2003). 'Security Council Deliberations: The Power of Better Argument' *European Journal of International Law* **14**: 437.

Johnstone, I (2008). 'Legislation and Adjudication in the UN Security Council: Bringing Down the Deliberative Deficit' *American Journal of International Law* **102**: 275.

Johnstone, I (2008). 'The Security Council as Legislature' in B Cronin and I Hurd (eds), *The UN Security Council and the Politics of International Authority.* Oxford, Routledge: 80.

Kahn-Freund, O (1974). 'On Uses and Misuses of Comparative Law' *Modern Law Review* **37**(1): 1.

Kanetake, M (2008). 'Enhancing Community Accountability of the Security Council through Pluralistic Structure: The Case of the 1267 Committee' *Max Planck United Nations Yearbook* **12**: 113.

Kant, I (1956). *Critique of Practical Reason.* Macmillan Publishing Co.

Kant, I (1959). *Foundation of the Metaphysics of Morals.* Charlottesville, VA, Michie Company.

Kelsen, H (1950). *The Law of the United Nations: A Critical Analysis of its Fundamental Problems.* E A Praeger, 1950.

Kennedy, D (1973). 'Legal Formality' *Journal of Legal Studies* **2**: 351.

Kennedy, D (1999). 'The Disciplines of International Law and Policy' *Leiden Journal of International Law* **12**: 9.

Kennedy, D (2007). 'One, Two, Three, Many Legal Orders: Legal Pluralism and the Cosmopolitan Dream' *NYU Review of Law and Social Change* **31**: 642.

Kennedy, D (2009). 'The Mystery of Global Governance' in J Dunoff and J Trachtman, *Ruling the World? Constitutionalism, International Law and Global Governance.* Cambridge, Cambridge University Press: 37.

Keohane, R O (2003). 'Global Governance and Democratic Accountability' in D Held and M Koenig-Archibugi, *Taming Globalization: Frontiers of Governance.* Cambridge, Polity: 130.

Ketchen, J C (2003). 'Revisiting Fuller's Critique of Hart—Managerial Control and the Pathology of Legal Systems: The Hart-Weber Nexus' *University of Toronto Law Journal* **53**(1): 1.

Kingsbury, B (1997). 'A Grotian Tradition of Theory and Practice?: Grotius, Law and Moral Skepticism in the Thought of Hedley Bull' *Quinnipiac Law Review* **17**: 3.

Kingsbury, B (2009). 'The Concept of "Law" in Global Administrative Law' *European Journal of International Law* **20**: 23.

Kingsbury, B and L Casini (2010). 'Global Administrative Law Dimensions of International Organizations Law' *International Organizations Law Review* **6**(2): 319.

Kingsbury, B and M Donaldson (2011). 'From Bilateralism to Publicness in International Law' in U Fastenrath et al, *From Bilateralism to Community Interest: Essays in Honour of Bruno Simma*. Oxford, Oxford University Press: 79.

Kingsbury, B et al (2005). 'The Emergence of Global Administrative Law' *Law and Contemporary Problems* **68**: 15.

Klabbers, J (2001). *An Introduction to Institutional Law*. Cambridge, Cambridge University Press.

Klabbers, J (2003). '(I Can't Get No) Recognition: Subjects Doctrine and the Emergence of Non-State Actors' in J Petman and J Klabbers, *Nordic Cosmopolitanism: Essays in International Law for Martii Koskenniemi*. Leiden, Martinus Nijhoff: 351.

Klabbers, J (2007). '*Kadi*. Justice at the Security Council' *International Organizations Law Review* **4**: 293.

Knop, K (2000). 'Here and There: International Law in Domestic Courts' *NYU Journal of International Law and Politics* **32**: 501.

Koh, H H (1997). 'Why Do Nations Obey?' *Yale Law Journal* **106**: 2599.

Kokott J and C Sobotta (2012). 'The *Kadi* Case—Constitutional Core Values and International Law—Finding the Balance?' *European Journal of International Law* **23**(4): 1015.

Korecki, L (2009). 'Procedural Tools for Ensuring Cooperation of States with the Special Tribunal for Lebanon' *Journal of International Criminal Justice* **7**: 927.

Koskenniemi, M (1995). 'The Police in the Temple: Order, Justice and the UN—A Dialectical View' *European Journal of International Law* **6**(3): 325.

Koskenniemi, M (1996). 'The Place of Law in Collective Security' *Michigan Journal of International Law* **17**: 455.

Koskenniemi, M (2006). 'Constitutionalism as Mindset: Reflections on Kantian Themes About International Law and Globalization' *Theoretical Inquiries in Law*: 9.

Koskenniemi, M and P Leino (2002). 'Fragmentation of International Law? Postmodern Anxieties' *Leiden Journal of International Law* **15**: 553.

Krisch, N (2006). 'The Pluralism of Global Administrative Law' *European Journal of International Law* **17**(1): 247.

Krisch, N (2008). 'The Open Architecture of European Human Rights Law' *Modern Law Review* **71**: 183.

Krisch, N (2009). 'Global Administrative Law and the Constitutional Ambition' *LSE Law, Society and Economy* Working Papers 10/2009.

Krisch, N (2010). *Beyond Constitutionalism: The Pluralist Structure of Postnational Law*. Oxford, Oxford University Press.

Ku, C and H K Jacobson (eds) (2003). *Democratic Accountability and the Use of Force in International Law*. Cambridge, Cambridge University Press.

Kumm, M (2004). 'The Legitimacy of International Law: A Constitutionalist Framework of Analysis' *European Journal of International Law* **15**: 907.

Kuthakas, C (2003). *The Liberal Archipelago*. Oxford, Oxford University Press.

Lacey, N (1992). 'The Jurisprudence of Discretion: Escaping the Legal Paradigm' in K Hawkins, *The Uses of Discretion*. Oxford, Clarendon Press.

Lauterpacht, H (1929). 'Decisions of Municipal Courts as a Source of International Law' *British Yearbook of International Law* **10**: 65.

Lefas, A (1978). 'A Comparison of the Concept of Natural Justice in English Administrative Law with the Corresponding General Principles of Law and Rules of Procedures in French Administrative Law' *Queen's Law Journal* **4**: 197.

Levit, J (2007). 'Bottom-Up International Lawmaking: Reflections on the New Haven School of International Law' *Yale Journal of International Law* **32**: 393.

Lilvoja, R (2008). 'The Scope of the Supremacy Clause of the United Nations Charter' *International and Comparative Law Quarterly* **57**: 583.

Lindseth, P (2005). 'Always Embedded Administrations: The Historical Evolution of Administrative Justice as an Aspect of Modern Governance' in C Joerges, B Stråth, and P Wagner, *The Economy as a Polity: The Political Constitution of Contemporary Capitalism*. London, Routledge Cavendish: 121.

Lopez, G et al (2009). *Overdue Process: Protecting Human Rights while Sanctioning Alleged Terrorists*, Fourth Freedom Forum, Kroc Institute for International Peace Studies, University of Notre Dame.

Loughlin, M (1992). *Public Law and Political Theory*. Oxford, Clarendon Press.

Loughlin, M (2003). *The Idea of Public Law*. Oxford, Oxford University Press.

Lowe, V (2001). 'The Politics of Law-Making: Are the Method and Character of Norm Creation Changing?' in M Byers, *The Role of Law in International Politics: Essays on International Relations and International Law*. Oxford, Oxford University Press: 207.

MacCormick, N (1993). 'Beyond the Sovereign State' *The Modern Law Review* **56**(1): 1.

MacCormick, N (1998). 'Risking Constitutional Collision in Europe' *Oxford Journal of Legal Studies* **18**: 517.

Maduro, M P (2003). 'Conrapunctual law: Europe's Constitutional Pluralism in Action' in N Walker, *Sovereignty in Transition*. Oxford, Hart Publishing: 501.

Maduro, M P (2009). 'Courts and Pluralism: Essay on a Theory of Judicial Adjudication in the Context of Legal and Constitutional Pluralism' in J Dunoff and J Trachtman, *Ruling the World? Constitutionalism, International Law and Global Governance*. New York, Cambridge University Press: 356.

Maher, G (1986). 'Natural Justice as Fairness' in N MacCormick and P Birks, *The Legal Mind: Essays for Tony Honoré*. Oxford, Clarendon Press.

Marossi, A Z and Bassett, M R (eds) (2015). *Economic Sanctions Under International Law: Unilateralism, Multilateralism, Legitimacy, and Consequences*. The Hague, TMC Asser Press.

Marriott, J (1927). *The Mechanism of the Modern State*. Oxford, Clarendon Press.

Mashaw, J (1985). *Due Process in the Administrative State*. New Haven, Yale University Press.

Mashaw, J (1997). *Greed, Chaos and Governance*. New Haven, Yale University Press.

McCorquodale, R (2004). 'An Inclusive International Legal System' *Leiden Journal of International Law* **17**: 477.

McCorquodale, R (2006). 'Beyond State Sovereignty: The International Legal System and Non-State Participants' *Revista Colombiana de Derecho Internacional*: 103.

McCorquodale, R (2006). 'The Individual and the International Legal System' in M D Evans, *International Law*. Oxford, Oxford University Press: 307.

McCrudden, C (2008). 'Human Dignity and Judicial Interpretation of Human Rights' *European Journal of International Law* **19**(4): 655.

McDougal, M S et al (1968). 'Theories About International Law: Prologue to a Configurative Jurisprudence' *Virginia Journal of International Law* **8**: 188.

Mégret, F and F Hoffmann (2003). 'The UN as a Human Rights Violator? Some Reflections on the United Nations Changing Human Rights Responsibilities' *Human Rights Quarterly* **25**: 314.

Mertus, J (2000). 'Considering Non-State Actors in the New Millennium: Toward Expanded Participation in Norm Generation and Norm Application' *NYU Journal of International Law and Politics* **32**: 537.

Michelman, F I (1977). 'Formal and Associational Aims in Procedural Due Process' in J R Pennock and J W Chapman, *Due Process.*. New York, New York University Press: 126.

Michelman, F I (2003–04). 'Reflection' *Texas Law Review* **82**: 1737.

Milanović, M (2009). 'Norm Conflict in International Law: Whither International Law?' *Duke Journal of Comparative and International Law* **20**(1): 69.

Mitrany, D (1948). 'The Functional Approach to World Organization' *International Affairs* **24**(3): 350.

Montesquieu, C (1979). *De l'esprit des lois*. Paris, Flammarion.

Montyon, A (1733–1820). 'Des agents de l'administration', Archives de l'Assistance publique de Paris, Fonds Montyon.

Moravcsik, A (1997). 'Taking Preferences Seriously: A Liberal Theory of International Politics' *International Organization* **51**: 513.

Moravcsik, A (2004). 'Is There a "Democratic Deficit" in World Politics? A Framework for Analysis' *Government and Opposition* **39**: 336.

Morgenthau, H (1993 (1948)). *Politics Among Nations: The Struggle for Power and Peace*. Boston, McGraw Hill.

Nardin, T (2000). 'International Pluralism and the Rule of Law' *Review of International Studies* **26**: 95.

Nettesheim, N (2007). 'UN Sanctions against Individuals—A Challenge to the Architecture of European Union Governance' *Common Market Law Review* **44**: 567.

Nijman, J (2004). *The Concept of International Legal Personality: An Inquiry into the History and Theory of International Law*. The Hague, TMC Asser Press.

Nonet, P and P Selznick (1978). *Law and Society in Transition: Toward Responsive Law*. New York, Octagon Books.

Note (1975). 'Specifying the Procedure Required by Due Process: Towards Limits on the Use of Interest Balancing' *Harvard Law Review* **88**: 1510.

Nye, J (2001). 'Globalization's Democratic Deficit: How to Make International Institutions More Accountable' *Foreign Affairs* **80**: 2.

O'Connell, M E (1999). 'New International Legal process' *American Journal of International Law* **93**: 334.

Ochoa, C (2007). 'The Individual and Customary International Law Formation' *Virginia Journal of International Law* **48**: 119.

Oosthuizen, G (1999). 'Playing the Devil's Advocate: The United Nations Security Council is Unbound by Law' *Leiden Journal of International Law* **12**: 549.

Orakhelashvili, A (2006). *Peremptory Norms in International Law*. Oxford, Oxford University Press.

Paust, J J (2011). 'NonState Actor Participation in International Law and the Pretence of Exclusion' *Virginia Journal of International Law* **51**(4): 977.

Pennock, J R and J W Chapman (1977). *Due Process*. New York, New York University Press.

Peters, A (2006). 'Compensatory Constitutionalism: The Function and Potential of Fundamental International Norms' *Leiden Journal of International Law* **19**: 579.

Peters, C J (2002). 'Participation, Representation, and Principled Adjudication' *Legal Theory* **8**: 185.

Petersmann, E-U (1997). 'How to Reform the United Nations System? Constitutionalism, International Law and International Organizations' *Leiden Journal of International Law* **10**: 421.

Pincoffs, E (1977). 'Due Process, Fraternity and a Kantian Injunction' in J R Pennock and J W Chapman, *Due Process*. New York, New York University Press: 172.

Pogge, T W (1992). 'Cosmopolitanism and Sovereignty' *Ethics* **103**: 48.

Postema, G J (1977). 'The Principle of Utility and the Law of Procedure: Bentham's Theory of Adjudication' *Georgia Law Review* **11**: 1393.

Postema, G J (1986). *Bentham and the Common Law Tradition*. Oxford, Clarendon Press.

Prost, K (2012). 'Fair Process and the Security Council: A Case for the Office of the Ombudsperson' in A M S D Frías, K Samuel, and N White, *Counter-terrorism: International Law and Practice*. Oxford, Oxford University Press: 409.

Rawls, J (1971). *A Theory of Justice*. Cambridge, Harvard University Press.

Reich, C A (1964). 'The New Property' *Yale Law Journal* **73**(5): 733.

Reich, J (2008). 'Due Process and Sanctions Targeted Against Individuals Pursuant to Resolution 1267 (1999). *Yale Journal of International Law* **33**: 505.

Reinisch, A (2001). 'Developing Human Rights and International Humanitarian Law Accountability of the Security Council for the Imposition of Economic Sanctions' *American Journal of International Law* **95**(4): 851.

Reinisch, A (2001). 'Securing the Accountability of International Organizations' *Global Governance* 7: 131.

Reisman, M (1993). 'The Constitutional Crisis in the United Nations' *American Journal of International Law* **87**: 83.

Reisman, M (1995). 'Assessing the Lawfulness of Nonmilitary Enforcement: The Case of Economic Sanctions' *American Society of International Law Proceedings* **89**: 350.

Reisman, W M (1993). 'The Constitutional Crisis in the United Nations' *American Journal of International Law* **87**: 83.

Roberts, A (2011). 'Comparative International Law? The Role of National Courts in Creating and Enforcing International Law' *International and Comparative Law Quarterly* **60**: 57.

Roberts, A (2001). 'Traditional and Modern Approaches to Customary International Law: A Reconciliation' *American Journal of International Law* **95**: 757.

Roberts, A and S Sivakumaran (2012). 'Lawmaking by Non-State Actors: Engaging Armed Groups in the Creation of International Humanitarian Law' *Yale Journal of International Law* **37**: 108.

Roberts, K (1995). 'Second-Guessing the Security Council: The International Court of Justice and Its Powers of Judicial Review' *Pace International Law Review* 7: 281.

Robinson, P (1996). 'The Criminal-Civil Distinction and the Utility of Desert' *Boston University Law Review* **76**: 201.

Robson, W A (1947). *Justice and Administrative Law: A Study of the British Constitution*. London, Stevens & Sons Limited.

Rosand, E (2004). 'The Security Council's Efforts to Monitor the Implementation of Al Qaeda/Taliban Sanctions' *American Journal of International Law* **98**: 745.

Rose-Ackerman, S (1994). 'American Administrative Law Under Siege: Is Germany a Model?' *Harvard Law Review* **107**: 1279.

Russell, R and J Muther (1958). *A History of the United Nations Charter*. Washington DC, Brookings Institution.

Sandel, M J (1982). *Liberalism and the Limits of Justice*. Cambridge, Cambridge University Press.

Saphire, R B (1978). 'Specifying Due Process Values: Toward a More Responsive Approach to Procedural Protection' *University of Pennsylvania Law Review* **127**: 111.

Sarooshi, D (2004). 'The Essentially Contested Nature of the Concept of Sovereignty: Implications for the Exercise by International Organizations of Delegated Powers of Government' *Michigan Journal of International Law* **25**: 1107.

Sarooshi, D (2005). *International Organizations and their Exercise of Sovereign Powers*. Oxford/New York, Oxford University Press.

Sarooshi, D (2008). 'The Role of Domestic Public Law Analogies in the Law of International Organisations' *International Organizations Law Review* **5**(2): 237.

Sax, J (1973). 'The Unhappy Truth About NEPA' *Oklohoma Law Review* **26**: 239.

Scanlon, T M (1977). 'Due Process' in J R Pennock and J W Chapman, *Due Process*. New York, New York University Press: 93.

Scanlon, T M (2003). 'Human Rights as a Neutral Concern' in T Scanlon, *The Difficulty of Tolerance: Essays in Political Philosophy*. Cambridge, Cambridge University Press: 113.

Schachter, O (1981). 'The Crisis of Legitimation in the United Nations' *Nordisk Tidsskrift Int'l Ret* **50**: 3.

Schauer, F (1995). 'Giving Reasons' *Stanford Law Review* **47**: 633.

Schlesinger, P and D Kevin (2000). 'Can the European Union Become a Sphere of Publics?' in E O Iriksen and J E Fossum, *Democracy in the European Union: Integration Through Deliberation*. London, New York, Routledge.

Security Council Report (2011). *Cross-cutting Report on the Rule of Law*. New York, SCR.

Simester, A P and A Von Hirsch (2006). 'Regulating Offensive Conduct through Two-Step Prohibitions' in A P Simester and A Von Hirsch, *Incivilities*. Oxford, Hart Publishing: 173.

Simma, B (1994). 'From Bilateralism to Community Interest in International Law' *Recueil des Cours de l'Académie de droit international de La Haye* **250**(VI): 217.

Simma, B (1995). 'The Contribution of Alfred Verdross to the Theory of International Law' *European Journal of International Law* **6**: 33.

Simma B (ed.) (2002). *The Charter of the United Nations: A Commentary*. Oxford, Oxford University Press.

Simpson, G (1996). 'Is International Law Fair?' *Michigan Journal of International Law* **17**: 615.

Simpson, G (2004). *Great Powers and Outlaw States*. Cambridge, Cambridge University Press.

Slaughter, A-M (2000). 'Agencies on the Loose? Holding Government Networks Accountable' in G A Bermann et al (eds), *Transatlantic Regulatory Cooperation*, Oxford, Oxford University Press: 521.

Slaughter, A-M (2004). *A New World Order*. Princeton, Princeton University Press.

Steele, J (2001). 'Participation and Deliberation in Environmental Law: Exploring a Problem-solving Approach' *Oxford Journal of Legal Studies* **21**(3): 415.

Steffek, J (2003). 'The Legitimation of International Governance: A Discourse Approach' *European Journal of International Relations* **9**: 249.

Steiker, C S (2001). 'Civil and Criminal Divide' in Joshua Dressler (ed), *Encyclopedia of Crime and Justice*. New York, Macmillan Reference USA: **1**.

Stein, E (2001). 'International Integration and Democracy: No Love at First Sight' *American Journal of International Law* **95**: 489.

Stein, T (2012). 'Too "Smart" for Legal Protection?: UN Security Council's Targeted Sanctions and a Pladoyer for another UN Tribunal' in H Hestermeyer et al, *Coexistence, Cooperation and Solidarity: Liber Amicorum Rüdiger Wolfrum.* Leiden, Martinus Nijhoff: 1539.

Stewart, R B (1975). 'The Reformation of American Administrative Law' *Harvard Law Review* **88**: 1667.

Summers, R (1974). 'Evaluating and Improving Legal Processes—a Plea for "Process Values"' *Cornell Law Review* **60**(1): 1.

Sunstein, C R (1985–86). 'Interest Groups in American Public Law' *Stanford Law Review* **38**: 29.

Sunstein, C R (1986). 'Factions, Self-interest, and the APA: Four Lessons Since 1946' *Virginia Law Review* **72**: 271.

Sunstein, C R (1988). 'Beyond the Republican Revival' *Yale Law Journal* **97**: 1539.

Tasioulas, J (1996). 'In Defence of Relative Normativity: Communitarian Values and the Nicaragua Case' *Oxford Journal of Legal Studies* **16**: 95.

Taylor, C (1989). *Sources of the Self: The Making of the Modern Identity.* Cambridge, Cambridge University Press.

Tomuschat, C (1983). 'Obligations Arising for States Without or Against their Will' *Recueil des Cours de l'Académie de droit international de La Haye* **241**(IV): 195.

Tomuschat, C (2009). 'The *Kadi* case: What Relationship is there between the Universal Legal Order under the Auspices of the United Nations and the EU Legal Order?' *Yearbook of European Law* 654.

Tribe, L (1988). *American Constitutional Law.* 2nd edn, Mineola, NY, Foundation Press.

Tully, J (1995). *Constitutionalism in an Age of Diversity.* Cambridge, Cambridge University Press.

Turner, S and S J Schulhofer (2005). *The Secrecy Problem in Terrorism Trials.* Liberty and National Security Project. Brennan Centre for Justice at NYU School of Law, New York, NYU.

Tyler, T (2006). *Why People Obey the Law.* New Jersey, Princeton University Press.

Tzanakopoulos, A (2010). 'Domestic Court Reactions to UN Security Council Sanctions' in A Reinisch, *Challenging Acts of International Organizations Before National Courts.* Oxford, Oxford University Press: 54.

Tzanakopoulos, A (2010). 'United Nations sanctions in domestic courts: from interpretation to defiance in *Abdelrazik v Canada (Minister of Foreign Affairs)*' *Journal of International Criminal Justice* **8**: 249.

Tzanakopoulos, A (2012). 'Human Rights and United Nations Security Council Measures' in E de Wet and J Vidmar, *Hierarchy in International Law: The Place of Human Rights.* Oxford, Oxford University Press: 42.

Tzanakopoulos, A (2012). 'Judicial Dialogue in Multi-level Governance: the Impact of the Solange Argument' in O K Fauchald and A Nollkaemper, *The Practice of International and National Courts and the (De-)Fragmentation of International Law.* Oxford, Hart Publishing: 185.

Van den Herik, L (2007). 'The Security Council's Targeted Sanctions Regimes: In Need of Better Protection of the Individual' *Leiden Journal of International Law* **20**: 797.

Van den Herik, L (2014). 'Peripheral Hegemony in the Quest to Ensure Security Council Accountability for its Individualized UN Sanctions Regimes' *Journal of Conflict and Security Law* **19(3)**: 427

Vandepoorter, A (2006). 'L'application communautaire des decisions du Conseil de Sécurité' *Annuaire francais de droit international* **52**: 102.

Verdirame, G (2011). *The UN and Human Rights: Who Guards the Guardians?* Cambridge: Cambridge University Press.

Verdross, A (1927). 'Le Fondement du Droit International' *Recueil des Cours de l'Académie de droit international de La Haye* **16**: 244.

Vereshchetin, V and R Müllerson (1990). 'International Law in an Interdependent World' *Columbia Journal of Transnational Law* **28**: 291.

von Bogdandy, A (2006). 'Constitutionalism in International Law: Comment on a Proposal from Germany' *Harvard International Law Journal* 47(1): 223.

Waldron, J (1999). *Law and Disagreement.* Oxford, Oxford University Press.

Walker, N (2002). 'The Idea of Constitutional Pluralism' *Modern Law Review* **65**(3): 317.

Walker, N (2003). 'Postnational Constitutionalism and the Problem of Translation' in J Weiler and M Wind, *European Constitutionalism Beyond the State.* Cambridge, Cambridge University Press: 27.

Walker, N (2008). 'Taking Constitutionalism Beyond the State' *Political Studies* **56**: 519.

Wallensteen, P et al. (2003). 'Making Targeted Sanctions Effective: Guidelines for the Implementation of UN Policy Options' in *Results from the Stockholm Process on the Implementation of Targeted Sanctions.* Uppsala University.

Walzer, M (1990). 'The Communitarian Critique of Liberalism' *Political Theory* **18**: 6.

Watson, G (1993). 'Constitutionalism, Judicial Review, and the World Court' *Harvard International Law Journal* **34**: 1.

Weil, P (1983). 'Towards Relative Normativity in International Law?' *American Journal of International Law* 77: 413.

Weiler, J (1999). *The Constitution of Europe: 'Do the New Clothes Have an Emperor?' and Other Essays in European Integration.*

Weiler, J (2004). 'The Geology of International Law—Governance, Democracy and Legitimacy' *Zeitschrift für ausländisches öffentliches Recht und Völkerrecht* **64**: 547.

Willis, G L (2011). 'Security Council Targeted Sanctions, Due Process and the 1267 Ombudsperson' *Georgetown Journal of International Law* **42**: 673.

Young, I M (2000). *Inclusion and Democracy.* Princeton, Princeton University Press.

Zamani, G and J Mazaheri (2015). 'The Need for International Judicial Review of UN Economic Sanctions' in Ali Z Marossi and Marisa R Bassett (eds), *Economic Sanctions Under International Law: Unilateralism, Multilateralism, Legitimacy, and Consequences.* TMC Asser Press: 219.

Zedner, L (2007). 'Seeking Security by Eroding Rights: The Side-Stepping of Due Process' in B J Goold and L Lazarus, *Security and Human Rights*. Oxford, Hart Publishing: 257.

Zürn, M (2000). 'Democratic Governance Beyond the Nation-State: The EU and Other International Institutions' *European Journal of International Relations* **6**: 183.

Zürn, M (2004). 'Global Governance and Legitimacy Problems' *Government and Opposition* **39**: 260.

Index